Freshwater Ecology

AQUATIC ECOLOGY Series

Series Editor

Prof. James H. Thorp
Kansas Biological Survey, and
Department of Ecology and Evolutionary Biology
University of Kansas
Lawrence, Kansas, USA

Groundwater Ecology
Janine Gilbert, Dan L. Danielopol, Jack A. Stanford

Algal Ecology
R. Jan Stevenson, Max L. Bothwell, Rex L. Lowe

Streams and Ground Waters
Jeremy B. Jones, Patrick J. Mulholland

Ecology and Classification of North American Freshwater Invertebrates, Third Edition
James H. Thorp, Alan P. Covich

Freshwater Ecology
Walter K. Dodds, Matt R. Whiles

Aquatic Ecosystems
Stuart E. G. Findlay, Robert L. Sinsabaugh

Tropical Stream Ecology
David Dudgeon

The Riverine Ecosystem Synthesis
James H. Thorp, Martin C. Thoms, Michael D. Delong

Freshwater Ecology

Concepts and Environmental Applications of Limnology

Second Edition

Walter K. Dodds and Matt R. Whiles

ELSEVIER

AMSTERDAM • BOSTON • HEIDELBERG • LONDON
NEW YORK • OXFORD • PARIS • SAN DIEGO
SAN FRANCISCO • SINGAPORE • SYDNEY • TOKYO
Academic Press is an imprint of Elsevier

Academic Press is an imprint of Elsevier
30 Corporate Drive, Suite 400, Burlington, MA 01803, USA
525 B Street, Suite 1900, San Diego, California 92101-4495, USA
84 Theobald's Road, London WC1X 8RR, UK

Notices
Knowledge and best practice in this field are constantly changing. As new research and experience
broaden our understanding, changes in research methods, professional practices, or medical
treatment may become necessary.

Practitioners and researchers must always rely on their own experience and knowledge in evaluating
and using any information, methods, compounds, or experiments described herein. In using such
information or methods they should be mindful of their own safety and the safety of others, including
parties for whom they have a professional responsibility.

To the fullest extent of the law, neither the Publisher nor the authors, contributors, or editors, assume
any liability for any injury and/or damage to persons or property as a matter of product liability,
negligence or otherwise, or from any use or operation of any methods, products, instructions, or
ideas contained in the material herein.

Library of Congress Cataloging-in-Publication Data
Dodds, Walter Kennedy, 1958–
  Freshwater ecology : concepts and environmental applications of limnology / Walter K. Dodds,
  Matt R. Whiles. — 2nd ed.
    p. cm.
  Includes bibliographical references and index.
  ISBN 978-0-12-374724-2 (hardcover : alk. paper) 1. Freshwater ecology—Textbooks. 2. Limnology—Textbooks.
  I. Whiles, Matt R. II. Title.
  QH541.5.F7D63 2002
  577.6—dc22   2010009285

British Library Cataloguing-in-Publication Data
A catalogue record for this book is available from the British Library.

For information on all Academic Press publications
visit our Web site at www.elsevierdirect.com

Contents

Preface

FOR THE STUDENT

This book was written for you. We obtained as much student input as possible by having student reviewers assess the text and the approach used in it. The idea for the text was based on teaching students who were not satisfied with the existing texts. Teaching aquatic ecology and limnology has made it clear to us that most students enter ecological sciences for practical reasons. They often are concerned about conservation of resources from a classical perspective (e.g., fisheries program) or from an environmental issue perspective. Most existing texts limit the applied aspects of aquatic ecology to a section at the end.

The aim of this text is to incorporate discussion of the issues as they arise when the basic materials are being covered. This allows you to see the applications of difficult topics immediately and, we hope, provides additional impetus for doing the work required to gain an understanding. We also attempted to use the broadest possible approach to freshwater ecosystems; scale and linkages among systems are important in ecology. Most students in ecological courses had some interest in the natural world as children. They spent time exploring under rocks in streams, fishing, camping, hiking, or swimming, which stimulated a love of nature. This book is an attempt to translate this basic affinity for aquatic ecosystems into an appreciation of the scientific aspects of the same world. It is not always easy to write a text for students. Instructors usually choose a text, giving the students little choice. Thus, some authors write for their colleagues, not for students. We tried to avoid such pressures and attempted to tailor the approach to you. We hope you will learn from the materials presented here and that they will adequately supplement your instructor's approach. When you find errors, please let us know. This will improve any future editions. Above all, please appreciate the tremendous luxury of being a student and learning. You are truly fortunate to have this opportunity and we are grateful for the time you take with this text.

FOR THE INSTRUCTOR

We hope this book will make your job a little easier. The chapters are short, mostly self-contained units to allow the text to conform to a wide variety of organizational schemes that may be used to teach about freshwaters. This will also allow you to avoid sections that are outside the scope of the course you are teaching. However, environmental applications are integrated into the text because we do not view the basic science and applications as clearly separate. We attempt to create instructional synergism by combining applied and basic aspects of aquatic ecology. Describing applications tends to stimulate student interest in mastering difficult scientific concepts. A variety of pedagogical approaches are used in an attempt to engage student interest and facilitate learning. These include sidebars, biography boxes, and method boxes. In the second edition we have also added advanced sections for areas where you might want the students to get a little more in depth but in areas that we often skip in our own lectures for a first class in freshwater ecology. We also include an appendix on experimental design in ecological science and a glossary because we have found many advanced undergraduates have little exposure to the practical side of how science is done. It is always difficult to know what to include and where to go into detail. Detailed examples are supplied to enforce general ideas. The choices of examples, perhaps not always the best overall, are the best we could find while preparing the text. Suggestions for improvements in this and any other areas of the text are encouraged and appreciated, and instructors using the first edition did just that. We apologize for any errors.

Why did we write this? In our experience, teaching limnology/freshwater ecology is more work than teaching other courses because of the breadth of subject, differential preparation of students, and associated laboratories but always seems to be the most fun. Of course it is fun; it is the best subject! We hope this book facilitates your efforts to transmit what is so great about the study of freshwater ecology.

WHY DID WE WRITE A SECOND EDITION?

Asking students to buy a new text rather than used requires a good reason to produce a new edition. In the near decade from the writing of the first edition, many new advances have occurred, and new topics have gained prominence with respect to environmental effects and hot areas of research interest. A second author (Matt R. Whiles) was added to broaden the perspective. In our attempt to cover these new areas we have added about 500 new and updated references, 50 new figures, 30 updated figures, color plates, two new chapters, and expanded the length of the text by roughly 40%. Particularly, we included more emphasis on wetlands and reservoirs than in the previous edition. Numerous errors were found in the first edition by dedicated students in Walter K. Dodds's classes (with a potential award of test points for each unique error) and instructors across the country who adopted the first edition. We have hopefully corrected these without introducing too many new ones.

Acknowledgments

We thank Dolly Gudder, who was involved in all aspects of the writing and compilation of this book, including proofing the entire text, drafting and correcting all the figures from the first edition, library research, writing the first draft of the index, and obtaining permissions. We are forever in her debt. Alan Covich provided extensive conceptual guidance and proofread the text; his input was essential to producing this work. Eileen Schofield-Barkley provided excellent editorial comments on all chapters. The fall of 1998 Kansas State University limnology class proofed Chapters 1 through 8 and 11 through 18. The Kansas State University limnology classes proofread all chapters (especially Michelle Let) and graciously field tested the text in draft form. The L.A.B. Aquatic Journal Club, Chuck Crumly, Susan Hendricks, Stuart Findlay, Steve Hamilton, Nancy Hinman, Jim Garvey, Chris Guy, and Al Steinman provided suggestions on the first edition. Early helpful reviews on book concepts were provided by James Cotner, David Culver, Jeremey Jones, Peter Morin, Steven Mossberg, Stuart Fisher, Robert Wetzel, and F. M. Williams. The anonymous reviews (obtained by the publisher) are also greatly appreciated, including those who completed a detailed survey on the prospect of a second edition. Many of the good bits and none of the mistakes are attributable to these reviewers. We appreciate the support of the Kansas State University Division of Biology, the Kansas Agricultural Experiment Station, and Southern Illinois University at Carbondale. This is publication 10-207-B from the Kansas Agricultural Experiment Station.

Thanks for corrections from the Freshwater Ecology classes at Kansas State University with particular thanks to Katie Bertrand and Andrea Severson. Thanks to Lydia Zeglin for help on describing molecular methods. Excellent suggestions for corrections of the first edition also came from Robert Humston, David Rogowski, John Havel, Sergi Thomas, Daniel Welsh, Erika Iyengar, Matt McTammany, Jack Webster, and Robert Humston. Kabita Ghimire produced the global distribution maps of freshwater habitats. Keith Gido, James Whitney, and Joe Gerken helped clarify Chapter 23. Thanks to Erika Martin, Joshua Perkin, and Kyle Winders for detailed review of the entire draft manuscript from the second edition. Thanks to Andy Richford who assisted in the development of the ideas for the second edition while he was editor at Elsevier.

Walter K. Dodds thanks his teachers over the years who guided him so well down the academic path: Ms. Waln, Steve Seavey, John Priscu, and especially Dick Castenholz and Eric Wickstrom. His students (Chris, Eric, Ken, Michelle, Mel, Randy, Nicole, Bob, Jon, Kym, Jessica, Justin, Alyssa, Kyle, Alex, and all the others) have kept asking the questions that fuel imagination. Walter's parents initiated his fascination for nature, and the encouragement of his siblings and in-laws kept him going. Dolly made it all possible.

Matt R. Whiles thanks F. E. Anderson, L. L. Battaglia, J. E. Garvey, J. W. Grubaugh, R. O. Hall, A. D. Huryn, K. R. Lips, R. Lira, and S. D. Peterson for valuable input and advice on this work. His early mentors, M. E. Gurtz, C. M. Tate, G. R. Marzolf, and J. B. Wallace provided guidance at critical points of his life and helped him realize that he could actually make a career out of his interests in ecology. His parents, Jim and Jane, and his sister, Wendy, tolerated and even supported his somewhat odd boyhood interests and laid the foundation for all of this. His wife, S. G. Baer, provided endless support, encouragement, and advice, and his recent and current graduate students (Amanda, Catherine, Checo, Dan, David, Eric, Jodi, Kaleb, Kim, Natalie, Therese) provided inspiration and tolerated him throughout this process.

Our children, Hannah, Joey, Sadie, and Rowland, put it all in perspective; the next generation is the reason this text includes environmental applications. We both deeply appreciate the support and love of our families.

Why Study Continental Aquatic Systems?

FIGURE 1.1
Crater Lake, Oregon.

Doi: 10.1016/B978-0-12-374724-2.00001-5

As is true of all organisms, our very existence depends on water; we need an abundance of fresh water to live. Although the majority of our planet is covered by water, only a very small proportion is associated with the continental areas on which humans are primarily confined (Table 1.1). Of the water associated with continents, over 99% is in the form of groundwater or ice and is difficult for humans to use. Human interactions with water most often involve fresh streams, rivers, marshes, lakes, and shallow groundwaters; thus, we rely heavily on a relatively rare commodity.

Why study the ecology of continental waters? To the academic, the answer is easy: because it is fascinating and we enjoy learning for its own sake. Thus, the field of *limnology*[1] (the scientific study of continental waters) has developed. Limnology has a long history of academic rigor and broad interdisciplinary synthesis (e.g., Hutchinson, 1957, 1967, 1975, 1993; Wetzel, 2001). One of the truly exciting aspects of limnology is the synthetic integration of geological, chemical, physical, and biological interactions that define aquatic systems. No limnologist exemplifies the use of such academic synthesis better than G. E. Hutchinson (Biography 1.1); he did more to define modern limnology than any other individual. Numerous other exciting scientific advances have been made by aquatic ecologists, including refinement of the concept of an ecosystem, ecological methods for approaching control of disease, methods to

Table 1.1 Locations, Amounts, and Turnover Times of Water Compartments in the Global Hydrologic Cycle

Location	Amount (Thousands km³)	Total %	% Inland Liquid Water	Approximate Residence Time
Freshwater lakes	125	0.009	1.45	75 years
Saline lakes and inland seas	104	0.008	1.20	
Rivers (average volume)	1	0.0001	0.01	5 months
Shallow and deep soil water	67	0.005	0.77	100 years
Groundwater to 4000 m depth	8,350	0.61	96.56	1,000 years
Icecaps and glaciers	29,200	2.14	—	12,000 years
Atmosphere	13	0.001	—	8.9 days
Oceans	1,320,000	97.3	—	3,060 years

(Data from Todd, 1970; Wetzel, 2001; Pidwirny, 2006; "The Hydrologic Cycle," Fundamentals of Physical Geography, *2nd Ed.)*

[1] The term "limnology" includes saline waters (Wetzel, 2001) and all other continental waters, but limnology courses traditionally do not cover wetlands, groundwater, and even streams. Thus, this book is titled Freshwater Ecology.

BIOGRAPHY 1.1 G. EVELYN HUTCHINSON

George Evelyn Hutchinson (Fig. 1.2) was one of the top limnologists and ecologists of the 1900s, and likely the most influential of the century. His career spanned an era when ecology moved from a discipline that was mainly the province of natural historians to a modern experimental science. In large part, he and his students were responsible for these developments. Born in 1903 in Cambridge, England, Hutchinson was interested in aquatic entomology as a youth and authored his first publication at age 15. He obtained an MA from Emmanuel College of Cambridge University and worked in Naples, Italy, and South Africa before securing a position at Yale University. He remained at Yale until the end of his career, and died in 1991.

Hutchinson's range of knowledge was immense, and his broad and innovative view of the world enriched his scientific endeavors. He was well versed in literature, art, and the social sciences, and published works on religious art, psychoanalysis, and history, including some of the most widely read and cited ecological works of the century. His four volumes of the *Treatise of Limnology* are the most extensive treatment of limnological work ever published. His writings on diversity, complexity, and biogeochemistry inspired numerous investigations. Hutchinson organized a research team to study the Italian Lake Ianula in the 1960s; this multidisciplinary approach has since become a predominant mode of ecological research. It is reported that he was always able to find positive aspects of his students' ideas and encouraged them to develop creative thoughts into important scientific insights. As a consequence, many of Hutchinson's students and their students are among the most renowned ecologists today.

Hutchinson earned many major scientific awards in his career, including the National Medal of Science. He wrote popular scientific articles and books that were widely distributed. He was a staunch defender of intellectual activities and their importance in the modern world. Hutchinson's mastery of facts, skillful synthesis, knack for asking interesting and important questions, evolutionary viewpoint, and cross-disciplinary approach make him an admirable role model for students of aquatic ecology.

FIGURE 1.2
G. Evelyn Hutchinson. *(Courtesy of the Yale Image Library).*

assess and remediate water pollution, ways to manage fisheries, restoration of freshwater habitats, understanding of the deadly lakes of Africa, and conservation of unique organisms.

Each of these advances will be covered in this book. We hope to transmit the excitement and appreciation of nature that come from studying aquatic ecology. Further justification for study may be necessary for those who insist on more concrete benefits from an academic discipline or are interested in preserving

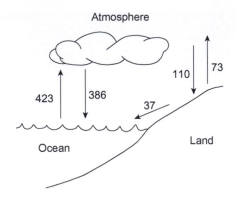

FIGURE 1.3

Fluxes (movements among different compartments) in the global hydrologic budget (in thousands of km³ per year). *(Data from Berner and Berner, 1987).*

water quality and aquatic ecosystems in the broader political context. There is a need to place a value on water resources and the ecosystems that maintain their integrity and to understand how the ecology of aquatic ecosystems affects this value. Water is unique and has no substitute. A possible first step toward placing a value on a resource is documenting human dependence on it and how much is available for human use.

Humankind would rapidly use all the water on the continents were it not replenished by atmospheric input of precipitation. Understanding hydrologic *fluxes*, or movements of water through the global *hydrologic cycle*, is central to understanding water availability, but much uncertainty surrounds some aspects of these fluxes. Given the difficulty that forecasters have predicting the weather over even a short time period, it is easy to understand why estimates of global change and local and global effects on water budgets are beset with major uncertainties (Mearns *et al.*, 1990; Mulholland and Sale, 1998). Scientists can account moderately well for evaporation of water into the atmosphere, precipitation, and runoff from land to oceans. This accounting is accomplished with networks of precipitation gauges, measurements of river discharge, and sophisticated methods for estimating groundwater flow and recharge.

The *global water budget* is the estimated amount of water movement (*fluxes*) between *compartments* (the amount of water that occurs in each area or form) throughout the globe (Fig. 1.3). This hydrologic cycle will be discussed in more detail in Chapter 4 but is presented here briefly to introduce discussion of how much water is available for human use. Total runoff from land to oceans via rivers has been reported as 22,000, 30,000, and 35,000 km³ per year by Leopold (1994), Todd (1970), and Berner and Berner (1987), respectively. These estimates vary because of uncertainty in gauging large rivers in remote regions. So, what are the demands on this potential upper limit of sustainable water supply?

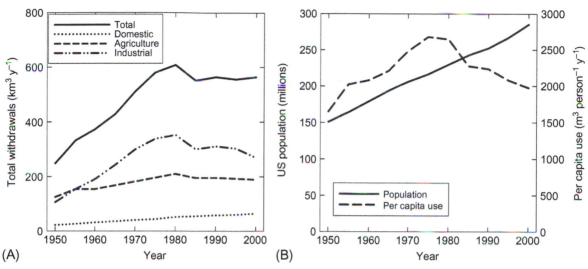

FIGURE 1.4

Estimated uses of water (A) and total population and per capita water use (B) in the United States from 1950 to 2000. Note that industrial and irrigation uses of water are dominant. Offstream withdrawals used in these estimates do not include hydroelectric uses. *(Data courtesy of the United States Geological Survey).*

HUMAN USE OF WATER: PRESSURES ON A KEY RESOURCE

People in developed countries generally are not aware of the quantity of water that is necessary to sustain their standard of living. In North America particularly, high-quality water often is used for such luxuries as filling swimming pools and watering lawns. Perhaps people notice that their water bills increase in the summer months. Publicized concern over conservation may translate, at best, into people turning off the tap while brushing their teeth or using low-flow showerheads or low-flush toilets. Few people in developed countries understand the massive demands for water from industry, agriculture, and power generation that their lifestyle requires (Fig. 1.4).

Some of these uses, such as domestic, require high-quality water. Other uses, such as hydroelectric power generation and industrial cooling, can be accomplished with lower quality water. Some uses are *consumptive* and preclude further use of the water; for instance, a significant portion of water used for agriculture is lost to evaporation. The most extreme example of nonrenewable water resource consumption may be water "mined" (withdrawal rates in excess of rates of renewal from the surface) from *aquifers* (large stores of groundwater) that have extremely long regeneration times. Such *withdrawal* is practiced globally (Postel, 1996) and also accounts for a significant portion of the United States' water use, particularly for agriculture (Fig. 1.5). In many

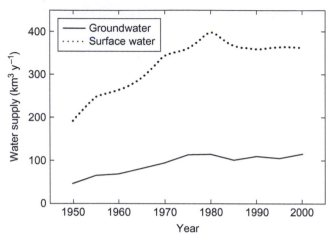

FIGURE 1.5

Amounts of surface and groundwater used in the United States from 1950 to 2000. Estimates include only withdrawals and not hydroelectric uses. *(Data courtesy of the United States Geological Survey).*

Table 1.2 General Ranges of Water Use with Varied Socioeconomic Conditions on a per-Capita Basis; Rates per Country Are Estimates for the Year 2000

Society	Range or Mean ($m^3 \, y^{-1} \, capita^{-1}$)
Irrigated semi-arid industrial countries	3,000–7,000
Irrigated semi-arid developing countries	800–4,000
Temperate industrial countries	170–1,200
China	431
Jordan	155
Switzerland	351
Turkmenistan	5,309
Uganda	9
United States	1,688

(After la Rivière, 1989; Falkenmark, 1992; Gleick et al., 2002)

cases use rates exceed the rate at which the aquifer is replenished, and the groundwater is over exploited. Other uses are less consumptive. For example, hydroelectric power "consumes" less water (i.e., evaporation from reservoirs increases water loss, but much of the water moves downstream).

How much water does humankind need? A wide disparity occurs between per capita water use in developed and less developed arid countries, particularly in semiarid countries in which surface water is scarce (Table 1.2). Israel is one of the most water-efficient developed countries, with per capita water use of $500 \, m^3$ per year (Falkenmark, 1992), about four times more efficient than the

United States. Increases in standard of living lead to greater water demands (per capita water use) if not accompanied by increases in efficiency of use.

The maximum total water available for human use is the amount that falls as precipitation on land each year minus the amount lost to evaporation. As mentioned earlier, the maximum amount of water available in rivers is 22,000 to 35,000 km^3 per year. However, much is lost to floods in populated areas or flows occurring in areas far removed from human population centers, leaving approximately 9,000 km^3 per year for all human uses (la Rivière, 1989). Humans cannot sustain use of water greater than this supply rate without additional supplies withdrawn from groundwater at rates greater than renewal, collected from melting ice caps, transported from remote areas, or reclaimed (desalinized) from oceans. These processes are expensive or impossible to sustain in many continental regions without tremendous energy input, as will be discussed later in this chapter.

Predicting future water use is difficult but instructive for exploring possible future patterns and consequences of this use. Total annual offstream withdrawals (uses that require removal of water from the river or aquifer, not including hydroelectric power generation) in the United States in 1980 were 2,766 m^3 per person and have decreased slightly since that time, mostly due to a decrease in total industrial use (Fig. 1.4). If all the people on Earth used water at the rate it is currently used in the United States (i.e., their standard of living and water use efficiency were the same as in the United States), over half of *all the water* produced by the hydrologic cycle would be used. Globally, humans currently withdraw about 54% of runoff that is geographically and temporally accessible (Postel *et al.*, 1996); if all people in the world used water at the per capita rates used in the United States, all the water available (geographically and temporally accessible) would be used.

On a local scale, water scarcity can be severe. Political instability in Africa is predicted based on local population growth rates and limited water supply (Falkenmark, 1992). Similar instabilities are likely to arise from conflicts over water use in many parts of the world (Postel, 1996). Large urban areas have tremendous water demands, and very strong effects on local and regional hydrology, generally decreasing water availability (Fitzhugh and Richter, 2004). For example, in some large arid cities major river flow is almost completely composed of sewage effluent, and some small streams that were ephemeral have become permanent because of runoff from lawn watering. In the arid southwestern United States, uses can account for more than 40% of the supply (Waggoner and Schefter, 1990). In such cases, degradation of water quality has substantial economic consequences.

The population of the Earth is currently over 6.7 billion people and is predicted to double over the next 43 years (Cohen, 1995). The global population will have increased more than 10% between the first and second editions of this book. Given the increase in human population and resource use (Brown, 1995), demand for water will only intensify (Postel, 1996). As the total population on Earth expands, the value of clean water will increase as demands escalate for a finite resource (Dodds, 2008). Population growth is likely to increase demand for water supplies, even in the face of uncertainty over climate in the future (Vörösmarty *et al.*, 2000). Increased efficiency has led to decreases in per capita water use in the United States since the early 1980s (Fig. 1.4). Efforts to boost conservation of water will become essential as water becomes more valuable (Brown, 2000).

Despite the existence of technology to make water use more efficient and maintain water quality, the ongoing negative human impact on aquatic environments is widespread. Most uses of water compromise water quality and the integrity of aquatic ecosystems, and future human impact on water quality and biodiversity is inevitable. An understanding of aquatic ecology will assist humankind in making decisions to minimize adverse impacts on our aquatic resources, and will ultimately be required for policies that lead to sustainable water use practices (Gleick, 1998). Accounting for the true value of water will help policy makers decide how important its use is, and students reading this book understand how water is a vital resource to learn about.

WHAT IS THE VALUE OF WATER QUALITY?

Accurate accounting for the economic value of water includes determining both the immediate benefit and how obtaining a particular benefit alters future use. Consumption and contamination associated with each type of use dictate what steps will be necessary to maintain aquatic ecosystems and water quality and quantity. Establishing the direct benefits of using water, including patterns and types of uses, is also necessary. Elucidation of benefits will allow determination of economic value of water and how its use should be managed.

Availability of water is clearly essential, but the quality of water also must be considered. Aquatic ecosystems provide us with numerous benefits in addition to direct use. Estimates of the global values of wetlands ($3.2 trillion per year) and rivers and lakes ($1.7 trillion per year) indicate the key importance of freshwaters to humans (Costanza *et al.*, 1997). These early estimates will increase as researchers are able to assign values to more *ecosystem goods and services* (Dodds *et al.*, 2008). Ecosystem goods and services are those provided by the ecosystem that are of value to humans. Freshwater aquatic ecosystems have the greatest value per unit area of all habitats. Estimates suggest that the greatest values of natural continental aquatic systems are derived from flood control, water supply, and waste treatment. The value per hectare is greater for wetlands, streams, and rivers than for any terrestrial habitats. The next section details how these values are actually assigned.

We explore values of aquatic ecosystems because monetary figures can influence their perceived value to society. Assigning values to ecosystems can provide evidence for people advocating minimization of anthropogenic impacts on the environment. Ignoring ecosystem values can be particularly problematic because perceived short-term gain often outweighs poorly quantified long-term harm when political and bureaucratic decisions are made about resource use. For example, restored wetlands may not be as valuable as conserved wetlands (Dodds *et al.*, 2008), so policy decisions should be made accordingly. The concept of *no net loss* when a constructed wetland can be substituted for a wetland that is removed for development may not capture all the values a natural wetland provides to society.

Quantification of some values of water is straightforward, including determining the cost of drinking water, the value of irrigated crops, some costs of pollution, and direct values of fisheries. Others may be more difficult to quantify. What is the value of a canoe ride on a clean lake at sunset or of fishing for catfish on a lazy river? What is the worth of the species that inhabit continental waters including nongame species?

What is the actual value of water? The local price of clean water will be higher in regions in which it is scarce. Highly subsidized irrigation water sells for about $0.01 per$m^3$ in Arizona, but clean drinking water costs $0.37 per$m^3$ in the same area (Rogers, 1986). Drinking water costs between $0.08 and $0.16 per$m^3$ in other areas of the United States (Postel, 1996). At the rate of $0.01 per$m^3$, and assuming that people on Earth use 2,664km^3 per year for irrigation (according to the Food and Agriculture Organization for 2001), the global value of water for irrigation can be estimated as $26 billion per year. This is probably an underestimate; in the 1970s in the United States, 28% of the $108 billion agricultural crop was irrigated (Peterson and Keller, 1990). Thus, $30 billion worth of agricultural production in one country alone could be attributed to water suitable for irrigation. Worldwide, 40% of the food comes from irrigated cropland (Postel, 1996). World grain production in 1995 was 1.7×10^{12} kg (Brown, 1996). In the 1990s, grain had a value of $0.50 per kilogram, and $340 billion per year came from irrigated cropland globally. This value has increased at least 15% by 2008.

Use of freshwater for irrigation does not come without a cost. Agricultural pesticide contamination of groundwater in the United States leads to total estimated costs of $1.8 billion annually for monitoring and cleanup (Pimentel *et al.*, 1992). Erosion related to agriculture causes losses of $5.1 billion per year directly related to water quality impairment in the United States (Pimentel *et al.*, 1995). This estimate includes costs for dredging sediments from navigation channels and recreation impacts, but it excludes biological impacts. These estimates illustrate some of the economic impetus to preserve clean water.

Table 1.3 Global Fisheries Production Relying on Freshwater in 2005		
Type	**Amount (Thousand Metric Tons)**	**Values (Million $ US)**
Carps, barbels, and other cyprinids	20,500	18,800
Tilapias and other cichlids	2,300	2,800
Miscellaneous	4,900	7,900
Sturgeon, paddlefish	21	71
River eels	266	1,000
Salmon, trout, smelt	2,143	9,891
Freshwater crustaceans	1,065	4,715
Freshwater mollusks	154	91
Total	31,349	45,268
(Food and Agriculture Administration, 2007)		

The economic value of freshwater fisheries, including aquaculture, worldwide is over $45 billion per year (Table 1.3). This economic estimate includes only the actual cash or trade value of the fish and crustaceans. In many countries, sport fishing generates considerable economic activity. For example, in the United States, $15.1 billion was spent on goods and services related to freshwater angling in 1991 (U.S. Department of the Interior and Bureau of the Census, 1993). In addition, 63% of nonconsumptive outdoor recreation visits in the United States included lake or streamside destinations, presumably to view wildlife and partake in activities associated with water (U.S. Department of the Interior and Bureau of the Census, 1993). Many of these visits result in economic benefits to the visited areas. Maintaining water quality is vital to healthy fisheries and healthy economies. Pesticide-related fish kills in the United States are estimated to cause $10 to 24 million per year in losses (Pimentel *et al.*, 1992). Finally, maintaining fish production may be essential to ensuring adequate nutrition in developing countries (Kent, 1987). Thus, the value of fisheries exceeds that of the fish. Managing fisheries clearly requires knowledge of aquatic ecology. These fisheries and other water uses face multiple threats from human activities.

Sediment, pesticide, and herbicide residues; fertilizer runoff; other nonpoint runoff; sewage with pathogens and nutrients; chemical spills; garbage dumping; thermal pollution; acid precipitation; mine drainage; urbanization; and habitat destruction are some of the threats to our water resources. Understanding the implications of each of these threats requires detailed understanding of the ecology of aquatic ecosystems. The effects of such human activities on ecosystems are linked across landscapes and encompass wetlands, streams, groundwater, and lakes (Covich, 1993). Management and policy decisions can be ineffective if the linkages between systems and across spatial and temporal scales are not considered (Sidebar 1.1). Effective action at the international,

SIDEBAR 1.1
Valuation of Ecosystem Services: Contrasts of Two Desired Outcomes

Ecosystem services refer to the properties of ecosystems that confer benefit to humans. Here, we contrast two types of watershed management and some economic considerations of each. The first case involves the effects of logging on water quality and salmon survival on the northern portion of the Pacific coast of North America, and the second involves water supply in some South African watersheds. The preferred management strategies are different, but both rely on understanding ecosystem processes related to vegetation and hydrologic properties of watersheds. When watersheds have more vegetation, particularly closer to streams, they have lower amounts of runoff and less sediment in the runoff. Removal of streamside vegetation is a major concern for those trying to conserve salmon.

Several species of salmon are considered endangered and the fish have direct effects on the biology of the streams in which they spawn (Willson *et al.*, 1998). Sport and commercial fisheries have considerable value on the northwest coast of North America. Dams that prevent the passage of adult fish and habitat degradation of streams are the two main threats to salmon survival in the Pacific coastal areas. Logging (Fig. 1.6), agriculture, and urbanization lead to degradation of spawning habitat. The main effects of logging include increased sedimentation and removal of habitat structure (logs in the streams). These factors both decrease survival of eggs and fry. Even moderate decreases in survival of young can have large impacts on potential salmon extinction (Kareiva *et al.*, 2000).

FIGURE 1.6
A clear cut across a small stream. *(Courtesy of the United States Forest Service).*

Economic analysis of efforts to preserve salmon populations includes calculation of the costs of modifying logging, agriculture, and dam construction and operation as opposed to the benefits of maintaining salmon runs. The economic benefits of salmon fisheries are estimated at $1 billion per year (Gillis, 1995). Costs of modifying logging, agriculture, and dam construction and operation probably exceed the direct economic value of the fishery.

The second case involves shrubland watersheds (fynbos) in South Africa that provide water to large agricultural areas downstream and considerable populations of people in urban centers and around their periphery (van Wilgen *et al.*, 1996). Introduced weed species have invaded many of these shrubland drainage basins (watersheds or catchments). The weeds grow more densely than the native vegetation and reduce runoff to streams. Also, about 20% of the native plants in the region are endemic and thus endangered by the weedy invaders.

Costs of weed management are balanced against benefits from increased water runoff. Costs associated with weed removal are offset by a 29% increase in water yield from the managed watersheds. Given that the costs of operating a water supply system in the watershed do not vary significantly with the amount of water yield, the projected costs of water are $0.12 per m^3 with weed management and $0.14 per m^3 without it. Other sources of water (recycled sewage and desalinated water) are between 1.8 and 6.7 times more expensive to use. An added benefit to watershed weed control is protection of native plant species. Thus, weed removal is economically viable.

The two cases illustrate how ecosystem management requires understanding of hydrology and biology. In the case of the salmon, vegetation removal (logging) is undesirable because it lowers water quality and reduces reproductive success. In the South African shrublands, removal of introduced weeds is desirable because it increases water yield. Knowledge of the ecology of the systems is essential to making good decisions.

---

federal, state, and local governmental levels, as well as in the private sector, is necessary to protect water and the organisms in it. Success generally requires a whole-system approach grounded with sound scientific information (Vogt *et al.*, 1997). Productive application of science requires explicit recognition of the role of temporal and spatial scale in the problems being considered and the role of the human observer (Allen and Hoekstra, 1992). Thus, we attempt to consider scale throughout the book. As discussed later, understanding of the mechanisms of problems such as nutrient pollution, flow alteration in rivers, sewage disposal, and trophic interactions has led to successful mitigation strategies. Many of our rivers are cleaner than they were several decades ago. Future efforts at protection are more likely to be successful if guided by informed aquatic ecologists interested in safeguarding our water resources.

ADVANCED: METHODS FOR ASSIGNING VALUES TO ECOSYSTEM GOODS AND SERVICES

An entire field of defining ecosystem goods and services has arisen as a branch of ecological economics. In this section we discuss how methods are actually assigned, and how this relates to freshwaters. A simple calculation can illustrate how valuable water is. About $9,000 \, km^3$ of water are available each year to humans through the global hydrologic cycle when and where we need it. It takes $2,443 \, J$ of energy to evaporate each gram of water. This means that $2.19 \times 10^{22} \, J$ of energy are required each year to move water from the ocean into the atmosphere. Humans currently use about $5 \times 10^{20} \, J$ of energy each year, mostly from fossil fuels. Thus it would take about 40 times our global energy use to evaporate all the water available, or 16 times to evaporate all the water we actually use. The average elevation on Earth is roughly $800 \, m$, and lifting the $9,000 \, km^3$ of water to that elevation would require about 10% of human energy use (this does not account for frictional loss in pipes or inefficiencies of pumps) that would be required if humans needed to pump the water from sea level to the areas they inhabit. Classical economics generally assumes that the water is simply available and assigns no value to it, even though to replicate the ecosystem service of providing fresh water from the ocean would require a substantial amount of the world's economy to provide if the global hydrologic cycle did not.

Several schemes have been constructed to value ecosystem goods and services. Probably the most widely agreed upon scheme was developed in conjunction with the Millennium Ecosystem Assessment (MEA, 2003). This scheme recognizes utilitarian (provisioning, regulating, and supporting aspects) and nonutilitarian values (ethical, religious, cultural, and philosophical aspects). Utilitarian values can be broken down into direct use values, indirect use values, and option values. Direct use values include consumptive (harvesting fish or aquatic plants) and nonconsumptive (water sports, recreation) use values. Indirect use values include water purification, water supply, and other ecosystem processes that benefit humanity. Option values are values that are not used currently, but may be used in the future (e.g., maintaining a fishery so your offspring can use it). Different values are determined with appropriate methodology. With respect to direct uses, consumptive values can be directly estimated by the market value of the item of interest.

Nonconsumptive values can be determined based on the amount of money people spend to partake in particular activities and surveys to gauge how much they would be willing to pay to maintain that activity (Wilson and Carpenter, 1999). For example, one can calculate how much property value would increase if a lake was cleaner, or how many more recreation days would be spent on a clean lake compared to a polluted lake coupled with the average amount of money spent per recreational visit. In an example of determining a relationship between an economic benefit and an ecosystem attribute, Michael et al. (1996) demonstrated that a 1-m increase in lake clarity translated into increased property values of $32 to $656 per meter of frontage. Thus, it can be established how much people are willing to pay for some aesthetic values.

Indirect uses can also be quantified. For example, if water pollution is of concern, the amount of money required to treat the water to make it usable would be quantified. In another example, flood regulation is an important property of some wetlands because they allow the water to spread out and cause less damage downstream. The cost of protecting downstream areas from floods if the wetlands were removed can be used to assign values for wetland flood protection.

The idea of quantifying ecosystem goods and services is controversial in the field of economics (e.g., Bockstael *et al.*, 2000). Critics argue that discounting (the lost cost of investment opportunity) is ignored. The argument is fundamentally philosophical, with proponents for conserving economic values of ecosystem goods and services arguing that benefits are accrued to society rather than individuals, so benefits are worth as much if they are used immediately or withheld as option values.

The second critique of valuation of ecosystem goods and services is that they do not have exact value (e.g., using willingness to pay survey methods). Economists who argue this are more comfortable assuming that values of ecosystem goods and services are externalities, or economic values that are not readily assigned a value. This critique is based on the argument that a poorly constrained estimate of value is worse than no estimate at all.

Nonutilitarian values are more difficult to quantify. Still, people are willing to pay to protect aspects of the environment that are important to them for ethical, religious, cultural, or philosophical reasons. For example, many people are willing to allot tax dollars to help protect parks and natural areas even if they are unlikely ever to visit those areas.

The fact remains that many aspects of freshwater environments are highly valued by humanity. Rivers and lakes are some of the most desirable places to live and recreate, and we have multiple dependencies on them as a source of water. As ecological economics matures as a field, more concrete values will continue to be assigned to freshwater ecosystem goods and services.

CLIMATE CHANGE AND WATER RESOURCES

There is now consensus among scientists that human-induced *climate change* is warming the planet (Intergovernmental Panel on Climate Change (IPCC), 2007), although some of the details remain uncertain. Knowledge of the basics of freshwater ecology will be essential if future scientists, such as the students reading this book, hope to be able to deal with the consequences of global climate change. The earth's warming climate is expected to negatively affect the quality and quantity of freshwater resources. Along with direct effects of warming temperatures, climate models forecast changes in regional precipitation patterns and overall higher variability in precipitation that will lead to increased frequency, magnitude, and unpredictability of flooding and droughts in many regions. Reductions in snow pack in mountain regions, which are already occurring at an alarming rate, will cause reduced stream flows. Warming will also result in earlier melting of snow pack, causing changes in the seasonal hydrology of receiving streams (Barnett *et al.*, 2008). Climate change will decrease freshwater availability in many areas because of increased losses to evaporation and human use, and warming water temperatures are expected to have both additive and synergistic effects with other stressors such as nutrient pollution and the spread of exotic species. Smaller bodies of water will likely be affected first by climate change because they have less thermal and hydrologic buffering capacity and are more highly influenced by local precipitation patterns (Heino *et al.*, 2009).

Warming temperatures will exacerbate current water pollution problems because increased water loss will reduce the volumes of water in lakes, streams, and wetlands, effectively concentrating pollutants and reducing the flushing of these materials (Whitehead *et al.*, 2009). Along with the obvious ecological consequences, water treatment costs will increase accordingly with increasing levels of pollutants. Freshwater habitats in many regions are already stressed from increasing nutrient inputs from human activities; this coupled with reduced water volumes and warming water temperatures will fuel growth of undesirable algae (Wrona *et al.*, 2006). Algal growths will lead to changes in community structure and reduced oxygen availability.

Predicting exact responses of different freshwater habitats to the combined effects of warming and other human impacts is difficult. For example, increased nutrient levels and water temperatures might initially increase biodiversity in a cold, oligotrophic lake, whereas similar changes to a eutrophic lake would likely result in a decrease in biological diversity because of reduced oxygen storage capacity (Heino *et al.*, 2009). Warming may also directly influence the availability of nutrients and other materials. For example, in arctic regions, predicted melting of upper layers of permafrost will liberate phosphorus, which can lead to cascading effects on the productivity of regional streams and lakes (Hobbie *et al.*, 1999).

Climate change will alter seasonal patterns in lakes and extend the ice-free season and period of summer stratification of lakes in higher latitudes. As temperatures warm, lake mixing patterns will change, with extended periods of summer stratification. Extended stratification will increase the probability of anoxia in cooler, deep water habitats that cold-water species need to survive.

Losses in biodiversity, and thus ecosystem function and stability, as a result of climate change and other human impacts are expected to be among the greatest in freshwater habitats (Xenopoulos *et al.*, 2005). Growth rate, adult size, and ultimately fecundity of most aquatic organisms are all influenced by temperature. Even subtle changes in average or maximum water temperatures can have measurable effects on the biota; in some cases these effects are sublethal, but can still be serious. For aquatic insects such as many mayflies, warming water temperatures just 2 to 3°C above optimal can greatly reduce the number of eggs produced by females (Vannote and Sweeney, 1980; Firth and Fisher, 1992), which has important implications for future mayfly populations, predatory fish production, and overall ecosystem health. Warming can also desynchronize life cycles and seasonal phenologies of consumers and resources, a pattern that has already been documented with some flowers and pollinators (Memmott *et al.*, 2007).

Global climate change has led us into a no-analog world, where predicting hydrology based on past patterns is difficult or impossible (Milley *et al.*, 2008). Thus, a mechanistic understanding of hydrologic process is necessary to predict the future availability of water in various regions as well as patterns of drought and flooding that will influence human society.

POLITICS, SCIENCE, AND WATER

Given the undeniable importance of water to human affairs, the fact that water issues are political is scarcely surprising. Globally, water could well become the first limiting resource for humanity (Dodds, 2008). More than 2 billion people on Earth live in areas where water is scarce (Oki and Kanae, 2006) and more than a billion lack adequate safe drinking water (Pimentel *et al.*, 2004). Locally, many communities are limited by water. Most major cities on Earth are located near water because people need an adequate supply, and rivers and lakes served as primary transportation avenues before road, rail, and air travel were available. Wetlands also have historic importance; the ancient Chinese built one of the world's great civilizations fueled for centuries with rice harvested from wetlands.

Now, particularly in arid regions, politics and water are deeply intertwined. In the arid western United States, water law is convoluted and water rights transcend most other property rights. Communities appropriate water from far away, and subsidize its use to encourage development. However, water is not just limiting in the United States; conflicts over water have occurred throughout recorded history (Gleick, 2008). Cut-off of water was used as a military tool as early as 2,500 BC, and has been used repeatedly to this day. Violent clashes have occurred over water repeatedly, including between Ethiopia and Kenya in 2006, between villagers in Kashmir India in 2002, between Chinese farmers in 1999 (with riot damages reaching $1 million), and in 1993 the Iraqi government started draining the Mesopotamian marshes to punish people living in the area. Political conflict over water across borders occurs in many places including India–Pakistan, Israel–Jordan, and Mexico–United States. The degree of conflict will only increase with more people vying for a finite amount of water and water pollution and exploitation decreasing the availability of clean water.

As we harm the quality of water, even less becomes available for use and it also becomes less suitable to support native biota. When water is a more limited commodity its market value swells, even more so as the economic values associated with ecosystem goods and services provided by water are increasingly recognized. The United States' Clean Water Act, the European Union's Water Framework Directive, India's system of Water Quality Standards, and the Law of the People's Republic of China on Prevention and Control of Water Pollution are just a few examples of how governments are attempting to protect water quality. Such protection requires sound scientific understanding of how aquatic ecosystems function and humans influence that function.

Managing aquatic resources successfully requires scientific understanding of freshwaters. Through scientific study of freshwaters, students may come to understand the beauty and complexity of a crucial part of the natural world. Part of studying freshwaters is learning the intrinsic value, in addition to the utilitarian value, of water. We hope that this text will serve as a resource for future stewards of our freshwaters.

SUMMARY

1. Clean water is essential to human survival, and we rely most heavily on continental water, including streams, lakes, wetlands, and groundwater.
2. The global renewable supply of water is about $39,000 \, km^3$ per year, and humans use about 54% of the runoff that is reasonably accessible. Thus, clean water is one resource that will be limited severely with future growth of the human population and increases in the standard of living. Local problems with water quality and supply may lead to political instability.
3. Economic analysis of the value of clean water is difficult, but factors to consider include the value of clean water for human use, the value of fisheries, and recreational use of aquatic habitats. The global benefits of these uses translate into hundreds of billions of dollars each year. Intangible benefits include preservation of nongame species and native ecosystems.
4. The study of the ecology of inland waters will lead to more sound decisions regarding aquatic habitats as well as provide a solid basis for future research.

QUESTIONS FOR THOUGHT

1. Why are you interested in studying aquatic ecology, and is such study important?
2. What is the difference between fluxes and compartments in water cycles, and what types of units are typically used to describe them?
3. What are some potential economic benefits to maintaining water quality?
4. What are the potential dangers of approaching conservation of aquatic resources from a purely economic viewpoint?
5. To whom does fresh water really belong?
6. List three trade-offs that are potentially involved in protecting native species in regulated rivers by attempting to mimic natural water discharge patterns.

Properties of Water

FIGURE 2.1

Bubbles formed in ice. This image reflects several properties of water. Surface tension forces the bubbles to be spherical, and gas that can dissolve in liquid water cannot do so in solid water. *(Courtesy of Steven Lundberg).*

Doi: 10.1016/B978-0-12-374724-2.00002-7

Unique physical properties dictate how water acts as a solvent and how its density responds to temperature. These physical properties have strong biological implications, and knowledge of water's characteristics forms the foundation for aquatic science. Physical properties of water are so central to science that they are used to define several units of measurement, including mass, heat, viscosity, temperature, and conductivity. Properties of water influence how it controls geomorphology, conveys human waste, links terrestrial and aquatic habitats, and influences evolution of organisms. In this chapter we explore how viscosity and inertia of water vary with scale, temperature, and relative velocity related to aquatic ecology. Movement of water is discussed in the final section, including how flowing water interacts with solid surfaces.

CHEMICAL AND PHYSICAL PROPERTIES

One of the many unusual properties of water is that it exists in liquid form at the normal atmospheric temperatures and pressures encountered on the surface of Earth (Table 2.1). Most common compounds or elements take the form of gas or solid in our biosphere (exceptions include mercury and organic compounds). *Polarity* of the water molecule and *hydrogen bonding* dictates the range of temperatures and pressures at which water occurs in a liquid state as well as additional vital distinguishing characteristics. Because oxygen atoms attract electrons, the probability is greater that electrons will be nearer to the oxygen than the hydrogen atoms. The angle of attachment (104.5°) between the two covalent bonds, one for each of the hydrogen atoms attached to the oxygen, means a slight positive charge near the hydrogen atoms, to one side of the molecule. This unequal distribution of charge leads to each water molecule exhibiting polarity. The negative region near the oxygen attracts positive regions near the hydrogen atoms of nearby water molecules resulting in hydrogen bonding (Fig. 2.2).

Table 2.1 Some Properties of Water

Property	Comparison with Other Substances
Density	Maximum near 4°C, not at freezing point, expands upon freezing
Melting and boiling points	Very high
Heat capacity	Only liquid ammonia is higher
Heat of vaporization	Among highest
Surface tension	High
Absorption of radiation	Minimum in visible regions; higher in red, infrared, and ultraviolet
Solvent properties	Excellent solvent for ions and polar molecules
	Increases for ions with increasing temperature, decreases for gases with increasing temperature

(Adapted from Berner and Berner, 1987)

Hydrogen bonding becomes more prevalent as water freezes but also occurs in the liquid phase (Luzar and Chandler, 1996; Liu *et al.*, 1996); without hydrogen bonding, water would be a gas at room temperature. When water freezes, the molecules form tetrahedral aggregates that lead to decreased *density*, mass per unit volume. Thus, pure ice has a density of $0.917\,g\,cm^{-3}$ at $0°C$, which is significantly less dense than liquid water at any temperature (Fig. 2.3A). Lower density as a solid than as a liquid is another unusual aspect of water; most compounds are denser in solid than in liquid phase. There are several forms of crystal ice with varied density, but the ecological relevance of these different forms is minor.

The density of liquid water, which is influenced by temperature and dissolved ions, can control the physical behavior of water in wetlands, groundwater, lakes, reservoirs, rivers, and oceans. Density also influences water flow and viscosity. Differences in density are important because less dense water floats on top of water with greater density. Such density differences can maintain stable layers. Formation of distinct stable layers is called *stratification*. Stratification is discussed in detail in Chapter 7 because it can control water movement and distribution of chemicals and organisms in lakes.

Maximum density of pure water occurs at $3.98°C$ (Fig. 2.3B, C). As water cools below $3.98°C$, hydrogen bonding begins to arrange the water molecules into the crystal structure of ice, leading to more space between the water molecules and a less dense fluid. As water is heated above $3.98°C$, it has a continuously greater decrease in density per degree temperature increase above $3.98°C$ (Fig. 2.3). Dissolved ions also increase water density. This density increase

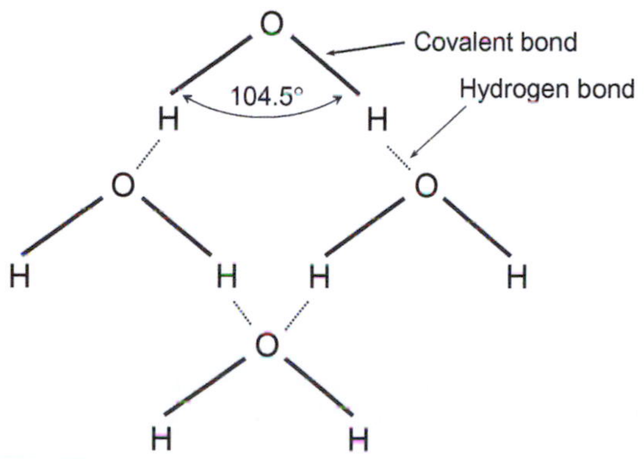

FIGURE 2.2
Schematic of hydrogen bonding among water molecules. The black lines represent covalent bonds; the dashed lines represent hydrogen bonds. This is an approximate two-dimensional representation. In water, three-dimensional cage-like structures are formed. In liquid water, these structures form and break up very rapidly.

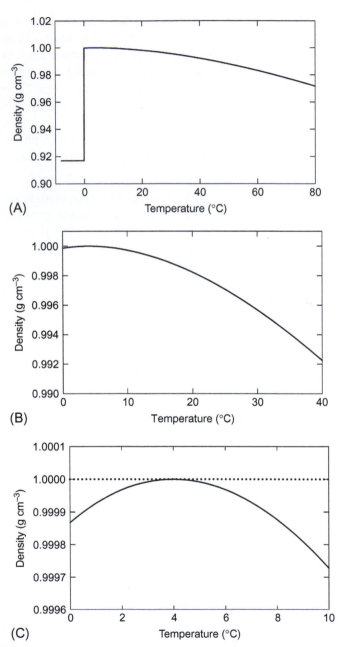

FIGURE 2.3

The density of water as a function of temperature from freezing to 80°C (A), from 0 to 40°C (B), and from 0 to 10°C to focus on the region of maximum density (C). At 0°C, ice forms with a density of 0.917 g per milliliter. *(Data from Cole, 1994).*

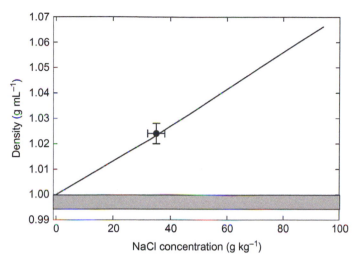

FIGURE 2.4

Comparison of density change caused by temperature to that changed by salinity. The range of variation in density with temperature is represented by the gray box. Seawater has an approximate salinity of 3.5% (indicated by the point bounded by error bars); saline lakes can exceed this value by many times. *(Data from Dean, 1985).*

that occurs at ionic concentrations found in some natural saline lakes can easily overcome or enhance temperature effects on stratification (Fig. 2.4).

Water is also one of the best *solvents* known and can dissolve both gases and ions. The solvent properties of water have greatly influenced geologic *weathering* of Earth's surface by dissolving ions from rocks. Weathering is the natural source to the biosphere of all nutrients that do not have gaseous form under normal conditions. Weathering also alters geomorphology. For example, about 20% of the continental land is *karstic terrain* (White *et al.*, 1995), a geological formation caused by rainwater dissolving limestone and leaving very rough land topography.

Most solids dissolve in water more readily as temperature increases (i.e., the rate at which a compound dissolves into water). For instance, this temperature effect on dissolved ions causes sugar to dissolve more quickly in hot than in iced tea. Most compounds, but not all, have higher solubility (saturation concentrations) at greater temperatures as well. Conversely, *solubility* of gases in water tends to decrease when temperature increases (see Fig. 12.7). Decreased solubility at high temperature is why effervescence is enhanced when a warm beer is opened compared to a cold one. Temperature's influence on gas solubility can have significant biological consequences; aquatic animals such as fishes are more likely to die of low dissolved oxygen (O_2)

Table 2.2 Heats of Fusion, Vaporizations, Heat Capacities, and Surface Tensions of Various Liquids

Substance	Heat of Fusion (cal g^{-1})	Heat Capacity at 25°C (cal g^{-1} °C^{-1})	Surface Tension at 20°C (dyne cm^{-1})	Viscosity (cp)	Heat of Vaporization (cal g^{-1})
Water	79.7	1.00	73	1.00	539.6
Benzene	30.3	0.41	40	0.65	94.3
Mercury	2.8	0.03	435	1.55	67.8
Oxygen	3.3	—	—	—	50.9

(Keenan and Wood, 1971; Weast, 1978)

stress when water temperatures are elevated because less dissolved oxygen is held in warm water and the animal's metabolic requirements for O_2 increase as temperature increases.

Additional properties of water include high heat capacity, heat of fusion (freezing), heat of vaporization, and surface tension. Water has a high *heat capacity*, that is, it takes a relatively large amount of energy to increase the temperature of liquid water. To illustrate, specific heat capacities (in calories required to change the temperature of 1 g of a substance by 1°C) are 1, 0.58, and 0.21 for water, ethanol, and aluminum, respectively. Similarly, *heat of fusion* and *vaporization* are high for water compared to other liquids (Table 2.2). A high heat capacity and heat of fusion means that a considerable amount of solar energy is required to heat a lake in the summer, and a long period of cold weather is required to freeze the surface of a lake. The slow heating and cooling of water is evident in the 5 to 7 day lag time that is often evident between air temperatures and stream water temperatures in many regions. High heat capacity buffers water against rapid changes in temperature. Thus, aquatic organisms generally do not experience the rapid temperature swings terrestrial organisms are subjected to (and generally are not exposed to temperatures <0°C or >100°C). Because they evolved in thermally buffered environments, many aquatic organisms are very sensitive to even small changes in temperature. For example, many aquarium fishes are not adapted to temperature swings and must be acclimated to new tanks slowly, so the rapid temperature change does not shock and kill them. Temperature sensitivity could also make many aquatic organisms highly susceptible to predicted climate changes (Firth and Fisher, 1992); global climate change is not only predicted to increase global mean temperatures, but also to increase climatic (temperature) variability as well.

A high heat of vaporization means that a considerable amount of energy is needed to evaporate water. Humans and other terrestrial animals take advantage of the heat of vaporization by perspiring; the evaporation of the moisture cools the skin (takes away energy). Surface waters are also cooled by evaporation. Because of its high heat capacity and heat of vaporization, water, in any form, is an important determinant of local, regional, and global climate. Global climate models must take heat of vaporization and heat capacity into account.

Another important aspect of water is *surface tension*. The high surface tension of water results from hydrogen bonding, which pulls water into a tight surface at a gas–water interface. Several organisms, such as water striders, take advantage of this surface tension to walk on the surface of water. Some lizards (*Basiliscus* and *Hydrosaurus*) also run across the surface of water using the support of surface tension (Vogel, 1994). Surface tension is also important to animals that live below the water surface; the respiratory siphons that many aquatic insect larvae use to breathe atmospheric air with and the air bubbles that many aquatic beetles and true bugs carry with them would not function properly without surface tension. These structures function because of the tendency of water molecules to stick together (*cohesion*) and to other surfaces (*adhesion*).

The influence of surface tension also comes into play when water droplets form spheres. Surface tension determines the size of the largest possible drop of water (Dodds, 2008). Most raindrops are 1 to 2 mm in diameter, but a few as big as 10 mm have been observed. Under microgravity (on the space shuttle) drops as large as 3 cm are stable. Finally, surface tension leads to *capillary action*, the ability of water to move up narrow tubes. Capillary action is central to forming the capillary fringe (the moist zone in sediments immediately above groundwater) because water creeps up the narrow spaces between sediment particles. Wetland plants with leaves above the water surface also use capillary action to move moisture up their stems to their leaves.

An interesting adaptation related to capillary action occurs in small surface-walking insects when they get out of the water. In areas where water meets solid objects, the surface tension and capillary action of the water causes a *meniscus* (an up-sloping surface where water meets a hydrophilic solid). This meniscus can be observed by looking closely at a clear glass of water at the edge where the surface of the water meets the glass. The meniscus is a very difficult barrier to overcome for very small organisms. However, they can position their bodies so that capillary action pulls them up the meniscus without their providing propulsive force from their legs (Hu and Bush, 2005). These animals essentially pull up on the water sloping up in front of them with their forelegs or front part of their body, and push down on the surface of the water in their middle, and capillary action moves them forward up the meniscus.

Finally, pressure increases with depth and solubility of gases is a function of pressure. Scuba divers must pay attention to high gas solubility at higher pressure because rapid ascent from depth allows gas bubbles to form from gas dissolved in the blood. This is known as the bends and can be fatal. Fish and other animals are also susceptible to the bends, and this is part of the reason why most fish caught in deep water will die even if released carefully. The fact that gases can dissolve at much higher concentrations under pressure explains how carbon dioxide can become supersaturated at depth in lakes (important in the Lake Nyos disaster explained in Chapter 13) and why fish that are subject to water released from deep waters in reservoirs can be harmed by the supersaturated gases (relative to atmospheric pressure). Pressure increases linearly with depth because the pressure is caused by the mass of water above the object.

$$P_x = P_{atm} \; \rho \; x \; g$$

where P_x is pressure at depth, P_{atm} is atmospheric pressure, x is the depth, ρ is density, and g is the force of gravity. In practical terms, a body is at approximately twice atmospheric pressure at a depth of approximately 10 meters below the surface; pressure will double again (4 times atmosphere) at 30 meters.

ADVANCED: THE NATURE OF WATER

The physical chemistry of water has advanced considerably because of the instrumentation that is available to study it. Properties of water are based in large part on how the individual molecules interact with their nearest neighbors. The nature of liquid water is determined by the hydrogen bond and van der Waals forces (Schmid, 2001). Apparently several types of hydrogen bonding are present simultaneously in liquid water, and there is rapid fluctuation between the types. Hydrogen bonds fluctuate on temporal scales between 10^{-14} and 10^{-11} seconds (Tokmakoff, 2007). The bonds are seemingly traded among water molecules (like handing off a baton in a relay race) rather than broken and reformed at a later time. The true nature of a hydrogen bond is still not yet completely understood (Tokmakoff, 2007). X-ray data currently suggest that in liquid water, most molecules are found in chains or rings formed by relatively strong hydrogen bonds within more diffuse networks of weaker hydrogen bonds (Wernet *et al.*, 2004). The regions flip back and forth from strong to weaker bonding, and regions of tight hydrogen bonding are maintained even at water temperatures close to boiling (Huang *et al.*, 2009).

The *hydrophobic* effect, the inability of some molecules to dissolve in water, is a function of the small size of the molecule and how the solutes interact with the hydrogen bonding. The exact mechanism behind this effect is still not well understood (Granick and Bae, 2008). The concept of hydrophobicity is extremely important in materials science but has obvious biological importance as well. For example, protein folding and dissolving depend upon this effect. The capillary effect also depends upon hydrophobicity. A more complex definition of polarity is also suggested by study of what will and will not dissolve in water (Schmid, 2001). The definition used by advanced physical chemists is more complex than what is taught in introductory chemistry classes. These properties may seem esoteric to aquatic ecology students, but they do determine how processes like weathering and transport of nonpolar pollutants occur.

RELATIONSHIPS AMONG WATER VISCOSITY, INERTIA, AND PHYSICAL PARAMETERS

Viscosity is the resistance to change in form, or a sort of internal friction. *Inertia* is the resistance of a body to a change in its state of motion. Water viscosity increases with smaller spatial scale, greater water movement, and lower temperature. Inertia increases with size, density, and velocity. These facts are underappreciated but are very biologically and physically relevant to aquatic ecology. Consequences of these physical properties include, but are not limited to, (1) why fish are streamlined, but microscopic swimming organisms are not; (2) why the size of suspended particles captured by filter feeding has a lower limit; and (3) why organisms in flowing water can find refuge near solid surfaces. Many aspects of these features of life in aquatic environments can be discussed conveniently using the *Reynolds number* (*Re*). This number can quantify spatial- and velocity-related effects on viscosity and inertia. The effects of viscosity and inertia and other properties of water on organisms have been described eloquently and in greater detail (Purcell, 1977; Denny, 1993; Vogel, 1994). Here, we attempt to describe water's physical effects briefly, using the Reynolds number as the basis of the discussion.

Relative viscosity increases and inertia decreases as the spatial scale becomes smaller. Viscosity increases because the attractive forces between individual water molecules become more important relative to the organism. Thus, the influence of individual water molecules is greatest when organisms are small or the space through which water is moving is small. We discuss the individual components of *Re*, *inertial force* and *viscous force*, and then provide an example calculation using these relationships. Mathematically, the ratio of inertia to viscosity is the Reynolds number:

$$Re = \frac{F_i}{F_v}$$

where F_i is the inertial force and F_v is the viscous force.

The equation for inertial force (F_i) is

$$F_i = \rho S U^2$$

where ρ is the density of the fluid, S is the surface area of the object, and U is the velocity of the fluid moving past the object (or the object moving through the fluid).

Inertial relationships can be stated in familiar terms: The faster the object, the greater the inertial force. A slowly pitched baseball does less damage to a batter than a $44\,\mathrm{m\,s^{-1}}$ (100 mph) fastball. Of course, larger objects have more inertia; a splash of water from a cup imparts less force than the splash from a bucket of water propelled at the same velocity, and a moderate sized wave can easily knock a person down.

The properties of inertia constrain aquatic organisms. For example, at small scales inertia is generally unimportant. A bacterium will coast 1/10th the diameter of a hydrogen atom if its flagellum stops turning (Vogel, 1994). In contrast, a large fish can coast many body lengths if it stops swimming. Low inertia at small scales also means that turbulence is less likely (i.e., individual parcels of water have less inertia).

The other part of the equation to calculate the Reynolds number is viscous force (F_v):

$$F_v = \frac{\mu S U}{l}$$

where μ is the *dynamic viscosity* of the fluid, a constant that describes the intrinsic viscosity of a fluid (e.g., corn syrup is intrinsically more viscous than water), and l is the length of the object.

Again, certain aspects of this relationship are intuitive. The viscous force can be thought of as a frictional force. Increasing velocity increases viscosity. Water feels more viscous to a person wading up a stream than one wading in a still pool. Dynamic viscosity related to properties of a fluid (i.e., μ) also may be important. Swimming completely immersed in tar would be much more difficult than swimming in water (although maybe not as difficult as it would seem as determined by comparing people swimming in water and syrup in television's "Myth Busters"). Surface area also influences viscous force. Pulling a large object (with a large frontal surface area) through water is more difficult (takes more force) than pulling one with a small frontal surface area. An object with a large frontal surface area and short length is more difficult to pull through water than one with a small frontal surface area and long length; pulling a boat sideways through the water is more difficult than pulling the same boat bow-first.

Viscous force is dependent upon spatial scale. One way to envision this force is to think of the hydrogen bonds in water as a spider web. A very small object such as a fly is heavily influenced by the web, and a larger object such as a human can barely perceive any resistance when walking through the web. Similarly, small particles experience substantial viscosity when moving through water and smaller particles take longer to settle out of water because they experience greater viscous force than do larger particles.

Environmentally related variation in dynamic viscosity can have major effects on aquatic organisms because dynamic viscosity (μ) is greater when temperature is colder (Fig. 2.5). Thus, it requires more energy for a fish to swim in cold than warm water, and it is more difficult for animals to filter out small particles at lower temperatures (Podolsky, 1994). Again, thinking about the spider web analogy helps here, if the water is colder, the hydrogen bonding is more prevalent. More hydrogen bonding would be equivalent to a stronger, more densely woven spider web that would be more difficult for a small object to pass through.

Effects of dynamic viscosity and viscous force are diverse and include constraints on (1) how aquatic organisms collect food, (2) how fast organisms swim (a bacterium with a cell length of 1 μm experiences viscous forces in water similar to a human swimming in tar), (3) when natural selection favors streamlined organisms, (4) how quickly particles settle in water, and (5) how fast groundwater flows. For example, when groundwater is moving through two sediment types that have the same surface area of channels through which fluid can flow but one has more 1μm diameter pores and the other has fewer

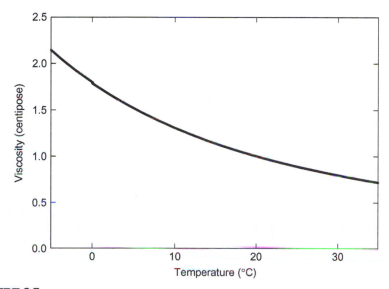

FIGURE 2.5

Viscosity as a function of temperature. Note that viscosity doubles when temperature drops from 30 to 0°C (i.e., a range of temperatures across seasons in temperate surface water). *(After Weast, 1978).*

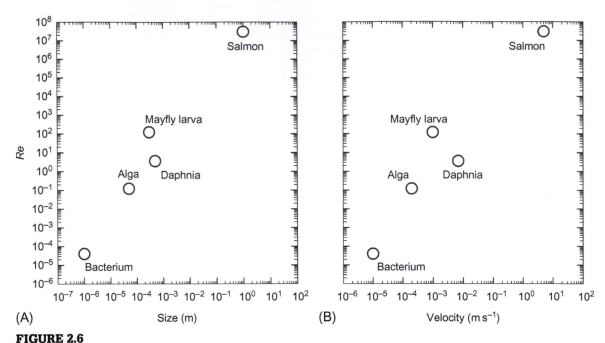

FIGURE 2.6

Reynolds number as a function of size (A) and velocity (B) for a variety of aquatic organisms. Note the log scales. *(Data from Vogel, 1994).*

pores of 5 μm diameter, the water velocity is slower through the sediment with the 1 μm pore diameter holes. The velocity is lower because the water is subject to greater friction while flowing through the smaller pores. If we put the equations for inertia and viscosity together and cancel, we get the equation for the Reynolds number:

$$Re = \frac{F_i}{F_v} = \frac{\rho\, S\, U^2}{\dfrac{\mu\, S\, U}{l}} = \frac{\rho\, U\, l}{\mu}$$

The Reynolds number is greater at large spatial scales (a large fish) than at small scales (a bacterium or protozoan) (Fig. 2.6). Calculations of this number (Example 2.1) reveal the wide variations in viscosity and inertia experienced between large and small organisms. A summary of the effects of scale related to *Re* is provided in Table 2.3. Reynolds numbers will be considered again when we discuss filter feeding by organisms in lakes and streams, microbial food webs, production of aquatic macrophytes, and flow of water in streams and groundwaters.

■ Example 2.1

Reynolds Number Calculations

Calculate viscous force (F_v), inertia (F_i), and Reynolds number (Re) for two cubes, one 1 µm and the second 1 cm on a side, each moving at 2 lengths per second, given that $\rho = 1 \times 10^6 \, \text{gm}^{-3}$ and $\mu = 1 \, \text{gm}^{-1} \text{s}^{-1}$. If turbulent flow is more likely to occur above $Re = 100$, which cube would have its hydrodynamic properties altered more by streamlining?

Parameter	Comment	1 cm Cube	1 µm Cube
l	length of cube	10^{-2} m	10^{-6} m
S	square of length	10^{-4} m^2	10^{-12} m^2
U	2 lengths per second	2×10^{-2} ms^{-1}	2×10^{-6} ms^{-1}
F_i	from equation	4×10^{-2} gms^{-2}	4×10^{-18} gms^{-2}
F_v	from equation	2×10^{-4} gms^{-2}	2×10^{-12} gms^{-2}
Re	F_i/F_v	200	2×10^{-6}

Note that inertia is very, very small for the small cube, leading to a much smaller Reynolds number. The 1 cm cube is above $Re = 100$, and would have its hydrodynamic properties altered more than the 1 µm cube if both were streamlined. ■

Table 2.3 Contrasting Effects of Scale and Reynolds Number on Aquatic Organisms

Parameter	Small Organism (<100 µm)	Large Organism (>1 cm)
Re	Low	High
Viscous force (F_v)	High	Low
Inertial force (F_i)	Low	High
Flow	Laminar or none	Turbulent
Body shape	Variable	Streamlined
Diffusion	Molecular	Transport (eddy)
Particle sinking rates	Low	High
Relative energy requirement for motility	High	Low

MOVEMENT OF WATER

At the very smallest scale, molecules move independently in the process called *Brownian motion*. The warmer the water, the more rapidly the molecules move. The average instantaneous velocity of individual water molecules is extremely rapid ($>100\,\mathrm{m\,s^{-1}}$), but because they continuously collide, individual molecules move from any location slowly ($50 \times 10^{-9}\,\mathrm{m\,s^{-1}}$; Denny, 1993).

Water flows (bulk movements of many molecules) on larger spatial scales and many methods have been used to measure water velocity (Method 2.1). Flow can be either laminar or turbulent. *Laminar flow* is characterized by flow paths in the water that are primarily unidirectional (parallel to each other). *Turbulent flow* is characterized by eddies, where the flow is not as unidirectional. Turbulent flow (mixing) decreases at small scales because viscosity dampens out turbulence as the Reynolds number decreases (below a value of

METHOD 2.1

Methods Used to Measure Water Velocity

The simplest, albeit somewhat inaccurate, way to measure water velocity on a large scale is to place an object that barely floats into moving water and measure the amount of time it takes to move a known distance. For example, an orange is often used because it floats just at the surface of the water and is a bright visible color. Experiments in our freshwater ecology classes confirm that apples and coconuts move at the same velocity as oranges. Unfortunately, this method is sensitive to wind speed, and surface water velocity is not indicative of velocity of the entire water column.

For flows in water >5 cm deep, propellers are often used to estimate velocity. The more rapidly water is moving, the more rapidly a propeller spins. Given an electronic method to count the revolutions per unit time of a propeller and suitable calibration constants, water velocity can be estimated. Electromagnetic flow meters measure the electrical current that is induced when a conductor is moved through a magnetic field. Because water is a conductor, a flow meter can be constructed to create a magnetic field, and the electrical current increases proportionally as water velocity increases.

For smaller scale (several centimeters or less) measurements of water velocity, other methods are more useful. Very small particles or dye can be suspended in flow, and movement can be timed along a known distance. Particle movement can be measured by photographing the moving particles using a flash of known duration. The length of the particle path on the photographs can be related to the flash duration, and water velocity can be calculated. Pitot tubes are small tubes with one end extending above the surface of the water and that have a 90° bend, which allows the open end of the tube that is under water to be positioned facing upstream. As water velocity increases, the pressure at the end of the tube increases, and the height of the water in the tube above the external water level increases proportionally to water velocity. Pitot tubes are inexpensive but are not sensitive to low water velocity and can foul easily. Head rods or head sticks use this same principle and are very inexpensive and easy to use. These devices are essentially meter sticks with blunt and sharp ends; the difference between depth readings taken with the blunt and sharp sides facing into the current is the hydraulic head and is proportional to the velocity. As with pitot tubes, simple equations are applied to field measurements to convert head to velocity.

Hot film, wire, or thermistors can be used to measure water velocity. These devices operate on the principle that moving water carries heat away from objects. Electronic circuitry can be used to relate water velocity to the cooling effect on the heated electronic sensor. These heat-based devices are most useful for biological and ecological investigations of the effects of water velocities on organisms, when low velocity and small spaces are most common.

Acoustic Doppler velocity (ADV) and laser Doppler velocity (LDV) methods have been used to measure water velocity. These methods rely on reflection of sound and light waves, respectively, from small particles in the flowing water. ADV and LDV are very useful because they allow determination of velocity in all three directions. They also allow very short-duration measurements (many per second) on very fine spatial scales (sub mm). Drawbacks include inability to sense velocity in confined spaces and high equipment costs.

approximately 1). Richardson (1922, p. 66) noted this cascade of turbulence into the viscous molecular realm in a more poetic way: "We realize thus that: big whirls have little whirls that feed on their velocity, and little whirls have lesser whirls and so on to viscosity."

Surfaces interact with flowing water. Water velocity slows, turbulence is damped, and flow becomes laminar near solid surfaces. Eventually, within a few microns of a surface, the primary movement of water is Brownian motion. The equation for F_v indicates that viscous force increases closer to surfaces (as spatial scale decreases). Thus, friction with the solid surface is transmitted more efficiently through the solution at small spatial scales near the surface (Fig. 2.7), water also flows more slowly, and flow becomes more laminar at the bottom and sides of stream channels and pipes or near any object that is moving through water. The outer edge of the region where water changes from laminar to turbulent flow is called the *flow boundary layer*.

The thickness of the flow boundary layer increases with decreased water velocity, increased roughness of the surface, increased distance from the upstream edge of an object, and increased size of the object (Fig. 2.8). The flow boundary layer is not a sharp, well-defined line below which no turbulence occurs, even though it is convenient to conceptualize the layer in such a fashion. Rather, the outer edge of the boundary layer represents a transitional zone between fully

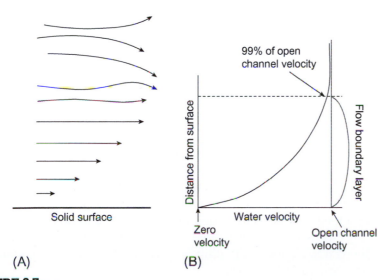

(A)   (B)

FIGURE 2.7

The concept of a flow boundary layer. (A) Arrows represent the velocity and direction of water flow. Inside the flow boundary layer, flow is approximately laminar and slows near the surface; outside the layer, turbulence increases. (B) The outer region of the flow boundary layer is where velocity is 99% of that in the open channel. Very close to the solid surface, water velocity approaches zero. *(Modified from Vogel, 1994).*

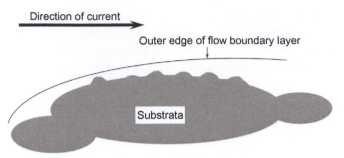

FIGURE 2.8

Schematic of thickness of the flow boundary layer as a function of surface roughness and distance from leading edge. Water is flowing from left to right. Picture the substrata as a rock in a stream. The thickness of the boundary layer increases with distance from the leading edge and is shallower over bumps and deeper over depressions.

turbulent flow and laminar flow. The relationships among flow boundary layer thickness and other factors have many practical aspects. For example, algal growth can have significant influences on hydrodynamic conditions several centimeters from the bottom and edge of a stream (Nikora *et al.*, 1997). Many aquatic animals and organisms can find refuge from high water velocities in cracks in rocks; this is because the cracks occur inside the flow boundary layer and have very slow water velocities. Aquatic invertebrates found in fast flowing waters often have characteristic flattened body forms that allow them to take advantage of the slower velocity microhabitats of the boundary layers.

The effects of scale on flow or movement through water have substantial practical implications. Very small objects experience laminar flow, and larger ones experience turbulent flow. Large organisms that swim through the water benefit from being *streamlined* (shaped to avoid turbulence); small organisms (bacteria sized) do not need to streamline (Fig. 2.9) because turbulence increases the force of drag. The breakpoint at which streamlining becomes useful is approximately 1 mm (depending on velocity and temperature). Greater water velocities also increase turbulence (Fig. 2.10). When a large organism moves through water, turbulent flow acts opposite to the motion of the organism and exerts a force that slows the organism. Scale related to flow is also important for organisms that live very close to large surfaces because they experience reduced water velocities (within approximately 0.1 mm, depending on flow velocity and surface geometry).

An interesting aspect of movement through water at a low Reynolds number is the effect a wall has on microbes swimming near solid surfaces. A 1 μm diameter bacterium swimming 50 body widths from a wall is slowed because friction with the wall is transmitted through fluid. Thus, when

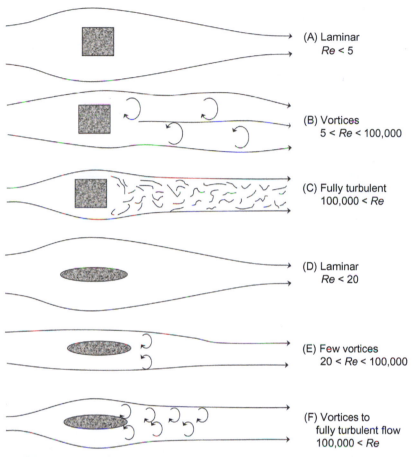

FIGURE 2.9

Patterns of flow behind two differently shaped solid objects at three different ranges of Reynolds numbers. When the Reynolds number is low, turbulence is minimal. Vortices start to form with increased Reynolds number; vortices and turbulence are more prevalent with the cubic object (B and C) than with the streamlined object (E and F). Compare to Figure 2.10.

microbes are observed swimming in a drop of water confined between a microscope slide and a cover slip, they are moving more slowly than if they were swimming freely. Also, an organism swimming in the interstices of fine sediments will have a lower velocity than it will in open water. In contrast, a fish swimming in a river is only modestly influenced by the solid surfaces it swims past. When viewing rapidly moving protozoa through a microscope, we can artificially increase the effects of the wall and viscosity by adding a cellulose solution that further increases the viscosity, slows the organisms, and makes them easier to observe.

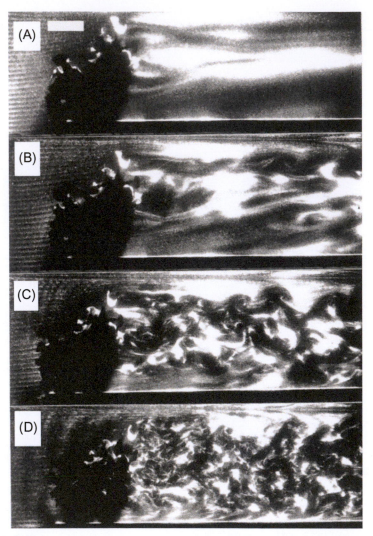

FIGURE 2.10
Water moving past an algal thallus at progressively higher velocities. Tracer particles allow visualization of turbulence. Velocities are 0.5 (A), 1.5 (B), 2 (C), and 3.5 cm s^{-1} (D). *(From Hurd and Stevens, 1997; reproduced with permission of the* Journal of Phycology*)*.

In contrast to very small organisms swimming, large organisms can take advantage of turbulence and vortices (Liao *et al.*, 2003). *Vortices*, spinning flows, are produced by an upstream swimming fish or can be shed by a stationary object in flow. Fish are able to move sinuously through these vortices and use less energy than if they swim through water without vortices flowing at the same average velocity. Very rapid and sensitive sensing of hydrostatic pressure is required to identify these vortices and react appropriately, a reminder that

Table 2.4 Overview of Some Additional Parameters Used to Describe Properties of Fluids

Parameter	Symbol	SI Units	Aquatic Ecological Relevance
Discharge	Q	$m^3 s^{-1}$	Downstream loading, variance characterizes flood and drought
Velocity	U	$m s^{-1}$	Any time an organism or the water it inhabits moves
Force	F	$N \, (kg \, m \, s^{-2})$	What causes object to change direction of motion
Pressure	p	$N \, m^{-2}$	Important in determining the solubility of gases, and the effect of depth on organisms
Shear stress	τ	$N \, m^{-2}$	Force on attached benthic organism
Dynamic viscosity	μ	$N \, m \, s^{-1}$	Intrinsic viscosity of a fluid
Density	ρ	$kg \, m^{-3}$	Alters stratification, sinking rates
Kinematic viscosity	ν	$m^2 s^{-1}$	Describes how easily fluids flow
Surface tension	σ	$N \, m^{-1}$	Capillary rise, spread of floating materials, ability to use water surface, size of bubbles

aquatic animals have a long evolutionary history and sense and experience water very differently from humans.

Reception of hydromechanical signals by small swimming crustaceans (copepods) is also related to friction transmitted through water at small scales. Copepods can sense small moving prey particles and move toward them, and they can sense predators (larval fish) and move away from them (Kiørboe and Visser, 1999). Copepods can react to moving predators slightly less than 1 cm away by making several long "jumps" away from the source of the hydrodynamic disturbance (Kiørboe et al., 1999). The actual response distances of the copepods to predators and prey are a function of velocity and relative size; complex models have been proposed to describe the effect (Kiørboe and Visser, 1999).

ADVANCED: EQUATIONS DESCRIBING PROPERTIES OF MOVING WATER

Engineers and hydrologists have developed a sophisticated body of knowledge of the mechanics and influences of moving water or organisms moving through water. The fundamental metrics of water movement and their ecological relevance (Table 2.4) and some equations that describe moving water and other important aspects (Table 2.5) are discussed in this section. Most of these equations can be related, in part, to the properties discussed with respect to Reynolds numbers.

Table 2.5 Some Numbers Used to Characterize Fluid Dynamics

Parameter	Symbol	Relevance for Aquatic Ecology
Reynolds number	Re	Allows scaling of viscosity and inertia[1]
Capillary rise	h	Calculates the height a liquid will move up through soil as a function of water density, degree of wetability of solid phase, and density of water[2]
Sinking velocity	U	Stokes law for small spherical objects[2]
Darcy's law	U	Velocity of fluid through porous media, groundwater flow[2]
Dalton's law	E	Rate of evaporation, for heat and water budgets
Drag		Amount of force required to move an object through a fluid[2]
Lift		Amount of lift a fin or attached organism experiences[2]
Froude number	Fr	Wave behavior at high velocity, floating objects moving across water surface[1]
Sherwood number	Sh	Advection to sheer across diffusion boundary layer[2]
Péclet number	Pe	Ratio of the convection-to-molecular diffusion of heat[2]
Schmidt number	Sc	Ratio of the convection-to-molecular diffusion of mass[2]

[1]Vogel, 1994
[2]Denny, 1993

Shear stress (tractive force per unit area) can be used to calculate the force on attached organisms or sediments on the bottom of a flowing river or stream.

$$\tau = \rho \frac{U}{l}$$

where τ = the shear stress, ρ is density of the fluid, U is velocity, and l is the distance from the object into the flowing water where full velocity is reached. The steeper the velocity gradient or the greater the density of the fluid, the more shear force is present. Organisms that live in fast flowing rivers are subjected to high shear stress, which has necessitated the evolution of adaptations such as flattened body forms, attachment hooks, or suction discs. Larval net-winged midges (family Blephariceridae) have six ventral suction discs located on their abdomen that allow them to persist in high shear stress habitats including cascades and waterfalls (Fig. 2.11).

Calculation of the sinking rates of particles in water provides a good example of the ramifications of properties of water and scale in aquatic habitats. In general, larger and denser particles sink more rapidly. This relationship can be calculated for small spherical objects using *Stokes law*:

$$U = \frac{2gr^2(\rho' - \rho)}{9\mu}$$

where U is velocity, g is gravitation acceleration, r is the radius of the sphere, ρ' is the density of the sphere, ρ is the density of the liquid, and μ is the dynamic viscosity. The relationship between sinking rate and density of spheres as a function of size can be seen in Figure 2.12. This relationship can be used to predict how long particles will remain suspended in water (Example 2.2). Denser, larger objects sink more rapidly.

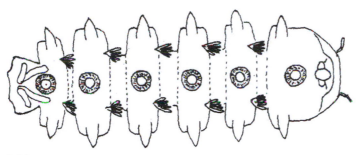

FIGURE 2.11

Ventral view of a larval net-winged midge (order Diptera, family Blephariceridae), showing the suction discs that allow them to live in high water velocities in streams. *(Reproduced with permission from Thorp and Covich, 2001).*

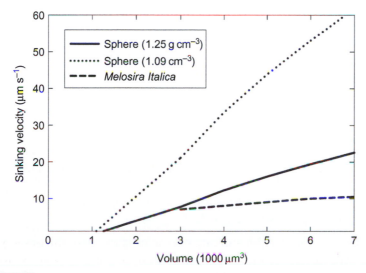

FIGURE 2.12

Sinking velocities of spherical particles with two different densities and of a filamentous diatom, *Melosira*, as a function of volume. The diatom can be found suspended in water and has a density of approximately $1.2\,\mathrm{g\,cm^{-3}}$ *(Data from Reynolds, 1984).*

■ Example 2.2

Calculating the Sinking Rates of Two Spheres through 10 m of Water

There are two spherical objects, one with a radius of 1 μm (about the size of a bacterium) and the other with a radius of 100 μm (the radius of a large spherical alga). Both have a density of $1100\,\text{kgm}^{-3}$. Assume the density of water is $998\,\text{kgm}^{-3}$, gravitational acceleration is $9.8\,\text{ms}^{-2}$, and viscosity is $0.001\,\text{kgm}^{-1}\text{s}^{-1}$. Calculate the time it would take for each object to sink through 10 m of still water. First, calculate the velocity of each sphere using Stokes equation. The small and large spheres have velocities of 2.2×10^{-7} and $2.2 \times 10^{-3}\,\text{ms}^{-1}$, respectively. At this velocity it would take the bacteria-sized sphere 1.44 years and the larger sphere 1.26 h to sink 10 m. Considering that there are few if any bodies of natural water that have no turbulent mixing, it is highly unlikely that a bacterium would ever sink out of the water. ■

Objects that deviate from spherical form can sink more slowly. *Melosira* is an alga that lives suspended in water. It has cylindrical cells and increases in size by adding the cylindrical cells end-to-end (like stacking barrels end-to-end). Thus, the volume of the colony can amplify with a much greater increase in surface area than if the colony was spherical. The relationship for viscous force (F_v) predicts that adding cells to a filament will lead to an increase in viscous force and thus a slower sinking rate relative to a sphere (Fig. 2.12). Slower sinking allows *Melosira* to stay in the lighted water column and grow.

Drag and *lift* are related aspects of life in moving water or life moving in water. The equation for drag is:

$$drag = \frac{1}{2}\rho\, U^2 S\, C_d$$

where ρ is the density of the fluid, U is the velocity of the fluid, S is the frontal area of the object, and C_d is the drag coefficient. As velocity, frontal area, and density of the fluid increase, overall drag increases. The value for C_d is determined empirically, but is a function of the streamlining of the object (when flow is turbulent) and will vary with U.

If a fluid moves more rapidly on one side of an object than the other, a force of lift is generated. This can be important both for organisms that swim, and for organisms attempting to stay attached to solid surfaces with water flowing past.

$$lift = \frac{1}{2}\rho\, U^2 S\, C_l$$

where the terms of the equation are almost identical to the equation for drag except the area term, S, is the projected area perpendicular to the direction of flow, and C_l is the coefficient of lift. Larger objects have more lift, as do those in fluid that is flowing at a greater velocity and those in denser media. A larger fin can have more lift. An organism with a larger surface area attempting to stay attached to the bottom of a stream needs to be attached more strongly than a smaller one assuming that both protrude the same distance into the stream channel (i.e., experience the same water velocity over the top of the organism).

The *Froude number* is useful for relating the propagation of waves through water to the velocity of the water or an object moving on its surface. This number describes forces when objects will move on the surface of the water (a duck or a beaver), when boats will hydroplane, and when rapidly flowing water will exhibit hydraulic jumps (e.g., standing waves that form in rapids). Standing waves form when the wave propagation is slower than the water velocity. The Froude number is calculated as:

$$Fr = \frac{2\pi U^2}{g\lambda}$$

where U is velocity, g gravitation force, and λ is wavelength.

Darcy's law allows for calculation of flow rates through porous media. This calculation is important in many areas of groundwater and soil science. For example, calculating the rate a pollutant will spread through groundwater requires calculation of water flow rates.

$$U = \frac{k_p \Delta p}{\mu l} = \frac{J}{S}$$

where U is velocity, k_p is permeability (depending on the size, shape, and volume of pores), p is pressure, μ is dynamic viscosity, l is length, J is flux, and S is surface area. The smaller the average size of the pores, the more viscous the fluid, and the lower the pressure- the lower the transport velocity. Pressure generation will be discussed more in the chapter on groundwater.

FORCES THAT MOVE WATER

Solar heating and *evaporation* of water are central to water movement. This solar energy input drives the hydrologic cycle by evaporating water from the ocean that is deposited subsequently as precipitation on land. When water flows downhill in rivers or groundwater, it releases the potential energy it gained against *gravity* when evaporated by the sun. The generation of hydroelectric energy illustrates the massive amount of power released as water flows back to

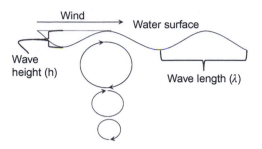

FIGURE 2.13

Schematic of wind-induced water movement in a lake illustrating the decrease in movement with depth from the surface, wave length, and wave height. The three-dimensional flow paths are more complex than in this simple diagram and will be discussed in Chapter 7.

the ocean. The erosive power of water is a direct effect of this release of potential energy. The Grand Canyon of the Colorado River is an impressive example of how much geological change the energy of flowing water can accomplish.

It takes a significant portion of the energy used by all humans on Earth to provide the water that is geographically and temporally available to humans via the hydrologic cycle. This calculation assumes the amount of energy to lift $9,000 \, km^3 y^{-1}$ to an average elevation of 1 km (Dodds, 2008). This cost, the high energy requirement for desalinization, and the energy required for horizontal movement (frictional losses) explain why ocean water is not a practical water source for most people on Earth. We discuss the physics of flowing water and its effects (hydrology) on diffusion in Chapter 3 and the physiography of rivers and streams in Chapter 6. Gravity plays a further role in water movement by causing density-induced currents that may be important in lakes (Chapter 7).

Wind causes surface waves and mixing in lakes, reservoirs, and ponds, and this water movement can have strong influences on organisms. Much of the water movement of a passing wave is simply a rotation of individual parcels of water. As depth increases, this rotation decreases (Fig. 2.13). Consequently, wind-induced mixing also decreases with depth. How wind partially controls the stratification and currents in lakes will be discussed in Chapter 7.

The *Coriolis effect* can influence large lakes (area greater than $100 \, km^2$) by causing rotational currents. The effect is an apparent deflection when a moving object is viewed from a moving point of reference. As applied to currents in lakes, it is not exactly a force that moves water; rather, the momentum of rotation of Earth and large-scale water currents resist this motion. For example, if you are standing on the surface of Earth looking north and watching a large flow of water moving toward you from north to south, Earth is rotating under the flow from west toward east, so the water would appear to move toward the west, while a current moving away from you would appear to move toward the east.

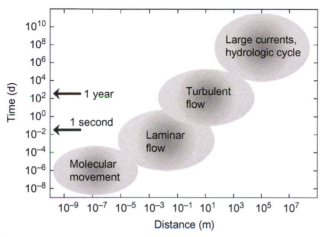

FIGURE 2.14

Schematic showing how water movement is related to spatial and temporal scales. The *x* axis ranges from the size of a protein to the size of Earth and the *y* axis from the time frame of molecular events to the age of Earth. This figure is not meant to imply that sharp boundaries exist between the adjacent regions. Rather, the regions should be viewed as fuzzy overlapping regions. Forces that drive the water movement are discussed in the text.

This effect is most evident in large-scale ocean currents, and rotation of large storms, but it can also cause counterclockwise currents in lakes in the Northern Hemisphere and clockwise rotation in the Southern Hemisphere.

Finally, organisms can move water on smaller spatial scales as a result of locomotion or attempts to change the amount of water movement. In some cases, the actions of animals can alter the movement of water through sediments and increase the exchange of materials between sediments and the water column. For example, clams and worms that live in sediments effectively mix mud and pore water, beavers and humans slow water in streams by building dams, and humans increase water movement with channelization of streams. The relationship of temporal and spatial *scale* and forces on types of water movement are summarized in Figure 2.14.

SUMMARY

1. The uniqueness of water is related in large part to the polarity of the molecule and associated hydrogen bonding.
2. Water is an excellent solvent. Ions generally become more soluble and gases become less soluble as water temperature increases.
3. Water is densest at 3.98°C. Ice is significantly (about 9%) less dense than liquid water. The variation of density with temperature is characterized by nonlinear relationships.

4. Density decreases above 3.98 °C, with a greater density decrease per degree temperature rise as temperature increases.

5. Dissolved ions can increase density. A part per hundred difference in salinity causes a greater density difference than a 50 °C temperature variation.

6. Hydrogen bonding of water leads to additional properties, including high heat of fusion, heat of vaporization, heat capacity, and surface tension, all of which are biologically very significant.

7. The Reynolds number (*Re*) is the ratio of inertia to viscous force and describes how the properties of water vary with spatial scale and water movement. Small organisms have low values of *Re* and little inertia relative to the viscous forces they experience; the opposite is true for larger organisms.

8. Biological activities, such as swimming, filter and suspension feeding, and sinking and many other aspects of aquatic ecology are constrained by properties of water that can be described by Reynolds numbers.

9. Water movement can be molecular, laminar, or turbulent. Molecular movement is also referred to as Brownian motion. Laminar flow is caused by physical processes that predominate at smaller Reynolds numbers and is more common close to solid surfaces or in the pores of sediments. Turbulent flow commonly occurs in open water at Reynolds numbers greater than 5.

10. Numerous equations have been developed by hydrologists and physicists to describe the properties of water, moving through water, and how these vary with temporal and spatial scale.

11. Evaporation and cooling of water vapor (the hydrologic cycle and gravity), wind, density differences, Coriolis effects, and activities of organisms (particularly animals) are all factors that can move water.

12. Water movements occur across a wide range of temporal and spatial scales and this forms part of the abiotic template to which aquatic organisms have adapted.

QUESTIONS FOR THOUGHT

1. Why is less of a temperature difference required for stratification (stable layers of different density) to occur in tropical lakes than in temperate lakes?

2. Why do bacteria generally sink more slowly than dead fish, even when both have approximately the same density?

3. If you wanted to simulate a 1 μm diameter spherical bacterium swimming at $10 \, \mu m \, s^{-1}$ with a 1 cm diameter sphere, how fast would you have to move the sphere to achieve the Reynolds number that the bacterium experiences?

4. Flow is slower in a small pipe than in a large pipe, given the same water pressure gradient. Explain this difference with respect to Reynolds number and viscous force.

5. Why are small swimming crustaceans not nearly as streamlined as larger fish?

6. Why might a lake with a large surface area mix more often and more deeply than a smaller lake?

Movement of Light, Heat, and Chemicals in Water

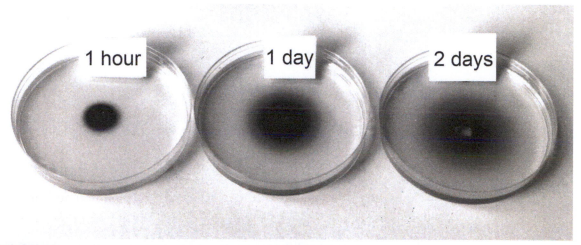

FIGURE 3.1

Diffusion of dye into a 0.5% agar solution in 10 cm diameter petri dishes as a function of time. The agar prevents turbulent mixing so the outward spread of the dye is indicative of the rate of molecular diffusion. Length of time since dye was added to a small depression in the middle of each agar plate is labeled.

Doi: 10.1016/B978-0-12-374724-2.00003-9

The movement of chemicals in water is a key factor to the survival and growth of most aquatic organisms and is central to the understanding of water pollution and material transport. Light is the ultimate energy source for most life on our planet. Without light, energy would not flow through ecosystems and most biogeochemical cycles would stop. Light heats water, which ultimately leads to stratification of lakes. The temperature and movement of water are tied closely to the rate at which chemicals move through water.

DIFFUSION OF CHEMICALS IN WATER

An insect detecting prey, algal cells acquiring nutrients, a microbe sensing its environment, fishes breathing, and a contaminant moving through an aquifer are all subject to the effects of diffusion of substances through water. Diffusion of heat and dissolved materials can be described in a similar fashion so the basics will be described in this section, and more specifically, the movement of heat will be described in the next section.

Diffusion of chemicals is affected by many factors, including the concentration gradient between two points (distance and concentration difference), *advective transport* of water (water currents that move the chemicals), temperature, size of molecules, the presence and structure of sediments or polymers excreted by organisms, and any direct movement of the chemicals by organisms. The rate of diffusive flux between two points is positively related to the difference in concentration and inversely related to the distance between the two points. The basic equation used to describe *chemical diffusion* is *Fick's law:*

$$J = D \frac{(C_1 - C_2)}{(x_1 - x_2)}$$

where diffusion *flux* (J, the amount of a compound diffusing per unit area per unit time) is a function of the *diffusion coefficient* (D, the intrinsic rate of diffusion independent of concentration and distance), the difference in concentration (C), and the distance (x) between points 1 and 2. The rate is greatest with large concentration gradients (difference between C_1 and C_2) and small distances ($x_1 - x_2$). Figure 3.2 provides a visual representation of this equation. The diffusion coefficient is a function of the fluid, size of the diffusing molecule (larger molecules diffuse more slowly), temperature, obstruction of diffusion by pore structure in sediments or other materials, and the rate of mixing of water. Fick's law can be applied to many biological problems related to diffusion, such as calculating the rate of spread of pollutants (Example 3.1).

Thermal movement of molecules (*Brownian motion*), as discussed in Chapter 2, causes them to mix randomly throughout the fluid and leads to *molecular diffusion*. The importance of Brownian motion is related inversely to spatial scale.

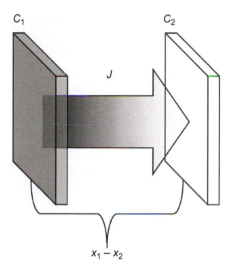

FIGURE 3.2

Schematic illustrating diffusion between two flat surfaces at different concentrations (C_1 and C_2). The rate of diffusion (J) is described by Fick's law (see text). The concentration at C_1 is greater than at C_2, so the net diffusion is toward C_2. Diffusion is less rapid as distance ($x_1 - x_2$) between the two planes increases and as the difference between the concentrations at the two planes ($C_1 - C_2$) decreases.

■ Example 3.1

Calculating Diffusion Flux into a Stream from a Groundwater Source

Nitrate (NO_3^-) is a common contaminant of drinking water that when present in excessively high concentrations leads to serious health problems. Groundwater is present below a feedlot and contains $100\,\text{mg liter}^{-1}$ NO_3^-–N (i.e., the number of milligrams per liter of nitrogen in the form of nitrate) at a distance of 10 m from a stream, where the concentration is $10\,\text{mg liter}^{-1}$ NO_3^-–N and the diffusion coefficient for NO_3^-, D, has been measured as $1.85 \times 10^{-3}\,\text{cm}^2\text{s}^{-1}$. Assume that D does not change from the feedlot to the stream. What is the daily flux of nitrate into the stream per square meter of stream bottom?

First, convert the nitrate concentration and distance into units consistent with the diffusion coefficient, so $100\,\text{mg liter}^{-1}$ NO_3^-–N $= 0.1\,\text{mg cm}^{-3}$, $10\,\text{mg liter}^{-1}$ NO_3^-–N $= 0.01\,\text{mg cm}^{-3}$, and $10\,\text{m} = 1{,}000\,\text{cm}$.

Then,

$$J = 1.85 \times 10^{-3}\,\text{cm}^2\text{s}^{-1} \times (0.1 - 0.01\,\text{mg cm}^{-3})/1{,}000\,\text{cm}$$
$$= 1.67 \times 10^{-7}\,\text{mg cm}^{-2}\text{s}^{-1}$$
$$= 0.14\,\text{g NO}_3^-\text{–N m}^{-2}\text{d}^{-1}$$

■

For example, many molecules randomly collide with a human swimmer from all sides, but because of the size of the human body relative to water molecules, the force is the same from all directions. A bacterium is small enough that more molecules probably will collide with it from one direction than another at any particular time, and thus it jiggles when viewed under a microscope. The probability of unequal collisions is greater the smaller the particle or ion, and the smaller relative difference in mass means that smaller molecules are more influenced by collisions with water molecules. Molecular diffusion is more rapid for smaller molecules (i.e., the diffusion coefficient, D, is greater) because small particles are more likely to be moved by collisions with water molecules. Thermal energy is a property of the average velocity of dissolved ions and water molecules. Consequently, molecular diffusion rates are greater as temperature increases (Fig. 3.3).

Molecular diffusion, in the complete absence of turbulent mixing, is so slow that it is not often observed directly at macro-scales (Fig. 3.1). For example, a cube of sugar would take days to dissolve completely and disperse evenly throughout a glass of water in the absence of any movement of the water. Cooling and convection currents are created as water cools at the surface from evaporation and drops to the bottom of the glass. These density currents cause the sugar to dissolve much more quickly in normal circumstances because the temperature of the glass is not controlled precisely. If a weak agar or gelatin solution was added to the glass, convection would be damped and diffusion would be slowed considerably.

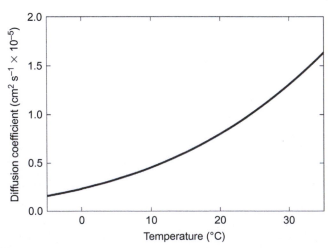

FIGURE 3.3

Effect of temperature on rate of diffusion of chloride.

Water currents often occur at spatial scales exceeding those of individual molecules, and such currents can dominate material transport. Molecular diffusion is many orders of magnitude slower than diffusion with water movement (variously referred to as *transport diffusion*, *advective transport*, or *eddy diffusion*). As discussed in Chapter 2, overall molecular movements are on the order of several nanometers per second, whereas velocity in streams and rivers can exceed 1 meter per second (a billion times more rapid). *Turbulent flow* causes transport diffusion and relatively high diffusion coefficients. Thus, if there is any appreciable movement of water, transport diffusion will dominate over molecular diffusion. With very small spatial scales, Reynolds number is small, viscosity is large, turbulence is not present, and molecular diffusion is more important. Transport diffusion overrides molecular diffusion in the water column of lakes and parts of wetlands, groundwaters, streams, and rivers. Very near solid surfaces or within fine sediments, molecular diffusion is the dominant mode of diffusion.

The ratio of molecular-to-transport diffusion is described by the dimensionless *Sherwood number*

$$Sh = \frac{Ul}{D}$$

where U is velocity, l is length, and D is the molecular diffusion coefficient. When Sh is large, advective transport dominates; when it is small, molecular diffusion predominates. However, D is very small in water ($\approx 10^{-5}\,\mathrm{m^2\,s^{-1}}$) so an organism must be very small ($<20\,\mu\mathrm{m}$) before molecular diffusion predominates (Denny, 1993). Thus, molecular diffusion rules for bacteria and microscopic invertebrates, but transport diffusion controls exchange of materials between the individual and the environment that most multicellular organisms inhabit.

In sediments, such as in groundwater or on lake bottoms, the rate of chemical diffusion is also influenced by the mean path length and size of the pores and channels within the sediment. Short path length and large channels lead to high permeability. When the channels are long or many dead-end channels exist, diffusion rates are slowed because molecules must take a longer path to diffuse between two points, and transport diffusion is limited because water velocity is slow. Determining mean channel length in sediment requires direct empirical measurements of diffusion of a tracer through the sediment.

Solutes can interact with sediments (both the inorganic particles and the organic materials) and further lower diffusion rates. For example, removing organic contaminants such as DDT and some other pesticides from groundwater is very difficult because they have low solubility (are hydrophobic) and are adsorbed onto the surface of the sediment particles, so they have low diffusion rates. Thus, bioremediation efforts (the use of microbes to clean groundwater) and other methods used to remove organic contaminants from groundwaters often employ detergents. Detergents increase diffusion rates by releasing hydrophobic molecules from sediments, increase the biological availability of organic contaminants, and speed the decontamination process.

Water velocity nears zero as a solid surface is approached, advective transport diffusion slows, and molecular diffusion predominates when water movement is sufficiently slow. The region near a solid surface where molecular diffusion predominates is called the *diffusion boundary layer*. This is similar to, but thinner than, the flow boundary layer discussed in Chapter 2. However, the same considerations apply to the thickness of diffusion and flow boundary layers; depressions in solid surfaces lead to thicker diffusion boundary layers, and thinner layers occur where a solid surface protrudes into flowing water and as water velocity and turbulence increase.

The diffusion rate is so much slower across the region where molecular diffusion predominates than regions with eddy diffusion that it represents a rate-limiting boundary layer to the passage of biologically active chemicals. Compounds that are required for metabolic activity must cross the diffusion boundary layer so metabolic rate can be limited by the thickness of this layer. The diffusion boundary layer constrains evolutionary pressures that shape chemically mediated interactions among organisms (Dodds, 1990; Brönmark and Hansson, 2000). For example, the dilution associated with transport diffusion makes it unlikely that biologically active compounds will be released into turbulent waters. The evolutionary cost of purposely releasing chemicals that are costly to synthesize is too great if chemicals are rapidly diluted to concentrations so low that they are ineffective.

Chemical diffusion rates are related to the size and geometry of organisms. *Surface area-to-volume* relationships are essential in the evolution of aquatic organisms. Large objects have a greater volume relative to their surface area compared to smaller objects. Geometry can affect relative rates of diffusion (of nutrients inward and metabolic waste products outward), which can limit metabolism. One way to conceptualize this is to think of the average distance of all points inside a sphere to the outside edge. On average, the distance to the edge of a larger sphere is greater than in a smaller sphere. Fick's law states that greater distance translates into less diffusion flux. Thus, metabolic processes in large organisms are more likely to be limited by diffusion.

Changes in form that increase surface area-to-volume ratios are one way to overcome limitations imposed by diffusion. Form is known to alter diffusion rates in small organisms suspended in water (Sidebar 3.1). Surface area-to-volume relationships also explain why large organisms (more than a few cells wide) have vascular systems. These systems move liquids through the organisms and

SIDEBAR 3.1

Does Diffusion Select for Morphology of Small Planktonic Organisms?

Factors crucial to survival of planktonic organisms (organisms suspended in water) include: (1) avoiding sinking into regions where light is insufficient for survival (Reynolds, 1994), (2) a surface area-to-volume ratio that maximizes nutrient and gas diffusion and increases competitive ability (Reynolds, 1994), and (3) development of defensive appendages that lower the probability of capture or ingestion by predators. The morphology (shape and size) of single- or few-celled planktonic organisms can influence all these factors.

Sinking can be affected by particle size (see formula for F_v and discussion of Stokes law in Chapter 2). Small particles experience high viscosity and sink more slowly. However, increasing surface area also increases viscosity F_v. Therefore, spines or projections (e.g., Figs. 9.8, 10.4, and 20.4) lower sinking rates.

Increasing the surface area with spines or projections also increases the surface area-to-volume ratio. Assuming that the spines or projections can allow materials to move through them to and from the cell, they also can increase the biologically active surface area. A greater biologically active surface area increases the diffusion rates of incoming nutrients and outgoing waste products. Thus, there can be natural selection for planktonic organisms to increase their biologically active surface area.

Sinking and diffusion rates may not be the only selective factors acting on morphology. Predation risk is also decreased by defensive appendages. Spines or projections can increase handling times and unwieldy organisms can clog mouthparts of planktonic predators. In conclusion, three selective pressures converge, and there are a wide variety of shapes of planktonic organisms. The relative importance of these three factors may vary with the organisms considered and their habitats. Considering the relationship between morphology and diffusion is still necessary to describe the selective pressures on shape of small planktonic organisms.

thus promote transport diffusion; otherwise, the relatively low rates of molecular diffusion would lower the maximum possible metabolic rate. Specialized organs associated with vascular systems such as the gills of fish and aquatic invertebrates have evolved to increase effective surface area and promote inward diffusion of O_2 and outward diffusion of CO_2. Insect larvae that live in high-velocity stream habitats can have reduced gills because the diffusion boundary layer is so thin at high water velocity, combined with the fact that larger gills increase drag. Such insect larvae can die if kept in still water because their respiration rate exceeds diffusion rates when advective transport is limited.

Movement of organisms can also increase movement of chemicals, although this movement is often not treated as a diffusion process. Animals in benthic sediments can disturb the sediments in which they live (*bioturbation*) and increase diffusion of materials by increasing turbulent mixing. Motile bacteria can absorb and then move organic contaminants through groundwater sediments more rapidly than expected, given permeability and water flow. Motile ciliates can increase O_2 transport through sediments up to 10 times above molecular diffusion (Glud and Fenchel, 1999). Migration of planktonic organisms from the bottom to the top of lakes (Horne and Goldman, 1994) and migration of fish upstream to spawn (Kline *et al.*, 1990) can result in significant redistribution of nutrients in aquatic systems. Emerging adult aquatic insects and amphibians can result in significant fluxes of nutrients from aquatic to terrestrial habitats (Gray, 1989; Nakano and Murakami, 2001; Regester *et al.*, 2008).

The ideas concerning movement of materials in water as a function of scale (distance and time) are similar to the concepts for water movement discussed previously (see Fig. 2.14). On the smallest scales, low *Re* numbers occur, turbulent mixing is not possible, and molecular diffusion dominates. Diffusion rates increase significantly at greater spatial scales because transport diffusion is likely to override molecular diffusion. Turbulent flow can occur at scales ranging from rivulets to large rivers and lakes. Finally, large currents in very large lakes and the hydrologic cycle move materials on continental and global scales.

LIGHT IN WATER

The interaction between light and water is important for at least three reasons: (1) light is needed for photosynthesis, (2) organisms with eyes or light sensors use light as a sensory cue, and (3) light heats water and ultimately leads to stratification in lakes (see Chapter 7). Feinberg (1969) provides a basic account of the properties of light and all limnologists should be familiar with the common names associated with wavelength ranges of the electromagnetic spectrum (Table 3.1).

Table 3.1 Names of Various Ranges of Wavelengths of Light and Their Biological Relevance

Name	Wavelength (m)	Biological Relevance
Radio	$2 \times 10^4 - 1 \times 10^{-4}$	Little effect
Infrared	$10^{-4} - 10^{-6}$	Heats water, some bacteria can use near infrared for photosynthesis
Visible red-orange	$7 \times 10^{-7} - 5.7 \times 10^{-7}$ (700–570 nm)	Visible, photosynthetically active, heats water
Visible green	$5.7 \times 10^{-7} - 4.9 \times 10^{-7}$ (570–490 nm)	Visible, photosynthetically active (some algae), heats water
Visible blue	$4.9 \times 10^{-7} - 4.0 \times 10^{-7}$ (490–400 nm)	Visible, photosynthetically active, heats water
Ultraviolet UVA	$4.0 \times 10^{-7} - 3.2 \times 10^{-7}$ (400–320 nm)	Mutagenic, cell damage
Ultraviolet UVB	$3.2 \times 10^{-7} - 2.8 \times 10^{-7}$ (320–280 nm)	Mutagenic, cell damage
Ultraviolet UVC	$2.8 \times 10^{-7} - 2.0 \times 10^{-7}$ (280–200 nm)	Mutagenic, cell damage, not present at significant levels in natural environments
Extreme UV, X-rays, gamma rays	$10^{-8} - 10^{-13}$	Mutagenic, cell damage, not present at significant levels in natural environments

Earth's atmosphere alters the intensity and composition of solar irradiance that reaches aquatic systems. Some atmospheric components remove specific wavelengths of light (Fig. 3.4). Others, such as dust and clouds, may scatter or absorb light less selectively. One of the most essential aspects of atmospheric influence on irradiance is the absorption of ultraviolet (UV) light by atmospheric ozone. Releases of chlorofluorocarbons (freons), bromide-containing compounds, and nitrous oxide have caused destruction of ozone in the upper atmosphere and have led to significant increases in the solar UV that reaches the surface, particularly at higher latitudes. Some of the consequences of increased UV to aquatic ecosystems will be discussed in Chapter 12. The absorption of light in the atmosphere is wavelength specific; the same is true once light enters water.

When light reaches the water surface, it can either be reflected or enter the water (Fig. 3.5). If the critical angle of the light reaching the atmosphere/water interface is too great, then it will all be reflected. The amount of light reflected is highly variable because it can be altered by waves on the water surface, the incident angle of the sun, the types of waves (e.g., whitecaps and size), and materials (e.g., dust, pollen, oils) on the surface of the water. If the surface is snow-covered ice, almost all the light is reflected away. Clear ice cover transmits most of the light that enters it.

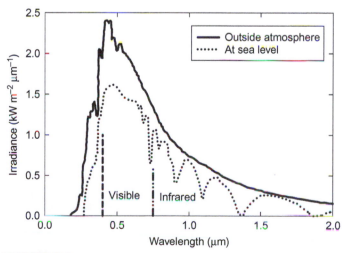

FIGURE 3.4
Spectral energy distribution of solar radiation outside Earth's atmosphere and inside the atmosphere at sea level. Note how the atmosphere changes the spectral distribution of light. *(After Air Force, 1960).*

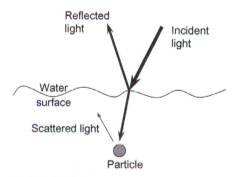

FIGURE 3.5
Schematic of light entering water, where it can be reflected back, scatter off of a particle, or be absorbed in the water column.

As light enters the water, *refraction* causes a bending of the light as it moves from less dense air to denser water. This refraction causes objects observed in water from above the surface to appear to be shallower or larger than they really are (contributing to "fish stories"). The refractive indices of water and ice are very similar, so only minor refraction light is expected as light passes between solid and liquid water in an ice-covered body of water.

Once light enters a parcel of water, it can be *absorbed*, *reflected*, or *transmitted*. Almost all light that is absorbed by water, suspended particles, or dissolved materials is converted to heat. Some of the light is reflected back into the atmosphere giving lakes their characteristic color as observed from above.

Biological activities related to light and water column heating are constrained by the amount of light that is transmitted to different depths. Light intensity decreases logarithmically with depth in a water column (*attenuation*). Attenuation rate is related to reflection and absorption by water, dissolved compounds, and suspended particles. In more productive water columns (*eutrophic*) with a large biomass of suspended photosynthetic organisms, or those with large amounts of suspended inorganic materials or high concentrations of dissolved colored materials, the water contains more material to absorb or reflect out the light; thus, it is not transmitted as far. In less productive (*oligotrophic*) lakes with low amounts of suspended particulate or dissolved colored material, light is transmitted to greater depths (Fig. 3.6).

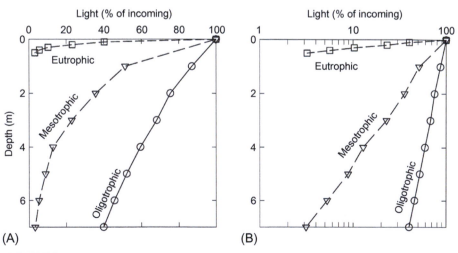

FIGURE 3.6

Light as a function of depth in three lakes—Waldo Lake (oligotrophic), Triangle Lake (mesotrophic), and a sewage oxidation pond (eutrophic), Oregon—plotted on linear (A) and log (B) scales. *(R. W. Castenholz, unpublished data).*

Light attenuation is *logarithmic* (Fig. 3.6), a physical characteristic that can be explained using the following example. Assume that 1/10 of the full sunlight entering the water column is transmitted to 10 m below a lake surface. If attenuation is constant, only 1/10 of the light remaining at 10 m is transmitted to the second 10 m (by 20 m); thus, only 1/100 remains. By 30 m only 1/1000 remains (1/10 of 1/100), and so on. The example is presented in units convenient to a log base 10 scale, but a plot of the light remaining with depth will be linear for a homogeneous water column using a logarithmic scale of any base. Various methods are used to measure light in water (Method 3.1) to characterize attenuation.

METHOD 3.1

Equipment Used to Measure Light in Aquatic Habitats

Many people who study aquatic habitats use a Secchi disk to estimate transparency. This simple method was first used in 1865 by an Italian astronomer, Father Pietro Angelo Secchi, to test ocean clarity. The Secchi depth is the depth at which a weighted, black-and-white disk, 20 cm in diameter, disappears from view. Black and white provides the maximum contrast regardless of the color of the light transmitted by the lake. The instrument provides the most consistent results in sunny, midday, calm water conditions off the shaded side of a boat or dock to minimize reflection off the water. Secchi depth corresponds to the depth at which approximately 10% of the surface light remains (Wetzel, 1983). The relationship between transmission of photosynthetically available radiation (or the extinction coefficient) and Secchi depth is complex and nonlinear because it depends on ambient light, scattering and absorptive properties of the water, and the measurer. However, Secchi depth can serve as an estimate of light attenuation (Fig. 3.7) and the method is inexpensive and easily performed by minimally trained observers. This method is useful for volunteer lake monitoring groups.

Several other methods are available for measuring light, depending on the reason for the measurements. If an investigator is interested in the total light that is heating the surface of a lake, then the total energy entering the lake should be measured. If photosynthesis is of interest, then the energy associated with wavelengths of light that support photosynthetic activity needs to be quantified. For

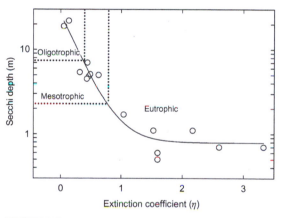

FIGURE 3.7

Secchi depth as a function of extinction coefficient (measured with a quantum meter, 400–700 nm) for 13 Oregon lakes. Boundaries between trophic states for Secchi depth set according to the classification of OECD (1982). (R. W. Castenholz, unpublished data).

other applications, specific wavelengths are of interest (e.g., determining the influence of UV-B radiation on biological activity).

To measure *irradiance*, it is necessary to correct for light that is coming at an angle to the sensor as opposed to a parallel beam coming straight toward the collector. The appropriate correction is a cosine curve, so *cosine collectors* are the sensors of choice to measure radiation from above.

In contrast, some biological processes such as photosynthesis are dependent on total light received from all directions. In this case, a *scalar* or *spherical* (360°) response collector is the sensor of choice (Kirk, 1994).

To measure the total energy entering a water body per unit area per unit time, a pyrheliometer is used, which compares the temperatures of a reflective metal surface and one that absorbs all incoming radiation. These measurements are useful for determining the heat budgets of water bodies.

A meter that is used to estimate light available for photosynthesis should measure the number of photons that are available to excite chlorophyll. These sensors measure photons between 400 and 700nm and the results generally are reported as *photosynthetically available radiation* or *photosynthetic photon flux density*. Units are in μmol quanta $m^{-2}s^{-1}$, or sometimes μEinsteins $m^{-2}s^{-1}$ (the Einstein is not an internationally recognized unit for a mole (6.02 $\times$ 10^{23}) of photons). Spherical sensors are often used for these measurements.

The relative absorption of specific wavelengths could also be of interest. In this case, selective filters can be fitted over sensors, or a spectral radiometer can be used. Spectral radiometers have diffraction gratings that allow photodetectors to sense the intensity of specific wavelengths of light.

Problems arise if light is measured in algal mats or sediments because very small sensors are needed (Jørgensen and Des Marais, 1988). In such cases, spherical tips on fiber optic collectors have been used (Dodds, 1992). Photodetectors or spectral radiometers can be used to analyze light collected by such sensors.

Several equations can be used to describe light attenuation, and coefficients of attenuation are useful indices for aquatic ecologists. If I_1 is the light intensity at depth z_1, and I_2 is the intensity at depth z_2, then the *attenuation* or *extinction coefficient* (η, identical to the absorption coefficient used in chemistry) can be calculated:

$$\frac{\ln I_1 - \ln I_2}{z_2 - z_1} = \eta$$

Alternatively, when the equation is rearranged and raised to the power of *e* to remove the natural logs,

$$I_2 = I_1 e^{-\eta(z_1 - z_2)}$$

Notice that the equations for light attenuation are based on natural logs (ln) and logs to the base 10 will give a different number that is not comparable to most published values without conversion. The exponential nature of the equation results describes a very rapid decline in light intensity with depth, and this concept is not intuitive to many beginning students. A more intuitive equation describes light in terms of transmission. Attenuation of light can also be presented simply in terms of *percentage of transmission* over a specific depth:

$$\frac{100\, I_2}{I_1(z_2 - z_1)} = percentage\ transmission\ per\ meter$$

where I_1 and I_2 are light intensity values measured at any two points (z_1 and z_2) 1 meter apart vertically in the water column.

■ Example 3.2

Use of Light Attenuation Equations

Calculate the attenuation coefficient and percentage transmission per meter at the surface of two lakes, one that is productive and one that is less productive. In both lakes, light is 1,500 μmol quanta m^{-2}s^{-1} at the surface. In the productive lake, the light is 1 mol quanta m^{-2}s^{-1} at 1 m, and in the unproductive lake it is 1,200 μmol quanta m^{-2}s^{-1} at 1 m.

Calculation	Unproductive	Highly Productive
I_0 (ln I), intensity at first depth	1,500 (7.31)	1,500 (7.31)
I_1 (ln I), intensity at second depth	1,200 (7.09)	1 (0)
% transmission m^{-1}	80	0.06
η	0.22	7.31

Note that transmission is greater and attenuation coefficient is less for the unproductive lake. ■

As mentioned previously, attenuation is a function of the sum of the absorption of light by the water, by the particles in the water, and by the compounds dissolved in the water. Attenuation coefficients are largest in eutrophic lakes and smallest in oligotrophic lakes (Example 3.2). As the productivity of a lake increases, so does the attenuation because lakes with greater productivity have more suspended particles and more dissolved organic compounds to absorb light. Other factors that can cause greater light attenuation include large *sediment* loads (typical of reservoirs in agricultural areas and shallow lakes) and lakes with high concentrations of humic compounds. *Humic compounds* result from decomposition of plant material and lead to brown-colored water with large absorption coefficients. Blackwater swamps, other wetlands, rivers, and some small lakes can have high concentrations of humic compounds.

Not only does total light intensity change with depth but also there is variation in the attenuation of relative amounts of different wavelengths. Pure water has maximum absorption in the visible wavelengths of red light and maximum transmission of blue (Fig. 3.8; color plate Fig. 1). Thus, clear (oligotrophic) lakes appear blue; the pure water in them absorbs green to red wavelengths and transmits blue (Fig. 3.8, color Fig. 2). There are always some suspended particles in lakes that reflect light back out of the surface into the eye of the observer. Blue wavelengths are more likely to be reflected back out before they are absorbed than are longer (redder) visible wavelengths.

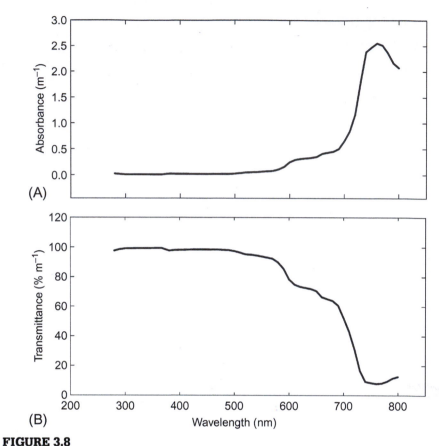

FIGURE 3.8

The absorption (A) and transmission (B) of light by pure water as a function of wavelength of light. *(Data from Kirk, 1994).*

Pigments of suspended algae (phytoplankton) absorb light in specific regions (Fig. 3.8; color plate Fig. 1). The most important of these is chlorophyll *a*, which absorbs blue and red light. As a lake becomes more eutrophic, more blue is absorbed and relatively more green is transmitted and reflected (Fig. 3.9). Thus, the spectral quality of light at depth is a function of the absorptive properties of the lake. Spectral transmission influences perceived colors of lakes and the objects within them (Sidebar 3.2).

Cyanobacteria (blue-green algae) and red algae have evolved pigments called *phycobilins* that allow them to use green light. This fact, combined with an understanding of optical properties including wavelength-specific attenuation, can be useful for describing some of the ecology of cyanobacteria. There is a greater relative transmission of green light compared to blue and red light

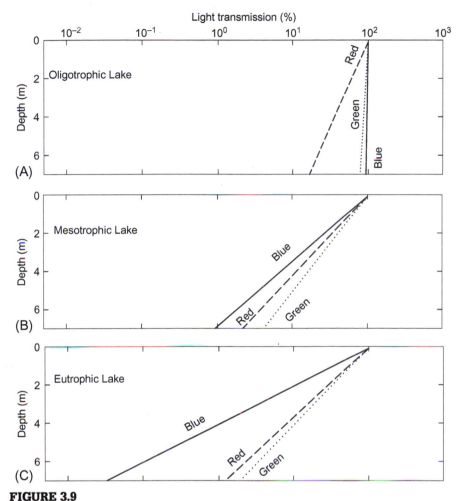

FIGURE 3.9

Light transmission as a function of color for an oligotrophic lake (Waldo Lake, 1984; A), a mesotrophic lake (Munsel Lake, 1984; B), and a eutrophic lake (Siltcoos Lake, 1983; C) in Oregon. *(R. W. Castenholz, unpublished data).*

in eutrophic lakes. At a meter or less of depth in a eutrophic lake, the light is poor in the red and blue wavelengths used by most photosynthetic organisms. Cyanobacteria use their specialized pigments to harvest green light and transfer it to chlorophyll *a* for photosynthesis. Dense populations of cyanobacteria are sometimes found well below the surface of lakes (Fig. 3.10). Algae that contain chlorophyll remove blue and red light in the water column above, but enough green light is transmitted to depth to allow photosynthesis for organisms such as red algae and cyanobacteria, which are adapted to use this wavelength range.

SIDEBAR 3.2

Why Are Lakes the Color They Are, and How Are Colored Fish Lures and Other Objects Perceived under Water?

The color of a lake can tell us much about its biological status. Several factors are involved in determining how an observer perceives lake color. Very unproductive (oligotrophic) lakes are a deep blue color because pure water transmits blue and absorbs green and red light. The blue light goes more deeply into the lake, but some of it is scattered by suspended particle material and returns to the surface. Scattered light that is blue is more likely to leave the lake than scattered green or red light.

In productive (eutrophic) lakes, blue and red wavelengths are absorbed by algal pigments (color plate Fig. 1), and green wavelengths travel relatively farther. The probability is greater that green light will be scattered back out of the lake than red or blue light, giving highly productive lakes their green color (color plate Fig. 2). These predictable relationships between lake color and productivity allow for large-scale assessments of lake productivity using remote sensing techniques and satellite images (Lillesand *et al.*, 1983).

Dissolved organic materials such as tannin and lignin can impart a brown color to a lake, pond, or stream. Observation of such lakes may reveal little suspended material, but the dissolved material absorbs all wavelengths of light, yielding a brownish color. Sediments can also color lakes. A lake that is very turbid with reddish clay will look red. High-altitude lakes near glaciers often contain very fine sediment particles (glacial flour) created by glacial action. These particles can lend a milky turquoise appearance to otherwise unproductive blue lakes.

Organisms can impart additional colors to lakes and ponds. Species of water ferns can be bright purple and reach very high densities on the surface of certain ponds. Photosynthetic bacteria can reach high densities, particularly in saline ponds, and yield purple, brown, yellow, or blue appearances. Dissolved metals can also color ponds. For example, high concentrations of copper can lead to metallic blue ponds or lakes.

The alteration of spectral quality with depth means that some wavelengths are never present in deeper waters, and colors are perceived differently there than under full sunlight. This may be important to the way that fish are able to perceive color. For example, under full sun, red fish lures appear red because they absorb green and blue light and reflect red. However, red light is not available deep in a lake. To a diver or fish some 5m deep in a mesotrophic lake a red lure would appear black. Fish lures appear to be different colors deeper in oligotrophic lakes than in eutrophic lakes (color plate Fig. 3). A white lure would appear blue 10m under the surface of an oligotrophic lake but green 1m deep in a eutrophic lake because of the predominant color of light found with depth in each lake. Certainly, the color patterns of many fish lures are designed to appeal to anglers rather than fish.

Aquatic ecologists should also remember these spectral properties in other contexts. For example, a deep-water fish may be colored bright red, but this actually may represent cryptic coloration when the fish is in deep water because the red appears black. Yet, if the fish moves into shallow water, the coloration may be visible and attractive to mates.

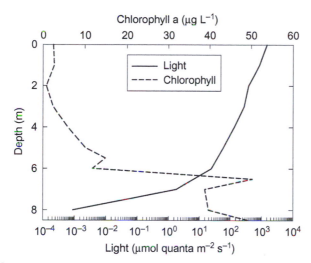

FIGURE 3.10

Profiles of chlorophyll a concentration and light with depth at Pottawatomie State Fishing Lake No. 2, Kansas. Deep chlorophyll peaks are attributable to the presence of large populations of cyanobacteria (*Oscillatoria*). The high biomass of algae occurs in a region with 1–0.001% of surface sunlight. Note how the attenuation of light increases (shallower slope of the light curve) because of the dense algal populations.

HEAT BALANCE IN WATER

The heat balance of water determines the water temperature, which in turn controls rates of biological activity. Heating, cooling, and movement of heat through water determine temperature of the water at any point in time. Changes in heating can lead to water currents (e.g., water sinks from the surface on cold nights or as it cools on the surface by evaporation). Heating and mixing also set up density differences that can lead to stratification and specific currents that occur in waters.

Heat energy in water is gained by solar radiation, diffusion of heat from the atmosphere, and warmer water flowing into a particular area (from precipitation, condensation, groundwater, streams, or rivers). Heat energy is lost by reflection, back radiation, evaporation, and flow of cooler water into an area. Change in temperature must be accounted for by the balance of heat into and out of any specific area. Solar, back radiation, and atmospheric interactions dominate heat budgets of most waters (Wetzel, 2001). Aquatic habitats such as wetlands with emergent vegetation may be influenced by interactions of the atmosphere with the emergent vegetation (e.g., modification of the atmospheric boundary layer, condensation on plant leaf surfaces, shading). Once heat enters a water body, distribution through the water begins by processes of diffusion and advection.

The movement of heat can also be viewed as diffusion. An equation similar to that presented for chemical diffusion (Fick's law) can be used to describe diffusion of heat. In this case, J_Q is heat flux per unit area, and the concentration difference is replaced by the temperature difference $(T_1 - T_2)$.

$$J_Q = K \frac{T_1 - T_2}{x_1 - x_2}$$

where J_Q is heat flux (Wm^{-2}), T_1 and T_2 are temperature at points x_1 and x_2, respectively, and K is the coefficient of thermal conductivity. Heat is transferred more rapidly across steep gradients of temperature (shorter distances or greater temperature differences).

As with molecular diffusion of dissolved materials, molecular diffusion of heat predominates when advective transport is minimal. Thus, when lakes stratify, they generally maintain that stratification until processes at the surface bring the surface layer temperature equal to that of the deeper layers. Stable stratification can result from heating of surface waters and the depth of the heated surface layer will be determined by how deeply light can penetrate and heat the water and how deeply the surface can be mixed by the wind or other water currents. Below the mixing depth the molecular diffusion of heat equilibrates disparate temperatures between the bottom and the top of the lake very slowly.

As with diffusion of chemicals, the relative rates of advective and molecular diffusion of heat can be captured with the *Péclet number*,

$$Pé = \frac{lU}{K}$$

where l is the characteristic length, U is the velocity, and K is the coefficient of heat conductivity.

Transport of hot or cold water by flow can redistribute heat rapidly. In contrast, heat is transferred more slowly by molecular collisions in the absence of transport. Thus, molecular diffusion of heat is considerably slower than transport diffusion of heat. Physical limnologists study distribution and diffusion of heat in lakes because they are related to stratification and mixing (Chapter 7).

The annual *heat budget* of a lake (actually not a "budget" but a heat storage capacity) indicates its heat content and how it changes over the course of a year. In temperate lakes, this budget is the total amount of energy required to increase the temperature of a lake from its winter minimum to its summer maximum. Not surprisingly, the annual heat budgets of tropical lakes are much smaller because their seasonal temperature differences are much smaller. Analyses of a large number of lake heat budgets (Wetzel, 2001) leads to several generalizations: (1) most heat gain occurs in the spring-summer period, (2) some periods lead to relatively large changes in heat content

(10%) in a few days, and (3) heat absorbed by sediments and heat of fusion of ice can have considerable influence on annual heat budgets.

In some ways, heat budgets of rivers and streams are more complex than those of lakes, particularly in smaller streams where groundwater temperatures have a large influence or where surface runoff (floods, snowmelt) can change heat input over periods of days. Rivers with substantial influence from reservoirs can further complicate heat budgets; shallow and wide reservoirs can artificially warm rivers. If reservoirs are deep and water is released from cool areas, they can artificially cool rivers, as is the case in the Grand Canyon, where the naturally warm Colorado River is cold because of reservoir releases.

SUMMARY

1. Diffusion of materials in water can be described by Fick's law. In this relationship, diffusion flux is related positively to a diffusion constant, the concentration gradient, and the reciprocal of distance between the diffusion source and location to which it is diffusing.

2. Diffusion constants can be affected by many factors. The most important is that moving masses of water result in diffusion constants many orders of magnitude higher than those for molecular diffusion in still water. Factors that increase diffusion constants also include increased temperature and bioturbation. Sediment properties, such as affinity of the diffusing molecules for the sediments and low permeability, can be associated with low diffusion constants.

3. Diffusion properties can constrain natural selection with respect to relationships among chemicals and organisms and body shape of organisms. This includes uptake of nutrients and interactions between organisms that are mediated by chemicals.

4. Quantity and quality of light entering water can be altered by the atmosphere, the surface of the water, and any ice or snow cover over water.

5. Light is attenuated logarithmically when it enters water.

6. Red light is absorbed directly by water, and blue light is transmitted deepest in oligotrophic lakes. Phytoplankton pigments absorb red and blue light. Cyanobacteria also can absorb green light.

7. Movement of heat can be described with equations similar to those used to describe diffusion of dissolved materials in water.

8. Heat balance accounting can characterize dynamics of temperature and stratification in aquatic ecosystems.

QUESTIONS FOR THOUGHT

1. Why would you expect transport diffusion to occur in a still glass of water, if evaporation was occurring at the surface, and how would you keep such transport diffusion from occurring?

2. Why do many small invertebrates have a heart but no system of blood vessels?

3. Calculate the surface area-to-volume ratios for a sphere, a cube, and a right circular cylinder (1 cm high), each with a volume of $1 \, cm^3$; relate these to diffusion of materials to cells of different shapes (r = radius, h = height, S = surface area, and V = volume; for a sphere $S = 4\pi r^2$, $V = 4/3\pi r^3$; and for a right circular cylinder $S = 2\pi r h + 2\pi r^2$ and $V = \pi r^2 h$).

4. Why might large rivers generally have greater attenuation coefficients than large lakes?

5. If objects at the surface of a lake are blue, black, red, or white, what color would they appear if they were viewed at 20 m depth in an oligotrophic lake?

6. Use the data plotted in Figure 3.6 to demonstrate that you obtain a straight line if the natural log (ln) of light is plotted against depth.

CHAPTER 4

Hydrologic Cycle and Physiography of Groundwater Habitats

FIGURE 4.1

A stream leaving a limestone cave on the South Island of New Zealand.

Identification of aquatic habitats is generally based on landscape geomorphology and hydrology. The hydrologic cycle describes the movement of water from the oceans into the atmosphere and across land. In combination with other geological processes, the hydrologic cycle determines many of the physical characteristics of aquatic habitats. This chapter provides some details about the hydrologic cycle and then discusses the physical geology of groundwaters. Chapters 5, 6, and 7 continue this theme with respect to wetlands, streams, rivers, lakes, and reservoirs. The reader should be aware of the many linkages among different types of aquatic habitats across the landscape, even though they are considered in separate chapters. Understanding of *physiography*[1] can provide a starting point for examining abiotic factors that drive aquatic ecosystems as well as the fact that different habitats grade into each other (e.g., what is the difference between a wetland and a pond)?

HABITATS AND THE HYDROLOGIC CYCLE

The definition of aquatic habitats can be based on geology and the *hydrologic cycle,* or the way water moves through the environment. Temporal and spatial variations in movement and distribution of water are called *hydrodynamics.* Parts of hydrodynamics were considered in Chapter 2; here we talk about how water moves across the surface of the land. Links between terrestrial and aquatic ecosystems, as well as links among different aquatic ecosystems, must be examined in order to understand how water moves across the surface of the continents. Chapter 1 included a brief description of a global water budget with respect to water availability to humans. This chapter provides a more detailed consideration of the hydrologic cycle and hydrodynamics. The importance of hydrology to ecology is reflected in the new field of study referred to as *ecohydrology*.

Aquatic habitats can be considered at a variety of spatial and temporal scales; the organism or process of interest dictates the scale chosen for study. For example, microbes can be influenced by proximity to a grain of sand, but ecosystem processes dominated by microbes can be altered by position in a watershed. Changes in small-scale microbial communities can occur over periods of hours, but changes in ecosystems can take decades to millions of years. An indication of the range of habitats and scales that can be used as a framework to link hydrodynamics and hydrology to aquatic ecology is presented in Table 4.1. Basic physical properties of water movement were discussed previously; now we explain how water moves across the landscape.

Weather patterns result in widely varied quantities of precipitation around the world (color plate Fig. 4). Precipitation, in combination with factors that

---

[1]Physiography is an older term for physical geography, and refers to the natural rather than social part of the geography of a region.

Table 4.1 Habitat Classification by Time and Distance Scales

Habitat	Time Range	Distance Range	Examples
Microhabitat	1 s–1 y	1 µm–1 mm	Fine particles of detritus, sand and clay particles, surfaces of biotic and abiotic solids in the environment.
Macrohabitat	1 d–100 y	1 mm–1,000 m	Riffles and pools in streams, rivers, and underground rivers. Logs, macrophyte beds, pebbles, boulders, position on lakeshore.
Local habitat	1 m–1,000 y	1 mm–100 km	Lake, regional aquifer, stream or riffle reach, shallow lake bottom.
Watershed	1 y–10^6 y	1 km–10,000 km	Areas feeding small streams to the basins of large rivers, including lakes, aquifers, and streams within boundaries.
Landscape	10 y–10^7 y	10 km–10,000 km	A mosaic of local habitats or watersheds.
Continent	10,000–10^9 y	>10,000 km	A composite of large drainage basins and aquifers.

This classification is only one of many ways to divide a natural continuum

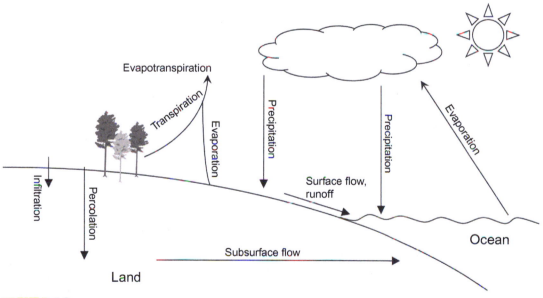

FIGURE 4.2
The hydrologic cycle.

influence return of water to the atmosphere, dictates how much water enters aquatic systems (Fig. 4.2). Precipitation can either be intercepted by vegetation or fall directly on nonvegetated surfaces. Water can return to the atmosphere by direct evaporation or by transpiration through plants. Transpiration and evaporation together are called *evapotranspiration*. If water is not lost to evapotranspiration, it infiltrates (flows down into) the soil or flows across its surface. Very heavy rain or impermeable surfaces can cause water flowing across the surface (*sheet flow*), which enters streams and can cause flooding.

Water that infiltrates can be stored as soil moisture (or as ice in polar or high-altitude areas). Soil moisture in incompletely saturated soils adheres to soil particles with capillary force. Water that is not stored *percolates* down through the soil layers into groundwater or flows through shallow surface soils to streams. This water is stored underground or eventually flows into rivers and streams (*runoff*).

Many factors combine to determine the amount of precipitation that ends up as runoff (*water yield*). The amount of precipitation required to cause runoff is greater when the temperature is high because potential evapotranspiration is high. Low temperatures and rates of evapotranspiration characterize polar regions. Thus, wetlands, lakes, and streams can form with moderate amounts of precipitation at high latitudes. Impermeable soils or surfaces have high water yield (including rocks, packed soils, and desert "pavements"). Dry soils can hold considerable amounts of moisture, so saturated soils will have greater water yield per unit precipitation. Human alterations to the landscape that compact soils and increase *impervious surfaces* (surfaces that do not allow infiltration) such as paving streets and parking lots can greatly alter patterns of surface flow and infiltration.

ADVANCED: PREDICTION OF AMOUNT AND VARIABILITY OF RUNOFF WITH GLOBAL CLIMATE CHANGE

Human activities can alter global patterns of precipitation and the hydrologic cycle in unpredictable ways. As Earth warms in response to increases of greenhouse gases (primarily CO_2), evaporation and precipitation will likely increase worldwide, but variability and distribution will also change. How such changes will influence local weather patterns is uncertain in many cases. Because freshwater habitats are influenced greatly by the balance between precipitation and evapotranspiration, predicting the impacts of the greenhouse effect and global change on specific habitats is complicated. The strongest effects will likely occur in areas that are currently arid or where precipitation is equal to or less than potential evapotranspiration (Schaake, 1990).

Predicting the amount of runoff as a function of precipitation is difficult enough when we assume that conditions are similar over the long run (e.g., when past trends can be used to predict the future). With global climate change, Earth is shifting to states that were not necessarily common previously; we are living in a "no analog" hydrologic world (Milly *et al.*, 2008). For example, many areas are supplied water by summer snow melt. Up to 60% of hydrologic change in the western United States from 1950 to 1999 may be related to human influence on climate (Barnett *et al.*, 2008) and is related to earlier snow melt and perhaps less snow accumulation.

On the whole, a warmer world means a more active (energetic) hydrologic cycle, so the total amount of precipitation falling should increase, as should the total amount of cloud cover. However, clouds and snow can amplify back-radiation of heat, making modeling difficult. A more active hydrologic cycle is also a more variable hydrologic cycle, creating variance that can make predictions more difficult. For example, 10 cm of rain that falls during five storms over a period of a month may lead to very different magnitude and frequency of runoff than 10 cm that falls in one storm over a month. The 10 cm falling in five storms may all evaporate before reaching groundwater or streams, and the 10 cm falling in one storm may run off very quickly in sheet flow.

Increased temperature coupled with increased CO_2 makes predicting the rate of evapotranspiration tricky. Evaporation should increase with temperature, as should transpiration, but plants can keep their stomata open less with increased CO_2 and still maintain substantial rates of photosynthesis. Thus, transpiration may actually decrease with increased availability of CO_2. More variable precipitation may alter vegetation as well. For example, more pronounced dry periods could lead to a greater probability of fire, leading to altered plant communities and a corresponding shift in whole-system transpiration.

Specific local conditions also make understanding the effects of global change on hydrology difficult. For example, local warming in the Himalayas linked to global climate change is accelerating the melting of glaciers. These glaciers and their moraines impound freshwater lakes. More rapid melting of glaciers increases the probability of catastrophic failures of these natural impoundments leading to destructive floods downstream. Such catastrophic flooding is expected to become more common in ice-covered regions such as Greenland.

Global circulation models are used to predict the influences of global warming and global climate change. These are very complex models that divide each column of atmosphere above each segment of land into a number of layers and model the interaction with the vegetation, soil, or ocean below. Each model gives somewhat different results based on varying base assumptions, requirements for data, and calculation approaches. The current general approach, as taken by the Intergovernmental Panel on Climate Change, is to look at what a wide variety of models predict (IPCC, 2007). Understanding the implications of such models for continental aquatic systems requires understanding the movement of water once it reaches terrestrial habits and moves into the aquatic realm.

MOVEMENT OF WATER THROUGH SOIL AND AQUIFERS

Water can either flow across the surface of soil (*sheet flow*) or move down into porous soils (*infiltration*). Sheet flow often ends up directly in stream channels (and can rapidly move materials including pollutants from terrestrial habitats), whereas infiltration can percolate to groundwater. As mentioned previously sheet flow mainly occurs during extremely high rates of precipitation or in areas with impermeable surfaces; most water that enters surface waters does so through groundwater. Several regions below the surface of soil that receive infiltration have been described (Fig. 4.3). The dry or moist sediments below surface soil layers form the *unsaturated zone* (also called the *vadose zone*). The depth of the unsaturated zone can vary from zero (where groundwater reaches surface water) to more than 100 m (in some deserts).

Capillary fringe is the area where groundwater is drawn up into the pores or spaces in the sediment by capillary action. This zone is generally 1 m or less above the *water table*, which is defined as the top of the region where virtually all the pore space is filled with groundwater. Below the water table is the groundwater habitat. Methods are available for sampling groundwater (Method 4.1), but the spatial extent (vertical and horizontal) of samples that can be collected is limited relative to those from surface water habitats. The extent of groundwater habitats and the adjacent capillary fringe and vadose zone often vary with precipitation patterns and can be difficult to define because they may not have distinct boundaries.

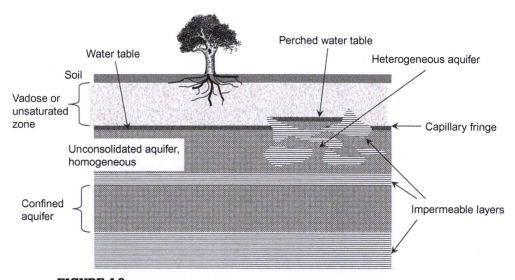

FIGURE 4.3
Various subsurface habitats.

METHOD 4.1

Sampling Subsurface Waters

Sampling groundwater and water in the vadose zone is more technically demanding than sampling streams or lakes. *Lysimeters* are used to sample water from the vadose zone (Fig. 4.4). This sampler has a ceramic cup on the end that absorbs water from the surrounding soil when it is placed under vacuum (Wilson, 1990).

Wells are generally used to sample groundwaters, but they can cover only a small part of the habitat. Shallow, temporary wells may be installed by hand where the water table is close to the surface and there are unconsolidated sediments. Deeper sampling requires well drilling machinery. When wells are drilled, samples of the pore water can be collected and sediments can be removed from the drilling apparatus. A split-spoon sample, in which the drill bit takes a core in its center as it cuts downward, is commonly used. The drill bit is then removed and split, and the core can be analyzed.

After a well is drilled, a casing is inserted through the length of the well with slots or screens placed in the region from which water is to be removed (Schalla and Walters, 1990). The outside of the well is then packed with sand fine enough to keep the sediments from the aquifer from entering and plugging the well when water is removed (Fig. 4.4). The fine packing materials used at the base of wells cause problems for researchers interested in groundwater animals because organisms larger than those able to pass though the packing material cannot be sampled. As a result, unless they have specially designed wells (e.g., such as those in Fig. 4.4 but without screens or fine packing materials), groundwater ecologists may miss significant components of the groundwater fauna. Material is packed in the hole outside the well casing above the slotted portion to form a seal. Otherwise, water could move vertically into the aquifer from the surface or between aquifer layers and contaminate the sample from the desired depth. Bentonite (a type of clay)

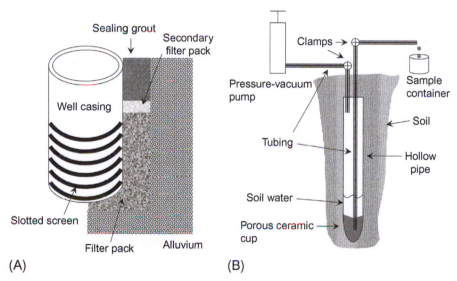

(A)   (B)

FIGURE 4.4

Some equipment used for sampling soil water and groundwater. (A) A slotted well casing packed with filtering materials allows sediment-free water to be sampled. (B) A vacuum sampler (lysimeter) relies on negative pressure to extract pore water from soil. The lysimeter is put under vacuum for sampling and then put under pressure to collect the sample as it flows out of the sampling tube.

is commonly used for this sealing because it is relatively chemically inert and swells when wetted.

Once a well is installed, it must be developed. Developing entails removal of a large volume of water and sediments in the water to ensure that the well flows clearly and supplies water representative of the aquifer. The well should be sampled regularly, with enough water removed so water will not become stagnant. During sampling, several volumes of water in the casing must be removed before the actual sample is collected to ensure that the water sampled is from the aquifer outside the well.

Various pumps and bailers are available for sampling groundwater. The type of analysis to be performed on the samples collected is determined before selecting the system. For trace metal analysis, pumps without metal parts that may contaminate samples are used. When organic materials are to be analyzed, pumps that do not use oil are essential because the oil may contaminate the water samples.

In this book we use *aquifer* to describe continuous groundwater systems, but some use the term only for groundwater reservoirs that are useful to humans. The aquifer is also referred to as the *phreatic* zone. A tremendous amount of water is stored below Earth's surface, but much of it is not available for human uses. A lot of the water is either too deep to access, unsuitable for human use because of high salinity, or tied up with mineral materials (many molecules and elements bind water strongly).

Water stays in aquifers for variable amounts of time. Some aquifers exchange with surface waters in weeks or months. These include karst aquifers with relatively large conduits for water to flow in (even underground "rivers" flow in some) and the shallow groundwaters connected to rivers in valleys with coarse substrata such as cobbles. On the other end of the spectrum are deep aquifers like the Ogallala in the central United States that hold water for thousands or even millions of years; water in these systems often is referred to as *fossil water*. Because of slow replenishment rates, fossil water is considered a nonrenewable resource.

Water can be dated by measuring its isotopic composition. A large spike of radioisotopes occurred during nuclear atmospheric testing from the mid-1940s to the 1960s. Tritium commonly is used as an indicator of surface water entering an aquifer from this time. Proportions of the ^{14}C isotope in dissolved substances in water can be used to determine age up to tens of thousands of years, and isotopes of other elements can age waters more than 1 million years old.

Much of the water flow from land into the world's oceans is from rivers. However, some areas, such as the southeastern United States, are characterized by large discharges of groundwater into marine waters (Moore, 1996). Likewise, the ecology of streams (Allan, 1995; Jones and Holmes, 1996; Brunke and Gonser, 1997), wetlands (Mitsch and Gosselink, 1993, 2007; Batzer and Sharitz, 2006), and lakes (Hagerthey and Kerfoot, 1998) is influenced by groundwater (Freckman *et al.*, 1997). Thus, knowledge of groundwater flows and processes is integral to the study of all aquatic systems.

Table 4.2 Representative Particle Sizes and Hydraulic Conductivity of Various Aquifer Materials

Material	Particle Size (mm)	Hydraulic Conductivity (m d^{-1})
Clay	0.004	0.0002
Silt	0.004–0.062	0.08
Coarse sand	0.5–1.0	45
Coarse gravel	16–32	150

(Bowen, 1986)

Soil texture and composition determine how rapidly water percolates into groundwater habitats. Impermeable layers, such as intact layers of shale or granite, do not allow water to flow deeper. In very fine clays or soils with large amounts of organic material, the rate of percolation can be very low. In contrast, gravel and sand allow for relatively rapid water flow (Table 4.2). Infiltration capacity partially determines the proportion of water that flows off the surface and the quantity that enters groundwater. The rate at which water percolates into groundwater is referred to as the rate of *recharge*.

Infiltration rate can have important practical consequences. For example, groundwaters can be contaminated when sewage sludge, mineral fertilizers, or pesticides are applied to cropland if infiltration rates are high enough and crop uptake insufficient so that contaminants enter groundwater. Thus, infiltration rate is an important aspect of determining fertilizer application levels (Wilson *et al.*, 1996). Extensive areas with impermeable surfaces associated with urban development can decrease flow into groundwaters, reducing recharge and increasing flooding problems when increased sheet flows occur, as discussed in Chapter 6.

Once water enters groundwater zones, permeability determines the potential rate of flow (*hydraulic conductivity*) and is variable and dependent on geology. Water will flow slowly in fine sediments and more rapidly where large pores and channels exist (e.g., in limestone aquifers with channels and unconsolidated sediments with large materials such as cobble). Hydraulic conductivity is partially dependent on the Reynolds number (see Chapter 2) because viscosity is high and flow is slow when Reynolds numbers are small (i.e., when sediment particles are small). *Darcy's law* (Table 2.5) can express the rate at which water moves through aquifers. This law states that the flow rate in porous materials climbs with increased pressure and decreases with longer flow paths. Darcy's law is used to mathematically describe the flow of groundwater and infiltration through the vadose zone (Bowen, 1986).

The amount of water that can be held in sediment is determined by its *porosity*, or the volume fraction of pores and fractures. If sediment is saturated with water, there are two components: (1) water that will drain away from the sediment (yield) and (2) water that is held in the sediment (retention). As pores get smaller, retention increases because surface tension has larger influence at smaller spatial scales. Greater flows often are found in more porous sediments because more porous materials tend to have more channels through which water can pass. For example, gravel and sand pack with relatively large spaces between the particles for water to flow through. This packing results in large connected channels. Exceptions to this relationship exist; high-porosity sediments may have a low hydraulic conductivity when a large proportion of the pores are dead ends and not involved in flow. Porosity may not be related directly to flow rates because of uneven distribution of pore sizes and the *tortuosity*, or average length of the flow path between two points, which varies as a function of type of material (Sahimi, 1995).

An example of high porosity but low conductivity material is carbohydrates excreted by microbes. These extracellular products have a high proportion of water and many microscopic pores, but allow little if any flow through them because the pores are small and the Reynolds number precludes flow at such high viscosity. Microbial secretions can decrease flow through sediments (Battin and Sengschmitt, 1999), making microbial ecology of groundwater and soils significant to hydrology. A specific example of this importance occurs when microbial growth plugs flows from wells or septic systems.

Samples of groundwater sediments and rocks are generally taken during well drilling to identify locations and depths that will optimize water yields. These samples are returned to the laboratory and analyzed for porosity and hydraulic conductivity. Given the expense of drilling wells, characterizing these properties, particularly when substrate heterogeneity is high, can be quite expensive.

Determining groundwater flow rates and directions can be very difficult. Measuring properties of groundwater is more complicated than measuring properties of any surface water because wells can only access a small portion of the habitat. Similarly, describing heterogeneity and the biota of groundwaters is beset with difficulties. Understanding groundwater hydraulics is essential to describing its connection with surface habitats. For example, much of the carbon in many shallow groundwater systems comes from terrestrial sources. If it takes hundreds or thousands of years for water to reach the habitat from the surface, then the habitat is likely to be extremely oligotrophic. In contrast, many karst aquifers have very brisk flow and contaminants from the surface such as nutrients and pesticides can spread rapidly through them.

Determining the velocity and direction of groundwater flow can be tricky. Several methods have been developed to measure flow direction and rate in groundwaters (Method 4.2). In all cases these methods are less straightforward

METHOD 4.2

Sampling Flow Rates and Directions in Groundwaters

Piezometers are used to determine the depth of water; these are tubes with open bottoms that are drilled into groundwater at various depths so the depth of water in the tube can be determined. Direction of flow is determined in both vertical and the horizontal directions. Nested piezometers (a group that is drilled to various depths in one location) are used to determine vertical flow direction, and piezometers spaced apart from each other can be used to determine horizontal direction of flow, as well as a vertical component.

The nest of piezometers is used to indicate pressure, with greater pressure leading to greater elevation of the top of the water inside the piezometer. If pressure differs between depths then water will flow from the region of higher to lower pressure. Nested piezometers indicate an upward flow if the elevation of the top of the water in the piezometer tube that penetrates the aquifer to the deeper point is greater than the elevation of the water in the shallower tube

(Fig. 4.5). The water is moving down if the elevation of the top of the water in the shallowest piezometer is greater than that in the deepest. If all elevations are the same, then the vertical flow is neither up nor down. If the flow is up, then water must be flowing into that location from somewhere else (either another part of the aquifer or through infiltration) and if the flow is down, then it must be replenished or the depth of the water table will decrease.

When piezometers are placed apart from each other, they can be used to estimate the local depth to the water table. If the elevation of groundwater at one site is lower than at another, and the two are hydraulically connected, then it is assumed that water is flowing from the higher to the lower site. The difference in elevation between the two sites is known as *hydraulic head*. Releasing a tracer (e.g., a dye or inert chemical) at the upper site and monitoring its appearance at the lower site can indicate water velocity directly.

Several additional methods can be used to assess flow rates through aquifers (Fitts, 2002; Kalbus *et al.*, 2006). Slug tests

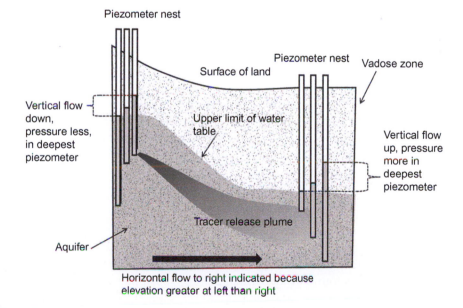

FIGURE 4.5

Ways to use small wells (piezometers) to estimate direction of groundwater flows. A nest (locally distributed) of piezometers can be used to determine vertical direction of flow, and spatially distributed piezometers can be used to determine lateral direction of flow.

consist of rapid water additions or removals from a well and monitoring of how long it takes the water level to reach its original depth. Equations can be used to analyze water depth data and relate them to flow. Longer periods of pumping can be used and the stable lower water table depth in the pumped well and nearby wells can be used to charac- terize flow rates. In tracer tests, water is added to the well with an inert tracer, and the rate that it is diluted out can be used to calculate flow rates. Flow meters can be installed in wells to determine precisely where in the depth profile the water is flowing from (to characterize heterogeneity of flow).

than measurement of surface water, and heterogeneous aquifers require sub- stantial sampling to understand their complexity.

GROUNDWATER HABITATS

Groundwater habitats can be divided into a variety of subhabitats (Fig. 4.6). For example, groundwater can flow through regions of continuous *homo- geneous* substrata (even distribution of permeable substrata such as sand, clay, or gravel) with little obstruction. Aquifers can occur where many alter- nate flow pathways exist because *heterogeneous* distribution of impermeable materials in the subsurface results in variable flow patterns and directions (e.g., aquifers with large rocks embedded in fine sediments or with patches of low hydraulic-conductivity material interspersed among high hydraulic- conductivity materials). Some rocks such as sandstone are permeable and have flow through the entire rock; others such as granite transmit water only along cracks or fractures in the rock and lead to very complex flow patterns. An aquifer between two impermeable layers is *confined*. Complicated groundwater flow patterns make determining the fate and source of waters difficult. Heterogeneous flow patterns are of concern when consider- ing groundwater contamination problems because such patterns interfere with assessing both the extent of the problem and attempts to clean up contaminants.

Groundwater is located worldwide, but the depth below the surface and the amount in the aquifer can vary across the landscape. In porous substrata, wells can yield a large amount of water. In some areas, such as those under- lain by solid rock, groundwater yields can be very low. Examination of the distribution of large aquifers demonstrates heterogeneity in the types of aqui- fers found in the United States (color plate Fig. 5). Some areas have extensive continuous aquifers (e.g., lower Mississippi Valley and High Plains) and oth- ers have sparser, localized aquifers (Rocky Mountain region). Groundwater in many of these aquifers is being depleted at rates faster than the rate of recharge. A famous example of this is the huge Ogallala or High Plains

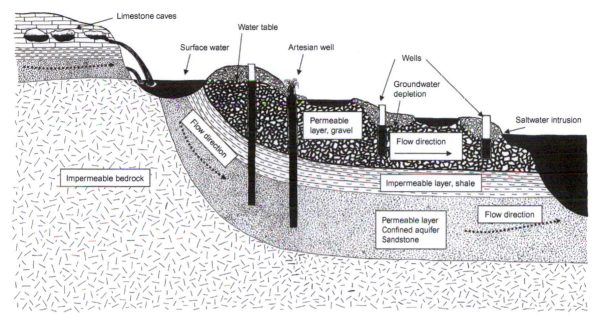

Limestone caves

Water table

Surface water

Artesian well

Wells

Groundwater
depletion

Saltwater intrusion

Flow direction

Permeable
layer, gravel

Flow direction

Impermeable bedrock

Impermeable layer, shale

Flow direction

Permeable layer
Confined aquifer
Sandstone

FIGURE 4.6

Water moves through groundwater and across the surface of the land in the hydrologic cycle. *(After Leopold and Davis, 1996; drawn by Sarah Blair, 1998).*

Aquifer (Sidebar 4.1). Groundwater depletion is commonly associated with irrigated land worldwide.

One of the major types of groundwater habitat is found in limestone regions with rough land surfaces called *karst* topographies (White *et al.*, 1995). Understanding specifics of karst aquifer hydrology is important for assessing the impacts of humans on groundwaters (Maire and Pomel, 1994). Large channels can form in these habitats because the water dissolves the limestone. If the water subsides, caves are left (Figs. 4.1 and 4.6). Pools and streams in limestone caves provide one groundwater habitat in which the geological formation allows humans to directly interact and sample the subsurface habitat. *Sinkholes*, depressions in the land's surface that are typical of karst regions, often are used for dumping household and farm wastes in rural areas. Sinkholes are closely connected to groundwater and cave systems, and they can be conduits for the rapid movement of pollutants into these undergrounds systems. There have been some rather spectacular occurrences of sinkholes forming rapidly and swallowing houses and other surface features (Fig. 4.9). Sinkholes can become lakes or wetlands in many areas. Hydrology

SIDEBAR 4.1
Mining the Ogallala Aquifer

The High Plains or Ogallala Aquifer stretches from Nebraska to the southern tip of Texas (Fig. 4.7). The aquifer underlies 450,000 km² and has an estimated thickness of up to 300 m and an estimated water volume of 4,000 km³. The volume is comparable to some of the larger lakes in the world. By the late 1980s the aquifer supplied 30% of all irrigation water in the United States (Kromm and White, 1992a).

Mean recharge rate is 1.5 cm per year, and withdrawal rates average about 10 times this rate. Between 1980 and 1994 large regions of the aquifer lost 10 m or more in depth. Precipitation to land above the aquifer is less than that required to support the crops that are irrigated from the aquifer (i.e., potential evapotranspiration exceeds precipitation). Annual withdrawals exceed the total annual discharge of the Colorado River, another important source of irrigation water in the United States (Kromm and White, 1992b). Some regions of the aquifer are very thick and can support withdrawals for decades. In many regions the water table has dropped far enough that it is not economically feasible to use the groundwater for irrigation (Kromm and White, 1992b) when grain prices are low and fuel prices are high. Water is being withdrawn at greater than sustainable rates, so the withdrawals can be referred to as "mining" the aquifer.

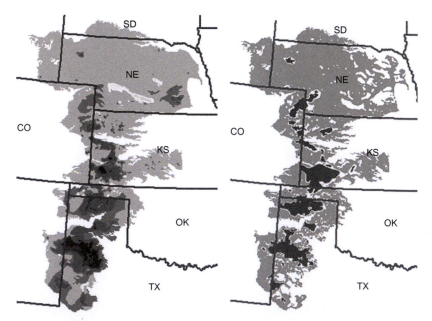

FIGURE 4.7
The extent of the High Plains Aquifer. In the left panel, zones of depletion >40 m depth before development to 1980 are shown in darkest colors. The most abundant gray color shows zones of no change, and the groundwater level has increased in the very light areas. In the right panel, the decline in depth between 1980 and 1997 is depicted, with the darkest colors indicating a >6 m decrease and the large gray areas indicating no significant change in depth. *(Data from the US Geological Survey).*

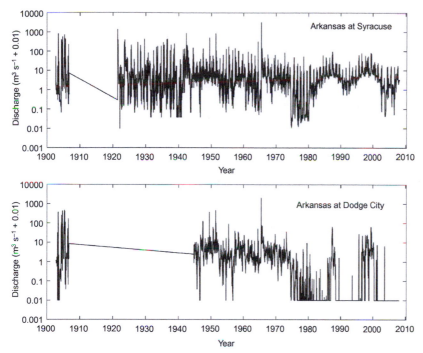

FIGURE 4.8

Discharge of the Arkansas River as it flows through a region of the High Plains Aquifer that has been used heavily for center-pivot irrigation since the 1960s. The Syracuse station is immediately upstream of the High Plains Aquifer and the Dodge City station is downstream. Note that the logarithmic scale of discharge is modified so that it reads 0.01 m^3 s^{-1} when the river is dry. The river downstream of the aquifer has flowed only during periods of flooding since the early 1970s; it was almost never dry prior to that. *(Data replotted from the US Geological Survey by the Central Plains Hydrologic Observatory).*

In addition to loss of economic uses, there are ecological impacts as the water table is drawn deeper underground. Depletion of the groundwater has caused decreased water supply and stream and river flow has disappeared in many regions (Kromm and White, 1992b). For example, the Arkansas River loses water to the aquifer because agricultural activity has lowered the water table, and now it only flows during floods (Fig. 4.8). Loss of flow has negative impacts on migrating waterfowl that use the river and decreases the ability of the river to dilute and remove pollutants. A number of small stream fishes in the region are endangered because most of the small streams no longer flow and perennial flow is not possible (Dodds *et al.*, 2004).

Conserving the remaining water makes good economic and ecological sense. It remains to be seen if more efficient irrigation technology and dryland farming will allow the same level of economic productivity as was made possible in the region during the past few decades by exploiting the High Plains Aquifer for irrigation water. As crop prices increase (for example, as occurred when the US government offered tax breaks for producing ethanol for fuel from corn), the incentive to pump deeper and deeper groundwater increases. When the cost of the energy to pump water up exceeds the profits obtainable from crops, dryland farming will replace irrigation unless subsidies are provided to farmers.

FIGURE 4.9
A sinkhole that formed in Bartow, Polk County, Florida. In May 22, 1967, it was 158 m long, 38 m wide, and 18 m deep. *(Photograph courtesy of the US Geological Survey).*

of karst aquifers is very complex, in part because it is difficult to predict the pathways of limestone dissolution (Mangin, 1994).

INTERACTION OF GROUNDWATERS WITH SURFACE WATERS

When groundwater impinges on the surface, a stream, lake, or wetland forms (Fig. 4.6). The dynamic zone of transition where both surface water and groundwater influences are found is referred to as the *hyporheic zone*. This zone forms a transitional habitat (*ecotone*) where there is a change between groundwater and surface water organisms. A hyporheic zone can be found between groundwater and wetlands, streams, or lakes (Gibert *et al.*, 1994).

The hyporheic zone represents the interface through which materials are exchanged between surface and groundwater. This zone may include interstitial water of sediments below lakes and wetlands, gravel bars in rivers, sand below streams, and many other benthic habitats in aquatic systems. Saturated fine sediments located on the fringes of water bodies are often referred to as the *psammon*. As with any habitat classification, the distinction between the hyporheic zone and groundwater is unclear because the zone is transitional, varies over space and time, and depends on whether material transport or habitats of organisms are considered (Gibert *et al.*, 1994). For example, hyporheic zones that are formed by river action can be quite complex because of erosion and deposition that naturally occur in the stream channel (Creuzé des Châtélliers *et al.*, 1994). These zones can be extensive (stretch many kilometers from the river) in larger valleys with substantial amounts of large cobble (e.g., glacial valleys with large amounts of rocky material in the unconsolidated sediments; Stanford and Ward, 1993). In the Flathead River system in Montana, nymphs of stoneflies (aquatic insects that have a terrestrial adult stage) have been found in well water drawn from over 2 km from the surface flowing river channel (Stanford and Ward, 1988). The ecological importance of hyporheic zones has become increasingly apparent to aquatic ecologists (Danielopol *et al.*, 1994; Gounot, 1994; Allan, 1995; Stanley and Jones, 2000).

Methods have been developed to quantify connections between hyporheic zones and surface water (Harvey and Wagner, 2000). These methods include those already discussed for determining groundwater flow rates as well as use of natural or added tracers to detect the source of water. The rate at which groundwater interacts with surface water can be biologically important. Streams are often fed by subsurface flow, and lakes and wetlands can also have extensive interaction with groundwater. There are several methods for determining the influence of groundwater on surface water (Kalbus *et al.*, 2006). Groundwater is often a different temperature from surface water, so temperature gradients can be used to assess rates of groundwater input. Relatively biologically conservative salts (their concentration is not influenced by organisms) are often dissolved in groundwater at different concentrations than in surface waters, and the input/dilution of these salts can be used to assess rates of interactions of groundwaters with surface waters. Similarly, radon concentrations are often greater in groundwaters relative to surface water, and these concentrations can be used to characterize inputs of groundwater.

SUMMARY

1. Freshwater habitats vary in scale from individual sediment particles to continental watersheds. The appropriate scale of investigation depends on the question being asked.
2. Precipitation falls unevenly across Earth and evaporates or is transpired (evapotranspiration) at different rates depending on a variety of factors, including global weather circulation patterns, geography, and landscape level influences. Water that is not lost to evapotranspiration either flows across the surface of the land to streams and rivers or infiltrates the soil to groundwater.
3. Characteristics of the medium between soil and groundwater alter the rate at which water flows below the surface and into aquifers. Generally, water flows more

slowly through fine-grained sediments. Once water enters an aquifer, the rate at which it moves through is also dependent on slope and the materials that make up the aquifer. Water flows very slowly between the pores of fine-textured sediments such as silts and clays or those with large amounts of organic materials, and relatively rapidly in coarse gravel or limestone with large channels and pores.

4. The hydrodynamics of groundwater dictate its use as a water resource, the ecology of its unique biota, and interactions with other aquatic habitats. Efforts to clean up groundwater pollution also depend on knowledge of groundwater and soil characteristics, particularly flow characteristics.

5. Groundwater interacts with surface water, and in many cases dictates important properties of that water. Many streams are primarily fed by groundwater, as are many lakes and wetlands. Surface water generally occurs where the top of the water table is above the elevation of the land surface. Hyporheic zones are at the interface between surface and groundwater habitats, and they can be extensive and ecologically significant components of many freshwater systems.

QUESTIONS FOR THOUGHT

1. Why would understanding the hydrodynamics of groundwater be important when a large oil spill occurs on land (e.g., leakage of a gasoline or oil storage tank)?

2. Darcy's law as presented in the text is one-dimensional (the equation in Chapter 2); why would a three-dimensional model be more useful?

3. Why might understanding the application of Darcy's law to sediments under a wetland be important when calculating a water budget for a wetland?

4. Why are groundwater temperatures one good indicator of annual air temperatures and ultimately regional warming?

5. How could temperature alter flow rates of groundwater?

6. What differences might you expect between the biota inhabiting hyporheic and groundwater habitats?

CHAPTER 5

Hydrology and Physiography of Wetland Habitats

FIGURE 5.1

A flock of wintering northern pintail ducks (*Anas acuta*) takes flight from the banks of a wetland in Northern Honshu, Japan. *(Photo courtesy of the US Geological Survey).*

Doi: 10.1016/B978-0-12-374724-2.00005-2

INTRODUCTION AND DEFINITION OF WETLAND HABITATS

Wetlands are critical habitats for many plants and animals, including numerous threatened and endangered species, and they provide vital and valuable ecosystem services such as flood control and the maintenance of water quality (van der Valk, 2006). Wetlands often receive and process large amounts of runoff from the landscape, and in some cases they are used to treat wastewater. In addition, wetlands are globally important in biogeochemical cycles and they are natural sources of methane (see Chapter 13), a trace gas that plays an important role in the regulation of climate (Schlesinger, 1997). Wetland sediments are valuable because they preserve a long-term record of environmental conditions, and sediments in peat bogs are mined for fuel and for use in gardens (Fig. 5.2). Many coal deposits that humans use for energy today were created from deposition in wetland habitats over millions of years. The study of wetlands is relatively new compared to that of lakes and streams because such study falls between the traditional disciplines of limnology and terrestrial ecology. However, there is now great interest in wetlands because of increasing recognition of their ecological, economic, and aesthetic value.

FIGURE 5.2

Peat moss mining from a bog in Ireland for use as fuel; face of the bank is approximately 1 m tall. *(Photo courtesy of Charles Ruffner).*

Many wetlands harbor high biological diversity, including numerous terrestrial and aquatic species, along with wetland specialists because wetlands are often *ecotones* (transitional habitats); they are frequently adjacent to both terrestrial habitats and streams, lakes, or oceans. Some wetlands such as ephemeral pools and acidic bogs are inhabited by a limited suite of species that are specially adapted to harsh conditions including short periods of inundation, low O_2 levels, or low pH. Although wetlands are often generalized as highly productive ecosystems, productivity is actually quite variable; estimates of net primary production in freshwater wetlands range from 800 to $4,000\,g\,m^{-2}\,y^{-1}$, and some tropical and temperate wetlands are among the most productive systems on the planet (Lieth, 1975).

One of the problems with studying and managing wetlands is defining them. Although distinct delineation is difficult between a wetland and a very shallow pond, or a slow, shallow side channel of a stream, the problem of finding a definition for wetland lies more in deciding what is a wetland versus what is terrestrial habitat. Wetland definitions and delineations are generally based on the plants that are present (often water-loving plants, called *hydrophytes*) or absent (flood intolerant species), and distinct soils with characteristics related to frequent inundation. The *hydric soils* of wetlands have characteristic colors and textures, often with dark, highly organic surface soils overlying gray mineral layers, and they are anaerobic even in the upper layers during the growing season. Some wetland soils are very nutrient poor, resulting in unusual assemblages of plants that are adapted to low nutrient conditions. Carnivorous plants such as pitcher plants and sundews, which lure and trap insects to gain nutrients (see also *Temporary Waters and Small Pools* in Chapter 15), can be abundant in nutrient poor wetlands.

What is legally considered a wetland has particular importance with respect to the requirements for wetland preservation. Policymakers are increasingly realizing the central importance of wetlands as wildlife habitat and key providers of valuable ecosystem functions. Environmentalists have pressed for more inclusive definitions of wetlands. Agriculturists, developers, and others want more freedom to develop and drain both seasonally wet regions and permanent wetlands. Consequently, numerous definitions of wetlands have been developed by scientists, policymakers, and others (Sidebar 5.1). However, there is still no single, indisputable, ecologically sound definition for wetlands because wetland types are so diverse (Sharitz and Batzer, 1999).

In many areas of the world, wetlands have been drained, filled in, or otherwise modified. In the United States, 70% of the *riparian* (near rivers) wetlands were lost between 1940 and 1980, and more than half of the *prairie potholes* (shallow glacial depressions in the northern high plains that form vital habitat for waterfowl) as well as the Florida Everglades have been lost since pre-European settlement (Mitsch and Gosselink, 2007). This degree of loss is typical for all types of wetlands in the United States. Twenty-two states have lost more

SIDEBAR 5.1
Definitions of Wetlands

Several definitions of wetlands have been chronicled by Mitsch and Gosselink (1993) and by the Committee on Characterization of Wetlands (1995). The definition used often depends on the requirements of the user.

United States Fish and Wildlife Service: Wetlands are lands transitional between terrestrial and aquatic systems where the water table is usually at or near the surface or the land is covered by shallow water. Wetlands must have one or more of the following three attributes: (1) at least periodically, the land supports primarily hydrophytes; (2) the substrate is predominantly undrained hydric soil; and (3) the substrate is nonsoil and saturated with water or covered by shallow water at some time.

Canadian National Wetlands Working Group: Wetland is defined as land having the water table at, near, or above the land surface or that is saturated for a long enough period to promote wetland or aquatic processes as indicated by hydric soils, hydrophytic vegetation, and various kinds of biological activity that are adapted to the wet environment.

Section 404 of the 1977 United States Clean Water Act: The term "wetlands" means those areas that are inundated or saturated by surface or groundwater at a frequency and duration sufficient to support, and that under normal circumstances do support, a prevalence of vegetation typically adapted for life in saturated soil conditions. Wetlands generally include swamps, marshes, bogs, and similar areas.

1985 United States Food Security Act: The term "wetland," except when such term is part of the term "converted wetland," means land that (1) has a predominance of hydric soils; (2) is inundated or saturated by surface or groundwater at a frequency and duration sufficient to support a prevalence of hydrophytic vegetation typically adapted for life in saturated soil conditions; and (3) under normal circumstances does support the prevalence of such vegetation.

1995 Committee on Wetlands Characterization, US National Research Council: A wetland is an ecosystem that depends on constant or recurrent, shallow inundation or saturation at or near the surface of the substrate. The minimum essential characteristics of a wetland are recurrent, sustained inundation or saturation at or near the surface and the presence of physical, chemical, and biological features reflective of recurrent, sustained inundation or saturation. Common diagnostic features of wetlands are hydric soils and hydrophytic vegetation. These features will be present except where specific physicochemical, biotic, or anthropogenic factors have removed them or prevented their development.

Some other words historically used to delineate wetlands or specific types of wetlands: The term wetland has been in regular use by scientists only since the mid-1900s. Terms used prior to this, or to indicate specific types of wetlands, include bog, bottomland, fen, marsh, mire, moor, muskeg, peatland, playa, pothole, reedswamp, slough, swamp, vernal pool, wet meadow, and wet prairie.

than half of their wetlands in the past 200 years (Fig. 5.3). The amount of loss and interconversion among wetland types has been great (2.5% lost over a period of 10 years), and it is driven primarily by agriculture. The creation of ponds, lakes, and reservoirs from wetlands is also an important aspect of human activities that has negative consequences for many wetland species (Table 5.1). Peatlands are under increasing pressure throughout the

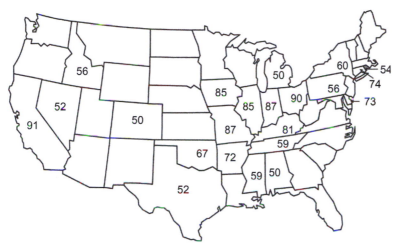

FIGURE 5.3

States that have lost 50% or more of their wetlands since 1780, labeled with percentage lost.
(Data from Dahl et al., 1991).

Table 5.1 Conversion and Losses of Various Wetland Types Including Agricultural and Urban Uses from the Mid-1970s to the Mid-1980s in the United States

Wetland Type	Amount in Mid-1970s	Amount in Mid-1980s	Total Loss (Gain)	Agricultural	Urban Land Use	Deepwater	Other	Conversion to Other Wetland Types
Swamps	223,000	209,000	−14,000	−4,000	−240	−200	−4,360	−5,200
Marshes	98,000	99,000	1,000	−1,500	−150	0	−350	3,000
Shrubs	63,000	62,000	−1,000	−1,000	0	0	−1,700	1,700
Ponds	22,000	25,000	3,000	900	0	0	1,800	300
Total	406,000	396,000	−11,000	−5,600	−390	−200	−4,600	−200

(Data from Dahl et al., 1991)
Deepwater Represents Conversion to Lakes or Reservoirs and Conversions are Placed in the Other Category if They are not to Agriculture, Urban, or Deepwater. Values are given in Thousands of km². Positive Values Indicate a Net Gain

world as a source of fuel and peat moss for gardening. In Southeast Asia, existing wetlands have been modified and many new artificial wetlands have been created to allow for rice culture (Grist, 1986); these *rice paddies* have fed billions of people over the centuries and formed the basis of one of the world's great civilizations, the Chinese dynasties. The decline in wetlands is global (Dugan, 1993); for example, large percentages of wetlands have been lost in the United States (54%), Cameroon (80%), New Zealand (90%), Italy (94%), Australia (95%), Thailand (96%), and Vietnam (99%).

WETLAND CONSERVATION AND MITIGATION

In response to decades of wetland destruction, and the realization of the ecological and economic values of wetlands, laws in the United States and other countries increasingly mandate protection of wetlands and, in cases where wetlands are "unavoidably" disturbed or destroyed, some form of compensation. Mitigation, or compensatory mitigation, is a procedure for offsetting wetland loss by creating another wetland, restoring or enhancing an existing wetland, or providing funding to conservation groups or agencies for wetland preservation. Wetland mitigation programs have slowed the net loss of wetland habitats, but there are questions regarding how closely some created or restored wetlands approximate natural systems in terms of their biological diversity and ecosystem functioning (Fig. 5.4) (Galatowitsch and van der Valk, 1995; Brooks *et al.*, 2005; Meyer *et al.*, 2008). Analysis of recovery of the value of ecosystem goods and services indicates that over 10 years, many wetland functions can be recovered in restored wetlands (Dodds *et al.*, 2008), but that conserving existing wetlands preserves the most value.

Section 404 of the United States Clean Water Act requires permits for discharging any fill materials into any US water bodies including wetlands. According to the Clean Water Act, any impacts to wetlands must first and foremost be avoided. Second, remaining impacts must be minimized. Finally, any unavoidable impacts must be compensated for through mitigation

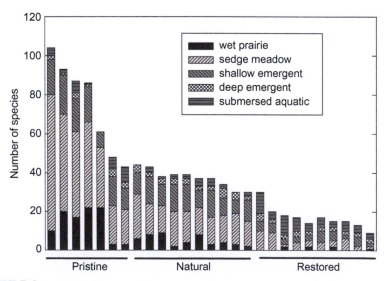

FIGURE 5.4

Species richness of five categories of wetland plants in pristine, natural (natural systems receiving agricultural runoff), and restored prairie pothole wetlands. The restored wetlands were formerly cropped and natural revegetation was allowed. *(Redrawn from Galatowitsch and van der Valk, 1995).*

procedures. The Food Security Act of 1985, often called "Swampbuster," further strengthened wetland protection in the United States; it includes a variety of provisions to discourage the conversion of wetlands for agricultural production. Under Swampbuster, farmers can be denied federal farm program benefits if they alter wetlands on their property.

Opportunities for mitigation procedures in the United States were expanded in the 1996 Federal Agricultural Improvement and Reform Act, which also began the process of developing wetland mitigation "banks." Wetland mitigation banks are wetlands that have been set aside or restored for the purpose of providing compensation for unavoidable impacts to other wetlands. The value of a bank is defined in compensatory mitigation credits. Mitigation banks represent third-party compensation, in that the responsibility for compensatory mitigation is assumed by someone other than the permit holder that impacts a wetland. Mitigation banking is becoming increasingly popular; the Society of Wetland Scientists wrote a position paper in 2004 endorsing mitigation banking as a sound practice that will help achieve the goal of no net loss of wetland habitats. However, most wetland scientists agree that site selection, design, monitoring procedures, and long-term management of mitigation banks and other forms of compensatory mitigation are still evolving and in need of further refinements.

In the European Union, the Water Framework Directive indirectly includes protection of wetlands. This directive requires protection and preservation of all surface waters with respect to species diversity and ecosystem functions. In contrast to the current US legal framework, small streams and isolated wetlands are considered connected to surface waters and are protected under this framework. Alternatively, human influence in Europe has been prolonged, and land-use intensive. Many European wetlands have been drained and lost forever.

On a global scale, the Ramsar Convention on Wetlands is one of the most successful conservation programs in the world. Named because it was originally signed in Ramsar, Iran, in 1971, the Ramsar Convention is an intergovernmental treaty that requires participating entities to designate and protect wetlands of international significance, as well as promote wetland conservation. As of 2009, there were 159 signatories to the Convention, with 1,847 wetland sites totaling 181 million hectares designated for inclusion in the Ramsar List of Wetlands of International Importance. Originally developed to protect critical habitats for migratory birds, the Ramsar Convention is the only international conservation program that explicitly targets wetlands.

WETLAND TYPES

Wetlands of various types are distributed worldwide (color plate Fig. 6), with particularly large areal coverage in northern Europe, northern North America, and South America. The processes that form these wetlands vary (Table 5.2) and

Table 5.2 Some Major World Wetland Types

Type	Description	Distribution	Geomorphology	Vegetation	Ecosystem Importance
Tidal salt marsh	A halophytic grassland or dwarf brushland on riverine sediments influenced by tides or other water fluctuations	Mid to high latitude, on intertidal shores worldwide	Form where sediment input exceeds land subsidence in regions with gentle slopes	Salt-tolerant grasses and rushes/periphyton	Highly productive, serve as nursery areas for many commercially important fish and shellfish
Tidal freshwater marsh	Wetland close enough to coast to experience tidal influence, but above the reach of oceanic saltwater	Mid to high latitude, in regions with a broad coastal plain	Area with adequate rain or river flow, with a flat gradient near coastline	High plant diversity including algae, macrophytes, and grasses	Highly productive, many bird species; often close to urban communities and susceptible to human impacts
Mangrove	Tropical and subtropical, coastal, forested wetland	25° north to 25° south worldwide	Forms in areas protected from wave action including bays, estuaries, leeward sides of islands, and peninsulas	Halophytic trees, shrubs, and other plants; generally sparse understory	Exports organic matter to coastal food chains, physically stabilizes coastlines; may serve as a nutrient sink
Freshwater marsh	A diverse group of inland wetlands dominated by grasses, sedges, and other emergent hydrophytes; includes important types such as prairie potholes, playas, and the Everglades	Worldwide	Widely varied	Reeds such as *Typha* and *Phragmites*; other grasses such as *Panicum* and *Cladium*; sedges (e.g., *Cyperus* and *Carex*); broad-leaved monocots (*Sagittaria* spp.); and floating aquatic plants	Wildlife habitat, can serve as nutrient sink
Northern wetland	Bogs and peatlands characterized by low pH and peat accumulation	Cold temperate climates of high humidity, generally in Northern Hemisphere	Forms in moist areas where lakes become filled in or where bay vegetation spreads and blankets; often a terrestrial ecosystem	Acidophilic vegetation, particularly mosses, but also sedges, grasses, and reeds	Low productivity system

(Continued)

Table 5.2 *Continued*

Type	Description	Distribution	Geomorphology	Vegetation	Ecosystem Importance
Deepwater swamp	Freshwater most or all of the season, forested	Southeast United States	Varied	Baldcypress-tupelo or pond cypress-black gum	Can be low nutrient or high nutrient; can serve as a nutrient sink
Riparian wetland	Wetland adjacent to rivers	Worldwide	In floodplains of rivers in regions with high water table	High diversity of terrestrial plants	Can provide key wildlife habitat and productivity particularly in more arid regions; can act as an essential nutrient filter

(After Mitsch and Gosselink, 1993)

generally are similar to the processes that form lakes (Chapter 7). The world's largest wetland complexes are found on all continents but Australia and Antarctica and in general are associated with large rivers or high latitude glacially influenced areas (Table 5.3). The traditional wetland classification system was developed as part of the National Wetlands Inventory (Cowardin *et al.*, 1979). Wetlands can be classified by geomorphology, hydrology, climate, nutrient input, and vegetation (Table 5.4). A classification system for wetlands has been proposed to allow assessment of wetland functions (Brinson *et al.*, 1994). Wetlands can be coarsely divided into two general types, inland and coastal. Four broad geomorphic classifications are riverine, depressional, coastal, and peatlands.

Coastal, or *fringe, wetlands* include *tidal marshes*, which can be freshwater or saline, and mangrove wetlands, which are saline and can be quite extensive along coastlines in the tropics. There are over 11 million hectares of coastal wetlands in the United States (Field *et al.*, 1991), but these systems are often highly modified or completely destroyed by coastal development. People are becoming increasingly aware of the economic and ecological significance of coastal wetlands, and lessons from hurricane Katrina in 2007 illuminated the importance of these systems as buffers from storms. Regions of New Orleans bordering intact coastal wetlands experienced less damage from Katrina than areas where wetlands had been drained and developed, and studies after the storm indicated that the overall destruction in the region would have been reduced if more extensive coastal wetlands had been present to buffer the storm (Costanza and Farley, 2007).

Tidal salt marshes are harsh habitats that are found throughout the world. These wetlands are brackish and influenced by ocean tides. In the United States,

Table 5.3 The Largest Wetland Complexes Globally

Rank	Location (Continent)	Description	Area (Thousand km²)
1	West Siberian Lowland (Eurasia)	Bogs, mires, fens	2,745
2	Amazon River Basin (South America)	Savanna and riverine forest	1,738
3	Hudson Bay Lowland (North America)	Bogs, fens, swamps, marshes	374
4	Congo River Basin (Africa)	Swamps, riverine forest, wet prairie	189
5	Mackenzie River Basin (North America)	Bogs, fens, swamps, marshes	166
6	Pantanal (South America)	Savanna, grassland, riverine forest	160
7	Mississippi River Basin (North America)	Bottomland hardwood forest, swamps, marshes	108
8	Lake Chad Basin (Africa)	Grass and shrub savanna, shrub steppe, marshes	106
9	River Nile Basin (Africa)	Swamps, marshes	92
10	Prairie Potholes (North America)	Marshes, meadows	63
11	Magellanic Moorland (South America)	Peatlands	44

(After Keddy et al., 2009)

Table 5.4 Some Examples of Expected Ecosystem Functions of Wetlands Based on Hydrodynamic Characterization

Primary Water Source	Climate	Geomorphological Aspects	Important Quantitative Attributes	Functions That Can Relate to Ecological Properties	Significance of Function or Maintenance of Characteristic
Precipitation	Humid	Poor drainage	Precipitation exceeds evapotranspiration during most times of year so soils waterlogged	Soil constantly waterlogged leading to peat formation and sediment anoxia	Low plant productivity related to anoxic sediment keeping plants from soil sources of nutrients, plants rely upon nutrients in precipitation only
Surface flow from flooding river	Mesic-humid	Floods occur at least annually	Frequency and height of floods and position of wetland an index of connectivity to river	Overbank flow creates influx of nutrients and moves sediments (changes physical structure)	Allows continued high production and high habitat heterogeneity
Groundwater influx	Mesic	Groundwater springs and seeps often at bottom of slopes or stream margins, some sediments must be permeable to allow influx	Aquifer permanence, yield of springs and seeps dominates hydrologic throughput	Groundwater supplies nutrients and flushes habitat, habitat often very stable	High plant production, stable plant community

Based on Brinson et al. (1994)

FIGURE 5.5

Researchers working on plant diversity plots in a *Spartina alterniflora* dominated tidal saltwater marsh on the Gulf Coast of Mississippi. *(Photo by Loretta Battaglia).*

these systems are usually dominated by *Spartina* grasses and *Juncus* rushes (Fig. 5.5). Wet and dry cycles, salinity, and large temperature fluctuations make these challenging habitats for biota and limit diversity, but high abundances of some types of crabs, snails, and other marine invertebrates live in these habitats and these, in turn, are critical food resources for waterbirds and other predators. Further inland, *freshwater tidal marshes* are influenced by tides but are not saline, and thus often have higher biological diversity and are more productive than their saline counterparts. A variety of plants, including grasses, sedges, cattails, and wild rice, are found in freshwater tidal marshes. Like their saline counterparts, these systems are critical foraging habitats for birds that feed on the abundant invertebrates, amphibians, and seeds.

Mangrove wetlands (or mangrove swamps) are coastal wetlands dominated by *halophytic* mangrove trees, which include 12 genera distributed across tropical and subtropical coastal areas. These wetlands develop in areas where wave action is minimal and sediments accumulate. Mangroves provide important protection from storms; it appears that there were far fewer fatalities from the

FIGURE 5.6
Above and below the water structure of mangrove trees. *(Photo courtesy of the US National Oceanic and Atmospheric Administration)*.

2004 South Asian tsunami in villages protected by mangrove wetland barriers than those where the mangroves had been removed for prawn farming.

Sediments in mangrove swamps are generally anoxic, and thus many species of mangrove trees such as black mangrove (*Avicennia germinanas*) have *pneumato-phores*, which are aerial roots that take up atmospheric oxygen for transport to the roots in the sediments. Mangrove swamps show characteristic gradients of species composition or zonation from the open water to land. The above- and below-the-water structure associated with mangrove trees provides important habitat for a variety of invertebrates and vertebrates, including commercially important shrimps and oysters (Fig. 5.6). Mangrove swamps and other types of coastal wetlands export much of their plant production to adjacent open water habitats as detritus, and these inputs are important fuel for marine food webs.

Wetlands known as *floating marshes* are common in some coastal areas. These systems are characterized by vegetation that forms thick mats of roots that, along with associated organic matter such as peat, float on the water. Because they develop on top of the water, the elevation of the plant community changes with fluctuating water levels. Within the normal range of water level fluctuations, floating marshes do not become inundated and thus are more resistant to flooding disturbances. Floating marshes in the Mississippi delta of the United

States are often dominated by maidencane (*Panicum hemitomum*), a grass that can grow over 2 m tall and forms thick, dense mats of root and rhizomes.

Extensive floating marshes form in Iraq near Basra above the confluence of the Tigris and Euphrates rivers. These marshes are formed on mats of cane material and serve as home to the "Marsh Arabs" who historically build reed houses on floating islands. The lush area has provided fish and wildlife to human inhabitants for all of recorded history. By the mid 1980s a third were drained by Iraq during the Iran–Iraq war to facilitate movement of supplies. Then Saddam Hussein drained more of the marshes that were serving as shelter for Iran-backed rebels. By the mid 1990s most of the marshes had been drained and most of the Marsh Arabs no longer lived in the area. Since then, some of the marshes have been restored, but it is not clear how well the ecosystem structure and function, or the historical way of life of the Marsh Arabs, will ever recover.

Inland wetlands include *depressional* and *fringe* formations such as *marshes* and *swamps*, *riverine* or *riparian wetlands*, and *peatlands*. Depressional formation processes are described more fully in the context of lakes in Chapter 7, and the formation of riverine wetlands is discussed in Chapter 6. In the United States, marshes and swamps are generally defined by the dominant vegetation; *marshes* are dominated by emergent herbaceous vegetation and *swamps* are dominated by trees. However, this generalization does not extend to some other parts of the world such as Europe and Africa, where wetlands dominated by herbaceous plants are often called swamps. Inland marshes and swamps are some of the most extensive wetlands on the planet and can vary tremendously in timing, frequency, and magnitude of inundation as well as morphology and size (Fig. 5.7). Extensive marshes such as the Florida Everglades, the fringe marshes associated with the Great Lakes, and the Pantanal in South America, are recognized globally as critical habitats for birds and other wildlife.

In the southeastern United States, extensive swamps such as the Okefenokee in southern Georgia and adjacent Florida and the Great Dismal Swamp of southern Virginia and North Carolina are dominated by trees such as baldcypress (*Taxodium distichum*), tupelo (*Nyssa* spp.), and others that are tolerant of saturated soils and frequent inundation. Trees growing in swamps often have different growth forms than even the same species growing in drier habitats. For example, baldcypress trees in swamps grow *knees* and *buttressed* bases (Fig. 5.8), which are absent or not as well developed on baldcypress trees growing in upland areas. Knees and buttressed bases may increase stability in the soft, saturated soils; baldcypress are one of the most resistant trees to windfall during storms. Buttressing is common among many species of trees found in wetland habitats.

Swamps and marshes are often associated with river *floodplains*. Much of the bottomland hardwood forests associated with large rivers in the southeastern United States (e.g., Mississippi), Africa (e.g., the River Nile, the Congo

FIGURE 5.7
An inland freshwater marsh in Louisiana with abundant floating and emergent vegetation including water lilies (*Nymphaea* spp.) and American lotus (*Nelumbo lutea*). *(Photo by Loretta Battaglia).*

River), and South America (the Pantanal) form globally important wetlands. In humid regions, floodplain forests can extend tens or even hundreds of kilometers (e.g., the Amazon basin) from the main channel of the river, whereas in arid climates they tend to be more constrained to narrow ribbons along the river. These *riparian* systems can range from ephemeral to nearly permanent inundation depending on local topography and connectivity to the river. Natural flood cycles facilitate the exchange of nutrients and energy between floodplain wetlands and the river channel (Junk *et al.*, 1989). These flood cycles are often disrupted or eliminated altogether by impoundments, levees, channelization, and other human modifications. When inundated, floodplain wetlands also serve as important spawning and nursery habitats for many invertebrates, fishes, and amphibians that cannot complete their life cycles in the main channel. In turn, through the process of *ichthyochory*, frugivorous fishes can serve as important dispersers of seeds they consume while foraging in inundated floodplain wetlands (Chick *et al.*, 2003; Correa *et al.*, 2007; Sidebar 23.1). The seeds of many other floodplain plants are dispersed passively through the water, a process known as *hydrochory*. As with fringe wetlands, riparian wetlands often show distinct gradients of vegetation types linked to soil moisture and elevation gradients.

FIGURE 5.8

A cypress swamp in Louisiana showing the buttressing and "knees" of the baldcypress trees. The water is covered with duckweed (*Lemna minor*), a floating aquatic plant that is abundant in wetlands. *(Photo by Loretta Battaglia).*

Riparian wetlands around the world are severely diminished and degraded for agricultural purposes because floodplain soils are fertile and in close proximity to water. However, in some tropical regions, extensive river-floodplains are still mostly intact. For example, parts of the Amazon River basin still have a natural, seasonal flooding regime. During the wet season, the river channel and inundated floodplain can be nearly 200 km wide along some reaches, and some floodplain forests remain inundated for over half the year. This system has served as a model for studies of the physical and biological interactions between rivers and riparian wetlands (Junk *et al.*, 1989).

The prairie pothole region of North America is an example of an extensive and critical marsh ecosystem formed through depressional processes (Fig. 5.9). Thousands of shallow wetlands called potholes were formed by retreating glaciers in this 776,000 square kilometer region extending from Iowa to central Alberta. The individual wetlands in this region vary from ephemeral to nearly permanent, and salinity is also highly variable. The wetlands can be high salinity in areas where they have no outlet and receive runoff every year that subsequently dries. This physicochemical diversity among the individual wetlands provides a template for a rich diversity of wetland species such as aquatic invertebrates

FIGURE 5.9
Prairie potholes in northwestern Minnesota. *(Photo courtesy of the US Fish and Wildlife Service).*

(Euliss *et al.*, 1999). Much of this region is heavily agricultural, and thus the majority of the wetlands were destroyed decades ago. Nonetheless, this wetland area is critical to migratory waterfowl and other flora and fauna in the region.

Various small, temporary water bodies can be classified as wetlands. These include ephemeral and intermittent vernal pools and other small wetlands. Although often small and inundated for short periods, these systems can harbor unique species and are critical habitats for many invertebrates and amphibians. The short hydroperiods of these systems limit the species that can inhabit them, and thus those species adapted to live in them are exposed to less predation pressure and less interspecific competition. Human modifications that alter the hydroperiods of these habitats often eliminate the specialist species that inhabit them. The ecology of temporary waters and small pools is discussed further in Chapter 15.

Peatlands are wetlands that accumulate decaying organic matter from mosses, sedges, and other wetland plants over time because the production of organic matter exceeds the rate of decomposition. Peat is a significant carbon storage that currently represents over one third of soil carbon on the planet (Gorham 1991), even though peatlands represent only 3% of the land area of the planet. As such, peatlands are generally regarded as carbon sinks that help ameliorate increasing carbon dioxide levels in the atmosphere. However, with warming temperatures and reduced water availability predicted for many regions as a result of climate change, these systems may shift to carbon dioxide sources because of increased decomposition, therefore exacerbating climate change.

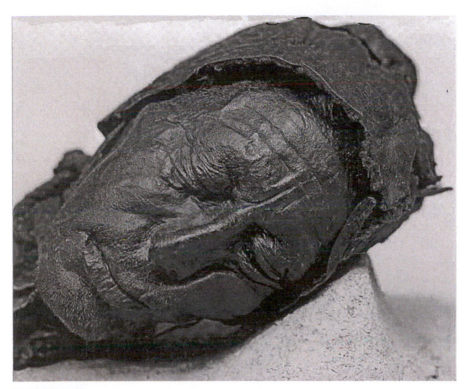

FIGURE 5.10
A bog body, the naturally mummified corpse of a man who lived during the fourth century BC. He was found in 1950 buried in a peat bog in Denmark. Low pH, cool temperatures, and anaerobic conditions result in remarkable preservation of soft tissues. *(Photo courtesy of Creative Commons).*

Bogs are a common type of peatland that have no significant water inflows or outflows and are thus *ombrotrophic*, meaning precipitation is the only water source. Bogs are nutrient-poor habitats inhabited by acid-tolerant plants such as sphagnum moss. Humic acids released during the slow decomposition of senescent mosses accumulate and make these systems acidic enough that processes such as decomposition are greatly suppressed. Although generally inhospitable habitats for many wetland animals, they are an excellent source of well-preserved remains of animals and humans from even thousands of years ago. The so-called "bog bodies" of Northern Europe are human bodies that are beautifully preserved in the cool, anoxic, acidic conditions of the bogs (Fig. 5.10; Sidebar 13.2).

Fens are similar to bogs in that they accumulate peat, but they are fed by additional sources of water such as groundwater and surface runoff and thus they are *minerotrophic*. The pH of the water in fens is usually neutral to alkaline, and nutrient concentrations are higher than bogs. As a result, fens tend to support a wider range of plant species, including sedges, grasses, spike rushes, reeds, and others. Fens with low nutrient availability and acidic water are known as *poor fens*, which are intermediate between fens and bogs; poor fens are generally dominated by short vegetation such as sedges. *Rich fens* have higher nutrient availability, are neutral or alkaline, and support more diverse plant assemblages. Fens and bogs often occur as the remnants of old lake basins that were originally created from glacial activity and gradually filled by sediments.

WETLAND HYDROLOGY

Hydrologic regimes of wetlands can range from highly variable to fairly constant. Hydrologic regime forms probably the most important abiotic template that influences wetland ecology (Wissinger, 1999). Important characteristics include permanence, predictability, and seasonality. For example, permanence controls the ability of large aquatic predators such as some fishes, amphibians, and larger invertebrates to inhabit a wetland. The range of permanence results in a *predator-permanence gradient*, because wetlands that are inundated for longer periods of time tend to harbor more predators. The presence or absence of these predators, in turn, structures the assemblages of lower trophic levels and can ultimately influence dynamics of basal resources such as plants and detritus. For example, many species of frog tadpoles cannot survive predatory fishes.

On a larger scale, diversity of hydrologic regimes among individual wetlands in a region can greatly influence wetland communities through the process of *cyclic colonization*, whereby individual wetlands that hold water longer are able to colonize less permanent wetlands when they become inundated (Batzer and Wissinger, 1996). Hydrologic diversity among individual wetlands in a landscape also enhances the diversity of plants and animals in a region.

Wetlands can receive any of three sources of water: precipitation, surface water, or groundwater. Hydrodynamic characteristics include fluctuations in water level and direction of water flow. Compared with lakes and streams, water loss by plant transpiration is usually more important to the hydrology of wetlands because they tend to be relatively shallow and densely vegetated. Tidal action can have a strong influence on hydrology of coastal wetlands such as coastal estuaries, and mangrove swamps, and tides ultimately result in bidirectional water flow. When rivers flow through wetlands, water moves unidirectionally. Hydrologic regimes of riparian wetlands are characterized by sporadic flooding with unidirectional flow, followed by extended periods of stagnation.

Conservation of wetlands clearly requires an understanding of hydrology. For example, in riverine wetlands hydrodynamic characteristics related to links to the river channels and geomorphology are important components of conservation (Bornette *et al.*, 1998a; Galat *et al.*, 1998). Hydrologic regimes are critical factors underlying the success of wetland restoration efforts (Hunter *et al.*, 2008; Meyer and Whiles, 2008).

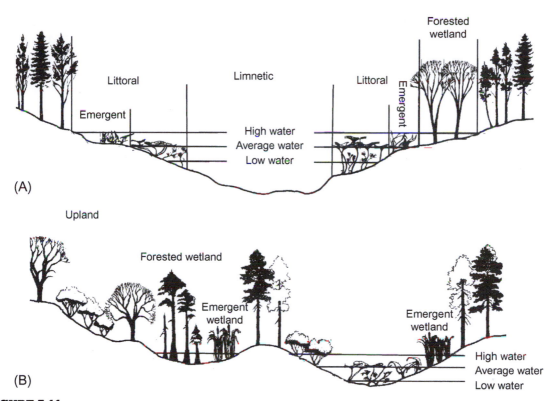

FIGURE 5.11
Classification of some subhabitats in two wetland types. Associated with still open water (A) and associated with a slow-moving stream (B). *(From Cowardin* et al., *1979).*

Climate interacts with hydrology to constrain hydrodynamics and the plants found in wetlands. Given the hydrogeomorphic properties, wetlands can then be classified by function (Table 5.4). Those wetlands that are fed by constant flows of groundwater may be influenced little by seasonal factors that control surface water flows. At the other extreme, seasonal wetlands such as playas or some riparian wetlands can fill during wet seasons and remain dry throughout the rest of the year.

Wetlands can be further categorized by nutrient input (eutrophic or oligotrophic), salinity, pH, and other components of water chemistry. Because of the variety of hydrogeomorphologic, climatic, and other factors, vegetation can range from dense forest to tundra or from macrophytes to trees (Table 5.4). A variety of different subhabitats can also occur within a wetland, depending on the degree and duration of inundation and the water depth (Fig. 5.11). In forested wetlands, hydrology, light availability, and subtle topographic gradients interact to influence forest composition and regeneration (Battaglia *et al.*, 2000).

Along with physical processes such as evapotranspiration, biotic factors can also influence wetland hydrology. Human activities obviously have strong influences on wetland hydrology. A large-scale example of this is the effort to manage the Everglades in Florida (Sidebar 5.2).

SIDEBAR 5.2
Managing Hydrology of the Florida Everglades

The Everglades and adjacent Big Cypress Swamp are parts of a large wetland area that covered more than $10,000 km^2$ of southern Florida prior to massive human modification during the past century (Gleason and Stone, 1994). This area is characterized by slowly flowing freshwater from the Kissimmee–Okeechobee–Everglades watershed. Movement of clean water through the wetlands is an ecosystem characteristic that is required to support the native flora and fauna.

Over the years, canals and dikes were used to drain large areas in the watershed for agriculture, development, and supplying water to the Miami metropolitan area. For example, in one decade in the late 1800s, millionaire Hamilton Disston drained 20,235 ha for agriculture. By 1917, four large canals (380 km total length) had been enlarged by the US Army Corps of Engineers. Because of these flood control and drainage practices, agricultural production and population increased dramatically in the region (Light and Dineen, 1994). In 1947, Everglades National Park opened, making official a desire to conserve at least part of the wetland.

Today, the South Florida Water Management District in large part controls the drainage systems erected during the past 100 years. The system is extremely complex and includes more than 700 km of canals, nine large pump stations, 18 gated culverts, and 16 spillways (Light and Dineen, 1994). These control structures must be managed to ensure delivery of freshwater for agriculture and Miami drinking water, to control flooding, and to provide enough clean water to maintain the ecological systems of the Everglades National Park. Variations in climate that need to be considered in this management include dry and wet seasons and extreme weather such as hurricanes (Duever et al., 1994). To further complicate matters, agricultural runoff has had detrimental effects on the native sawgrass, the input of nutrient-enriched water to sensitive areas must be managed (see Chapter 18), and animal communities respond variably to different management approaches (Rader, 1999).

Plans and actions to mitigate problems associated with altered hydrology and pollution in the Kissimmee–Okeechobee–Everglades are varied. The largest restoration project (up to the 1990s in North America) involves reversing the effects of channelization in the Kissimmee River. Biological effects of this restoration are discussed in Chapter 22. In other parts of the watershed, land is being purchased and water running off of agricultural areas is being treated to assist with nutrient removal. Nutrient pollution problems and solutions are described in Chapter 18.

The tremendous economic stakes (billions of dollars) conflict with preservation of what is left of the natural environment. Given the large tax base of the region, this has led to a situation in which numerous hydrologists, modelers, and aquatic ecologists, among others, are paid by local governments to make decisions that minimize the human impact on the Everglades while attempting to maximize the human benefits. In 2000 there were 68 separate projects included in the Comprehensive Everglades Restoration Plan. In 2009, the US government put over $200 million into hydrologic restoration. The problems are complex, but if they are not solved in the near future, the Everglades may be lost forever (Harwell, 1998). Although much important scientific information has been generated, still more is needed to provide the basis for rational management decisions about this important system.

Animals such as beavers, alligators, and some other freshwater vertebrates can greatly alter wetland hydrology and other biotic and abiotic features of wetlands. Organisms like these that strongly alter their environments are known as *ecosystem engineers*. Beavers have altered the geomorphology of entire valleys (Naiman *et al.*, 1994) and are thus one of the more commonly cited examples of ecosystem engineers. Alligators in the Everglades construct holes that keep the wetland from becoming completely dry during times of low precipitation and these deeper areas serve as refugia for fishes, snails, turtles, and other aquatic species.

Completely terrestrial animals can also influence and even create wetlands; American bison wallow in certain areas that, through repeated use, become small, depressional wetlands that serve as important habitats for some plants, invertebrates, and amphibians (Knapp *et al.*, 1999; Gerlanc and Kaufman, 2003; see also *Temporary Waters and Small Pools* in Chapter 15).

WETLANDS AND GLOBAL CHANGE

Wetlands may be greatly influenced by global change because water levels in many wetlands are the product of a delicate balance between rates of precipitation and evapotranspiration. Changes in either of these processes may directly affect input and output of water or may indirectly affect wetland water levels by altering depths to groundwater. Predicted warmer temperatures in many regions will produce higher evapotranspiration rates, which can lower water levels even if precipitation rates remain constant. Given the complex and variable hydraulic characteristics across the different types of wetlands, the magnitude and direction of the changes in these habitats are not easy to predict.

Predicted warming temperatures and reduced water levels are expected to affect important ecosystem processes in wetlands. Drier, warmer conditions enhance decomposition rates and thus oxidative losses, altering the balance of systems such as bogs that normally accumulate organic carbon. This is expected to create a positive feedback, whereby production of the greenhouse gasses CO_2 and CH_4 are enhanced because of increased decomposition, exacerbating the process of climate change. Further, changes in the hydrologic regimes of wetlands will also disrupt the life cycles of wetland animals, many of which are adapted to specific hydrologic patterns of the wetlands they inhabit (Batzer and Wissinger, 1996; Whiles *et al.*, 1999; Wissinger, 1999). Rising sea levels associated with warming temperatures will greatly influence coastal wetlands; predictive models based on the Intergovernmental Panel on Climate Change (IPCC, 2007) estimates for 2,100 indicate the area of coastal salt marshes will decline by 20% to 45%, resulting in significant losses of ecosystem services such as nutrient uptake and flood control (Table 5.5).

Table 5.5 Predicted Changes in Tidal Marsh Areas and Associated Plant Biomass and Nutrient Uptake and Processing on the Atlantic Coast Along Georgia Using The Mean (52 cm) and Maximum (82 cm) Estimates of Sea Level Rise From The Intergovernmental Panel on Climate Change Special Report on Emissions Scenarios (Meehl *et al.*, 2007)

Coastal Wetland Type	Change in Area (km²)		Change in Plant Biomass (t yr⁻¹)		Change in N Sequestration (t yr⁻¹)		Change in Potential Denitrification (t yr⁻¹)	
	52 cm	82 cm	52 cm	82 cm	52 cm	82 cm	52 cm	82 cm
Tidal freshwater	+1	−32	+1,400	−44,800	+8	−262	+7	−211
Brackish	+41	−4	+70,200	−6,800	+307	−30	+184	−18
Saltwater	−226	−496	−225,100	−494,000	−542	−1,188	−384	−843
Cumulative	−184	−532	−153,500	−545,600	−227	−1,480	−193	−1,072
Overall % change	−11%	−33%	−8%	−28%	−4%	−23%	−4%	−25%

(Data are from Craft et al., 2009)
Changes in Plant Biomass and N Sequestration and Potential Denitrification are Based on Mean Values for The Three Types of Wetlands. Negative Values Indicate a Reduction in Area and/or Process

SUMMARY

1. Wetlands are distributed worldwide, provide important habitat for wildlife, provide vital ecosystem services such as flood control and maintenance of water quality, and are important regulators of greenhouse gases.

2. The types of wetlands that have been described are extremely variable and generally defined by their morphology and formational process, the length of time they contain water, their vegetation, and the degree of marine influence. The geology of wetlands varies in different parts of the world, and there is no dominant process that leads to wetland formation worldwide.

3. Rice paddies constitute a human-created type of wetland that feeds a large portion of the world's human population.

4. The hydrology of many wetlands has undergone major changes due to human activity. Many wetlands have been drained and lost, and others are compromised severely. Thus, wetlands are among the most endangered habitats throughout the world.

5. Wetland protection efforts in the United States and globally have steadily increased over the past few decades. These efforts generally focus on conservation of existing wetland habitats or, if impacts are inevitable, mitigation procedures. Wetland mitigation involves the conservation or restoration of wetland habitats in order to compensate for modification of wetlands.

6. Wetlands can be coarsely divided into coastal and inland systems; there are four broad geomorphic classifications including riverine, depressional, coastal, and peatlands. Coastal wetlands can be heavily influenced by tides, whereas the hydrology of riparian wetlands is influenced by rivers. Depressional wetlands can

be influenced solely by local precipitation or various combinations of precipitation and surface and groundwater inputs.

7. Hydrology has a pervasive influence on wetland processes and communities. Biotic interactions such as predation are generally more important in wetlands with long hydroperiods. At the landscape scale, hydrologic diversity among individual wetlands facilitates regional biodiversity and colonization processes.

8. Wetlands are particularly susceptible to predicted climate changes because they are the product of a delicate balance between precipitation and evapotranspiration. These changes may further alter biogeochemical cycles and climate because globally wetlands store tremendous quantities of carbon.

QUESTIONS FOR THOUGHT

1. Do you know of any local wetlands that are endangered or have been drained in your lifetime?
2. How can temporary wetlands in arid habitats be extremely important to wildlife?
3. How might wetland mitigation and restoration procedures be improved?
4. Why do extensive wetlands exist in the high Arctic, even though annual precipitation is similar to that in many temperate or tropical deserts?
5. When new wetlands are created to replace those destroyed by development, what ecosystem goods and services might be easy to replicate and which ones might be difficult to re-create?

Physiography of Flowing Water

FIGURE 6.1

Salt Creek Falls, Oregon.

DOI: 10.1016/B978-0-12-374724-2.00006-4

Rivers and streams are central to life. Small streams are dominant interfaces between all other aquatic habitats and land. Streams and rivers move materials from land to sea through lakes and estuaries, forming a vital link in global biogeochemistry. Streams and rivers have been well characterized by hydrologists because of interest in flooding, erosion, and water supply (for a basic description of river geology and hydrology, see Leopold, 1994; for practical aspects of special interest to ecologists, see Gordon *et al.*, 1992). To comprehend the importance of streams in aquatic ecology, it is necessary to understand their morphology and geology. In this chapter, we discuss ways to describe streams, characteristics of stream flow, geology, and how streams move materials.

CHARACTERIZATION OF STREAMS

We first describe characterization using watershed features such as discharge, number of upstream branches, and area. Then we describe how streams are classified with respect to water velocity and changes in discharge. Finally, classification by vegetation in the watershed is explored. One way to characterize a stream is by the size of its drainage area. *Drainage area* is the land area that is drained by all the tributary streams above a chosen point in the main channel. In North America *watershed* is synonymous with drainage area, but Europeans use the word *catchment*; the word basin is also used by some. Watershed in European terminology is the boundary of the catchment (i.e., the ridge that divides catchments). Even though the ridge "sheds" water and the basin "catches" it, we use the American definition of watershed.

Globally, more permanent streams and rivers occur in regions where there is more precipitation (color plate Fig. 6). A greater number of intermittent streams are found in drier regions (color plate Fig. 6). The *discharge*, or volume of water passing through a channel per unit time, is approximately related to the area of the watershed (Fig. 6.2). The amount of discharge produced per unit area has an approximate upper bound. Many drier areas have lower discharge per unit area than wetter areas, creating a 100-fold variance in the relationship between watershed size and discharge for a watershed of any given size. The world's largest rivers occur in large drainage basins with significant amounts of precipitation. A list of the 15 largest rivers in the world is presented in Table 6.1. The Nile is the longest river at 6,758 km, but it historically ranks only thirty-sixth in discharge because it drains a relatively arid landscape. Now, the Nile ranks even lower because in many years it is almost dry by the time it reaches the Mediterranean Sea because of such heavy use of the river.

Stream order is another way to characterize streams. The most common method for ordering streams is the *Strahler classification system*, often used by ecologists

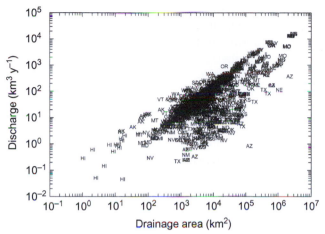

FIGURE 6.2

Discharge as a function of area for a large number of watersheds in the United States. Letters are the abbreviations for the states from which the data were obtained. *(Data courtesy of the US Geological Survey).*

Table 6.1 Average Discharge of the 15 Largest Rivers of the World, in Order of Rank

Rank	River	Country	Drainage Area (km²)	Length (km)	Av. Ann. Disch. (m³ s⁻¹)	Rank Order Length	Rank Order Drainage Area
1	Amazon	Brazil	5,950,000	6,597	176,000	3	1
2	Congo	Congo	3,700,000	4,586	41,000	8	2
3	Yangtze	China	1,940,000	5,744	33,100	5	9
4	Orinoco	Venezuela	980,000	2,735	23,000	20	16
5	La Plata	Uruguay	3,110,000	3,894	22,000	14	4
6	Brahmaputra	Bangladesh	930,000	2,896	20,000	17	19
7	Yenisei	Russia	2,610,000	5,937	20,000	4	6
8	Ganges	India	1,000,000	2,510	19,000	24	13
9	Mississippi	United States	3,210,000	6,693	18,000	2	3
10	Lena	Russia	2,490,000	4,312	16,000	10	7
11	Mekong	Indochina	790,000	4,248	16,000	11	23
12	Irrawaddy	Burma	430,000	2,011	14,000	26	—
13	Ob	Russia	2,450,000	5,567	12,000	6	8
14	Tocantins	Brazil	910,000	2,639	11,000	21	20
15	Amur	Russia/China	1,850,000	4,344	11,000	9	10

(After Leopold 1994)

when describing basic stream characteristics (Fig. 6.3). In this method, the smallest streams are assigned first order. Order only increases when two streams of the same order join. Modifications to this method of stream ordering have been proposed, each with their own benefits and drawbacks (Allan, 1995; Allan and Carillo, 2007). Following are some of the practical problems with determining and using stream order: (1) determination of the smallest permanent stream in a network is often difficult; (2) maps cannot always be relied on for accurate hydrological information because blue line and dashed blue line features (permanent and intermittent flowing waters) are not determined consistently; and (3) stream order does not always correlate closely with discharge,

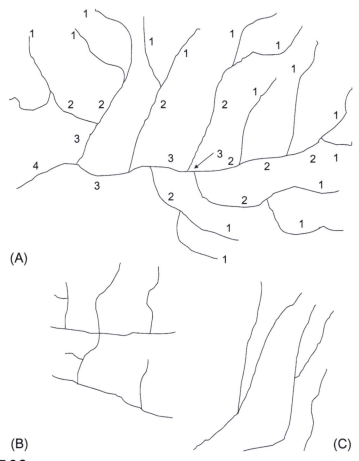

(A)

(B)   (C)

FIGURE 6.3

The Strahler method of stream ordering on a dendritic stream network (A). Order increases only when two streams of equal order meet. Other types of drainage patterns include rectangular, which may be found in Karst systems (B), and parallel, which occur mainly in deeply eroded areas (C). *(Modified from Strahler and Strahler, 1979).*

water chemistry, or other important abiotic factors, although low-order streams have some predictable differences from higher orders.

More recently, computer-based tools have been used for more accurate assessments of stream order and drainage patterns. Geographical Information Systems (GIS) can be coupled with detailed digital elevation maps to identify stream networks in predefined watersheds. The mapping methodology can also be combined with models of water yield specific to soil types and different regions to allow prediction of where there should be flowing water. To a lesser degree, these tools have some of the same problems as traditional maps, but they can provide fast, automated estimates of order for many streams in a region. Stream order will continue to be used as a primary method to characterize streams at hierarchical levels of organization, but students should not put too much emphasis on the exact order of a stream.

In general, a greater number of low-order streams occur in a watershed. Although streams of higher order have a greater length per stream, the total length of low-order streams is often greater (Fig. 6.4). The high relative abundance of small streams means that processes that occur where small streams interact with land or groundwater dominate interactions between aquatic and terrestrial systems. Stream drainage systems can also be classified by the pattern of the stream channels (Fig. 6.3). Various patterns develop in response to geological factors and can give rise to very different segment lengths of each order. Patterns range from highly reticulated to almost linear in shape.

Another useful way to characterize stream hydrology is by discharge and water velocity. Discharge is related to, but different from, water velocity. Discharge is a volume of water passing through a channel per unit time, whereas *water velocity* is the speed of water at any point in a channel (also referred to as current). Flow is a general term for movement that can mean discharge, water velocity, or both. We have discussed the strong biological effects of water velocity on organisms in Chapter 3 and will describe effects of both discharge and water velocity in streams in subsequent chapters.

Discharge is related to water velocity through the cross-sectional area

$$Q = VA$$

where Q = discharge, V = velocity, and A = cross-sectional channel surface area. Cross-sectional area = average width * average depth. The way these features are measured varies with stream size. In larger streams and rivers it is reasonably accurate to measure average depth and width and multiply them to get cross-sectional area. Average water velocity can be determined with numerous measurements across the channel with depth. An alternative is release of an inert tracer like sodium chloride or a dye at a known rate and

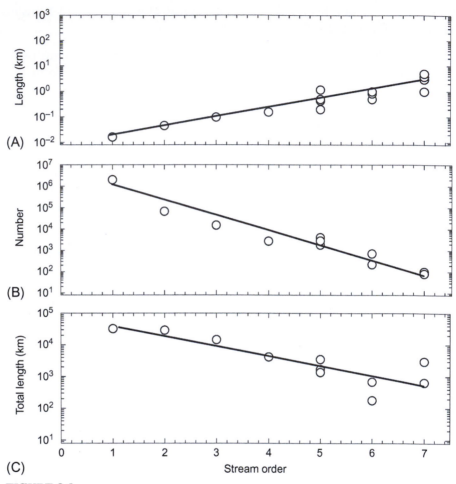

FIGURE 6.4

Relationships between stream order, average lengths of each order (A), number of streams of each order (B), and total length of streams of each order (C) for several watersheds in the southwestern United States. On average, streams of lower order are shorter, but they can also be more numerous; individual watersheds can have a greater total length of low-order streams. *(Data from Allan, 1995, and Leopold et al., 1964).*

calculation of discharge from the amount of dilution. For example, a dye tracer can be used to calculate discharge when added at a constant rate

$$Q_{stream} = Q_{dye}[C_{dye}]/[C_{stream}]$$

where Q_{stream} = discharge of the stream, Q_{dye} = pumping rate of the dye, $[C_{dye}]$ = concentration of the dye, and $[C_{stream}]$ = concentration of dye in the stream.

FIGURE 6.5
A weir used to measure discharge from water height, viewed from downstream. There is a depth sensor and a lateral pipe system to log water depth continuously in this particular weir. The depth sensor and data logger are housed in an enclosure behind the bison to the left.

This method requires adequate mixing of the dye with the streamwater. It is a better method in small streams where it is more difficult to determine average depth and make accurate velocity measurements. Another common method for measuring discharge in a small stream is with a *weir*, which is a constructed constriction in the stream channel with a known geometry (Fig. 6.5). In this case water height can be related to discharge by making repeated measures of depth and discharge at different depths and constructing a *rating curve*, which is used to predict discharge from depth.

A plot of discharge against time is called a *hydrograph*. Discharge can vary from fairly constant in rivers fed primarily with groundwater (Figs. 6.6A and 6.6B) to intermittent in headwater streams of drier regions (Figs. 6.6C and 6.6D). Steep watersheds and intense storms can lead to great variability in discharge (Fig. 6.6E). Damming changes the natural hydrograph in several ways. Generally, the magnitude and duration of floods are moderated and low discharge periods may be rarer (Fig. 6.7). If the dam is used for power generation, there can be fluctuations in downstream discharge relative to changes in power demand throughout the day, but much of the seasonal variation in flow can be decreased.

Streams can be classified as *perennial*, flowing all the time or at least at all times except extreme droughts; *intermittent*, flowing some of the time and receiving water from groundwater; or *ephemeral*, flowing rarely and not receiving input

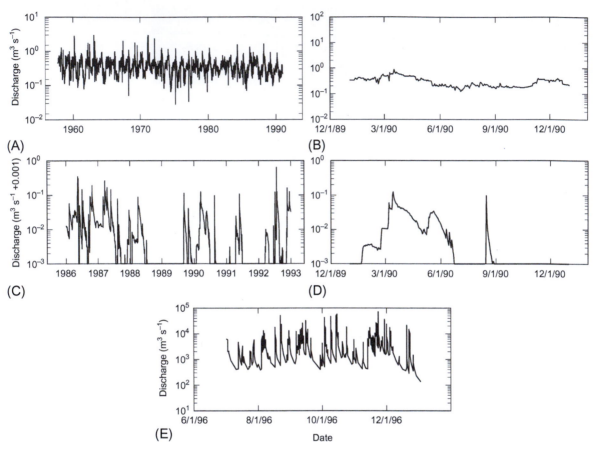

FIGURE 6.6

Hydrographs from three river systems plotted on log scales. The Niobrara River in Nebraska is mostly spring fed and shows relatively little variation in discharge among (A) and within years (B; note only about a 10-fold difference in each year, whereas two or three orders of magnitude are covered in the remaining hydrographs). Kings Creek in Kansas is a small, intermittent, prairie stream, with alternating periods of wet and dry over the years (C). A typical year in Kings Creek includes both times of no flow and floods (D; note 0.001 = 0 discharge in C and D). A stream in a steep watershed (Slaty River on the west coast of New Zealand) that experiences frequent rainstorms exhibits approximately weekly floods (E). *(Data from A and B courtesy of US Geological Survey; data from C and D courtesy of Konza Prairie Long-Term Ecological Research project; and data from E courtesy of Barry Biggs and Maurice Duncan).*

from groundwater. Small, headwater ephemeral stream channels are sometimes difficult to identify except during wet periods when actively flowing channels develop as soils become saturated. This dynamic nature of headwater stream networks is related to the concept of *variable source area hydrology*, which is the expansion and contraction of saturated areas in a watershed during wet and dry periods (Hewlett and Hibbert, 1967).

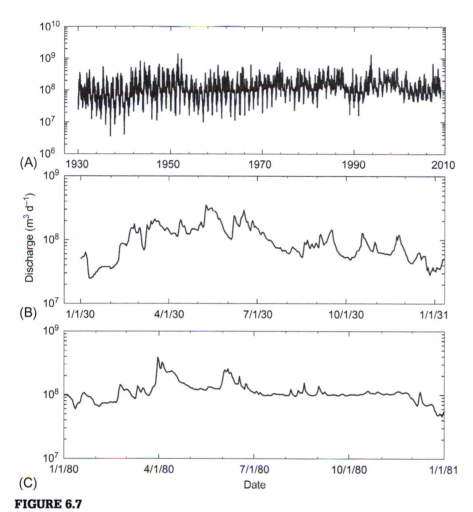

FIGURE 6.7

Discharge of the lower Missouri River. Prior to the 1950s, discharge was more variable than after dams were installed (A). A typical year before regulation (B) reveals a period of very low discharge in the winter, a spring peak, and a gradual decrease after early summer. After regulation (C), discharge is about the same throughout much of the year, except in winter when it is allowed to decrease after barge traffic halts. *(Data courtesy of the US Geological Survey).*

Fluctuation in water discharge can characterize streams and it can have a profound influence on the community structure of stream organisms (Poff and Ward, 1989). In this approach, discharge variability, flooding patterns, and extent of drying are used to create a classification system (Table 6.2). Classification based on discharge patterns is one approach that can be used by stream ecologists to describe stream characteristics and relate them to the organisms that inhabit them.

Table 6.2 A Method of Classifying Streams by Discharge Patterns and Relationship to Aquatic Communities

Drying Frequency	Flood and Discharge Frequency/Predictability	Stream Type	Effect on Biota
Often	Rare–frequent	Harsh intermittent	Strong
Low	Frequent	Intermittent flashy	Strong
Low	Infrequent	Intermittent runoff	Strong
Rare	Frequent unpredictable floods, low discharge predictability	Perennial flashy	Strong
Rare	Frequent predictable floods, low discharge predictability	Snow and rain	Strong–intermediate
Rare	Infrequent floods, low discharge predictability	Perennial runoff	Strong–intermediate
Rare	Infrequent floods, high discharge predictability	Mesic groundwater	Weak
Rare	Infrequent predictable floods, high discharge predictability	Winter rain	Seasonally strong
Rare	Infrequent predictable floods, high discharge predictability	Snowmelt	Seasonally strong

(After Poff and Ward, 1989)

Streams can also be characterized by their surrounding landscape and the associated vegetation. Thus, scientists speak of desert streams, forest streams, or arctic streams. This classification method can be useful because the terrestrial vegetation in the landscape that the streams drain may drive the biological processes that occur in streams. For example, many stream invertebrates rely on leaves and woody debris from terrestrial vegetation for food. Also, the vegetation can alter the stream channel morphology. For example, replacement of natural riparian forest with pasture can lead to narrower streams because grasses and other pasture vegetation encroach on the stream channel (Davies-Colley, 1997). When streams are characterized by vegetation, most runoff occurs from tropical evergreen forests. Grassland and temperate forests are important sources of runoff as well (Table 6.3). Such classifications can be used to quantify global fluxes of materials such as carbon, which can be linked to terrestrial vegetation types (Meybeck, 1993).

STREAM FLOW AND GEOLOGY

Important links occur between groundwater and streams. Most streams are fed by groundwater for the majority of the time. Consider the regularly flowing headwater streams with which you are familiar; it rarely rains hard enough

Table 6.3 Calculated Areal Cover of Seven Vegetative or Cover Classes and Associated Runoff, Counts of Perennial and Intermittent River, and Combined Vegetation Types. The Seven Classes are Ordered by Runoff

Vegetation Class	Coverage		Runoff					Rivers	
	$km^2 \times 10^6$	%	$km^3 y^{-1}$	%	Water Yield $(m y^{-1})$	Perennial (Counts)	%	Intermittent (Counts)	%
Broadleaf evergreen forest	13.4	9.0	14,663	29.8	1.09	16,269	10.4	407	1.1
Grasslands–wooded grasslands	42.2	28.4	13,709	27.9	0.33	39,676	25.4	15,085	39.2
Temperate forests, seasonal forests	28.7	19.3	9,438	19.2	0.33	57,985	37.1	3,069	8.0
Cultivated land	13.3	8.9	4,377	8.9	0.33	14,820	9.5	4,141	10.8
Ice	15.9	10.7	3,838	7.8	0.24	1,446	0.9	0	0.0
Tundra	7.1	4.8	2,235	4.5	0.32	20,142	12.9	6	0.0
Shrub–desert	27.9	18.8	909	1.8	0.03	5,968	3.8	15,747	40.9
Total	148.4	100	49,169	100		156,306	100	38,455	100

(From Dodds 1997; reprinted by permission of the Journal of the North American Benthological Society)

for water to flow across the surface of the land (*sheet flow*). Rather, infiltration through soil and subsurface sediments feeds into the stream to maintain flow. This constant level of discharge in streams is called *base flow*. Streams with increased discharge from groundwater as they move downstream are called *gaining* streams. Streams that lose flow because they are at a level above groundwater are referred to as *losing* (influent) streams. Streams can be gaining or losing over shorter reaches depending upon slope and groundwater flow paths.

Increased or prolonged rain events can cause rapid increases in discharge, called *floods* or *spates* (Fig. 6.8). These events often occur randomly, but a probability can be calculated that an individual event of a specific magnitude will occur given a certain amount of time (Fig. 6.9). Thus, when hydrologists speak of a 10-year flood, they are referring to an event that on average will occur once every 10 years. In other words, in any year there is a 1 in 10 chance of such a flood, regardless of whether such a flood has not occurred for the past 20 years or whether one occurred the previous year. Severity of floods is related to watershed characteristics as well as to the intensity of precipitation events. Where runoff is more rapid and infiltration is less, floods are often more severe. Channelization and increases in impermeable surfaces (pavement and buildings) associated with urbanization cause increased flooding (Figs. 6.8 and 6.9).

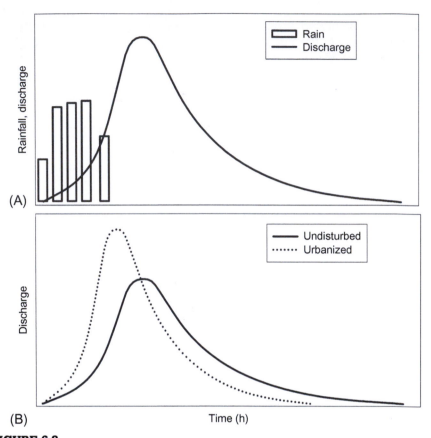

FIGURE 6.8

A hypothetical hydrograph of a storm event with precipitation and runoff in a natural area (A) and hypothetical comparison of watershed responses before and after urbanization (B). *(After Leopold, 1994).*

Water velocity also varies within stream channels. As you move down a small stream, you will usually find some shallow areas where the influence of the bottom can be seen at the surface of the flowing water. These turbulent, shallow areas are called *riffles*. Deep areas with relatively low water velocity are called *pools*, and areas with rapidly moving water but a smooth surface are called *runs* (Figs. 6.10 and 6.11). Pools tend to accumulate fine sediments and runs have coarser substrata. A section of river with several runs, pools, and riffles is called a *reach*. Several reaches together and sometimes the length between major tributaries are called *segments*.

The distance downstream between successive riffle and pool areas is approximately five to seven river widths in many streams (Leopold, 1994). The alternating riffle and pool pattern is absent when the bottom material consists of

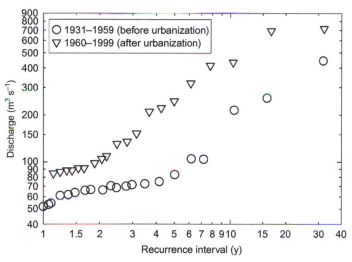

FIGURE 6.9

Flood frequencies plotted as recurrence intervals as a function of discharge for all recurrence intervals more than 1 year for the Seneca Creek watershed in Maryland before and after urbanization *(peak discharge data from the US Geologic Survey; computed as in Leopold, 1994).*

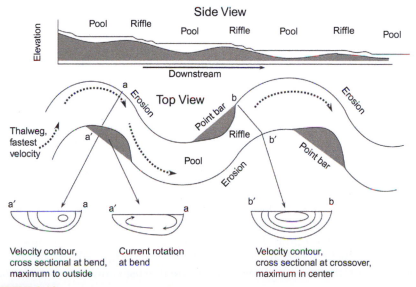

FIGURE 6.10

Conceptual diagrams of stream geomorphology. (Top) The side view is a cross-sectional lengthwise view showing pool and riffle sequence. (Middle) The top view shows a meandering stream, the thalweg (line of maximum velocity), and zones of erosion and deposition (point bars). (Bottom) The water velocity contours (cross-sectional across the channel) show how the maximum velocity is outside of the bend and the lateral current direction. When the thalweg crosses the channel, the maximum velocity is in the center of the channel.

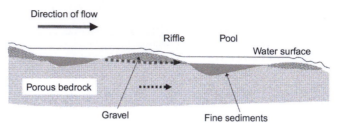

FIGURE 6.11

Cross-sectional diagram of a stream showing the riffle pool sequence, accumulation of fine sediments, and water flow through the shallow subsurface (hyporheic).

fine sand or smaller particles, is present when it is gravel and small rocks, and occurs sometimes when large boulders make up the stream bottom. In steep mountainous regions with large rocks, pools alternate with small falls. These cascades depend on geographical structure and do not exhibit the typical riffle and pool sequence. The riffle and pool sequences can be tied to stream flow and meandering patterns.

A cross-section of the straight portion of a stream shows that the water velocity is often maximum in the center (termed the *thalweg*) and minimum near the sides and bottom (Fig. 6.10). In areas where the stream or river is curved, the maximum velocity occurs nearer to the outside of the bend. Furthermore, because the water on top is moving more rapidly than the water on the bottom of the bend, the direction of flow tends to move downward and cut into the outside bank. Higher velocity and down cutting lead to erosion on the outside bend. The slower water velocity on the inside of the bend allows deposition on the inside of the curve and formation of a point bar (Fig. 6.10) and these processes lead to stream meandering.

Streams *meander* (wander in "S"-shaped patterns) unless they are constrained by outcrops or bedrock or conditions are conducive to forming braided channels. Water flowing across surfaces of glaciers, currents flowing in oceans, and rivers flowing into reservoirs or oceans, can all meander. The process is a self-organizing procedure that can be characterized by the contemporary mathematical tools of fractal geometry (Stølum, 1996). Meandering is characterized by erosion and deposition, which exaggerate the meander over time (Fig. 6.10). In general, erosion occurs on the outside of the curve in the downstream portion of the bend because sediment concentrations are low in the water and it is cutting down into the bank. Water moves along the bottom of the channel downstream and toward the inside of the bend. Because the water has the highest concentration of sediment as it moves toward the inside of the bend and it slows and drops the sediments as it reaches the inside of the bend, it forms a *point bar*. A point bar is the zone of deposition on the inside of the curve. Eventually, the meander

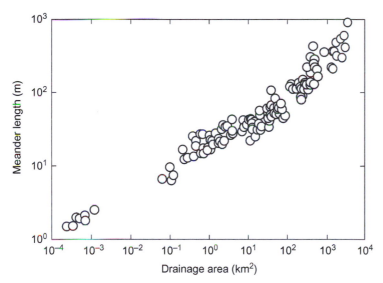

FIGURE 6.12

Relationship between drainage area and meander length. *(Redrawn from Leopold et al., 1964).*

cuts itself off. In a small stream the cut-off meander can lead to a short-lived pond, in a large river it forms an *oxbow lake*. Meander formation occurs in similar fashion in rivers of all sizes; the wavelength (the distance to meander out, back, out the other way, and back again) averages about 11 times the channel width, and the radius of curvature of a channel bend is generally about one-fifth of the wavelength (Leopold, 1994). Thus, meander size is a function of discharge (Fig. 6.12).

The idea that there are regular relationships between stream width, distance between pools and riffles, and meander length is captured by the theory of a dynamic equilibrium (Leopold, 1994). This concept is useful because it describes not only the dynamic nature of rivers and streams, but also that there may be an ideal morphology that dissipates energy most efficiently. Humans alter the flow and morphology of rivers and deviation from this general idealized morphology may be problematic; a stream may not bend to the wishes of engineers and eventually regain its morphology via dynamic equilibrium, in ways that humans do not want. Restoration of streams needs to consider the dynamic nature of stream channels or it is doomed to failure; even concrete channels eventually give way to the natural process of streams.

A detailed review of river meandering suggests that a variety of factors influence the nature of meandering (Callander *et al.*, 1978). The *sinuosity* is the distance that water travels between two points divided by the direct distance between the two points. A straight channel has a sinuosity value of 1. Meandering rivers have sinuosity values of 1.3 or greater. Sinuosity increases with slope, but to a point, where the river channel can abruptly change to braided flow. Sinuosity also increases with the ratio of mean width to depth and as the percentage of silt and clay in the river increases.

In addition to meandering, some rivers also flow in a *braided* fashion. This pattern generally occurs when water flows in broad sheets across noncohesive sediments such as sand and when flow is variable. Individual channels combine and split, form and disappear, sometimes over relatively short periods of time. Braiding is a basic physical process that can be modeled using multiple cells where sediment is transported from one cell to the next (Murray and Paola, 1994).

Meandering over time and deposition of materials by a river leads to a *floodplain* that is relatively flat across the river valley (Fig. 6.13A). Floodplains are often inundated seasonally and provide numerous wetlands and other types of habitats for many species of plants and animals. Over geological time, a river moves back and forth across its valley. Any traveler on an airplane can observe the related landscape heterogeneity; always ask for a window seat. Rivers like the Platte River of the US Great Plains, which have highly variable flows and sand substrata, or rivers on the plains below mountains such as the Canterbury Plains on the South Island of New Zealand, become highly braided, and this natural process is associated with the formation of a great diversity of freshwater habitats within the floodplain. Human activities that alter flows and modify channel structure greatly reduce the habitat heterogeneity of these systems.

The aggregate effect of these different aspects of channel structure, in addition to the effects of fallen trees and large rocks, leads to a highly heterogeneous system on the scale of tens to hundreds of meters (Fig. 6.13B). The degree of heterogeneity in the floodplain includes raised levees that are deposited naturally along stream channels and old scoured oxbows. In large rivers, depressions caused by river actions may form important wetlands, ponds, or lakes. This heterogeneity alters response to floods and greatly influences the ecology of rivers and the riparian zone's function as an interface between terrestrial and aquatic habitats (Naiman and Décamps, 1997).

An active natural river creates a habitat mosaic for organisms to inhabit along its margins (Lorang and Hauer, 2007). The flowing channel is referred to as the *fluvial zone*. The zone where the river regularly floods is considered parafluvial, and the remaining floodplain the orthofluvial. Movement of materials in a network can cause an in-stream habitat mosaic as well. For example, where side channels enter a river, the geomorphology of sediment loading creates an alluvial fan as one stream enters a larger one. The fan has effects that cascade upstream leading to lower water velocity and a wider stream channel (Benda *et al.*, 2004).

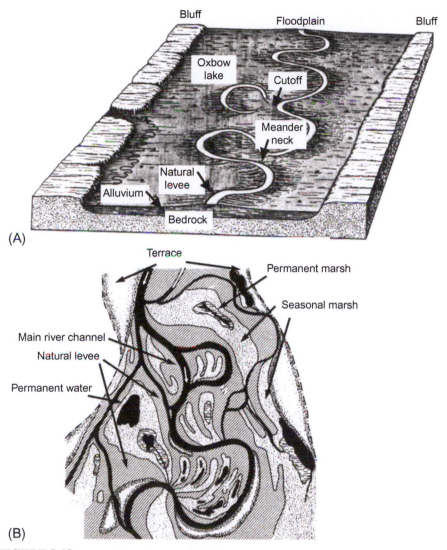

(A)

(B)

FIGURE 6.13
(A) General features of a floodplain (from Strahler and Strahler, *Elements of Physical Geology*, Copyright ©1979, reprinted by permission of John Wiley and Sons, Inc.) and (B) a diagram of heterogeneity of a tropical floodplain (*from Welcomme, 1979; reprinted by permission of Addison Wesley Longman Ltd.*).

Few relatively pristine river systems remain. Damming, channelization, and excessive water use have resulted in major human impacts on the geomorphology of rivers throughout the world (Sidebar 6.1). Other organisms, such as beavers, hippopotamuses, crocodiles, and elephants, alter channel morphology and riparian areas (Naiman and Rogers, 1997), but not to the degree that humans have.

SIDEBAR 6.1
Human Impacts on Rivers and Streams from Damming, Channelization, and Flood Control Measures

Many rivers have been channelized, and riparian (streamside) vegetation has been removed to allow rapid boat travel, increased drainage, and agricultural and urban expansion. This has dramatic influences on the shoreline habitat. For example, a 25-km stretch (as the crow flies) of the Willamette River in western Oregon had 250 km of shoreline in 1854 (Fig. 6.14), but human activity decreased it to 64 km by 1967 (Sedell and Froggat, 1984). There was a concurrent loss of at least 41% of the riparian wetlands during this time period (Bernert *et al.*, 1999). This removal of virtually all slow-moving portions and straightening of meanders is common in rivers in areas where humans live. Additional measures, such as creating projecting short structures that move flow from

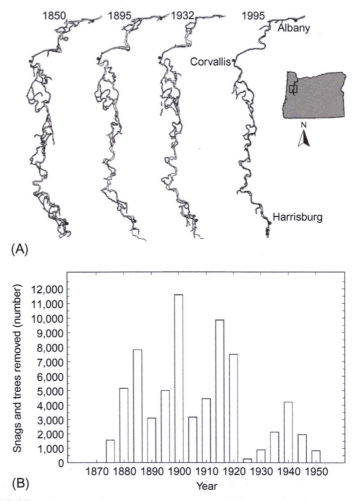

(A)

(B)

FIGURE 6.14

Changes in the Willamette River related to channelization from 1850 to 1995 (A, image courtesy of Ashkenas, Gregory, and Minor, Oregon State University) and (B) snag and tree removal *(data from Sedell and Froggatt, 1984).*

the edge to the middle of the channel (wing dykes), are common and further alter river discharge. Such alteration has been well documented on the Missouri River (Hesse *et al.*, 1989). In addition, removal of large instream obstructions, such as logjams and stumps (often called snagging operations), to facilitate navigation is common and this destroys vital habitat for aquatic organisms. Many large river systems are stable substrate limited and much of the fish and invertebrate diversity and productivity is associated with accumulations of large woody debris (e.g., Benke *et al.*, 1984). Even historic factors come into play in human alterations of rivers and streams; the more than 10,000 mill dams that were erected in the Northeast United States from the 1700s to early 1900s could have substantially altered stream geomorphology by burying wetlands in sediments and creating deeply incised stream channels throughout the region (Walter and Merritts, 2008).

Flood control measures on many large rivers include levees to contain high discharge. When floods do occur, the levees constrain the water, making it move faster and deeper in the main channel instead of spreading out across the floodplain and flowing with a lower average velocity as it would naturally. If a flood does breach the levy suddenly, it causes considerable damage because of the rapid current velocity when the levee breaks. In addition, such levees constrain flows and act like dams to upstream regions that do not have levees. In the Mississippi River basin, mean annual flood damage has increased by 140% during the past 90 years. This increase is probably attributable to increases in numbers of levees and removal of riparian wetlands (Hey and Philippi, 1995). Now, reestablishing floodplains and riparian wetlands on large rivers is beginning to be seen as a viable flood control strategy (Opperman *et al.*, 2009).

Landowners often channelize smaller streams flowing across their property so they can develop or plant crops closer to the edge of the stream and drain their land more quickly. This channelization causes the water to move downstream at higher velocity, increasing erosion in the channelized reach and upstream. Channelization also causes the stream to have stronger flooding impact downstream, leading more landowners to channelize their stream banks. Removal of riparian vegetation creates an even worse situation because natural retention of sediment and slowing of floodwater does not occur, thus increasing the severity of floods.

Human alteration of stream and river hydrology has effects across all spatial scales. At the smallest scales (Paragamian, 1987), it removes habitat for aquatic organisms. At intermediate scales, alterations of stream hydrology are associated with increased erosion and more severe flooding. At the largest scales, human activities lead to global changes in the transport of materials by rivers. The economic impact of all these human influences is likely very great. Effective management requires understanding of dynamics of natural rivers (Poff *et al.*, 1997).

Attempts at restoration of hydrology and channel complexity has become common (Bernhardt *et al.*, 2005). Unfortunately such restorations are rarely effective (Alexander and Allan, 2007) and often not based on appropriate design standards (Palmer *et al.*, 2005). Complete restoration of large rivers impacted by humans is unlikely, but partial rehabilitation may be possible (Gore and Shields, 1995).

A new trend of dam removal seems to be developing with the hope that some of the more harmful and less useful impoundments can be removed. At least 467 documented dam removals have occurred in the United States since 1912 for environmental, safety, economic, or other reasons (American Rivers, 1999). These removals have had some beneficial effects, but some of the degradation from dams to aquatic habitats is irreversible (Middleton, 1999) and dam removal may even have some negative aspects such as allowing upstream movements of exotic species and the release of large quantities of sediments stored behind the dam (Orr *et al.*, 2008). Removal effects on macroinvertebrates can even extend upstream (Pollard and Reed, 2004). Analysis of a small dam removed from a Pennsylvania stream indicated that most effects of the removal downstream were gone once the sediments that were washed from the old reservoir had been flushed from the downstream channel, but some effects lingered a year afterward (Thomson *et al.*, 2005).

Most large rivers have been altered significantly from their natural state. Globally, 172 of 292 large river systems are impacted by dams (Nilsson *et al.*, 2005). For example, 70% of the discharge from the 139 largest rivers in North America, Europe, and the former Soviet Union is affected by irrigation, diversion, or reservoirs. Most of the unaffected river systems that remain in these regions are in the far north (Dynesius and Nilsson, 1994). In the United States, it is estimated that there are 2.5 million dams (National Research Council, 1992). Reservoirs cause sediments to settle from rivers, and rivers downstream become "sediment starved" and significantly more erosive. Natural sandbars are less likely to form in a starved river. In an extreme case, construction of the Aswan Dam led the River Nile to become sediment starved. The lack of sediment, in turn, has led to erosion of the Nile Delta, where the River Nile enters the Mediterranean, and a subsequent loss of valuable agricultural land that has served as Egypt's breadbasket for millennia (Milliman *et al.*, 1989).

Reservoirs can also alter the riparian habitat by interfering with flooding; in dry, sandy rivers they allow establishment of more riparian vegetation and reduce the width of the channel (Friedman *et al.*, 1998). Dams also fragment riparian habitat, leading to distinct changes in plant communities by altering patterns of dispersal and recruitment (Nilsson *et al.*, 1997; Jansson *et al.*, 2000). Perhaps surprisingly, effects of reservoirs and other disturbances can extend to organisms upstream (Pringle, 1997). For example, interruption of salmon runs reduces the transfer of nutrients, into small streams via spawning salmon that eventually die in the stream channel.

Reservoirs can also disrupt natural temperature patterns. When relatively cool rivers flow into wide shallow reservoirs and the reservoirs release their water downstream, it can lead to artificially higher water temperatures in the river below. Alternatively, deep reservoirs can always have cold deep water (see Chapter 7 for an explanation of temperature stratification). If water is released from deep in the reservoir, artificially low temperatures can occur. For example, the Colorado River is naturally a warm river that would not normally support salmonid populations. However, there are now large populations of trout in the Grand Canyon because cold water is always released from the upstream Glen Canyon dam. Unfortunately, these altered conditions and the predatory trout are harming endangered populations of native fishes.

ADVANCED: CLASSIFICATION OF RIVER AND STREAM TYPES AS A RESTORATION AND MANAGEMENT TOOL

River restoration has become important as people have come to realize that alteration of hydrology, water chemistry, and biology of rivers has unintended consequences in urban (Bernhardt and Palmer, 2007) and other areas

(Bernhardt *et al.*, 2005). Such efforts require restoration of the natural hydrology and understanding of how the dynamic equilibrium of geomorphology can be restored to lead to long-term stability of the system. Restoration targets need to consider geomorphology, life history of species that are being managed for (Jansson *et al.*, 2007) and desired ecosystem functions (Groffman *et al.*, 2005). While morphology is often restored, it is less likely that flow dynamics are restored (Kondolf *et al.*, 2006). Still, one key to restoring rivers and streams is restoration of their morphology, and this requires classification of existing natural and altered morphology.

Classification of river systems and how their morphology has been modified is central to restoring river systems. One of the most popular river classification systems is the Rosgen system. The main objectives of this system are to (1) predict river behavior from existing morphology, (2) develop specific hydraulic and sediment relationships appropriate for the type of stream, (3) provide a way to assess reference reaches, and (4) provide a common framework for characterizing stream morphology and condition. The first level of classification characterizes the channel and valley shape, slope, and pattern (e.g., confined or wide valley). The second level of classification is a determination of field classification of stream type. For example, width, depth, channel entrenchment, sinuosity, slope, and substratum type in the stream (e.g., sand, gravel, cobble) are surveyed. This description of the Rosgen system is very cursory, and a two-week training course is generally required to learn the classification methodology and system; it is a complex, multitiered method for surveying streams and has been adopted by a number of government agencies.

The Rosgen method is somewhat controversial. Criticisms (Simon *et al.*, 2007, 2008) refuted by Rosgen (2008) are that the method is descriptive and does not appropriately account for temporal evolution of stream morphology based on current geomorphologic principles. The critique suggests that peer-reviewed publication is required to establish the method as scientific and that some restoration projects have failed because of inappropriate application of the classification method. Rosgen (2008) asserts that the critique improperly represents his classification method and its application. Those adopting the Rosgen approach should be aware of this controversy.

MOVEMENT OF MATERIALS BY RIVERS AND STREAMS

Rivers can have a very long geological time of existence. There are many cases where mountain ranges have rivers that cut through them. This generally means that the river is older than the mountains, and the river could erode down faster than uplift could raise the mountains. Rivers are the primary source of terrestrial

Table 6.4 Average Chemical Composition of River Water Throughout the World

Attribute	Current Concentration	Natural Concentration	Pollution	% Increase
Ca^{2+}	14.7	13.4	1.3	9%
Mg^{2+}	3.7	3.4	0.3	8%
Na^+	7.2	5.2	1.3	28%
K^+	1.4	1.3	0.1	7%
Cl^-	8.3	5.8	2.5	30%
SO_4^{2-}	11.5	6.6	4.9	43%
HCO_3^-	53.0	52.0	1.0	2%
SiO_2	10.4	10.4	0.0	0%
Total dissolved solids	110.1	99.6	10.5	11%
Dissolved nitrogen	21.5	14.5	7.0	32%
Dissolved phosphorus	2.0	1.0	1.0	50%

(From Berner and Berner, 1987, and Meybeck, 1982). Concentrations in mg L^{-1}

materials to the oceans and over time they can move mountains to the bottom of the ocean.

Rivers carry materials dissolved from land to the sea. In addition, they move larger particles by *erosional processes*. Movement of dissolved materials (Fig. 6.15) to the sea has been altered by human activities (Table 6.4). Dissolved materials in rivers have increased, particularly nutrients such as nitrogen (Vitousek, 1994), phosphorus, and sulfur, because human activities such as burning of fossil fuels and use of fertilizers have increased the actively cycling phases of these materials. Ultimately, the increased transport of nutrients in rivers may alter productivity of coastal marine systems (Downing *et al.*, 1999); increased export of nutrients from agricultural regions of the central United States to the Mississippi River and ultimately the Gulf of Mexico are the primary factor contributing to the large "dead zone" (hypoxic, or low oxygen, area that can cover >22,000 km^2) in the Gulf of Mexico (Rabalais *et al.*, 2002). Streams and rivers play a part in the global carbon cycle by moving dissolved and suspended organic materials (e.g., woody debris and leaf fragments) from terrestrial habitats to the sea (Table 6.5). In this context, tropical rain forests, where a large proportion of continental runoff originates, are extremely important. Desert and semiarid habitats are not as important because of their low runoff (Table 6.3). Understanding global cycles of materials and the effects of global change requires knowledge of how rivers move materials through the environment.

Table 6.5 Total Organic Carbon Export by Rivers in Different Terrestrial Environments

Environment	Average Total Carbon Export (g m^{-2} y^{-1})	Total Carbon Load (10^{12} g C y^{-1})
Tundra	0.6	4.5
Taiga	2.5	39.6
Temperate	4.0	88.0
Tropical	6.5	241
Semi-arid	0.3	4.6
Desert	0	0
Total	3.8	378

(Meybeck, 1982)

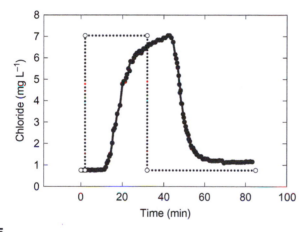

FIGURE 6.15

Movement of a pulse of chloride through a stream channel with considerable areas of slack flow (transient storage zones). In a perfect channel, the pulse would be square as it moves downstream. *(After Webster and Ehrman, 1996).*

More and larger particles can be moved downstream as discharge increases (Fig. 6.16). Mobile material can be divided into two categories, suspended load and bed load. The *suspended load* is the fine material that is suspended in water under normal flows. This suspended load can also be referred to as *turbidity* or *total suspended solids*. The *bed load* moves along the bottom by sliding, rolling, and bouncing. It never moves more than several particle diameters above the bottom. The relative movement of both types of particles and the size of the particles that can remain suspended depend on the water velocity and turbulence (Fig. 6.17). Streams become turbid after a rain. This turbidity

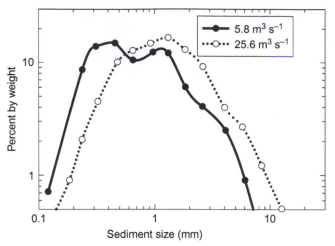

FIGURE 6.16

Movement of particles as a function of particle size for maximum and minimum flows in the East Fork River. Note that larger particles move more readily at greater discharge rates. *(Reprinted by permission of Harvard University Press from Leopold, 1994, © 1994 by the President and Fellows of Harvard College).*

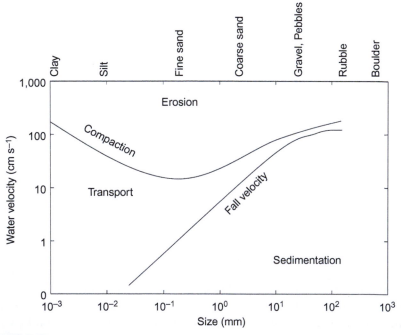

FIGURE 6.17

Transport and erosion of particles as a function of water velocity. Note that the lines delineating transitions between erosion, transport, and sedimentation represent fuzzy rather than abrupt transitions. *(After Allan, 1995 and Morisawa, 1968).*

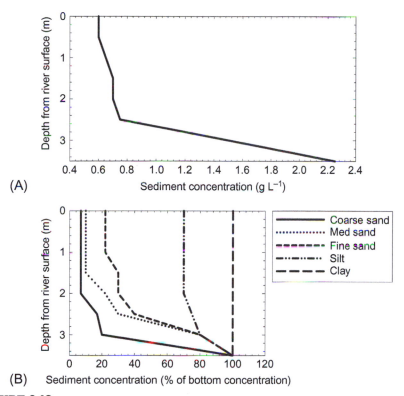

FIGURE 6.18

Distribution of particles with depth in the Missouri River expressed as total particle concentration (A) and as percentage concentration of bottom sediments for different sized particles (B). *(Data plotted from Wilber, 1983).*

is partially attributable to increased sediment inputs from land, but it is also related to the higher water velocity that can re-suspend and transport materials from within the channel.

Turbidity is not evenly distributed in rivers. The total concentration of suspended materials is higher near the bottom (Fig. 6.18A). However, the trend of greater concentration with depth does not hold with the finest particles (such as clay), which remain in suspension throughout the water column. The effect becomes more pronounced with larger particles such as sand (Fig. 6.18B).

Understanding the movement of solid materials can be very important to understanding the ecology of streams. Floods cause erosion and can alter habitat. Some species of fish and invertebrates are dependent upon entrained organic particles as a food source (Wallace and Merritt, 1980), and some cannot survive when the concentration of suspended particles is too great. Movement of rocks on the bottom of the streams can also have direct ecological effects on the organisms that live on or under those rocks. On land, a rolling stone gathers no moss; in a stream, a rolling stone will not gather much algae or many invertebrates (McAuliffe, 1983).

In rivers with contaminated sediments, transport of those sediments during high flows can lead to pollution events because many pollutants such as pesticides and metals adsorb to the surface of sediment particles. For example, in the Clark Fork River in Montana, years of mining have led to contaminated sediments in the basin (Johns, 1995). Each time it rains hard, local fishermen hold their breath; massive trout kills result if enough toxic sediment is resuspended (see Sidebar 16.4). Another way that streams are seriously impacted by sediments is from "hill topping" or mountain top removal coal mining (Fig. 6.19). With this mining technique, the top of a mountain containing coal is removed, and the waste rock is disposed of by filling the stream valleys skirting the mountain; streams under the fill are obviously obliterated. Movement of dissolved and suspended toxins into the areas downstream from the fills has negative effects on the animals in the streams (Hartman *et al.*, 2005).

Movement of materials is also a key aspect of erosion. Very small particles such as clays are less susceptible to erosion than slightly larger particles (Fig. 6.17).

FIGURE 6.19
Hill top, or mountain top removal mining for coal. *(Image courtesy of Ohio Valley Environmental Coalition, Vivian Stockman).*

Due to the relationship between particle size and transport, gravel is often found in riffles and fine sediments are deposited in pools. Under base flow, riffles are erosional habitats and pools are depositional habitats. This may lead the reader to ask, why are pools deep and riffles shallow? The answer is that pools become erosional habitats, whereas riffles are depositional during floods. So, floods scour out pools and deposit the materials in riffles and at baseflow the materials are moved slowly from riffles into pools.

Lateral movement of materials is also important in forming habitat for organisms. For example, when a tributary enters a river it tends to deposit materials leading to damming of the main channel. This damming slows flow in the river above the tributary and can lead to a more rapid elevation drop below the tributary (Fig. 6.20).

Retention and transport of particulate materials can be important in describing the long-term effects of sediment pollution. For example, sediments that enter stream channels from erosion caused by watershed disturbance can be retained for decades, prolonging the recovery time from sediment pollution events (Trimble, 1999). The length of time that sediments can be stored in stream networks is an active area of hydrologic research.

FIGURE 6.20
The Grand Canyon at Hance Rapids. Note the stream entering from the left deposits materials that cause a pool behind the confluence, and a steep elevation drop immediately downstream leading to the rapids. Also note the river downcutting through sediment layers. *(Photograph courtesy of the US Geological Survey, by L. Leopold).*

The effects of change in temporal and spatial scales on stream habitats can be linked to processes of erosion and habitat change, leading to a hierarchical classification of stream habitats (Fig. 6.21). Such classification provides a useful framework with which to approach the links between river hydrology and aquatic ecology and a template for the interaction of organisms with habitat (Gregory *et al.*, 1991). Consideration of scale is a vital component for

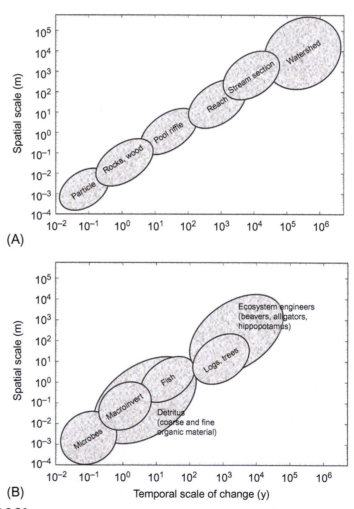

FIGURE 6.21

Hierarchical representation of the geomorphological changes in stream channels (A) and related biotic processes (B). *(After Alan, 1995; Frissell et al., 1986; Mitsch and Gosselink, 1993).*

understanding patchiness in stream ecosystems (Stanley *et al.*, 1997); scale controls chemical transport (Dent *et al.*, 2001) and may be useful in the study of water quality (Hunsaker and Levine, 1995).

ADVANCED: CHARACTERIZING THE MOVEMENT OF DISSOLVED MATERIALS IN RIVERS AND STREAMS

Dissolved materials move down streams as a function of discharge and exchange with the biotic and abiotic components of stream channels. If a chemical is added in a defined pulse to a stream, the pulse will become less coherent as it moves downstream (Fig. 6.15). Streams have areas with relatively slow velocity and others with high water velocity, which causes spreading of the pulse. In addition, dissolved materials can interact with the sediments or biota as they flow downstream. Departures from the ideal transport of dissolved materials (i.e., a coherent pulse moving through a frictionless, inert channel) can be used to investigate the retentive properties of the channel and the effect of the hyporheic zone (Webster and Ehrman, 1996). The hyporheic zone can play a significant role in retention of nutrients (Hill and Lymburner, 1998).

Characterizing the influence of slower regions of flow including groundwaters, backwaters, pools, and even water in algal mats requires release of tracer substances and modeling to match observed transport patterns. Two major processes, *advection* and *dispersion*, describe the pulse of dissolved materials as they flow downstream. Advection is the average velocity of the stream. Dispersion is the diffusion of material (mainly via turbulent transport) from the main pulse. Regions of slower flow are referred to as *transient storage zones*. As a pulse of material moves downstream, some of it moves into the slower zones of flow and is gradually released back into the main channel. If a solute is released in a coherent pulse, downstream the pulse broadens out, and the extended tail after the main pulse is from material that is rereleased from transient storage zones into the main channel. The equations describing these processes are complex partial differential equations, and in practice are not solved mathematically, but can be modeled with readily available software (Bencala and Walters, 1983; Runkel, 2002).

The common models assume that there is an average cross-sectional surface area of the main channel (A) and a second compartment that is connected to some degree to the main channel that has the cross-sectional area A_s. The computed value of A_s/A then is an indicator of the relative size of the two compartments. With larger values of this ratio, more transient storage area and more dispersion are expected. The parameter α indicates a coefficient of exchange between this *transient storage zone* and the main channel, and dispersion is noted by the factor D. From these parameters a number of additional characteristics can be calculated, including the average distance a molecule travels downstream before entering a transient storage zone, the average amount of time a molecule resides in the transient storage zone, and the Péclet number (Pe), which indicates the relative importance of advection and dispersion.

This two-compartment model describes solute dynamics relatively well. However, it is clearly a simplification. Some transient storage zones exchange very quickly with the main stream, and others much more slowly. Thus, there is a range of different transient storage zone types that can occur in a single stream reach.

Characterizing the movement of dissolved materials has practical applications. As more streams become urbanized, researchers are striving to understand the implications of the hydrologic changes associated with waters that flow through areas of dense human population. Nutrient uptake is impaired in urban streams, and assessing these changes requires characterization of hydrology and movement of dissolved materials (Meyer *et al.*, 2005). Self-purification of rivers by processes, such as denitrification that removes nitrate (see Chapter 14), can be enhanced in transient storage zones.

SUMMARY

1. Permanent rivers are more common where precipitation is greater; intermittent rivers are characteristic of regions with low or sporadic precipitation.
2. Rivers can be classified by watershed area, stream order, variations in discharge over time, and vegetation.
3. Stream habitats can be characterized into riffles, pools, runs, and falls. These habitats are mostly scale independent.
4. Streams naturally meander; the water flow patterns alter erosion and deposition in such a way that meanders increase in size until they eventually close off and form oxbow lakes. This meandering is scale independent and occurs in rivers and streams regardless of size.
5. Humans have drastically altered the morphology of stream and river channels throughout the world with activities such as channelization, construction of dams, construction of levies, and alterations of flow from water use.
6. Rivers transport dissolved materials and are thus important in global biogeochemical cycles. The average concentrations of sulfur, nitrogen, and phosphorus have increased 30% to 50% because of human activities.
7. Erosion and transport of silt and larger materials on river bottoms are influenced by discharge and particle size.

QUESTIONS FOR THOUGHT

1. Why might some regions in deserts serve as runoff sinks (i.e., have more water flowing in than out)?

2. How can levees act as dams to upstream areas during floods and increase erosion in the channel where they are present?

3. Why does channelization increase the movement of bed load during floods?

4. How may the amount of suspended materials in streams alter the rates of photosynthesis of organisms attached to the stream bottom?

5. How might vegetative characteristics in a watershed relate to stream flow in terms of total amount and in terms of how many floods occur per year?

6. How does frequency and predictability of flooding relate to the possibility that stream organisms are adapted to flooding?

Lakes and Reservoirs: Physiography

(A)

FIGURE 7.1

Satellite images of the Great Lakes (A) and Smithville Reservoir (Missouri) (B). The reservoir is 16 km long. Note the dendritic pattern of the reservoir and the relatively smooth shorelines of the glacially formed Great Lakes. The numerous black dots around the reservoir are farm ponds. *(Images from the US Geological Survey).*

Doi: 10.1016/B978-0-12-374724-2.00007-6

(B)

FIGURE 7.1
Continued

Lakes of all sizes provide us with fisheries, recreation, drinking water, and scenic splendor. Having a clean lake nearby increases property values. Large lakes (Table 7.1) have played a part in the history, economy, and culture of many nations. Lakes also provide excellent systems for ecological study. The boundaries of the lake community and ecosystem often appear distinct, the water well mixed, and the bottom relatively homogeneous, making lakes tractable systems for ecologists. Much effort has been made to study the physical and biological aspects of lakes (e.g., some of those in Table 7.2 have been studied intensively for over a century) and to manage pollution. The foundation of these studies is an understanding of the geomorphology of the lakes. Different lake morphologies give rise to different levels of productivity and physical effects of water retention, circulation, currents, and waves. For example, the fates of toxins and nutrients in lakes depend partly on lake circulation, which is a function of lake physiography. In this chapter, we describe formation, lake morphometry, the process of stratification, and water movement in lakes.

FORMATION: GEOLOGICAL PROCESSES

What is a lake? We define a *lake* as a very slowly flowing or nonflowing (*lentic*) open body of water in a depression and not in contact with the ocean.

Table 7.1
Some Properties of the 10 Largest Lakes by Either Depth, Area, or Volume, Globally Arranged by Maximum Depth

Lake	Continent	Formation	Mixis	Area (km²)	Max. Depth (m)	Mean Depth (m)	Volume (km³)	Length (km)	D_L	Retention (y)
Baikal	Asia	Tectonic	Meromictic	31,500	1,741	730	23,000	2,200	3.4	323
Tanganyika	Africa	Tectonic	Meromictic	34,000	1,470	572	18,940	1,900	3.1	5,500
Caspian	Asia/Europe	Tectonic	Meromictic	436,400	946	182	79,319	6,000	2.6	Sink
Nyasa	Africa	Tectonic	Meromictic	30,800	706	273	8,400	1,500	2.7	
Issyk Kul	Asia	Tectonic	Meromictic	6,200	702	320	1,732	760	2.8	305
Great Slave	N. America	Glacial	Dimictic	30,000	614	70	2,088	2,200	3.6	
Crater	N. America	Volcanic	Monomictic	55	608	364	20	35	1.3	4.9
Matano	Asia	Tectonic		164	590	240	39	80	1.8	
Toba	Asia	Volcanic-tectonic	Monomictic	1,150	529	216	249	100		
Hornindalsvatn	Europe	Glacial	Dimictic	508	514	237	12	65	2.6	
Great Bear	N. America	Glacial	Dimictic	29,500	452	81	2,381	2,100	3.3	124
Superior	N. America	Glacial	Monomictic	83,300	307	145	12,000	3,000	2.9	184
Michigan	N. America	Glacial	Monomictic	57,850	265	99	5,760	2,210	2.6	104
Huron	N. America	Glacial	Monomictic	59,510	223	76	4,600	2,700	3.1	21
Victoria	Africa	Tectonic	Polymictic	68,800	79	40	2,700	3,440	3.7	23
Aral	Asia	Tectonic	Meromictic	62,000	68	16	970	2,300	2.6	Sink

Data in this and the Following Table are from Several Sources Including Gasith and Gafny (1990), Herdendorf (1990), Hutchinson (1957), and Horne and Goldman (1994). Sink Indicates Lake is in a Closed Basin

Table 7.2 Selected Lakes of Historical or Research Interest that are Not Included on Table 7.1

Lake	Continent	Formation	Mixis	Area (km²)	Max. Depth (m)	Mean Depth (m)	Volume (km³)	Length (km)	D_L	Retention (y)
Biwa	Asia	Tectonic	Monomictic	618	46.2	45	28	46	2.6	5.4
Erie	N. America	Glacial	Monomictic	25,820	64	21	540	1,200	2.1	3
Eyre	Australia	Tectonic	Polymictic	0–8,583	4	2.9	23		4.9	Sink
Geneva	Europe	Glacial	Monomictic	580	310	153	89	70		
Kinneret	Asia	Tectonic	Monomictic	1.7	43	26	4.3	2.2	1.16	7.3
Loch Ness	Europe	Glacial	Monomictic	56.4	230	133	7.5	39	3.2	2.8
Mendota	N. America	Glacial	Dimictic	39.8	25.3	12.8	0.47	9.1	1.6	4.6
Ontario	N. America	Glacial	Monomictic	18,760	225	91	1,720	1,380	2.8	8
Tahoe	N. America	Tectonic		499	501	249	124	125	1.6	700
Titicaca	S. America	Tectonic	Monomictic	7,700	280	106	820	176	3.46	1,343
Vanda	Antarctica	Glacial	Amictic	5.2	67	33.8	0.17	5.6	2.28	75 (ablation*)
Windermere (both basins)	Europe	Glacial	Monomictic	14.3	64	23	0.39	17	1.2	

*Ablation is evaporation directly from the ice that covers the lake

This definition includes saline lakes but excludes estuaries and other mainly marine embayments. The distinction between a small shallow lake or pond and a wetland is not clear, and neither is that between a very slow, wide spot in a river and a lake or reservoir with high water throughput; all aquatic habitats occur across a continuum of physical attributes, such as depth and water velocity.

Permanent lakes are common where more precipitation occurs and where geology allows for formation of water-retaining basins (color plate Fig. 6). Some areas have geological histories that result in more lakes. For example, if we compare the distribution of wetlands (color plate Fig. 6) to the distribution of freshwater lakes, relatively more lakes than wetlands occur in northern North America, and wetlands are relatively important in northern Asia and northeast Europe. Intermittent lakes (those that dry sometimes) are distributed sparsely throughout the world, with greater numbers in arid areas (color plate Fig. 6). The western United States, South Australia, India, central Asia, and central Africa all have high numbers of intermittent lakes.

Humans have made many lakes and ponds and, in doing so, have greatly altered the distribution of lentic habitats. Most regions inhabited by humans with few natural lakes and even moderate precipitation have significant numbers of ponds and reservoirs. At least 2.6 million small impoundments are found in the conterminous United States, and they account for about 20% of the lentic water in the country (Smith *et al.*, 2002). Similar patterns occur in Europe and all other developed areas on Earth. The large number of rivers in the Northern Hemisphere that have been impounded to form reservoirs was discussed in Chapter 6.

More small than large lakes exist in the world. On a regional level, this has important consequences for the chemistry and productivity of lakes (Hanson *et al.*, 2007) because small lakes have biological and physical characteristics that are distinct from larger lakes in the same region. However, the sum of the area of lakes of each size globally is fairly constant, with the few very large lakes having a major impact on total area (Fig. 7.2). A variety of geological processes lead to the formation of these lakes (Table 7.3). Hutchinson (1957) described these processes in detail; we give a brief version here.

Tectonic movements of Earth's crust (Fig. 7.3) form some of the largest and oldest lakes. For example, warping of Earth's crust formed the Great Rift Valley in Africa and gave rise to Lakes Edward, Albert, Tanganyika, Victoria, Nyasa, and Rudolf. This group contains some of the oldest, deepest, and most ecologically and evolutionarily interesting lakes on Earth. Although small tectonic lakes are more numerous than large ones, large tectonic lakes cover an area greater than that covered by the small ones on a global scale

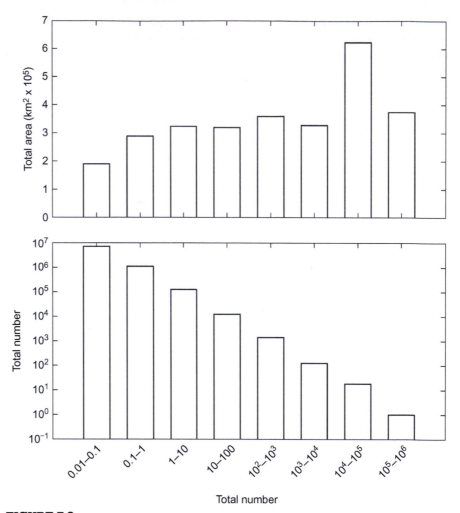

FIGURE 7.2
Global numbers and total areas of lakes by surface area size class. *(Data from Meybeck, 1995).*

(Fig. 7.4). *Graben* lakes are tectonic lakes formed where multiple faults allow a block to slip down and form a depression. Lake Baikal of Siberia, the deepest and oldest lake on Earth, is a graben lake. About 3 km of sediment has accumulated on the bottom of Lake Baikal over 16 million years (Fig. 7.5). Tectonic movements also form *horst* lakes. In this case, the blocks tilt and leave a depression that can be filled by water (Fig. 7.3).

Damming by natural processes can form lakes. Examples of these processes include landslides (Fig. 7.6), lava flows, drifting sand dunes, and glacial

Table 7.3

Some Ways that Lakes Form and Essential Characteristics of Each Type. Many More Types are Possible (e.g., Hutchinson, 1957)

Lake Type	Formation Process	Essential Characteristics and Examples
Tectonic	Basin formed by movement of Earth's crust Graben: A block slips down between two others. Horst: diagonal slippage.	Can be very old and very deep. Lake Baikal, Asia; Lake Tanganyika, Africa.
Pothole or Kettle	Formed when ice left from retreating glacier is buried in till (solid material deposited by glacier) and then melts.	Small lakes/wetlands. Prairie pothole region in Alberta and North and South Dakota.
Moraine	Glacial activity deposits a dam of rock and debris.	Narrow, fill valley.
Earthslide	Movement of earth dams a stream or river.	Similar to reservoirs. Quake Lake, Montana.
Volcanic–Caldera	Volcanic explosion causes hole that is filled with water.	Often round and deep. Crater Lake, Oregon.
Dissolution Lake	Limestone dissolves and lake forms.	Small, steep sides.
Oxbow	River bend pinches off, leaves lake behind.	Shallow, narrow, may be seasonally flooded.

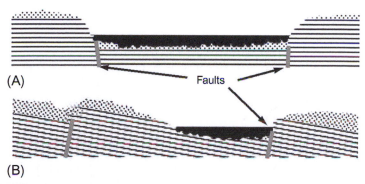

FIGURE 7.3

Diagrams of two modes of lake formation associated with tectonic processes. (A) Graben, a block drops below two others. (B) Horst, blocks tip and a lake forms along a single fault line.

moraines. In addition, beaver ponds, damming by excessive plant growth, flows of rivers at deltas, glacial ice dams, and pools formed at the edges of large lakes by shore movement are classified into this general category. These lakes usually are not large, but some exceptions exist. Lake Sarez is a

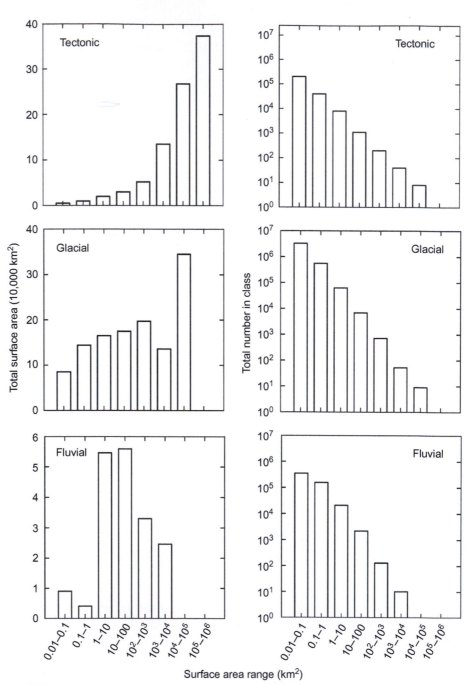

FIGURE 7.4

Global numbers and total areas of lakes of different geological origins by surface area size class. Note, there is one tectonic lake in the largest size class. *(Data from Meybeck, 1995).*

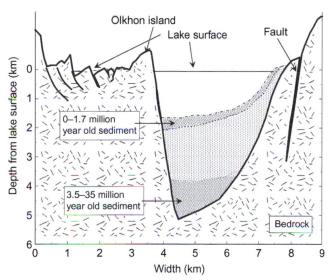

FIGURE 7.5

Cross section of Lake Baikal, the south basin, in the region of maximum depth (1637 m). *(Redrawn from Belt, 1992; and Mats, 1993).*

FIGURE 7.6

Earthquake Lake forming after a large earthquake in 1959 in Montana caused a massive landslide damming the Madison River. This picture taken before the lake was full shows debris still in the water. *(Image courtesy of the US Geological Survey).*

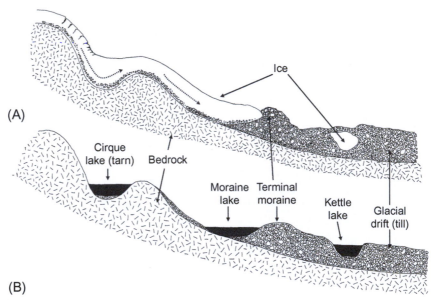

FIGURE 7.7

Formation of some types of glacial lakes. (A) A cross section of a glacier moving down a valley. (B) After the glacier has retreated, it leaves cirque, moraine dammed, and pothole lakes. *(Drawn by Sarah Blair).*

large landslide lake that formed in Russia starting with an earthquake in 1911. The mound formed by the landslide is 574 m high and the lake is 500 m deep. There is concern that the dam will break in another earthquake and the 56 km long lake will cause catastrophic flooding downstream. Other landslide lakes can create considerable danger to people downstream because of the possibility of catastrophic failure. On 12 May 2008, an earthquake registering 8.0 on the Richter scale killed at least 69,000 people in Sichuan China and caused a landside that created the large, unstable Tangjiashan Lake on the Jianjiang River. Engineers have worked hard to create a stable spillway, but the lake may still breach in flood conditions, leading to even more mortality related to the earthquake.

Glacial activity is responsible for the formation of many lakes in temperate regions and for formation of more lakes than any other process (Fig. 7.7). Areas far enough north or south historically to have had glaciations tend to have far more lakes than those closer to the equator. Several processes associated with glacial activity lead to lake formation. Glaciers scour as they move down valleys. The ice flow of these glaciers creates basins. Lakes occur where glaciers have scoured more deeply, leading to the formation of *cirque* (also called tarn) lakes in the "amphitheaters" at the heads of the valleys. The glacier forms chains of *paternoster* lakes as it flows farther down the valleys (Figs. 7.7 and 7.8).

FIGURE 7.8
Paternoster lakes (a string of glacial lakes) in a snowy mountain valley in the Cascades, Skymo Lake.
(Photograph courtesy of the US National Park Service).

Glacial scour can lead to formation of extremely large lakes. The Great Slave Lake in Canada was carved to a depth of 464 m below sea level by the massive weight of the continental ice sheet and is the deepest lake in North America. The Laurentian Great Lakes of North America (e.g., Superior, Huron, and Erie) were partially formed by glacial action. *Fjord* lakes such as Loch Ness (home of a legendary creature) are long glacial lakes formed in steep valleys. One of the strangest lakes associated with glaciers is the gigantic lake that has recently been described below the ice in Antarctica (Sidebar 7.1).

As glaciers move, they entrain rocks and sediments into the ice. Where glaciers melt at the edges and front, they deposit these materials. As the glaciers retreat they leave this material, called *glacial till*, behind. If large blocks of ice remain in this till, they melt and eventually leave lakes, ponds, or wetlands called *kettles* or potholes (Fig. 7.7, Fig. 5.9). This process formed the many lakes and ponds that provide vital habitat to waterfowl in the northern prairies of North America, although many have been filled for agricultural purposes.

If the forward flow of a glacier is approximately equal to its backward melting rate, a wall of material is formed called a *terminal moraine*, which can

SIDEBAR 7.1
A Large Lake beneath the Ice in Antarctica

In 1974 and 1975, an airborne radio-echo survey of Antarctic ice depths led to the discovery of a lake under the ice. The ice sitting on the lake's surface is flat relative to the surrounding ice on land, and remote satellite measurements of ice elevation determined the size of the lake to be close to the area of some of the North American Great Lakes (Kapitsa *et al.*, 1996). The lake is estimated to cover about 15,000 km^2, is 125 m deep, and rests below about 4 km of ice. Preliminary calculations suggest that the residence time of the water in the lake is tens of thousands of years, and that the lake basin is about 1 million years old. However, there is significant water exchange between the lake and the ice sheet (Siegert *et al.*, 2000). Scientists have taken cores through the ice sheet to 3950 m depth (about 120 m above the lake). The ice at 3310 m is about 420,000 years old and was formed by refrozen lake water (Jouzel *et al.*, 1999). Analyses of the refrozen lake water from the ice cores indicate the presence of a microbial community (Priscu *et al.*, 1999; Karl *et al.*, 1999) and some of these bacteria may still be viable (Karl *et al.*, 1999). Data suggest that these lakes below the Antarctic ice cap are pressurizing and may quickly release their contents under the ice into the ocean leading to a rapid drop in ice elevation (Wingham *et al.*, 2006). Sampling the lake without contaminating it will be technically difficult but is certain to yield interesting results.

impound water flow and lead to formation of lakes. Materials deposited along the sides of glaciers form lateral moraines, which can also dam streams and create lakes. Glacial lakes tend to be smaller than tectonic lakes, but a few very large glacial lakes (e.g., the North American Great Lakes) make up a considerable area when considered on a global basis (Fig. 7.5).

The release of large volumes of water from behind glacial ice dams is a catastrophic mode of lake loss followed by downstream formation of smaller lakes. Some of these outbursts happen on a moderate scale now; huge ones occurred during the last ice age. These outbursts occurred as a result of pooling of glacial water as large ice sheets receded, followed by collapse of the ice dam. Such outbursts created massive floods that scoured out existing lakes and created new lakes below any spillways that existed in the channels. Kehew and Lord (1987) suggest that such outbursts established the courses of most major rivers in the midcontinental United States and Canada. Lake Missoula was formed in western Montana behind the retreating ice sheet and was responsible for several massive floods downstream in the Columbia River basin. The lakes in the Grand Coulee in eastern Washington State are below the spillways where the floods dug out massive quantities of sediment and deeply incised the channels.

Volcanic activities can lead to formation of lakes. Explosions of volcanoes or pockets of steam can leave behind depressions in craters that fill with water. These lakes are called *caldera* or *maar* lakes. An example of a volcanic lake is the exceptionally clear and deep Crater Lake in Oregon (Fig. 1.1). The lake was formed when the volcano Mount Mazama exploded about 6000 years ago. This eruption must have been a catastrophic event for Native Americans living in the region at the time because several meters of ash were deposited throughout western North America. The resulting crater filled with water and a subsequent eruption formed a volcanic cone in the lake known as Wizard Island. Similarly, Lake Yellowstone in Wyoming formed as the result of a very large volcanic explosion, and such an explosion could occur again in this location leading to devastation in the western United States and severe global cooling because of ash ejected into the atmosphere.

Water can dissolve sedimentary rocks and lead to depressions that form lakes. In karst regions, *sinkholes* form where limestone is dissolved and the cavity collapses (Fig. 4.9), and in cases where the bottom is sealed after the collapse, a lake can form. Sometimes these sinkholes simply drop down below the level of groundwater and a lake or pond is created at the bottom of a very steep-walled hole. Such pools, called *cenotes*, are common in Yucatán in Mexico; the most famous is probably the Cenote of Sacrifice in Chichén Itzá where numerous Mayan artifacts, including gold objects, have been found as well as human bones from sacrificial offerings. Sinkholes can also occur where old subterranean salt deposits are dissolved or where sandstone is washed away. These dissolution lakes are generally small.

Activities of rivers can form *fluvial* lakes, including oxbow lakes where meanders pinch off (see Chapter 6). A levee lake is another fluvial type that is formed next to rivers where periodic floods scour and fill depressions parallel to river channels. The lowland areas surrounding the Amazon River contain many of these lakes. They are connected to the Amazon during times of high flow. Fluvial lakes provide an important habitat for many organisms and are involved intimately with the ecology of the river. Fluvial lakes tend to be smaller than either glacial or tectonic lakes and are less abundant globally than the other two lake types (Fig. 7.4).

Additional processes that can form lakes include erosion by wind (*aeolian* lakes), crater formation by meteoric impacts, and formation of depressions by alligators (*Alligator mississippiensis*) or bison (*Bos bison*), dam building by beavers (*Castor* spp.) and accretion of corals leading to isolation from the ocean and lakes in the centers of coral atolls.

LAKE HABITATS AND MORPHOMETRY

A lake can be divided into several subhabitats or zones. Lake habitats in general are referred to as *lentic* or *lacustrine* (i.e., habitats with deep, nonflowing waters). The open water of a lake, particularly the water column above sediments that do not receive enough light to maintain benthic photosynthetic organisms, is the *pelagic* zone. The *profundal* zone is the benthic habitat below the pelagic waters. The profundal zone is influenced by materials that settle from the pelagic waters and usually has sediment composed of fine silt or mud. The shallow zone of a lake, where enough light reaches the bottom to allow the growth of photosynthetic organisms, is referred to as the *littoral* zone. The relative occurrence and extent of these different subhabitats is determined by the size and shape of the lake.

Morphometry, or the shape and size of lakes and their watersheds, is one of the first ways that people classify lakes. The *bathymetric map* (a depth-contour map of a lake bottom) of a lake provides important information on geomorphologic properties (Fig. 7.9). Generally, the first measurement made is of the area of the lake (A) and the second is of depth (z). The maximum depth (z_{max}), mean depth ($\bar{z}$), and volume (v) are also of interest. The volume is the product of area and mean depth:

$$v = A\bar{z}$$

In general, lakes with a shallow mean depth are more productive than deeper lakes. Greater productivity of shallow lakes is a consequence of wind mixing

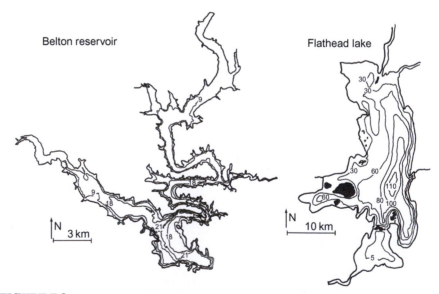

FIGURE 7.9
Bathymetric maps of Belton Reservoir, Texas (left), and Flathead Lake, Montana (right). The reservoir is dendritic and shallow, with the deepest portion near the dam (lower right). Flathead Lake was formed by a combination of tectonic and glacial processes and is deep with a regular shoreline and shallow outlet (lower left).

the nutrients up from the bottom more readily, more extensive littoral habitat for primary producers that use the lake bottom, and other morphometric considerations.

If the volume of a lake and the amount of water entering and leaving the lake are known, then the *retention time* or *water residence time* of the water in the lake can be determined. The average retention time (R) can be generally calculated as follows if volume (v) and water loss (L) from evaporation and outflow are known:

$$R = \frac{v}{L}$$

This equation assumes inflow and outflow are equal. However, in some cases it is more difficult to calculate water retention because closed basins gain water and increase in volume until evaporation balances out inputs. Water retention time can vary widely from several hours for a small pond with a large inflow to thousands of years for very large lakes. Water residence time is important in determining the residence time of pollutants in a lake, how quickly the biota can be washed out, and the general influence of tributaries entering a lake. For example, Lake Tahoe, a beautiful, clear lake in the western United States, has about a 700-year residence time (Table 7.2) so it is very sensitive to nutrient pollution (see Chapter 18).

Another important aspect of morphometry is the irregularity or degree of convolution of the shore. An index used to quantify this is called the *shoreline development* (D_L). This index compares the minimum possible circumference of the lake, given a specific surface area (i.e., a circle), to the actual circumference of the lake and its surface area. A value of 1 for shoreline development is a circle and a larger value means that the shoreline is highly dissected. High shoreline development is generally related to small values of mean depth and mode of formation, and is indicative of a high degree of watershed influence. Shoreline development is calculated as follows:

$$D_L = \frac{L}{2\sqrt{\pi A_0}}$$

where L is the length of the shore, and A_0 is the surface area of the lake.

Lakes with high shoreline development are often naturally productive relative to those with low shoreline development (Example 7.1). The idea of a dissected shoreline resulting in increased productivity leads to consideration of the watershed of a lake as a determinant of productivity. A lake with a relatively large watershed will have more land area to contribute nutrients.

■ Example 7.1

Compare Morphology of Two Lakes—Crater Lake and Milford Reservoir

Crater Lake, Oregon, and Milford Reservoir, Kansas, are lakes of contrasting properties, despite the fact that they are similar in surface area. Crater Lake (Fig. 1.1) is a deep oligotrophic lake in the crater of a large volcano; its scenic grandeur has earned it national park status. Milford Reservoir is a typical midwest US lake; it is eutrophic but widely used for recreation, including a vibrant fishery. Much of the difference in levels of productivity in these two systems can be related to their contrasting morphometric properties, but the level of agricultural activity in the watershed of Milford Reservoir is also much greater. The following are vital characteristics of the two lakes:

	Crater Lake	Milford Reservoir
Area (km^2)	55	65
Mean depth (m)	364	7.4
Watershed area (km^2)	81	64,465
Shoreline length (km)	42	22
Inflow (m^{3}s^{-1})	4.3	27.2

Calculate volume, water replacement time, and shoreline development. Also calculate the ratio of the area of the watershed to the lake volume and speculate about how these features may alter trophic state.

Volume for Crater Lake = area × mean depth = 55 × 0.364 = 20.3 km^3
Volume for Milford Reservoir = 65 × 0.0074 = 0.48 km^3
Water replacement for Crater Lake = volume/discharge = 20.3/0.136 = 150 years
Water replacement for Milford Reservoir = 0.48 km^3/(0.858 km^{3}y^{-1}) = 0.56 years
Shoreline development for Crater Lake = D_L = 42/(2$\sqrt{\pi 55}$) = 1.6
Shoreline development for Milford Reservoir = D_L = 122/(2$\sqrt{\pi 65}$) = 4.3
Watershed area/lake volume for Crater Lake = 4.0
Watershed area/lake volume for Milford Reservoir = 134,000

All these parameters but one suggest that the watershed will have a much greater influence on the water in Milford Reservoir and that nutrients and light should be greater in Milford Reservoir. The only caveat is that the water replacement is so slow in Crater Lake that once a nutrient enters the system, it could be recycled for some time. The high shoreline development index, watershed area to volume, and low mean depth suggest that Milford Reservoir should indeed be more eutrophic than Crater Lake.   ■

Such a lake is likely to be more productive than a lake with a small watershed. Land use practices also play a major role in determining nutrient inputs.

UNIQUE PROPERTIES OF RESERVOIRS

Reservoirs are common features of today's landscape, so understanding how they differ from natural lakes is important. Damming can form natural lakes, and presumably reservoirs are not much different, except that natural lakes usually do not release deep waters downstream, which is sometimes the case with reservoirs. Occasionally, the capacity of natural lakes is increased and outflow is regulated by adding a dam.

Unlike natural lakes, reservoirs are deep near the dam and generally become shallower near the deltas of the rivers that feed them. Reservoirs are often limited by the surrounding topography, so they have a smaller mean depth than many natural lakes. However, some high-dam reservoirs are 200 to 300 m deep. Shallow mean depth can lead to increased mixing and associated suspended solids, as well as decreased likelihood of stratification.

Reservoirs fill the drainage basins of rivers and streams, and each arm of a reservoir moves up into a former stream channel. Thus, a typical reservoir has a dendritic or tree-like shape (Figs. 7.1 and 7.9). A dendritic shape results in a high value for the shoreline development index. Shallow mean depths and high shoreline development indeces indicate that many reservoirs are very productive, unless turbidity limits light for photosynthetic production.

Larger reservoirs often have more lake-like biological and physical characteristics as one moves from the inlet rivers toward the dam. In the shallow inlets, river flow dominates, but as the water velocity and turbulence decreases with greater width and depth, a lentic character is established. Thus, species commonly associated with rivers are gradually replaced with those more commonly found in lakes as one moves closer to the dam.

Rivers deposit sedimentary material as velocity slows, creating river deltas at the upstream end of reservoirs. Reservoirs fill with sediments from the top down toward the dam. All reservoirs eventually fill with sediments, with the useful life of some reservoirs only a few decades. Controlling upstream sediment inputs can prolong the life of a reservoir.

Reservoirs have global importance. Even though the amount of sediment erosion from the land has increased worldwide, the amount of sediment that reaches the oceans has decreased because sediments settle in reservoirs, and there are ever more reservoirs, particularly in tropical areas (Syvitski *et al.*, 2005). Organic materials also settle in reservoirs leading to significant retention of organic carbon (Friedl and Wüest, 2002) that would otherwise be exported to the ocean.

Most lakes have relatively stable water levels because their outflow stream is at a single elevation, except in cases where outflow is severely constricted and flooding can cause increases in lake elevation. In contrast, most reservoirs have the capacity to raise and lower water levels as a practical consequence of their function of water storage and power generation. Thus, a relatively unnatural condition occurs where large areas of shoreline are inundated for part of the year. Often, water level is tightly prescribed to allow for recreational boating on the reservoir or to meet downstream navigation requirements. This mode of reservoir operation leads to a ring of vegetation that cannot withstand inundation around the reservoir at its maximum elevation; a barren area is often seen below this ring when water level is low. The level of alteration can affect the ability of submerged plants to survive, and thus the nature and extent of the littoral zone can increase sediment erosion problems, as well as influence the spawning behavior and success of fishes in the reservoir.

STRATIFICATION

Factors influencing density of water that were discussed in Chapter 2, and heating effects of light discussed in Chapter 3, have profound effects on mixing in lakes. These effects influence the biogeochemistry, biology, and physical geology of lakes. A primary factor creating stratification of lakes is the difference in density resulting from temperature or salinity variation. The classical understanding of lake stratification is based on consideration of cold-temperate lakes, so this seasonal sequence of stratification is considered first.

During the early spring in a cold-temperate lake, the water is *isothermal*, or approximately the same temperature from top to bottom (Fig. 7.10). An isothermal lake can be completely mixed by wind, leading to *spring mixing*. The entire lake will continue to mix as long as the wind continues to blow. The surface of the water is warmed by solar energy as the spring season progresses. Surface waters of the lake heat the most because infrared radiation (heat) is absorbed quickly with depth. If you have ever swum in cold water on a calm, sunny spring day, you are familiar with the phenomenon of the top several centimeters of the water being much warmer than the deeper water. Such stratification is only temporary because wind can mix a shallow layer of warm water into the lake.

When a series of calm, warm days occurs, the lake stratifies. Surface waters of the lake heat enough so that the wind cannot completely mix the warmer, less

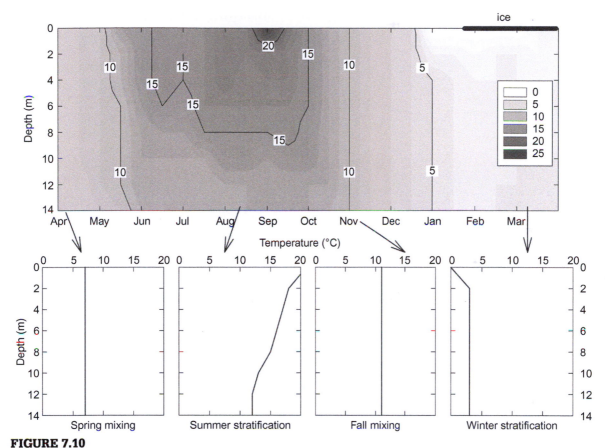

FIGURE 7.10

(Top) A depth contour plot of lake temperature over the course of a year in a dimictic cold-temperate lake (Esthwaite Water, an English lake). The thick black line at the top right corner of contour plot indicates ice cover. (Bottom) Two-dimensional representations of the temperature versus depth at each phase of stratification. *(Data from Mortimer, 1941).*

dense surface water into the cooler water below. The top of the stratified lake is called the *epilimnion*. The zone of rapid temperature transition just below the epilimnion is the *metalimnion* or *thermocline*. The bottom of the lake is at fairly constant temperature and is called the *hypolimnion* (Fig. 7.11). The stratified layers will stay distinct until a prolonged period of cool weather occurs. The period with distinct layers is called *summer stratification*. Prolonged summer stratification is a combined function of the very slow rate of diffusion of heat across the metalimnion because turbulent mixing no longer occurs, and because of the continued heating of the epilimnion. Because there is minimal mixing across the metalimnion, no eddy diffusion of heat occurs; only molecular diffusion occurs. Slow rates of molecular diffusion were discussed in Chapter 3.

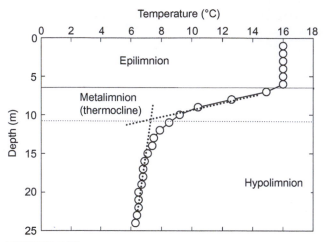

FIGURE 7.11

Temperature as a function of depth for Triangle Lake, Oregon, on October 1, 1983, and positions of epilimnion, metalimnion, and hypolimnion. *(Data from R. W. Castenholz).*

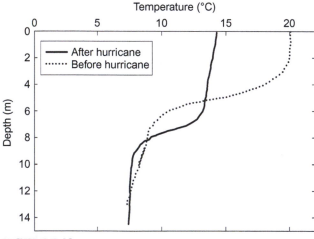

FIGURE 7.12

Stability of the thermocline in Linsley Pond, Connecticut, before and after a hurricane on September 21, 1938, with wind speeds up to 100 km h^{-1}. *(From G. E. Hutchinson, A Treatise on Limnology, Vol. 1, Geography, Physics and Chemistry. John Wiley & Sons, Inc., 1957. Reprinted by permission of John Wiley & Sons, Inc.).*

The epilimnion is very stable relative to the ability of the wind to mix a lake. There can be some mixing of the top of the hypolimnion (*entrainment*) with extreme winds, but even hurricane-force winds will not fully mix a well-stratified lake (Fig. 7.12) because the mixing power of the wind declines exponentially with depth (see next section). Stratification will break down only when the autumn weather can cool the epilimnion to approximately the same temperature (*isothermal*) as the hypolimnion. Cool air coupled with continued heat losses from surface evaporation decreases the temperature of the surface. Cooled surface water is denser than the water immediately below, so it sinks. The wind can mix the lake once the entire lake is isothermal, and the *fall mixing* period begins. The lake will potentially continue to cool and mix until formation of an ice cover on the surface of the lake. In very deep lakes, the wind still cannot mix the bottom waters with the top, even if the lake is isothermal.

The surface of the lake can freeze when the temperature of the entire lake is below 3.9°C. If the lake is warmer, cool water from the surface will continue to sink and mix with the less dense water below it. If the lake temperature is 3.9°C, the water is at its densest, so cooler water will sit on the surface of the lake as long as the wind does not mix it. The surface of the lake can freeze if there is a cold, calm night. Low-density ice and cold low-density water will sit on the surface of the lake once the surface has frozen. No more wind reaches the surface of the water once ice forms, so no more mixing occurs. Winter stratification is the second period during the year when a cold-temperate lake does not mix (Fig. 7.10) and it lasts as long as cold weather maintains the ice cover.

Duration of ice cover provides one of the best long-term records of the effects of global warming on freshwater systems. Detailed data on dates of formation and breakup of ice cover for many lakes and rivers for the past 100 years are available, and records exist for Lake Suwa in Japan almost continuously since 1450. These data suggest that freeze dates have shifted 5.8 days later over the past 100 years and breakup dates 6.5 days earlier (Magnuson *et al.*, 2000).

Much terminology is used to describe the mixing regimes of lakes (Table 7.1). Lakes that mix twice a year are called *dimictic*. Lakes that only mix once during the year are called *monomictic* and are common in temperate and subtropical regions where winters are not cold enough to freeze lake surfaces but are cool enough to allow the lakes to become isothermal. Lakes that never mix are called *amictic* or *meromictic*. *Polymictic* lakes stratify or mix several times a year and are found mainly in tropical regions.

Several conditions cause meromictic lakes. Seasonal temperature regimes can be constant enough (mainly in tropical areas) that lakes rarely mix. The temperature difference in tropical lakes does not need to be as great to form a stable stratification as in the temperate zone because the water temperatures are higher and a greater relative difference in density occurs for each degree difference in water temperature (Fig. 2.3). For example, the density difference is greater between 20 and 25°C than between 10 and 15°C water.

Some lakes rarely if ever mix if they contain dissolved compounds in the hypolimnion that stabilize density layers. Salinity differences can lead to extremely stable stratification. In this case, more saline water can sit below cooler surface waters when the salinity-caused density difference is greater than the temperature-related differences. Several conditions can cause such salinity differences. In tropical lakes that are stratified for long periods of time, nutrients enter the surface waters from rivers. These nutrients enter the biomass of the planktonic food web, and when organisms die they sink. The sinking organisms slowly release dissolved nutrients as they decompose and a portion is transported to the hypolimnion. Slowly, the salinity of the hypolimnion increases and the stratification is stabilized. Saline spring or groundwater inputs can also stabilize lakes. Some of the dry-valley lakes in Antarctica have such stable layers (see Chapter 15).

In arid regions, evaporation can lead to increases in dissolved salt concentrations. Fresh river water flowing into a lake will remain on top of the denser saline water. A fresh surface lake is a common occurrence in closed basins where saline lakes form. Such was the case in the Dead Sea, where a stable stratification was maintained for nearly 300 years until water diversions for human uses reduced inflow to the lake and the dilute surface layer disappeared in the 1970s (Gavrieli, 1997).

Fjord lakes can also have saline waters below freshwater. In this case, a glacial valley is formed where the glacier carves the valley to an elevation below sea level. As the glacier recedes, saline marine water floods the valley. The floor of the valley rebounds from the weight of the glacier and if a raised portion exists at the end of the valley (e.g., a terminal moraine) the saline ocean water can be isolated. Freshwater flows on top of the saline water and a lake forms with saline water on the bottom and freshwater on the top.

The biological and biogeochemical effects of stratification on lake organisms are strong. Molecular diffusion rates that dominate movement of dissolved materials across the metalimnion are slow enough that a significant depletion of O_2 in the hypolimnion will lead to anoxia during the summer. In turn, O_2 loss from the hypolimnion means that biogeochemical cycling and lake productivity are altered. Anoxia and biogeochemistry are discussed in detail in Chapters 12 through 14. Given the very complex chemical and physical characteristics in many stratified lakes, several methods for sampling lake waters from different depths have been developed (Method 7.1).

METHOD 7.1

How Do Limnogists Sample Water and Sediments from Lakes?

The main consideration in determining how to sample water from a lake is what information is sought. If the information can be obtained with a submersible sensor (such as for temperature, dissolved oxygen, pH, and conductivity) it is easiest to take measurements without removing samples. However, for many chemical and biological parameters, water must be removed from known depths with a minimum of contamination or turbulence.

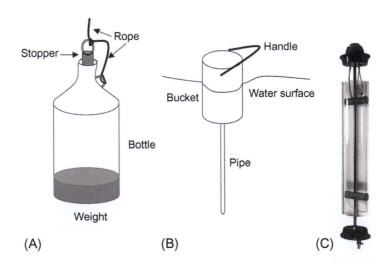

FIGURE 7.13

(A) A simple water sampler made from a weighted bottle and stopper; (B) a sampler that collects water from depth by displacing water at the surface; (C) a Kemmerer sampler. Only samplers (such as type C) that close at depth are suitable for collecting dissolved gas samples. *(Photograph courtesy of Wildlife Supply Company).*

Several devices are available that allow water to be sampled from depth. Many different types of limnological equipment are lowered into the water on ropes or cables. The devices are generally triggered with a weight that is dropped down the line to the sampling equipment. This weight is called a *messenger*.

Van Dorn bottles and Kemmerer bottles are devices used to remove water from lakes (Lind, 1974). The Van Dorn bottle consists of a tube with covers over each end. The messenger releases a catch so the stretched rubber connectors can pull the covers onto the tube, sealing the water into it. Kemmerer bottles operate similarly, but rather than using rubber connectors to pull ends onto a tube, gravity is used (Fig. 7.13). Other devices include pumps to remove water from depth, bottles with strings attached to stoppers that can be unplugged at depth, and pipes that allow water to flow up to containers that displace surface waters (Fig. 7.13).

Lake substrata and associated sediments, the waters contained in them, and organisms in them, can also be sampled with devices called *dredge* or *grab samplers* that snap shut when a messenger is dropped down an attached rope (Fig. 7.14). Alternatively, depending on the type of sample and information that is desired, various types of coring devices can be used; these range from small, hand-held corers to large, boat-mounted oceanographic drilling devices, depending on the specifics of the habitat and type of sample needed. A variety of suction or hydraulic dredge samplers also have been developed for collecting lake and ocean sediments. Suction samplers require a power source

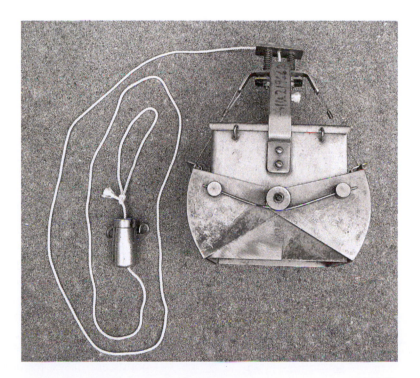

FIGURE 7.14

An Ekman grab sampler for sampling sediments from the bottom of a lake. These types of grab or dredge samplers come in a variety of shapes and sizes. Most have a spring-loaded trigger (visible on the top of the sampler), which is tripped by a messenger (cylindrical weight on the rope) that is sent down the attached rope; when the messenger trips the trigger, the sampler snaps shut around a parcel of sediment. *(Photograph by S. Peterson).*

and, like coring devices, they can range from handheld samplers to large, boat-mounted machines.

The choice of device for water or sediment sampling depends on the type of sample that is required, and there are many factors to consider before sampling. Toxic materials generally should be avoided and if chemical analysis on metals is to be done, metal samplers should not be used. The violent closure of some samplers can harm organisms that are susceptible to pressure shock. Zooplankton may avoid an opaque sampler more than a clear one because of their predator avoidance behaviors.

WATER MOVEMENT AND CURRENTS IN LAKES

Movement of wind is generally the main cause of waves across lakes, although motorboat activity can cause significant wave action. Wave action is important partially because it is associated with surface mixing and erosion of the shoreline. Lakeshore erosion can lead to habitat destruction and large financial losses associated with property damage; many environmental engineering firms specialize in controlling erosion. Wave action can influence which species can successfully inhabit the different depths of the shallow benthic zone (littoral zone). The two main determinants of wave height are the strength (speed and duration) of the wind and the length of lake on which the wind acts. The influence of the wave also is dependent on the geometry and materials that make up the shoreline. A long shallow beach can dissipate the force of breaking waves across a larger area and is less prone to erosion than is a steep lake shoreline.

The length of lake on which the wind acts is called the *fetch* (Fig. 7.15). The longer the fetch, the higher the waves (Fig. 7.15). A perfectly round lake would be affected similarly by wind from any direction. On an irregularly shaped lake, certain wind directions lead to the largest waves. In smaller ponds and lakes, features such as hills or trees can prevent the wind from making very large waves. For example, deforestation can lead to a deeper epilimnion (by about 2 m in some small Canadian lakes) because of increased mixing during spring

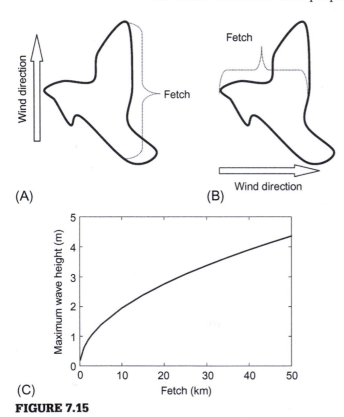

FIGURE 7.15

How fetch of an irregularly shaped lake varies with wind direction (A, B) and relationship between maximum wave height and fetch (C). *(Equation for C from Wetzel, 1983).*

warming (France, 1997a). Even though the processes of surface wave formation are the most apparent to human observers, the wind causes other water movements in lakes that can also be important.

As wind moves water on the surface of a lake forward, it must be replaced by water from below. This process leads to spiral circulation patterns called *Langmuir circulation* cells (Fig. 7.16). As wind pushes the surface water forward, it must be replaced by deeper water, and this process sets up zones of upwelling and downwelling water. The lines of upwelling and downwelling occur parallel to the direction of the wind, and set up spiral movement of water parcels. The spirals rotate in alternating directions compared to the ones next to them. Floating materials aggregate at the water surface along the downwelling lines and form streaks in the same direction as the wind is blowing. These circulation cells are several meters wide, and streaks can commonly be observed on lakes and reservoirs under a constant wind.

In addition to the smaller scale waves and Langmuir cells, movement of water can also occur within the whole lake's volume. When a sustained wind occurs, it causes water to pile up on the downwind side of the lake, and when the wind suddenly ceases the surface of the lake can rock. This rocking of a

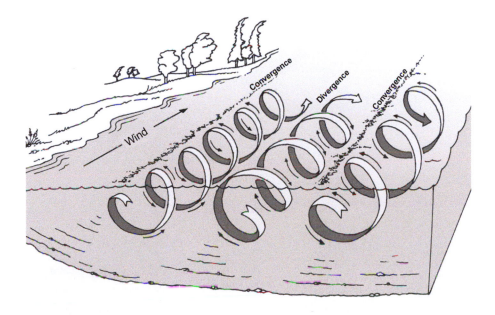

FIGURE 7.16
Langmuir circulation cells on a lake. Note zones of convergence where spiral currents meet and go down into the lake. *(Image by Pete Else).*

lake's entire surface is called a *seiche*. The surface seiche in Lake Erie can be several meters leading to destructive storm surges and stranded watercraft. Seiches can have periods of a few hours to days, and can also be formed by earthquakes.

An interesting phenomenon, an internal wave, occurs in stratified lakes that are subjected to a sustained unidirectional wind. The force of the wind causes the water in the epilimnion to move across the lake to the downwind side (Fig. 7.17) and the depth of the epilimnion is greater downwind than upwind. Under extreme winds, the hypolimnion can come to the surface on the upwind side. When the wind ceases, the less dense water of the epilimnion moves back across the surface of the lake, and the hypolimnion moves back toward its original position. Like a pendulum, the surface of the hypolimnion rocks back farther than its original position. Thus, an *internal seiche* is created where the surface of the lake appears still, but the plane that forms the top of the hypolimnion continues to oscillate for hours or days after the wind ceases. Aside from the intrinsic elegance of the seiche as a physical phenomenon, this type of water movement has an important biological implication.

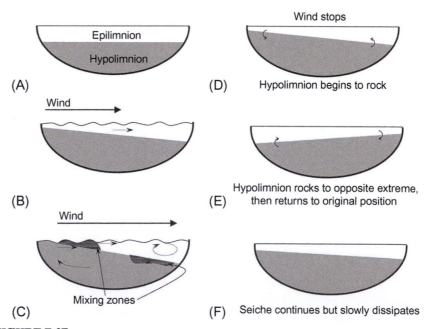

FIGURE 7.17

Formation of an internal seiche and entrainment associated with wind. Dashed arrows show water flow. (A) The lake under calm conditions; (B) the wind deepens the epilimnion on the right; (C) a strong wind mixes some of the epilimnion with the hypolimnion; (D) the wind stops and the hypolimnion begins to oscillate; and (E and F) the amplitude of the seiche diminishes over time.

Even though the hypolimnion is very stable, seiches can lead to a moderate amount of mixing of hypolimnetic and epilimnetic water. The movement of this water up to the epilimnion leads to entrainment. Entrainment causes nutrient-rich water from the hypolimnion to reach the epilimnion (Fig. 7.17C), causing stimulation of primary production. Nutrient mixing can be significant biologically because the mixing rate far exceeds the rate of molecular diffusion that usually predominates between the hypolimnion and the epilimnion. Seiches can also influence rooted plants and benthic invertebrates by altering temperature and nutrient regimes. The lake model exercises discussed by Wetzel and Likens (1991) are highly recommended for students who want a clearer understanding of the processes of stratification and seiches.

Very large lakes, as discussed in Chapter 3, can also be subjected to Coriolis forces causing a counter-clockwise rotation in the Northern Hemisphere. Other forces that move water in lakes can include human-caused disturbances, earthquakes, and floods.

SUMMARY

1. A variety of processes form lake basins, including tectonic, glacial, fluvial, volcanic, and damming processes. Glacial lakes are the most numerous worldwide, but some of the largest, deepest, and oldest lakes are formed tectonically. Fluvial lakes can be very important to riverine ecology.

2. Lake basin morphology is described with various parameters, including mean depth, area, maximum depth, volume, shoreline development (D_L), and watershed area relative to lake surface area. Shallow lakes with large watersheds and highly dissected shorelines are generally the most productive.

3. Reservoirs are unique habitats that have characteristics of both streams and lakes. Depending upon how they are constructed and managed, reservoirs can have profound impacts on upstream and downstream habitats and communities.

4. Waves are greatest where the wind has the longest length of lake (fetch) on which to act.

5. Stratification can alter the water circulation in lakes and thus alter biogeochemical, ecosystem, and community properties. Mixing can occur often (polymictic), once a year (monomictic), twice a year (dimictic), or rarely (amictic or meromictic), depending on climate and type of stratification.

6. Thermal stratification occurs when warm surface water sits above denser, cooler waters. The warm surface layer of a thermally stratified lake is the epilimnion, the zone of steep temperature transition is the metalimnion, and the deepest stable zone is the hypolimnion.

7. High concentrations of dissolved substances can also lead to stratified layers in lakes. Such chemically driven stability can exceed temperature-driven stability

because density differences can be greater than are possible with natural temperature differences.

8. Wind causes Langmuir circulation patterns, which lead to streaks of floating material on the water surface but also mix the lake to depth.

9. A sustained wind that suddenly stops can cause oscillation of the lake surface (an external seiche) or the hypolimnion (internal seiche). This rocking can lead to breakdown of stratification. Mixing of deeper waters into the surface is called *entrainment*.

QUESTIONS FOR THOUGHT

1. Why is it sometimes difficult to assign a single geological explanation to a lake's origin?

2. Why is a lake with a high value for D_L likely to have smaller waves than a lake of comparable surface area with a D_L close to 1?

3. Why do more lakes occur farther from the equator?

4. In which order (from greatest to least) should lakes be ranked with respect to the overall average ratio of maximum depth divided by the mean depth: tectonic, glacial, and fluvial?

5. Langmuir circulation cells concentrate particles slightly denser than water below the surface of the water: Where will these particles be concentrated and why?

6. Under what conditions would thermal stratification lead to anoxia in the hypolimnion?

7. Explain why some rivers flowing into lakes flow down into the hypolimnion, some flow across the surface, and others flow into the metalimnion.

Types of Aquatic Organisms

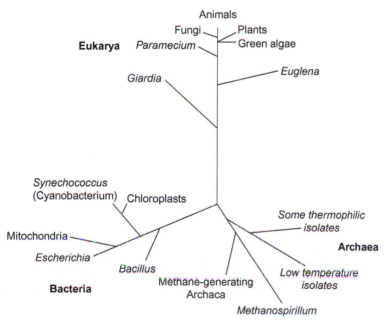

FIGURE 8.1

A simple phylogenetic diagram based on rRNA data. *(Data combined from Fenchel and Findlay, 1995, and Pace, 1997).*

167

Doi: 10.1016/B978-0-12-374724-2.00008-8

Correct identification of freshwater organisms is essential to understanding their ecology. Plants, animals, and microbes interact with the environment to alter water quality and perform ecological "services," such as decomposition and nutrient cycling. Identifying species in food webs is an essential part of managing fisheries. The presence, absence, and overall diversity of invertebrates, microbes, algae, plants, and fishes can be used to indicate pollution problems through various biological assessment procedures. Such assessment relies upon the identification of pollution-sensitive and pollution-tolerant taxa to indicate water quality. Proper identification of these organisms is required for biological assessment techniques that use species diversity as an indicator of pollution. Some species of algae are toxic, so identification of these species may be important in maintaining safe water quality. Tracking the invasion and influence of pest organisms also requires taxonomic expertise. Finally, endangered species can have a major influence on management decisions, again requiring taxonomic information. This chapter provides very basic taxonomic principles, introduces different ways to classify organisms, and outlines how they survive and the habitats they frequent. This chapter also provides background for the following two chapters on microbe, plant, and animal groups found in freshwaters, and it introduces some essential terminology related to habitat and species interactions.

THE SPECIES CONCEPT

Biologists consider the species the fundamental unit of taxonomic division, and the scientific system for naming organisms is based on distinguishing species. Species are then classified into ever-broader groups (Table 8.1). A traditional definition of biological species is "a genetically distinctive group of populations whose members are able to interbreed freely under natural conditions and are reproductively isolated from all members of other such groups" (McFadden and Keeton, 1995). However, many aquatic organisms (especially the microbes, but also many plants and animals) do not reproduce sexually. Other species may be able to reproduce with organisms that are considered to belong to different species (e.g., many species of trout are able to hybridize). Bacteria have no sexual reproduction but can pass genes within or among different taxonomic groups. Even for those organisms that do reproduce sexually, it is often difficult to test if two individuals will interbreed successfully or to determine the degree of genetic similarity. We have limited data on the reproductive biology of most aquatic organisms in their habitat except for some game fish and emergent wetland plants. Finally, there is no objective measure of how "genetically distinctive" an organism must be from another before it qualifies as its own species.

Table 8.1 Botanical and Zoological Naming Scheme for Organisms, Name Endings Associated with Each Taxonomic Group, and Examples of Naming of a Cattail and a Beaver

Taxonomic Classification (Sub- or Supergroup)	Botanical Name Endings	Example	Zoological Name Endings	Example
Super kingdom or domain	-a	Eukarya	-a	Eukarya
Kingdom	-ae	Plantae	-a	Animalia
Phylum or division	-phyta	Anthophyta	-a	Chordata (subphylum Vertebrata)
Class	-opsida	Liliopsida	-a	Mammalia
Order	-ales	Typhales	-a (-formes, birds; -oidea, some mammals)	Rodentia
Family (subfamily, tribe, subtribe)	-aceae	Typhaceae	-idae	Castoridae
Genus		*Typha*		*Castor*
Species (subspecies)		*latifolia*		*canadensis*

In practice, systematists differentiate among most nonbacterial species on the basis of morphological characteristics (although molecular techniques are becoming more common). The operational species definition uses an older, formal definition of a species as "a group of organisms that more closely resemble each other, with respect to their physical appearance (morphology), physiology, behavior, and reproductive patterns, than they resemble any other organisms" (McFadden and Keeton, 1995). However, no hard and fast line delineates the amount of morphological differentiation that is necessary for organisms to be considered distinct species. Difficulty arises because of the natural morphological variation found in the same species (Fig. 8.2), especially those living in different environments. Thus, a large sample of individuals from different environmental conditions is often useful to ensure correct identifications. Generally, systematists that specialize in a specific group of organisms have clearly defined the characteristics used to differentiate among distinct species.

The lack of a single system to define species that works well across all taxonomic groups leads to a utilitarian species definition for those who are not taxonomists: A species may be considered distinct if the majority of the systematists studying the group of organisms agree that it is a distinct species.

FIGURE 8.2

Eichhornia azurea with different morphology of floating versus submersed leaves within the same plant. *(Courtesy of Steve Hamilton).*

This approach allows aquatic ecologists to use systematic data without information about sexual reproduction of the species they are studying or other complex methods of analysis. The taxonomic identity can be communicated within the bounds of the best current scheme of identification, providing a solid basis for ecological information. A reliable scientific name can be used effectively in searching for biological information on a species, particularly with electronic search engines increasing such abilities.

The most recent taxonomic methods use biological molecules, such as DNA, RNA, lipids, and proteins, to distinguish species. An excellent discussion of these and other taxonomic techniques can be found in Graham and Wilcox (2000). One of the most successful methods relies on the RNA found in ribosomes for taxonomic determinations (Method 8.1). We are reaching a time where only a small tissue sample or a sample of a few cells is required for rapid definitive identification of many species.

Some widely adopted approaches to classifying microbial species are assuming distinct species if DNA sequences vary by more than 30% or if ribosomal rRNA sequences vary by more than 5% (Buckley and Roberts, 2006). These approaches to classifying species are made purely for convenience and 30% or 5% are mostly subjective levels of variation upon which to define species. The methods are empirical and not based on a theory of microbial evolution (e.g., there is no indication that a genome is a coherent group of genes over evolutionary time at some arbitrary degree of rRNA similarity). As of now, there is not enough known about evolution of microbial genomes to apply a theoretical definition of species. The currently used methods do allow investigators to define the species complement of an environment that contains many species, a large proportion of which have not yet been cultured and characterized.

METHOD 8.1

Using Ribosomal RNA Analysis for Identification of Organisms

Ribosomal ribonucleic acid (rRNA) sequence comparison has been used increasingly as a taxonomic tool, particularly for microorganisms but also for taxonomically difficult macroscopic organisms. Several RNA molecules occur in each ribosome (the site for protein synthesis in the cell). The 16S rRNAs in bacteria and 18S rRNAs in eukaryotes are generally used for taxonomic comparisons. There are several advantages of the use of rRNA. The molecules are found in all living organisms and have some sequences that are very conservative, meaning they retain some sections that all organisms have in common. The rRNA molecules also contain regions that change more rapidly over evolutionary time and can be used for detailed analysis at the species or subspecies level. Finally, molecular methods have developed in which small amounts of rRNA from natural populations can be amplified and analyzed.

The technique is based on comparing sequences of rRNA molecules from different organisms. The greater the time since evolutionary divergence of species, the more the sequences will vary. The rRNA (Fig. 8.3) has specific sections that are required for the function of the molecule and others that are less essential. The sections of the rRNA that are essential to protein synthesis are highly conserved (i.e., their nucleotide sequence varies little among species) because mutations in these regions are usually fatal. Such mutation is a very quick evolutionary dead end. Thus, a taxonomist needing to make kingdom-level comparisons would choose a highly conserved section of the rRNA, and one distinguishing among closely related species would choose a section that accumulates mutations more rapidly.

Comparison of rRNA sequences is particularly useful for microbes for which taxonomy is difficult because of the small degree of morphological variation. In cases in which the technique can be compared to the taxonomic trees generated by morphology (e.g., cyanobacteria), reasonably good agreement occurs between results from conventional techniques and RNA analysis (Giovannoni *et al.*, 1988; Taylor, 1999).

When the rRNA sequence is known for a particular species, this information can be used for *in situ* identification of species. A complementary DNA strand is synthesized

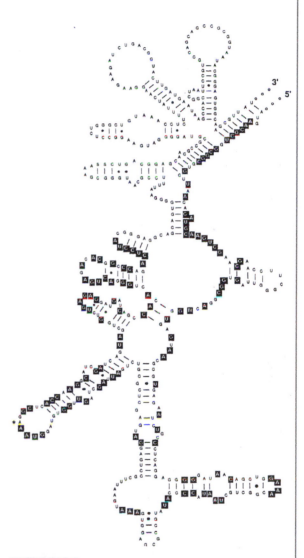

FIGURE 8.3

A map of a 16S rRNA molecule from an oligotrophic planktonic marine bacterium. Shaded characters indicate where the same nucleotides are in the same position in Eukarya. Related bacteria occur in freshwater (Bahr *et al.*, 1996). *(From Giovannoni et al.; reprinted with permission from* Nature 345, *60–62, 1990, Macmillan Magazines Limited).*

that matches a unique section of the DNA that codes for the rRNA, and a label such as a fluorescent or radioactive molecule is attached to the end of the complementary strand. The labeled probe then will attach selectively to target organisms in the natural environment with the complementary sequence and can be used for rapid *in situ* identification.

Use of molecular techniques based on rRNA will likely become more common and more feasible for aquatic ecologists. Scientists first successfully determined RNA sequences from environmental samples taken from open ocean (Giovannoni *et al.*, 1990) and hotspring organisms (Ward *et al.*,1990). Now, the method has become relatively rapid and commonly used by aquatic microbial ecologists. Interestingly, many of the rRNA sequences found in natural environments are from organisms that have not been grown successfully in the laboratory. The rRNA identity can

be used in a variety of ways. For example, a known rRNA sequence was used to label a bacterium from the natural environment. Following labeling, "optical tweezers" (lasers used to manipulate microscopic particles) were used to isolate and culture individual marked cells of known bacterial species (Huber *et al.*, 1995). In the future, scientists will use related techniques for microbes and other groups whose taxonomic identity is difficult to establish *in situ*. For example, aquatic ecologists rarely identify the larvae of chironomid midges to species because identification is very time-consuming and requires considerable experience. A rapid molecular method for nonsystematists would be very helpful in water quality studies using midge larvae as biological indicator species. The ability to put a sample with a complex community of organisms and rapidly have the numbers and kinds of organisms present in that sample determined is within reach.

ADVANCED: MOLECULAR METHODS FOR ASSESSING MICROBIAL DIVERSITY IN NATURAL ENVIRONMENTS

Microbial ecologists have developed a variety of techniques to assess biodiversity of microbes in their natural environment and their activities. Molecular tools are now making it possible to determine the species that are present in the natural environment and some of their rates of gene expression. Gene sequences for small subunit ribosomal RNA (SSU rRNA, 16S rRNA) can be extracted from the environment from very small samples. All organisms use ribosomes to make proteins, and the genetic code for ribosomal structure is very similar ("highly conserved") across the whole tree of life. The approach is to find a short segment of ribosomal sequence that is in common with all microbial species that are desired and use this to "prime" a polymerase chain reaction (PCR) to exponentially increase the number of copies of the targeted rRNA genes, which then can be sequenced with standard methods.

The PCR reaction is based on breaking (denaturing) DNA strands in two with heat and resynthesizing the complementary strand to each half of the two-stranded molecule. Repeated cycles of this breaking and synthesis exponentially increase the number of copies from a very small sample. While there may be problems with the generality of the method including taxonomic biases, contamination, and the lack of truly universal primers with sequences common to all organisms (Forney *et al.*, 2004), this approach is very useful.

After the pool of 16S rRNA "identity" genes has been amplified from a mixed environmental sample, fingerprinting methods can be used to assess diversity. For these methods, small subunit rRNA genes are amplified with PCR from a complex community and then they are separated based on their sequence composition, creating a "fingerprint" of the taxonomic composition of the sample. These are generally accomplished with denaturing gradient gel electrophoresis (DGGE) or terminal restriction fragment length polymorphism (T-RFLP). Denaturing gradient gel electrophoresis depends on a gel that is made containing a chemical gradient that increasingly denatures the molecules as they migrate through a charged gel. Since different sequences of DNA are differentially susceptible to denaturation, they migrate different distances through the gel. The resulting bands of DNA can be stained to reveal a ladder-like pattern that reflects diversity.

Terminal restriction fragment length polymorphism uses restriction enzymes (enzymes that split DNA at particular sequences) to make DNA fragments that are labeled with a fluorescent molecule and separated by gel electrophoresis. The phylogenetic variation in DNA sequences causes different lengths of fragments. Restriction polymorphism also is analyzed with a gel that creates a pattern of bands, a "fingerprint," that reflects the taxonomic composition of a diverse microbial community.

The combination of targeted PCR of environmental genes and fingerprinting techniques can be used to separate functional genes as well. The genes chosen can be very broad genes in common with all organisms (e.g., genes for DNA synthesis) or some restricted to a taxonomic or functional group (e.g., genes for photosynthetic machinery). Instead of investigating diversity of all microbes in a certain phylogenetic group, the functional gene approach investigates diversity of all microbes with a certain metabolic function (e.g., the ability to use a particular form of nitrogen). Taking this approach further, the abundance of expressed RNA molecules can be analyzed as an indicator of relative gene activity in a sample.

One general problem with molecular analyses of phylogenetic diversity is that bacteria tend to have lateral transfer of some genes. This lateral transmission means that "housekeeping" genes (like ribosomal RNA) that are essential to cell function and are more difficult to transfer among organisms are the best to approximate traditional phylogenetics. This transfer can occur along several routes including viral, plasmid exchange, and incorporation of dissolved DNA by chance. An example of the problem would be choosing a gene to pull one taxonomic group out of a diverse community (e.g., a gene that codes for the ability to break down a specific organic compound) if that ability has transferred to widely differing types of bacteria.

On the horizon now is the ability to sequence all DNA collected from a specific environment, an approach generally termed *metagenomics*. This allows for determination of the total genomic diversity (a somewhat different concept than the idea of total species diversity). In this case DNA is chopped (partially digested) with enzymes into lengths that are readily sequenced. Since each bacterium has one circular genome, if the digestion is incomplete, there should be overlapping sequences from the same organism. These can be assembled to recreate the entire sequence of the genomes of microorganisms in the environment. The ability to handle huge amounts of data and rapidly sequence massive numbers of base pairs has just begun to allow this approach to be applied in natural habitats. These methods can also be applied to eukaryotic species and are showing great promise in determining taxonomic patterns in groups that are traditionally difficult or time-intensive to assign taxonomic identity to.

MAJOR TAXONOMIC GROUPS

Three major groups of organisms have been proposed at the broadest level of classification: the Eukarya (eukaryotes), the Bacteria, and the Archaea (Woese *et al.*, 1990). The Bacteria and Archaea were known formerly as the Prokaryota. Before microscopic and chemical recognition of the unique cellular composition of bacteria, organisms were classified into animals (mobile) or plants (sedentary and green). After light microscopy became established, the classification dividing organisms into animal and plant groups became difficult because many microbes are photosynthetic, motile, and exhibit simple behaviors (e.g., attraction to light or food). The observation of diverse microbial lifestyles blurred traditional distinctions between animals and plants. Electron microscopy allowed definitive differentiation between organisms with complex inner architecture (eukaryotes) and those with more simple cells (then called prokaryotes). Recently, analysis of rRNA and other biological molecules has revealed that the Archaea split from the Eukarya shortly (relative to the 4-billion-year-old Earth) after they diverged from the Bacteria (Fig. 8.1). Such analyses have also revealed that the Bacteria, Archaea, and Eukarya should be assigned to super kingdoms or domains, not to the traditional kingdoms (Woese *et al.*, 1990). If Eukarya is assigned a kingdom-level designation, then it retains the name Eukaryota. The most basic tree of life means that the term prokaryote has little meaning in modern systematics because it includes the widely divergent Archaea and Bacteria. The interpretation of rRNA data has recently been called into question, however, because of possible transfer of genetic material among organisms over evolutionary time (Williams and Embley, 1996; Doolittle, 1999). The issue remains to be resolved, but at least the idea of three domains allows an appreciation of the tremendous diversity of organisms in the Eukarya, Bacteria, and Archaea.

Even in the eukaryotic groups, where morphology is often distinct, phylogenetic relationships do not always follow traditional taxonomic lines (Fig. 8.4).

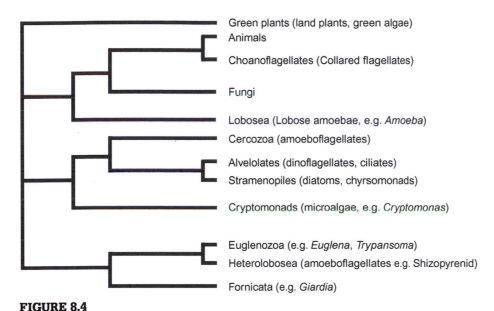

FIGURE 8.4

A phylogenetic diagram of relationships of some major groups of eukaryotic organisms found in freshwaters. *(Basic relationships derived from the Tree of Life Web project, http://www.tolweb.org/tree/).*

For example, the group protozologists use to classify amoebae actually falls into three widely divergent groups, and those groups are more different from each other than are humans and fungi. Molecular data, combined with the fact that there is lateral transfer of genes, and lack of sequence information for many organisms lead to many phylogenetic questions that currently have no definitive answers. Phylogeny is currently a field in a state of rapid flux, so many relationships are provisional. Students should also realize that this is a very exciting time to be involved in phylogenetic research, as the most basic evolutionary questions of origins of species are being solved in this field on a daily basis.

One aspect of the most basic issues in origin of diversity is the concept that eukaryotic cells formed organelles by ingesting and appropriating bacterial cells. The cells were enveloped and became endosymbiotic, living inside of the Eukaryotic hosts (Sapp, 1991). This idea is called the *serial endosymbiosis theory* and was developed through the work of Dr. Lynn Margulis. Most biologists accept this as a theory for the origin of chloroplasts and mitochondria. Mitochondria and chloroplasts found in eukaryotes were initially cells of purple photosynthetic bacteria and cyanobacteria (blue-green algae), respectively. This evolutionary innovation occurred several billion years ago and vastly increased the complexity of organisms. Molecular evidence strongly supports independent origins of mitochondria and chloroplasts. The numerous less tightly integrated intracellular associations that exist in aquatic ecosystems provide partial support for the serial endosymbiosis theory. For example, inclusion of photosynthetic algae in protozoa and small animals is functionally similar to chloroplasts in plants. Such associations are found in freshwaters and will be highlighted in this book, including *Chlorella* in *Hydra*, *Paramecium bursari*, and *Spongilla*.

There are six or more kingdoms of organisms if one accepts molecular taxonomy that indicates the three major domains of organisms. The kingdoms include the Plantae, Animalia, Fungi, and Bacteria. The group that has been classified as Protista likely will be divided into several kingdoms, given the broad spread in molecular trees (Figs. 8.1, 8.4) and to maintain a classification scheme consistent with evolutionary origins (Hickman and Roberts, 1995), but the classification of this group is currently in flux (Patterson, 1999). The Archaea probably contains two kingdoms as well (Woese *et al.*, 1990).

One point that becomes clear when analyzing the molecular evidence is that the divergence within the Bacteria and Archaea exceeds that of the Eukarya. Thus, the diversity of morphology and behavior that is so evident in the plants and animals has arisen recently. Molecular diversification is the forte of the Bacteria and Archaea.

CLASSIFICATION OF ORGANISMS BY FUNCTION, HABITATS, AND INTERACTIONS

For the aquatic ecologist, taxonomy of organisms is based not only on phylogenetic relationships but also on their functional roles in communities and ecosystems. Such classifications include how the organisms acquire carbon, what habitat they occupy, and how they interact with other organisms. These concepts are discussed here because they are used in the following chapters to characterize organisms.

Organisms can be *autotrophic* (self-feeding) and rely on carbon dioxide (CO_2) as the primary source of carbon to build cells. The other option is to be *heterotrophic* (other-feeding) and acquire carbon for cells from organic carbon (Table 8.2). Some organisms are able to use both autotrophic and heterotrophic processes to obtain carbon.

Most autotrophs use light as an energy source to reduce CO_2 to organic carbon and are classified as *photoautotrophic* (i.e., photosynthetic organisms). Some microbes are able to use chemical energy, as opposed to light, as a source of energy to reduce CO_2 to organic carbon; these organisms are *chemoautotrophic*. The chemoautotrophs are less important to most carbon budgets

Table 8.2 Classification of Organisms by Energy Source and Nutrient Requirements

Carbon Source	Mode	Energy Source/ Electron Donor	Electron Acceptor	Organisms (Example)
CO_2	Photoautotroph	Light/H_2O	O_2	Cyanobacteria/eukaryotic algae/macrophytes
CO_2	Photoautotroph	Light/H_2S	Organic C	Green and purple sulfur bacteria (anaerobic)
CO_2	Photoautotroph	Light/H_2, organic C	Organic C	Purple nonsulfur bacteria (anaerobic)
CO_2	Chemoautotroph	Reduced inorganic compounds (e.g., NH_4^+, H_2S, Fe^{2+}, H_2)	O_2	Bacteria (e.g., *Nitrosomonas*, iron bacteria, hydrogen bacteria)
Organic C	Heterotroph	Organic C	Organic C	Fermentative bacteria (anaerobic)
Organic C	Heterotroph	Organic C	O_2	Aerobic bacteria, protozoa, and animals
Organic C	Heterotroph	Organic C	Oxidized compounds such as NO_3^-, SO_4^{2-}, Fe^{3+}	Anaerobic bacteria that respire organic C

(After Yanagita, 1990). See Chapters 9, 10, 12, 13, and 14 for further discussion of types of organisms. Not all types are shown

than photoautotrophs, but they dominate some unusual environments and play a key part in several nutrient cycles (see Chapter 14).

Organisms that acquire carbon from other living organisms (predation, herbivory, parasitism, etc.), dissolved or particulate organic compounds, or dead organisms (decomposers and carrion eaters) are heterotrophic. Heterotrophs that decompose organic carbon are sometimes called *saprophytes* or *detritivores*.

A variety of additional classifications are used to describe the functional roles of organisms in aquatic food webs (*functional feeding groups*; Cummins, 1973). Organisms that sieve particles from the water column are called *filterers*; those that build nets or have morphological features that filter particles out of flowing waters are *passive filterers*, whereas those that actively pump water or create currents are *active filterers*. Organisms that acquire their nutrition from small organic particles in the benthos are called *collectors*. *Shredders* break up larger organic materials like decaying leaves for their nutrition, and *scrapers* remove biofilms from hard benthic substrata.

Functional feeding groups are somewhat similar to *guilds*, which are often used in terrestrial studies of consumers and communities, except that the guild concept is based on organisms using the same resource in the same fashion (Root, 1967). Members of the same functional group may use different resources. For example, some shredders may feed on decomposing leaves, whereas other shredders may feed on wood or living aquatic plants, but they all shred large organic materials. Functional and guild analyses can be more relevant than taxonomic considerations for relating organisms to ecosystem processes like nutrient cycling and energy flow.

Consumers are often classified further by their position in the food web. For example, *grazers* or *herbivores* (primary consumers) eat algae, plants, or sometimes bacteria (primary producers). *Carnivores* or secondary consumers eat other animals, and *top carnivores* eat animals but are generally eaten by no larger animal. Thus, classification schemes based on modes of obtaining nutrition are one of many ways to classify organisms.

Additionally, organisms may be classified by the habitat they occupy and some of the special terminology for this purpose is presented in Table 8.3. Such classification can be useful because it allows an investigator to make predictions about abiotic and biotic conditions important to organisms. For example, an epilithic alga (living on rock) in a rapidly moving stream may

Table 8.3 Some Terms Used to Classify Aquatic Organisms by Habitat

Habitat	Description
Benthic	On the bottom
Emergent	Emerging from the water
Endosymbiotic	Living within another organism
Epilithic	On rocks
Epigean	Above ground
Epipelic	On mud
Epiphytic	On plants
Epipsammic	On sand
Hygropetric	In water on vertical rock surfaces
Hyporheic	In groundwater influenced by surface water
Lentic	In still water
Littoral	On lake shores, in shallow benthic zone of lakes
Lotic	In flowing water
Neustonic	On the surface of water
Pelagic	In open water
Periphytic (Aufwuchs, biofilm, microphytobenthos)	Benthic, in a complex mixture including algae
Profundal	Deep in a lake
Symbiotic	Living very near or within another organism
Stygophilic	Actively use groundwater habitats for part of life cycle
Stygobitic	Specialized for life in groundwater
Torrenticole	Adapted to live in swiftly moving water

experience a relatively high water velocity. An epiphytic alga on a macrophyte may have competitive or facilitative interactions with the macrophyte. This basic terminology is necessary to understand the literature on aquatic ecology.

Organisms can also be classified by how they interact with other organisms (*interspecific interactions*). Many different types of interspecific interactions are possible, and we adhere to the interaction scheme shown in Table 8.4. There are *direct interactions* and *indirect interactions*. Direct interactions occur between individuals of two species and involve no others; indirect interactions are mediated by other species. The terms for interaction types presented here are not all standard, but they allow for a very general classification scheme. *Exploitation* is a general term for an interaction that harms one species and helps another. This term is not widely accepted yet, but includes interactions that may not be formally considered *predation* or *parasitism*. For example, an epiphyte that harms a macrophyte but receives benefit from living on its leaves is exploiting the plant. *Mutualism* is used to denote any positive reciprocal interaction. Others have used various terms to denote mutualism, including symbiosis, synergism, and protocooperation. *Symbiosis* refers to organisms that live close together (but not to how they are interacting), and synergism and protocooperation have not received widespread use outside of studies of animal behavior.

Of all the interaction types found in macroscopic ecological communities, *commensalism* (positive on one, none on the other) and *amensalism* (negative on one, no effect on the other) are likely the most common, followed by exploitation and then competition and mutualism (assuming that positive interactions are as likely as negative interactions; Dodds, 1997b). In general, commensalism and amensalism have received almost no attention in the ecological literature; predation has received the most, followed by competition and mutualism. It is up to future ecologists to study amensalism and commensalism more intensively. These interactions will be discussed in detail in Chapters 19 through 22.

Table 8.4 Classification of Interactions between Two Species (A and B)

Effect of A on B	Effect of B on A	Name of Interaction
Positive	Negative	Exploitation (includes predation and parasitism)
Negative	Negative	Competition
Positive	Positive	Mutualism
None	Positive	Commensalism
None	Negative	Amensalism
None	None	Neutralism

ORGANISMS FOUND IN FRESHWATERS

Freshwater habitats contain representatives of most of the groups of organisms on Earth. Several books are available to introduce students to freshwater organism diversity in general (e.g., Reid and Fichter, 1967; Needham and Needham, 1975), and there are also numerous readily available guides covering various taxonomic groups (Table 8.5). Detailed taxonomic analyses of many groups (e.g., genus and/or species level identifications of algae or invertebrates) often require keys available in specialized books or in the primary literature, and there are no universally accepted taxonomic keys for problematic groups such as some water mites and some groups of protozoa.

The Archaea and Bacteria are difficult to distinguish unless they can be brought into culture and metabolic characteristics can be used as taxonomic characteristics (Holt *et al.*, 1994). Algae are the primary autotrophs in many aquatic ecosystems and are well represented in freshwaters (South and Whittick, 1987; Graham and Wilcox, 2000; Wehr and Sheath 2003). The classification and ecology of microbes are discussed in Chapter 9. The incredible diversity of the algae and protozoa make observation of aquatic samples under the microscope fascinating; we can imagine the excitement that the first microscopes generated when they revealed an entirely new world.

Taxonomy of the Eukarya is relatively well defined, at least at the family levels for most groups. Protozoa are common in all freshwater habitats and can often be identified to the family or genus level if a good microscope is available. All major phyla of invertebrates, with the exception of the Echinodermata, have some freshwater species, and some groups such as the insects and arachnids are much more diverse in freshwaters compared to

Table 8.5 Some Initial Resources for Identifying Various Groups of Aquatic Organisms

Group	References
General	Reid and Fichter (1967), Needham and Needham (1975)
Algae	Prescott (1978), Dillard (1999), Wehr and Sheath (2003)
Protozoa	Jahn *et al.* (1979), Lee *et al.* (2000) Thorp and Covich (2001)
Nonvascular plants	Conrad and Redfearn (1979)
Aquatic plants	Riemer (1984), Cook (1996), Borman *et al.* (1997)
Aquatic invertebrates	Lehmkuhl (1979), Thorp and Covich (2001), Simth (2001)
Aquatic insects	Merritt *et al.* (2008), McCafferty (1988)
Reptiles and amphibians	Conant and Collins (1999), Stebbins (2003)
Fish	Eddy and Underhill (1969), Page and Burr (1991)

marine environments. Students are often surprised to learn that some groups of organisms that they traditionally associate with marine habitats, such as the sponges and jellyfish, can be found in abundance in some freshwaters.

Some freshwater invertebrates, such as the aquatic insects and arachnids, are secondarily aquatic in that their evolutionary paths involved marine ancestors that evolved into terrestrial forms, which subsequently colonized and adapted to freshwater habitats. Others, such as the freshwater cnidarians and decapods, evolved directly from marine ancestors. Secondarily aquatic forms often have morphological or physiological features that reflect their terrestrial ancestry. For example, many aquatic insects use air bubbles or breathing tubes to breathe atmospheric air, even though they spend the bulk of their time underwater.

Many beginning students find the invertebrates the most fascinating of freshwater organisms; take a child to a stream and turn over rocks and you will see how early such fascination can begin. Invertebrates are abundant, diverse, and ecologically important in just about every habitat on the planet, and freshwaters are no exception. The diversity of body forms, life cycles, and behaviors among the invertebrates is astonishing. Invertebrates constitute the bulk of animal abundance and diversity in freshwater habitats. Because of their abundance, diversity, and relatively fast individual and population growth rates, invertebrates are ecologically very important in freshwater habitats, as well as in other systems (Wallace and Webster, 1996; Thorp and Covich, 2001). Identification of many invertebrate groups to the species level, particularly for immature forms (e.g., aquatic fly larvae and mite larvae), is difficult, but taxonomic keys are available for adult and immature forms of most invertebrate groups. Major groups of invertebrates are discussed in more detail in Chapter 10, and aspects of their ecological significance in Chapters 19 through 21.

Vertebrates are also abundant and diverse in freshwater habitats, although the bulk of this diversity is accounted for by one group, the bony fishes (superclass Osteichthyes). Aside from the fishes, most vertebrates associated with freshwater habitats or are only partially aquatic (e.g., birds and mammals) or have biphasic amphibious life cycles (e.g., frogs and salamanders). Identification of vertebrates is generally easier than invertebrates because fewer, larger, and better-studied organisms are represented, although some groups of smaller-bodied fishes (e.g., some minnows and darters) and immature forms of amphibians can be challenging and may require an expert for accurate species-level identification. Many freshwater vertebrates have been assigned common names and are already familiar to students. The major groups of freshwater vertebrates are introduced in Chapter 10, and fish ecology and fisheries management is discussed in Chapter 23.

As with vertebrates, many plants in aquatic systems have been well characterized. Emergent wetland species generally are included in traditional plant taxonomic references (Smith, 1977). For the more obscure mosses and liverworts, identification is more difficult, and even some larger groups such as the sedges can be problematic to identify. Truly aquatic plants are only moderately diverse; a good introduction to their ecology is provided by Riemer (1984), and major groups of freshwater plants are discussed in Chapter 9.

SUMMARY

1. Several species definitions are available. The most utilitarian approach is to define a species by criteria established by the taxonomists of a particular group.
2. Traditional taxonomic schemes have distinguished among organisms using behavior, metabolic characteristics, and morphology. Recently, molecular techniques have been used.
3. Traditional taxonomic classifications at the broadest level (e.g., kingdom and phylum) are probably not completely natural, and more research is necessary to untangle these evolutionary relationships.
4. Bacteria and Archaea are two groups with the greatest amount of metabolic diversity. Behavioral and morphological diversity are greatest in the Eukarya.
5. Organisms can be classified by their mode of obtaining nutrition and by the habitat they inhabit in addition to their evolutionary relatedness.
6. Organisms that use CO_2 as their primary carbon source are autotrophic; those that use organic carbon are heterotrophic. Autotrophic organisms include those that obtain energy from light (photoautotrophic) and chemicals (chemoautotrophic). Heterotrophic organisms include predators, detritivores, and organisms that live on dissolved organic compounds.
7. Organisms can be classified by their direct interactions (competition, mutualism, exploitation, commensalism, amensalism, and neutralism) with other organisms.
8. Most groups of organisms found on Earth include species that live in freshwater habitats, some of which evolved directly from marine ancestors and others that are secondarily aquatic and evolved from terrestrial species.

QUESTIONS FOR THOUGHT

1. Why might legislation designed to prevent extinction of species require a precise definition of a species?
2. Why did the inclusion of mitochondria and chloroplasts in cells of Eukarya represent a sudden large increase in complexity of cells and how does this observation damage the creationists' "blind watchmaker" arguments?
3. Why do freshwater invertebrates have fewer species as a whole than marine invertebrates?
4. Why does lateral transfer of genetic material among widely disparate organisms (e.g., plants and bacteria) cause difficulties for molecular taxonomists?

5. Can you think of more specialized habitats than those listed in Table 8.3?

6. Is taxonomy fixed when a species is described or do perceived taxonomic relationships among organisms change over the years as more information becomes available?

7. Evolution can occur at much more rapid rates than previously thought (over a few generations); how could this influence taxonomy?

8. How would morphological plasticity interfere with taxonomic identification?

Microbes and Plants

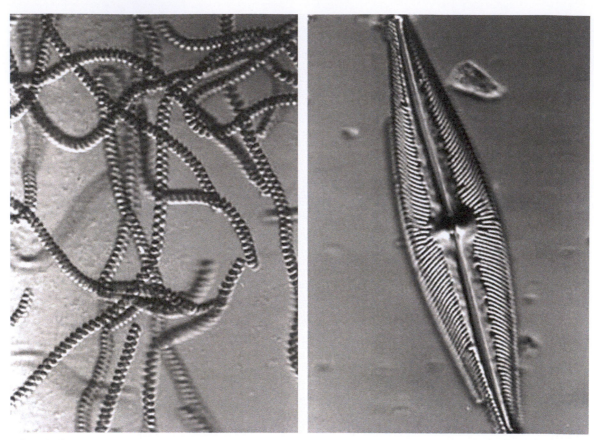

FIGURE 9.1

Micrographs of a spiral cyanobacterium (*Spirulina*) and the frustule (silicon shell) of a diatom (*Navicula*). *Spirulina* spirals are 5 μm wide *Navicula* is 30 μm wide.

Primary producers capture much of the energy that flows through freshwater food webs, and microbes are responsible for the bulk of the biogeochemical transformational fluxes (including decomposition and nutrient recycling) in aquatic systems. Some of this freshwater biogeochemistry (e.g., production of methane by wetlands) has importance on a global scale because methane and carbon dioxide are important greenhouse gases. Knowledge about these organisms is essential for those involved with water quality issues as well as general ecological studies. In this chapter, we consider the microbes and plants found in freshwaters. This placement of microbes and plants into a single chapter is certainly an artificial classification; the organisms considered here span taxonomic groups from viruses to complex plants and include *Bacteria, Archaea,* and *Eukarya*. With the advent of molecular techniques, taxonomy, particularly for the microbial world, is in a state of rapid flux. For practical purposes we mostly follow traditional taxonomic protocol, because many determinations are based on obvious morphological characteristics, and this method is readily available to researchers who are not specialists in taxonomic identification via molecular methods. Still, where the traditional taxonomy does not match molecular methods at very deep phylogenetic levels, we point out the differences.

VIRUSES

All organisms have viruses. We consider that viruses are not really organisms because they cannot survive without a host and are not capable of basic metabolic function. Nonetheless, they are important in population dynamics of aquatic organisms, aquaculture, and public health (Table 9.1). Taxonomy of viruses is possible with modern molecular methods and this is a rapidly developing field, both with respect to epidemiology and viruses that are common in the environment but do not infect humans, their crops, or their livestock. Viruses also transfer genetic material among microorganisms, so they can be important in issues related to release of genetically engineered microorganisms into the environment. Over longer time periods, the lateral transfer of genetic material, mediated in part by viruses, causes difficulty in taxonomy because genes or groups of genes "jump" across phylogenetic lines.

Viruses are very small particles (25–350 nm) that remain in suspension because they are too small to sink relative to turbulence and Brownian motion at very small scales. Particles that look like viruses that have been isolated in the laboratory are commonly seen in lakes when scanning electron microscopy is used (Fig. 9.2). It is difficult to determine if particles that look like viruses are actually infectious and to determine their host. Generally, an assay for the numbers of infectious units per volume water based on exposure of organisms is the definitive test for active viruses. A series of dilutions containing viruses is made and organisms are exposed to the dilution series. The number of infections caused by the various dilutions is used to calculate the number of active viruses. This protocol is used in testing for viruses that infect humans (with a surrogate infective host) as well as those that infect other organisms. Molecular probes and labeled antibodies are available for the more common human infectious agents, and can be synthesized for any virus for which the RNA or DNA sequence is known.

Viruses are classified as DNA or RNA, by occurrence of single- or double-stranded nucleic acids, and by the molecular weight of nucleic acids. Classification of the capsid (protein coat around the nucleic acid), including the number of subunits, the shape or symmetry, and where in the host cell the capsid is assembled, are also important. Some viruses also have a lipid

Table 9.1 Some Organisms Causing Human Diseases That Can Be Transmitted by Water or Waste Water

Group	Organisms	Disease/Symptoms
Bacteria	*Salmonella* spp.	Typhoid fever, paratyphoid fever, gastroenteritis
	Shigella spp.	Gastroenteritis- dysentery
	Vibrio cholerae	Cholera
	Escherichia coli	Gastroenteritis
	Leptospira icterohaemorrhagiae	Weil's disease
	Campylobacter spp.	Gastroenteritis
	Yersinia enterocolitica	Gastroenteritis
	Mycobacterium spp.	Tuberculosis/respiratory illness
	Legionella pneumophila	Legionnaire's disease/acute respiratory illness
Virus	Hepatitis A	Liver disease
	Norwalk agent, Rotaviruses, Astroviruses	Gastroenteritis
	Poliovirus	Polio
	Coxsackievirus	Herpangia/meningitis, respiratory illness, paralysis, fever
	Enteroviruses (68–71)	Pleurodynia/meningitis, pericarditis, myocarditis
Protozoa	*Giardia lamblia*	Diarrhea, abdominal pains, nausea, fatigue, weight loss
	Entamoeba histolytica	Acute dysentery
	Acanthamoeba castellani, Naegleria spp.	Meningioencephalitis
	Balantidium coli	Dysentery
	Cryptosporidium spp.	Dysentery
Helminths	Nematodes (*Ascaris lumbricoides, Trichuris trichiura*)	Intestinal obstruction in children
	Hookworms (*Necator americanus, Ancylostoma duodenale*)	Hookworm disease/gastrointestinal tract
	Tapeworms (*Taenia* spp.)	Abdominal discomfort, hunger pains
	Schistosoma mansoni	Schistosomiasis (liver, bladder, and large intestine)

(After Bitton, 1994)

or lipoprotein coat. Characterization schemes differ among viruses of plants, animals, and microorganisms. Animal viruses are recognized by the diseases they cause, plant viruses by the disease and plant species that serve as host, and microbial viruses by the organism they infect. Additionally, prions (infectious proteins) and viroids (naked RNA) may be significant parasites, although little is known about their importance in nature. The ecology of

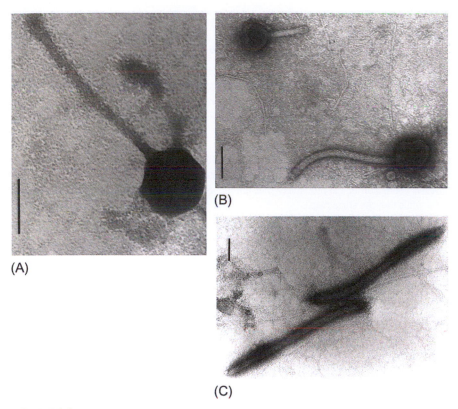

(A)

(B)

(C)

FIGURE 9.2
Electron micrographs of aquatic virus-like particles from two high-mountain lakes. Scale bars = 100 nm.
(Reproduced with permission from Pina et al., 1998).

virus assemblages in their natural environment is complex as they evolve very rapidly to stay ahead of defenses evolved by their hosts (Anderson and Banfield, 2008).

Viruses can be specific for one species or strain of organisms or more widely infective. Those of the greatest interest to humans cause disease, and many of these can be transmitted in water (Table 9.1). Viruses that infect unicellular organisms are often fatal if an infection proceeds because the reproductive virus lyses (bursts) the cell. Understanding the dynamics of microbial communities requires knowledge of how viruses are transmitted.

A successful virus in an aquatic habitat must make contact with the correct type of host cell. The virus must remain active long enough to randomly encounter the appropriate host. The spread of viral infections is greater when the density of host cells in the environment is high and the length of time that a virus can remain viable outside the host is long. Many things may inactivate viruses when outside of their hosts, including UV light, absorption onto cells

SIDEBAR 9.1
Survival of Human Pathogenic Viruses in Groundwater

Pathogenic viruses can enter groundwaters through many different sources. Some of the most common sources are land disposal of sewage, overflow from septic systems, and livestock waste. Leachate from solid waste landfills also can contain viruses. For example, human infections were traced to viral contamination of groundwater in Georgetown, Texas (coxsackievirus and hepatitis A), and Meade County, Kentucky (hepatitis A; Lipson and Stotzky, 1987), and 1500 people were infected with Norwalk virus from a contaminated spring in Rome, Georgia (Bitton, 1994). Contamination of drinking water wells with hepatitis A, polio, or enteroviruses has been documented throughout the world. Knowing how long these viruses can remain infective is important to allow estimation of the probability that groundwater flows will move them to drinking water wells while they are still active.

Infective viruses have been demonstrated to travel over 50m (depth) from septic tanks into drinking water wells. Controlled studies have demonstrated movement up to 1.6km horizontally through soils (Gerba, 1987). Clearly, viruses could travel even greater distances in aquifers with rapid water velocity, such as karst systems or alluvium with coarse cobbles (Sinton *et al.*, 1997). For example, poliovirus was demonstrated to move at least 20m in a cobble aquifer with less than 1% virus mortality (Deborde *et al.*, 1999).

Factors that influence the movement of viruses into and through groundwater include the rate of water flow through the sediment, the retentive properties of the sediments, and the survival time of the virus. Factors that lower viral survival times in sediments include high temperatures, microbial activity, drying, lack of aggregation with other particles, and low organic matter. Inactivation of viruses can be very rapid, but poliovirus can remain active for up to 416 days in sandy soils (Sobsey and Shields, 1987).

Understanding the hydrology of soils and sediments is necessary to assess the problems that may be related to sewage contamination of groundwaters. Aquatic microbial ecologists are only beginning to elucidate the mechanisms of deactivation of viruses related to microbial infections. Because little is known about community dynamics of groundwater microbes, this is a potentially valuable and exciting field for future study.

or remains of cells that are not proper hosts, and predation by microflagellates that can ingest very small particles. Inorganic particles can enhance viral survival by limiting the previous factors but can also lower infection rates because tight association with the inorganic particles lowers the probability of contact with a host cell. Some of these factors influence how long viruses can survive in groundwater, which can be an important public health issue (Sidebar 9.1).

ARCHAEA

The *Archaea* are single-celled organisms with no nucleus or organelles. They reproduce asexually by fission or budding and no species are known to produce spores. Analysis of ribosomal RNA sequences has led to a classification of these organisms as separate from the Bacteria (Fig. 8.1). Archaea are probably more closely related to the Eukarya than the Bacteria, because they have similar genes for transcription and translation to the Eukarya.

The Archaea originally were thought to predominate mainly in extreme environments including anaerobic waters, hot springs, and hypersaline environments such as salt lakes. Molecular methods have revealed that Archaea are common in all environments, although not as numerically dominant as in some extreme habitats. These organisms are common in the ocean and probably in many freshwater habitats as well. The study of Archaea and their importance in freshwater habitats, as well as the general phylogeny of the group, is a rapidly developing and very exciting area of aquatic microbial ecology.

The Archaea are broadly diverse and include members that can create methane from carbon dioxide and hydrogen (methanogens), some that can use organic carbon (particularly in very cold habitats with low carbon availability), and some groups that are capable of photosynthesis, but do not generate oxygen gas. Others can use inorganic compounds such as ammonium or sulfide as a source of energy and are chemoautotrophic. Some groups such as the methanogens have global biogeochemical importance, especially those populations found in wetlands. The identification of species or strains of archaebacteria generally is based on metabolic characteristics (Table 9.2) and molecular analyses.

BACTERIA

The *Bacteria* are single-celled organisms without organelles or a nucleus. They are ubiquitous and may have greater active biomass (protoplasm) than any other group of organisms on Earth (Whitman *et al.*, 1998), although that distinction may also belong to Archaea. Bacteria are in every habitat and they regulate flows of energy and nutrients through aquatic ecosystems.

The Bacteria are the most metabolically diverse group of organisms on Earth. These unicellular organisms include heterotrophs that obtain energy from oxidizing organic carbon, predators, and parasites. Bacteria also can be autotrophic (chemoautotrophy and photoautotrophy) and *mixotrophic* (a mixture of heterotrophic and autotrophic activities). Some of the most crucial biogeochemical fluxes mediated by these organisms will be discussed in Chapters 12 through 14.

Some of the most important human pathogens regularly transmitted by water are bacterial (Table 9.1), and bacteria may be pathogens to many aquatic plants and animals. The human pathogens are not restricted solely to developing countries. Outbreaks of waterborne illnesses occur in parts of the world with a high standard of living (Young, 1996). Finally, bacteria are the most common organisms used in bioremediation, which is the process of using organisms to clean up (break down) pollution (see Chapter 16).

Table 9.2
Major Groups of Bacteria from *Bergey's Manual* (Holt et al., 1994). The Number of Described Genera is only Included as a Rough Guide to Relative Diversity in the Group.

Major Group	Group	Genera Represented in Aquatic Systems (# of Described Genera)	Importance
Gram-negative with cell wall	Spirochetes	*Spirochaeta* (8+)	Free living in aquatic waters, some pathogens
	Aerobic/microaerophilic, motile, hilical/vibriod	*Campylobacter, Bdellovibrio* (16)	In freshwaters, some denitrifiers, includes the predatory *Bdellovibrio*
	Nonmotile curved	*Ancylobacter* (8)	In freshwater
	Aerobic/microaerophilic rods and cocci	*Azotobacter, Puedomonas, Francisella, Legionella* (84)	Aerobic nitrogen fixation (e.g., *Azotobacter*), some disease organisms, very diverse group
	Facultative anaerobic rods	*Escherichia, Vibrio* (45)	Contains many waterborne disease organisms
	Anaerobic straight, curved, and helical rods	*Thermotoga* (47)	Common from anoxic muds and animal intestinal tracts, also a number of thermophiles and halophiles
	Dissimilatory sulfur reducing	*Desulfomonas* (18)	Reduces oxidized sulfur compounds to sulfide
	Anaerobic cocci	*Megashaera* (4)	Mainly animal parasites
	Rickettsias and chlamydias	2 subgroups	Parasites
	Anoxygenic photosynthetic	*Rhodospirillum* (28+)	Able to use sulfide as electron donor for photosynthesis
	Oxygenic photosynthetic	*Cyanobacteria* (37)	Important photosynthetic and nitrogen fixers
	Aerobic chemilithotrophic	*Thermothrix, Nitrobacter* (28)	Important biogeochemically, including nitrifiers, iron oxidizers, and sulfur oxidizers
	Budding or appendaged	*Caulobacter* (25)	*Caulobacter* indicative of oligotrophic conditions
	Sheathed	*Clonothrix* (7)	Some important in iron and manganese cycles

	Gliding, nonphotosynthetic, nonfruiting		
	Myxobacteria	Beggiatoa (28)	Mostly aquatic, some important in sulfur cycling
		Polyangium (12)	Decomposers, predominantly in soils but some in freshwaters
Gram-positive with cell walls	Cocci	Streptococcus, Trichococcus (24)	Some pathogens, also found in aquatic habitats
	Endospore forming	Bacillus (10)	Widespread species, some sulfur oxidizers
	Nonsporing regular rods	Lactobacillus (8)	Widespread, some fish pathogens
	Nonsporing irregular rods	Microbacterium (36)	An artificial group, some pathogens, mainly in soil, some aquatic thermophiles
	Mycobacteria	Mycobacterium (1)	Widely distributed in soil and water, some pathogens
	Actinomycetes	Dactylosporangium, Streptomyces (49)	Fungi-like morphology, important in decomposition in soils
No cell wall, Bacteria	Mycoplasmas	Mycoplasma (6)	Smallest known self-reproducing organisms
Archaea	Methanogens	Methanobacterium (18)	Generates methane
	Sulfate reducers	Archaeoglobus (1)	Deep sea vents
	Extreme halophiles	Halobacterium (6)	Require at least 1.5 M NaCl for growth
	No cell wall	Thermoplasma (1)	Found in mine waste
	Extreme thermophiles	Sulfolobus (14)	Optimum growth from 70–105°C

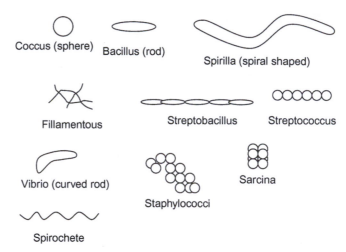

FIGURE 9.3
Some bacterial morphologies.

The methods used to determine bacterial groups are often based on morphology or simple metabolic characteristics and likely do not accurately represent evolutionary relationships among all bacteria. A common identification scheme is presented in *Bergey's Manual of Determinative Bacteriology* (Holt *et al.*, 1994), but this scheme is gradually being replaced by molecular methods. Classically, differentiation among bacterial species or strains is based on reaction to staining compounds, morphology, motility, production of extracellular materials, color, and metabolic capabilities. Metabolic capabilities have the greatest utility to discriminate among taxa because the other attributes vary little among species. Most bacteria are only 1 or 2 μm in diameter, but larger and smaller examples exist and a modest variety of morphologies occur (Fig. 9.3). The major groups of bacteria, based on standard bacteriological techniques, are presented in Table 9.2.

The number of species of bacteria is unknown, partially because of a fundamentally different species concept. Most vertebrates and many of the plant species have been described, but less than 1% of bacterial species have been described (Young, 1997). Microbial diversity has received limited study because of difficulties associated with identification in natural samples. Furthermore, obtaining representative samples in some habitats, such as groundwaters, is difficult (Alfreider *et al.*, 1997). The gut of each aquatic invertebrate species could harbor several unique microbial species. Each milliliter of water or gram of sediment has species that have never been cultured, so it is possible that there are more species of Bacteria than any other type of organism.

Early analysis of intensively studied hot spring communities suggests that only a fraction of the viable bacteria in any habitat can be cultivated successfully with current techniques (Ward *et al.*, 1990). In a study of Octopus

Spring in Yellowstone National Park, rRNA sequences were obtained from natural samples. Few of these sequences matched known sequences, even though numerous microbes have been cultured from the same study site and hot springs likely contain simple microbial communities. Subsequent work with substantially greater data sets and more efficient methods has verified the original finding by Ward *et al.* (1990) that only a portion of the bacteria in an environment have ever been cultured.

Zwart *et al.* (2002) searched published rRNA data and found 34 species clusters that are exclusively or mostly found in freshwater; 23 of these clusters included no cultured organisms. A survey of lakes indicated that larger lakes had more species (distinct rRNA bands) than smaller lakes, that lakes near each other tended to have fairly unique groups of bacteria, and that there were 10 to 100 dominant species in each lake (Reche *et al.*, 2005).

Cyanobacteria (Bluegreen Algae or Cyanophytes)

The *Cyanobacteria* are a particularly important group of photosynthetic bacteria. Aquatic ecologists are very concerned with the cyanobacteria because of their tremendous impact on water quality; they form *blooms* or extremely high cell densities in eutrophic waters. Many researchers have adopted the more modern terminology "cyanobacteria" rather than "bluegreen algae" to clearly delineate the fact that this group is bacterial. The taxonomy of these organisms has been studied more completely than that of other bacteria because most are large and morphologically distinct under the light microscope.

Most cyanobacteria are O_2-producing photosynthetic bacteria. Fossils similar to extant cyanobacteria are the oldest known records of life (Schopf, 1993). Cyanobacteria have been successful for billions of years and are currently able to exploit some of the most extreme habitats on Earth, including very cold, very hot, and extremely saline environments.

Cyanobacteria are found in most habitats and can range from $1\,\mu m$ in diameter to several $100\,\mu m$ (Fig. 9.4). Shape ranges from simple spheres to complex branching structures. Specialized cells include *akinetes* (resting cells) and *heterocysts* (the site of most nitrogen fixation). Nitrogen fixation is the acquisition of gaseous N_2 into cellular nitrogen and will be discussed more thoroughly in the context of nitrogen cycling (see Chapter 14).

Proteinaceous vacuoles called *gas vesicles* lend buoyancy to the cyanobacteria and lead to formation of surface scums under calm conditions. These surface scums can be up to $1\,m$ thick, can have very objectionable odors, and can render a lake virtually useless for swimming and other forms of recreation. The gas vesicles give the cyanobacteria a competitive advantage in eutrophic conditions by allowing them to compete well for light at the surface and shade out the phytoplankton below. Halting vesicle synthesis allows cells to sink to deeper, nutrient-rich waters.

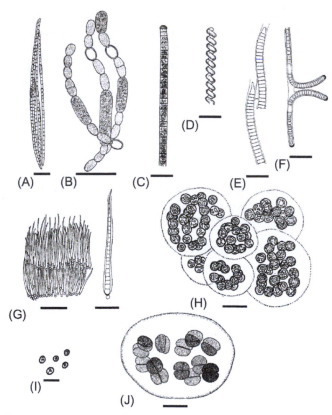

FIGURE 9.4

Selected genera of cyanobacteria, with length of scale bar: (A) *Aphanizomenon*, 30 μm; (B) *Anabaena*, 20 μm; (C) *Oscillatoria*, 20 μm; (D) *Spirulina* 10 μm; (E) *Phormidium* (distinguish from C by mucous sheath), 20 μm; (F) *Scytonema* (with false branching), 30 μm; (G) *Rivularia*, habit view 50 μm, single trichome 25 μm; (H) *Microcystis*, 20 μm; (I) picocyanobacteria (indeterminate genus), 3 μm; (J) *Chroococcus*, 10 μm. *(A, B, C, G, and H reproduced with permission from Prescott, 1982).*

Cyanobacteria are excellent competitors for light because they have *phycobilins*, pigments that absorb light in the green region (where chlorophyll does not absorb; see color plate Fig. 1). The ability to use green light allows some species of cyanobacteria to inhabit very deep waters and still remain photosynthetically active. The ecological consequences of phycobilins were first introduced in Chapter 3 (Fig. 3.10).

Cyanobacteria are often difficult for herbivores to consume, partly because of their gelatinous coatings and also because many of them produce toxins (Sidebar 9.2). These toxins have likely evolved to limit grazing, but are very broad spectrum and may also have adverse effects on fish, humans, crop plants, and many other organisms. Cyanobacteria also synthesize compounds

SIDEBAR 9.2
Cyanobacterial Toxins

Cyanobacteria are among several groups of toxic primary producers that can be found in freshwater. Cyanobacteria produce several general types of toxins, neurotoxins, hepatotoxins, and cylindrospermopsin. These toxins can be responsible for a variety of problems, including illness of humans who drink water containing the toxins, death of dialysis patients dialyzed with water containing the toxins, dermatitis from skin contact, potential long-term liver damage from contaminated water supplies, and animal deaths from drinking water containing cyanobacterial blooms (Falconer, 1999; Codd *et al.*, 1999a, 1999b). Twenty-five genera containing 40 species of cyanobacteria have been confirmed to have members that produce toxins (Codd, 1995; Carmichael, 1997). These numbers will increase as additional taxa are surveyed and new toxins are identified.

The neurotoxins produced by cyanobacteria act very rapidly (also known as very rapid death factors) and are responsible for the deaths of domestic animals that drink from water containing the toxins (Carmichael, 1994). The neurotoxins are lethal at very dilute concentrations; the notorious toxin dioxin is 10 to 60 times less toxic than the cyanobacterial aphantoxin (Kotak *et al.*, 1993). The neurotoxins include anatoxin-a, anatoxin-a(s), saxitoxin, and neosaxitoxin (the first two are unique to cyanobacteria). Some cyanobacterial genera containing species that are known to produce neurotoxins include *Anabaena*, *Aphanizomenon*, and *Oscillatoria*. It is difficult to know if a species is producing a toxin in a particular lake because different strains of each species can produce different amounts of toxins.

Cylindrospermopsins inhibit protein synthesis. This group of toxins has been described more recently. Cylindropermopsins are potentially problematic because they can be found in appreciable concentrations dissolved in natural waters. These toxins can harm the liver, kidneys, small intestine, and white blood cells (Falconer and Humpage, 2006).

Hepatotoxins kill animals by damaging the liver, including the associated pooling of blood. These toxins are in a family of at least 53 related small peptides. There is concern that these compounds lead to increased rates of liver cancer (Carmichael, 1994). Microcystins are found in the serum of people exposed to the toxins for more than 50 days after initial exposure (Hilborn *et al.*, 2007). The Canadian government implemented a recommended water quality guideline of 0.5 μg liter^{-1} microcystin-LR (the most common hepatotoxin) as a result of this threat, and other countries will likely follow suit (Fitzgerald *et al.*, 1999; Codd *et al.*, 1999b). The World Health Organization recommends a maximum level of 1 μg liter^{-1} microcystin-LR and a number of European countries as well as Australia, New Zealand, and Brazil have adopted regulations based on this recommendation.

Genera with species known to produce hepatotoxins include *Microcystis* and *Nodularia*. These genera pose a threat to drinking water quality because they commonly form large blooms in nutrient-rich drinking water reservoirs during summer. In the treatment of algal blooms in lakes, methods that lyse the cells and release toxins should be avoided (Lam and Prepas, 1997). Copper treatments commonly used on algal blooms release most toxins present within 3 days, but lime (calcium hydroxide) will remove algae without immediate release of toxins (Kenefick *et al.*, 1993). The toxins are remarkably stable once they enter drinking water and can be removed only by chlorination and activated charcoal. Chlorination of drinking water rich in organics may be problematic because it may form chlorinated hydrocarbons (known carcinogens). Methods for controlling cyanobacterial blooms will be discussed in Chapter 18.

Given the intense blooms of cyanobacteria that can form in some lakes, the ecological importance of these toxins in terms of ecosystem and community properties is likely underappreciated.

The cyanobacterial toxins are known to affect food crop (bean) photosynthesis when they are present in irrigation water (Abe *et al.*, 1996). They can also modify zooplankton assemblages (Hietala and Walls, 1995; Ward and Codd, 1999), reduce growth of trout (Bury *et al.*, 1995), kill turtles (Nasri *et al.*, 2008), interfere with development of fish and amphibians (Oberemm *et al.*, 1999), have caused massive flamingo die offs (Krienitz *et al.*, 2003), and presumably can affect other organisms. However, some animals may actually prefer water containing toxic algae even though it is toxic to them (Rodas and Costas, 1999). The toxins can also be bioconcentrated by clams (Prepas *et al.*, 1997), mussels, aquatic insect larvae, and oligochaetes (Chen and Xie, 2008).

A note of caution should be made related to cyanobacterial toxins. Some companies provide dietary supplements made from cyanobacteria (bluegreen algae). A screening of 87 samples of cyanobacterial nutritional supplements found toxins in 85 of them. Even *Spirulina*, once thought to be nontoxic, has some strains capable of producing toxins (Dittman and Wiegan, 2006). If the genera in the product have strains known to produce toxins, verification that tests for cyanobacterial toxins are conducted routinely (Schaeffer *et al.*, 1999) with negative results is advisable before any of the products are consumed.

Predicting when toxic blooms will occur has remained elusive. An extensive survey of 241 lakes in the Midwestern United States measured numerous chemical and morphological factors. Only 50% of the variance in toxin concentrations could be accounted for, and many relationships predicting toxins were nonlinear (Graham *et al.*, 2004). Even very large lakes, such as the Great Lakes of North America, are susceptible to cyanobacterial toxin problems (Watson *et al.*, 2008).

Other groups of algae (the dinoflagellates and the diatoms) have toxic species or strains but cause problems more rarely in freshwaters. Cases of fish poisoning have been related to dinoflagellate blooms (similar to the marine red tide) in freshwater lakes or reservoirs. The factors that lead to blooms of these toxic algae are poorly understood.

---

with "earthy" or "musty" flavors and odors that can lead to undesirable qualities in drinking water and poor taste of fish inhabiting waters with high population densities of cyanobacteria (Schrader *et al.*, 2005).

Benthic species of cyanobacteria can be found in diverse habitats, including wetlands, streams, and temporary waters, and are somewhat more common in oligotrophic waters (in contrast with planktonic species that are more commonly found in eutrophic waters). We discuss benthic cyanobacteria in hot springs and hypersaline habitats, including their motility and behavior, in Chapter 15.

Cyanobacteria tend to do well in warmer waters, and some species are capable of photosynthesis at higher temperatures than any eukaryotic algae. Given that cyanobacterial blooms are undesirable, an additional concern is that such blooms are more probable with warmer weather. With predicted global warming, cyanobacterial problems will probably increase (Jöhnk *et al.*, 2008).

PROTOCTISTA

The protists include a wide variety of organisms from single celled to multicellular, including algae (other than cyanobacteria) and the heterotrophic protozoa.

Together, the various groups of protists are responsible for much of the primary production and nutrient recycling that occurs in aquatic habitats. Traditionally these were split into the protozoa (heterotrophic and unicellular) and the protophyta (autotrophic) groups, but the difference between many of these is a matter of absence or presence of chloroplasts, with all other characteristics the same. These diverse and often elegant organisms are often the first microbes students see and have doubtlessly inspired numerous careers in microbiology and aquatic ecology. Those whose imaginations have not yet been captured may want to examine the excellent photomicrographs of algae by Canter-Lund and Lund (1995). As discussed in Chapter 8, our understanding of the phylogenetic relationships among protists is in a state of flux. We present the following groups roughly in conventional rather than modern phylogenetic terms, mainly because this is how taxonomic books that students can use are organized.

Eukaryotic Algae

A wide variety of species of algae are found in freshwaters, but only the most common will be discussed here. The *algae* are defined as nonvascular eukaryotic organisms that are capable of oxygenic photosynthesis and contain chlorophyll *a*. Some algae may be considered protozoa because they have colorless forms that survive by ingesting other organisms. Table 9.3 and the following text summarize the characteristics of selected groups of eukaryotic algae.

Rhodophyceae, the Red Algae

Red algae are abundant in marine habitats, but are relatively rare in freshwaters, where they are usually found in lotic habitats. For example, *Batrachospermum* (Fig. 9.5C) is a red alga found in streams and springs throughout the world. The algae are red because of their phycoerythrins, which impart a red hue. Phycoerythrins are phycobilin pigments similar to those found in the cyanobacteria that allow red algae to use bluegreen wavelengths of light. Because of their ability to use wavelengths of light that are less absorbed by others and thus penetrate farther, red algae are often abundant in heavily shaded or deep habitats.

Chrysophyceae, the Golden Algae

The Chrysophyceae, sometimes called *golden algae*, are common components of the plankton in oligotrophic lakes. They have two flagella and, interestingly, most species are able to shift between photosynthesis and ingesting smaller organisms or particles for food. Many members of the Chrysophyceae are thus considered facultative heterotrophs.

Dinobryon is a common genus that forms chain-like colonies (Fig. 9.5). The large size of *Dinobryon* probably makes them difficult for herbivorous zooplankton to consume. *Dinobryon* are common in the summer in lakes when zooplankton

Table 9.3 Characteristics of Major Groups of Freshwater Algae

Group (Common Name)	Dominant Pigments	Cell Wall	Habitats	Approximate # of Species (% Freshwater)	Ecological Importance
Cyanobacteria	Chl *a*, phycobilins	Peptidoglycan	Oligotrophic to eutrophic, benign to harsh environments	1,200–5,000 (50%)	Some fix nitrogen, some toxic, floating blooms characteristic of nutrient-rich lakes
Rhodophyceae (red algae)	Chl *a*, phycobilins	Cellulose	Freshwater species in streams	1,500–5,000 (5%)	Rare in freshwaters except *Batrachospermum* in streams
Chrysophyceae	Chl *a*, chl *c*, carotenoids	Chrysolaminarin	Freshwater, temperate, plankton	300–1,000 (80%)	*Dinobryon*, a common dominant in phytoplankton
Bacillariophyceae (diatoms)	Chl *a*, chl *c*, carotenoids	Silica frustule	Plankton and benthos	5,000–12,000 (20%)	An essential primary producer, both in freshwaters and globally
Dynophyceae	Chl *a*, chl *c*, carotenoids	Cellulose	Primarily planktonic	230–1,200 (7%)	Some toxic, some phagotrophic, involved in many symbiotic interactions
Euglenophyceae	Chl *a*, chl *b*	Protein	Commonly in eutrophic waters, associated with sediments	400–1,000	Can be phagorophic, indicative of eutrophic conditions
Chlorophyceae (green algae)	Chl *a*, chl *b*	Naked, cellulose, or calcified	Oligotrophic to eutrophic, planktonic to benthic	6,500–20,000 (87%)	Very variable morphology, very important primary producers. Filamentous types in streams, unicellular in plankton
Charophyceae	Chl *a*, chl *b*,	Cellulose, many calcified	Benthic, still to slowly flowing water	315 (95%)	Often calcareous deposits

(After South and Whittick, 1987, and Vymazal, 1995)
See Figures 9.4–9.6 and 9.8 and 9.9 for representative genera and some morphological characteristics.

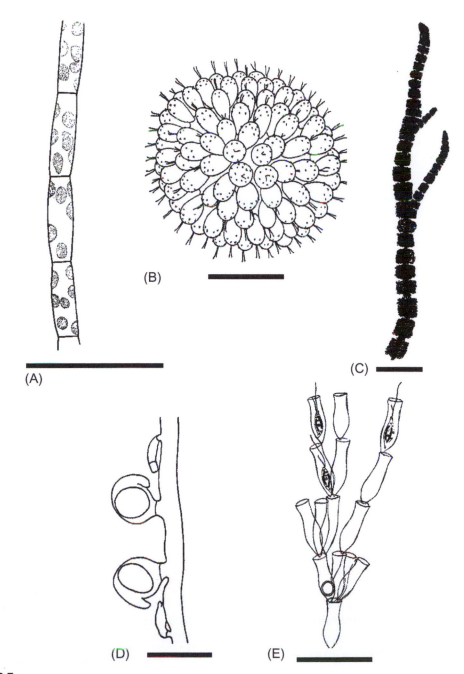

FIGURE 9.5

Selected algal genera, with scale bar length: (A) *Tribonema* (a xanthophyte), 40 μm; (B) *Synura* (a chrysophyte), 50 μm;
(C) *Batrachospermum* (a red alga), 1 cm; (D) *Vaucheria* (a xanthophyte), 200 μm; and (E) *Dinobryon* (a chrysophyte), 20 μm.
(From Prescott, 1978, 1982, reproduced with permission of The McGraw-Hill Companies).

populations are elevated. It is interesting that the colony is able to swim in a directed fashion well enough to stay in the photic zone of lakes, given that it is made up of individual cells inside of shells (loricas) that are not in contact with each other. Like diatoms, members of the Chrysophyceae contain significant quantities of silica, but it is in the form of small scales outside the plasma membrane rather than serving as a solid wall that surrounds the cell.

Bacillariophyceae, the Diatoms

The diatoms are extremely important primary producers in lakes, streams, and wetlands. They are often dominant in plankton *tows* during the spring in oligotrophic–mesotrophic lakes and in the benthic zone of lakes, streams, and wetlands year round.

The key defining characteristic of diatoms is the silica opalescent–glass cell wall called the *frustule*. The silica frustule makes them one of the few groups of organisms (along with the chrysophytes) that can be limited by silicon. This frustule has two halves, and the halves fit together to make an elongate, pennate (Fig. 9.6), or circular centric form. Centric forms are common in the plankton, and pennate forms are common in the benthic habitats. The frustules may be attached to form chains or filaments of many cells. The frustules are resistant to dissolution, so they may remain in the sediments for some time. This attribute makes them a valuable tool in forensic medicine with respect to drowning victims (Sidebar 9.3).

Diatoms are also useful in paleolimnological studies because they sink and accumulate in the sediments and leave a record of the community structure of planktonic diatoms. Sampling the sediments by sectioning with depth and identifying diatom frustules in each depth layer can produce a story of which diatom assemblages dominated in the lake over long periods of time. Changes in diatom assemblages over time can then be related to changes in climate, water chemistry, nutrient availability, and other factors that influence diatom assemblages. Isotopic analyses are used to date sediments from various depths so that ecological changes inferred from diatom frustules can be linked to specific time periods. These techniques have been used to examine the history of lakes, such as long-term temporal patterns of salinity and trophic state, and were used to show that acid precipitation was the result of industrialization (Flower and Battarbee, 1983). Ruth Patrick, one of the leading environmental researchers in the United States, has made diatoms and their use in environmental studies her specialty (Biography 9.1).

Dinophyceae, the Dinoflagellates

The dinoflagellates are commonly found in lakes and occasionally in streams. They are unicellular and free swimming, and they are subsequently found in

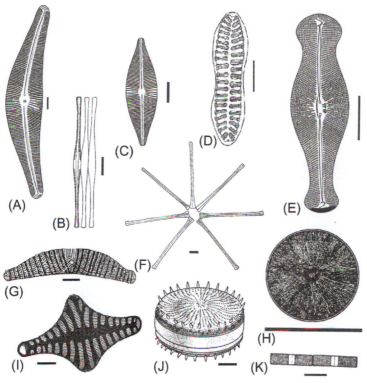

FIGURE 9.6

Common genera of diatoms with scale bar length: (A) *Cymbella*, 10 μm; (B) *Navicula*, 10 μm; (C) *Surirella*, 10 μm; (D) *Gomphonema*, 10 μm; (E) *Epithemia*, 10 μm; (F), *Asterionella*, 10 μm; (G) *Fragilaria*, 10 μm; (H) *Melosira*, 10 μm; (I) *Staurosirella*, 3 μm; (J) *Stephanodiscus*, 2 μm; (K) *Coscinodiscus*, 10 μm. *(A, B, F, H reproduced with permission from Patrick and Reimer, 1966, 1975; D and K reproduced with permission from Prescott, 1982).*

SIDEBAR 9.3
Diatoms in Forensics

Diatoms can be used as a tool to determine if drowning is a cause of death and where a drowning or other crimes occurred. When a person dies by drowning, one of the last things he or she does is take a breath of water. The water enters the lungs and bursts some of the alveoli (site of contact between blood and atmosphere). Diatoms in the water enter the bloodstream and are circulated through the body until the heart stops. When a forensic scientist searches for the cause of death, a tissue sample can be digested in strong acid or with enzymes and the diatom frustules will remain behind (Timperman, 1969). If diatom frustules are in the tissues (particularly the bone marrow), the person likely died from drowning (Ludes *et al.*, 1996). If the person was not breathing when he or she entered the water (i.e., he or she was already dead), no diatom frustules will be found deep within the organs. The technique is useful enough that the establishment of routine monitoring programs for diatoms has been recommended in areas where frequent drowning cases occur (Ludes *et al.*, 1996).

Diatoms can be used to establish location of drowning or trace suspects to a particular place because specific diatoms occur in known areas. In one case, a Finnish man was assaulted and thrown into a ditch. Five years later, the corpse was discovered and diatoms found in the lungs and bone marrow were the same species as those occurring in the ditch. Investigators concluded that the death was caused by drowning in the ditch where the body was found (Auer, 1991). In another case, a group of teenagers assaulted two boys who were fishing in a pond and attempted to drown them. The boys escaped and the teenagers were apprehended. Investigation confirmed that the teenagers had been at the pond because the residue found on their shoes contained the same diatom community as the pond mud (Siver *et al.*, 1994).

BIOGRAPHY 9.1 RUTH PATRICK

Dr. Ruth Patrick (Fig. 9.7) is one of the leading diatom systematists in the world. She has used her taxonomic expertise to extend the general theory of how aquatic microorganisms colonize new habitats and for nearly 50 years has assessed the condition of stream ecosystems from the structure of biological communities. Her publication list spans 62 years and includes over 200 works, 143 as sole author and 38 as the first author; she is also coauthor of the authoritative monograph on diatom systematics in North America. Still active in her 90s, Patrick recently completed the third book of a series on rivers and estuaries. She has exhibited a continued dedication to pollution control in aquatic systems, where she has pioneered the use of diatoms as indicators of chronic pollution.

Patrick has served as president for major scientific societies and has served on committees for several US presidents, Congress, the National Academy of Sciences, and others. Patrick was the recipient of the prestigious National Medal of Science, conferred by President Clinton. This is added to a long list of awards that includes 25 honorary doctorates and election to the National Academy of Sciences.

Given the time period that her career spanned, she overcame tremendous obstacles to become a leading scientist when women were not typically scientists, an environmentalist when few were concerned about human impacts on the environment, and only the twelfth woman in 100 years elected to the National Academy of Sciences. A hallmark of Patrick's career has been her insistence on making a positive difference. Her

FIGURE 9.7
Ruth Patrick.

father, who allowed her to climb onto his lap to look through a microscope when she was 4 or 5 years old, guided her. Patrick notes that her father would get up from the dinner table every night and say, "Remember, you must leave this world a better place." She has.

the phytoplankton. Species can also be found in wetlands and ponds. They can have cellulose plates or armor covering their body (Fig. 9.8). One flagellum encircles the cell, and another trails behind. Many members of this group are able to ingest other organisms. Some have no photosynthetic pigments, and some exist as predators, ingesting smaller cells.

Some dinoflagellates have complex life cycles and are able to assume a variety of forms, including spores, ameboid forms, and flagellated cells (Burkholder and Glasgow, 1997). In addition, some species of dinoflagellates ingest small unicellular algae and use them as chloroplasts. The dinoflagellates form a group that does not fit comfortably in the old classification system of plants or animals, because there are closely related autotrophic and heterotrophic species.

The toxic dinoflagellate *Pfiesteria piscicida* has caused concern recently. This organism is found in estuaries and has caused fish kills in the Chesapeake Bay. *Pfiesteria piscicida* may harm humans and swimming advisories are publicized when the organism is known to be present. Nutrient pollution transported via freshwaters to the estuary probably exacerbates blooms of this toxic alga (Burkholder and Glasgow, 1997), although there is some controversy over the causes of blooms and the biology of the organism.

Euglenophyceae

The euglenoids can be pigmented or colorless and thus can be autotrophic or heterotrophic. Pigmented members of this group (hence the "phyceae" indicating "plant") often have pigments similar to those of the green algae, but are always unicellular and generally motile. Molecular analyses suggest they are not closely related to the green algae, in spite of their similar pigment content (Fig. 8.4). They are found most commonly in eutrophic situations, including shallow sediments. Euglenoids are capable of ingesting particles. A flexible protein sheath covers the cell, and ameboid cell movement can occur. Additionally, many cells have a single flagellum that can be used for locomotion. Characteristic features include a red photosensitive spot in one end and numerous chloroplasts in the cell (Fig. 9.8).

Chlorophyceae and Charophyceae, Green Algae and Relatives

These algae range from simple single-celled organisms to complex multicellular assemblages (Fig. 9.8). They are found in all surface aquatic habitats from damp soil and wetlands to the benthic zone of rapidly flowing streams and the plankton of large lakes, and they are the most diverse freshwater algae group.

Some species are found mainly in oligotrophic habitats, whereas others are common in eutrophic habitats. Unicellular types are most common in lake plankton but can also be found in benthic habitats. There is a tremendous

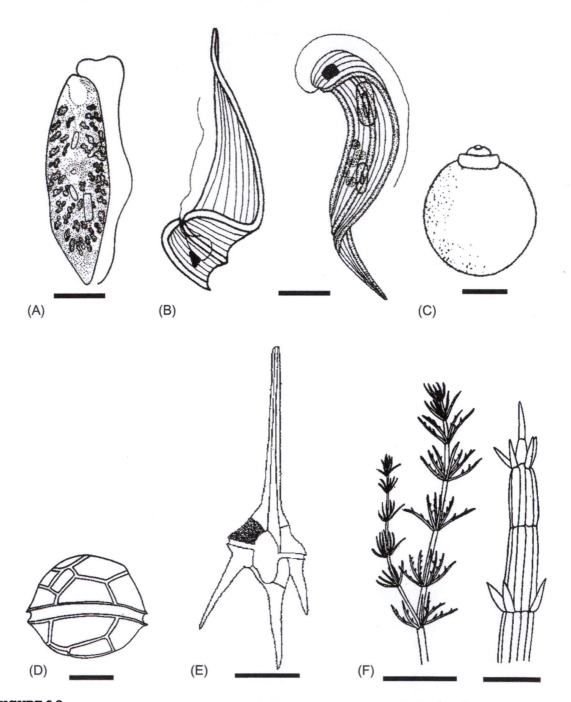

FIGURE 9.8

Selected algal genera, with scale bar length: (A) *Euglena* (a euglenophyte), 20 μm; (B) *Phacus* (a euglenophyte), 20 μm; (C) *Trachelomona* (a euglenophyte), 20 μm; (D) *Peridinium* (a dinoflagellate), 20 μm; (E) *Ceratium* (a dinoflagellate), 20 μm; and (F) *Chara* (a charophyte) large view 2 cm, close-up 500 μm. *(Reproduced with permission from Prescott, 1982).*

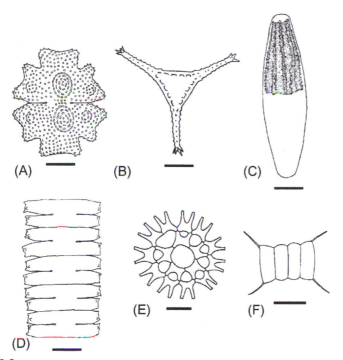

FIGURE 9.9

Some colonial and single-celled green algae and scale bars: (A) *Euastrum*, 30 μm; (B) *Staurastrum*, 25 μm; (C) *Netrium*, 35 μm; (D) *Spondylosium*, 25 μm; (E) *Pediastrum*, 20 μm; (F) *Scenedesmus*, 20 μm. *(Reproduced with permission from Prescott, 1982, and Wehr and Sheath, 2003).*

variety of unicellular morphologies (Fig. 9.9). Species with filamentous morphologies (Fig. 9.10) are generally attached to the benthic substrata in streams and lakes. One species, *Basicladia*, is most commonly found attached to the backs of aquatic turtles. Filamentous green algae are often the most obvious algae in nutrient-enriched streams and along shorelines of nutrient-enriched lakes, with massive populations that degrade water quality and hinder recreational activities observed in some cases.

The Charophytes (stoneworts) are related closely to the Chlorophyceae, but are more complex (Fig. 9.8). The stoneworts are likely the evolutionary precursors to land plants. These plant-like algae are called stoneworts because they precipitate calcium carbonate on the exterior of their cell walls and feel rough, or stone-like to the touch. They are not vascular, but have multicellular reproductive structures more like land plants than the other algae. The charophytes can sometimes cause problems because of immense biomass that impedes water flow or navigation on rivers. Charophytes may also be an important component of more productive wetlands. *Chara* can be abundant in the benthic zone of some oligotrophic lakes. Many species of charophyte

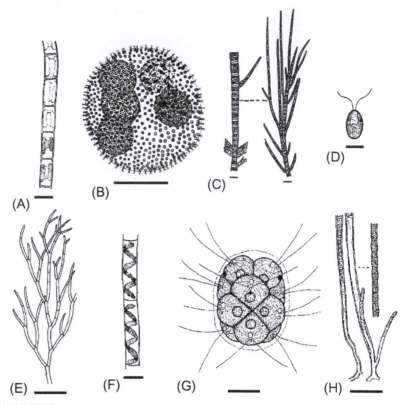

FIGURE 9.10
Common genera of filamentous and flagellated green algae, with scale bar length: (A) *Ulothrix*, 20 µm;
(B) *Volvox*, 10 µm; (C) *Stigeoclonium*, 20 µm; (D) *Chlamydomonas*, 10 µm; (E) *Cladophora*, 50 µm;
(F) *Spirogyra*, 20 µm; (G) *Pandorina*, 10 µm; and (H) *Basicladia*, 30 µm. *(Reproduced with permission
from Prescott, 1982, and Wehr and Sheath, 2003).*

are sensitive to nutrient enrichment and distribution of the stoneworts has
been used to indicate nutrient pollution. Species found only in very clean
waters may be endangered and some species are now protected in Europe.

Additional Algal Groups
Additional groups of algae are found in freshwaters, including the Cryptophy-
ceae, the Tribophyceae, and the Phaeophyceae. Members of these groups can
be important in freshwaters. However, detailed description is left to phycology courses and the comprehensive phycological texts (South and Whittick,
1987; Graham and Wilcox, 2000: Wehr and Sheath, 2003).

Protozoa
Protozoa are found in nearly every habitat on Earth, and can be very abundant
and diverse in freshwaters. This group includes autotrophs and heterotrophs;
the "zoa" presumably denotes motility. Many types are heterotrophic and survive

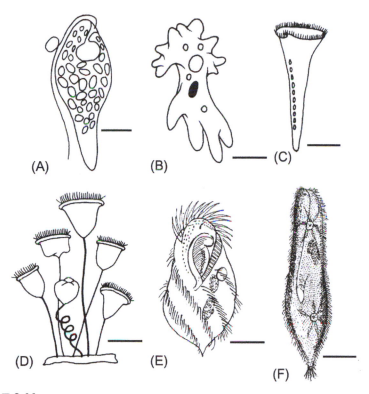

FIGURE 9.11
Selected protozoa and size of associated scale bars: (A) *Khawkinea*, a zooflagellate, 20 μm; (B) *Amoeba*, 50 μm; (C) *Stentor*, a solitary ciliate, 200 μm; (D) *Vorticella*, a colonial ciliate 75 μm; (E) *Hypotrichidium*, a ciliate, 30 μm; (F) *Paramecium*, a ciliate, 60 μm. *(Reproduced with permission from Thorp and Covich, 1991b).*

by ingesting particles or absorbing dissolved organic carbon. They are very important predators of bacteria in aquatic environments. There are also many ecologically and economically important parasites in this group. Some of the smallest protozoa are only slightly larger than bacteria and can ingest virus-sized particles. The largest are visible to the unaided eye. This is a very diverse group (Fig. 9.11), and includes the most complex single-celled organisms known.

Life histories are generally simple. Sexual reproduction is widespread but not universal. Many protozoa form cysts that are resistant to environmental extremes, allowing them to persist in harsh habitats and quickly exploit moisture when it becomes available. Other morphological variations among life cycle stages can also occur, such as differentiation between forms that search for food and those that consume it (Taylor and Sanders, 2001).

Various classifications of the protozoa have been proposed (Taylor and Sanders, 2001), and molecular analyses indicate that there should be several phyla of protozoa (Figs. 8.1, 8.4). Members of the protozoa include organisms from the entire lower portion of the Eukarya part of the phylogenetic tree. For the sake of simplicity, we will discuss the major functional groups of protozoa, following the groupings of Taylor and Sanders (2001).

The flagellated protozoa (those with one or more flagella) can be divided into two major functional groups. The phytoflagellates include groups that are autotrophic, heterotrophic, and mixotrophic, some of which were previously discussed (e.g., dinoflagellates, chrysophytes, euglenoids, and flagellated green algae). The zooflagellates (Fig. 9.11) are all heterotrophic. The phylum Zoomastigina in this group includes several important human parasites (e.g., trypanosomes and *Leishmania*) and many free-swimming forms. The heterotrophic nanoflagellates are very small zooflagellates that are often the most important consumers of pelagic bacteria and can serve a vital role in nutrient cycling.

The ameboid protozoa, including phyla Karyoblastea and Rhizopoda, constitute the largest group of described species of protozoa. The amoeboid protozoa include those that move by protoplasmic flow and use of pseudopodia (extensions of the protoplasm; e.g., *Amoeba*, Fig. 9.11B). The flowing movements of amoebae illustrate this process. These groups of protozoa are more often associated with benthic habitats and sediments than open freshwaters, but some floating forms exist. Some amoeboids live in shells (called *tests*) that they secrete or construct from sediments. Two species from two genera (*Naegleria and Acanthamoeba*) can cause meningitis in human swimmers. Many of the amoeboid protozoa are also important microbial predators.

The ciliated protozoa (phylum Ciliophora, including 10 classes) are characterized by having more than four cilia, which are primarily used for locomotion. Nuclear dualism is a unique feature of the ciliates; individuals have one or more macronuclei and one or more micronuclei. This group contains the familiar *Paramecium* and other free-swimming genera (Fig. 9.11F). Some members of this group are floating or sessile predators with ciliated larval stages for dispersal. The common attached protozoan, *Vorticella*, is also a member of the Ciliophora (Fig. 9.11C). Most of the ciliates feed on bacteria or other protozoa, but a few species feed on algae and detritus. *Balantidium coli*, an intestinal parasite of pigs and occasionally humans, is the only member of the Ciliophora known to cause human disease.

FUNGI

Fewer species of fungi occur in aquatic habitats than in terrestrial habitats. Nonetheless, they are very important in the degradation of detritus (leaf litter and woody debris) that enters streams and lake margins, and as a source of nutrition for detritivores. Fungal colonization softens detritus and increases its nutritional value. Without the activity of fungi and bacteria, detrital carbon sources would probably remain unavailable to many invertebrates (Arsuffi and Suberkropp, 1989; Suberkropp and Weyers, 1996).

Aquatic Fungi

More than 600 species of fungi occur in freshwaters (Wong *et al.*, 1998), including those of the Labyrinthulomycetes (slime molds), Ooprotista (algal fungi), Ascomycetes (sac fungi), Basidoiomycetes (club fungi), and Deuteromycetes (imperfect fungi). Most fungi are saprophytes, meaning they live on dead organic matter. The systematics of the fungi are based on their reproductive features and morphology, except for the Dueteromycetes, in which no reproductive structures have been found. As molecular techniques are applied to fungi, the taxonomy of the Deuteromycetes will probably be resolved more satisfactorily.

Of the groups of fungi, only the Oomycota (formerly the Phycomycetes) are predominantly aquatic. In general, they are unicellular. The Oomycota include many parasitic species that are pathogens of planktonic algae, small animals, and the eggs of crustacean larvae and fish (Rheinheimer, 1991). The Chytridiomycota (chytrids) live in aquatic habitats and moist soils.

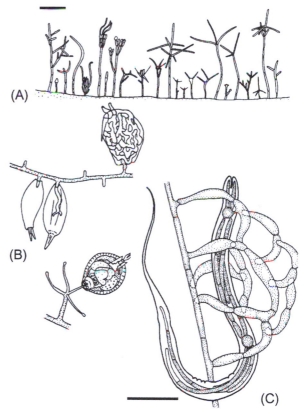

(A)

(B)

(C)

FIGURE 9.12
Selected aquatic fungi: (A) Aquatic deuteromycetes, (B) zoophagous species with trapped rotifers, and (C) *Arthrobotrys oligospora* and a trapped nematode. *(Reproduced with permission from Rheinheimer, 1991).*

Although members of this group are considered fungi, they have a motile zoo-spore with one whip-like posterior flagella, more like the protists. The chytrids can be important parasites and the chytrid *Batrachochytrium dendrobatidis*, which is discussed in Chapter 11, is the primary cause of many of the catastrophic amphibian declines occurring around the planet (Lips *et al.*, 2006, 2008).

The Ascomycetes and Deuteromycetes (particularly the aquatic Hyphomycetes) are often abundant on decaying leaves and wood (Fig. 9.12). The Hypho-mycetes are divided into the Ingoldian fungi (with branched or radiate conidia) and the helicosporous (with helical conidia) fungi (Alexopolus *et al.*, 1996). Yeasts (in the group Ascomycetes) can be found in rivers and lakes, particularly in polluted waters. "Sewage fungus," associated commonly with organic-rich pollution, is actually composed predominantly of sheathed bacteria, not fungi.

Fungi are not abundant in pristine groundwater because of low concentrations of organic matter (Madsen and Ghiorse, 1993). Likewise, pristine spring water rarely has significant numbers of fungi. Fungi may be locally abundant where pollution or natural inputs enrich groundwater with organic compounds, but when the groundwaters are anoxic, the fungi are less abundant than in oxic water.

An interesting mode of nutrition for some aquatic fungi is predation on rotifers or nematodes. The fungi that prey on rotifers have sticky appendages that trap the organisms and then rapidly grow into them. The nematode trapping fungi inhabit soils and aquatic sediments and can form a net of loops or snares that trap the nematodes as they crawl through (Fig. 9.12).

Fungi often associate with plant roots (mycorrhizae), particularly in terrestrial habitats. However, these fungi are important in some wetlands, where they can alter the dominant plant species, depending upon wetland hydrology. Experiments with plants in northeastern US calcareous fens suggests that the importance of mycorrhizal fungi decreases with increased inundation (Wolfe *et al.*, 2006).

Aquatic Lichens

Lichens (a symbiotic partnership between a fungus and an alga) are never found in groundwaters, and only rarely in the benthic zone of lakes and rivers. However, fairly dense growths of the lichen *Dermatocarpon fluviatile* can be found in benthic habitats of streams and some lakes. Lichens can be abundant in wetlands, particularly those in northern temperate, boreal, or polar regions. In one northern European wetland, lichens and mosses were responsible for 9% of the carbon input to the bog (Mitsch and Gosselink, 1993). Lichens on rocks near the waterline of lakes have received some study (Hutchinson, 1975). Lichen species change with distance above the water as their tolerance of submergence decreases.

The taxonomy of lichens traditionally is based on external morphology. The morphological differentiation among lichens requires determining if the form is foliose (leaf-like), fruticose (finger-like projections), or crustose (appressed to a solid surface). Molecular techniques are also being used to distinguish lichen species.

PLANTAE

Plants dominate in many shallow waters. Plants in water are called *macrophytes*. They are the dominant organisms in wetlands and many lake margins and streams. They can play essential roles in biogeochemistry and ecology. For example, macrophyte beds can provide important spawning habitat and shelter for small fishes. Macrophytes can also be the dominant photosynthetic organism in small or shallow lakes. Thus, they can even be central to lake food webs. Riparian plants are very important in biogeochemistry because they intercept materials (e.g., pollutants, nutrients, and sediments) that are entering aquatic habitats from the terrestrial ones. Furthermore, much of the diversity of plants in many terrestrial landscapes is associated with riparian zones or wetland areas. Aquatic plants can cause nuisance conditions, and there are aquatic or riparian species that are invasive and cause problems.

Nonvascular Plants

Bryophytes (mosses and liverworts) are abundant in some freshwaters. Traditionally, they received little study relative to their importance in some systems, but appreciation for their importance is increasing (Arscott *et al.*, 1998). Bryophytes can be important components of streams, particularly in alpine and arctic regions, where they can be productive and provide important structural habitats for invertebrates (Bowden, 1999). Aquatic mosses can be divided into three orders (Hutchinson, 1975): the Sphagnales, the Andreales, and the Bryales. The Sphagnales and Bryales have numerous aquatic representatives, and the Andreales has few. The Sphagnales contains only one genus, *Sphagnum*.

Species of the genus *Sphagnum* are often a dominant component of the vegetation in the shallow acidic waters of peat bogs and can be very important in many high-latitude wetlands. The total global biomass of *Sphagnum* is greater than that of any other bryophyte genus (Clymo and Hayward, 1982). Carbon deposition in these peat bogs may be important in the global carbon cycle. The moss promotes acidic conditions because microbial breakdown of organic material produced by *Sphagnum* produces organic acids. The acidity leads to a stable dominance by the acid-tolerant moss and slows breakdown of organic material. Thus, peat accumulations are significant in the bogs where *Sphagnum* dominates. Peat and bogs were discussed in more detail in Chapter 5.

The Bryales includes several interesting aquatic genera, including *Fontinalis*, which is found to 120 m deep in Crater Lake, and *Fissidens*, which has been found to 122 m deep in Lake Tahoe (Hutchinson, 1975). Thus, among plants, the mosses are found in some of the deepest habitats.

Vascular Plants

The angiosperms, the true flowering plants, are the dominant aquatic vascular plants, with representatives of both monocots and eudicots. In addition, some of the ferns and fern allies are associated with aquatic habitats. Of the angiosperms, the monocots are relatively more important in aquatic habitats than they are in terrestrial habitats (Hutchinson, 1975). Vascular plants represent

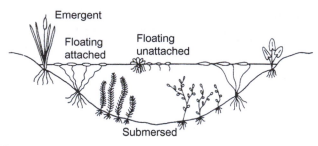

FIGURE 9.13

Growth habit types of aquatic plants. *(Reproduced with permission from Riemer, 1984).*

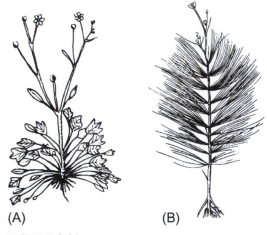

(A)   (B)

FIGURE 9.14

Two morphologies of the buttercup: *Ranunculus polyphyllus* growing on land (A) and submersed (B). *(From G. E. Hutchinson,* A Treatise on Limnology, *Vol. 3, 1975. Reprinted by permission of John Wiley & Sons, Inc.).*

important structural habitat for other organisms (e.g., fishes, invertebrates, and epiphytes), food for some herbivores and detritivores, and, as with algae, their photosynthesis and respiration can greatly influence the local chemical environment, including availability of dissolved oxygen.

Aquatic ecologists tend toward classifications of plants based on functional roles and habitats (Fig. 9.13). Distinguishing aquatic plants on the basis of morphology can be difficult because there is tremendous variation in morphology of plant structures within genera. Also, leaf size and shape can change appreciably in the same species grown under different environmental conditions (Fig. 9.14) or even in the same plant above and below water (Fig. 8.2).

The traditional categories include (1) *floating unattached* macrophytes (roots not attached to substratum; Fig. 9.15), (2) *floating attached* plants (leaves floating at the surface, and roots anchored in the sediments), (3) *submersed* plants (entire life cycle, except flowering, under water; generally attached to sediment; Fig. 9.16), and (4) *emergent* (growing in saturated soils up to a water depth of 1.5 m and producing aerial leaves; Fig. 9.17).

A wide variety of plant groups have given rise to aquatic species; some of the representative genera are listed in Table 9.4. Trees associated with wetlands are important in defining wetland types (Table 9.5). The genera of submersed plants tend to be confined to aquatic habitats, but emergent genera also have many representatives in terrestrial habitats (Hutchinson, 1975). The aquatic plants form a vital part of the ecosystem, but some have become serious invaders or pests. Invasion of North American wetlands by purple loosestrife (*Lythrum salicaria*) (Sidebar 9.4) is one of many examples of nuisance exotic plants that are altering freshwater ecosystems.

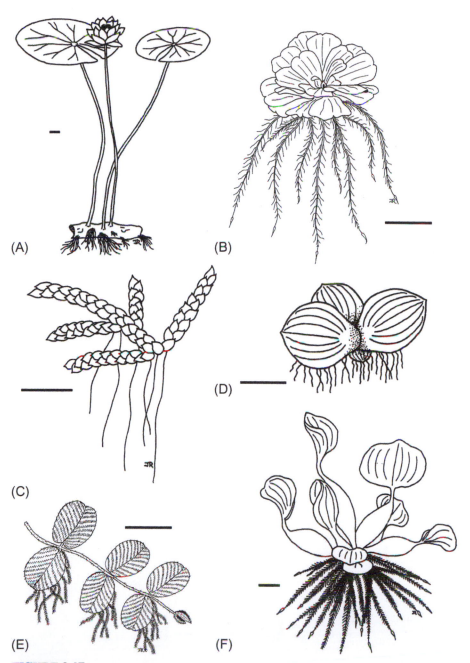

FIGURE 9.15

Some species of floating attached (A), and floating unattached (B–F) macrophytes: (A) *Nymphaea*, water lily; (B) *Pistia*, water lettuce; (C) *Azolla*, water velvet; (D) *Spirodela polyrhiza*, duckweed; (E) *Salvinia*, water fern; (F) *Eichhornia crassipes*, water hyacinth. Scale bar = 2 cm. *(Reproduced with permission from Reimer, 1984).*

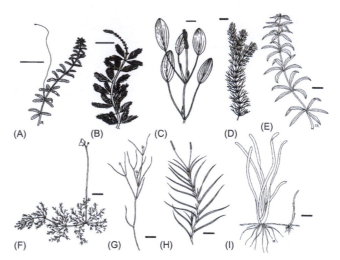

FIGURE 9.16

Some species of submerged aquatic plants: (A) *Elodea canadensis*, waterweed; (B) *Myriophyllum spicatum*, water milfoil; (C) *Potamogeton nodosus,* pondweed; (D) *Ceratophyllum,* coontail; (E) *Hydrilla verticillata*, hydrilla; (F) *Utricularia*, bladderwort; (G) *Potamogeton pussilus*, pondweed, compare with C; (H) *Potamogeton robbinsii*, pondweed, compare with C and G; and (I) *Vallisneria americana*, wild celery; Scale bar = 4 cm. *(Reproduced with permission from Reimer, 1984).*

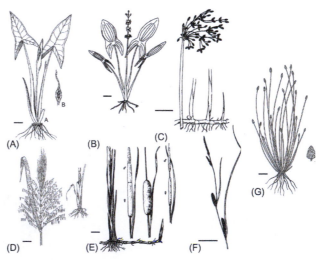

FIGURE 9.17

Some emergent aquatic plants: (A) *Peltandra virginica*, arrow arum; (B) *Sagittaria latifolia*, arrowhead; (C) *Scirpus validus*, great bulrush; (D) *Zizania aquatica*, wild rice; (E) *Typha*, cattail; (F) *Carex lanuginose* wooly sedge; (G) *Eleocharis obtuse*, spikerush. Scale bar in A, B, D, and E 10 cm; C, F, and G 5 cm. *(Reproduced with permission from Riemer, 1984).*

Table 9.4 Some Common Genera of Aquatic Plants, with Their Diversity, Habit, and Distribution

Habit	Genus	Common Name	Species	Habitat/Distribution/Comments
Emergent	*Oryza sativa*	Rice	1	Tropical temperate food crop, most important in world
	Peltandra	Arrow arum	3	Shallow water eastern North America
	Sagittaria	Arrowhead	20	New world, tuber edible by waterfowl
	Scirpus	Bulrush	150	Worldwide, mostly North America
	Sparganium	Bur reed	20	Temperate and arctic Northern Hemisphere
	Typha	Cattail	10	Worldwide, common in monospecific stands in swamps, marshes, and along streams
	Phragmites	Giant reed	3	Worldwide
	Pontederia	Pickerelweed	5	New world, shallow muddy areas
	Juncus	Rushes	225	Northern Hemisphere, few submerged species
	Carex	Sedges	1000	Worldwide, common in damp areas to shallow waters
	Eliocharis	Spikerush	200	Worldwide, may be completely submerged, but will be sterile if so
	Zizania	Wild rice	2	North America, is a prized food
Floating attached	*Nuphar*	Spatterdock, yellow water lilies	25?	Large yellow flowers
	Brasenia	Water shield	1	Scattered worldwide
	Nymphaia	White water lilies	40	Worldwide, introduced in many places, large flowers, mostly white
Floating unattached	*Lemnaceae*	Duckweed	30	Worldwide, can cover small pond surfaces
	Salvinia	Water fern	12	Tropical
	Eichornia crassipes	Water hyacinth	1	Tropical and subtropical, one of the worst weeds in the world, also used for tertiary sewage treatment
	Pistia stratiodes	Water lettuce	1	Tropical and subtropical, can be a pest
	Azolla	Water velvet		Worldwide, contain N fixing cyanobacteria, may be important in traditional rice culture, can be red or purple on surface of water
Submersed	*Utricularia*	Bladderwort	150 (30 aquatic)	Seed-like bladders trap and digest aquatic animals
	Ceratophyllum	Coontail	30	Worldwide
	Elodea	Elodea	17	North and South America, but introduced elsewhere as an escapee of aquaria
	Cabomba	Fanwort	7	Tropical to temperate New World
	Hydrilla verticillata	Hydrilla	1	Similar to *Elodea*
	Najas	Naiad	50	Worldwide
	Potomogeton	Pondweeds	100?	Worldwide, found in most types of fresh surface waters, important food sources for wildlife, some pest species
	Myriophyllum	Water milfoil	40	Africa, some species introduced, pests elsewhere
	Vallisneria	Wild celery	10	Worldwide, warm areas

(After Riemer, 1984)

Table 9.5 Some Important Trees Associated with Wetlands in North America

Common Name	Scientific Name	Distribution
Black spruce	*Picea mariana*	Boreal wetlands
Tamarack	*Larix laricina*	Boreal wetlands
Red maple	*Acer rubrum*	Temperate wetlands
Northern white cedar	*Thuja occidentalis*	Northeast temperate North America
Atlantic white cedar	*Chamaecyparis thyoides*	Southeast United States
Cypress	*Taxodium* spp.	Southeast United States, deepwater swamps
Tupelo	*Nyssa aquatica*	Southeast United States, deepwater swamps
Cottonwood	*Populus* spp.	Riparian wetlands
Willow	*Salix* spp.	Riparian wetlands
Red mangrove	*Rhizophora* spp.	Brackish tropical waters
Black mangrove	*Avicennia* spp.	Brackish tropical waters

SIDEBAR 9.4

Invasion of Wetlands by Purple Loosestrife

Purple loosestrife (*Lythrum salicaria*) is a perennial plant that is invading many North American wetlands. It is an emergent plant with beautiful purple flowers. Purple loosestrife can reach some of the highest levels of biomass and annual production reported for freshwater vegetation (Mitsch and Gosselink, 1993). Unfortunately, the plant is a poor food source for most waterfowl, and large stands with a high percentage of cover possibly lower the numbers of nesting sites for ducks and other water birds as well as provide additional cover for predators. Purple loosestrife can outcompete native plants and lower biodiversity (Malecki *et al.*, 1993). In wet areas that are used for hay, it lowers the forage value. However, Anderson (1995) suggested that the effects of purple loosestrife have been overestimated and more research should be done to quantify its impacts on native ecosystems.

Lythrum salicaria occurs naturally in Europe from Great Britain to Russia (Mal *et al.*, 1992). It was introduced to eastern North America in the ballast of ships and as a medicinal and decorative plant (Malecki *et al.*, 1993). Loosestrife has since become a serious pest species around the Great Lakes of North America and has spread across Canada and the United States to the west coast. Introductions have also occurred in Australia, New Zealand, and Tasmania (Mal *et al.*, 1992).

Several control strategies have been attempted (Malecki *et al.*, 1992), including herbicides (which also harm other wetland species), physical removal, burning, manipulation of water levels to favor native species, use of native insects, and introduction of exotic insects to control the plant. There are concerns that insects introduced as control agents will harm native species. After tests for host specificity, three insect species were released in the United States as control agents (Piper, 1996). Their efficiency in loosestrife control is not yet known. Certainly, an understanding of wetland ecology is crucial to assessing the impact of and control options for this exotic invader.

SUMMARY

1. Viruses are common in freshwater habitats and have important consequences in terms of diseases of aquatic organisms and human health, as well as population dynamics of microorganisms.

2. Archaea are important in extreme habitats and for some types of biogeochemical cycling, particularly the formation of methane. They are probably more important in most aquatic habitats than appreciated previously.

3. The biomass, metabolic diversity, and species diversity of Bacteria probably exceed those of any other group of organisms on Earth. Understanding the role of bacteria is central to attempts to understand the aquatic environment, including organic carbon decomposition, energy flow, and nutrient cycling.

4. Cyanobacteria are a significant ecosystem component of many lighted aquatic habitats. They can cause problems when they form large blooms. Many varieties produce toxins and foul the water.

5. Algae constitute an important and diverse group found in freshwaters and form the basis of many aquatic food webs. The importance of algae also includes an intimate role in water quality. Diatoms are used in paleolimnology to document historical biological patterns over thousands of years.

6. Protozoa are a very diverse and taxonomically problematic group. Protozoa are generally the main consumers of bacteria in aquatic systems.

7. Fungi are responsible for much of the degradation of particulate organic material that occurs in freshwaters, and some groups are also commonly parasitic on aquatic organisms.

8. The Bryophytes can be important in some shallow aquatic habitats. Formation and maintenance of high-latitude wetland communities by *Sphagnum* in peat bogs are of global significance.

9. Plants can be classified as submerged, floating, or emergent, and they represent important structural habitat for other organisms. Vascular plants are dominant contributors to organic matter in many shallow aquatic ecosystems. The types of flowering plants present define many wetlands.

QUESTIONS FOR THOUGHT

1. Do more types of viruses exist than species of organisms on Earth?

2. Should separate taxonomic definitions of species be used for microbes than those that are used for animals?

3. Why are there no known fish-pollinated aquatic angiosperms?

4. How does stability of the benthic substrata partially determine if benthic systems are dominated by microalgae or by macrophytes?

5. Should molecular taxonomy methods be more useful for aquatic angiosperms or unicellular algae?

6. Why are floating leafed macrophytes relatively rare on large lakes?

7. Why are bryophytes relatively more important in alpine and high-latitude freshwater habitats?

Multicellular Animals

FIGURE 10.1
The caddisfly larva, *Psychoglypha sub-borealis*, and the snail, *Vorticifex effusa*. The caddis larva is 1 cm long, and the snail is 0.5 cm long. Both are from Mare's Egg Spring, Oregon.

In this chapter we discuss animals, some of the most fascinating organisms found in aquatic habitats because of their diversity and behavior. Many animals play critical roles in freshwater ecosystems (Wallace and Webster, 1996; Vanni, 2002). They can be important indicators of the health of freshwater ecosystems because they integrate stresses, and sensitive species are generally not found in polluted habitats (Palmer *et al.*, 1997; Covich *et al.*, 1999). Food webs are a major pathway of energy flow through ecosystems and generally are dominated by animals. We have already discussed the economic importance of commercial and sports fisheries (Chapter 1). This chapter considers each group of organisms approximately in order of evolutionary origins.

INVERTEBRATES

Phylum Porifera

Sponges often are thought of as marine animals but they can be abundant in freshwaters. There are about 25 species of freshwater sponges in North America and about 300 species worldwide (Frost, 1991). They can be found in a variety of lentic and lotic habitats, and some species have small ranges, whereas others are widespread. Sponges are among the most primitive animals and obtain nutrition by filtering particles from the water.

Sponges feed on particles ranging in size from several hundred micrometers to smaller than bacteria ($<1\,\mu m$). They feed selectively, and they do not digest unsuitable particles and eject them instead. Many species of sponges have algal endosymbionts that provide photosynthate and give them a bright green color. Most of these green species harbor a green alga, *Chlorella*, but some

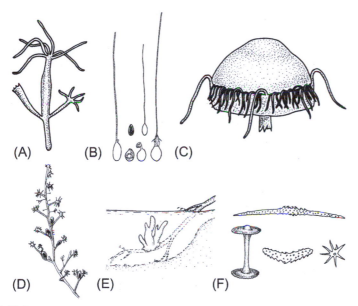

FIGURE 10.2

Example freshwater cnidarians and structures (A–D) and sponges and structures (E–F). Organisms and structures and their approximate lengths are as follows: (A) *Hydra*, ~5 mm; (B) discharged and undischarged *Hydra* cnidocysts, the microscopic stinging structures that contain venom; (C) *Cordylophora* colony, 20 mm; (D) *Craspedacusta* medusa, ~1 cm; (E) a sponge colony growing on a stick, 20 cm; (F) spicules made of silicon from several species of sponges (about 50 μm each). *(A, C–F reproduced with permission from Thorp and Covich, 2001; B from Smith, 2001).*

contain algae from other classes. Freshwater sponges are important food for some predators, including spongillaflies, insects whose larvae are specifically adapted to pierce and suck fluids from sponge cells.

Macroscopically, colonies can appear as round clumps, as flattened encrusting bodies, or as finger-like growths (Fig. 10.2E). All freshwater sponges are composed of collagen and siliceous spicules (Fig. 10.2F). Sponges have no multicellular organs, but they do have a variety of specialized cells, including epithelial cells, flagellated cells that pump water through a canal system in the sponge, and digestive cells that break down ingested particles and transport nutrients to various parts of the sponge. Sponges reproduce both sexually and asexually. Asexual reproduction can range from simple fragmentation to formation of specialized resistant stages called gemmules that are resistant to adverse conditions such as drying and can aid in survival in marginal habitats as well as serve as a dispersal stage.

Phylum Cnidaria

The Cnidaria (formerly called Coelenterata), which includes jellyfish, anemones, corals, and hydroids, are mostly marine, but a few species of small jellyfish (Fig. 10.2C), polyps, and colonial hydrozoa occur in freshwaters, all of which are members of the class Hydrozoa (Slobodkin and Bossert, 1991; Smith 2001). These animals are radially symmetrical, and all have cnidocytes, specialized cells that contain cnidocysts (nematocysts), which can fire variously modified filaments to capture prey and may contain strong toxins (Fig. 10.2B). Most cnidarians exhibit very obvious alternation of generations between sessile (hydroid or polyp) and free-swimming (medusa) forms. The cnidarians can reproduce asexually through budding or fission, or sexually. Freshwater species often have a drought-resistant stage that allows for widespread dispersal and persistence in marginal habitats.

Hydra (Fig. 10.2A) may be the most commonly observed freshwater cnidarian occurring in streams, wetlands, and lakes. These small polyps are 1 to 20 mm long with 10 to 12 tentacles crowning a tubular body. *Hydra* species can float in the plankton or neuston, but are most commonly observed attached to hard substrata or macrophytes in benthic habitats. *Hydra* can move across solid surfaces by "cartwheeling"—attaching tentacles, releasing the posterior end, and flipping it over the body to reattach on the other end. They can also move vertically in the water column by secreting a gas bubble that allows them to float to the surface. These small predators feed on a variety of smaller invertebrates such as copepods and cladocerans, which are paralyzed with nematocysts.

As noted for sponges, *Hydra* can also appear bright green from the endosymbiotic green alga, *Chlorella*. This relationship is usually optional for *Hydra*. The alga produces photosynthate in lighted habitats but is ejected or digested during an extended period of low light. *Hydra* species with algal symbionts can have higher growth rates in the light than those without algae (Slobodkin and Bossert, 1991).

Cordylophora (Fig. 10.2D) are the familiar colonial polyps that can be very abundant in brackish and freshwater habitats. Large colonies resemble moss on rocks and can reach up to 70 mm above the substrata. *Cordylophora* are unique among the cnidarians in that there is no true medusiod generation.

A unique jellyfish, *Mastigias*, occurs in a saline, marine-influenced lake in Palau in the West Caroline Islands (Hamner *et al.*, 1982). The lake is stratified and has an anoxic, high nutrient hypolimnion. This medusa contains an endosymbiotic dinoflagellate. The medusa moves down at night to follow the copepods it consumes for food and it moves up during the day to allow the endosymbionts to photosynthesize. During the day, the medusae migrate up to 1 km horizontally to maximize exposure to light. Similarly, in temperate, freshwater lakes the freshwater jellyfish *Craspedacusta* (Fig. 10.2C) can be important in linking the lower and upper strata of water during short periods of time. Their role in food webs can be substantial when they aggregate in swarms of more than 1000 medusae m^{-3} (Angradi, 1998; Spadinger and Maier, 1999). *Craspedacusta* have been found in freshwater habitats throughout the world, and they are somewhat unusual among the freshwater cnidarians in that the medusa is the more obvious, commonly observed life stage.

Phyla Platyhelminthes and Nemertea

The Platyhelminthes includes three classes: the Turbellaria (free-living flatworms), the Trematoda (flukes), and the Cestoda (tapeworms). This is a diverse phylum that includes some species that reproduce only sexually, others that reproduce only through asexual means including budding and fission, and some that can reproduce either way. The phylum Platyhelminthes includes groups of significant ecological and economic importance.

The Turbellaria (Figs. 10.3H–10.3J) are common in freshwaters, with about 400 species found throughout the world (Kolasa, 1991). Most turbellarians are predators or scavengers of dead animals, but some smaller forms are omnivores that feed on detritus, microbes, and living or dead smaller invertebrates. Some predaceous forms use secreted mucus to entrap and subdue their prey. Turbellarians do not have an anus or closed circulatory system, but they do have an intestine and a ciliated epidermis over the entire body. The Turbellaria are divided into two groups—the microturbellarians, with about 300 species, and the macroturbellarians, or Tricladida. Microturbellarians usually have cosmopolitan distributions, whereas the triclads are distributed less widely. Microturbellarians can be found in rivers, ponds, lakes, and subsurface habitats. *Microstomum* and a few other types of microturbellarians have the interesting adaptation of ingesting *Hydra* and retaining their undischarged cnidocysts, which are kept on the body surface and serve as a defense (Smith, 2001).

Triclads in the families Dugesiidae and Planariidae are probably the best known turbellarians to beginning students, as they are common in freshwater habitats and common subjects of study in introductory biology laboratories. These free-living flatworms are distinguished by the extension of a muscular pharynx from the mouth that ingests food. Caves and underground waters have many unique and endemic triclad species; diversity of triclads is particularly high in karst regions.

The flukes represent one of the major groups of animal parasites of humans. Many flukes have complex and fascinating life cycles that include multiple hosts; many rely on freshwater hosts such as snails or fishes for at least part of their life cycle. One of the more well-known flukes, *Schistosoma*, is the cause of schistosomiasis. This parasite uses pulmonate snails as the intermediate host, so understanding the epidemiology of this parasite requires knowledge of the ecology of snails. Humans are a definitive host of *Schistosoma*, meaning the flukes reproduce within them and eggs are then passed in feces. Thus,

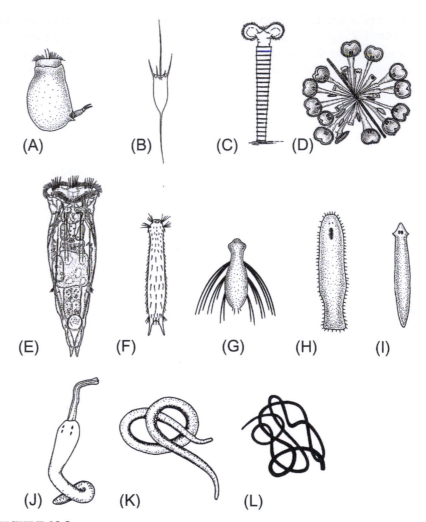

FIGURE 10.3

Representative rotifers (A–E), gastrotrichs (F–G), flatworms (H–J), and a nemertean (K), a nematode (L), and a nematomorphan (M). Organisms shown and their approximate lengths are as follows: (A) *Gastropus*, 0.3 mm; (B) *Kellicottia*, 0.6 mm; (C) *Limnias*, a tube-building rotifer, ~0.15 mm; (D) a colony of *Sinantherina*, colony diameter 2 mm; (E) *Epiphanes*, 0.6 mm; (F) *Chaetonotus*, 0.4 mm; (G) *Stylochaeta*, 0.4 mm; (H) *Macrostomum*, 2 mm; (I) *Girardia* (*Dugesia*), 10 mm; (J) *Protostoma* with proboscis extended, ~20 mm; (K) a nematode, 1 mm; (L) a horsehair worm, Nematomorpha, 10 cm. *(A, B, D–H, and K, L reproduced with permission from Thorp and Covich, 2001; C and J from Smith, 2001).*

management of snails and sewage treatment are both central to controlling schistosomiasis. Schistosomiasis is one of the most significant and devastating human diseases on the planet; an estimated 200 million people are infected worldwide, and about 20 million of those have severely debilitating infections. Swimmers' itch is caused by a related fluke that is normally parasitic on birds—in this case, the specialized cercariae fluke larvae burrow into bathers' skin, leaving an irritating, but usually harmless rash.

The nonsegmented proboscis worms (Fig. 10.3J) in the Nemertea are distinguished from the Turbellaria by having an anus and closed circulatory system. Twelve freshwater species and many more marine species have been described. All known species are benthic predators, and freshwater forms are generally found in warm, shallow, heavily vegetated habitats (Kolasa, 1991; Smith, 2001). Nemerteans actively hunt; when a prey item is encountered, it is repeatedly stabbed with stylets located in the proboscis and further subdued with sticky mucus secretions from the proboscis. Nemerteans are generally small, but large forms can exceed 100 mm in length.

Phylum Gastrotricha

The gastrotrichs can be tremendously abundant in freshwaters (10,000–100,000 m^{-2}) but are poorly studied (Strayer and Hummon, 1991). About 250 species in 22 genera have been described from the freshwaters of the world (Smith, 2001). Morphologically, they are about 50 to 800 µm long and bowling pin shaped. Gastrotrichs usually have a distinct head with sensory appendages and a cuticle that is covered with scales or spines (Figs. 10.3F and 10.3G).

Gastrotrichs are found mainly in benthic habitats and are common in the shallow waters including littoral zones of lakes and ponds, as well as many wetland habitats including temporary pools. Gastrotrichs lay two types of eggs, tachyblastic and opsiblastic eggs; the latter has a heavier shell and can withstand drying, heat, and freezing for years, allowing them to persist in ephemeral habitats. Gastrotrichs are among the few animals that can withstand extended anoxia and concomitant exposure to sulfide (Strayer and Hummon, 1991). Gastrotrichs feed on bacteria, protozoa, algae, and detritus. Most genera have cosmopolitan distributions.

Phylum Rotifera

About 2,000 species of rotifers occur in freshwaters, and members of the phylum generally have cosmopolitan distributions. Rotifers can be found in all freshwater habitats. They are more diverse in fresh than marine waters, and some species inhabit saline lakes (Wallace and Snell, 1991).

Rotifers are small (60–250 µm long) and distinguished by a ciliated head region (corona) that moves water in a circular fashion. Rotifer body forms range from worm-like to vase-shaped and some are colonial (Figs. 10.3A–10.3E). They have a well-developed digestive system that includes a mastax to grind food, a stomach, an intestine, and an anus. Rotifers have a small brain (around 15 cells) and eyespots that allow them to respond to environmental stimuli with moderately complex behavior. Rotifers can also have a "foot" that extends ventrally and several appendages called "toes" that can be used for movement or attachment.

Some rotifers reproduce only sexually, others asexually, and yet others mostly by asexual parthenogenesis with occasional sexual reproduction. Amictic generations reproduce without any recombination of DNA. Sexual reproduction often results in formation of a thick-walled, highly resistant resting egg. Thus, sexual reproduction may be a strategy to avoid undesirable environmental conditions. It is noteworthy that the bdelloid rotifers are the largest metazoan taxon completely without sexual reproduction, and males have never been seen (Welch and Meselson, 2000). This verifies that sexual reproduction is not a requirement for multicellular animals to maintain stable species over the tens of millions of years the bdelloid rotifers have been a distinct group. The bdelloid rotifers are capable of horizontal gene transfer so evolution can occur in the absence of sexual reproduction (Gladyshev *et al.*, 2008).

Several species of rotifers are adapted to drying, and in some cases can survive decades of desiccation and revive within minutes to hours after rewetting. Such species are important animal components of temporary waters and are also well adapted for dispersal.

Many species of rotifers are omnivorous filter feeders. These species use their cilia to actively filter large volumes of water. Some benthic species act as "sit and wait" predators that engulf prey when it swims near (Wallace and Snell, 1991). Not all captured particles are consumed; rotifers feed selectively and reject unsuitable food particles. Rotifers are important consumers of bacteria in aquatic habitats and are therefore important links between the microbial loop and higher trophic levels. Rotifers are a major food source for zooplanktivores, including many small crustaceans and insects, but they are generally too small for planktivorous fish to capture. Some species produce long spines in response to predation risk; consequently, body form can vary considerably (Fig. 10.4). Other species have adaptive behavioral responses to avoid predators such as making rapid jumps when touched.

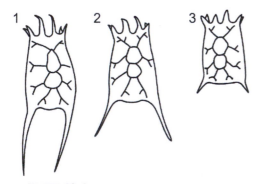

FIGURE 10.4

Change in body form (cyclomorphosis) of the planktonic rotifer *Keratella quadratica* in successive generations in laboratory culture. *(Reproduced with permission from Hutchinson, 1967).*

Phylum Nematoda

Nematodes or roundworms are found in freshwaters, soils, and marine habitats. Probably because of their small size and complex taxonomy, they have not received much attention in freshwater habitats, but their importance in freshwaters and all other habitats is undisputed. Two thousand freshwater species were reported in 1978 (Pennak, 1978), about 7% of the total number of species (Abebe *et al.*, 2008). Nematodes are usually benthic but can be found in most aquatic habitats, including those as extreme as hot springs and snowmelt pools (Poinar, 1991). Nematodes often reach densities of 1 million m^{-2} in freshwater habitats.

The nematodes are nonsegmented, worm-like, cylindrical in cross-section, and possess a complete alimentary tract and a body cavity (Fig. 10.3K). They also have a well-developed nervous system, excretory system, and musculature. Most freshwater nematodes are less than 1 cm in length. Reproduction is parthenogenic in some species, and sexual reproduction can also occur.

A wide variety of feeding strategies occur among the nematodes. Some feed on detritus, some feed on algae, many feed on aquatic plants, and some species are carnivorous. Predatory nematodes may be the biggest consumers of other species of nematodes in their natural habitats. Nematodes can be important parasites of humans, other animals, and plants. In freshwater habitats, many species are endoparasites in other invertebrates as well as vertebrates. Species that are parasitic on humans use mosquitoes, black flies, or other dipterans as intermediate hosts. The nematode that causes river blindness (onchocerciasis) is transmitted by black flies (Simuliidae); the World Health Organization estimated that in 2008, 17.7 million people were infected, 270,000 were blind, and about a half million had visual impairment, most living in Africa.

Phylum Nematomorpha

Members of the Nematomorpha are known as horsehair worms or gordian worms (Fig. 10.3L) and are parasites. Some species are parasitic on humans, but invertebrates and other vertebrates serve as hosts. The free-living adults are several centimeters to 1 m long and about 3 mm wide. They often appear like thin sticks or thick hairs and nonliving until they start to move. Their typical life cycle includes adults that reproduce sexually and lay eggs. The eggs hatch, and the larvae then encyst on vegetation where they may be ingested by a host animal (commonly an invertebrate, although not always an aquatic host; grasshoppers and crickets are common hosts). Sometimes, this host is engulfed by another predator, and the parasitic larva infects the predator. The host must be in contact with water for the mature adult to emerge from the body cavity. At this stage of development, the horsehair worm can be almost as large as the host. The parasite may alter the behavior of the host and increase the chance that it ends up in water (Thomas *et al.*, 2002). Although reproductive, the adult never ingests food and has no functional digestive tract.

Phylum Mollusca

The freshwater mollusks include two classes, the Gastropoda (snails and limpets) and the Bivalvia (clams and mussels). Mollusks are widespread, conspicuous, and often abundant. They are soft-bodied, unsegmented animals. Their body has a head, a muscular foot, a visceral mass, and a mantle that often excretes a calcareous shell. They form an important part of the biodiversity and food webs of many aquatic ecosystems.

The gastropods constitute the most diverse class of the phylum Mollusca, with about 75,000 species of marine and freshwater snails worldwide. Freshwater snails are very diverse in North America, with about 500 species present. The gastropods have a univalve (one-piece) shell and a file-like radula that is used to scrape surfaces while feeding (Brown, 1991). Shell geometry can range from simple and conical as in the freshwater limpets (Fig. 10.5F); to spiral and flat (*Planorbella*; Fig. 10.5E) or spiral and elevated (Fig. 10.5D). Most snails reproduce sexually, but a few are parthenogenic. Many freshwater species are hermaphroditic; hermaphroditic species usually reproduce sexually by exchanging sperm with other individuals, although self fertilization has been documented in some. Reproductive activity ranges from once per year to continuous, depending on the species and the geographic region.

Snails generally feed on detritus, periphyton and biofilms, macrophytes, and occasionally carrion. They usually prefer periphyton to macrophytes (Brönmark, 1985), and they can be major consumers of periphyton in aquatic systems. In the process of grazing, snails can greatly alter the composition and structure of periphyton assemblages and can even stimulate growth by reducing light limitation and excreting nutrients and mucus that fertilize periphyton and biofilms (Steinman, 1996). Despite the protection of their shell, there are many important predators of snails. Some, such as crayfish and some species of sunfish, crush the shells. Other predators, such as leeches, flatworms, and some aquatic insect larvae, invade the shells (Brown, 1991). Because shells get thicker and stronger as snails grow, predation by predators that crush the shells, such as crayfish, is often size-dependent, which makes smaller individuals more vulnerable (Crowl and Covich, 1990). Freshwater snails are intermediate hosts for some important parasites such as the human lung fluke (*Paragonimus*) and blood flukes (*Schistosoma*).

The freshwater bivalves are characterized by a shell with two halves and enlarged gills with long ciliated filaments used for filter feeding (McMahon, 1991). Most of the native North American bivalves burrow in sediments. The bivalves can be found in the benthic zone of streams, lakes, and rivers. Several bivalve species have recently invaded North America, including the Asiatic clam *Corbicula fluminea* (Fig. 10.5B), the zebra mussel *Dreissena polymorpha* (Fig. 10.5C), and the quagga mussel *Dreissena bugensis*. The zebra mussel has caused significant economic and ecological damage (Sidebar 10.1).

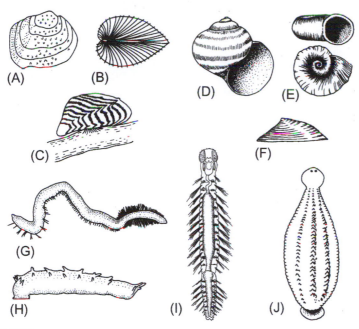

FIGURE 10.5

Some representative mollusks (A–F) and annelids (G–J). Organisms shown and their approximate lengths are as follows: (A) a unionid mussel, *Quadrula*, 7 cm; (B) the Asiatic clam, *Corbicula*, 3 cm; (C) a zebra mussel, *Dreissena* on a stick, 3 cm; (D) *Pomacea*, 4 cm; (E) *Planorbella*, 3 cm; (F) the freshwater limpet, *Ferrissia*, 4 cm; (G) *Branchiura*, 10 cm; (H) *Ceratodrilus*, 3 mm; (I) *Aeolosoma*, ~6 mm; (J) *Placobdella*, 16 mm. (A–H and J reproduced with permission from Thorp and Covich 2001; I reproduced with permission of Smith, 2001).

SIDEBAR 10.1
Invasion by the Zebra Mussel

The zebra mussel was established in the Volga drainage in Europe from its native Caspian Sea drainages before 1800. It spread from the Volga throughout most of the major river systems in Europe through the canal transportation system following introduction to some far western European countries in the early 1800s (Hutchinson, 1967). This invasion represents the first major expansion of its range caused by human activity. In 1986, a ship with water from a European port, probably taking on cargo at the St. Claire River, dumped ballast water and released the zebra mussel into North America. Since that time, the population has increased its distribution to cover much of the Mississippi drainage (color plate Fig. 7). This explosive spread has occurred partly because females can produce more than 1,000,000 eggs each reproductive cycle that give rise to

FIGURE 10.6
Zebra mussels foul a current meter that has just been removed from a lake. *(Image courtesy of the US National Oceanic and Atmospheric Administration).*

easily transportable veliger larvae (de Vaate, 1991; Sprung, 1993) and partly because of the many ways the mussel can be transported. Carlton (1993) reports two natural transport methods (currents and animals other than humans), and 20 human-caused movements, including water traffic, fisheries activities, and navigation. Ultimately, this species probably will spread through much of North America (Strayer, 1991). These mussels attach tightly to any solid surface with byssal threads, can reach tremendous densities (Fig. 10.6), and overgrow native species. This invader has many potential effects on food webs and ecosystems (Table 10.1).

The Hudson River provides a good case study for examining the effects of the zebra mussel on ecosystem properties (Strayer *et al.*, 1999). The zebra mussels were first observed in the Hudson River in 1991 and by 1993, densities were high enough that the mussels filtered the entire water column every 1.2 to 3.6 days. The adult mussels are filter feeders and can remove suspended particles

Table 10.1 Some Community, Food Web, and Ecosystem-Level Effects of the Zebra Mussel, *Dreissena polymorpha*

Effect	Citation
Lower phytoplankton biomass, greater water clarity	Makarewics *et al.*, 1999; Holland *et al.*, 1995; Fahnenstiel *et al.*, 1995a; Heath *et al.*, 1995
Decreased dissolved oxygen from mussel metabolism	Effler *et al.*, 1998; Caraco *et al.*, 2000
New food source for waterfowl; effective conduit for biomagnification of organic toxicants	Mazak *et al.*, 1997
Increased variability in phytoplankton biomass and phosphorus concentrations	Mellina *et al.*, 1995
Increased abundance and depth of macrophytes related to decreased turbidity	Skubinna *et al.*, 1995
Declines in native unionid bivalves and clams	Nalepa, 1994; Strayer *et al.*, 1998
Increases in periphyton biomass and productivity related to increased water clarity, about equal to phytoplankton decreases	Lowe and Pillsbury, 1995; Fahnenstiel *et al.*, 1995b
Increased numbers of small planktonic heterotrophic bacteria, decreased numbers of predatory protozoa	Cotner *et al.*, 1995; Findlay *et al.*, 1998
Increased nitrogen regeneration rates	Gardner *et al.*, 1995; Holland *et al.*, 1995
Decreases in planktonic protozoa	Lavrentyev *et al.*, 1995
Moderate effect on cladoceran grazing rates in open, deep waters of a large lake	Wu and Culver, 1991

including phytoplankton, small zooplankton, and detritus. Amounts of all these suspended particles decreased concomitantly with mussel increases. This consumption ultimately may impact fish by lowering the amount of food available to animals that feed on suspended particles and serve as food for piscivorous fish. The increase in zebra mussels has been accompanied by a decrease in native bivalves, particularly native unionid mussels. Two of the species of unionid mussels will likely be extirpated in the Hudson River. Dissolved phosphorus and water clarity have increased since the invasion of the mussels, probably leading to increases in macrophytes. The various ecosystem effects can be viewed as positive or negative. For example, an increase in water clarity may be good, but the effects are obviously disastrous for some native species.

Zebra mussels have caused chronic problems in water intakes in both Europe and the United States (Kovalak *et al.*, 1993). Layers of attachments up to 30 cm thick can clog pipes and screens. Removal efforts are time-intensive and costly, usually involving dewatering (shutting down and drying) water systems and cleaning with high-pressure water hoses (Kovalak *et al.*, 1993). Chemical controls such as treatment with chlorine and other toxic chemicals are effective in confined areas (e.g., water pipes), but toxic when released to the environment. The economic impact was estimated to be $5 billion in the Great Lakes region alone by the year 2000 (Ludyanskiy *et al.*, 1993), and recent surveys put actual costs at a minimum of $10 million per year (O'Neill, 1997). Biological controls are being investigated, but no inexpensive method is available that has been widely adopted. Biological controls may be somewhat effective; crayfish (Perry *et al.*, 1997), waterfowl (de Vaate, 1991), and some fish prey upon zebra mussels (in particular introduced species, but also native drum, sunfishes, and redhorses; French, 1993). There is no good way currently to control zebra mussels at the ecosystem level.

FIGURE 10.7
A red-eye bass (*Micropterus coosae*) "attacking" the lure of a freshwater mussel (*Lampsilis cardium*).
(A) View of the gravid gill that serves as a lure. After the fish bites the mantle (C), the glochidia are
released in a cloud and the fish rapidly leaves (D). The mussel is about 6 cm long. *(From Haag and
Warren, 1999; images courtesy of Wendall Haag).*

Unionid mussels are the most diverse group of bivalves found in freshwaters
of North America (Fig. 10.5A). Unionids are unique in that they produce specialized larvae called glochidia, which are important for dispersal. These larvae attach to host fish species, mainly in the gill region (but some attach to
fins), and encyst for 6 to 60 days. After encystment, the larvae drop off the
host, settle into the substrata, and develop into sedentary adult forms with
shells. Some adult female mussels attract potential hosts with elaborate lures
that are muscular extensions of the mantle. These lures may look like small
fish or other probable prey items, and when potential host fish attack them
the glochidia are released forcibly (Fig. 10.7). The unionid mussels are also
unique because of their long life span. Individuals can live from 6 to 100
years; other bivalves live less than 7 years. The long life span and infrequent
reproduction make unionids vulnerable to human impacts on streams. Of the
297 native species and subspecies of North American mussels, 19 are extinct,
62 are federally listed in the United States as endangered or threatened, and
130 need further study to determine their conservation status. These species
are linked to many ecological processes through their role as filter feeders
(Vaughn and Taylor, 1999). Conservation of mussels and other species is discussed in Chapter 11. At one time, unionids were harvested in large quantities

FIGURE 10.8
Pearl shell buttons and a *Megalonaias nervosa* mussel shell from the Mississippi River that was drilled for buttons. *(Photograph by J. W. Grubaugh).*

for the button industry (Fig. 10.8), and they are still the basis of a historically important pearl and shell fishery that continues at modest levels.

Phylum Annelida

Annelids are segmented worms with a tubular body and a specialized digestive system with a terminal mouth and anus. Their body cavity has thin transverse septa that delineate the segments. They generally reproduce sexually by cross-fertilization and are often hermaphroditic, but many also reproduce asexually by budding. The freshwater annelids include the oligochaetes, the leeches, and several other less diverse groups.

Aquatic oligochaetes (Fig. 10.5G) are similar to their terrestrial analogs (earthworms). They usually have four bundles of chaetae (hairs) on each segment (Brinkhurst and Gelder, 1991). Most of these worms burrow through sediment in lotic and lentic habitats and ingest organic particles, but some are important algal feeders or predators. The aquatic oligochaetes have a thinner muscular layer around their body, so fisherman use terrestrial oligochaetes (earthworms) for bait more often as they will not disintegrate on a hook. Some oligochaetes, particularly the tubificid (Naididae) worms such as *Tubifex*, are highly resistant to low O_2 concentrations and high levels of organic pollution. Thus, they are used as indicator species of polluted waters and can be components of biotic indices to assess ecosystem health.

Oligochaetes can also be vectors for important parasites such as whirling disease, which infects trout in North America, Europe, South Africa, and New Zealand.

Members of the annelid class Polychaeta are generally found in marine habitats, where they can be very abundant and tremendously diverse in size and form. In contrast, only a few species of polychaetes are found in freshwater habitats, including *Aelosoma* (Fig. 10.5I). Members of the genus *Aelosoma* are found in freshwater habitats throughout the world, including open water habitats of large lakes, stream sediments, and groundwater habitats, where they feed on fine organic particles. Many polychaetes are predators or omnivores, but some have specialized appendages for filter feeding.

Leeches (Hirudinae) are mostly predators that feed on midge larvae, amphipods, oligochaetes, and mollusks, but some are parasites that feed on the blood of vertebrates, including humans. Species of leeches that use blood are being investigated for the pharmacological value of the anticoagulants used during feeding (Davies, 1991; Salzet, 2001). Leeches are distinguished by several characteristics, including dorsal–ventral flattening (Fig. 10.5J), an oral sucker and usually a posterior sucker, usually 34 true body segments, and a muscular body. Leeches are commonly found in shallow, warm waters, including slow-moving streams and rivers, lakes, and wetlands (but generally not acid peat bogs). Some species live in moist terrestrial habitats and can be quite large, reaching nearly 30 cm in length.

Leeches can sense moving animals in the water and swim quite actively toward them. When parasitic leeches attach to prey for a blood meal, they attach with the posterior sucker and explore for a suitable feeding spot with the anterior end. The oral sucker is then attached, three painless cuts are made with the jaws, and anticoagulants are injected. The leech eats its fill and then drops off the host. Frequent meals are not necessary, and specimens have been kept alive without feeding for more than 2 years (Smith, 2001).

Members of one annelid order, the Branchiobdellida (sometimes called crayfish worms; Fig. 10.5H), are exclusively commensals and parasites of crustaceans, particularly crayfish. Branchiobdellids range from 1 to 10 mm in length, and an individual crayfish can harbor hundreds of individuals representing multiple species. Most branchiobdellids feed on organic materials and smaller organisms on the surface of their host, or food particles from the host's feeding, and actual parasitism in this group is rare.

Bryozoans: Phyla Entoprocta and Ectoprocta

Bryozoans are generally *sessile* (attached to substrata) colonial invertebrates that use ciliated tentacles to capture suspended food particles. This group

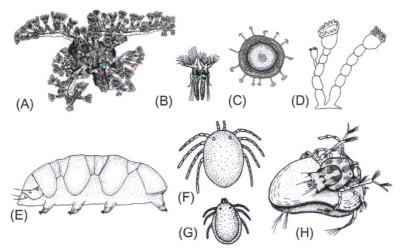

FIGURE 10.9

Some representative bryozoans and structures (A–D), a tardigrade (E), and water mites (F–H). Organisms and structures shown and their approximate lengths are as follows: (A) *Hyalinella* colony, ~2 cm wide; (B) *Hyalinella* zooids, 2.5 mm; (C) bryozoan statoblast, 1.4 mm; (D) *Urnatella* colony, 5 mm; (E) heterotardigrade, 0.3 mm; (F) generalized adult water mite, 1 mm; (G) generalized larval water mite, 0.5 mm; (H) a water mite feeding on an ostracod. *(A, B, and H reproduced with permission of Smith, 2001; C–G reproduced with permission from Thorp and Covich, 2001).*

is primarily marine, with more than 4,000 species worldwide, about 50 of which are freshwater species (Pennak, 1978). Bryozoans are composed of many individual zooids, each of which is approximately tubular and has a crown of tentacles (Figs. 10.9A–10.9D). This group actually includes two distantly related phyla, the Entoprocta (Fig. 10.9D), which have zooids with external segmented stalks and lack a coelom, and the Ectoprocta (Figs. 10.9A and 10.9B), in which external stalks are absent and a coelom is present. Bryozoans can reproduce sexually or asexually through budding and fission, and individual zooids are hermaphroditic. Statoblasts (Fig. 10.9C), the small, resistant structures of the Ectoprocta that are formed through budding, are important for dispersal and surviving harsh conditions. The Bryozoa are generally restricted to warm water and can be found in both still and running waters (Wood, 1991). They require solid substrata such as rocks or wood for attachment. The colonies can reproduce asexually by formation of encapsulated dormant buds and can also reproduce sexually once a year. The bryozoa are fairly resistant to predation. Large, gelatinous *Pectinatella* colonies, which can exceed the size of a softball, are a common and spectacular sight in some lakes and reservoirs in the southeastern United States, where they are sometimes called "dinosaur snot." Bryozoans can be very abundant in lentic habitats, and in some cases can be a nuisance because they clog pipes in water treatment facilities.

Phylum Tardigrada

Water bears (tardigrades) are microscopic animals with a cosmopolitan distribution (Nelson, 1991); about 350 species are reported worldwide. They have a bilaterally symmetrical body with four body segments, each with a pair of legs (Fig. 10.9E). Most adults are tiny, ranging from 250 to 500 μm long. Some tardigrade species reproduce through parthenogenesis, while others have sexual reproduction; many are hermaphroditic.

Tardigrades are well known for their ability to withstand drying, even for periods up to 4 to 7 years with no apparent biological activity, a characteristic related to their unusual habitat preferences. The tardigrades are rarely found in the plankton; aquatic species mainly inhabit benthic habitats in which they can reach very high densities, notably in the capillary water in wet sand of beaches. Most species are semi-aquatic and are associated with droplets of water and water films on terrestrial mosses and liverworts. They are some of the only animals found in extreme habitats, such as Arctic and Antarctic lakes, streams, and ice sheets (McInnes and Pugh, 1998; Pugh and McInnes, 1998).

Phylum Arthropoda

Arthropods are one of the most successful groups of organisms on the planet and they are well represented in all continental surface waters. They are important components of aquatic biodiversity and central to ecosystem function. They are characterized by a chitinous exoskeleton and stiff-jointed appendages (including legs, mouthparts, and antennae). There are three subphyla common in freshwaters: the Chelicerata (class Arachnida—water mites and aquatic spiders), the Uniramia (insects and Collembola), and the Crustacea (e.g., crayfish, shrimp, amphipods, isopods, mysids, fairy shrimp, Cladocera, and Copepoda).

Class Arachnida

Water mites are a taxonomically diverse and poorly studied group including several families. More than 5000 species of water mites have been described worldwide; about 1,500 are estimated to occur in North America, but only half of them are named. They can be extraordinarily diverse, with as many as 75 species representing 25 genera m^{-2} in plant beds in eutrophic habitats, and 50 species from 30 genera in a single stream riffle (Smith and Cook, 1991).

Water mites have a mouth region and a body that has a fused cephalothorax section (head and thorax) and an abdomen (Figs. 10.9F–10.9H). Six pairs of appendages are present; the last four pairs, the legs, are the most conspicuous. The appendages can have setae, or spines, which are used as characteristics for identification.

Water mites inhabit a variety of benthic habitats, including springs, riffles, interstitial habitats, lakes (with mostly benthic but a few planktonic forms), and temporary pools. The majority of water mites are carnivorous or parasitic (primarily on aquatic insects). More sedentary species may feed on carrion or possibly detritus.

Among the true spiders, no North American species are completely aquatic. However, several species live near water and are able to run on the water surface and even dive beneath it. Fishing spiders, members of the genus *Dolomedes*, are common in and along lake and stream margins where they are active predators on aquatic insects and small vertebrates such as small fish and amphibians. They trap air in fine hairs on their bodies and subsequently can remain submerged up to 30 minutes. The European water spider *Argyroneta aquatica* (Clerck) spends its entire life under water. These spiders keep an air bubble around their body for breathing and build silk underwater air containers. These containers are used for shelter and nests (Schütz *et al.*, 2007). Many spiders, particularly members of the genus *Tetragnatha* (long-jawed spiders) build their webs close to water on emergent vegetation and other structures to catch adult aquatic insects. Spiders can be important consumers of insects in riparian habitats and wetlands.

Subphylum Insecta and Collembola (Uniramia)

Many insect orders contain aquatic or semi-aquatic species (Merritt *et al.*, 2008). Insects are found in most freshwater habitats, where they are arguably the best studied group of freshwater invertebrates. The majority of the aquatic insects spend most of their immature lives in the water, and the adults emerge from the aquatic environment to mate and disperse. The insects have three major body regions (head, thorax, and abdomen), one pair of antennae, compound eyes (as adults), and specialized mouthparts. The thorax has three segments, each with a pair of legs, and each leg is divided into five parts. The immature forms of the aquatic insects are variously referred to as *larvae*, *nymphs*, or *naiads*. The term larva is generally applied to immature forms or groups with complete metamorphosis, and nymph is generally used to describe immature stages of those with incomplete metamorphosis, although opinions on this terminology vary among zoologists. Some characteristics of the orders of insects with aquatic or semi-aquatic species are summarized in Table 10.2. Each of the orders will be discussed in turn.

The Collembola, or springtails, are no longer considered insects by most taxonomists. They are small, eyeless, and wingless arthropods (usually less than

Table 10.2 Characteristics of Orders of Aquatic Insects and Collembola

Order	Common Name	Approximate Number of Aquatic and Semi-Aquatic Species	Primary Habitats	Functional Feeding Groups
Collembola	Springtails	50	Lentic, shallow	Collectors
Ephemeroptera	Mayflies	2,250	Lentic and lotic	Scrapers, collectors, few predators
Odonata	Dragon flies and damselflies	5,500	Lentic and lotic	Predators
Orthoptera	Grasshoppers and crickets	~100	Mostly lentic	Shredders
Plecoptera	Stoneflies	2,140	Lentic mainly	Shredders, collectors, predators
Trichoptera	Caddisflies	7,000	Lentic and lotic	Predators, scrapers, collectors
Megaloptera and Neuroptera	Fishflies, alderflies, and spongillaflies	300		Predators
Hemiptera	Bugs, plant hoppers, and others	3,800	Lentic and lotic	Predators
Lepidoptera	Aquatic caterpillars	100	Lentic and lotic	Shredders, scrapers
Coleoptera	Water beetles	5,000	Lentic and lotic	Predators, scrapers, collectors, shredders
Hymenoptera	Ants, bees, and wasps	100	Lentic and lotic	Parasites
Diptera	Flies and midges	>30,000	Lentic and lotic	Predators, scrapers, collectors, shredders

6 mm long) that possess a characteristic ventral tube (collophore) that functions in respiration and osmoregulation and can also be adhesive (Fig. 10.10A). The name springtail refers to the ability of some collembolans to jump using a specialized bifurcate structure on the abdomen called a furcula. They differ from insects by having only six abdominal segments and the mouthparts are withdrawn into a pouch in the head capsule. Most species are terrestrial or semi-aquatic and occur in lentic habitats, but they can also be abundant along stream margins. Collembola are commonly found in the neuston, where their hydrophobic integuments allow them to rest on the surface tension. Their biology and ecology in freshwaters are not well known, but they are generally considered primary consumers that feed on detritus, algae, and biofilms.

Mayflies (Ephemeroptera) are often dominant components of the aquatic invertebrate communities of streams (where they can provide an important food source for fishes) and in the benthic zone of some lakes. Mayfly nymphs are distinguished from other aquatic insects by long filaments on their posterior end (generally three) and the presence of conspicuous gills on the abdominal segments (Figs. 10.10B and 10.10C). The abdominal gills of mayfly nymphs range from plate-like, to feathery, to long filaments. Another unique

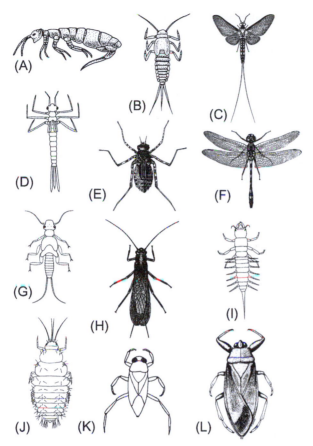

FIGURE 10.10

Some representative aquatic insects. (A) a semi-aquatic springtail (Collembola), 1 mm; (B) *Baetis* mayfly nymph (Ephemeroptera), 1 cm; (C) adult *Hexagenia* mayfly (Ephemeroptera), ~3 cm; (D) a damselfly larva (Odonata), *Calopteryx*, 1 cm; (E) a dragonfly nymph, *Macromia* (Odonata), ~5 cm; (F) an adult dragonfly, *Macromia*, ~8 cm; (G) a stonefly nymph, *Isoperla* (Plecoptera), 0.7 cm; (H) an adult stonefly, *Clioperla*, ~3 cm; (I) an alderfly larvae, *Sialis* (Megaloptera), 2 cm; (J) a spongilla fly larvae, *Climacia* (Neuroptera), 0.5 cm; (K) an adult backswimmer, *Notonecta* (Hemiptera), ~2.5 cm; (L) an adult giant water bug, *Lethocerus* (Hemiptera), ~7 cm. *(A, B, and J reproduced with permission from Thorp and Covich, 2001; C, E, F, H, and L reproduced with permission from Borror et al., 1989; D, G, I, and K reproduced with permission of Hilsenhoff, 1991).*

feature of mayflies is the subimago stage. This "extra step" in the mayfly life cycle follows the larval stage and precedes the adult stage. The subimago stage is the only winged pre-adult stage known in insects. Most mayfly nymphs feed on periphyton or biofilms, but some filter organic particles from the water and a few are predaceous. Mayfly nymphs generally crawl about on the substratum, but some species are rapid swimmers and others construct burrows. Mayflies are diverse in well-oxygenated, unpolluted streams and are often used as indicators of good water quality and ecosystem health. Adult mayflies do not feed and live for only a few days while they attempt to reproduce. Massive emergences of millions of mayfly adults in spring and summer in some regions are a spectacular site; densities can be so high that they stop traffic and are detected with Doppler weather radar.

The Odonata (dragonflies and damselflies) are fantastic predators as aquatic nymphs and terrestrial adults. About one-third of the larvae are lotic and two-thirds are lentic (Hilsenhoff, 1991). The nymphs can be distinguished from other aquatic insects by the long, hinged labium that has been modified to eject rapidly and seize moving prey. Nymphs have large compound eyes and short antennae (Figs. 10.10D–10.10F) and move about the substrata by crawling; some stalk their prey, which can include small fish and tadpoles along with other invertebrates. To escape predators, nymphs can forcefully eject water from the rectum as a form of jet propulsion. Adults are able to fly at 25 to 35 km per hour, are significant predators of mosquitoes and other insects, and are viewed as beneficial insects. Adult odonates are some of the most familiar aquatic insects and many have beautiful colors and patterns (Fig. 10.10F).

The order Orthoptera (grasshoppers, crickets, and others) does not include any truly aquatic forms, but some species are adapted for a semi-aquatic existence. Semi-aquatic crickets and grasshoppers are generally either found in association with the emergent aquatic plants that they feed on, or some are associated with damp marginal substrata. Many have the ability to swim on the surface of the water. Some tropical forms can swim under the water surface, remain submerged for a few minutes, and lay their eggs in underwater portions of plants.

The Plecoptera (stoneflies) are important in streams as food for fish and other vertebrates. Some species are predators of other invertebrates, while others are important leaf shredders, which facilitate the decomposition and recycling of materials. Stoneflies are associated primarily with pollution-free, cool, highly oxygenated running waters. This preference for clean habitats has led to their use in biotic indices for stream water quality and biotic integrity. The nymphs are somewhat similar in body form to mayflies, but can be distinguished from other aquatic insect nymphs by the presence of two long cerci (appendages) on the posterior end of the abdomen and their elongate, often flattened bodies (Figs. 10.10G and 10.10H). The gills of stonefly nymphs are generally less conspicuous than those of mayflies or, if they are obvious, they usually appear as small tufts or finger-like projections.

The order Hemiptera includes the suborders Heteroptera (true bugs; Figs. 10.10K and 10.10L) and Homoptera (plant hoppers, aphids, and others), most of which are terrestrial. About one-third of the aquatic or semi-aquatic species live on the water surface and two-thirds live in the water. All breathe air, so those that live underwater generally use siphons or air bubbles to breathe. This group is well represented in lentic and lotic habitats, particularly in highly vegetated areas, and they are tolerant of low dissolved oxygen concentrations in the water. The Hemiptera are distinguished by mouthparts that are modified to form a sucking and piercing beak, which is used to pierce plant tissues in herbivorous forms or feed on the blood of animals in carnivorous species. This group includes some of the most familiar aquatic insects, including the giant water bugs and water scorpions (big enough to prey on small fish and tadpoles), the water boatmen (often seen in shallow littoral zones), and the gerrids (water striders).

The order Neuroptera used to include what is now considered a separate order, the Megaloptera (fishflies, dobsonflies, alderflies, and hellgrammites). Neuropterans are mostly terrestrial, except for spongillaflies, which live in and feed on freshwater sponges (Fig. 10.10J). Megalopteran larvae have seven or eight pairs of lateral filaments and large mandibles (Fig. 10.10I). All megalopteran larvae are predators, mostly on other invertebrates. Larval forms with abdominal gills generally inhabit cool, oxygen-rich streams, but some species with respiratory siphons are found in wetlands and other lentic habitats. Some megaloptera larvae (*Corydalus*, the hellgrammites) can be more than 6 cm long and live for 2 to 5 years before emerging.

Caddisflies (Trichoptera; Figs. 10.1 and 10.11D–10.11F) are best known for their ability to build cases (Fig. 10.12), retreats, and nets that they construct using silk and materials from the environment such as sand grains or parts of leaves. Some species are free living, highly mobile, and lack cases or nets. Most species occur in running waters, but some occur in lakes and wetlands. Along with the mayflies and stoneflies, a diverse caddisfly community indicates a clean stream or river. The free-living caddisflies are mostly predators, those that build cases are primarily herbivores on periphyton or eat dead leaves, and species that spin nets are filter feeders. The net-spinning caddisflies use stream flow to filter organic materials from the water column onto their nets. Adult caddisflies look like moths, which reflects the close relationship between caddisflies and the Lepidoptera (butterflies and moths), along with the ability of the larvae to produce silk. Like mayflies, caddisflies can reach very high densities in some freshwater habitats. Although massive emergences of adults can be a nuisance to humans, emergences can signal anglers that fish may be actively foraging.

The order Lepidoptera (butterflies and moths) is not diverse in freshwaters, but they can be abundant in some stream and lentic habitats. Those that live in streams generally graze on diatoms and periphyton, whereas those that live in lakes and wetlands often feed on aquatic vascular plants and may bore

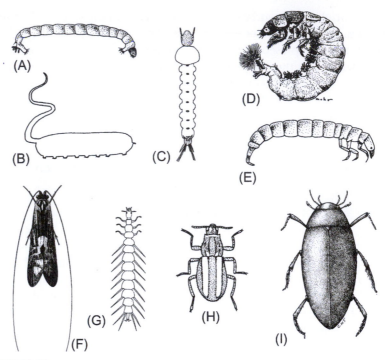

FIGURE 10.11
Some representative aquatic insects. (A) a midge larva, *Chironomus* (Diptera) 0.5 cm; (B) a syrphid fly larvae, *Eristalis* (Diptera) with respiratory siphon extended, ~5 cm; (C) a mosquito larvae, *Anopheles* (Diptera), ~1.5 cm; (D) a hydropsychid caddisfly larvae, *Hydropsyche* (Trichoptera), ~2 cm; (E) a caddisfly larva, *Polycentropus*, 1.5 cm; (F) an adult caddisfly, *Macronemum*, ~2.5 cm; (G) a whirligig beetle larva, *Dineutus* (Coleoptera), 1 cm; (H) a riffle beetle *Stenelmis*, 0.5 cm; (I) a hydrophilid beetle, *Hydrophilus* (Coleoptera), ~2.5 cm. *(A–C, E, and H reproduced with permission from Thorp and Covich, 2001; D, F, and I reproduced with permission of Borror et al., 1998; G reproduced with permission of Hilsenhoff, 1991).*

into stems and leaves. Larvae are similar in form to terrestrial caterpillars, but some forms that live in streams have prominent filamentous gills. Like the caddisflies, this group has the ability to produce silk, which is used to build retreats in stream-dwelling forms and cases in lentic forms. Like their terrestrial counterparts, aquatic caterpillars pupate and emerge as aerial adults.

The Coleoptera (beetles) species with aquatic larval and/or adult stages (Figs. 10.11G–10.11I) represent only about 3% of this mostly terrestrial order. However, there are so many species of beetles in the world that they are significant components of the biodiversity of both lentic and lotic habitats. The group includes the riffle beetles (Elmidae, with aquatic larvae and adults), the water pennies (Psephenidae larvae, attached to rocks), the whirligig beetles

(Gyrinidae, with adults found on the surface), and the predaceous diving beetles (Dytiscidae) and water scavenger beetles (Hydrophilidae), both of which have aquatic larvae and adults.

Some adult Coleoptera, along with some Hemiptera, transport air bubbles or air films (plastrons) underwater by means of specialized hairs. Underwater, these air bubbles can function as a physical gill, extracting dissolved oxygen from the surrounding water as oxygen levels decline in the bubble. The aquatic larvae of some Curculionidae and Chrysomelidae are equipped with specialized spines that pierce plant tissues and extract dissolved oxygen from submerged portions of emergent macrophytes. Because of the many adaptations for using atmospheric oxygen, coleopterans are often abundant and diverse in the poorly oxygenated waters of wetlands. Many coleopterans are important predators, whereas others are adapted to scrape periphyton, graze aquatic plants, or use detritus.

The order Hymenoptera (bees, wasps, ants) does not include any truly aquatic species, but some small wasps parasitize aquatic insects, and in some cases the adult wasp actually swims or crawls underwater to lay eggs in the host. Most "aquatic" Hymenoptera parasitize larval or pupal stages of aquatic insects such as aquatic flies, caterpillars, and caddisflies. The nearly featureless larvae are endoparasites in the host, and like their terrestrial counterparts, most are fairly host-specific. One spider wasp (family Pompilidae) specializes on the semi-aquatic spider, *Dolomedes*. These wasps can crawl underwater, where they hunt the spiders as they crawl on the surface or dive. Once stung, the paralyzed spider is carried to a nest on the shore where the female wasp lays eggs on it (Bennett, 2008).

Flies and midges (Diptera) constitute a large group, to which about 40% of all aquatic insects belong (Hilsenhoff, 1991). This group is dominated by the family Chironomidae (Fig. 10.11A), which includes about one-third of all species of aquatic Diptera, and can reach densities of well over 10,000 individuals m^{-2} in productive freshwater habitats. The Diptera also includes many aquatic larvae with adults that are nuisance species and important disease vectors, including the mosquitoes (Culicidae, Fig. 10.11C), black flies

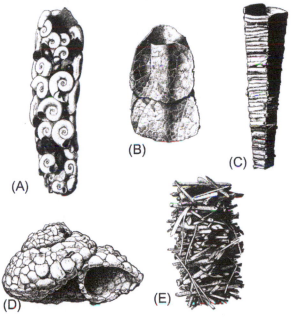

(B)

(C)

(A)

(D)

(E)

FIGURE 10.12

Caddisfly larvae cases illustrating diversity of materials used and form of construction, all about 1 cm long. (A) *Philarctus*, (B) *Clostoeca*, (C) *Brachycentrus*, (D) *Helicopsyche*, and (E) *Platycentropus*. (Reproduced with permission from Wiggins, 1995).

(Simuliidae), biting midges (Ceratopogonidae), and horseflies and deerflies (Tabanidae). Some mosquitoes are very important vectors of diseases such as Dengue fever and malaria in many parts of the world, and black flies transmit Onchocerciasis (river blindness). Although some are considered pests, this group includes members that are a primary food source for fishes, waterfowl, and other aquatic and riparian predators. This group also includes the phantom midges (Chaoboridae), which are important predators on zooplankton in many lakes. Dipterans show a great degree of diversity and adaptation to different types of aquatic habitats. The "blood worms," which are actually chironomid midge larvae, have hemoglobin in their blood that imparts a red coloration and allows them to survive in low-oxygen sediments. *Eristalis* (Syrphidae, Fig. 10.11B) larvae have elongated respiratory siphons that allow them to breathe atmospheric oxygen and thrive in anaerobic waters such as sewage lagoons. Shorefly larvae (Ephydridae) can be abundant in extreme environments, such as hot springs and saline lakes.

Subphylum Crustacea

The subphylum Crustacea is a mostly aquatic group with abundant and diverse freshwater forms. This group includes some species that are critical in food webs and ecosystem processes; cladocerans and copepods are key primary consumers in many lakes and ponds, and decapods (crayfish and others) are important omnivores in benthic food webs of many rivers, lakes, and wetlands. The Crustacea exhibit a wide diversity of body forms, but are characterized by respiration through gills or across the body surface, a hard chitinous exoskeleton that is often reinforced with calcium, two pairs of antennae, and most body segments with paired and jointed appendages. The crustacean taxonomic groups considered here are the Ostracoda, Copepoda, Branchiopoda, Decapoda, Mysidaceae, Isopoda, Amphipoda, and Bathynellacea.

The Ostracoda (seed shrimp; Fig. 10.13F) are benthic species that are covered by a carapace often composed of chitin and calcium carbonate that is modified into two shells or valves that cover the body. Ostracod shells are particularly well preserved in sediments; paleolimnologists make use of them as indicators of ancient environments and these fossils are the oldest known microfauna (Delorme, 1991). Ostracods (sometimes spelled ostracodes) are found in most aquatic habitats, including temporary pools (even in very small pools in the bracts of bromeliads), ponds, lakes, rivers, and groundwaters. Superficially, they resemble small bivalves, but the body concealed by the shells has the typical crustacean array of variously modified segmented appendages. They can be found in just about all types of freshwater habitats, including hot springs and hypersaline habitats. Ostracods mostly eat detritus and algae; a few are scavengers or predators. Ostracods rarely exceed 3 mm in

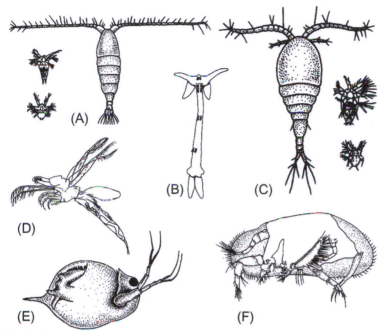

FIGURE 10.13

Some representative Crustacea. (A) calanoid copepod and nauplii, adult 0.5 mm; (B) the parasitic copepod, *Lernea*, ~7 mm; (C) cyclopoid copepod and nauplii, adult 0.5 mm; (D) the cladoceran, *Leptodora*, 3 mm; (E) *Daphnia*, 0.5 mm; (F) the ostracod, *Candona* with left valve carapace removed, 0.5 mm. *(A and C–F reproduced with permission from Thorp and Covich, 2001; B reproduced with permission of Smith, 2001).*

total length, but they can reach tremendous densities in some benthic habitats and may serve as an important food source for predators that specialize on benthic microfauna. Sexual and asexual (parthenogenesis) reproduction are evident in this group, with some species capable of both modes.

The Copepoda is a group of microcrustaceans that can be important components of zooplankton in both freshwater and marine systems, but most freshwater species are found associated with the benthic zones of streams, wetlands, and groundwaters. About 500 freshwater species have been described worldwide (Williamson, 1991). They are characterized by a cylindrical body (Figs. 10.13A–10.13C) that is 0.2 to 4 mm long, with numerous segmented appendages on the cephalothorax and abdomen. They have conspicuous first antennae and a single, simple, anterior eye. Most copepods are drab in coloration, but some forms are brilliant shades of red or orange. The copepods can be herbivorous, detritivorous, carnivorous, or parasitic. Parasitic forms such as *Lernea*, a common ectoparasite on fish, are highly modified and do not resemble free-living forms whatsoever (Fig. 10.13B).

Reproduction in copepods is usually sexual. The fertilized eggs hatch into a larval stage called a *nauplius*. Some copepods can produce eggs that are resistant to drying, and copepods are often among the first invertebrates to appear in freshly inundated habitats such as intermittent streams and wetlands. Six naupliar stages are followed by six copepodite stages (the last being the adult stage). Plankton tows from lakes often reveal numerous nauplii. Behavior of the copepods can be complex, with predator avoidance, mating behaviors, and foraging behaviors all well developed.

The Branchiopoda includes the Cladocera (water fleas) and the fairy, tadpole, brine, and clam shrimps. This order is generally composed of small crustaceans with flattened, leaf-like legs. They occur in a wide variety of freshwater habitats but usually are not abundant in flowing waters. There are about 400 species worldwide (Dodson and Frey, 1991). The fairy, tadpole, and clam shrimps are relatively large crustaceans and they are susceptible to predation by fish and large invertebrates. Many have desiccation-resistant stages that allow them to live in temporary pools where predators are scarce. Brine shrimp *(Artemia salina)* are tolerant of very saline waters and can establish dense populations in lakes that are too saline to contain fish. Brine shrimp are commonly used as food for aquarium fish and in aquaculture.

Cladocerans (Figs. 10.13D and 10.13E) are often extremely important zooplankters in lakes, and they can occur in wetlands and lotic environments as well. Thus, cladocerans are frequently studied aquatic invertebrates. Many methods have been developed to sample the cladocerans and other zooplankton from lakes (Method 10.1). Cladocerans have a small, central compound eye and a carapace that is used as a brood chamber. Adults range from 0.2 to 18 mm in length, and body forms in this group are quite variable. The thorax has four to six pairs of legs that beat continuously and create water currents that bring in algae, protozoa, and detritus. The food is filtered onto the many fine setae on the legs and then moved toward the mouth. In planktonic species, the large, second antennae are used for swimming, and the animals move with a jerky motion. The jerky motion leads to the common name "water fleas" for the common cladoceran *Daphnia*.

Reproduction in many cladocerans is parthenogenic for most of the year. Thus, females are encountered most often in nature. At times, often when conditions are suboptimal for growth, males and sexual females are produced, and sexual reproduction occurs. The fertilized eggs are encased in a resistant *ephippium* that can remain dormant until conditions once again are favorable for growth and reproduction. There are several juvenile instars that generally appear similar to adults.

METHOD 10.1

Sampling Zooplankton

A variety of zooplankton nets and sampling methods have been devised. Techniques depend on the species of interest. Zooplankton populations are patchy over space and time, so it is necessary to sample at a variety of depths in a lake to obtain adequate population estimates and to ensure that all species are captured.

For very small species such as rotifers, which are able to pass through larger nets, samples are collected with a Van Dorn or similar sampler (see Chapter 7), preserved, and concentrated by settling or use of a fine filter in the laboratory. For larger species, more water must be processed and a variety of net configurations can be used with standardized mesh sizes.

Details for obtaining a volumetric sample are presented in Wetzel and Likens (1991). One approach is to filter a volume of water through a net in the lake; another to remove a set volume of water from the lake and filter it at the surface. Metered nets often are used for obtaining quantitative zooplankton samples. These nets have a flow meter on the mouth and are towed horizontally at a set depth and then retrieved. The meter indicates the volume that was sampled. Alternatively, vertical tows can be taken. Nets are available that close when a messenger is sent down the line that is used to pull the net through the water. These nets can be used to sample vertically over a fixed range of depths. If the area of the mouth of the net and the distance sampled are known, then the volume of water filtered can be calculated.

Several techniques are available for removing water from a lake and then filtering it. In all cases, the collection devices are clear to reduce the chance that predator avoidance behaviors related to evading large, dark objects do not decrease the amount of zooplankton collected. One method is to take a Van Dorn sample and pour the water sample through a filter to concentrate the zooplankton. Another method involves lowering a box that encloses a known amount of water at a specific depth. The box has an exit net where the water flows out after it is raised above the surface. This device is called a Schindler trap.

Samples are preserved, generally with formalin, and returned to the laboratory for counting and identification. Quick preservation is necessary to prevent predatory species from consuming their prey before the sample is transferred to alcohol for counting and possible longer-term preservation.

A change in body form across seasons called *cyclomorphosis* is one aspect of cladoceran biology that has received considerable attention and is also evident in rotifers. The variation in shape is diverse (Fig. 10.14) and can be related to temperature, varied predation pressure, other abiotic factors, or synergistic effects of several factors (Yurista, 2000). Cladocerans exhibit a variety of interesting behaviors including swarming, predator avoidance, and mating and feeding selectivity. The most obvious behavior is probably vertical migration in lakes, with the populations moving deep during the day to avoid sight-feeding predators and moving to the surface at night to feed on phytoplankton.

The Decapoda includes some of the most familiar crustaceans such as crayfishes, crabs, and shrimps. Most decapod species are found in marine or estuarine habitats, but hundreds of freshwater crayfish, shrimp, and crab species have been described, and they can be very abundant and diverse in the tropics. Decapods live in both lentic and lotic environments; some species have evolved to live in caves and groundwaters, and others live in swamps and intermittent wetlands. They are the largest and longest lived crustaceans in most freshwater systems. The bodies of shrimp and crayfish are more or less cylindrical (Figs. 10.15A and 10.15C) with a carapace that encloses at least a branchial chamber. Chelae, or pincers, are well developed in most adult decapods and are the feature most commonly associated with this group. Although many are drab colored or even pigmentless (e.g., many cave-dwelling species), some crayfish and freshwater shrimps are brightly colored. Freshwater decapods can provide a food source for humans; therefore, aquaculture of freshwater decapods such as crayfish and shrimps is an important economic activity in some regions and is a multimillion dollar industry worldwide. Some freshwater decapods can be serious pests in rice fields, where they consume plants and disturb the water (Schmitt, 1965). Many of the freshwater shrimps in the tropics are amphidromous; adult females release planktonic larvae that drift downstream to estuaries where they further develop before migrating back into freshwater streams (March *et al.*, 1998). Impoundments have serious negative impacts on these ecologically and economically important crustaceans (Benstead *et al.*, 1999).

Crayfish are omnivorous and can be dominant consumers (Hobbs, 1991). Freshwater shrimp are more often grazers or detritivores, but some, such as the large *Macrobrachium* shrimps, are predators. The

FIGURE 10.14

Cyclomorphosis of adults of the cladoceran *Daphnia retrocurva* over a season in Bantam Lake, Connecticut, during 1945. Only body shape was traced. *(From Brooks, 1946).*

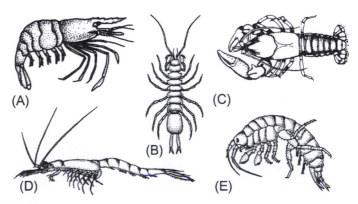

(A)   (C)   (B)   (D)   (E)

FIGURE 10.15

Some representative Crustacea. (A) a cave shrimp, *Palaemonias*, 2 cm; (B) the isopod, *Caecidotea*, 2 cm; (C) the crayfish, *Cambarus*, 10 cm; (D) the opossum shrimp, *Mysis*, 1 cm; (E) the amphipod *Gammarus*, 1 cm. *(Reproduced with permission from Thorp and Covich, 2001).*

freshwater decapods can be important prey for fishes and other large animals. Some species of crayfish have become nuisance invaders and may outcompete native species. All decapods were thought to reproduce sexually, in one form or another, until the recent discovery of the marbled crayfish in aquaria in Germany. Marbled crayfish (or Marmorkrebs) reproduce through parthenogenesis (Scholtz *et al.*, 2003), and, although they are well known in the aquarium trade, the natural range and ultimate origins of this crayfish remain virtually unknown. The rapid reproduction of this animal, and its widespread adoption as an aquarium pet, has led to considerable concern about its spread as an invasive species, with escapees becoming established in Madagascar and Europe.

Isopoda (isopods) are known commonly as pill bugs or sow bugs in terrestrial habitats. There are many marine and terrestrial species, but freshwater forms can also be found. The isopods are strongly dorsal–ventrally flattened (Fig. 10.15B) and range in length from 5 to 20 mm. Many of the species in North America can be found in springs, spring brooks, groundwaters, and streams, and those found in subterranean habitats often lack pigments. A few are found in the littoral zones of lakes. Most isopods are thought to be detritivores and scavengers because they are often observed eating dead aquatic animals. The isopods *Caecidotea* (formerly *Asellus*) and *Lirceus* have received some study because of their roles in processing organic matter and as an indicator of water quality in both surface and subsurface waters.

The Amphipoda (scuds and sideswimmers) are similar in size to the isopods, but they are flattened laterally (Fig. 10.15E). There are about 800 freshwater species worldwide (Smith, 2001). Amphipod eyes usually are well developed (except in the subterranean species). The amphipods are omnivorous and often considered scavengers. They are mainly nocturnal benthic species and can be present in densities up to $10,000\,m^{-2}$. Amphipods are very important as fish food. The amphipod *Hyalella* is cultured widely and is one of the most commonly used test organisms for freshwater toxicology studies. Freshwater amphipods, as well as isopods, reproduce sexually.

The Mysidaceae (opossum shrimps; Fig. 10.15D) and Bathynellaceae are usually minor components of the freshwater fauna. The mysid shrimps are planktonic zooplanktivorous species that are several millimeters long. Twenty-five species occur in freshwaters, whereas 780 occur in marine habitats. Mysids can be an important food source for fishes and strong competitors of other planktivores in some temperate lakes. The bathynellids are microscopic to macroscopic invertebrates that can be found in the hyporheic zones of some streams and other groundwaters. Because of their cryptic habitat, they often have been ignored, and new species are actively being described.

PHYLUM CHORDATA, SUBPHYLUM VERTEBRATA

The fishes are considered the most important freshwater animals by many people. Other types of vertebrates also live in or rely on freshwater habitats. These organisms are covered well in most biology curricula, and we will only briefly discuss them here.

Fishes

Fishes are the most diverse aquatic vertebrates, with more than 24,000 described species, and this number may rise to 28,000 as taxonomy progresses (Moyle and Cech, 1996; Table 10.3). Slightly less than half of these species are found in freshwater and more than 3,000 can be found in Amazonian waters alone. Fishes provide the major economic impetus to conserve and protect many freshwaters because of their importance as food and for recreation. A considerable amount of research has been conducted on their ecology. The major groups of fishes that can be found in freshwaters are discussed here. The lampreys (Fig. 10.16A) are the most primitive and are jawless; the jawed fishes include the Chondrichthyes (sharks and rays) and the Osteichthyes (bony fishes). Of these groups, the bony fishes are by far the most diverse in contemporary fresh and marine waters (Matthews, 1998).

Fishes have a variety of adaptations that allow them to specialize in various ecological roles and to display complex behavior. Body form is variable (Figs. 10.16 and 10.17). The streamlined bodies of active predators such as the salmon (Fig. 10.17E) allow them to pursue prey. Lie-and-wait predators with pointed snouts (Fig. 10.17B), torpedo-shaped bodies, and large posterior fins are able to generate sudden thrust and acceleration. Fishes that live near the surface and obtain food from it generally have upturned mouths, flattened heads, and large eyes (Fig. 10.16B). Bottom fishes generally have flattened bodies and often special adaptations, such as the lack of a swim bladder or the presence of sensory barbels or paddles for sensing prey in benthic habitats (Figs. 10.16E and 10.17C). Deep-bodied fish have narrow bodies and large fins (often spiny; Fig. 10.16D), and they are adapted to maneuvering in tight quarters such as dense macrophyte beds. Eel-like fish (Fig. 10.16C) have reduced fins and elongated bodies, and they are adapted to move through narrow spaces, soft sediments, or holes.

Fishes also have extensive sensory adaptations. The ability for chemoreception is acute; chemical concentrations as low as 10^{-13} moles per liter can stimulate behavioral responses (Moyle and Cech, 1996). Sounds and vibrations can also be sensed by the inner ear and lateral line system. This includes the ability to sense remote objects by their hydromechanical signals. Fishes also have a series of pit organs that allow sensing of weak electrical currents generated by prey. Many fishes have well-developed eyes that facilitate sight feeding and sexual displays.

Table 10.3 Some Orders of Fishes That Have Freshwater Representatives

Order	Common Name of Freshwater Species	Main Region of Dominance	Approximate Number of Freshwater Species
Petromyzontiformes	Lampreys	Worldwide, coastal	40
Myliobatiformes	SRrays	Coastal	
Lepidosireniformes	Lungfishes	AF, SA	
Polypteriformes	Bichirs	AF	10
Acipenseriformes	Sturgeons, paddlefishes	PA, NA, AS	28
Lepisosteiformes	Gars	NA	7
Amiiformes	Bowfin	NA	1
Osteoglossiformes	Elephantfishes, bonytongues	AF	200
Anguilliformes	Eels	Worldwide	26
Clupeiformes	Herrings, shads	Worldwide, some freshwater	80
Gonorynchiformes	Milkfishes	AF, AS	29
Cypriniformes	Minnows, carps, algae eaters, suckers, loaches	Worldwide	2,600
Characiformes	Characins	SA, AF	1,300
Siluriformes	Catfishes	Worldwide	2,280
Gymnotiformes	Knifefishes	SA	62
Esociformes	Pikes, mudminnows	NA, PA	10
Osmeriformes	Smelts, galaxiids	Worldwide, AU	71
Salmoniformes	Salmon, trout, whitefishes, chars, graylings	NA, PA	66
Atheriniformes	Silversides, rainbow fishes, blueeyes	AU, AS	160
Beloniformes	Needlefishes, halfbeaks	AS	45
Cyprinodontiformes	Top minnows, killifishes, pupfishes	Worldwide	805
Synrannchiformes	Swamp eels, spiny eels	AF, AS, SA, US	87
Scorpaeniformes	Sculpins	NA, PA	62
Perciformes	Basses, perch, sunfishes, darters, cichlids, gobies, gouramis	Worldwide	2,200

(From Moyle and Cech, 1996; Matthews, 1998).
Regions are African (AF), Neotropical (Central and South America, SA), Asia (Southeast Asia and peninsular India, AS), Palearctic Region (northern Eurasia, PA), Nearctic Region (North America, NA), and Australian Region (AU).

Freshwater fishes can be divided into three geographic types. A few marine fishes can enter freshwaters for extended periods of time and may be found in the lower regions of coastal streams. Obligatory freshwater species must inhabit freshwaters for at least part of their life cycles and many cannot tolerate marine waters for any significant length of time. *Diadromous* fishes spend part of their life cycles in marine systems. These species include fishes that

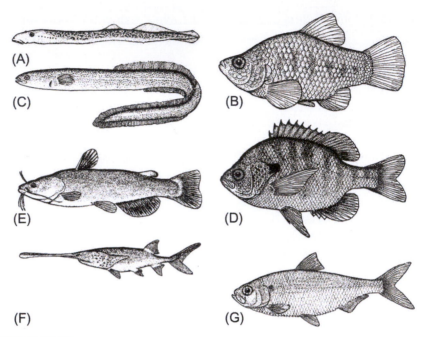

FIGURE 10.16

Some representative fishes. (A) lamprey, *Petromyzon*, 60 cm; (B) desert pupfish, *Cyprinodon*, 7 cm; (C) eel, *Anguilla*, 1 m; (D) bluegill, *Lepomis*, 20 cm; (E) bullhead, *Ictalurus*, 50 cm; (F) paddlefish, *Polyodon*, 2.5 m; and (G) alewife, *Alosa*, 40 cm. *(Reproduced with permission from Eddy and Underhill, 1969).*

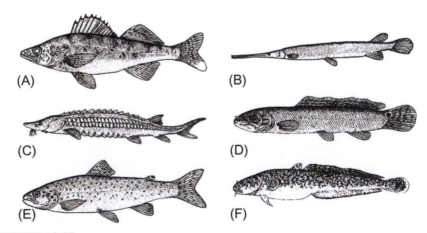

FIGURE 10.17

Some representative fishes. (A) walleye, *Stizostedion*, 90 cm; (B) gar, *Lepisosteus*, 1.5 m; (C) sturgeon, *Scaphirhynchus*, 20 cm; (D) bowfin, *Amia*, 60 cm; (E) salmon, *Salmo*, 1 m; and (F) burbot, *Lota*, 90 cm. *(Reproduced with permission from Eddy and Underhill, 1969).*

are *catadromous* and move from freshwater to saltwater to spawn (e.g., some eels) and those that are *anadromous* and move from saltwater to freshwater to spawn (e.g., salmon).

Tetrapods

Fishes gave rise to several groups of vertebrates that are prominent inhabitants of some freshwater systems. These include the amphibians, reptiles, birds, and mammals. The total diversity of these species is not high relative to that of the fishes or the arthropods, but they draw public interest. Several species such as beavers have extensive ecosystem effects.

Amphibians are divided into three groups: the salamanders (Urodela or Caudata), the caecilians (Gymnophiona), and the frogs and toads (Anura). The anurans account for 4,100 of the more than 4,600 species of amphibians (Pough *et al.*, 1998). Many of these species spend part or all of their time in freshwaters, particularly as eggs or larvae. Adult amphibians are predators but some larvae, such as many tadpoles, consume algae and detritus. Amphibians can be associated with lentic and lotic habitats. Many species of tadpoles are very susceptible to predation by fish and other large predators, although some are toxic or distasteful and are resistant to predation (Alford and Richards, 1999). Amphibians are consumed by a wide variety of animals, including fishes, other amphibians, reptiles, birds, and mammals, and even some of the larger invertebrates. Many amphibians are sensitive to environmental change and are thus good indicators of habitat disturbance and pollution. Amphibians are among the most imperiled groups of animals on the planet; catastrophic declines in amphibian diversity have been documented in many parts of the world (see Sidebar 11.2).

Aquatic reptiles, birds, and mammals are familiar to most people. A list of some of the important or charismatic species (Table 10.4) includes some that can be abundant in wetlands, rivers, and lakes. Some of these species can alter hydrology or nutrient cycles. Others are of interest because of their rarity or unusual nature. Most of these species are associated with shallow habitats, such as the littoral zone of lakes, ponds, rivers, streams, and wetlands.

Table 10.4 Some Reptiles, Birds, and Mammals of Interest in Freshwater Ecosystems

Taxonomic Group	Common Name	Habitat/Trophic Position	Comments
Testudines	Turtles	Rivers, wetlands, ponds, lakes/predators and herbivores	About 260 species worldwide
Eunectes	Anaconda	Rivers, wetlands, ponds/predators	Largest snake in world, up to 10 m long
Nerodia and *Natrix*	Water snakes	Rivers, wetlands, ponds, lakes/predators	
Crocodilians	Crocodiles, alligators, caimans	Rivers, wetlands, ponds, estuaries/predators	All 21 species endangered or threatened
Podicipediformes	Grebes	Lakes, estuaries/predators	21 species
Pelecaniformes	Pelicans, cormorants, anhingas	Lakes, estuaries, marine/predators	
Anseriformes	Swans, geese, ducks, screamers	Lentic and lotic waters/predators and herbivores	Temporary ponds and wetlands important to many species
Phoenicopterifromes	Flamingos	Shallow lagoons and lakes/filter feeders	Five species, all tropical
Coconiiformes	Herons and storks	Shallow water/predators	
Falconiformes	Osprey, hawks, eagles	Surface waters/predators	Mainly take fish from surface
Gruiformes	Cranes and rails	Shallow waters/predators	
Cinculus	Dipper	Mountain stream/insectivore	Dive into clear, running mountain streams to forage
Charadriiformes	Shorebirds and gulls	Shallow waters/predators	
Gaviiformes	Loons	Lakes/predators	
Ornithorhynchus anatinus	Duck-billed platypus	Lakes, rivers/predators	An Australian monotreme
Noctilio	Fishing bat	Lakes, rivers, wetlands/predators	
Phoca siberica	Baikal seal	Lake Baikal/predators	Only freshwater seal
Lontra	River otter	Streams, lakes/predators	
Delphinidae	Dolphins	Coastal rivers/predators	Asia and South America
Platanistidae	River dolphins	Large rivers in Asia and Indopacific/predators	Generally blind, locate prey by echolation
Sirenia	Dugong, manatee	Tropical, estuarine, coastal rivers/herbivores	
Hippopotamus amphibious	Hippopotamus	Lakes, rivers/herbivores	
Castor	Beaver	Lakes, rivers, wetlands/herbivores	Northern Hemisphere
Myocastor coypus	Nutria	Rivers, wetlands/herbivores	South American native introduced into Europe and North America

SUMMARY

1. The major groups of animals in freshwaters are the Porifera, Cnidaria, Turbellaria, Nemertea, Gastrotricha, Rotifera, Nematoda, Mollusca, Annelida, Bryozoa (Entoprocta and Ectoprocta), Arthropoda, and Chordata.
2. The taxonomy of the smaller organisms is less completely resolved than that of larger organisms.
3. Sexual reproduction is often sporadic or nonexistent in more primitive organisms. It is usually required in larger organisms such as the vertebrates, with notable exceptions.
4. Body form can vary with season or exposure to predation in several groups, including the rotifers and cladocerans. A seasonal change in body form is called cyclomorphosis.
5. The Crustacea and Insecta are the most diverse animal groups in freshwater systems and have adapted to all major aquatic habitats.
6. Immature aquatic insects are particularly diverse in rivers and streams.
7. Fish assume an important role in aquatic food webs. Their body shape is related to their place in the food web and their habits.
8. Many of the mammalian, amphibian, reptilian, and avian species that use freshwater habitats are endangered; some have become extinct.

QUESTIONS FOR THOUGHT

1. What constrains Crustacea from reaching the size of humans?
2. What groups of invertebrates do you think are most likely to contain undescribed species? Why?
3. Why do most aquatic insects leave the water to mate?
4. Why has behavior evolved to be more complex, but biochemistry less complex, in animals in comparison to microorganisms?
5. Why are aquatic insects rare and not diverse in marine habitats relative to freshwater habitats?
6. Why are animal assemblages of lakes generally less diverse than those in streams?
7. How have animals adapted to subsurface habitats?
8. Why is lifespan tied to the utility of various freshwater animals for use in biotic assessment indices?
9. Why are there more species of Crustacea than mammals and birds?
10. Why do fish that are ambush predators have a different morphology than piscivorous species that forage in the pelagic zone?

CHAPTER 11

Evolution of Organisms and Biodiversity of Freshwaters

FIGURE 11.1

(A) A groundwater-dwelling isopod (*Caecidotea tridentata*).

Doi: 10.1016/B978-0-12-374724-2.00011-8

FIGURE 11.1

(B) A green heron (*Butorides striatus*). *Image courtesy of Steve Hamilton.*

Describing patterns of distribution of organisms and factors that control the distributions is central to the study of ecology and historically has been the focus of much limnological study. An entire field, conservation biology, is now dedicated to the science of conserving organisms. We are living in a time of unprecedented rates of human-caused extinction, particularly in freshwater habitats (Cairns and Lackey, 1992; Jenkins, 2003). Efforts to understand, manage, or conserve species require knowledge of distribution of organisms in their habitats, how they evolved, and how they maintain populations. In addition to causing many species to go extinct and threatening many more, people are introducing exotic species into habitats at very high rates. Species introductions can lead to direct negative effects on humans and often harm the ecosystems that exotic species invade.

In this chapter, we discuss measures of diversity and how and why diversity varies among and within habitats, and we describe some particularly spectacular cases of high diversity. Extinctions caused by humans and some of the associated consequences are also discussed.

MEASURES OF DIVERSITY

The simplest measure of diversity is *species richness*, the total number of species found in an area. Another important aspect of diversity is how equally represented the numbers of each species are, which is called *evenness* or *equitability*. For example, consider two habitats, both with 10 species. If one habitat has approximately even numbers of each species, it will have a high evenness. If the other is dominated by one species and has one or a few representatives of the rest, it has the same species richness but lower evenness. This can be visualized in Figure 11.2. Ponds B and F have the same richness, but pond F has a greater evenness. Some indices include both evenness and richness.

One commonly used measure of diversity that includes both evenness and richness is the Shannon index,[1] H':

$$H' = -\sum_{j=1}^{S} p_j \ln(p_j)$$

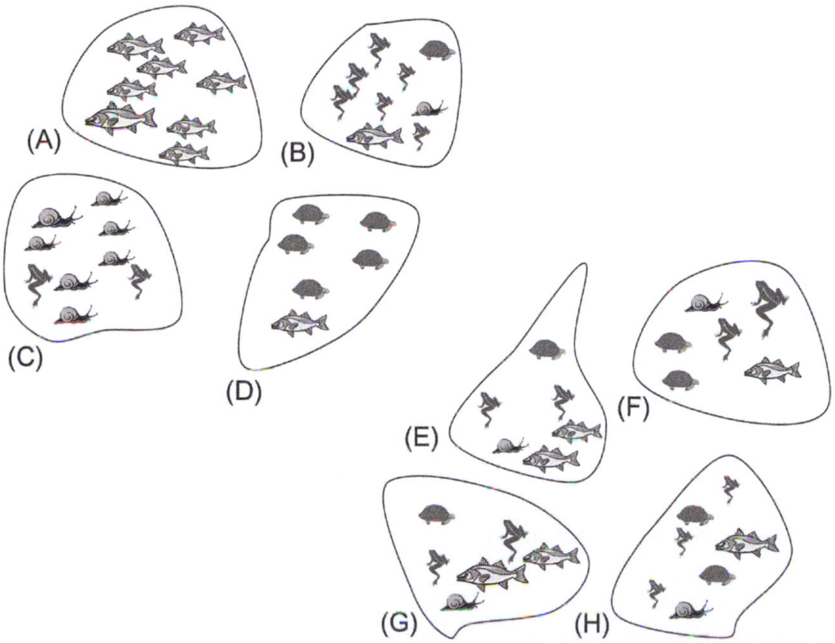

FIGURE 11.2

A diagram of diversity in two sets of ponds. When A–D are considered versus E–H, both have approximately the same overall diversity. When α diversity (within-habitat diversity) is measured, ponds in the A–D group have lower diversity than those in the E–H group. When β diversity (between-habitat diversity) is measured, ponds in the A–D group have higher diversity than those in the E–H group.

---

[1] Often incorrectly referred to as the Shannon-Weiner or the Shannon-Weaver index.

where there is an assemblage of organisms with S species and p_j is the proportion of species j (i.e., the number of individuals of species j divided by the total number of organisms in the assemblage). The summation sign at the beginning of the equation means that the proportions of all the species in the community are summed. The equation is most often calculated with the natural log, but others use $\log 10$ or $\log 2$. Here, we use the natural log, but care should be taken when comparing results to other studies to be certain that the index was calculated the same way. H' increases with more species and with greater evenness. One index used to indicate evenness is to divide H' by the diversity at maximum evenness for a given number of species. The following equation can be used to create an index of evenness using this approach:

$$E = \frac{H'}{\ln S}$$

where E is the evenness, and S is the number of species. For this index to be comparable among different communities, they need to be extensively sampled because if they are not, the total number of species is not known (Pielou, 1977). How to calculate diversity and evenness is demonstrated in Example 11.1.

■ Example 11.1

Calculating Diversity

Samples of aquatic insect larvae are taken from both pool and riffle stream habitats. Find and calculate the mean species richness, the Shannon diversity index, and the evenness for both habitats.

Sample	Species	#m^{-2}	p_j	$p_j \ln p_j$	H'	E	S
Pool	Perlesta placida	10	0.24	−0.34			
	Zealeuctra claasseni	20	0.49	−0.35			
	Stenonema femoratum	5	0.12	−0.25			
	Baetis spp.	6	0.15	−0.28			
	Total pool	41			1.22	0.88	4
Riffle	Perlesta placida	17	0.21	−0.33			
	Zealeuctra claasseni	18	0.23	−0.34			
	Stenonema femoratum	5	0.06	−0.17			
	Baetis spp.	3	0.04	−0.13			
	Tipula spp.	21	0.26	−0.35			
	Pseudolimnophila spp.	16	0.20	−0.32			
	Total riffle	80			1.64	0.91	6

The diversity in the riffle is higher, because more species are present and evenness is greater. Statistical experimental designs including replication followed by statistical techniques such as Analysis of Variance should be used to determine if these differences are indeed significant (see Appendix). These data are presented only to demonstrate calculation of diversity.   ■

Within-habitat diversity (α *diversity*) and between-habitat diversity (β *diversity*) are additional aspects of diversity. These two aspects link spatial scale and diversity. Each habitat has a variety of subhabitats. For example, a wetland may contain habitat types ranging from open water to damp soil, and each habitat has its characteristic species. Figure 11.2 illustrates how spatial arrangement can be related to determinations of diversity. The two groups of four ponds (A–D and E–H) each have the same number of total species and approximately the same evenness if all the ponds in each group are considered. However, each individual pond in the first group (A–D) has lower diversity (evenness and richness) than each pond in the second group. The ponds in the first group have low within-habitat diversity (α diversity) but high between-habitat diversity (β diversity).

We have considered only three indices of diversity here—species richness, Shannon diversity, and evenness. Ecologists have used several other indices of diversity throughout the years. Some mathematical benefits and drawbacks of the various indices are discussed by Pielou (1977) and Magurran (2004) but are beyond the scope of this text. Once diversity is estimated, the next issue to be considered is the major patterns of β diversity.

One approach to classification of habitats is to use the distributions of organisms (β diversity) and abiotic characteristics, termed the *ecoregion* concept. Ecoregions are relatively large areas of land or water that contain a geographically distinct assemblage of natural communities. For example, the freshwater habitats of North America have been divided into 76 ecoregions from eight major regions based on biological distinctiveness with regard to fish, amphibians, crayfish, unionid mussels, and aquatic reptiles (Abell *et al.*, 2000). It is not yet clear how well these ecoregions extend to insects, plants, and algae. Given that freshwaters can be biologically distinct, the next issue we will consider is the processes that lead to observed distributions of diversity.

TEMPORAL AND SPATIAL FACTORS INFLUENCING EVOLUTION OF FRESHWATER ORGANISMS

The ultimate source of biological diversity is evolution. Two main factors influence the evolution of new species: time available for evolution and reproductive isolation. These factors, coupled with what Hutchinson (1959) called the mosaic nature of the environment (what we now call spatial and temporal aspects of habitat heterogeneity), have resulted in millions of species. We discuss these factors with regard to lakes, streams, groundwaters, and wetlands. Consideration of spatial and temporal scale is important in describing the biological processes. Specific examples of situations in which evolution has resulted in high numbers of species will be described. Some short-term determinants of biodiversity will be explored in the next section.

One of the most striking demonstrations of the importance of evolution to biological diversity is the high number of species found in geologically ancient habitats that have had ample time for new species to arise. Typically, such habitats contain many *endemic* species, those species with a restricted distribution. Additionally, such habitats provide some of the clearest examples of *adaptive radiation* (evolution of many species from a single or few founder species).

Ancient lakes contain a large proportion of freshwater biodiversity (Cohen, 1995). These tectonic lakes have unique assemblages of vertebrates and invertebrates compared to the Great Lakes of North America (Fig. 11.3). The Great Lakes were recently covered by ice during the last ice age, and thus are young in an evolutionary sense. This high degree of endemism occurs even though the surface area (i.e., potential habitat) of the Great Lakes is almost 10 times greater than that of the largest of the ancient lakes. The effect of habitat area

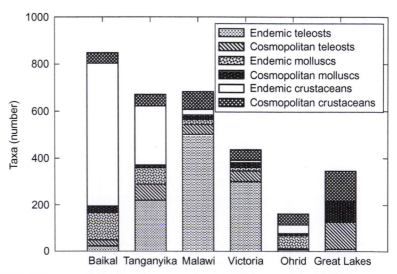

FIGURE 11.3

Number of endemic species for some large lakes of the world. The area of the North American Great Lakes is approximately 10 times greater than that of any of the other lakes shown. *(Reproduced with permission from Cohen, 1995).*

will be discussed in the next section, but we expect more species in a greater area, so the relative diversity of old tectonic lakes relative to the Great Lakes is even more remarkable.

Lakes that have existed continuously for millions of years generally have been subjected to changes that lead to geographic isolation. Variations in water level, formation, and isolation of various subbasins and fluvial processes can allow for divergence and evolution of new species (Brooks, 1950). Divergence and evolution of species can occur because of reproductive isolation of populations. Reproductive isolation allows organisms to evolve to be different enough that they become distinct species that can no longer interbreed if their populations become contiguous again.

Lake Baikal, located in Russia, may be the most studied of the ancient lakes. It has hundreds of endemic crustacean species, 291 of which are amphipods. Figure 11.4 demonstrates a small part of the diversity in body form that can be found in these amphipods; this diversity in body form is consistent with differences in rRNA sequences (Sherbakov *et al.*, 1998). In addition, there are 86 endemic species of turbellarians, 98 endemic mollusks, and 29 endemic species of fish (of which 26 species are sculpins (Cottoidei); Sherbakov, 1999). Eighty to 100% of the species in these groups are endemic. Slightly less than half of the 679 species of diatoms are endemic (Brooks, 1950). As with most large lakes, this incredible biodiversity is threatened by human-caused pollution (Galaziy, 1980).

The African Rift has formed a deep depression in the continent of Africa that has held lakes for millions of years. Lake Tanganyika is one of these and has been in

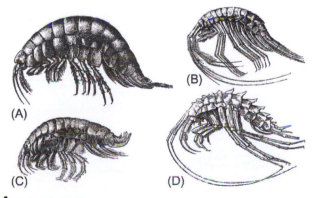

FIGURE 11.4

A variety of Gammarids from Lake Baikal, demonstrating diversity of body form: (A) *Ommatogammarus albinus*, body length up to 25 mm; (B) *Abyssogammarus sarmatus*, body length up to 63 mm; (C) *Crypturopus pachytus*, body length up to 18 mm; (D) *Garjajewia cabanisi*, body length up to 80 mm. *(Reproduced with permission from Kozhov, 1963).*

existence for 9 to 12 million years. It is renowned for its large number of endemic fishes, dominated by the perch-like tropical cichlid fishes. Greenwood (1974) reported that the family Cichlidae is represented by 126 species, all of which are endemic. An additional 67 species of fishes from other families are found in this African lake, 47 of which are endemic. Coulter (1991a) reported 287 species and subspecies of fish, 219 of which are found only in Tanganyika. The Coulter and the Greenwood numbers do not exactly match because of changes in taxonomy and inclusion of subspecies by Coulter.

Lakes Victoria and Malawi, also found in the African Rift, have more species of Cichlidae than Lake Tanganyika but not more endemic species in other families, and they have fewer endemic genera. The variety of adaptations that have evolved in the fish from Rift Lakes is astounding (Stiassney and Meyer, 1999). Various species of fish in this one family live by eating zooplankton, phytoplankton, gastropods, benthic algae, macrophytes, detritus, or fish. Among the piscivores, some eat the scales of other fish and have evolved to look like their prey so they are not detected as predators (Fryer and Iles, 1972). The diversity of body form in just the one family, Cichlidae (Fig. 11.5), can exceed that seen in all families found in some temperate habitats.

Behavioral diversity in the Cichlidae in these lakes is also high (Hori *et al.*, 1993). Wide variation in coloration (Axelrod, 1973) is probably related to

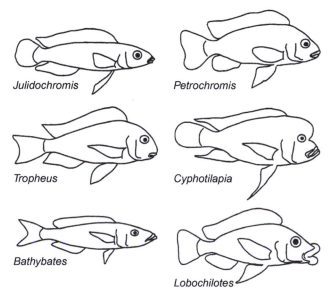

FIGURE 11.5

Some of the many varieties of fishes in the family Cichlidae from Lake Tanganyika. *Julidochromis*, omnivorous; *Petrochromis*, herbivorous; *Tropheus*, herbivorous; *Cyphotilapia*, ambush predator, gastropods; *Bathybates*, piscivorous; *Lobochilotes*, insectivorous. *(From Gillespie et al., 2001).*

sexual selection that leads to sympatric speciation (i.e., evolution of a species within one habitat; Seehausen and van Alphen, 1999). Many cichlids exhibit parental care, which ranges from guarding eggs and fry from predators to incubating eggs and protecting fry in the oral cavity (Keenleyside, 1991). Two species can brood in the same region and mutually defend against predators. In another case of evolved cooperation, mixed groups of predators can cooperate to increase success (Nakai, 1993). These complex behavioral adaptations can arise only in a diverse animal assemblage.

Tanganyika also supports a high proportion of endemic animals in other groups, including 12 of 20 leech species, 37 of 60 gastropods, 33 of 69 copepod species, 74 of 85 ostracods, and 22 of 25 decapod species (Coulter, 1991b). As in Lake Baikal, the millions of years this lake has existed has allowed for diversification of many animal groups. Unfortunately, the biological diversity of Tanganyika is increasingly threatened by overfishing for food and the aquarium trade, as well as other human impacts such as deforestation in the watersheds that drain into the lake, and the inevitable pollution that human activities cause.

An obvious question when considering the various groups that have produced large numbers of endemic species in ancient lakes is the following: Why does a particular group diversify (Gillespie et al., 2001)? For example, the teleost fishes are very diverse in the African Rift Lakes like Tanganyika, but the amphipods have the greatest number of endemic species in Baikal. Amphipods occur in the African lakes, and teleosts occur in Baikal, so why not extensive speciation in both groups in both lakes? Chance plays a large part in evolution. If the right group already happens to be present and already is fairly diverse, it may be predisposed to amplify the existing variation when conditions occur that are favorable to evolution of new species. The evolution of greater diversity can then be accelerated in particular related groups. For example, the diverse open water fish assemblages found in Lake Malawi (one of the African Rift Lakes) probably arose from multiple independent lines of species that inhabit the shoreline (Fryer, 2006). Thus, a diverse shoreline community led to a diverse community in the pelagic (open water) areas of the lake. Likewise, the diverse cichlid fish assemblages found in the rivers of southern Africa probably arose in a now dry lake (Joyce et al., 2005).

Certain biological traits may favor rapid speciation in a group. The gastropods provide an example of some characteristics in a specific group that can lead to evolution of high numbers of species. Gastropod taxa that are depth tolerant, brood their young, and have poor dispersal ability can diverge more rapidly. All these characteristics increase the chances of reproductive isolation and, consequently, evolution of independent lineages, as is known from study of evolutionary mechanisms (Michel, 1994).

River basins can also have a continuous existence over geological time, leading to evolution of high numbers of unique species. The Mississippi River drainage has been geologically stable, the river has flowed continuously, and little Late Cenozoic extinction has occurred (e.g., in the last 10 million years). Much of the diversity in North American fishes has radiated from this drainage basin (Briggs, 1986); there are 300 fish species found in Tennessee alone (Etnier, 1997). The mussel assemblages in the superfamily Unionacea have evolved 227 native species in North America (McMahon, 1991), most of which are found in the Mississippi River basin. However, many of these mussel species are now endangered by human activities.

Geographic isolation is another factor that leads to high levels of endemism. It has long been known that islands have relatively high numbers of endemic species (but lower total numbers of species). Islands can be thought of as an extreme case of geographic isolation. Such isolation occurs in continental habitats as well; many freshwater habitats are essentially islands in a terrestrial "sea." Geographic isolation into different habitat types has been demonstrated to lead to species divergence in three species of sticklebacks (*Gasterosteus* spp.), but divergence did not occur where isolated habitats were similar (Rundle *et al.*, 2000). Dry landscapes can lead to more isolation of aquatic habitats and greater divergence of amphipod species (Thomas *et al.*, 1998) and leeches (Govedich *et al.*, 1999). Nine species of ostracods in the genus *Elpidium* evolved in Jamaica (Little and Hebert, 1996). These ostracods inhabit the small pools of water that form in bromeliads (terrestrial epiphytic plants). Thus, segregation of habitats that leads to evolution of new species can occur over very small spatial scales.

Groundwaters in aquifers with channels large enough to allow movement of animals have given rise to many unique endemic species. The organisms found in groundwaters can be accidental wanderers from surface waters, can occur mostly in groundwaters and have some adaptations to the subsurface life (*stygophile*), or can be specifically adapted to life in groundwaters (*stygobite*) (Gibert *et al.*, 1994). Adaptations to life in groundwater habitats include loss of eyes and pigmentation, increased length of sensory appendages, slowed metabolic rates, long life histories, and production of fewer and larger eggs.

Apparently, dispersal ability is low in groundwater fauna (Strayer, 1994), probably contributing to endemism. For example, little recovery of some North American hypogean invertebrates occurred following the last glaciation. In some cases, the species found in groundwaters can vary over small spatial scales. Diversity of ostracods in the shallow groundwater (hyporheos) near the Sava River (Croatia) varies considerably across a 135 m transect moving away from the river channel (Rogulj *et al.*, 1994). Studies of subterranean crustacean species found in subsurface streams in a Virginia karst aquifer indicated the importance of spatial scale (Fong and Culver, 1994; Culver and Fong, 1994). These authors demonstrated unique species assemblages that varied at scales from individual rocks to differences between pools and riffles. Greater differences were found among different branches of the streams, and the greatest differences were found among adjacent subsurface drainage basins.

Rivers with extensive hyporheic systems in very permeable or channelized aquifers can have a unique biota associated with the subsurface water (Marmonier et al., 1993; Dole-Olivier et al., 1994; Popisil, 1994; Ward and Voelz, 1994; Hakenkamp and Palmer, 2000; Boulton, 2000). In a fascinating series of studies on groundwater communities in a gravel-bed river system, numerous aquatic insect larvae and other invertebrates were documented up to 2 km from the river channel (Stanford and Ward, 1988). Some of these species are unique stoneflies that spend their entire nymphal life cycle underground and then return to surface water habitats to emerge as adults and mate (Stanford and Gaufin, 1974; Gibert et al., 1994). Hyporheic habitats may be constant enough to provide refugia for ancient taxa of cladocerans over evolutionary time (Dumont, 1995).

The data available on biodiversity of hypogean animals suggest that a tremendous number of undescribed species exist. The difficulties encountered in sampling these organisms, coupled with the documented spatial variation in distributions of the animals, make it highly probable that species will be missed unless areas are sampled intensively. Despite what we do not know about the distribution of these species, many are highly susceptible to human-caused pollution of groundwaters. Thus, the use of groundwater invertebrates as bioindicators of contamination has been suggested (Malard et al., 1994), particularly as indicators of low dissolved oxygen conditions (Malard and Hervant, 1999).

Some temporary pools are also centers of endemism. Such pools are isolated from other aquatic habitats and require special adaptations in the form of desiccation resistance. The vernal pools of California are examples of this because they provide habitat for highly endemic plant communities. In addition, when 58 of these vernal pools were sampled, 67 species of crustaceans were found and about half these were endemic, mostly Anostraca, the fairy shrimp (King et al., 1996). Such pools with endemic plant and crustacean communities have been described in dry habitats throughout the world (Thorne, 1984; Belk, 1984).

Several cases of endemism have yet to be explained. For example, benthic copepods show high levels of endemism in South American wetlands (Reid, 1994). However, it is not known why such endemism occurs in these wetlands and not in those of North America, such as the vernal pools in California in which the Anostraca are diverse.

SHORT-TERM FACTORS INFLUENCING LOCAL DISTRIBUTION OF SPECIES

Numerous factors, including species interactions, productivity, species intro-ductions, habitat type, and colonization, control how many and what types of species occur in a specific environment. Here, we discuss colonization and habitat type. Later, we focus on species introductions. Species interac-tions are discussed in Chapters 19 through 22, and productivity is discussed in Chapters 17 and 18. Factors related to colonization are dispersal ability, competitive ability, and distance from sources of colonists when new habitats form. Several areas of ecological theory relate to these concepts.

Habitat type and diversity are interrelated. The habitat serves as a template on which evolution can occur. Organisms must specialize in a type of habitat to some degree to be evolutionarily successful, and habitats are partitioned by scale and other physical attributes. For example, a lake has several distinct types of habitats that are dominated by different groups of organisms (Fig. 11.6). Changes in these habitats can influence biodiversity. For instance, decreases in woody debris on lakeshores associated with residential development are expected to lead to decades-long decreases in biodiversity (Christensen *et al.*, 1996). Similarly, decreases of benthic habitat diversity in streams lead to decreases in macrophyte diversity (Baattrup-Pedersen and Riis, 1999), and macrophytes can control heterogeneity of sediments (Schneider and Melzer, 2004). Within taxonomic groups, different species also specialize in specific habitat types, which has been well documented for many plants and animals, even those as simple as rotifers (Pejler, 1995). Scale also can be an important

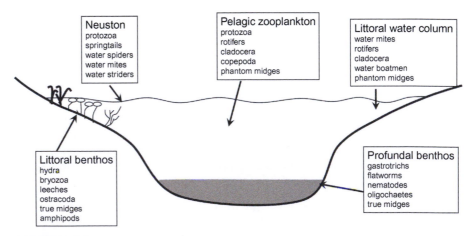

FIGURE 11.6

Representative invertebrate groups as a function of habitat in a small lake. *(Redrawn from Thorp and Covich, 1991c).*

determinant of diversity patterns, with microbes being associated with the smallest particles and animals such as beavers actually creating and responding to changes at the watershed level. Figure 6.21 illustrates how scale can affect biodiversity in streams and rivers.

How habitats are linked to each other is also a central determinant of biodiversity. Interfaces among habitat types are regions with high numbers of species. High diversity at edges occurs because species that are adapted to both habitat types can occur there (species ranges overlap), in addition to species that specifically exploit the transitional zone. The relationships among biodiversity and riparian zones of streams are a prime example of high diversity associated with an ecosystem interface (Tockner and Ward, 1999). Maintenance of riparian zones can preserve both diversity and ecosystem function in a watershed (Naiman *et al.*, 1993; Spackman and Hughes, 1995; Patten, 1998) and in the stream (Vuori and Joensuu, 1996). Likewise, hyporheic zones as an edge between groundwater and streamwater should be managed when biodiversity of river systems is of concern (Stanford and Ward, 1993).

Dispersal ability is also important in determining distributions of species and biodiversity. Widely distributed freshwater species (cosmopolitan species) often have the following characteristics: (1) life history stages that can survive transport (often desiccation resistance); (2) production of large numbers of individuals in the transportable life stage form; and (3) the ability to survive, including competing successfully and avoiding predation, in new habitats (Cairns, 1993).

Some species are distributed very broadly. For example, many protozoa have cosmopolitan distributions; about 8% of all protozoan species ever described on Earth can often be found in a single sample (Fenchel *et al.*, 1997). Protozoa have all the characteristics listed in the prior paragraph and thus are very successful at colonizing new habitats. Consequently, identification keys for protozoa that have been written anywhere in the world are useful in most other locations (Cairns, 1993).

Some of the natural modes of species transport include swimming or being moved by currents through connected habitats; transport on the feet or in the guts of animals (including insects, birds, and mammals); windborne dispersal; and having life stages specialized for dispersal (e.g., many amphibians and aquatic insects). The possibility of transport of algae by waterfowl in their guts has been documented (Proctor, 1959, 1966), and is probably common on the local scale (Figuerola and Green, 2002). A few larger zooplankton species can be transported by wind, but are not as likely to be transported by waterfowl (Jenkins and Underwood, 1998).

Resting stages of organisms are particularly important in some cases—not only in dispersal but also in the ability to respond to long-term changes in the environment. For example, copepods produce eggs that can remain viable but physiologically inactive in sediments for long periods of time (*diapause*). This long-term survival is an adaptation to changing environmental conditions and allows the copepods to become established in environments that only sporadically support a reproductive population (Hairston, 1996). Hairston *et al.* (1995) studied sediments of two small freshwater lakes in Rhode Island. They dated the sediments of the lakes with isotopes and hatched eggs taken from sediments of different ages. In one lake, the mean age of eggs that hatched was 49 years, and the maximum was 120 years. In the second lake, the mean was 70 years and the maximum was 332 years.

A consistent ecological pattern relating spatial scale and diversity is the positive relationship between the size of an island of habitat or area (A) and the number of species that are found in that habitat (Preston, 1962). If S is the number of species, then

$$S = cA^z$$

or, in a form easier for calculations,

$$\log S = \log c + z(\log A)$$

where c is a constant measuring the number of species per one unit area, and z is the slope of the line on a plot of the log of the area versus the log number of species. The relationship between size of habitat and area holds well for many types of organisms. The species–area relationship applies broadly (Rosenzweig, 1995) and includes diatoms colonizing glass slides (Fig. 11.7),

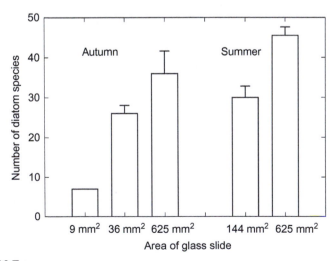

FIGURE 11.7
Number of diatom species on glass plates of various surface areas after 1 week in Ridley Creek. *(Data from Patrick, 1967).*

invertebrates on stones in streams (Fig. 11.8A), and fishes in lakes (Fig. 11.8B). This relationship can be used to estimate the effects of habitat reduction on species diversity (Example 11.2).

Spatial considerations also determine diversity of a given habitat. The distance from a source of colonists paired with the area of the habitat can be used to estimate a relative expected diversity (MacArthur and Wilson, 1967). Although the specifics of these models are somewhat complex to treat here, they include

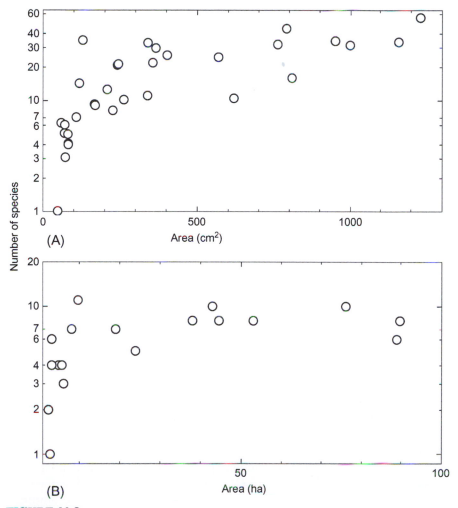

FIGURE 11.8

Number of invertebrates as a function of stone size in an Australian stream (A; from Douglas and Lake, 1994) and number of fish as a function of lake size in small Wisconsin lakes (B; data from Tonn and Magnuson, 1982).

■ Example 11.2

Calculating Expected Numbers of Species after a Habitat Reduction

A developer wants to decrease the area of a small lake in Wisconsin from 90 ha to 10 ha. Given that data from Tonn and Magnuson (1982) can be used to calculate values of $c = 2.211$ and $z = 0.3325$ for lakes of similar size, calculate the expected number of fish species for each size of lake.

For a 90 ha lake $\log S = \log 2.211 + 0.3325 * \log(90) = 0.994$,

so $\quad\quad\quad\quad S = 9.9$ species of fish

For a 10 ha lake $\log S = \log 2.211 + 0.3325 * \log(10) = 0.677$,

so $\quad\quad\quad\quad S = 4.8$ species of fish

Thus, we can expect that half the fish species will be supported in the smaller lake. This calculation ignores other possible factors such as what type of habitat will be present after the lake is altered and the variance in the predictive relationship (fairly high). It also does not tell us which species will survive. However, such relationships are useful for illustrating the idea that the size, as well as the quality of habitat, is an important parameter in conserving biodiversity. ■

two important points related to biodiversity and success of invaders. First, lowering the connectivity of habitats (i.e., sources of natural colonists) can disrupt the natural diversity. Second, human transport of organisms can vastly reduce the effective isolation of some habitats. Ramifications of such transport of organisms will be discussed later.

Succession, the sequence of species inhabiting newly formed habitat, is a dominant and long-standing ecological principle. The central idea related to diversity is that species that occur early in succession reproduce rapidly and disperse widely. Species that occur late in succession are better competitors and displace the early colonists. Diversity is low early in succession because only the early colonists are present. Diversity is also low late in a successional sequence because only the competitive dominants survive. Diversity is maximal in the intermediate successional stages because both early and late successional species can be found. Despite these general patterns, chance can play a large part in the actual species that become established in a successional sequence in new aquatic communities (Jenkins and Buikema, 1998). Specific examples of succession will be discussed in Chapter 22.

The *dynamic equilibrium model* of species diversity suggests that natural disturbance, such as drying or flooding, can decrease diversity if it exceeds the ability of species to recover. However, it can increase diversity if it removes dominant species that outcompete others (Huston, 1994). This theory will be discussed more thoroughly in Chapter 22.

The *intermediate disturbance hypothesis* (Connell, 1978) relates diversity to disturbance frequency and intensity. This hypothesis suggests that maximum diversity is found at intermediate levels of natural disturbance. At high rates of disturbance, only early successional species, which are generally good colonizers but poor competitors, are able to survive. At very low levels of disturbance, late successional species, which are strong competitors, dominate. At intermediate levels of disturbance, both types of species can be found, and diversity is maximized. The problem with the intermediate disturbance hypothesis is that it provides no *a priori* way to establish relative disturbance intensity among communities. This hypothesis will be discussed in greater detail in Chapter 22.

Disturbances by humans can also cause variations in species diversity (color plate Fig. 8). Assessing changes in types and numbers of species present forms the basis of several indices of biotic integrity. Use of such indicators is described in Chapter 16.

GENETICS AND POPULATIONS OF SPECIES

Genetic diversity, within and among species, is a key component of biological diversity. Using modern molecular techniques, ecologists can now analyze sediment and water samples for microbial genetic diversity by amplifying and sequencing DNA extracted and amplified from environmental samples. Within species, genetic diversity is critical for long-term viability because it determines an organism's ability to adapt to dynamic environments. Genetic diversity determines the ability of populations and species to survive adverse periods (e.g., prolonged drought or climate changes), disease outbreaks, and other causes of mass mortality, because more genetically diverse populations are more likely to include individuals with differential susceptibility to these stressors.

Genetic diversity includes the proportion of polymorphic loci across the genome (gene diversity), numbers of individuals within a species with polymorphic loci (heterozygosity), and the number of alleles per locus. Ultimately, mutation is the source of genetic diversity, and it is maintained by continued mutations, emigration and immigration, and environmental variability. Environmental variability can influence genetic diversity at multiple scales over relatively short time periods. For example, genetic diversity of the European freshwater snail, *Radix balthica*, decreased locally in response to a severe drought (within a single study site; α diversity), but genetic diversity across sites (β diversity) simultaneously increased at larger scales (Evanno *et al.*, 2009).

These patterns of α and β diversity were the same for snail species diversity within and across the multiple study sites.

Conservation efforts for endangered species are increasingly focused on maintaining or enhancing genetic diversity because it is inherently linked to their survival (Vrijenhoek, 1998). The same is true for fish and invertebrate aquaculture programs, where genetic diversity is key to long-term viability and productivity. Conservation efforts for rare species strive to keep population sizes above the *minimum viable population*, which is the lowest number of individuals required for a population to survive in the wild. Within a population, only some members actually contribute genes to succeeding generations, and these individuals constitute the *effective population size*, which can be a fraction of the actual population, depending on a variety of demographic factors and life history specifics.

Below the minimum viable population size, a population is no longer expected to survive stochastic (random or chance) environmental or demographic events. Accurate estimates of minimum viable population size are difficult, and they vary greatly depending on the life cycles and natural histories of individual species of interest. Using life history information and a criterion of less than 10% probability of extinction in 100 years, minimum viable population size for the threatened Rio Grande cutthroat trout was estimated at 2,750 individuals (Cowley, 2008). A recent review of published estimates of minimum viable population sizes for a variety of terrestrial and aquatic endangered species found a median value of 4,169 individuals using a standardized population persistence probability of 99% (Traill *et al.*, 2007).

The *extinction vortex* (Gilpin and Soulé, 1986) suggests that as populations become smaller, their susceptibility to extinction accelerates exponentially. Because of both environmental and demographic stochasticity, smaller populations have much greater probabilities of going extinct. As populations decline, they experience degraded genetic diversity and individual fitness, and numbers of individuals fluctuate widely as a function of environmental variability (Fagan and Holmes, 2006). The extinction vortex illustrates the "slippery slope" of extinction and the importance of maintaining viable population sizes for effective conservation efforts.

ADVANCED: POPULATIONS AND THEIR SPATIAL DISTRIBUTION

When there are numerous populations with limited gene flow among the populations, the group of subpopulations is referred to as a *metapopulation* (Gilpin and Hanski, 1991). In metapopulations, multiple populations exchange individuals of the same species through dispersal and, in the process, enhance and maintain genetic diversity. The concept is closely related to evolution. Very distinct metapopulations may eventually diverge and form two species. The metapopulation concept is central to conservation because it describes how individual populations can make up the entire population of a species. The extinction vortex described earlier can apply to small subpopulations, and without dispersal from outside, some subpopulations are destined for extinction.

The metapopulation concept is also related to source-sink population dynamics. If a group of subpopulations makes up a metapopulation, some of them may go extinct and require input from adjacent populations to persist. These are the *sink populations*. Other populations can occur in favorable habitats and supply individuals to other populations. These are *source populations*.

Although metapopulation models are applicable in some cases (Harrison, 1991), less is known about their application in freshwaters. Simple metapopulations were tested against fish assemblage distributions across 10 sites in the Cimarron River, Oklahoma (Gotelli and Taylor, 1999). These authors suggested that metapopulation models may need to account for variability in colonization and dispersal rates that more simplistic models do not. This suggests that the metapopulation concept can be important, and spatial characteristics of population structure are essential to understanding population dynamics, but empirical application of simple models should be approached with caution. Perhaps simpler, more spatially structured populations, such as those found in a series of wetlands or ponds, would be more likely to exhibit classic metapopulation characteristics.

In a concept related to metapopulations, *metacommunities* are groups of communities spread across a landscape that interact through dispersal and exchanges of multiple species (Holyoak and Leibold, 2005). The applicability of the metacommunity is not well characterized for ecology, as the concept is relatively new, and it is more difficult to establish community structure. Population structure can be indicated with genetic tools, so the metapopulation concept is more straightforward.

The Controversy Over Endemism and Microbial Species

There are two major views of microbial diversity: the first is that "everything is everywhere and the environment selects" (Whitfield, 2005); the second is that there is dispersal limitation of microbes. The first view is supported by a large amount of work on protozoa. These studies have been able to find few if any unique species of protozoa, and taxonomic keys from one part of the world work as well in another distant location (Finlay, 2002). The fact that microbes have such huge populations suggests that even if the probability of transport of any individual cell is low, the probability that at least one cell out of billions will be transported is high. Extreme weather events like hurricanes and tornadoes can move water and soil high into the atmosphere. Birds migrate long distances, some from the North to the South Poles, and are known to transport organisms. Furthermore, sediments taken from ponds can yield species of protozoa that are not present in the natural habitat, but can be induced to grow by altering environmental conditions such as temperature, salinity, and food sources.

Still, there are microbes that do not have good dispersal stages. Diatoms are an example of this. Diatoms are easily identified taxonomically because of the distinctive ornamentations on their frustules. A substantial number of endemic freshwater species are found in New Zealand (Kilroy *et al.*, 2007), a location that is geographically very distant from many other freshwater sources. Although this high endemism appears related to a geographic isolation, these diatoms simply might not have been found elsewhere because of incomplete taxonomic coverage, or that the distinctive ornamentation is not indicative of distinct species. The second argument, that the diatoms are not really different species than those found in other parts of the world, could be settled with molecular analyses such as *DNA barcoding*, which looks at differences in mitochondrial DNA to assess phylogenetic relationships. Barcoding techniques often are used to identify *cryptic species*, species that are morphologically similar, but genetically different enough to be considered distinct species.

The argument between the two worldviews has yet to be settled. A recent study of 16 regional data sets from North America and Europe demonstrated that assemblages of lake diatoms responded slowly to changes in pH. These data suggest that dispersal rates are slow enough that it might be possible for genetic differentiation and endemism to develop (Telford *et al.*, 2006). As ever more habitats are sampled, and molecular techniques allow for distinction of morphologically similar species, a definitive answer should emerge as to what groups have truly global populations and what other groups are geographically isolated.

INVASIONS OF NONNATIVE SPECIES

Invasion by species introduced by humans is probably the most permanent form of pollution. Dispersal of species is one of the less recognized forms of global change instigated by humans (Vitousek *et al.*, 1996). Once an invader becomes established, eradicating it is almost impossible. We do not know of any examples of eradication of an aquatic pest species once it has become established in a large area (although some species may be successfully controlled to low numbers). Understanding the ecology of ecosystems is essential to managing invasive species. As humans increase speed and frequency of global travel and move more materials across international boundaries, the probability for transport of undesirable species increases dramatically. Humans have created a world without borders for many species (Mack *et al.*, 2000) and rates of human-caused species invasions are far greater than natural rates, although natural invasions of species influence how evolution of new species relates to community formation. Numerous case studies illustrate potential problems associated with unwanted species introductions; we will present only a few here.

Why do species invade successfully? Predicting which species will invade and the effects of the invasion is difficult (Fuller and Drake, 2000). A sound knowledge of the natural history of the invader and the community it has entered allows for the best predictive assessment of the influence of an invading species. Qualitative techniques may be used to assess the risks associated with freshwater species introductions (Li *et al.*, 2000).

Moyle and Light (1996) proposed a conceptual model of species invasions and establishment based on case studies of species of invading fish. Their points appear useful and are generalized here with additional observations provided by Kolar and Lodge (2001). As with any conceptual ecological model, exceptions exist, and the model is presented mainly as a method to approach the problem. First, most invaders fail to establish. Most failed invasions are never documented; thus, the success rate of invasions is not well known. Second, most successful invaders are integrated without major effects on the ecosystem or community, although some have major effects. Third, all aquatic systems can be invaded. Fourth, major community effects are observed most often in low-diversity systems, including island and highly disturbed habitats. Fifth, top predators that invade successfully are more likely to have strong community effects than successful invaders at lower trophic levels. Sixth, species must have physiological and morphological characteristics suited to the environment to invade successfully. Seventh, invaders are most likely to become established when native assemblages are disturbed. Natural or anthropogenic disturbance increases susceptibility to invasion. Eighth, success of invaders can depend on environmental variability. Invasion and establishment in an extremely harsh and variable environment may be difficult unless invaders are preadapted to the variation (see the sixth point). Variation can also play a role in the seventh point. Ninth, very stable systems may be susceptible to invasion, although the data are weak for this generalization. Tenth, the greater the number of invading individuals and times they are introduced, the greater the probability they will become established. Finally, species with a history of being invasive are most likely to invade other habitats.

The previous model does not follow more traditional treatments of *community assembly theory* that depend on the sequence of invasion and species interactions.

Many treatments of invasion success have revolved around competition and predation as determinants (Roughgarden, 1989). For example, early attempts at understanding success of invasions led to suggestions that alien species are more likely to colonize cropland and disturbed habitats, and by extension, more diverse communities are less susceptible to invasions (Elton, 1958). This idea of diverse communities being resistant to invasion has since been challenged (Moyle and Light, 1996; Levine, 2000). Still, disturbance and alteration of habitat can influence the success of species invasions. For example, reservoirs can harbor lentic species in river networks that historically had little lake habitat, they can receive many introduced fish species, and these fish that were rare or absent historically can invade rivers and streams upstream from the reservoirs (Falke and Gido, 2006).

The Great Lakes of North America have experienced a tremendous number of invasive species (Sidebar 11.1, Table 11.1), including the zebra mussel (*Dreissena polymorpha*), which has had considerable economic and ecological

---

SIDEBAR 11.1
Invaders of the North American Great Lakes

The Great Lakes have a long history of invasions of exotic organisms caused by humans (Table 11.1). As of 1990, more than 139 alien species had become established, including plants, fish, invertebrates, and algae (Mills *et al.*, 1994). By 1998, 145 alien species had been documented (Ricciardi and MacIsaac, 2000), and by 2009, the National Center for Research on Aquatic Invasive Species reported 181 nonnative species in the Great Lakes. Some of these introductions have had massive economic impacts, such as the effect of the zebra mussel on water intakes for municipalities and industries and the crash of the large lake trout fishery caused by the sea lamprey. Many of the other introductions have had moderate impacts, or impacts that do not directly affect humans in an economic sense.

Routes of introduction include intentional releases, movement through the aquarium trade and bait buckets, release of ship ballast water taken from other areas, and connection of the Great Lakes with the Atlantic Ocean by a system of shipping canals (the Erie Canal and the St. Lawrence Seaway).

Attempts are being made to limit the number of new introductions. For example, ships from overseas that originate in freshwater ports are required to exchange ballast water while at sea to avoid additional introductions of freshwater species. However, species that are tolerant to salinity changes will not be kept out by these methods (Ricciardi and MacIsaac, 2000). It is likely that new species will continue to be transported into the Great Lakes. Control of new species introductions has been attempted since the first edition of this book was written (data from 1999), yet the number of invasive species continues to climb. The voluntary measures and the mandated open ocean ballast change regulations have had little influence to date, and the rates of invasion continue to increase (Holeck *et al.*, 2004). The damage associated with some of these species and the existing nonnative organisms will continue to be considerable.

Table 11.1 Some Invaders of the North American Great Lakes

Organism	Year	Source	How Introduced	Effects
Sea lamprey, *Petromyzon marinus*	~1830	Atlantic	Shipping canals	Decreases native lake trout
Purple loosetrife, *Lythrum salicaria*	1869	Europe	Ship ballast	See Sidebar 9.4
Alewife, *Alosa pseudoharengus*	1873	Atlantic	Shipping canals	Suppresses native fish species; new prey for salmon
Chinook salmon, *Oncorhynchus tshawytscha*	1873	Pacific	Intentional	New piscivore; important sport fish
Common carp, *Cyprinus carpio*	1879	Europe	Intentional	Destroys habitat for waterfowl and fish
Brown trout, *Salmo trutta*	1883	Europe	Intentional	New piscivore; important sport fish
Coho salmon, *Oncorhynchus kisutch*	1933	Pacific	Intentional	New piscivore, important sport fish
White perch, *Morone americana*	~1950	Atlantic	Shipping canals	Competes with native fish
Eurasian watermilfoil, *Myriophyllum spicatum*	1952	Eurasia	Not known	Competes with native plants
European ruffe, *Gymnocephalus cernuus*	1986	Europe	Ballast water	Competes with native fish and eats eggs
Zebra mussel, *Dreissena polymorpha*	1988	Europe	Ballast water	Biofouling; outcompetes native species

(After Mills et al., 1994, and other sources)

impacts. The spread of the zebra mussel is one of the more pronounced examples of the effects that a nonnative species can have, in terms of both explosive invasion and economic consequences (see Sidebar 11.1).

Many other introductions of nonnative species to freshwaters have occurred (Table 11.2) and many of these were intentional. Introduction of fish species has led to homogenization of fish communities; many of the same species are found across the continent now (Rahel, 2000). Hopefully, this practice of purposely stocking exotic species will decrease in popularity as it becomes clear that unintended consequences are common with species introductions.

Climate change will probably increase the numbers of invasive species in aquatic habitats (Rahel and Olden, 2008). Species intolerant of colder water temperatures will be able to spread to higher latitudes as these regions warm. These invaders will likely be mobile, weedy species.

Table 11.2 Some Species Introductions Not Covered Elsewhere in the Text

Species	Source	Habitat Introduced To	Effects	Citation
Trout, *Salmo* spp.	Intentional introduction	New Zealand lakes and streams	Extinction of native galaxiid fishes	Flecker and Townsend, 1994; Townsend, 1996
Carp, *Cyprinus carpio*	Intentional introduction	North American lakes and streams	Habitat destruction, increased turbidity, competition with native fishes	Lever, 1994
Asiatic clam, *Corbicula fluminea*	Probably deliberate	Sandy sediments of lakes, rivers, and streams in North America	Possible competitive effects on native bivalves, high rates of filtration	Strayer, 1999; McMahon, 2000
Water hyacinth, *Eichornia crassipes*	Intentional introduction	Tropical and subtropical rivers worldwide	Interferes with boat traffic and waterflow; outcompetes native macrophytes	
Opossum shrimp, *Mysis relicata*	Intentional introduction	Flathead Basin	Collapse of salmon fishery	Spencer *et al.*, 1991
Daphnia lumholtzi	Unknown, probably human caused	Southeastern and south-central United States	Possibly more resistant to predation by zooplanktivores	Havel and Hebert, 1993; Kolar and Wahl, 1998
Bythotrephes cederstroemi	Ballast water from Europe	Northeast United States	Possibly more resistant to predation by zooplanktivores	NRC, 1996
Crayfish, *Pacifastacus leniusculus*	Intentional introduction	Europe	Transmission of disease and competitive removal of native crayfish	Vorburger and Ribi, 1999

This list contains only a small portion of the introduced species that have entered freshwaters (e.g., Benson, 2000)

EXTINCTION

Man has been reducing diversity by a rapidly increasing tendency to cause extinction of supposedly unwanted species, often in an indiscriminate manner. Finally we may hope for a limited reversal of this process when man becomes aware of the value of diversity no less in an economic than in an esthetic and scientific sense.

—Hutchinson (1959)

Humans likely will cause extinctions of at least half of all the approximately 10 million species on Earth in the next 50 to 100 years (May, 1988), despite the fact that the problem has been recognized clearly for more than 50 years. May calculated that worldwide extinction rates are currently 1 million times greater than rates of evolution, and the majority of taxonomists and ecologists understand that we are causing the sixth mass extinction event over the 3.5 billion years life has existed on Earth (Pimm *et al.*, 1995; Wake and Vredenburg, 2008).

In an even bleaker assessment of the situation, it has been argued that area-diversity relationships (discussed previously) can be used to estimate the ultimate effects of habitat destruction, and that in the long term the habitat destruction caused by humans to date will result in the loss of 95% of Earth's species (Rosenzweig, 1999). Freshwater habitats are probably the most impacted by humans. Because they integrate the landscapes they drain, aquatic systems suffer corresponding effects from all extensive terrestrial disturbances. Humans live near water and discharge their wastes into it. Control of hydrology by humans destroys habitats for many species. Consequently, losses of freshwater species have been substantial (Warren *et al.*, 2000; Jenkins, 2003; Strayer, 2006), and the outlook for sensitive aquatic species is bleak (Folkerts, 1997).

Rates of species extinction caused by humans far exceed the rates of evolution of freshwater organisms. More than 300 species of endemic fishes may have evolved in Lake Victoria, Africa, during the past 12,000 years (Johnson *et al.*, 1996), one of the highest known rates of species evolution ever documented. Many of these species have disappeared during the past two decades from overfishing, introduction of nonnative predators, and eutrophication (Barel *et al.*, 1985; Kaufman, 1992; Seehausen *et al.*, 1997). About 200 of these species disappeared in a single decade, primarily because of introduction of the predatory Nile perch, *Lates niloticus* (Goldschmidt *et al.*, 1993). The Victoria fishes may represent one of the most rapid speciation events, but species accrued 500 times more slowly than humans destroyed them.

Researchers generally assume that endemic species are most likely to go extinct and deserve the most attention for conservation. However, some cosmopolitan species also could go extinct because they generally rely on habitats that are temporarily available for colonization and act as fugitives that move across the landscape from one habitat to another. The massive alteration of habitats associated with human activities may cause extinction of such species, and this may have a broader geographic effect on community formation and ecosystem function than extinction of localized endemics (Cairns, 1993).

Many endangered species are aquatic or require aquatic habitat. In 2009 for the United States, the US Fish and Wildlife Service listed hundreds of aquatic species or subspecies on their threatened and endangered list, including 138 fishes, 72 bivalves, 38 snails, 25 amphibians, and 22 crustaceans. All these numbers have increased, by about 30%, since 1999. In addition, mammals and birds that require aquatic habitat are listed, such as the whooping crane (*Grus americanus*), the Florida panther (*Puma concolor coryi*; the only remaining habitat for this animal is wetland), and the piping plover (*Charadrius melodus*). Globally, the IUCN (World Conservation Union Red List, searched August 2009) lists 16,924 vulnerable or endangered freshwater species. It lists 260 species as already extinct. Many more may be endangered, but because they are inconspicuous they have not been listed (Strayer, 2001).

Many wetland plants, particularly those associated with the vernal pools of California, are also listed as endangered and threatened. About half of all the crayfish species in the United States and Canada are threatened. Pressures on native crayfish include habitat destruction, limited natural ranges, and introduced competitors (Taylor *et al.*, 1996).

Some of the most endangered fishes are the very large species found mostly in tropical waters (Stone, 2007). The largest freshwater fishes can be up to 500 kg mass, and 700 cm long (the Chinese paddlefish *Psephurus gladius*). Of the 20 largest species, six are endangered. This does not include the endangered sturgeon species that move between fresh and salt water. Most of these species are endangered by overharvesting, but pollution and habitat fragmentation (by dams) are also of concern. Interviews with elders who have fished the waters where these giants live for decades confirm that numbers and sizes of these massive fishes are decreasing.

Just because a group of species is not listed does not mean that no species in the group are threatened with extinction. We know much less about the small organisms than the large organisms. New species of aquatic insects, protozoa, algae, and bacteria are described daily, whereas new species of mammals and plants are described more rarely. It is clear that pollution, dewatering, and habitat destruction have negative effects on all freshwater organisms from fish to bacteria. Thus, small organisms may not be listed as endangered simply because their existence or the state of their population in nature is not known.

Many species have been *extirpated* (locally extinct). For example, the pearly mussels (Unionacea) are among the most endangered groups of animals on Earth. These mussels are long-lived and inhabit rivers and are thus influenced by any disturbance occurring in the watershed above (Strayer *et al.*, 2004). Of the 44 species of unionid mussels ever found in Kansas, four are extinct, six are endangered statewide, and an additional 16 are listed as threatened (Obermeyer *et al.*, 1997). Not all these species are rare or extinct in other parts of their historical range. The numbers of extinct, endangered, and threatened mussel species are similar in all states throughout the Mississippi drainage. Between 34% and 71% of the mussel species found in the southeastern United States are threatened or endangered. Thirty-six species that are thought to be extinct in the United States were found in the Tennessee River basin alone (Neves *et al.*, 1997). Likewise, some western states have high proportions of threatened endemic species. For example, in the Great Basin, Klamath, and Sacramento basins, 58% of the native fishes are endemic. Essentially all the native fish species are threatened in Nevada (Warren and Burr, 1994).

Ongoing amphibian declines are resulting in mass extirpations and extinctions of much of the amphibian diversity on the planet (Collins and Crump, 2009; Sidebar 11.2). Many of the amphibian species that we are losing inhabit

SIDEBAR 11.2
Global Amphibian Declines

Amphibian populations are declining around the planet at alarming rates (Collins and Crump, 2009). Recent estimates indicate that 43% of all amphibian species have experienced some degree of population decline, 33% are globally threatened, and over a hundred may be extinct (Stuart *et al.*, 2004). Further, most of these losses have occurred in the past three decades. The causes of some amphibian declines are still not entirely understood, but climate change, ultraviolet-B radiation, fungal pathogens, pesticides, habitat destruction, overexploitation, or combinations of these factors have all been implicated (Kiesecker *et al.*, 2001; Davidson, 2004; Pounds *et al.*, 2006; Lips *et al.*, 2006). In cases of severe habitat destruction and pollution, declines of amphibians and other freshwater organisms are not surprising. However, in many regions, amphibians are disappearing from seemingly undisturbed habitats such as high mountain cloud forests in the tropics, national parks in the United States, and other protected areas around the world. Scientists originally labeled these declines "enigmatic," but recent investigations have linked many of these mysterious events to the chytrid fungus *Batrachochytrium dendrobatidis* (Bd) (Skerratt *et al.*, 2007; Lips *et al.*, 2008). Since its discovery in the 1990s, Bd has now been associated with amphibian declines on every continent inhabited by amphibians. Bd causes the disease chytridiomycosis, which spreads rapidly through populations and can be lethal to most species under appropriate conditions. In many cases infected frogs become lethargic, slough their skin, and die, but otherwise show few symptoms.

Sudden, catastrophic amphibian declines associated with chytridiomycosis have been documented in remote, mountainous regions of Central America (Fig. 11.9), resulting in extirpations of about half the species and an overall 50% decrease in the abundance of amphibians (Lips *et al.*, 2006). Recent

FIGURE 11.9

Jars of frogs killed by the chytrid fungus *Batrachochytrium dendrobatidis* in a national park in Panama, and a dead *Bufo haematiticus* lying in a stream. The white substance on the toad's parotoid glands is a poisonous secretion that the toad released as it was stressed and dying. *(Photographs courtesy of Scott Connelly and Forrest Brem).*

extinctions of some African, Australian, Mexican, and South and Central American frogs have now been linked to chytridiomycosis. Unfortunately, chytridiomycosis outbreaks are most devastating in cool, moist habitats at high altitudes, where amphibian species diversity and endemism are highest. Studies in Central America indicate that species more closely associated with streams are most susceptible (Lips *et al.*, 2003; Brem and Lips, 2008), and thus the ecological impacts of these losses may be greatest in streams (Whiles *et al.*, 2006). Scientists have now linked many amphibian declines to chytridiomycosis, but there are still unanswered questions about the disease, such as how it spreads, how it survives in areas once amphibians have declined, and why it is so lethal to so many species in so many habitats.

In particular, there are still no definitive answers for why chytridiomycosis suddenly appeared in different parts of the planet. Scientists agree that it is an exotic species in some areas, but there is no consensus on where it originated and why it is now killing amphibians in so many regions. Some studies implicate climate changes that have made the disease more virulent and stressed amphibians, but there is also evidence that Bd may have simply spread around the planet with the global transport of amphibians or other animals and plants. Two species, the African clawed frog (*Xenopus laevis*) and the American bullfrog (*Lithobates catesbeianus*), have received considerable attention because they have been moved around the planet by humans in large quantities, they are very hardy, and they appear capable of carrying the disease with them (Weldon *et al.*, 2004; Daszak *et al.*, 2004). Massive quantities of *X. laevis* were shipped from Africa around the world for decades for use in human pregnancy tests, scientific studies, and the pet trade, and today there are established feral populations of *X. laevis* in Europe, the United States, and South America. Large numbers of American bullfrogs have also been exported from their native habitats in eastern and central North America for food, and bullfrogs have also established feral populations around the world. Recent studies further suggest that humans may be responsible for moving Bd and other diseases through the amphibian trade. Importation data for live amphibians into three major US ports of entry (Los Angeles, San Francisco, and New York) indicated that nearly 28 million amphibians of various species entered the country from 2000 to 2005. Testing with polymerase chain reaction (PCR) techniques indicated that over half of these imported amphibians were infected with Bd and about 9% with ranaviruses, another disease agent linked to amphibian declines (Schloegel *et al.*, 2009).

Habitat destruction may well be the most important factor in global declines. Most of the world's amphibian species reside in tropical areas. In particular tropical rain forests are very speciose, with hundreds of species found in the Amazonian basin alone. The rate of rainforest clearing is high and continues unabated.

This extinction event is unprecedented in geographic or taxonomic scope in recent history. There are undoubtedly ecological consequences beyond the inherent tragedy of this catastrophic loss of biodiversity. Loss of the amphibians will result in declines in predators of amphibians and changes in the structure and function of freshwater habitats (Whiles *et al.*, 2006; Connelly *et al.*, 2008).

Regardless of the ultimate causes for extinctions there are now some focused efforts to save remaining populations and species by establishing captive breeding programs, as well as researching ways to stop the spread of chytridiomycosis. If habitat destruction is not stopped, the only place to see many of these species may be in zoos.

freshwater habitats for at least part of their lives, and declines are homogenizing amphibian species diversity (Smith *et al.*, 2009).

WHAT IS THE VALUE OF FRESHWATER SPECIES DIVERSITY?

Processing of human wastes by aquatic microbes has been estimated to be worth more than $700 billion per year worldwide, and bioremediation of hazardous waste will save 85% of the costs over nonbiological methods of treatment (estimated at $135 billion worldwide over the next 30 years; Hunter-Cevera, 1998). The economic benefits of fisheries were discussed in Chapter 1, and biodiversity is important to the growing aquaculture industry. Species diversity provides genetic diversity and animals and plants for culture. Biological diversity is also important because of the reliance of aquaculture systems on healthy biological systems (Beveridge *et al.*, 1994). Inherent with species diversity is biochemical diversity, which has already provided myriad medicines and other important chemicals that benefit humans: this is still a rich area of scientific exploration.

Organisms can contribute to ecosystem structure and processes in a variety of ways, and thus biological diversity is important for sustained ecosystem function and integrity. Vital ecosystem processes such as decomposition and nitrogen cycling are governed by various species of microbes, plants, and animals in aquatic systems, and thus a reduction in the numbers of species in a system can be expected to negatively influence these processes at some point, as we will see in the remainder of this text.

Another reason for concern over extinction of species is that the species we live with are analogous to the canary in the coal mine; when the bird dies, it is time to get out. When numerous species go extinct, this could indicate that the ability of Earth to support life is endangered (Lawton, 1991).

Although important considerations, basing arguments for conserving species purely on economic and utilitarian grounds is dangerous. Many species may exist that have no economic value and are not essential to ecosystem function. Many species are not attractive to the general public, unless they have big eyes or flashy coloration. Do these species deserve to go extinct? It is common in public debate over listing a species as endangered to hear the question, "Aren't people more important than animals?" We suspect what is usually meant in this case is that an individual believes his or her right to make money the way he or she wants is more important than the right of a species to exist. If the immediate past is any gauge, species will continue to be driven to extinction, regardless of their value.

SUMMARY

1. Several aspects of diversity are used by ecologists, including the number of species (richness), evenness, within-habitat (α) diversity, and between-habitat (β) diversity.

2. Evolution ultimately drives diversity. Aquatic habitats that have been in existence for long periods of time, such as some tectonic lakes and old river basins, tend to have high diversity and many endemic species. Spatial segregation of habitats also influences species evolution.

3. Over short-term timescales (decades and centuries), factors other than evolution may influence diversity, including size of habitat, connectivity of habitats, habitat disturbance, sources of new species, species interactions, productivity, and species introductions.

4. Genetic diversity within populations and species is important for long-term survival. Conservation efforts and aquaculture programs are increasingly focused on maintenance of genetic diversity.

5. Population size is a critical factor in the survival of species. As populations decline below the minimum viable population size, the probability of extinction increases exponentially.

6. Spatial arrangement of subpopulations, metapopulation structure, is important for evolution and conservation.

7. Human-caused species introductions are common forms of pollution in aquatic environments. Some of these invading species have serious economic and ecological impacts on the communities that are invaded. Most successful invasions are difficult or impossible to reverse.

8. Humans are causing extinction of existing species at rates from hundreds to millions of times greater than the evolution of new species is occurring. Many of us alive today will likely witness the disappearance of half or more of the biodiversity on Earth, including many species in aquatic habitats.

QUESTIONS FOR THOUGHT

1. Why are temperate lakes generally relatively species poor?
2. What are the evolutionary advantages of very long diapause periods?
3. Are there more species of large than small animals?
4. Should conservation be based on local diversity (e.g., if a species is rare in one state and common in another, should it be conserved where it is already rare)?
5. What is the benefit of an ecosystem-based method of conservation?
6. Should DNA samples of endangered species be stored?
7. Will captive breeding programs allow us to preserve amphibian diversity?
8. How might genetic diversity be different than species diversity, particularly in microbial species?
9. Why might long-lived species appear to be doing OK because there is a large number of large adults but then abruptly go extinct?

CHAPTER 12

Aquatic Chemistry and Factors Controlling Nutrient Cycling: Redox and O_2

FIGURE 12.1

The macrophyte *Elodea* with O_2 bubbles from photosynthesis.

Doi: 10.1016/B978-0-12-374724-2.00012-X

Chemistry controls physiology, biogeochemistry, behavior of pollutants, and many aspects of how organisms interact with their environment and each other. The way chemicals are transformed and move through the environment is an essential aspect of ecosystem function. Nutrients often control primary production, and understanding nutrient cycling is central to understanding the influence of nutrient pollution. On a larger scale, nutrient cycling is tied to global-scale patterns of climate and production. Understanding nutrient cycling requires careful study of the concept of oxidation–reduction state (redox). The dynamics of the element oxygen and how it reacts with other compounds central to life are related closely to redox, so both are discussed in this chapter. In following chapters, additional nutrients (carbon, nitrogen, phosphorus, sulfur, and iron) are discussed. The material in this chapter forms a conceptual foundation for treatment of all biogeochemical discussions in this book as well as those on toxins, saline systems, and chemical interactions among organisms.

CHEMICALS IN FRESHWATERS

Among the chemical parameters of interest to the aquatic ecologist, some of the most important are the concentration of all inorganic ions (charged molecules) and concentrations of specific ions and gasses dissolved in solution. The total amount of dissolved and suspended organic material (carbon containing) and concentrations of specific organic molecules are also of interest. The chemistry of natural waters is an extremely complex field of study (Stumm and Morgan, 1981; Brezonik, 1994).

Chemical materials in aquatic systems can either be *dissolved* or *particulate*. The distinction between dissolved materials and particulate materials is arbitrary and commonly based on their ability to pass through a filter of a specified size. For example, dissolved metals and dissolved organic carbon are generally defined based on methods to analyze water passed through a 0.45-μm filter, but total dissolved solids are determined using material that passes through a filter with slightly greater than 1 μm retention (Eaton *et al.*, 1995). A more natural definition scheme has been proposed (Gustafsson and Gschwend, 1997) in which chemicals are divided into three classes: dissolved, *colloidal* (particles not settled by gravity), and *gravitoidal* (particles that will settle) (Fig. 12.2). A colloidal particle is defined specifically as any particle whose movement is not affected significantly by gravitational settling and that provides a location where chemicals can escape from the aqueous solution. This definition scheme has not been broadly adopted since its proposal because definitions based on analytical methods are more practical. Still, the scheme allows more realistic consideration of the physical and chemical state of materials associated with natural waters.

Total mass of ions is estimated as *total dissolved solids* (total mass of material dissolved in water, often noted as TDS), *salinity*, or *conductivity*. Conductivity

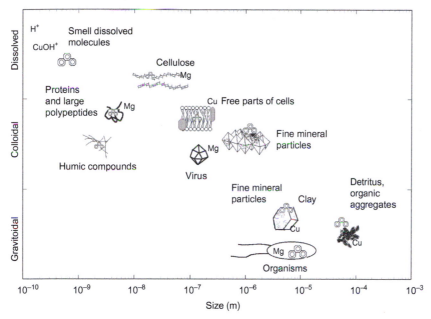

FIGURE 12.2

Chemical associations with dissolved, colloidal, and gravitoidal particles. The association is depicted with three trace substances, inorganic copper and magnesium ions, and the three-ringed organic compound phenanthrene. Dissolved chemicals depicted include a variety of organic chemicals. Colloidal particles include membrane pieces, viruses, inorganic precipitates, and aggregations of organic chemicals. Gravitoidal particles include clays, planktonic cells, and larger aggregates of organic and inorganic materials. *(Redrawn from Gustafsson and Gschwend, 1997).*

and salinity are commonly used because they are easy to measure. Conductivity is proportional to the relative amount of electricity that can be conducted by water. The more dissolved ions present, the higher the conductivity. Conductivity can be correlated approximately to system productivity because high nutrient waters have high conductivity, but other factors including concentration of nonnutrient salts also influence conductivity. For example, saline lakes have extremely high conductivity, but do not necessarily have correspondingly high productivity if nutrients such as nitrogen and phosphorus are low. The value for total dissolved solids does not always exactly correlate to that of conductivity. Only ionic compounds are included in conductivity, whereas uncharged molecules (such as many dissolved organic compounds) also contribute to total dissolved solids. Also, various salts lead to different conductivity. For example the ions ferrous iron (Fe^{2+}) and ferrous lead (Pb^{2+}) have the same charge, but iron and lead have molecular weights of 55.85 and 207.2 grams per mol, respectively, roughly a four-fold difference. Salinity is the mass of dissolved salts per unit volume. Complete chemical analysis is necessary to determine true salinity, but conductivity is generally used as a surrogate.

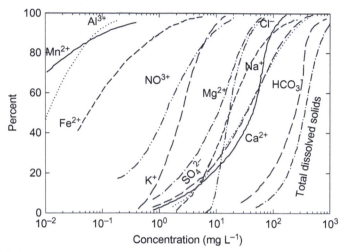

FIGURE 12.3

Cumulative curves showing the frequency distribution of various ions in water running off of land. Each line shows the percentage of water samples with less than the corresponding concentration. *(Redrawn from Davis and De Weist, 1966).*

As water moves through terrestrial ecosystems, materials are dissolved or *weathered* from the land. Chemical weathering releases dissolved matter, whereas mechanical weathering releases particulate matter that may react to form dissolved matter at some point. Mechanical weathering can include freezing and thawing as well as erosional activity of wind and water that moves particles against each other. Microbes can enhance rates of weathering (Bennet *et al.*, 2001), and their activity requires water and time. Thus, the total concentration of dissolved matter is related inversely to the amount of runoff because the higher the runoff, the less time water has to dissolve ions. However, the relationship between runoff and total dissolved solids is variable because of differences in geomorphology, geology of the parent material (e.g., the relative abundance and solubility of the ions in the native sedimentary and igneous rocks and soils), and area of runoff. The amount of dissolved materials associated with a set amount of precipitation can vary over an order of magnitude as a result of differences in geomorphology or the composition of the parent material (Fig. 12.3). Furthermore, some ions, in particular nutrients such as nitrate, can decrease in concentration as they are assimilated (incorporated into biomass) by terrestrial and aquatic biota. Therefore, the relative abundance of ions in water flowing from terrestrial habitats also depends on chemical interactions with biota. Much of the material that enters watersheds probably does so through smaller rivers and streams (Hynes, 1975; Alexander *et al.*, 2000).

Solubility and relative abundance of elements in Earth's crust lead to some general patterns of average relative abundance of dissolved ions in river waters. For example, about 75% of Earth's crust is composed of silicate, but it is relatively insoluble so it does not dominate in natural waters. Aluminum (abundant but not soluble) and manganese (low abundance and solubility) are found at low concentrations relative to calcium or sulfate. Concentrations of some ions are constant across many rivers. Silicate is very abundant, but with relatively low weathering rates. Silicon has moderate biological importance (in some plants and diatoms) but generally concentrations are constant. Concentrations of other elements (e.g., sodium and iron) vary over many orders of magnitude, depending on regional geology (Fig. 12.3). Concentrations of nitrate and phosphate vary over several orders of magnitude because plants have high affinity for these nutrients when they limit primary production, but agricultural fertilization or weathering of natural deposits can lead to very high concentrations. Variance in nitrogen and phosphorus concentration is large because biological demands can lead organisms to deplete concentrations to very low values in relatively pristine waters, and human activities have boosted nutrient concentrations far above natural levels in many rivers.

Solubility of compounds in waters can be characterized by solubility constants. Solubility of most compounds increases with temperature, but several important compounds ($CaCO_3$ and $CaSO_4$) become less soluble at higher temperatures. Equilibrium solubility constants indicate what concentration of a material will dissolve in water when it is saturated. When concentrations are below saturation, the material will continue to dissolve, above saturation the compound will precipitate.

Concentration of hydrogen ions (protons) is also central to maintaining biological activity as well as defining the chemical nature of the environment. This concentration, or acidity, is expressed as pH. The pH is a logarithmic scale, corresponding to the following equation:

$$pH = -\log_{10}\{H^+\}$$

where $\{H^+\}$ is the hydrogen ion activity, which is closely related to concentration expressed in moles per liter. Hydrogen ions interact with other ions at all but the most dilute concentrations, so concentration does not exactly correspond to hydrogen ion activity. The smaller the pH value is, the greater the hydrogen ion activity. On this log scale, each change of one unit of pH corresponds to a 10-fold change in hydrogen ion activity. Pure water has a pH of 7.0 (1.0×10^{-7} mol H^+ liter^{-1}). Vinegar and beer have a pH of about 3, stomach acid has a pH of 2, and household ammonia has a pH of about 11. The actual range found in most aquatic ecosystems is near neutrality (several orders of magnitude in activity; pH $= 7 \pm 1$). Acid precipitation causes devastation in many aquatic systems because it both lowers the pH several orders of magnitude (a hundred- to thousand-fold increase in H^+ activity), and it mobilizes and increases solubility of toxic metals. This problem will be discussed in Chapter 16. Increases in global atmospheric CO_2 concentrations are also increasing hydrogen ion activity as carbonic acid forms when the gas is dissolved in water. Runoff from mining waste can also be very acidic. Some natural habitats such as acid springs (pH < 1) or alkaline lakes (pH > 10) have extreme pH values. The value of pH is easy to determine because reliable hand-held probes are readily available, easy to use, and inexpensive.

Alkalinity or *acidity* indicate the capacity of water to react with a strong acid and base, respectively, and are measured routinely to determine the suitability of domestic water and the relative content of inorganic carbon. *Hardness* is the sum of magnesium and calcium ions present and an indicator of the ability of water to precipitate soap and is also of particular interest for domestic water supplies. Hard water causes scaling as well as interfering with the ability of soaps and detergents to clean properly. Scaling can reduce the efficiency of hot water heaters, and some vegetables lose flavor when cooked in hard water. For these reasons, hard water is not desirable for domestic uses and many water supplies are treated to reduce hardness. Water treatment plants typically use chemical precipitation methods to remove hardness, and in individual homes, ion exchange resin systems are common. Many companies claim that magnetic treatment can lower hardness of water. There is currently no credible scientific evidence to back up the viability of these products based on magnetism, nor an adequate scientific explanation for the mechanism of magnetism lowering water hardness.

Additional bulk chemical parameters that are commonly measured as indicators of water chemistry or quality include color, taste, odor, suspended solids, and *turbidity*. Turbidity is measured as the light absorption or the light scattering of water. Light absorption is often influenced by suspended particles such as clays or suspended algae, but other factors can influence absorption of light such as tannins and lignins from decomposition of organic material. Turbidity is significantly correlated with total suspended solids. Total suspended solids (TSS) are determined by filtering water onto a preweighed filter and then drying and weighing the filter. Material can be filtered onto non-combustible glass fiber filters and heated at high temperature (e.g., 450°C) to burn off organic material. The loss in mass following combustion can be used to calculate total suspended organic material and total suspended inorganic material.

These bulk parameters are traditionally the first indicators used to characterize general water quality. Other dissolved materials will be discussed in more detail in this and other chapters. Concentration of specific dissolved ions, other dissolved compounds, and suspended particulate material can be analyzed by a wide variety of methods (Method 12.1) and these methods continue to evolve as technology develops.

METHOD 12.1

Analysis of Concentration of Specific Dissolved Ions and Other Dissolved and Particulate Materials in Natural Waters

Many methods are used to detect the concentrations of specific ions in natural waters; only a few will be discussed here. For a more definitive treatment of a wide variety of methods, refer to *Standard Methods for the Determination of Water and Wastewater* (Eaton *et al.*, 2005). Probably the simplest method of analysis is the use of ion-specific probes. The probes are calibrated in solutions with known ion concentrations. The calibrated probes can be immersed in the unknown solution and the concentration of the ion read directly from the meter. Most biologically important ions are found in concentrations too low for analysis with ion-specific probes in unpolluted habitats; thus, other methods of analysis are required.

Colorimetric methods are used commonly to analyze concentrations of specific ions. These methods are based on formation of a specific colored compound from the ion of interest. The concentration of the colored compound can then be determined with a spectrophotometer (a machine that measures the absorption of light at specific wavelengths). Beer's Law (the Beer-Lambert Law) states that the relationship between absorption of light by a colored compound and concentration of that compound is linear at intermediate concentrations.

$$A = eCl$$

where A = absorbance, e = extinction coefficient, C = concentration of a chemical that absorbs light, and l is the distance the light needs to travel through the solution. Beer's Law allows the laboratory worker to make a standard curve from which the unknown concentration in natural waters can be calculated directly. Autoanalyzers are used routinely in laboratories that need to analyze a large number of water samples. These machines automatically take a small sample, add the appropriate chemical mixtures for a colorimetric reaction to occur, measure absorption at the appropriate wavelengths, and use a computer program to calculate concentration automatically from absorption.

Ion chromatography is also used in many laboratories. This technique is based on the idea that dissolved ions will pass at different rates through specific materials. A column filled with an appropriate material is used, and a carrier solution is passed through the column. A small amount of the sample is added to this carrier solution, and the sample is carried through the column. The time needed for specific ions to pass through is called the retention time and is known for ions of interest. Ions in the sample are detected after they pass through the column and are separated according to their retention times. A nonspecific detector for ions (e.g., a sensitive conductivity meter) can be used to estimate the amount of material passing out the end of the column. The method is rapid and can be used to analyze a large number of ions simultaneously. However, it is generally less sensitive than colorimetric methods.

Atomic absorption spectrometry is used to estimate concentrations of dissolved and particulate metals. This method is based on the idea that individual elements emit light at very specific wavelengths when their electrons are excited. This is the same concept used to detect the elemental composition of distant stars. For the analysis, the water sample is injected into a chamber, where it is subjected to high energy that causes excitation of electrons. The intensity of the light at the wavelength specific to the ion is directly proportional to the amount of the ion. The signal is compared to a standard concentration, thus giving the concentration of the element in the sample.

Bulk dissolved and particulate organic carbon concentrations are usually analyzed by conversion to CO_2 followed by analysis of concentration of this gas. Some controversy exists regarding the efficiency of different methods (e.g., UV, high temperature, or persulfate oxidation digestions) used to decompose organic carbon to CO_2 (Koprivnjak *et al.*, 1995). Given the complex chemical composition of dissolved and particulate organic carbon, it is not surprising that analysis is difficult. Several methods are available for analyzing particulate materials. One involves degradation of particulate material into dissolved ionic forms and then analysis by methods mentioned previously for dissolved ions. Another method involves combustion at high temperature and analysis of the resulting gasses (e.g., organic nitrogen is converted to N_2 gas). As with organic carbon, problems can occur with the efficiency of degradation.

Organic materials can be analyzed more specifically with gas or liquid chromatography and by a number of advanced analytical methods such as spectral fluorometry, mass spectrometry, and other techniques. These methods rely on a principle similar to ion chromatography as discussed earlier; molecules of different mass or charge pass through separation columns at various specific rates. With appropriate detectors, very small concentrations (nanomoles per liter) can be sensed. This ability to sense vanishingly small concentrations is important because many toxins are highly active even at very small concentrations. If the output of a chromatograph is hooked to a mass spectrometer, the exact mass of compounds that are separated by the chromatography can be determined to identify specific compounds more exactly.

REDOX POTENTIAL, POTENTIAL ENERGY, AND CHEMICAL TRANSFORMATIONS

Some fundamental chemical principles are required to understand chemical interactions and transformation. Students are encouraged to review their basic chemistry. Some key concepts are chemical equilibrium, activation energy, free energy, and solubility constants. These concepts will be mentioned, but in-depth discussion is beyond the scope of this text. Some concepts, such as redox potential, will be discussed in more detail because experience has shown that students require substantial review on this topic to understand chemical interactions and transformations.

The relative availability (concentration) of electrons for chemical reactions in solution is referred to as *oxidation–reduction potential* or *redox potential*. This parameter is important because it quantifies the general chemical environment of the water. This aspect of chemistry determines the potential energy requirements or yields during biotic or abiotic processes that transform chemical compounds in the environment. Stated differently, redox potential is a large determinant of what chemical reactions will occur without an input of energy, what reactions will require energy, and what products will be favored in the environment. Even though the concept of redox can be difficult for some students to grasp, it is worth the effort because it provides an understanding of the way that nutrients cycle and the behavior of pollutants in the environment (Christensen *et al.*, 2000). Building a rational framework based on principles of chemistry is essential to comprehending the cycles that underlie ecosystem function.

Redox potential of natural systems is simple to measure with electrodes that assess availability of transferable electrons relative to the availability of electrons in hydrogen gas. Such sensors read in millivolts, with low values (e.g., below 100 mV) denoting large numbers of transferable electrons and high values of redox denoting few transferable electrons. Even though redox potential is easy to measure, prediction of an exact redox potential for natural aquatic habitat based on chemical composition is difficult because of the myriad of chemical compounds that co-occur, because the different compounds may

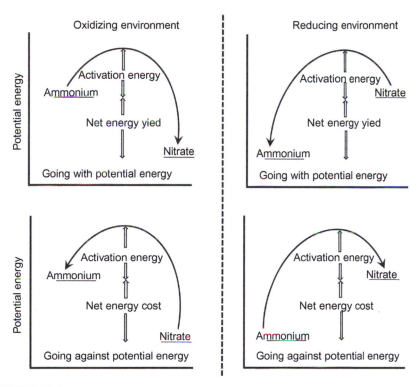

FIGURE 12.4

Representative Gibbs free energy diagrams for ammonium and nitrate ions in oxidizing and reducing environments. The activation energy is the energy required to move from a state of high to low potential energy, and the energy yield is what is released when the transformation occurs. The activation energy plus the energy yield are required to accomplish a transformation from low to high potential energy.

not be at equilibrium, and due to the complex interactions among many dissolved organic and inorganic chemicals.

The central idea that links the concept of redox potential to biogeochemical cycling is that chemical compounds have *potential energy* when they have a redox significantly different from their surrounding environment. *A compound can have a high potential energy if it is a low redox compound in a high redox environment or a high redox compound in a low redox environment.* This is a difficult concept for many students to grasp and bears close thought. Gibbs free energy diagrams are the best way to visualize the potential energy (Fig. 12.4). The redox potential of different chemical transformations can be plotted relative to each other (Fig. 12.5). The energy yield of a chemical transformation is the distance between the head and tail of the arrow on Figure 12.5. The transformation will yield energy if the redox potential of the solution is less

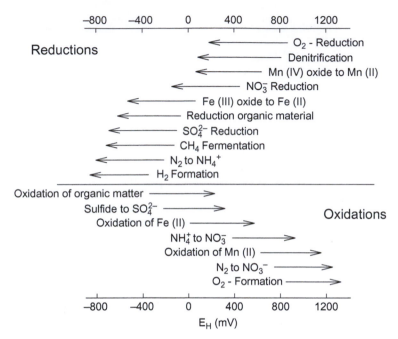

FIGURE 12.5

Microbe-mediated chemical transformations plotted to show energy yield as the difference between the tail and the head of the arrow and redox potential required to complete transformation. *(Redrawn from W. Stumm, and J. J. Morgan,* Aquatic Chemistry: An Introduction Emphasizing Chemical Equilibria in Natural Waters. *1981. John Wiley & Sons, Inc. Reprinted by permission of John Wiley & Sons, Inc.).*

than the head of the arrow for the reductions and greater than the head of the arrow for the oxidations.

It is important to distinguish the total potential energy to be gained or lost during conversion among chemical forms, in addition to determining the energy required for the reaction to occur. The energy required to make the conversion is called the activation energy (Fig. 12.4). The reaction will not occur rapidly if the activation energy needed is high. For example, ammonium can exist in an oxidized solution containing dissolved oxygen without spontaneously combining with the dissolved oxygen gas and converting to nitrate, even though ammonium has a significantly higher potential energy than nitrate under oxidizing conditions. The activation energy of this conversion is too high for the reaction to proceed at significant rates under normal conditions found in natural aquatic environments.

One of the primary determinants of redox is the concentration of dissolved oxygen gas (O_2). O_2 concentration has been considered a master variable in determining biogeochemistry and redox because of the O_2 molecule's tremendous affinity for electrons. When O_2 is present, redox values must be high—generally in excess of 200 mV. This high redox potential indicates an oxidizing environment, with very few available electrons that allows only specific chemical reactions (those that release electrons) to proceed without a net input of energy.

Iron concentrations in a dimictic lake are an example of how O_2 concentrations regulate redox potential to control the concentration of a chemical in the aquatic environment (Fig. 12.6). In this case, ferrous iron (Fe^{2+}), the reduced form of iron, is soluble, but ferric iron (Fe^{3+}), the oxidized form of iron, forms an insoluble precipitate in water. Ferrous iron converts readily to ferric iron in the presence of O_2 (i.e., the activation energy for the reaction is low) and then precipitates and settles to the sediments if O_2 is present through the water column and mixing rates are low. When the lake is mixed fully and in contact with the atmosphere, iron concentrations are low throughout the water column. As O_2 is depleted from the deeper stratified layers, the dissolved iron concentration increases as ferrous iron diffuses out of the sediment. In this case, low redox potential and high dissolved iron concentrations are correlated closely.

Organisms can promote chemical reactions that would not otherwise occur by lowering the activation energy, and in some cases providing energy to force a chemical reaction that goes against potential energy. This promotion of chemical reactions is accomplished with catalytic enzymes that lower activation energy. In the previous example, in which ammonium is stable in aquatic habitats containing O_2, microorganisms can lower the activation energy required to oxidize ammonium to nitrate. This reaction releases energy because nitrate has a lower potential energy than ammonium in the presence of O_2. Bacteria can direct this energy toward cellular growth; the process is called nitrification and will be discussed in greater detail in Chapter 14.

Organisms can drive chemical reactions against potential energy (create more energetic chemical compounds). Such reactions require more input of potential energy than is stored in the products. Photosynthesis is an excellent example of a reaction that goes against potential energy; CO_2 is transformed to sugar (with a higher potential energy) using the energy of sunlight to accomplish the energy-requiring transformation. Many forms of nutrient assimilation require energy for uptake and transformation to forms that can be used in biochemical pathways. For example, nitrate requires more energy to use than ammonium and nitrogen gas more energy than nitrate (see Chapter 14) but if nitrogen is limiting, then the energy will be used to transform the nutrients so they can be used.

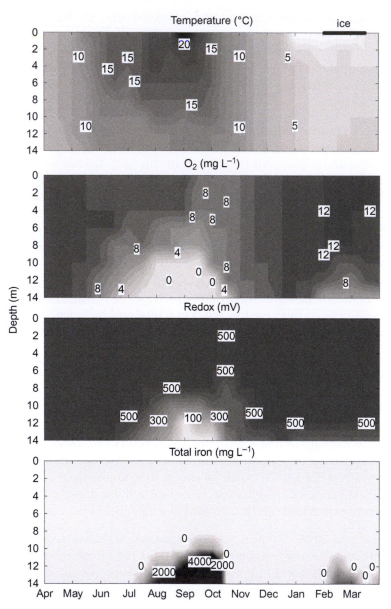

FIGURE 12.6

Temporal patterns of temperature, O$_2$, redox, and total iron in Esthwaite Water (an English lake) over a year. This type of figure is common for representing time series in lakes. The contours represent the boundaries of the values, with depth on the *y* axis and time on the *x* axis. *(Redrawn from Mortimer, 1941).*

OXYGEN: FORMS AND TRANSFORMATIONS

The element oxygen can be found in many forms in the natural environment, including water. The predominant form in the atmosphere is oxygen gas, O_2, at about 21% of atmospheric gas. Oxygen is found in numerous compounds in combination with many other elements. Oxygen is a major component of organic compounds and biologically relevant inorganic compounds. As a result, it is important to understand the behavior and distribution of oxygen in the natural environment in order to appreciate its impact on aquatic ecosystems.

The amount of O_2 dissolved in water is a function of many factors, including metabolic activity rates, diffusion, temperature, and proximity to the atmosphere. This amount can be expressed in several related concentration units, including mg liter^{-1}, mol liter^{-1}, and percentage saturation. The percentage saturation is the concentration of O_2 relative to the maximum equilibrium concentration for that solution. *Dissolved oxygen* (DO) refers to the O_2 dissolved in water (as opposed to oxygen that is part of other chemical compounds).

The *saturation concentration* of O_2 is determined as the equilibrium concentration when pure water is in contact with the atmosphere for an extended period of time. The amount of O_2 that can be dissolved in water is a function of temperature; the lower the temperature, the greater the concentration of O_2 under equilibrium conditions (Fig. 12.7). In addition, the greater the atmospheric pressure, the greater the saturation O_2 concentration. Atmospheric pressure is

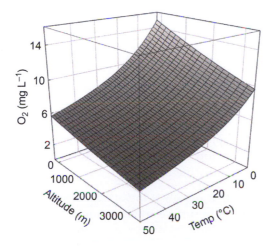

FIGURE 12.7

Saturation concentrations of dissolved O_2 as a function of temperature and altitude. The equation that describes the curve is: $\ln(O_2) = 2.692 - 1.27 \times 10^{-4}$ (alt) $- 6.15 \times 10^{-10}$ (alt)$^2 - 0.0286$ (temp) $+ 2.72 \times 10^{-4}$ (temp)$^2 - 2.09 \times 10^{-6}$ (temp)^3, where $O_2 = $ mg L^{-1}, alt is altitude in meters, and temp is temperature in °C. *(equations modified from Eaton et al., 1995).*

a function mainly of altitude (Fig. 12.7), but variation with barometric pressure can also be important. A 10% change in barometric pressure at one location is extreme, but air pressure on a very tall mountain (5,400 meters) can be half the value observed at sea level. Finally, O_2 saturation increases with increasing water pressure (i.e., depth in the water) and decreases with increasing salinity.

A point of confusion for some students is the concept that O_2 concentrations can exceed the level of saturation. If O_2 becomes highly supersaturated, it will form bubbles and come out of solution. However, at concentrations several-fold greater than saturation, O_2 can remain dissolved and slowly equilibrate with the atmosphere. Likewise, when O_2 is below saturation in solution, it will slowly come to equilibrium with exposure to the atmosphere. Both *supersaturation* and *subsaturation* are common in natural waters because they can be caused by photosynthesis and respiration, respectively.

The presence or absence of O_2 is an important aspect of aquatic ecosystems because it determines whether and what type of organisms can live in a given ecosystem. Habitats that have any O_2 in the water are referred to as *oxic* or *aerobic*, and those without detectable O_2 are *anoxic* or *anaerobic*.[1] The absolute concentration of O_2 is also important, and several methods have been developed to measure it (Method 12.2).

The main cause of biological consumption of O_2 in most environments is aerobic respiration. All organisms must metabolize, and the oxidation of organic carbon with molecular O_2 gives a greater energy yield than does oxidation using other molecules. The general reaction is

$$CH_2O + O_2 \rightarrow CO_2 + H_2O + \text{chemical energy}$$

where CH_2O is a general stoichiometric representation of the formula for sugar (not the structure of formaldehyde), and chemical energy is in the form of ATP. However, the ratios of C to H and O vary in different organic compounds (e.g., phospholipids have relatively low oxygen concentrations relative to sugars). The previous formula should be considered only an approximate representation of respiratory metabolism.

In addition to the O_2 that dissolves in surface water from the atmosphere, a significant amount of O_2 in surface environments is contributed by photosynthesis. A generalized equation for oxygenic photosynthesis is

$$CO_2 + H_2O + \text{light energy} \rightarrow CH_2O + O_2$$

---

[1] The older terminology (aerobic and anaerobic) is less precise than the newer terminology (oxic and anoxic) because the prefix aer- refers to air, which also contains other gasses.

METHOD 12.2

Measuring O_2 Concentration

The most commonly used method for measuring O_2 concentration in the field is an O_2 electrode. The electrode has a gold-plated cathode and a reference electrode (anode). When a voltage is applied across the cathode and the anode, the cathode reacts with O_2 and causes an electrical current to flow. The more O_2 that reacts with the cathode, the higher the current that registers on the meter. A higher current corresponds with a higher O_2 concentration. This type of electrode must be calibrated regularly for variations in temperature and atmospheric pressure. A more recently introduced electrode takes advantage of a fluorescent compound that fluoresces more brightly when exposed to more O_2. The compound is impregnated in a membrane. The membrane is excited optically and the resulting wavelength of fluorescence is detected. The signal is proportional to the O_2 concentration. These two types of electrodes are useful because they can be constructed with long lead wires and lowered into lakes or wells to assess *in situ* O_2 concentrations. These electrodes can also be attached to a data logger in a submersible housing and used to monitor dissolved O_2 unattended over days or weeks. Very small-scale cathode tips have been constructed to allow for determination of O_2 with <0.1 mm spatial resolution (Revsbech and Jørgensen, 1986).

The other older method of measurement of O_2 is Winkler titration. This method has been used for many years and is very reliable. In this method, reagents are added that cause iodine to react with O_2 molecules, forming iodate, IO_3^-. The iodate is then titrated back to iodine, and the amount of titrant necessary to react with the iodate is directly proportional to the O_2 concentration (Eaton *et al.*, 1995). The Winkler method is more time-consuming but is often more accurate and more precise than use of standard O_2 electrodes. The Winkler method requires collecting samples without introducing O_2. Van Dorn samplers (see Chapter 7) and other methods provide techniques for collecting samples and not exposing them to the atmosphere.

Oxygen is difficult to detect at very low concentrations (Fenchel and Finlay, 1995), so it is problematic to determine whether or not an environment is strictly anoxic. Strict anoxia is required for some biogeochemical processes. Some organisms are extremely sensitive to low O_2 concentrations and must live in stringently anoxic conditions. Specialized dyes and sample handling methods are used to sample for strict anoxia.

The photosynthesis equation is essentially the reverse of the equation for respiration. Again, the equation for photosynthesis is a general equation because the exact composition of the organic molecules (represented as a general stoichiometry for sugars, CH_2O, in the equation) varies depending on the biological molecules being synthesized.

Respiration and photosynthesis are primarily responsible for maintaining a constant concentration of O_2 in Earth's atmosphere. At smaller scales, O_2 concentrations are not constant. When photosynthesis dominates, O_2 concentrations exceed saturation, but when respiration dominates, a habitat could become anoxic in the absence of any O_2 input from the atmosphere.

PHOTOSYNTHESIS

Photosynthesis is the process that provides the energy to run most ecosystems. Light, temperature, and nutrients all control photosynthetic rates. Light is generally the predominant factor over the short term.

Some of the terminology used to describe photosynthetic processes must be explained initially. Because respiration and photosynthesis are both occurring simultaneously, even within individual photosynthetic organisms (another point of confusion among beginning biology students), we can distinguish between net primary production (*NPP*) and gross primary production (*GPP*) as follows:

$$NPP = GPP - respiration$$

Net primary production is the photosynthesis that occurs in excess of the respiratory demand. *Gross primary production* is the total amount of photosynthesis that occurs before the losses to respiration are accounted for. The equation relating *NPP*, *GPP*, and respiration is used frequently in aquatic ecology (Example 12.1). The value for *NPP* (positive, negative, or close to zero) determines if O_2 is being supplied or consumed by organisms in the environment.

■ Example 12.1

Calculating Net and Gross Photosynthetic Rate and Respiration from Lake Water Samples

An experiment was done to determine photosynthetic rate of phytoplankton in a lake at midday. Water was sampled, and the initial O_2 content was determined to be $8.0\,mg\,liter^{-1}$. Three clear and three dark bottles were filled with this lake water and suspended in the lake at the depth of collection for $1\,h$, then the water in the bottles was analyzed for O_2 content. Water in the clear bottles had 9.5, 10, and $10.5\,mg\,O_2\,liter^{-1}$ and in the dark bottles had 7.7, 7.5, and $7.3\,mg\,O_2\,liter^{-1}$. Calculate net and gross photosynthetic rate and respiration.

Net photosynthetic rate refers to the photosynthesis that occurs in excess of respiratory demand and is calculated by the increase in O_2 in the light bottles compared to the initial O_2 concentration. The average final concentration was $10\,mg\,O_2\,liter^{-1}$, so the net photosynthetic rate $= (10 - 8\,mg\,O_2\,liter^{-1})/1\,h = 2\,mg\,O_2\,liter^{-1}h^{-1}$.

The respiration calculation is similar; the rate is the difference between the dark bottle and the initial O_2 concentration. Thus, respiration rate $= (8 - 7.5\,mg\,O_2\,liter^{-1})/1\,h = 0.5\,mg\,O_2\,liter^{-1}h^{-1}$. Note that when respiration is calculated in the same way as net photosynthesis, a negative flux rate (negative rate of O_2 production) is obtained. Finally, gross photosynthetic rate is net photosynthetic rate + the oxygen consumed by respiration. Gross photosynthetic rate $= 2 + 0.5\,mg\,O_2\,liter^{-1}h^{-1} = 2.5\,mg\,O_2\,liter^{-1}h^{-1}$. This value can also be calculated by subtracting the concentration value for the dark bottles from that of the light bottles and dividing by time. ■

When the production of an entire ecosystem (all primary producers and heterotrophs in an area) is being spoken of, the terms *net ecosystem production* (*NEP*), *gross ecosystem production* (*GEP*), and *ecosystem respiration* (*ER*) are often used and the same equation applies:

$$NEP = GEP - ER$$

Photosynthesis–irradiance (P–I) relationships describe the effects of light on photosynthetic rate. Several parameters are generally used to describe this relationship (Fig. 12.8A). These P–I parameters include the respiration rate (O_2 consumption in the dark), the *compensation point* (where gross production equals respiration), α (the initial slope of the line), P_{max} (the maximum photosynthetic rate), and β (a parameter describing the deleterious effects of high light or *photoinhibition*). Understanding this curve provides initial insight into how light alters photosynthetic rates and the strategies that photosynthetic organisms can use to compete successfully in their environment.

Organisms that live in low-light habitats (e.g., deep in lakes, in a small shaded stream flowing through a forest, or under the dense canopy of a forested wetland) have several characteristics allowing them to survive and compete that can be described by P–I curves. They have a relatively low compensation point so respiration is equal to photosynthetic O_2 production at very low light. Such organisms also have a rapid increase in photosynthetic rate as light increases (a steep α). These organisms tend to be photoinhibited at relatively low irradiance (Fig. 12.8B).

The light field can vary spatially and temporally. Clouds, vegetation, and waves cause variation in the light intensity reaching photosynthetic organisms (as discussed earlier in Chapter 3). Some species of algae are adapted (have appropriate evolutionary characteristics) to compete more effectively in a variable light field (Litchman, 1998). Steep spatial light gradients (e.g., from full light to less than 1% sunlight over cm or mm) can occur in natural environments; periphyton and sediments attenuate light rapidly in benthic habitats. In lakes light is attenuated less rapidly unless turbidity is high. Organisms must acclimate to spatial light gradients, and different light conditions may favor different species. Primary producers can either have permanent physiological characteristics (be adapted) to specific light conditions, or can alter their physiology (acclimate) to maximize production under various light conditions.

Mixing also influences the light regime of phytoplankton in lakes. If turbulent mixing is deep enough, the average light experienced can be too low to support growth. The mixing depth below which growth does not occur is called the *critical mixing depth*. Phytoplankton are most likely to be mixed below the critical depth during winter when a lake is not stratified and ambient light is low. High levels of inorganic turbidity also lead to lakes where phytoplankters are mixed below the critical mixing depth.

Organisms that live in intense sunlight near the water's surface have special characteristics that allow them to succeed in this harsh habitat. The strongest effect of high light is photoinhibition caused by the direct or indirect damaging effects of high-intensity light on the molecules in the cells. The most likely

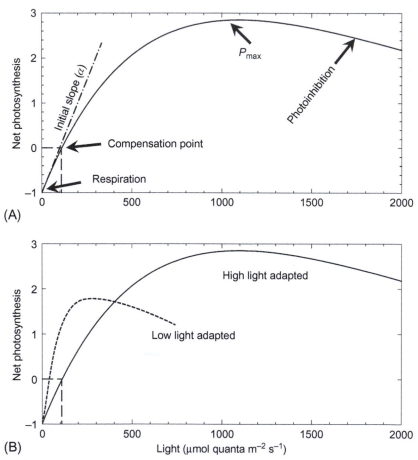

FIGURE 12.8

Diagram of a representative relationship between net photosynthetic rate and irradiance (A) and comparison of high-light- and low-light-adapted species (B). The low-light species has a steeper α, lower compensation point and P_{max}, and greater photoinhibition. Irradiance is in photosynthetically available radiation, and photosynthetic rate is in arbitrary units. A single species can acclimate to light in a similar fashion.

site of damage in photoinhibited cells is in the photosynthetic apparatus at photosystem II (Long *et al.*, 1994). Most studies of photoinhibition in the natural environment have been on phytoplankton (Long *et al.*, 1994). Photosynthetic organisms can protect themselves from high light by synthesizing special protective pigments and, if they are motile, by moving into areas with lower irradiance. Carotenoids with a red color are often synthesized, and these compounds can be concentrated by zooplankton and fish and lead to an orange or red color of their tissues (commonly seen in high altitude lakes where irradiance is intense). Photosynthetic organisms that are acclimated to intense light have low values for α as well. The compensation point of organisms acclimated to high light occurs at relatively high irradiance. Thus, an organism that is acclimated to high light will not compete well in low light. The inhibitory influence of ultraviolet radiation on photosynthetic rates is of particular concern given the increased amounts reaching Earth's surface with the thinning of the atmospheric ozone layer (Sidebar 12.1).

SIDEBAR 12.1
The Influence of UV Radiation on Aquatic Photosynthetic Organisms

Human activities have led to significant decreases in concentrations of ozone (O_3) in the stratosphere, allowing more UV radiation to penetrate the atmosphere and causing substantial biological effects (Bancroft *et al.*, 2007). The two types of UV radiation that are of concern are UV-A (320–400 nm) and UV-B (280–320 nm). They have increased significantly at high and low latitudes, with the greatest increases near the poles. Even though phase-out of chemicals that cause ozone depletion began seriously in the early 1990s, UV levels have continued to rise with the largest Antarctic ozone hole (as of this writing) occurring in 2006. UV is expected to gradually decrease over the next 50 years (Madronich *et al.*, 1995) but it could take considerably longer for Earth's atmosphere to recover. Understanding the ecological effects of this increased UV irradiance requires knowledge of how it influences photosynthetic organisms (Häder, 1997).

The first question is how much are the primary producers exposed to the UV? A variety of atmospheric factors can alter incoming UV, including seasonal variation in O_3 depletion, amounts of UV-scattering particulate material (including pollutants) in the air, increased cloudiness, and altitude. Yearly variation in incident UV irradiance can be considerable (Leavitt *et al.*, 1997). Once UV enters water, the depth of the water and concentration of UV-absorbing compounds control the amount that ultimately reaches aquatic organisms. Dissolved organic carbon is the predominant material that absorbs UV, and rapid attenuation occurs with moderate dissolved organic carbon concentrations (i.e., more than several mg C liter^{-1}). Given these considerations, another important determinant of UV exposure is the position of organisms in the aquatic habitat. Organisms inhabiting shallow waters, such as those in wetlands, open streams, and littoral regions of lakes, may have very high exposures. The epilimnia of lakes, particularly those in high altitudes, may receive high levels of UV. Finally, species of primary producers can protect themselves by synthesizing mycosporine-like amino acids (in the diatoms), scytonemen (in cyanobacteria), or

Table 12.1 Some Possible Effects of Increased UV on Primary Producers

Effect of Increased UV	Reference
Harms cyanobacteria	Bebout and Garcia-Pichel, 1995; Castenholz, 2004
Decreases photosynthetic rates of phytoplankton and periphyton in high altitude tropical lake	Kinzie *et al.*, 1998
Decreases nitrogen uptake rates of plankton	Behrenfeld *et al.*, 1995
Lowers populations of consumers of the producers	Bothwell *et al.*, 1994; Häder *et al.*, 1995
Damage to DNA	Jeffrey *et al.*, 1996
Damage to ability of cyanobacteria to fix nitrogen	Kumar *et al.*, 1996
Minimal effects with increased dissolved organic carbon	Morris *et al.*, 1995
Damages photosynthetic apparatus	Nedunchezhian, 1996
Harms stream mosses	Rader and Belish, 1997a
Selects for tube building or mucopolysaccharide-producing diatoms	Rader and Belish, 1997b
Has selective effects on competitive ability of different periphyton species	Francoeur and Lowe, 1998; Vinebrooke and Leavitt, 1999; Tank and Schindler, 2004; Hodoki, 2005
Alters response to nutrient enrichments	Bergeron and Vincent, 1997
Decreases growth of *Sphagnum* in a bog	Gehrke, 1998; Robson *et al.*, 2003
Alters phytoplankton species composition	Laurion *et al.*, 1998; Xenopoulos and Schindler, 2003
Exacerbates the biotic effects of acid precipitation on lakes	Yan *et al.*, 1996

flavenoids (in green algae and higher plants). However, synthesis of these compounds costs the plants energy and may lower overall production (Karentz *et al.*, 1994).

Given all these considerations, the question still remains: What is the effect of increased UV on aquatic primary production? A variety of influences occur from increased UV on primary producers (Table 12.1). However, the most complicated of the UV effects are related to community interactions and ecosystem influences (Karentz *et al.*, 1994). In artificial stream experiments, natural levels of solar UV drastically reduced grazing midge larvae. In these experiments, UV had a negative direct influence on the periphyton over weeks, but the release from grazing pressure with higher UV eventually led to more luxuriant periphyton growth over months (Bothwell *et al.*, 1994). In contrast, experiments on grazing snails did not support a general prediction that UV always relaxes grazing pressure (Hill *et al.*, 1997), but production of grazing tadpoles may be negatively affected (Licht, 2003). In lakes, UV can have a variety of effects on microbial components of the food web, and complex interactions can occur among UV, dissolved organic C, and nutrient supply (Bergeron and Vincent, 1997; De Lange *et al.*, 2003). In addition, all primary producers have viruses that may infect and destroy cells. With increased UV the survival of viruses is decreased, and consequently rates of transmission and cell mortality are lower. More research is necessary if we are to understand the influence of UV on aquatic communities and ecosystems, especially in the context of climate warming and lake acidification. It has been suggested that both of these processes will lower dissolved organic C, leading to increased UV penetration into aquatic ecosystems (Schindler *et al.*, 1996).

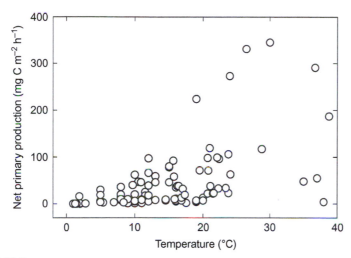

FIGURE 12.9

Relationship between photosynthetic rate and temperature based on 94 measurements of stream epilithon production in 14 studies. *(Reproduced with permission from DeNicola, 1996).*

Other factors in addition to light influence photosynthetic rate. Temperature has a distinct influence (Fig. 12.9). As with other metabolic rates, the rate of photosynthesis approximately doubles with each 10°C increase in temperature up to a species-specific threshold (DeNicola, 1996). Above this threshold, further increases in temperature harm the photosynthetic organisms and lower the rate; eventually, with a great enough increase in temperature, death occurs (see Chapter 15). The amount of nutrients available also has an influence on photosynthetic rates. Nutrient-starved cells will lower their rates of photosynthesis when nutrients become available so cellular metabolism can be directed toward acquisition of nutrients (Lean and Pick, 1981), but ultimately (after a day or two), nutrients will stimulate photosynthetic rate as cells synthesize the required molecular machinery from the newly available nutrients.

Water velocity can alter photosynthetic rates (Fig. 12.10). As the diffusion boundary layer thickness decreases with greater water velocity, the transport of material across the layer increases. The influx rate of CO_2 across this diffusion boundary layer can be an important determinant of photosynthetic rate of macrophytes (Raven, 1992), as can diffusion controls over the influx of nutrients. High levels of dissolved O_2 can inhibit photosynthesis. Increased water velocity can increase transport of O_2 and other potentially inhibitory chemicals. Therefore, increases in water velocity often increase photosynthetic rates. However, very high velocities may stress photosynthetic organisms and actually lead to decreases in photosynthetic rates (Stevenson, 1996).

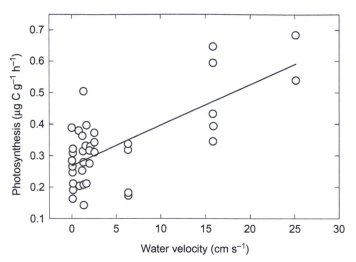

FIGURE 12.10
Relationship between water velocity and photosynthetic rate of the benthic cyanobacterium *Nostoc*. *(From Dodds, 1989, with permission of the* Journal of Phycology).

Organic compounds, particularly herbicides that target molecular photosynthetic machinery, can have strong negative effects on photosynthetic rates. Chemicals designed to control macrophytes often act by interfering with photosynthesis (Murphy and Barrett, 1990). Organic chemicals that lower photosynthetic rates can also occur naturally; macrophytes can release organic compounds that inhibit photosynthetic rates of epiphytes that grow on their surface (Dodds, 1991).

Determining rates of primary production over extended periods of time in entire ecosystems is difficult given the large observed variation in photosynthetic rates and the wide variety of factors that influence rates. Measurements must be repeated under an array of conditions to achieve accurate estimates of rates of primary production. In addition, a variety of methods are available for photosynthetic rate measurements (Method 12.3).

RESPIRATION

Respiration is the primary route of O_2 consumption in most environments. Other biotic and abiotic activities can lead to consumption of O_2 such as oxidation of ammonium, sulfide, or iron (discussed earlier with respect to iron, and in more depth in Chapter 14), but these rates are usually low relative to respiratory consumption in most environments. Similar factors can influence respiration rates.

METHOD 12.3

Techniques for Measuring Photosynthetic Rates in Freshwaters

Several methods have been developed to measure photosynthetic rates, with the best method often determined by the system of interest and the question being asked. Two general classes of methods are available—those that make use of chemical tracers and those that measure bulk change in dissolved products or substrates of photosynthesis. Numerous scientific articles and books have been published on various methods; Wetzel and Likens (1991) present one of the clearest accounts for lakes and Bott (2006) for streams.

Changes in dissolved O_2 concentration form the basis of the most widely used methods. For this technique, a sealed system containing the photosynthetic organisms normally is used, and the O_2 change in light versus dark enclosures is compared following a known incubation time. This technique requires moderately high biomass and is difficult to apply to oligotrophic phytoplankton because O_2 production and consumption rates are low relative to background O_2 concentrations. However, if great care is taken, O_2 methods may still be suitable for oligotrophic lakes (Carignan et al., 1998).

Several enclosed systems have been used to determine photosynthetic rates with O_2 exchange. Attempts are generally made to mimic in situ conditions when rates are measured. Glass bottles are commonly used for phytoplankton studies. UV-transparent plastic or quartz bottles are used if the investigator wants to account for UV effects. It is necessary to duplicate water movement found in the natural habitat for investigation of photosynthetic rates of benthic algae. Enclosed recirculating systems constructed of transparent plastic are generally used for these measurements (e.g., Dodds and Brock, 1998).

Open-water methods are also used for measuring O_2 production and consumption. Odum (1956) pioneered such methods for use in streams. The technique is based on measuring the increase in O_2 as light increases or the decrease in dark. The exchange rate with the atmosphere must be known for these measurements; this exchange rate is difficult to measure. Although equations exist to estimate aeration rates, their application in small streams is questionable. Trace gases such as propane can be used to estimate this

exchange rate (Marzolf et al., 1994, 1998). With the whole-stream O_2 method, the stream communities remain under their natural conditions. The production and consumption of CO_2 can be used to measure respiration and photosynthesis, respectively, in a fashion similar to that outlined previously for O_2. The main problem with this approach is that background concentrations of CO_2 and associated dissolved forms (discussed in Chapter 13) are considerably greater than even dissolved O_2 concentrations. Measuring changes in CO_2 is primarily useful for emergent plants or macrophytes with very high biomass.

For any methods that track changes in dissolved O_2, results can simply be expressed as change in O_2 per unit time, or converted to actual carbon fixed by photosynthesis or released by respiration using the following relationship:

$$mgC = mgO_2 * \frac{1}{PQ} * \frac{12}{32}$$

where PQ is the *photosynthetic quotient* (moles of O_2 released during photosynthesis per moles CO_2 incorporated = 1.2 for algae at moderate light levels; Strickland and Parsons, 1972), 12 is the atomic weight of C, and 32 is the molecular weight of O_2.

Photosynthetic organisms also consume protons (increase pH) while they photosynthesize. Some investigators have used this fact to estimate photosynthetic rates in the natural environment. This method is not highly sensitive and may not work well in waters that are resistant to pH changes (buffered).

The radioactive isotope of carbon ($^{14}CO_2$) has seen broad application in measurements of photosynthetic rates. If the ratio of $^{14}CO_2$ to ambient unlabeled $^{12}CO_2$ is known, then the rate of uptake of radioactive carbon into plant carbon can be used to calculate total photosynthetic rate. This approach is useful in very oligotrophic waters where O_2 methods fail. However, the practical difficulties of using radioactive isotopes (in laboratory and particularly in field settings) hamper this method. Also, separating net from gross photosynthetic rate with $^{14}CO_2$ techniques is difficult because some of the carbon fixed by photosynthesis can be respired immediately. Careful planning is necessary before $^{14}CO_2$ methods are employed.

Recently, a fluorescence method, pulse amplitude-modulated fluorescence, has been applied to measure relative photosynthetic rates. This method relies on the biophysics of photosystem II and how it responds to pulses of saturating light over time. The problem with this method for determination of absolute photosynthetic rates of complex assemblages is that different species of primary producers respond differently to fluorescence (Juneau and Harrison, 2005), and it should be used mainly for short-term comparisons of relative physiological responses.

All organisms must respire. If there is a lot of living biomass, then respiration rates will be substantial in an ecosystem. A common error of beginning students is to discount the fact that photosynthetic organisms respire. An example of the importance of respiration by producers occurs when algal blooms lead to anoxia because the algae continue to respire at night or during cloudy periods when they are not photosynthesizing.

Quality of organic carbon available for consumption is a primary determinant of rate of respiration. Easily metabolized molecules such as simple sugars can lead to greater rates of respiration than a source that is less bioavailable (such as cellulose). Details of carbon use are discussed in Chapter 13. Other nutrients such as nitrogen and phosphorus can increase availability of carbon, biomass of heterotrophic organisms, and ultimately respiration rate.

Respiration rates are also controlled by temperature; some of the earliest physiological observations were on temperature effects on respiration. Homeotherms (warm-blooded animals) generally have higher rates of respiration than poikilotherms (cold-blooded animals). The Van't Hoff-Arrhenius equation describes the general relationship between temperature and rate of metabolism:

$$R = e^{-E/kT}$$

where R = rate, E is the activation energy, k is Boltzmann's constant, T is absolute temperature (degrees Kelvin), and e is the mathematical constant approximately equal to 2.718.

Exchange rates with the environment can also alter rates of respiration; diffusion controls the rate of transport of O_2 and other compounds required for respiration. As with photosynthesis, any other compound that influences physiology can alter respiration rate (such as toxins, extreme pH, salinity). Animals can have rapid variation in respiration rates based upon their levels of activity.

CONTROLS OF DISTRIBUTION OF DISSOLVED OXYGEN IN THE ENVIRONMENT

As mentioned previously, the distribution of O_2 over time and space in aquatic habitats is a function of O_2 transport (influx and efflux) as well as production by photosynthesis and consumption by respiration. Given the natural variation in the rates of these different processes, and differences driven by the relative inputs of organic C, habitats can be either anoxic or oxic. In this section, we describe spatial and temporal variations of O_2 in lakes, sediments, groundwaters, and small particles.

Measurement of vertical patterns of dissolved O_2 in lakes and relation to thermal stratification is a common exercise in limnology courses. Such measurements illustrate the processes leading to production and consumption of dissolved O_2 and the biological importance of density stratification in lakes. Movement of O_2 across the metalimnion is slow because it depends mostly on molecular diffusion. In addition, the hypolimnion is usually deep enough that little light reaches it. With little or no photosynthesis, respiration predominates in the hypolimnion as organic carbon rains down from above in the form of settling planktonic cells and other organic particles. Given a high enough rate of carbon input and associated heterotrophic activity, O_2 can be consumed completely during a summer season of a dimictic or monomictic lake (Fig. 12.11A).

The hypolimnia of eutrophic lakes tend to lack O_2 by the end of summer and oligotrophic hypolimnia of monomictic or dimictic lakes retain at least some O_2. The greater deficit of O_2 in eutrophic lakes occurs because the amount of organic carbon sinking per unit time into the hypolimnion from the epilimnion is greater in eutrophic lakes than in oligotrophic lakes. Understanding anoxia of hypolimnia is important because taste and odor problems in drinking water become more acute in anoxic conditions and as trophic state increases (see Chapter 18) and because the presence or absence of O_2 can determine distributional patterns of organisms. For example, fish will generally avoid an anoxic hypolimnion.

O_2 can also disappear from the hypolimnion of an amictic lake, even if the lake is oligotrophic. Such anoxia occurs because the O_2 is depleted gradually in the hypolimnion over a long period of time. As planktonic organisms die and sink into the hypolimnion, they introduce organic carbon, and this rate exceeds the downward diffusion of O_2. Lake Tanganyika in Africa (discussed in Chapters 7 and 11) is an example of an oligotrophic lake with an anoxic hypolimnion; even though the lake is 1,470 m deep, only approximately the top 100 m are oxygenated.

In some situations, dissolved O_2 can exceed saturation with respect to the atmosphere in lakes. When there is a high biomass of phytoplankton at the metalimnion, an O_2 peak may result. This deep chlorophyll maximum can be associated with substantial photosynthesis and O_2 concentration can build in the metalimnion where mixing is limited.

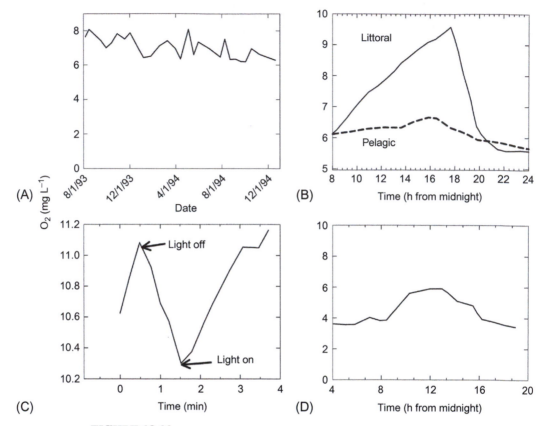

FIGURE 12.11

Temporal variation in O_2 in Kansas groundwater (A), the pelagic and littoral zones of an Indiana lake (B), a periphyton assemblage (C), and a stream (D). In panel C, the periphyton was brought into the laboratory, and an O_2 microelectrode was placed 100 μm into the mat under lighted conditions. The light was turned off and back on where the arrows indicate. *(Data in (A) courtesy of Konza Prairie Long-Term Ecological Research Site; data in (B) from Scott, 1923; C, original data; D, Odum, 1956).*

Significant daily changes in the O_2 concentration in the epilimnion of a lake related to the balance between photosynthesis and respiration can also occur. Even though there can be considerable exchange with the atmosphere, during relatively calm days an O_2 supersaturation can build up in the epilimnion, particularly in littoral zones (Fig. 12.11B). This observation of relatively high O_2 production in the littoral highlights the importance of benthic primary producers in lakes, a subject that Robert Wetzel has researched extensively (Biography 12.1).

When lakes mix completely, dissolved O_2 is blended throughout. In some cases, when lakes are hypereutrophic and wind (mixing) and light are low

BIOGRAPHY 12.1 ROBERT WETZEL

Robert Wetzel (Fig. 12.12) had a major influence on the field of aquatic ecology. One of his many contributions was to demonstrate that primary production by periphyton and macrophytes in lakes is often greater than that by phytoplankton. Many researchers incorrectly assumed that phytoplankton photosynthesis dominated in lakes because of the large volume of water containing phytoplankton (see Chapter 24). In addition to establishing the importance of benthic producers, his research documented the fate and cycling of carbon in ecosystems. Perhaps his greatest influence has been through his textbook, *Limnology* (Wetzel, 1983), which has been used to train innumerable aquatic ecologists. The most recent edition,

FIGURE 12.12
Robert Wetzel.

Limnology: Lake and River Ecosystems (2001), is the essential reference text for advanced limnologists.

Dr. Wetzel became interested in ecology because an inspiring high school biology teacher took the time to show him the virtues of nature. From this start, he became one of the most respected aquatic ecologists in the world. He published over 400 papers and 23 books, and received numerous awards and honors. He was deeply involved in international limnological pursuits, and was an elected member of the Danish, Russian, and Hungarian National Academies of Science.

When Wetzel was interviewed for the first edition of this book, he said he found an academic career extremely rewarding, and he deeply appreciated the freedom to satisfy his intellectual curiosity. He had adventures related to limnology as well. For example, he recalled a time sampling alone on Borax Lake in a rubber raft when a youth on shore with a rifle decided to try and shoot the raft out from under him. Fortunately, he survived that experience.

Wetzel cautioned new students of aquatic ecology not to put too much stock in simple explanations and to remember that biology is a complex and sophisticated field of study. He suggested that an important area for future study is the regulation of growth and the productivity of aquatic organisms by chemically mediated signals among them (Wetzel, 1991).

Robert Wetzel died on April 18, 2005 at the age of 68. His academic career was inspired by that of his hero, G. E. Hutchinson. He carried on the tradition of synthesis and intellectual rigor established by Hutchinson and became a giant in the field of limnology.

(under extended cloudy conditions or early in the morning), the heterotrophic demand for O_2 can be great enough to cause the water to become anoxic. An ice cover can also lead to anoxia by limiting photosynthesis and O_2 transport into lakes. Anoxia can cause fish kills in lakes and streams (Sidebar 12.2).

Anoxic conditions are very common in sediments of lakes, streams, and wetlands, even if there is ample O_2 in the water above. Anoxia develops because the sediments retard mixing, and diffusion is generally molecular. Furthermore, sediments serve as a store of organic carbon, and heterotrophic activity is

SIDEBAR 12.2

Fish Kills Result from Anoxia in Streams and Lakes

When organic carbon is high and exchange with the atmosphere is low, habitats can temporarily become anoxic or hypoxic (low O_2, generally below $2\,mg\,L^{-1}$) and fish and other animals die. This condition occurs when total respiratory demand of all organisms in the ecosystem exceeds input of photosynthetic and atmospheric O_2. High temperatures often exacerbate the problem because the metabolic rates are greater and O_2 solubility is lower in warmer water (Cooper and Washburn, 1949).

Fish kills can occur during the summer in eutrophic lakes. Kills occur in hypereutrophic lakes when a highly productive system experiences a series of calm, cloudy days. Under these conditions, algal blooms have high total respiration rates and little input of atmospheric O_2 occurs, leading to anoxia and fish kills. Such summer kills can be common in many areas and similar kills can occur year-round in the tropics.

Fish kills also occur in lakes in the winter when an ice cover prevents O_2 transport into the water. If snow covers the ice, light transmission and photosynthetic O_2 production are low. However, a eutrophic lake contains a significant amount of biomass, so respiration continues and O_2 concentrations decrease. Fish kills occur in such situations unless the fish can find an inlet stream or O_2 is bubbled into the lake.

Fish kills from anoxia are relatively common in many areas. For example, in the state of Missouri, from 1970 to 1979 there were more than 40 known winter kills and at least 100,000 fish deaths. During the same period, there were about 20 summer kills resulting in the death of more than 200,000 fish (Meyer, 1990). Very large-scale fish kills have been attributed to anoxia in Lake Victoria, Africa. In this case, storms suspend sediments and wash organic material from surrounding wetlands into the lakes. The resulting anoxia kills tons (hundreds of thousands) of fish (Ochumba, 1990). These kills are problematic because local people rely on the fish for food.

Hypoxic conditions can exacerbate other stresses that fish experience such as high temperature and exposure to toxins (either man-made or resulting from toxic algal blooms). Thus keeping O_2 concentrations at reasonable levels in naturally oxic waters is key for maintaining the biotic integrity of fish assemblages.

Sewage is released untreated into rivers and streams throughout the world. Such releases occurred in the United States and Western Europe until the 1970s, when environmental laws were enacted requiring reductions in the amount of organic carbon (biochemical oxygen demand) in sewage discharges. When the load of organic carbon in untreated sewage stimulates respiration and consumes O_2 at a rate in excess of that which can be replenished by exchange with the atmosphere, a river can become anoxic and the fish die. Such problems have become rare in developed countries since municipalities have been required to lower the organic carbon in the sewage that they release. Consequently, fish species that are less tolerant of low O_2 are becoming reestablished in areas where they were absent for many years.

relatively high. Even though a highly active photosynthetic community can lead to supersaturated O_2 concentrations at the surface of lighted sediments, anoxic conditions are often found within depths of millimeters or centimeters below the sediment surface (Fig. 12.13B).

Plants rooted in anoxic sediments must often cope with a lack of O_2 for their roots. Thus, aquatic plants can either transport O_2 to their roots or exhibit fermentative metabolism. Wetland plants often have specific adaptations to living in saturated soils including O_2 transport to roots. Vascular systems that transport O_2 down to the roots also serve to transfer methane, CH_4, an important greenhouse gas, to the atmosphere (as discussed in Chapter 13).

In sediments that are exposed to light, there is invariably a photosynthetic community associated with the sediment surface. The production of these communities is generally high, leading to very steep gradients in O_2 over depth (Fig. 12.13B) or time (Fig. 12.11C). The shape of the O_2 curve with depth in sediments is similar to the shape of the O_2 curves with depth from hypereutrophic lakes, but the vertical scale is in millimeters instead of meters. Distribution of O_2 across these sediments can be an important factor controlling biogeochemical cycling. Thus, factors that alter O_2 distribution, such as animal burrows, can have strong ecosystem effects.

O_2 concentration in streams can also vary over time (Fig. 12.11D) allowing for *in situ* measurement of photosynthetic rates (Method 12.2). O_2 variation over time can provide an index of the relative degree of system productivity, with large diurnal swings in O_2 occurring in the more productive systems. Information on O_2 dynamics can be used in ecosystem analyses or as a management tool to assess effects of river pollution (Auer and Effler, 1989). In rivers that receive a great deal of untreated sewage, O_2 can disappear completely (Sidebar 12.2).

Variation occurs in O_2 over time in wetlands. This fact has been used to estimate rates of primary production and respiration in the Everglades (McCormick *et al.*, 1997). The variation of O_2 was found to be extreme in some areas that receive phosphorus pollution in the Everglades. The water column was completely anoxic at night and fully saturated during the day.

Groundwaters can be oxic or anoxic depending on the relative supply of organic carbon to heterotrophs, the scale that is considered, the residence time of the water in the aquifer, and the dissolved O_2 concentration in the incoming water. Many pristine aquifers are oxygenated, but human activities can lead to anoxia. However, row-crop agriculture actually may decrease organic C input into aquifers leading to higher O_2 concentrations (Fig. 12.14A). Septic systems, feedlots, spills of organic chemicals, and subsurface disposal of sewage effluent can lead to anoxic groundwaters (Madsen and Ghiorse, 1993). When groundwater is deep enough, it is generally anoxic because of very low supply of O_2, high temperatures associated with geothermal heating, O_2 consumption by organisms, reactions with inorganic chemicals, and long turnover times. If groundwaters become anoxic, biogeochemical cycling is altered and many

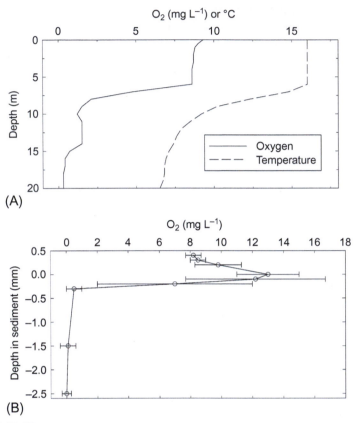

FIGURE 12.13

O_2 profiles from the stratified Triangle Lake, Oregon, on October 1, 1983, mid-morning (A) and from the South Saskatchewan River in an active algal mat, midday on June 9, 1993 (B). Note the lack of O_2 in the hypolimnion and the possible deep photosynthetic activity (at 10–14 m) that causes a slight increase in O_2 in panel A and the supersaturating O_2 concentration at the sediment surface in panel B. *(Data for panel A courtesy of R. W. Castenholz; data for B from Bott et al., 1997).*

species of invertebrates and microbes cannot exist in the aquifer. Thus, biodiversity of groundwaters can be controlled in part by oxygen.

Additional small-scale anoxic habitats exist that are important in a variety of aquatic environments. Organic materials may have anoxic zones associated with them. For example, decaying leaves may have anoxic zones at their surface and inside individual leaves or entire aggregates of leaves, even though they are in completely oxygenated waters (Fig. 12.14B). The digestive systems of many animals are anoxic, and a distinct microflora forms in these locations. The role of animal digestive tracts in biogeochemical cycling has not been well researched. In lighted sediments, the algal community that is present can cause great variations in O_2 concentration with time. Upon darkening,

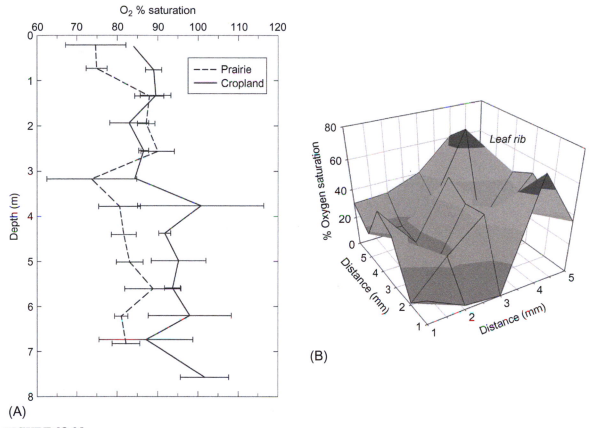

FIGURE 12.14

O_2 profiles in unconsolidated sediments and groundwater below cropland and prairie (A) and an O_2 concentration map at the surface of a leaf particle collected from a groundwater habitat (B). Statistical analysis showed that the O_2 concentration was significantly lower for prairie than under cropland (panel A). The groundwater was at 4.2 m in the prairie profile and 5.3 m in the cropland profile. *(A, reprinted from Dodds et al., 1996a, with permission from Elsevier Science, 1996; B, reproduced with permission from Eichem et al., 1993).*

respiration rapidly consumes O_2; there is a measurable decrease within 1 s (Fig. 12.11C). Organisms living in such a habitat must be adapted to rapid changes in O_2 concentration.

SUMMARY

1. Materials in water can be dissolved, colloidal, or particulate (gravitoidal).
2. Conductivity (total dissolved ions) and pH are chemical properties of water that are important descriptive parameters used by aquatic ecologists because they can control the distribution and activity of organisms.

3. Redox potential is another important parameter that controls chemical and biochemical processes in aquatic ecosystems. The way chemicals are transformed in the environment is determined partially by the redox potential of the environment. Redox potential is an estimate of the relative concentration of the available electrons in the environment. O_2 concentration is a primary determinant of redox potential.

4. Potential energy drives chemical and biochemical processes. It is present when the redox potential of the reactants is very different from the redox potential of the surrounding environment. A chemical reaction will release energy if the products of the reaction have less potential energy than do the reactants.

5. Organisms can promote chemical reactions by lowering the activation energy with enzymes. In the case of reactions that go against potential energy, another source of energy must be present. For example, organisms can shunt chemical energy into a chemical reaction. Photosynthetic organisms convert light into chemical energy and use this chemical energy to drive carbon fixation (convert CO_2 to sugar).

6. A chemical reaction will not occur spontaneously if the activation energy is too large, regardless of the potential energy that will be released.

7. The presence of O_2 is very important because of its role in redox and respiration. The concentration of O_2 varies with space and time in aquatic environments. Habitats with O_2 are called oxic or aerobic, and those without O_2 are called anoxic or anaerobic.

8. Photosynthesis produces O_2 in aquatic environments, and aerobic respiration consumes it.

9. Photosynthetic rates are controlled mainly by the amount of light, temperature, and nutrient availability. Organisms are able to acclimate to low light. High light (especially UV) can cause photoinhibition, and adaptations such as special protective pigments can be used as protection.

10. Respiration rates are mainly controlled by carbon availability and quality, temperature, and nutrient availability.

11. The balance of photosynthesis and respiration, the degree of contact with the atmosphere, and transport processes determine actual O_2 concentration.

12. Some common anoxic habitats include the hypolimnia of eutrophic lakes; organic-rich sediments in lakes, streams, and wetlands; organic-rich groundwaters; digestive tracts of animals; decaying vegetation in water; and particles with high rates of associated microbial activity.

QUESTIONS FOR THOUGHT

1. Why are nutrients that are taken up by cells often preferred in a reduced form, even when organisms inhabit oxidized environments? (Hint: Think of the conditions under which life evolved.)

2. Why do midge larvae that live in the profundal benthos of lakes often turn bright red when brought to the surface?

3. Why are aquifers in karst regions often oxic?
4. Why is oxidation of organic carbon by O_2 more efficient than anaerobic respiration?
5. Dense periphyton mats can float off of the bottom of a lake during the day, float to the surface, and then sink again at night. Why might this happen?
6. Winter fish kills from anoxia can occur in Arctic lakes that are not very productive and do not freeze to the bottom. Why?

Carbon

FIGURE 13.1

Lake Nyos, Cameroon, Africa. This is the site of a catastrophic CO_2 release that killed 1,700 people in 1986. *(Image courtesy of George Kling).*

Carbon cycling is central to the way ecosystems operate from local to global scales. Carbon can be organic (combined with oxygen and hydrogen) or inorganic (e.g., as carbon dioxide). Organic carbon is the currency of energy exchange in aquatic ecosystems. Understanding carbon cycling is central to understanding food webs and how aquatic communities are structured and supported. Inorganic carbon in water is involved in the bicarbonate equilibrium, which is connected intimately to pH control and responses to acid precipitation. Methane production by wetlands is one example of the direct impact of the carbon cycle of freshwater ecosystems on global biogeochemistry and the greenhouse effect. In this chapter, forms of organic and inorganic carbon and fluxes of carbon in the environment (including the carbon cycle) are discussed.

FORMS OF CARBON

Inorganic Carbon

Inorganic carbon is found in the atmosphere, primarily in the form of carbon dioxide (CO_2), where the concentration in 2009 was approximately 380 ppm, and was increasing by about 2 ppm per year. The concentration has been constantly increasing since the industrial revolution, intensifying the greenhouse effect. The greenhouse effect (increases in atmospheric CO_2, other gases, temperature, and associated climate change) undoubtedly will influence aquatic ecosystems (Hutchin *et al.*, 1995; Tobert *et al.*, 1996; Magnuson *et al.*, 1997; Megonigal and Schlesinger, 1997); such effects are discussed in more detail in Chapter 1 and elsewhere in the book.

When CO_2 is dissolved in water, it can exist in a variety of forms, depending on pH. The forms are *carbon dioxide, carbonic acid, bicarbonate,* and *carbonate*. The sum of the concentrations of all these forms is the inorganic carbon concentration and is signified as ΣCO_2. Under most conditions in aquatic systems, CO_2 is rapidly converted to carbonic acid so they will be considered the same. The chemical conversions among these forms are referred to as the *bicarbonate equilibrium*. Understanding this series of chemical reactions is necessary to comprehend how aquatic ecosystems are buffered against changes in pH and how CO_2 becomes available for photosynthesis (Butler, 1991). The bicarbonate equilibrium can be represented as

$$CO_2 + H_2O \Leftrightarrow H_2CO_3 \Leftrightarrow H^+ + HCO_3^- \Leftrightarrow 2H^+ + CO_3^{2-}$$
$$\text{carbon dioxide} \quad\quad \text{carbon acid} \quad\quad \text{bicarbonate} \quad\quad \text{carbonate}$$

The $\Leftrightarrow$ symbol indicates an equilibrium reaction. Adding or taking away chemicals from any part of the reaction can force the reaction to compensate. For example, if acid (H^+) is added to a bicarbonate solution, the equilibrium is weighted too heavily to the right-hand side of the equation, so the

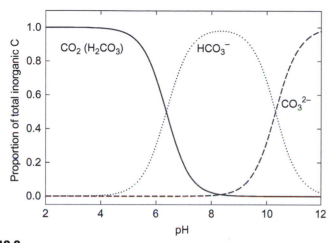

FIGURE 13.2

The relative concentrations of inorganic compounds involved in the bicarbonate equilibrium as a function of pH.

bicarbonate will convert spontaneously to carbonic acid or carbon dioxide. This is demonstrated easily by adding an acid such as vinegar to a solution of the sodium salt of bicarbonate (baking soda). Adding the acid will cause production of CO_2 as the equilibrium is reestablished. Because the CO_2 gas has a limited solubility in acidic water, it will bubble out. Thus, as the pH changes so do the relative amounts of bicarbonate, carbonate, and carbonic acid (Fig. 13.2).

Increased atmospheric pressure allows greatly increased amounts of CO_2 to be dissolved in solution. Thus, carbonated beverages stored under pressure (such as soda, beer, and champagne) lose CO_2 when opened. These beverages must be slightly acidic, so the equilibrium is forced to the side of CO_2. A similar, but catastrophic, release of CO_2 had disastrous consequences in the African Lake Nyos (Sidebar 13.1).

The equilibrium of inorganic carbon also explains the acid-neutralizing or *buffering* capability of bicarbonate and carbonate. Protons form covalent bonds with carbonate and bicarbonate and ultimately with hydroxyl ions if CO_2 is produced. The protons are not free, so pH changes only a small amount relative to the concentration of H^+ added to solution as long as there is carbonate or bicarbonate to react with. Once these forms are converted to CO_2 then pH can decrease rapidly with further additions of acid. Systems with a significant amount of dissolved bicarbonate (e.g., limestone watersheds) are able to resist the effects of acid precipitation. An opposite response occurs to addition of base (OH^-). The OH^- ions associate with the H^+ ions so the

SIDEBAR 13.1
The Lake Nyos Disaster

One thousand seven hundred people died on August 21, 1986, near Lake Nyos (Fig. 13.1) in Cameroon, Africa. At about 9:30 PM, people in the area heard a loud rumbling. One survivor reported viewing a mist rising off the lake and a large water surge (subsequently demonstrated to have washed up to 25 m high on the southern shore). Some survivors reported smelling an odor like rotten eggs, experiencing a warm sensation, and then losing consciousness. When they awoke 6 to 36 h later they were weak and confused, and many of their family members were dead. Thousands of livestock and other animals succumbed as well (Kling *et al.*, 1987). Research teams visiting the area later pieced together a picture of what caused this disaster. The deaths resulted from a catastrophic release of more than a million metric tons of CO_2 from the lake; CO_2 is heavier than air, and the large amount released from the lake filled the valley around the lake, displacing O_2 and suffocating people and animals.

Before the CO_2 was released, something caused CO_2 bubbles to start coming out of solution, leading to mixing of the lake. Once this process began, large amounts of CO_2 were rapidly released. The lake went from clear blue to reddish brown; the iron dissolved in the hypolimnion was mixed up and oxidized forming a rusty precipitate at the surface of the lake. Evans *et al.* (1993) suggested that volcanic eruption, alteration in limnological factors, or both initiated the gas release. The limnological explanation revolves around the idea that seasonal mixing allowed for the CO_2 to degas rapidly after a threshold CO_2 concentration was reached in the hypolimnion. Evidence for this hypothesis includes the observation that a similar gas release killed 37 people at nearby Lake Manoun during the same season two years earlier (Kling, 1987).

FIGURE 13.3
Water jet in Lake Nyos upon installation of a CO_2-releasing pipe into the hypolimnion. *(Photograph courtesy of Michel Halbwachs).*

A buildup of CO_2 has been documented in this amictic tropical lake since the disaster (Evans *et al.*, 1993) and could lead to hypolimnetic saturation again within 140 years. However, saturation may be reached near the bottom in less than 20 years (Evans *et al.*, 1994). The source of CO_2 is most likely volcanic activity occurring below the lake. The lake is stratified, with a chemocline (transitional zone of a lake with stratification stabilized by salinity) currently at about 50 m depth. Thus, mixing of CO_2-rich waters with the atmosphere to relieve high concentrations deep in the lake is prohibited by limnological factors. Relatively high concentrations of dissolved CO_2 can build up at depth because of the high pressure under the water.

Several actions were contemplated to avoid future catastrophic releases. The final solution was to use a natural pump-lift pipe to remove excess CO_2 from the hypolimnion. This ingenious solution placed a pipe down into the hypolimnion with the other end above the surface of the water. Water is pumped up from below, and as it decreases in pressure, dissolved CO_2 bubbles out of solution and causes an airlift pump (Fig. 13.3) that continues pumping water to the surface with no further energy input (Halbwachs *et al.*, 2004). The fountain, when this pipe was installed, shot 50 m into the air. About 20% of the gas is methane, and now plans are under way to harvest this gas and use it as fuel. Never before has the field of physical limnology had such direct involvement in a human health issue. Careful future monitoring will be necessary to avoid repeating the disaster. Currently, there is also concern that the natural dam of volcanic rock that impounds the lake is weakening and could collapse. Rapid downstream flooding would probably lead to thousands more deaths. Research is under way to find how to stabilize the dam.

---

equilibrium balances by moving toward the bicarbonate side. Alternatively, as CO_2 is dissolved in water it causes decreases in pH. Thus rain is naturally slightly acidic, and increases in atmospheric CO_2 are causing acidification of surface waters.

The acid- and base-neutralizing capacity of bicarbonate and the predominance of bicarbonate ions in many systems have led to using *alkalinity* and *acidity* titrations to estimate ΣCO_2. For alkalinity titrations, acid can be added with little initial change in pH. After all the bicarbonate and carbonate have reacted with the added acid, further additions cause proportionally greater decreases in pH per unit of acid added. The alkalinity is the amount of acid needed to cause these greater decreases in pH, which can easily be measured with pH sensors or dyes while a sample is titrated with a known concentration of acid.

The dependence of the bicarbonate equilibrium on pH yields plots that can be used to calculate the relative concentrations of each of the forms of inorganic carbon when the pH is known (Fig. 13.2). Such data are useful because CO_2 is the form of inorganic carbon required for photosynthesis. Many photosynthetic organisms can convert bicarbonate to CO_2, but CO_2 is still the most easily used form. Thus, knowing the alkalinity and pH of a solution allows an investigator to calculate the amount of inorganic carbon that is immediately available for photosynthesis.

Calcite ($CaCO_3$) is an important precipitate of the bicarbonate equilibrium. The precipitate can form spontaneously when CO_2 is removed from solution (as a way to balance the equilibrium). This precipitation can occur when photosynthesis removes CO_2, when physical factors remove CO_2 (e.g., degasing of spring waters), or when organisms such as mollusks build their shells (Wetzel, 2001). The resulting precipitate can build up impressive concretions of whitish calcium bicarbonate on the stems of the green alga *Chara* (hence the name *stoneworts*), in terraced outflows of hot springs, and in some benthic habitats with photosynthetic microorganisms. Mammoth Hot Springs in Yellowstone National Park is composed of massive carbonate terraces.

Organic Carbon

Organic carbon (carbon compounds bonded with hydrogen, usually oxygen, and sometimes other elements) takes a tremendous variety of forms. The broadest classifications are *dissolved organic carbon* (DOC) and *particulate organic carbon* (POC). Stream ecologists further divide the particulate fractions into *fine particulate organic matter* (FPOM) and *coarse particulate organic matter* (CPOM). A dividing line of $0.45\,\mu m$ has been proposed for the difference between DOC and FPOM, and a line of 1 mm has been proposed for the division between FPOM and CPOM (Wallace *et al.*, 2006). Further divisions of the particle sizes have been used by some investigators (e.g., ultrafine particles or large debris). CPOM and FPOM contain both living and dead organic material, though nonliving material derived from terrestrial plants often dominates in streams, and living material dominates the plankton of many lakes.

The amount of organic carbon in ecosystems that is biologically available is particularly important in habitats dominated by heterotrophic organisms (e.g., groundwaters, sediments, and forested streams) but is important in all aquatic environments (Dodds and Cole, 2007). Given the huge number of organic compounds that can be synthesized by organisms, and all the possible pathways of degradation leading to by-products, even determining the total organic chemical content of water can be difficult. Some of the organic carbon compounds may be very resistant to degradation, and others may have high biological availability (Wetzel, 2001), so simply lumping them together may not make ecological sense unless the aim is to characterize total export from or storage by an environment. However, one common method to estimate the total available organic carbon for heterotrophs is based on the concept of *biochemical oxygen demand* (BOD) or the total demand for oxygen by chemical and biological (respiration) oxidative reactions. Change in concentration of BOD is a cornerstone of understanding the effects of sewage effluent on aquatic ecosystems (Method 13.1).

METHOD 13.1

What Is Biochemical Oxygen Demand and How Is It Measured?

Biochemical oxygen demand (BOD) is a simple and practical indicator of the total organic content that is available to organisms plus any chemicals that spontaneously react with O_2. The procedure is straightforward: Water is incubated in sealed bottles, and the decrease in O_2 over time is monitored. If all the O_2 is consumed over the time of the incubation, the original water sample must be diluted and analyzed again. During the incubation, naturally present heterotrophic organisms use O_2 to respire the organic carbon that is biologically available and any chemicals that spontaneously react with O_2 (e.g., sulfide) will also do so, allowing analysts to assess total BOD in wastewaters (Eaton *et al.*, 1995).

When sewage is released into natural waters, it provides heterotrophs with additional organic carbon and creates a demand for dissolved O_2. High sewage influx leads to anoxic conditions in the waters, particularly during summertime low-flows. During summer low-flow, dilution is at a minimum, and with high temperatures dissolved O_2 concentrations are low. Thus, regulations for the degree of sewage treatment are based on the amount of BOD released into the receiving waters. A permissible loading rate is calculated based on dilution, aeration rate in the receiving water, and BOD concentration of the sewage effluent. BOD measurements are often made daily to assess the efficiency of sewage treatment facilities.

The dissolved pool of organic carbon can be divided into two major classes—humic and nonhumic substances. *Humic compounds* are large-molecular-weight compounds and lend a brownish color to water. The nonhumic fraction includes sugars and other carbohydrates, amino acids, urea, proteins, pigments, lipids, and additional compounds with relatively low molecular weights. Such a classification is an ecological classification because nonhumic substances are generally broken down by heterotrophic processes to yield humic compounds (Stumm and Morgan, 1981).

The humic substances are produced as the by-products of the breakdown of still larger molecular weight compounds such as *celluloses, tannins,* and *lignins.* These compounds are resistant to microbial use (except see Lovley *et al.,* 1996). Tannins and lignins leach from bark and leaves of plants. Humic compounds can be classified into three groups—the *humic acids* (soluble in alkaline solutions and precipitate in acid), the *fulvic acids* (remain in solution in acidic solutions), and the *humin* (not extractable by acid or base) (Stumm and Morgan, 1981). The compounds can be formed from a variety of different source molecules, ultimately derived from either terrestrial or aquatic sources (Reemtsma and These, 2005).

Both the humic substances and the tannins have the following important features (Stumm and Morgan, 1981): (1) They attach to many other organic substances (e.g., they can be important in transport and fate of organic pollutants), (2) they form complexes with metal ions (which can be particularly important in keeping iron in solution), (3) they form colloids including large organic flocs, (4) they color the water brown or tan (absorbing light that could fuel photosynthesis) when in high concentrations such as in blackwater swamps, and (5) they are resistant to biological degradation.

Advanced analytic methods are allowing better characterization of carbon compounds found in natural waters. A study of dissolved organic matter in Lake Ontario using nuclear magnetic resonance indicated that the majority of the compounds were aliphatic (did not contain carbon rings), with heteropolysaccharides, carboxyl-rich and acyclic molecules (Lam *et al.*, 2007). Methods such as size exclusion chromatography, mass spectrometry, and scanning fluorescence spectroscopy all allow analyses of specific types of molecules to determine characteristic patterns that serve as "fingerprints" for dissolved organic matter. For example, fluorescence spectroscopy can separate general classes of dissolved organic carbon by microbial or terrestrial origin (McKnight *et al.*, 2001).

TRANSFORMATIONS OF CARBON

Photosynthesis and aerobic respiration were discussed in Chapter 12. These two fluxes are central to carbon cycling on a global scale and in surface freshwater habitats. There are also bacteria that are capable of *anoxygenic photosynthesis* (photosynthesis with no O_2 production) that will be discussed later. In general, the complexity in carbon cycling transformations lies in anaerobic cycling and in use of complex organic compounds in oxic and anoxic habitats.

Organisms have evolved the ability to use most types of organic molecules (the exception so far being many human-synthesized plastics). This evolution is evident in the rapid appearance of strains of microbes able to use novel organic carbon compounds produced by humans and released into the environment. The process of *in situ* bioremediation is based on this phenomenon (see Chapter 16). Similarly, organisms are perpetually participating in an evolutionary arms race, with one species evolving chemical defense or chemicals to lower activities of competitors, and other species evolving resistance to the chemicals.

Most naturally occurring compounds can be broken down, and numerous metabolic pathways exist given the millions of distinct organic compounds that can be found in the environment. The general metabolic approach to deal with large organic compounds (e.g., proteins, cellulose, tannins, and fatty acids) is to modify the compound by energy-requiring hydrolysis reactions into a more readily used compound that can be metabolized with energy-yielding reactions (Fig. 13.4). Thus, cellulose is broken down into

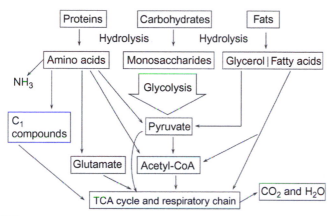

FIGURE 13.4
A general diagram of aerobic breakdown of organic carbon by organisms. *(Modified from Rheinheimer, 1991).*

its component sugars at an energy cost, but metabolism of the sugars provides energy in excess of this initial cost. Complex organic compounds can be degraded in the cell, or enzymes can be excreted outside the cell (*exoenzymes*) to break down larger organic carbon compounds into compounds that can be taken up (Sinsabaugh *et al.*, 1991). A full understanding of carbon cycling will require elucidation of cycling of complex organic carbon compounds (Hobbie, 1992; Wetzel, 2001).

Advanced: Breakdown Rates of Organic Carbon

Empirical measurements of organic carbon decay suggest that mass generally declines with an exponential relationship:

$$M_t = M_0 \, e^{-kt}$$

where t = time, M_t and M_0 are mass at times t and 0, respectively, and k is the decay coefficient. This relationship is common because the most labile organic material is used first, and the more recalcitrant material, with slower absolute decomposition rates, is used later. Generally, these rates are determined by incubating organic materials such as leaves, wood, dead alga, or macrophytes in natural environments and determining the change in organic mass over time (e.g., Benfield, 2006).

When organic material of just one type is plentiful, rates should be constant over time (e.g., when there is a high concentration of glucose the breakdown rate is roughly a constant mass per unit time). When concentrations become low, then rates decrease because of decreased uptake rates of materials. When organic materials such as leaves enter aquatic ecosystems, the simple sugars,

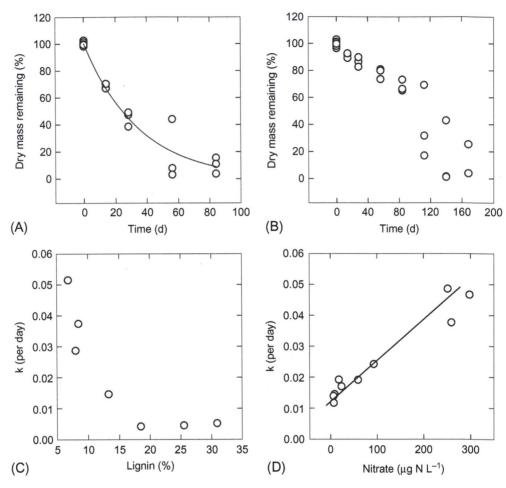

FIGURE 13.5

Breakdown of leaves in streams. (A) Breakdown of alder follows a typical exponential decay pattern, (B) decomposition of beech leaves increases following a flood (at around 80 days), (C) decomposition rates (k) of various species as a function of lignin content, and (D) rates as a function of nitrate concentration in the stream. *(Data for B, and D from Gessner and Chauvet, 1994; data from C from Suberkropp and Chauvet, 1995).*

proteins, and starch are used very quickly and the more complex structural materials such as cellulose and lignin break down more slowly. Bacteria and fungi colonize the leaves rapidly and begin to break them down (Gessner and Chauvet, 1994). These organisms have the enzymes necessary to break down cellulose and lignin. As discussed later in Chapter 19, animals that eat low quality organic materials, like senescent leaves, are actually deriving much of their nutrition from the microbial community associated with decay of these materials (Cummins, 1974).

Different species of terrestrial vegetation invest variable amounts in recalcitrant compounds in their leaves to protect against herbivory and other mechanical damage. Rapidly growing species often have leaves that break down very quickly and others have much tougher leaves that are more difficult to degrade (Fig. 13.5).

Other factors such as temperature and inorganic nutrients can also control rates of decay of organic matter. Leaves and wood are very low in nitrogen and phosphorus, and nutrients can stimulate decomposition rates (Gulis *et al.*, 2004). A whole-stream nutrient enrichment experiment in an Appalachian Mountain stream documented that leaf breakdown rates were greater with nutrient addition, with concurrent increases in the microbial respiration rates and ultimately greater biomass of macroinvertebrates associated with leaves (Greenwood *et al.*, 2007). Human land use can alter rates of leaf breakdown; in agriculturally influenced streams increased nutrients expedite breakdown rates and in urbanized streams more frequent disturbance increases fragmentation of leaves leading to quicker decomposition (Paul *et al.*, 2006).

Oxidation of Organic Carbon with Inorganic Electron Acceptors Other Than O_2

Organic carbon can release the most energy to organisms if it is oxidized with O_2. In the absence of O_2, the next best energy yield is from other electron acceptors (such as nitrate, sulfate, and oxidized iron) to oxidize organic C. The relative efficiency of these oxidations depends on the oxidation state of the compound. In other words, O_2 is the most oxidized compound abundant in the natural environment that organisms can use to react with the reduced organic carbon, so the greatest amount of energy can be obtained by aerobic respiration. Other compounds are used in order of redox (see Fig. 12.5). It is possible that for short periods an oxidation of carbon that is not the most efficient can occur (e.g., a bacterium that can only use nitrate to oxidize organic carbon may continue to do so even in the presence of O_2). If the transformation is less efficient, the organism that relies on the less efficient mode of oxidizing carbon will ultimately be outcompeted, unless it is able to switch to the more efficient mode. In variable environments, conditions can change rapidly so an organism that does not rapidly change metabolic strategy can be successful, even if it is using a less efficient mode of respiration for part of the time. For example, it may be more energetically expensive to completely turn off the biochemical machinery for anaerobic respiration than to retain it until the next anoxic event.

In anoxic habitats, this series of organic carbon oxidations can be distributed across a redox gradient (across millimeters or centimeters in sediments, micrometers or millimeters in decaying organic particles, or meters at the

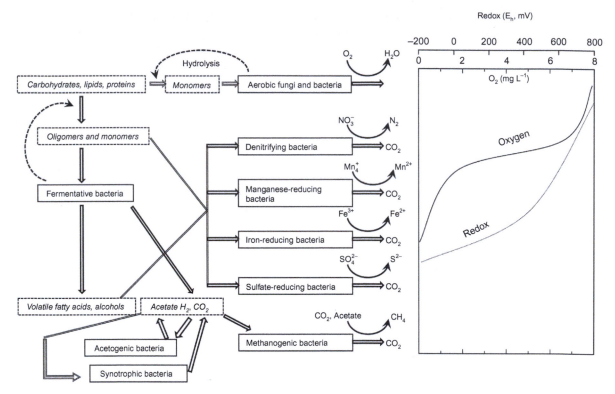

FIGURE 13.6

A general diagram of breakdown of organic material across a redox gradient. Organisms are shown in solid boxes, and chemical pools are shown in dashed boxes. Curved arrows with solid lines indicate alternative electron acceptors for oxidation of organic carbon, and the preferred redox of each of these transformations approximately corresponds with the redox curve drawn on the right. *(Redrawn from Westermann, 1993).*

interface of an anoxic hypolimnion). The O_2 is used first where contact with the atmosphere occurs. After O_2 is used up, then each successive type of oxidant is used up according to the maximum potential energy yield, leading to areas with progressively lower redox (Fig. 13.6). Thus, NO_3^- is used first, followed by Mn^{4+}, Fe^{3+}, and SO_4^{2-}. All these oxidations are more efficient than acetogenesis and methanogenesis, which will be discussed in following sections.

Fermentation

In addition to oxidizing organic compounds with inorganic electron acceptors, heterotrophic organisms in anoxic environments can use organic carbon by *fermentation*, a process of rearranging the organic molecules to yield more simple organic and inorganic compounds (e.g., acetate, ethanol, CH_4, CO_2, H_2, and H_2O) and energy. A wide variety of these reactions occur because

Table 13.1 Some Representative Fermentative, Methanogenic, and Acetogenic Transformations That Occur in Anoxic Communities

Reaction	Name	Comment
Fermentation		
Glucose→2 ethanol + 2 CO_2	Ethanol fermentation	Formation of alcohol
Glucose→2 lactate + 2 H^+	Lactate fermentation	
Glucose→ethanol + acetate + CO_2 + H_2	Mixed acid fermentation	Produces variable amounts of products
Glucose→butyrate + 2 CO_2 + 2 H_2	Butyrate fermentation	
3 lactate→2 propionate + acetate + CO_2	Propionate fermentation	Gives Swiss cheese flavor
Alanine + 2 glycine→3 acetate + 3 NH_3 + CO_2	Paired amino acid fermentation	Important when proteins being broken down
Acetogenesis		
2 CO_2 + 8 H^+→acetate + 2 H_2O	Heterotrophic acetogenesis	
2 CO_2 + 4 H_2→acetate + 2 H_2O	Autotrophic acetogenesis	
Methanogenesis		
CO_2 + 4 H_2→CH_4 + 2 H_2O	Autotrophic methanogenesis	Uses CO_2 and H_2 as a source of energy
Acetate→CH_4 + CO_2	Acetoclastic methanogenesis	Disproportionation of acetate

Acetogenic and Methanogenic Processes Require Very Low Redox (Fig. 12.5). The Listed Processes Represent a Small Proportion of the Fermentative Processes That Can Occur in Anoxic Communities

there is such a tremendous number of different structures of organic molecules that can be used; examples are presented in Table 13.1. Many of these reactions are of enormous commercial benefit (e.g., fermentation to produce alcohol), but they are also central to the carbon flux of anoxic aquatic habits.

A general feature of these fermentation processes is that many yield organic acids. These acids lower pH and generally decrease rates of further degradation. No individual species of fermenter is able to metabolize organic polymers (such as cellulose, proteins, and lipids) completely to CO_2 and H_2. In contrast, individual species are able to degrade polymers to CO_2 and H_2O in the presence of O_2 (Fenchel and Finlay, 1995). Thus, complex communities of "syntrophic" microorganisms are required to continue energy cycling in anoxic systems. Individual species from these complex groups of heterotrophic anoxic microbes cannot grow in isolation without a supply of very specific metabolic substrates, so they tend to "cooperate" to break down organic materials.

As complex organic compounds produced by terrestrial plants are degraded in the absence of O_2, humic compounds are formed. Degradation of the

---

SIDEBAR 13.2
Acid Bogs in Archeology, Paleolimnology, and Palynology

Bogs led to preservation of organic materials for thousands of years. The most spectacular finds of archeological interest are preserved human bodies (see also Chapter 5 and Fig. 5.10). Several circumstances characteristic of carbon cycling in bogs are required for preservation of bodies: (1) The water must be deep enough to prevent carrion-eating animals from consuming the body when it is first deposited; (2) the water must remain anoxic to inhibit microbial growth and oxidation (preserved materials will decay within days when reexposed to O_2); (3) sufficient concentrations of tannic acids must be present to cause tanning of the skin (in nonacidic waters, only the bones are preserved); and (4) water must be cold, generally below 4°C, to inhibit microbial growth. Above this temperature, flesh rots and acids attack and degrade the bones (Coles and Coles, 1989). Bodies 2,000 years old have been found so well preserved that the color of the hair and eyes could be determined, as well as the last meal eaten. Many human artifacts preserved in bogs have been uncovered as well.

Scientists also use the preservation of organic materials in anoxic bogs to indicate changes in plant communities over time. Sediments can be dated, and preserved pollen, pigments, or other plant parts can be used to indicate the local plant community. Such information may be useful for detecting environmental trends over centuries or longer time periods (Taylor and Taylor, 1993). For example, isotopic composition of sedges and mosses preserved in peat bogs can be used to estimate atmospheric CO_2 content with a temporal resolution of about a decade (White *et al.*, 1994). These data are useful to climate modelers who are interested in rapid climate change events that have occurred during the past 12,000 years.

---

organic compounds leads to lowered pH and creation of phenolic and acidic organic compounds that are inhibitory to further microbial activity. Lowered pH encourages precipitation of the humic compounds. As these build up in the sediments, further degradation slows. For the most part, humic substances are difficult for microbes to degrade, but they can be used slowly by some microbes (Lovley *et al.*, 1996). Anoxic conditions lead to slower breakdown of organic materials because of reduced efficiency of carbon oxidation and inhibition of microbial activity by metabolic by-products. Accumulation of organic compounds under anoxic and acidic conditions has led to formation of peat bogs. These wetlands are sites of organic accumulation, and meters of material can accumulate over many years. Such bogs are distributed worldwide, notably in boreal regions of the Northern Hemisphere. Similar processes lead to the formation of blackwater swamps. Humic compounds in blackwater swamps cause the water to appear brown or black in color, and the acidity allows buildup of organic carbon. Peat bogs are important in archeology and paleolimnology (Sidebar 13.2).

Methanotrophy
Methane and carbon monoxide are commonly produced in fermentation reactions. These compounds are simple but energetically inefficient for most

microorganisms to metabolize. Some specialized aerobic bacteria, the *methylotrophs*, can harvest energy by oxidizing simple compounds containing methyl groups and carbon monoxide. Bacteria that specifically oxidize methane are referred to as *methanotrophs*. These bacteria are central to the global carbon cycle because their activity is a major reason that methane does not build to large concentrations in Earth's atmosphere. Methanotrophs are generally found living in close proximity to anoxic habitats from which there is a constant diffusive flux of methane.

Methanogenesis, Acetogenesis, and Disproportionation of Acetate

Methane can be produced under anoxic conditions by Archaea in a process called *methanogenesis* in extremely low redox conditions. CO_2 actually has more potential energy than the reduced methane at low redox (Figs. 13.6 and 12.5). Some of the reactions that produce methane are presented in Table 13.1. Microbes can use CO_2 and H_2 or acetate to produce methane. Using acetate is also referred to as a disproportionation, because one of the two carbons ends up as methane, and the other as CO_2 (i.e., proportioned into two different redox states).

Rates of methane and CO_2 formation occasionally can be so great that gas bubbles containing methane are released directly from anoxic sediments (*ebullition*). Biogas production as an alternative energy source is driven by methanogenesis, as was formation of natural gas deposits stored underground across geological time. Methane is a greenhouse gas, and current increases of methane concentration in Earth's atmosphere related to human activities are evident. Anoxic habitats are common in wetlands, and freshwater wetlands contribute significantly to the global methane budget (Sidebar 13.3).

Microbe-mediated reactions similar to methanogenesis include acetogenesis (making acetate). Acetate production can be more important than methane production in some conditions, particularly when pH is lower. Samples of these reactions are also presented in Table 13.1.

A CONCEPTUAL INTRODUCTION TO NUTRIENT CYCLING

In this section, we introduce how to diagram a complete nutrient cycle, rather than considering individual fluxes specific to individual elements. Such cycling is a key feature of all compounds used by organisms. All the fluxes that occur in an environment make up the *cycle*. An account of the relative magnitude of the fluxes is called a *budget*. If any chemical form in a cycle is not recycled, it builds up in the environment and eventually becomes unavailable

SIDEBAR 13.3
Global Emission of Methane Related to Wetlands

Global methane concentrations have more than doubled in the past 200 years. The absolute level is less than that of CO_2 in the atmosphere, but the rate of increase is greater. Increasing concentrations of methane cause concern because methane absorbs about 20 times as much heat per molecule as CO_2 and acts as a greenhouse gas. While some details of the increase are not well established (Schlesinger, 1997), it is clear that human activities put a significant amount of methane into the atmosphere. Such human activities include burning, gas and coal production, methane release from landfills, and ruminant production.

Wetlands are predominant sources of methane, even more so when rice paddies (agricultural wetlands) are considered (Table 13.2). Wetlands and rice paddies make up 54% of the total methane produced globally. Methane escapes wetlands because anoxic processes lead to methanogenesis. Most of the methane produced is intercepted and oxidized by methanotrophic bacteria and this is not accounted for in the budget in Table 13.2. However, the stems of plants form a conduit from anoxic sediments to the atmosphere, bypassing the populations of methanotrophs found at the oxic-anoxic interface in the sediments (Joabsson *et al.*, 1999).

Understanding carbon cycling in wetlands will aid future efforts to predict atmospheric methane concentrations. For example, it is unknown what will happen to methane production in high latitudes if the permafrost thaws and significant numbers of new wetlands are formed. It has also been demonstrated that increased CO_2 can stimulate methanogenesis in wetlands (Megonigal and Schlesinger, 1997), complicating the relationships among climate change, CO_2, and methane. Satellite data indicate that roughly half the methane emissions occur in the tropics and currently about 2% in the Arctic. About 7% of the rise in atmospheric methane between 2003 and 2007 could be attributed to increased emissions from warming of mid-latitude and Arctic wetlands (Bloom *et al.*, 2010).

Table 13.2 Global Sinks and Sources of Methane

Sources and Sinks	Methane Flux ($10^{12}\,g\,CH_4\,y^{-1}$)	Percentage of Total
Sources		
Natural wetlands (microbial)	115	21
Freshwaters (microbial)	5	1
Termites (microbial)	20	4
Oceans and geological (microbial and abiotic)	20	4
Human activities (burning, landfills, gas and coal mining)	230	43
Rice paddies (microbial)	60	11
Grazing animals (microbial)	85	16
Total sources	535	
Sinks		
Reaction with OH and loss in atmosphere	475	88
Soil microbes	30	6
Increase in atmosphere	30	6
(After Schlesinger, 1997)		

to all organisms. Evolution has led to organisms that can metabolize most naturally occurring organic carbon compounds (including novel compounds synthesized by humans). Such adaptations can occur in days or weeks in the case of most novel organic compounds. Likewise, almost any inorganic chemical with potential energy in an environment can be converted by organisms to yield energy. Thus, we have complete nutrient cycles.

Nutrient cycles have general features that can be used to build a conceptual framework of the way that materials move through the environment. Some of this framework is built on the requirements of organisms and some on the constraints that redox exerts on chemical reactions. The general features make it easier for students to understand nutrient cycles and ultimately to link the cycles together in a larger, more comprehensive view of aquatic ecosystems.

All organisms require nutrients, and they can acquire them as dissolved inorganic forms, organic forms, or both. For example, people obtain carbon in the form of organic molecules, but plants can obtain carbon from either CO_2 or organic carbon sources. This acquisition of nutrients is called *assimilation*, and generally occurs regardless of the redox state of the environment. Likewise, organisms excrete inorganic nutrients; the general name for this is *remineralization* or *regeneration*. Remineralization of organic carbon to CO_2 is respiration or fermentation in the carbon cycle. When reduced inorganic chemicals are in an oxidized environment (e.g., an oxic region), microbes can oxidize them, and metabolic energy can be gained. Energy is gained because the flux goes with potential energy. Similarly, oxidized compounds can release energy when reduced in an anoxic environment.

Here, we introduce a method for diagramming nutrient cycles to assist the reader in understanding and remembering complete nutrient cycles. It is important to keep in mind the idea of potential energy in environments of differing redox (see Chapter 12). A generalized nutrient cycle is presented to illustrate the point (Fig. 13.7). In this diagramming method, the oxidized inorganic forms are listed from right to left, oxic processes are placed in the top half of the diagram, and anoxic processes are placed in the bottom half of the diagram.

The boundary between anoxic and oxic that passes through the horizontal center of the diagram can be considered similar to the dividing line between an oxic epilimnion and an anoxic hypolimnion, the difference between an oxic sediment surface and an anoxic portion deeper in the sediments, or the difference between the oxic habitat outside a decaying leaf and the anoxic habitat inside. Assimilation and heterotrophy generally occur regardless of the presence of O_2, so they are placed across the oxic/anoxic boundary.

In general, processes that yield energy move from left to right on the top (oxic) part of the diagram and from right to left on the bottom (anoxic) part

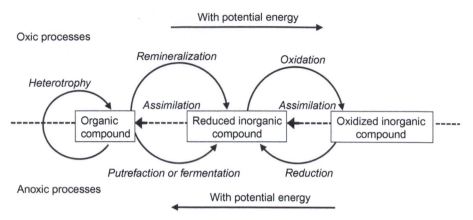

FIGURE 13.7

Diagram of a hypothetical nutrient cycle. This will be the general format used to represent nutrient cycles. Oxic processes are above the center line and anoxic processes are below. Those that move on the center line are required, independent of O_2 concentration. Inorganic forms are listed from left to right, from reduced to oxidized. Thus, transformations are generally occurring with potential energy if they move from left to right in the top half of the diagram or from right to left in the bottom half of the diagram.

of the diagram. Some processes move against potential energy. For example, photosynthesis uses carbon dioxide to synthesize organic carbon in oxic environments. Each of the nutrient cycles has its own idiosyncrasies related to the different chemical properties of the individual nutrient and the conventions of researchers, but this general diagramming method should assist in learning the generalities across nutrient cycles.

THE CARBON CYCLE

The carbon cycle is diagrammed using the technique presented for general nutrient cycles (Fig. 13.8). CH_4 is listed separately from other organic forms because of its crucial role in global carbon cycling. Also, photosynthesis is an assimilatory flux, but the processes of oxic and anoxic photosynthesis are different, so they are separated in this chart. All processes occur in freshwater, and freshwaters have global importance in several portions of the carbon cycle.

The global importance of methane production in wetlands has already been discussed (Sidebar 13.3). Wetlands are also globally important in the process of carbon deposition or *sequestration*. Most of the coal on Earth formed from trees that died and accumulated in wetlands and other freshwater systems and were buried. Large amounts of peat are stored in tundra regions, and one of the concerns is how much of that peat will enter the global atmospheric stores of CO_2 and CH_4 as global warming melts permafrost that is preserving large amounts of peat in the northern regions of the planet.

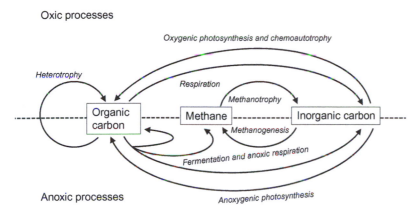

Oxic processes

Anoxic processes

FIGURE 13.8

A diagram of the generalized carbon cycle.

Carbon transport through freshwaters is also of global importance. Far more heterotrophic activity fuels food webs in marine systems than previously thought, and many marine ecosystems are net heterotrophic (Dodds and Cole, 2007). Extensive impoundment of water in reservoirs has altered now much organic carbon is transported from terrestrial to aquatic systems; about 20% of global carbon transport from rivers into the ocean is now trapped in reservoirs (Syvitski *et al.*, 2005). Deforestation, urbanization, and hydraulic modification all have the potential to influence carbon transport.

ADVANCED: HYDROGEN AND CARBON CYCLING

Hydrogen gas (H_2) production and anaerobic carbon cycling are closely linked. Syntrophic assemblages of bacteria use hydrogen cycling coupled with carbon cycling to complete fermentation cycles (Fenchel and Finlay, 1995). For example, pyruvate, ethanol, or butyrate fermentation pathways can lead to excretion of H_2 as a way to dissipate reducing power (electrons). If H_2 builds up, then the fermentation can proceed no farther. However, if methanogens are present to consume excess H_2 or other pathways occur (such as using sulfate to oxidize hydrogen gas), then fermentation can proceed. The relationship between hydrogen-consuming and hydrogen-producing organisms forms the basis of syntrophy.

Whereas syntrophic associations are primarily composed of Archaea and Bacteria, some Eukarya are capable of fermentation that yields H_2. Some protists have hydrogenosomes that convert pyruvate to acetate and H_2. This source of H_2 is probably only moderately important in aquatic ecosystems in most cases.

Control over methanogenesis by sulfate is also mediated by H_2. In this case sulfate reducers (covered in more detail in the next chapter) oxidize acetate and H_2 while converting sulfate to sulfide. These organisms are better competitors for acetate and H_2 because redox relationships dictate greater energy yield occurs when sulfate (used by sulfate reducers) rather than CO_2 (used by methanogens) is used as an oxidant. Essentially, coupling of the anoxic C cycle with H_2 dictates that methanogenesis will only occur at very low rates when sulfate is present in the environment. Maintaining a balance of electrons and the wide variety of chemical interactions that microorganisms can mediate leads to complex interactions among the various nutrient cycles; considering hydrogen gas is an important part of understanding these anoxic interactions.

SUMMARY

1. Inorganic carbon is found in the form of carbon dioxide (CO_2) gas in the atmosphere at about 380 ppm (in 2009), with continuous increases of about 2 ppm per year. This gas readily dissolves in water, where it can take the form of CO_2, carbonic acid (H_2CO_3), bicarbonate (HCO_3^-), and carbonate (CO_3^{2-}).

2. The equilibration between the different ionic forms of inorganic carbon is called the bicarbonate equilibrium. One of the most important factors driving equilibrium is the pH of the water. Under low pH, CO_2 is formed; under high pH, the equilibrium moves toward carbonate. The equilibration can buffer natural waters against changes in pH.

3. Organic carbon can be dissolved or particulate; the particulate fractions traditionally are divided into fine and coarse components. Dissolved organic material can be divided into humic and nonhumic substances. Nonhumic substances generally have low molecular weight and are metabolized easily by aquatic microbes. Humic substances have high molecular weight and are more resistant to breakdown.

4. Organic carbon is produced by oxygenic (O_2-producing) and anoxygenic photosynthesis, as well as by chemoautotrophic microorganisms. Organic carbon is oxidized to CO_2 by a variety of metabolic pathways, including aerobic respiration, anaerobic respiration using oxidized inorganic compounds to release energy from organic molecules in the absence of O_2, and fermentations.

5. Methanotrophic organisms rely on oxidation of methane as a primary energy source and generally are situated close to anoxic habitats where methane is produced.

6. Methanogenic organisms produce methane from CO_2 and H_2 or from acetate under extremely low redox conditions. These organisms in aquatic habitats have a significant input to the global atmospheric methane pool.

QUESTIONS FOR THOUGHT

1. Why could global warming alter deposition rates of organic carbon in peat bogs?
2. Should wetlands be preserved even though some are a net source of greenhouse gases?

3. Why can the microlayer at the surface of lakes be important in determining influx and efflux rates of lake water CO_2?

4. Do the processes of manganese reduction and iron reduction by bacteria have to occur in regions of different redox in the environment?

5. Why is CO_2 more likely to limit production of benthic organisms than phytoplankton?

6. Why is it important to know the pH when conducting experiments to measure photosynthetic rate with $^{14}CO_2$?

7. Why is alkalinity generally high in limestone watersheds?

CHAPTER 14

Nitrogen, Sulfur, Phosphorus, and Other Nutrients

Doi: 10.1016/B978-0-12-374724-2.00014-3

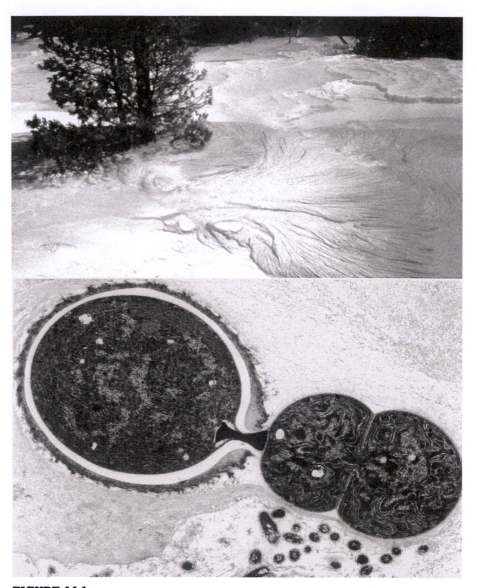

FIGURE 14.1
Streamers composed of the sulfur-oxidizing bacterium *Thermothrix* at Mammoth Terrace, Yellowstone National Park (*courtesy of R. W. Castenholz*) and a transmission electron micrograph of a heterocyst (the site of nitrogen fixation in *Nostoc* and other cyanobacteria) attached to a smaller dividing vegetative cell with a diameter of approximately 8 μm. (*Micrograph courtesy of N. J. Lang*).

Other nutrients in addition to carbon are central to the way that aquatic ecosystems function. Some forms of these nutrients, such as sulfide, ammonium, and nitrate, can cause water quality problems if they are present in large quantities. In addition, primary production can be limited by one or more of these nutrients (most commonly phosphorus and nitrogen), so knowledge of their forms and cycling is an important part of understanding ecosystem function. In this chapter we discuss cycling of nitrogen, sulfur, phosphorus, iron, and silicon. The nutrient cycles are all interrelated; none stands alone. Thus, the chapter concludes with a discussion on how the cycles interact and revisits how redox controls nutrient cycling.

NITROGEN

Nitrogen Forms

The most common form of nitrogen (N) in the biosphere is N_2 gas. The atmosphere is composed of about 78% N_2. Water generally contains N_2 as a dissolved gas. Even though N_2 is less soluble in water than O_2, the higher atmospheric concentration of N_2 leads to dissolved concentrations at equilibrium with the atmosphere similar to those observed for O_2. The N_2 molecule is very difficult for organisms to use directly because it has a triple covalent chemical bond that requires a significant amount of energy to break. Thus, forms of organic and inorganic nitrogen other than N_2 are referred to as *combined nitrogen.*

The two most important forms of dissolved inorganic N in natural waters are *nitrate* (NO_3^-) and *ammonium* (NH_4^+). Ammonium is the ionic form found in neutral to acidic waters. Under high pH conditions the ion is converted to *ammonia* gas (NH_3), which can move between the atmosphere and the water and also is toxic to aquatic organisms. Roughly less than 10% of ammonia is in gaseous form below pH 8, and at pH 9 about half the ammonia is in the gas form. *Nitrite* (NO_2^-), an additional form of dissolved inorganic nitrogen, occasionally is found in significant concentrations in natural water, especially when sewage is present, and can be problematic because of its toxic nature. The sum of nitrate, nitrite, and ammonium is often referred to as *dissolved inorganic nitrogen*. Other forms of inorganic N that also are dissolved in water, such as N_2 gas, are generally not considered because they are not as biologically available. Nitrous oxide (N_2O) is a gas that is also found dissolved in low concentrations in many waters and in the atmosphere. Nitrous oxide is known as laughing gas but also absorbs heat more efficiently than CO_2 and is an important contributor to the greenhouse effect.

Organic N can take many forms, including amino acids, nucleic acids, proteins, and urea. *Urea* may be particularly important because it is excreted by many organisms (although most aquatic organisms excrete far more ammonium) and may be a crucial form in nitrogen cycling. Organic N in aquatic habitats can be in the form of *dissolved organic nitrogen* and *particulate organic nitrogen*; usually a $0.45\,\mu m$ filter is used to separate dissolved from particulate.

Nitrogen Fluxes

Nitrogen must be assimilated by organisms in some form because it is a required component for many biological molecules such as proteins. Many animals can assimilate nitrogen only in the form of organic molecules, such as amino acids and nucleic acids in tissues of other organisms they consume. Enzymes that cleave proteins are excreted into the surrounding water or associated with cells (Billen, 1991) because whole proteins cannot be taken across the cell membrane and must be broken into amino acids for uptake. Primary producers and bacteria generally use nitrate, nitrite, or ammonium, which they can take up from the water. The assimilatory pathways for nitrogen acquisition in such organisms require the nitrogen to be in the form of ammonium in the cell (Fig. 14.2). Thus, the enzyme nitrate reductase converts nitrate to nitrite and nitrite reductase transforms nitrite to ammonium, which can then be assimilated. These reductase enzymes require energetically costly reducing equivalents in the form of NADH, NADPH, or ferridoxin. Nitrate reductase requires molybdenum to function properly, so a limited supply of molybdenum may lead to an inability to use nitrate. In oxic environments, ammonium has higher potential energy than nitrate, so energy is required to

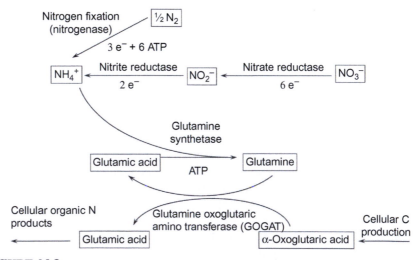

FIGURE 14.2

Nitrogen assimilation. This figure illustrates that nitrogen must be assimilated in the form of ammonium, and energy requirements for assimilation are $N_2 > NO_3^- > NO_2^- > NH_4^+$.

convert nitrate to ammonium before assimilation. Many aquatic bacteria and primary producers prefer ammonium because it requires less energy to use, and affinity for ammonium is relatively high.

Many bacteria, including some cyanobacteria (Young, 1992) and some Archaea, have the capacity to assimilate N_2. This capacity is known as *nitrogen fixation*. The transformation does not occur spontaneously because it requires extremely high activation energy. Conversion of N_2 to ammonium is accomplished with the enzyme nitrogenase, which requires molybdenum as an essential component. The process is one of the most energetically expensive metabolic reactions, requiring at least six ATP molecules and three electrons (reducing equivalents) for each ammonium produced. Another important property of nitrogenase is that it is inactivated by O_2. Organisms either have to inhabit anoxic habitats to fix nitrogen or they have to protect the enzyme from exposure (Bothe, 1982).

Some groups of cyanobacteria have formed specialized cells called *heterocysts* to protect nitrogenase from O_2 (Figs. 14.1 and 14.3). These cells are clearly unique in appearance under the microscope and have a variety of adaptations that allow for nitrogenase activity in the heterocysts and photosynthetic O_2 evolution in adjacent cells. The adaptations include high respiratory rates in heterocysts to consume O_2, thick gel or mucilage around the heterocysts to retard inward diffusion of O_2, and loss of O_2 evolution in photosynthesis with retention of cyclic photophosphorylation (i.e., generation of ATP by photosystem I) in heterocysts (Haselkorn and Buikema, 1992). Other groups

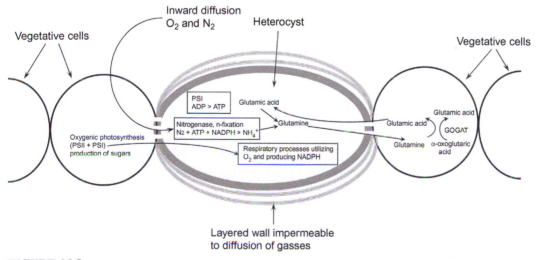

FIGURE 14.3

A diagram of cyanobacterial vegetative cells, a heterocyst, adaptations to protect nitrogenase from deactivation by O_2, and mode of N transport from the heterocyst into vegetative cells.

of cyanobacteria have no heterocysts but form aggregations in which O_2 is depleted. Heterocysts and patterns of aggregation allow low O_2 concentrations while promoting high-energy availability through respiration for the costly process of nitrogen fixation (Paerl and Pinckney, 1996). Molecular control over nitrogen fixation is well documented (Böhme, 1998) and the chemical signaling controlling heterocyst formation has been described (Yoon and Golden, 1998). Chemical signaling controlling heterocyst formation is particularly interesting because it is the first documented case of intercellular signaling with a peptide among the Bacteria.

Some wetland and riparian species of plants are associated with nitrogen-fixing bacteria. Some species of alder (*Alnus*) trees can harbor symbiotic bacteria that fix N_2. The fixed nitrogen can enter aquatic habitats through leaching or when leaves fall into the water. Some wetlands can also have significant populations of nitrogen-fixing cyanobacteria, which may be free-living or associated with other organisms. For example, the aquatic fern *Azolla* has endosymbiotic, nitrogen-fixing cyanobacteria. *Azolla* growth is encouraged in traditional rice culture; the field is drained to release the fixed nitrogen and then flooded and planted with rice. Cyanobacterial crusts associated with sediments in wetlands can also fix substantial amounts of nitrogen.

Nitrogen can also be fixed in the atmosphere when lightning produces enough energy to cause N_2 and O_2 to combine and form nitrate. Thus, rainwater naturally contains very small amounts of nitrate. Additional nitrogen is found in rainfall and particulates in the atmosphere that are suspended from terrestrial systems. Burning of fossil fuels introduces nitrogen oxides into the atmosphere. In aquatic systems with low amounts of nitrogen, atmospheric deposition can be a significant source of nitrogen pollution. Nitrogen is also fixed from the atmosphere during the industrial production of fertilizers. This process has approximately doubled worldwide rates of nitrogen fixation and led to widespread nitrogen contamination of waters (Vitousek, 1994; Vitousek *et al.*, 1997).

When ammonium is in water with dissolved O_2, it has potential energy relative to the more oxidized forms nitrite and nitrate. Some chemoautotrophic bacteria are able to use this stored potential energy with the process of *nitrification*. This two-step process yields energy to allow for carbon fixation:

$$\underset{\text{ammonium oxidation}}{NH_4^+ \rightarrow NO_2^- + \text{energy}} \underset{\text{nitrite oxidation}}{\rightarrow NO_3^- + \text{energy}}$$

Different genera of bacteria are responsible for ammonium oxidation (e.g., *Nitrosomonas*) and nitrite oxidation (*Nitrobacter*). Nitrification is a dissimilatory process because the nitrogen form changes but it is not assimilated.

Nitrification rates are minimal in anoxic habitats because nitrate has higher potential energy than nitrite and ammonium at low redox. Ammonium and nitrite oxidation are important processes for several reasons: (1) nitrification forms a vital link in the nitrogen cycle (the only other processes that produce nitrate are weathering of rocks (generally occurring slowly) and lightning), (2) nitrite formed by this process can have negative health implications, and (3) nitrifying bacteria compete for ammonium with primary producers. Previously, researchers had assumed that nitrifying bacteria occur only where ammonium is high. However, these bacteria occur commonly in oligotrophic, low ammonium environments and are able to compete successfully for ammonium with other microbes (Ward, 1996).

Denitrification forms a key link in the nitrogen cycle because it leads to the conversion of inorganic combined nitrogen (nitrate) to the relatively unavailable N_2 gas. This process has been heavily studied, given its importance in agriculture and water quality (Payne, 1981). Denitrification uses nitrate as an electron acceptor in a fashion analogous to the use of O_2 for aerobic respiration. The forms of nitrogen are converted in the sequence

$$NO_3^- \rightarrow N_2O \rightarrow N_2$$

During this process, organic carbon is converted to CO_2 and cellular energy. The process of denitrification is less efficient than aerobic respiration because nitrate is less oxidized than O_2 and yields less free energy when reacting with organic carbon.

Finally, excretion of nitrogen by organisms is an important flux. If an organism ingests more nitrogen than is needed, the excess must be excreted. If water is limiting, animals convert nitrogenous waste into urea because ammonium requires more water to produce and is toxic at high concentrations. However, production of urea requires more energy than ammonium. Ammonium is often the primary form of nitrogen that is excreted by aquatic organisms because water is rarely limiting, although some organisms (some fishes) excrete urea. Urea excretion is more important when water ammonium concentrations are high or in alkaline waters where ammonium ion is converted into toxic ammonia gas. When ammonium is remineralized, it is referred to as *ammonification*. The study of ammonification is particularly important in fish culture because excessive ammonium produced by high densities of fish can be toxic (Jana, 1994).

Nitrate is generally the predominant form of dissolved inorganic nitrogen in oxidized waters, and ammonium is the predominant form in anoxic waters. The absolute amounts of each are highly variable in many freshwaters (see Fig. 12.3; nitrate). The biotic processes that lead to reduced and oxidized forms of ammonium (e.g., nitrification, remineralization, and denitrification) can vary over time and space in freshwater habitats. Also, soils have a greater affinity for nitrate than ammonium, but different soils have varying degrees of affinity. Rainwater generally has more nitrate than ammonium, and nitrogen concentrations are geographically patchy, dependent in part on human activities. Thus, absolute concentrations of dissolved inorganic nitrogen can be highly variable.

In a lake with an anoxic hypolimnion, ammonium dominates the hypolimnion, and nitrate is mainly confined to the epilimnion during stratification (Fig. 14.4). Several processes drive this spatial pattern of nitrate and ammonium concentrations: In the epilimnion, nitrification transforms ammonium

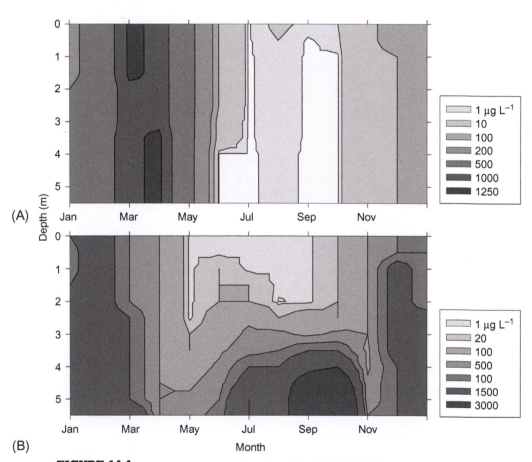

FIGURE 14.4

Distribution of nitrate (A) and ammonium (B) in hypereutrophic Wintergreen Lake, Michigan, as a function of depth and time. Ice cover occurred from January to March. Darker colors represent higher concentrations. Contours are reported in μg liter^{-1}. *(Reproduced with permission from Wetzel, 1983).*

to nitrate in the presence of O_2, but in the hypolimnion nitrate is removed by denitrification under anoxic conditions, and nitrogen continues to be excreted in the form of ammonium. A similar pattern is observed in oligotrophic lakes and wetlands, except that high ammonium is generally confined to the anoxic zone within the sediments.

High ammonium in streams is often associated with input of anoxic groundwaters or pollution (Duff and Triska, 2000). The temporal variation of nitrate in streams can be large, given the complex interactions between precipitation and groundwater sources (Fig. 14.5), whereas concentrations in groundwaters

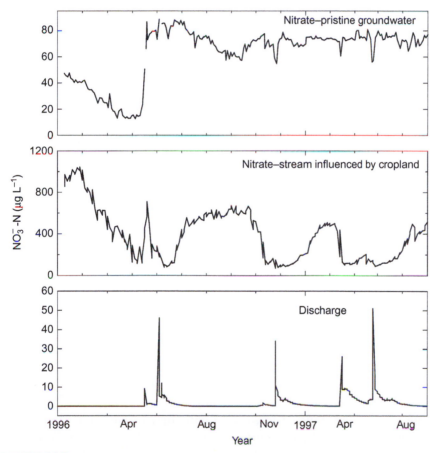

FIGURE 14.5

Concentrations of nitrate in groundwater flowing from undisturbed prairie (top, solid line), nitrate in Kings Creek, Kansas (middle, a stream influenced by groundwater that passes under cropland), and discharge in the same stream (bottom). High nitrate is related to input of groundwater from below fertilized cropland that dominates flow during periods of low discharge, and these concentrations are substantially greater than found in pristine groundwater. *(Data courtesy of Konza Prairie Long-Term Ecological Research site).*

may be more stable. Understanding processes that control nitrate in streams generally requires understanding nitrogen dynamics in the surrounding terrestrial ecosystem. For example, riparian zones decrease inorganic nitrogen and increase the ratio of ammonium:nitrate in watersheds of the Amazon (McClain *et al.*, 1994).

Advanced: Alternative Pathways of Anoxic N Cycling and Nitrous Oxide Production by Freshwater Environments

Several pathways compete with denitrification and other processes that create N_2, yielding ammonium instead. Under some conditions, nitrate can be reduced to ammonium instead of N_2 in a process called *dissimilatory nitrate reduction* (DNRA). Several alternative pathways lead to the conversion of nitrate to ammonium including a fermentative pathway where nitrate is used to oxidize organic carbon and an alternative pathway where nitrate is used to oxidize sulfide to yield energy. DNRA can account for 0 to 60% of the total of nitrate removal that is not accounted for by assimilation (Burgin and Hamilton, 2007).

Several pathways in the nitrogen cycle remove nitrogen from the ecosystem by conversion to N_2 in addition to denitrification. For example, nitrate can be reduced by oxidizing reduced forms of iron or manganese. Nitrite can also react with ammonium in a process referred to as *anaerobic ammonium oxidation (anammox)*.

All of these processes form N_2 gas and are potential removal routes for nitrogen from ecosystems. These pathways have only recently been described and it is uncertain how important they are in freshwaters; in marine waters anammox can be responsible for up to 60% of N_2 production (Burgin and Hamilton, 2007). Given the relative paucity of information on DNRA, anammox, and other pathways that yield N_2 gas in freshwaters, it is an exciting time to be researching the nitrogen cycle.

Nitrous oxide is an important greenhouse gas; even though it is present in trace amounts in the atmosphere, it has a half-life of a century in the atmosphere, about 300 times the heat absorption of carbon dioxide per molecule, and is currently increasing rapidly in concentration. In addition, nitrous oxide contributes substantiality to the depletion of atmospheric ozone and to subsequent increases in UV radiation. Rivers and estuaries could account for a considerable portion of nitrous oxide production that is stimulated by human use of nitrogenous fertilizers (Kroeze *et al.*, 2005). The general idea is that nitrous oxide is lost during the processes of denitrification or nitrification. Conditions under which these losses occur in aquatic

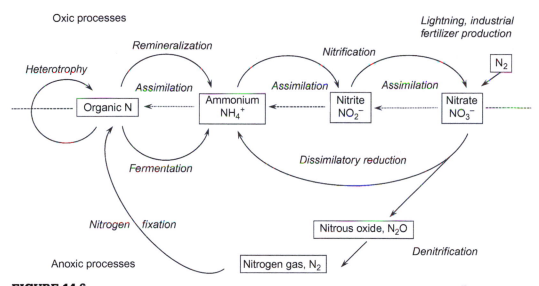

FIGURE 14.6
A conceptual diagram of the nitrogen cycle.

environments are not completely understood because they have not been studied extensively.

Nitrogen Cycle

The nitrogen cycle (Fig. 14.6) can be represented in the format provided for carbon in Chapter 13, divided into oxic and anoxic components, with more oxidized forms to the right of the diagram, and fluxes that go with potential energy occurring from left to right on the top of the diagram and from right to left on the bottom of the diagram. Anoxic habitat could be the hypolimnion of a stratified eutrophic lake; sediments in a lake, stream, or wetland; or a decaying organic particle as discussed in Chapter 13 for the carbon cycle. This generalized cycle illustrates some of the complexities of the nitrogen cycle related to varied redox potential of the different forms of inorganic nitrogen.

Understanding the complex fluxes in the nitrogen cycle is crucial because nitrogen can limit primary production in lakes, streams, wetlands, ground-waters, riparian zones, and marine habitats. In addition, nitrate and nitrite can be toxic (Sidebar 14.1), and understanding the nitrogen fluxes and factors controlling them may allow for mitigation of some pollution problems. For example, if denitrification can be maximized, then nitrate pollution that enters surface waters will return to the atmosphere as benign and inert N_2 gas.

SIDEBAR 14.1

Problems with Excessive Nitrate Contamination

The United States and many other countries have enacted limits on the amount of nitrate that can be present in drinking water. In the United States, the allowable level is 10 mg NO_3–N liter^{-1}. The primary reason for this limit is methemoglobinemia, a medical condition that occurs because nitrite binds more strongly to hemoglobin than O_2. This condition can result when water with large amounts of nitrite or nitrate is consumed. In infants, denitrifying bacteria produce nitrite. The condition is most often fatal in infants who are fed with formula made with high-nitrate water. Methemoglobinemia causes the infant's skin to appear blue and it suffocates, and thus is often called *blue baby syndrome*.

Nitrate is also converted to carcinogenic nitrosamines in the stomach. A relationship has been documented between gastric cancer rates and fertilizer use (Fig. 14.7). The relationship is only a correlation; other factors such as lifestyle, pesticide and herbicide use, and diet may explain the correlation. However, the interdependence of cancer and nitrate should be investigated more thoroughly.

The problem of nitrate contamination is widespread and worst in highly agricultural areas. In France, 266 groundwater sources were sampled, and 98% of those exceeded the recommended levels of 10 mg N liter^{-1}. Worldwide, 10% of rivers exceed 9 mg N liter^{-1} (Meybeck *et al.*, 1989). Some of the large contributors to nitrate in groundwater and streams include agricultural fertilization and feedlots (confined livestock facilities). Some sewage treatment plants release large amounts of nitrate as well.

Once an aquifer is contaminated with nitrate, treating the problem is very difficult. One possibility is to fertilize with organic carbon to cause respiratory consumption of O_2 by microbes and

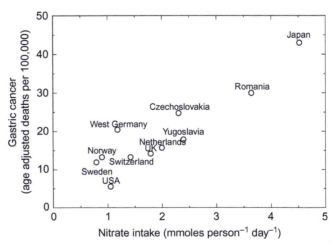

FIGURE 14.7

Correlation between nitrate intake and rates of gastrointestinal cancer. *(After P. E. Hartman. 1983. Reprinted by permission of Wiley–Liss, Inc., a subsidiary of John Wiley & Sons, Inc.).*

promote denitrification. The nitrogen then is lost as N_2. The problem with this treatment is that it leads to anoxic groundwater. Anoxia can result in other problems, such as high iron content and bad tastes and odors related to sulfide and other chemicals.

Perhaps the best way to control the problem of nitrate contamination is to do so at the source. Options include lining livestock feedlots and treating all runoff before it reaches the groundwater or streams, only adding as much fertilizer to crops as is necessary, leaving intact riparian buffer zones (Hedin *et al.*, 1998), better design of septic drain fields, and treating sewage to remove nitrogen (discussed in later chapters). All the treatment and control options require a basic understanding of nitrogen cycling and cycling of other nutrients.

SULFUR

The sulfur cycle is more complex than either the nitrogen or the carbon cycle. The complexity is driven by the different redox states in which sulfur can occur, and is of interest because it illustrates that organisms have evolved to use most of the possible common compounds in the biosphere that have potential energy. The sulfur cycle is also important because some water quality problems revolve around sulfide contamination. Importantly, sulfur is tightly coupled to the inorganic metal cycles such as iron and manganese and, thus, indirectly to phosphorus.

Sulfur Forms

Inorganic sulfur can take a wide variety of forms. Some of the forms that occur include:

$$S^{2-} \qquad S^{o} \qquad S_2O_3^{2-} \qquad SO_4^{2-}$$

sulfide   elemental sulfur   thiosulfate   sulfate

In order of increasing redox (more oxidized) these are S^{2-}, S^{o}, $S_2O_3^{2-}$, and SO_4^{2-}. Hydrogen sulfide can be dissolved in water as an ion (S^{2-}) under basic conditions or the ion is converted to a gas (H_2S) under acidic conditions. Hydrogen sulfide gas is toxic in high concentrations but can be detected by odor in exceptionally small concentrations. Hydrogen sulfide is partially responsible for the distinctive smell (an odor reminiscent of rotten eggs) of anoxic sediments in lakes and wetlands.

Sulfur is found in many organic molecules and is part of some amino acids that are central to protein structure. Dimethylsulfide is a significant gaseous product of phytoplankton metabolism that forms an important component of the global sulfur budget.

Fundamental Sulfur Transformations

Numerous transformations are possible because of the many redox states that sulfur can take. A few of these transformations are illustrated in Table 14.1.

Table 14.1 Some Sulfur Transformations

Equation	Conditions	Name	Classification
$H_2S + \frac{1}{2} O_2 \rightarrow S^\circ + H_2O$	Oxic	Sulfide oxidation	Abiotic and dissimilatory biotic
$S^\circ + H_2O + 1\frac{1}{2}O_2 \rightarrow H_2SO_4$	Oxic	Elemental sulfur oxidation	Abiotic and dissimilatory biotic
$4NO_3^- + 3S^\circ \rightarrow 3SO_4^{2-} + 2N_2$	Anoxic	Inorganic sulfur oxidation	Dissimilatory biotic
$2CO_2 + 2H_2S + 2H_2O + light \rightarrow 2(CH_2O) + H_2SO_4$	Anoxic	Photosynthetic sulfur oxidation (anoxygenic photosynthesis)	Biotic
$CH_3COOH + 2H_2O + 4S^\circ \rightarrow 2CO_2 + 4H_2S$	Anoxic	Acetate oxidation	Dissimilatory, reduction
$2(CH_2O) + H_2SO_4 \rightarrow 2CO_2 + 2H_2O + H_2S$	Anoxic	Anaerobic respiration, sulfate as the electron acceptor	Dissimilatory, reduction
$4H_2 + H_2SO_4 \rightarrow 4H_2O + H_2S$	Anoxic	Anaerobic hydrogen respiration	Dissimilatory, reduction
$S_2O_3^{2-} \rightarrow SO_4^{2-} + S^{2-} + H^+$	Anoxic	Disproportionation	Dissimilatory
$SO_4^{2-} \rightarrow S^{2-} \rightarrow cysteine$	Oxic/anoxic	Assimilation	Biotic
$H_2S + Fe^{2+} \rightarrow FeS + H_2$	Anoxic	Iron pyrite formation	Abiotic, spontaneous

Note: Others occur, but this table partially illustrates the complexity of the sulfur cycle

The general fluxes of heterotrophy and remineralization (aerobic or anaerobic) are present, as for all other nutrient cycles.

Reduced forms of sulfur can combine with O_2 with a net release of potential energy (Table 14.1). *Abiotic oxidation* occurs spontaneously, but more slowly than *biotic sulfur oxidation*, so the biotic oxidation of reduced sulfur compounds by chemoautotrophic organisms is generally confined to areas at the interface between oxic and anoxic habitats where the supply of reduced sulfur compounds is high, and there is not much time for abiotic oxidation as the reduced sulfur compounds have just diffused into oxic conditions. In unique cases, the inflow of sulfide-rich water into caves with airspace or water containing O_2 can lead to an entire cave ecosystem supported by sulfur-oxidizing bacteria (Sidebar 14.2).

Oxidized sulfur can be used as an electron acceptor to respire organic carbon in a process similar to denitrification. This *dissimilatory sulfur reduction* is a primary source of sulfide found in anoxic sediments and waters. Sulfate forces the redox to remain moderately high and can indicate conditions under which methanogenesis will not occur.

Several unique transformations exist in the sulfur cycle. One is called *disproportionation*, an anoxic transformation in which thiosulfate is converted to sulfate and sulfide, yielding energy (Table 14.1). Another transformation is the use of sulfide as an electron donor for *anoxygenic photosynthesis*, which yields sulfate in a process analogous to using water as an electron donor for oxygenic photosynthesis. The production of sulfate in anoxic habitats leads to the possibility of complete autotrophic and heterotrophic components and a

---

SIDEBAR 14.2

A Romanian Cave Ecosystem Fueled by Sulfide-Oxidizing Primary Producers

In 1986, access was obtained to a groundwater ecosystem within Movile Cave, which has a considerable influx of springwater rich in H_2S. The cave has a terrestrial fauna with 30 obligate cave-dwelling invertebrates, of which 24 are endemic. The aquatic fauna includes 18 invertebrates, with nine endemic species. Air pockets in the cave have floating microbial mats (composed of bacteria, fungi, and five nematode species; Riess *et al.*, 1999) and other mats are attached to walls. Mats are floated by bubbles of methane, presumably produced by methanogenic bacteria. Scientists hypothesized that sulfide-oxidizing bacteria form the base of the food web in these caves.

Subsequent studies demonstrated that the microbial communities were able to assimilate ^{14}C-labeled bicarbonate into microbial lipids, verifying that the assemblages growing in the cave were chemoautotrophic. The ratios of stable isotopes of carbon ($^{13}C/^{12}C$) and nitrogen ($^{15}N/^{14}N$) were used to ascertain that the carbon fixed by microbial autotrophs was entering the food webs. This analysis demonstrated that the microbial biofilms, the invertebrate consumers of the microbes, and the invertebrate predators had isotopic ratios more similar to each other than to those found in nearby surface aquatic or terrestrial samples (Sarbu *et al.*, 1996). Furthermore, isotopic signatures were consistent with the hypothesis that invertebrates subsisted on microbes or other invertebrates that feed on microbes in the cave. Analyses of the microbial community by molecular methods suggested that the methane bubbles in the microbial mats support methanotrophic microbial biomass (Hutchens *et al.*, 2004). This unusual cave ecosystem and a similar one recently discovered in Israel (Por, 2007) are unique, but perhaps more widespread than now known because of the difficulty of finding them.

---

complete sulfur cycle with no O_2 present. Such processes may have been central to early life in the anoxic habitats of primordial Earth.

Sulfide can combine with metals to form *pyrites*, the most common being iron pyrite (fool's gold). These precipitates have very low solubility under anoxic conditions and can represent an important loss of iron to aquatic ecosystems. The precipitate is black and gives anoxic sediments their characteristic black color. Sulfur-oxidizing bacteria can hasten breakdown rates of pyrite when it is exposed to O_2 because they can use it chemoautotrophically. This feature is taken advantage of by some mining where metals are bound to pyrites; bacteria that oxidize the pyrite compounds are mixed with the ore, and the metals are released into solution as the pyrite is oxidized. Regardless of whether this oxidation occurs biotically or abiotically, the end product is sulfuric acid, so acid mine waste can be a substantial pollution problem in surface waters influenced by mine drainage.

Advanced: More Complex Sulfur Transformations

Like methane, N_2, and nitrous oxide, dimethyl sulfide is a gaseous by-product of organisms. Gas can enter the atmosphere and lead to a net loss of sulfur from some ecosystems. The significance of this production has not been well studied in freshwaters; it is of interest to those who link global sulfur budgets to marine primary producers (Malin and Kirst, 1997), and production by fermentations in freshwater sediments may be important as well (Lomans *et al.*, 1997; Yoch *et al.*, 2001). Dimethylsulfide is ultimately oxidized to sulfate in the atmosphere. The gas is important globally because it provides nuclei for formation of clouds and could help regulate global climate.

Bacteria and Archaea have evolved a dazzling array of ways to cycle sulfur, several of which have already been discussed in this chapter. For example, a variety of oxidized compounds, including nitrate, Fe^{2+}, and Mn^{2+} can be used by chemoautotrophs to oxidize reduced sulfur compounds such as sulfide and elemental sulfur. An interesting adaptation to oxidizing sulfur has been documented for the bacterium *Beggiatoa* (Hinck *et al.*, 2007). These large filamentous gliding bacteria have large vacuoles and can use them to store nitrate that is taken up in oxic areas in hot springs. The *Beggiatoa* can then glide down into the anoxic sediments where sulfide concentrations are high and use the nitrate to oxidize the sulfide and generate energy.

Additionally, elemental sulfur can be used to oxidize acetate, sulfate can react with hydrogen gas to yield energy, and denitrification can occur using elemental sulfur and nitrate. Sulfur plays a pivotal role in the cycling of other compounds because it removes iron from solution, which would otherwise precipitate phosphate. Sulfide can precipitate with many other metals; and when sulfate is present, methanogenesis is not energetically favorable.

Sulfur Cycle

The cycling of sulfur is studied mainly because of interactions with other nutrients. The sulfur cycle in aquatic systems is indirectly important to ecosystem productivity because sulfur is rarely a limiting nutrient for primary producers; the requirement is low and the levels available are generally relatively high. The sulfur cycle can be diagrammed using the method presented previously (Fig. 14.8).

PHOSPHORUS

Phosphorus is the key nutrient that limits primary production in many aquatic habitats. Nutrient pollution and use by organisms are discussed extensively in Chapters 17 and 18. Unlike carbon, nitrogen, and sulfur, phosphorus is mainly found in only one inorganic form (phosphate), so most research has centered on organic transformations and interaction of other inorganic chemicals with phosphate.

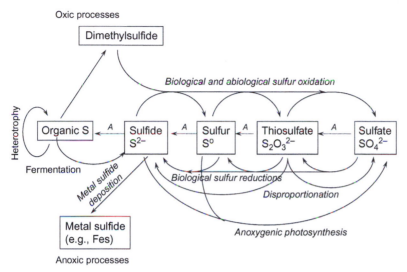

FIGURE 14.8

A conceptual diagram of the sulfur cycle. A = assimilation.

Phosphorus Forms

Phosphate (PO_4^{3-}) is a dominant form of inorganic phosphorus in natural waters, but concentrations are often below detection in pristine waters (about $1\,\mu g\,liter^{-1}$). Determining the precise level of phosphate is difficult because standard methods of analysis also detect a variable and poorly defined group of phosphate compounds (Rigler, 1966). Cells can also store phosphate as a polymer (polyphosphate). Organic phosphorus occurs in a variety of organic compounds including nucleic acids and lipids. The crudest classification is dissolved organic phosphorus (DOP) and particulate phosphorus (PP). The sum of DOP, PP, and phosphate is total phosphorus (TP). Phosphorus is used in cells for nucleic acids, phospholipids, and other compounds.

Phosphorus Transformations

The interaction of phosphate with iron is important in determining the availability of phosphorus in many aquatic systems. Phosphate will precipitate with some metals, including ferric iron (Fe^{3+}, an oxidized form of iron). The precipitation occurs only in the presence of O_2, and the complex dissociates under anoxic conditions. Precipitation of ferric phosphate leads to the deposition of phosphorus in sediments when surface water is oxygenated. When the precipitate settles into an anoxic zone (such as an anoxic hypolimnion of a lake) the phosphate dissociates. Transport processes such as eddy diffusion, which move dissolved materials, then move the phosphate. Thus, phosphate is brought back up to the surface when fall mixing breaks down an anoxic hypolimnion.

Another important precipitation of phosphate occurs with calcium to form calcium phosphate (apatite). This reaction is favored at high alkalinity, high phosphate, and high pH (Stumm and Morgan, 1981). Carbonate-rich waters commonly will exhibit these conditions. Photosynthesis by periphyton or macrophytes will raise pH and will encourage precipitation of apatite. This process of deposition can strip phosphorus from the water column and lead to long-term sequestration of phosphorus in the sediments. The process is part of the reason for strong phosphorus deficiency in the Everglades (Dodds, 2003a). Over geological time the process can lead to formation of marine sedimentary rocks that are very rich in phosphate. Phosphate that is mined for fertilizers and industrial uses mostly originates from this source.

Phosphate is assimilated at very low concentrations by cells under phosphorus-limited conditions. Evidently, billions of years of natural selection under phosphorus-limiting conditions have created a selective pressure for very efficient uptake mechanisms. Uptake of nutrients will be discussed in Chapter 17. This uptake results in extraordinarily low (nanomoles per liter) concentrations of phosphate, and these concentrations are far below those normally detected by conventional methods (Hudson and Taylor, 2005). Natural concentrations of phosphate in very phosphorus-limited waters are also far lower than conventional reagent-grade ultrapure laboratory water; organisms are better at removing phosphate from water than are standard water purification methods created by humans.

Organisms have *phosphatase* enzymes that cleave dissolved organic phosphorus compounds to liberate phosphate. The phosphatases are common inside of cells as they are necessary for normal cell function. Toxic microcystins synthesized by cyanobacteria inhibit phosphatase activity and are thus toxic to most organisms.

Phosphatases can also be excreted outside the cell (extracellular) or be associated with the exterior cell surface (Chróst, 1991; Olsson, 1991). These enzymes increase the availability of phosphate to cells, so the excretion of extracellular phosphatases increases when phosphorus becomes scarce. The ubiquitous nature of these compounds in natural waters leads to rapid turnover of many organic phosphorus compounds. Assays for phosphatase activity are used to gauge the degree of phosphorus deficiency in aquatic habitats; under limiting conditions organisms tend to produce ample amounts of phosphatase.

Heterotrophy results in rearrangement of organic phosphorus compounds as in all other cycles (Fig. 14.9). Many animals cannot use phosphate directly, but must consume organic phosphorus to meet their needs (e.g., lipids, nucleic acids). Organisms excrete excess phosphorus as phosphate or organic phosphorus in both oxic and anoxic environments.

In contrast to other nutrient cycles, phosphate cannot serve to oxidize organic carbon (e.g., denitrification and dissimilatory sulfate reduction). Reduced phosphorus compounds are extremely toxic, and natural phosphate concentrations are low, which explains why organisms have not evolved the ability to use phosphate as an electron acceptor in anaerobic respiration.

Given the importance of phosphorus in nutrient pollution, numerous phosphorus budgets have been well documented. Chapter 18 explores the process of quantification of the effects of phosphorus in lakes and describes wetlands as systems that remove phosphorus from water.

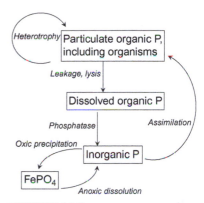

FIGURE 14.9
A diagram of the phosphorus cycle.

SILICON, IRON, AND OTHER TRACE NUTRIENT CYCLES

Silicon cycles drive the dominance of diatoms in many surface waters and can be a primary determinant of algal community structure in lakes and streams. Iron is intimately tied to the sulfur and phosphorus cycles in addition to being a water quality variable of concern on its own. Manganese, molybdenum, and many other nutrients have their own cycles as well.

Silicon

Aluminosilicate minerals are the most abundant in Earth's crust (Schlesinger, 1997). However, silicate is generally not the major dissolved ion in natural water because of the limited solubility of silicon-containing compounds (Stumm and Morgan, 1981). Microbes contribute to weathering rates of silicates and this can be important in chemistry and geology of groundwaters (Hiebert and Bennet, 1992). Silicon can take several forms, including *silica* (SiO_2) and *silicic acid* (H_4SiO_4). Silicic acid is generally the form dissolved in waters. The greatest concentrations are in waters draining volcanic watersheds, and the lowest from limestone watersheds.

Clays are composed mainly of silicon-containing compounds and are a primary component in turbid aquatic systems and sediments. Clays can be very important as sites for microbial activity and areas where chemicals can adsorb. For example, clays can adsorb and hold organic pollutants, making it difficult to clean spills in groundwaters. Sand, composed mainly of silicates, also forms a major component of the benthic substrata of many aquatic systems.

Diatoms rely on silicon as a component of their specialized cell walls called *frustules*, where it is deposited as opal. The opal is relatively insoluble, so it sinks to the sediments and persists after the diatom dies. The burial of these frustules allows for paleolimnological determinations of the diatom species present over time in lakes. Diatomaceous earth, which is soft, sedimentary rock formed from ancient accumulations of diatom frustules, is used widely in water filtration systems, as a mild abrasive, and as a natural insecticide. Emergent grasses also use silicate as a structural material and to defend against herbivores, and freshwater sponges make their spicules from silicate for similar reasons.

Silicon can become depleted in the epilimnion of both eutrophic and oligotrophic lakes during summer. Depletion occurs because of the relatively slow cycling of silicon, its incorporation into diatom frustules, and subsequent sinking of frustules before silicon can be redissolved. Dissolved silicon concentrations build near the sediments during summer and winter stratification (Fig. 14.10) because diatom frustules sink to the sediments and slowly dissolve. The temporal and spatial variation in silicon availability may be important in the successional processes of phytoplankton, as diatoms deplete silicon seasonally (Fig. 14.11).

In contrast, concentrations of silicon in rivers are remarkably constant over time and across continents (Wetzel, 2001); thus, periphytic diatom communities in flowing waters are less likely to be influenced by seasonal changes in silicon concentrations. However, increased nitrogen loading to world rivers is apparently stimulating diatoms, leading to greater silicon uptake and lower concentrations of silicate exported by rivers to marine habitats (Turner *et al.*, 2003).

Iron

Iron is a key element in many biological molecules, including electron transport proteins, hemoglobins, enzymes used in synthetic pathways for chlorophyll and proteins, and enzymes used for nitrate assimilation, photosynthesis, and other essential metabolic processes. The total demand for iron is not high relative to nitrogen, carbon, and phosphorus because iron serves as a cofactor in these reactions (one or a few iron atoms for each entire protein). However, iron is required as a minor nutrient and also serves as a key link among many biogeochemical interactions, such as the phosphorus and sulfur cycles.

Iron occurs as a dissolved ion in several forms. Under high redox (oxic conditions), iron takes the form of *ferric* (Fe^{3+}) ions. At low redox iron is in the *ferrous* (Fe^{2+}) form. O_2 converts ferrous to ferric iron by accepting electrons from the ion, but the process occurs over many minutes (Stumm and Morgan, 1981). Ferric ions react with hydroxyl ions to form the flocculent precipitate ferric hydroxide ($Fe(OH)_3$) under oxic conditions and near-neutral pH. Those who obtain water from anoxic wells with iron-rich groundwater can directly observe this precipitation. The water appears clear when it is withdrawn, but

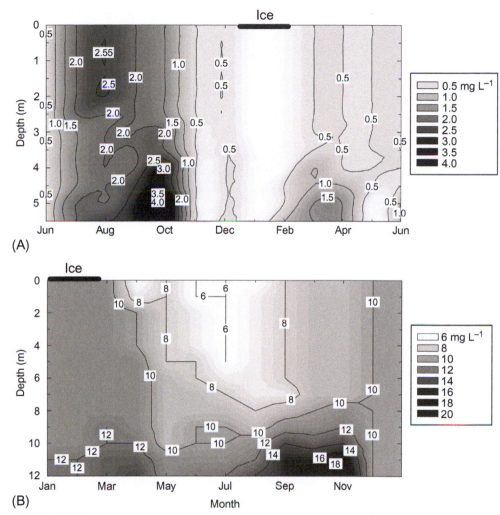

FIGURE 14.10

Concentration of silica as a function of depth and time in hypereutrophic Wintergreen Lake, Michigan (A), and oligotrophic Lawrence Lake, Michigan (B). Concentrations are given in mg liter^{-1}, with darker contour fills corresponding to greater concentrations. *(Reproduced with permission from Wetzel, 1983).*

exposure to O_2 converts the ferrous iron to ferric iron, which then forms a flocculent precipitate as ferric hydroxide; a clear glass of water looks a bit rusty after a few minutes. Modern water treatment plants often aerate drinking water and allow the ferric hydroxide to settle in a pool if anoxic groundwater rich in iron is used as a source.

Inorganic iron concentrations vary seasonally in stratified lakes, with especially pronounced variation in concentrations related to anoxic events such

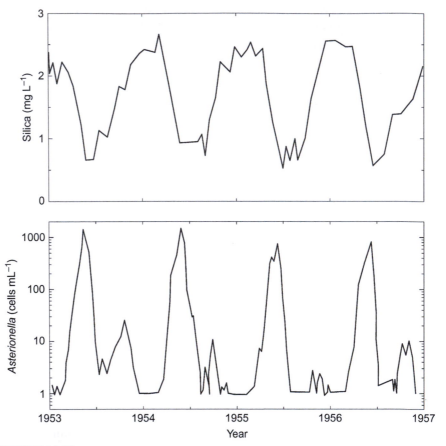

FIGURE 14.11

The relationship between epilimnetic silicon and biomass of the diatom, *Asterionella*, in Lake Windermere, England. Note how decreases in dissolved silica correspond with high densities of diatoms. *(Data from Lund, 1964).*

as summer stratification in eutrophic lakes (see Fig. 12.6). These patterns are tied closely to the redox of the lake.

Higher concentrations of iron can be maintained in oxic solutions by the actions of *chelators*. These compounds briefly bind iron and other metals and do not allow precipitation. Most chelators in natural waters are organic compounds, although inorganic compounds can also serve as chelators. Some organic chelators of iron are produced by algae and excreted into solution to keep iron available for uptake. Other substances, such as humic materials formed by plant decomposition, are also chelators.

Large concentrations of humic materials found in natural water can complex with iron so successfully that they make iron unavailable to organisms.

Thus, some dystrophic wetlands and lakes are unproductive not only because of high concentrations of humic materials that absorb light but also because they make iron less available. At intermediate concentrations, these chelators make iron more available by interfering with precipitation, but they are detrimental at high concentrations.

Ferric iron can also precipitate phosphate, as mentioned previously. These flocculant precipitates will settle to the sediments in an oligotrophic lake or wetland. If the hypolimnion is anoxic as expected in a eutrophic lake, the $FePO_4$ precipitate will dissociate into phosphate and ferrous iron. This process is a key interaction between elemental cycles and has a major impact on algal production.

Iron pyrite forms when ferrous iron reacts with sulfide. This precipitate has limited solubility and forms black deposits in anoxic sediments. Thus, in wetlands or lakes with O_2 in the water column immediately above the sediments, the first few centimeters of sediment are light colored, but as the deeper anoxic portions are reached, the dark pyrites color the sediments dark brown or black. This reaction and its indirect mediation of the phosphorus cycle can influence responses of management of nutrient pollution (Sidebar 14.3).

SIDEBAR 14.3
Sulfur and Iron Dynamics Increase Wetland Phosphorus and Associated Growth of Noxious Plants

Large areas of The Netherlands are composed of peaty lowlands. These wetland areas are fed with river water to compensate for groundwater withdrawals for agriculture (Smolders and Roelofs, 1993). The inflow of river water was hypothesized to cause deterioration in water quality (increases in turbidity, filamentous algae, and duckweed) and a shift away from characteristic emergent wetland plants. Attempts were made to halt this deterioration by stripping incoming river water of phosphate, but the water quality did not improve.

Further investigation revealed that the sulfate and iron contents in the river water were altering the biogeochemical cycling. The river water entering the wetlands was stripped of phosphate, high in sulfate, and low in iron. The sulfate diffused into the anoxic sediments where it was reduced by microbes to sulfide. Increased sulfide concentrations in the anoxic sediments led to greater precipitation of ferrous iron as iron pyrite into the sediments. Because the iron was removed from the system, it could no longer precipitate with phosphate in the oxic waters of the wetland. The lowered amount of phosphate precipitation allowed higher concentrations of phosphorus to be maintained in the surface waters and led to stimulation of algal growth. Concurrent increases in the sulfide levels in the pore waters combined with low iron concentrations were hypothesized to harm the native macrophytes via sulfide toxicity and growth limitation by low iron.

This European case study is an excellent example of the importance of understanding nutrient cycling and how interactions among the cycles influence primary producers. The management of the system appeared to be simple, but complex interactions among nutrient cycles conspired against the feasibility of a "dilution solution to pollution."

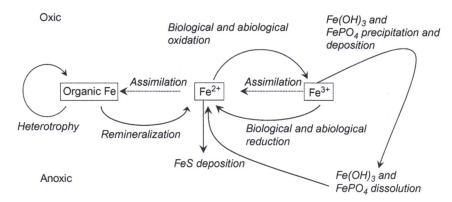

FIGURE 14.12
A conceptual diagram of the iron cycle.

Ferrous iron has potential energy in the presence of O_2 and can serve as an energy source for microbes able to oxidize it before spontaneous conversion occurs. This reaction is confined to areas with O_2 that are close to an anoxic habitat that provides ample iron in the reduced (ferrous) form. This oxidation is one route for dissociation of iron pyrite (FeS) in oxic environments.

Iron-oxidizing bacteria can cause problems in groundwater wells in anoxic habitats high in iron. In these situations, O_2 dissolved in the well water moves down the well. The aquifer immediately outside the bottom of the well has a zone in which O_2 and Fe^{2+} are found together. The iron oxidizers can use this potential energy source to incorporate CO_2 and grow. Bacterial growth can eventually plug the well intake screens and the porous materials surrounding the well intake. Removing the bacterial growth can be difficult, and measures such as back-flushing wells provide only temporary help.

Iron cycling (Fig. 14.12) is similar to nitrogen, carbon, and sulfur cycling because the concepts of redox and potential energy can be used to order and understand the processes. The iron cycle was one of the first nutrient cycles to receive extensive study (Mortimer, 1941).

CYCLING OF OTHER ELEMENTS

Cells need a number of minor elements including copper, manganese, selenium, zinc, and molybdenum. Sodium, potassium, chloride, and boron are also required. Most of these are found in low concentrations. These elements can also act as pollutants at greater concentrations.

Minor nutrient cycles, such as molybdenum, selenium, and copper, are also much less studied probably because even though organisms require these micronutrients, the requirement is small and the chances of these elements limiting system production are relatively low (but see Goldman, 1972). Potassium can limit plant production in some wetlands. An exception to the idea that minor nutrients limit productivity is the case of molybdenum limitation in Castle Lake, California. In this case, molybdenum is used for nitrogen assimilation. Alder trees in the watershed scavenge the molybdenum, creating apparent molybdenum limitation in the lake (Goldman, 1960). Limitation can be overcome by supplying nitrogen in the form of ammonium that does not require molybdenum to use, unlike nitrate, which requires a molybdenum-containing nitrate reductase before assimilation can occur.

Manganese cycling is similar to iron cycling in many ways (Wetzel, 2001). Reduced manganese can be combined with O_2 by chemoautotrophs to yield energy. Oxidized manganese can be used as an electron acceptor for anaerobic respiration. There are some differences between the iron and manganese cycles. For example, iron oxidation is spontaneous in the oxic waters of lakes, but manganese oxidation requires the high pH and redox associated with photosynthesizing organisms (Richardson *et al.*, 1988). Manganese nodules are occasionally found in freshwaters. These nodules are formed when bacteria oxidize reduced iron and manganese (Atlas and Bartha, 1998; Chapnick *et al.*, 1982).

Selenium and copper are both required for proper cell function, but they are far more commonly studied as pollutants. Some agricultural practices can lead to selenium occurring at toxic levels, and some forms of mine waste can lead to copper pollution.

GRADIENTS OF REDOX AND NUTRIENT CYCLES AND INTERACTIONS AMONG THE CYCLES

Redox is a major factor controlling biological and abiological nutrient transformations as stressed in previous discussions of biogeochemical cycling. Metabolic activities that predominate in a specific environment can be predicted in part by the redox potential of that system. Redox decreases with increasing depth in anoxic sediment or across the metalimnion of a eutrophic stratified lake. Changes in redox also occur over time, giving rise to an orderly sequence of preferred nutrient transformations. The more oxidized a compound is, the more energy that can be generated when using it as an electron donor to oxidize organic carbon with anaerobic respiration. The order that compounds can serve as electron acceptors (their relative redox state) was presented in Figure 13.6.

Redox gradients are sites of high rates and diverse types of metabolic activities. These microbial "hot spots" occur because of the dependence of many aquatic microbial geochemical processes on either reduced chemicals in oxidized environments or oxidized chemicals in reduced environments. Given Fick's law of diffusion, the steeper the gradient, the greater the diffusive flux. Thus, redox gradients are locations where reduced compounds are diffusing rapidly into oxidized conditions and vice versa.

The complex interrelationships of nutrient cycling in the vicinity of a redox gradient is illustrated by considering the hypothetical distribution of some of these activities and populations of microbes (Fig. 14.13). The fact that multiple chemicals interact with each other and are transformed by organisms leads to substantial complexity. Additional complexity arises from the variety of fermentative processes that can occur and the fact that many different microbial strains or species can be responsible for each of the processes.

Gradients of redox can occur in a variety of habitats (discussed in Chapter 12). Consideration of redox adds more complexity to the view of biogeochemical cycling because of the series of redox potentials. In the absence of O_2, redox gradients still provide a habitat gradient with different metabolically driven fluxes depending on the amount of potential energy that can be extracted. Because many of these organisms have a long evolutionary history, natural selection dictates that organisms relying on inefficient metabolic pathways will be outcompeted and are unlikely to dominate in areas where other processes are more efficient.

One of the influences of humans is to alter the natural distribution of redox heterogeneity. For example, urbanization can lower heterogeneity, and decrease anoxic carbon-rich habitats leading to lower rates of denitrification. Restoration of stream channel heterogeneity can increase the rates of denitrification by creating numerous small low-redox areas that are conducive to denitrification (Groffman *et al.*, 2005).

The complex interactions among organisms can be thought of as constrained by redox but also as tied together by organic carbon because carbon provides energy and materials for organism growth. For example, chemoautotrophic organisms, such as those that oxidize ammonium, sulfide, or ferrous iron, use the energy obtained to assimilate CO_2. Assimilatory processes can provide new nutrients to a habitat, as in the case of nitrogen fixation. The redox series of metabolic interactions discussed previously is in large part a series of increasingly less efficient ways for organisms to oxidize organic carbon to CO_2 and release energy.

Some unexpected interactions can occur among nutrient cycles that make sense based on redox. For example, some wetlands can be protected from eutrophication by nitrate pollution (Lucassen *et al.*, 2004) because nitrate has a higher redox and is used as an electron acceptor before oxidized Fe^{3+}.

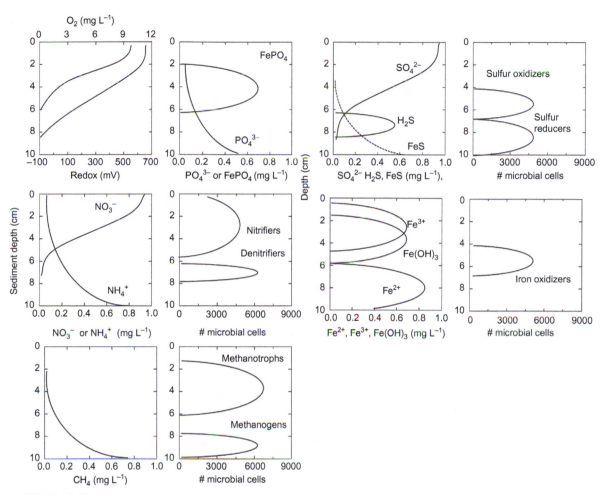

FIGURE 14.13

Relationship among redox gradients, dissolved oxygen, nutrient concentrations, and functional groups of microorganisms responsible for biogeochemical fluxes. This figure illustrates the steep gradients that occur at oxic/anoxic interfaces, and how such interfaces are a hot spot for biogeochemical activities.

Oxidized iron can then precipitate the phosphate. The phosphate is deposited in the sediments and is less available to wetland plants. Understanding this effect requires a thorough understanding of redox and interactions among chemicals in freshwaters.

In addition, key links occur between the cycles that are mediated by inorganic chemical reactions. These include interactions between ferric iron and phosphate, ferrous iron and sulfide, and calcium and phosphate.

SUMMARY

1. The major forms of inorganic nitrogen are N_2 gas, nitrate, nitrite, and ammonium. Organic nitrogen occurs in many forms, including amino acids, proteins, nucleic acids, nucleotides, and urea.

2. The major fluxes in the nitrogen cycle include denitrification (using nitrate to oxidize organic C yielding N_2), oxidation of ammonium to nitrate by chemosynthetic bacteria, assimilation of ammonium, fixation of N_2 by bacteria, and excretion of ammonium by heterotrophs.

3. Forms of inorganic sulfur include sulfide, thiosulfate, sulfate, elemental sulfur, and metal sulfides. Organic sulfur is found in proteins and amino acids.

4. The most important fluxes in the sulfur cycle are biological and abiological sulfur oxidation, biological sulfur reduction (a form of anoxic respiration), production of sulfide by fermentation, disproportionation, and metal pyrite precipitation and deposition.

5. Phosphorus is a key element that determines the productivity of many aquatic ecosystems and can be found in the forms of phosphate and organic phosphate. Organic phosphate can be cleaved to phosphate by phosphatase enzymes produced by organisms. Phosphate forms a low-solubility precipitate with ferric iron in the presence of O_2 that can cause its removal from oxic environments.

6. Silicon is a vital component of the cell walls of diatoms and can be a key factor in controlling composition of phytoplankton communities. It is redissolved slowly; thus, when frustules sink out of the photic zone in lakes, it takes months for the silicon to become available again. Under some conditions, frustules accumulate over time and form silica-rich sedimentary rock.

7. Iron occurs as ferric and ferrous ions in oxic and anoxic habitats, respectively. It can also occur as a metal pyrite (FeS) in anoxic habitats and a flocculent precipitate $(Fe(OH)_3)$ in oxic habitats. Iron is an important component of many proteins, including those for electron transport, nitrate assimilation, and chlorophyll synthesis.

8. Bacteria in oxic environments can oxidize ferrous iron. Ferric iron combines with OH^- or PO_4^{3-} to form low-solubility precipitates. Chelators can keep ferric iron in solution (prevent it from precipitating with OH^-). Chelators are organic molecules that form complexes with iron.

9. Gradients of redox that occur in natural waters allow for complex populations of microbes that are capable of a wide variety of nutrient transformations to form in localized hot spots. There is a predictable order in which each process will occur across the redox gradient, given the relative potential energy of each chemical reaction at each redox point.

10. Nutrient cycles do not occur in isolation. Complex interactions occur among all of them, in part because organisms have similar nutrient requirements. Interactions also occur because different chemicals interact with each other in the absence of organisms.

QUESTIONS FOR THOUGHT

1. How might acid precipitation dominated by sulfuric acid increase the availability of phosphate, aside from pH effects?

2. Why are isolated chemosynthetic communities, such as those found where sulfide enters oxic cave waters or methane enters sediments, still ultimately dependent on photosynthetic organisms?

3. When people control eutrophication, if only nitrogen is removed, cyanobacteria can dominate. Why?

4. What was Earth like before oxygenic photosynthesis and how did the sulfur cycle likely drive the redox processes of ecosystems in aquatic habitats at that time?

5. Why are areas downstream from hyporheic zones where water reenters streams often sites of very high algal production?

6. Why do sewage treatment plant operators generally want to avoid denitrification in ponds designed to settle solids?

7. How might phosphate-rich water be treated with iron to remove phosphate (including O_2 concentrations and forms of iron used)?

8. Why does silicon often disappear from the epilimnion of lakes more rapidly than phosphorus or nitrogen?

9. Why is it difficult to reduce the effects of excessive phosphorus pollution in lakes when O_2 disappears from the hypolimnion?

Unusual or Extreme Habitats

Doi: 10.1016/B978-0-12-374724-2.00015-5

FIGURE 15.1

Lake Frxyell (the lake is under the flat ice in the center of the valley and is about 5 km long), a permanently ice-covered lake in Antarctica (*top, courtesy of Dale Anderson*), and tufa pillars, which are several meters high, at hypersaline Mono Lake, California. (*Bottom, courtesy of Mark Chappell*).

Abiotic extremes create unusual environments that are engaging to aquatic ecologists because of their novelty. Microorganisms that can live in almost boiling water and animals that can live in near-freezing water are fascinating. In addition to the academic interest, these habitats can provide insight into how organisms may tolerate pollution and other human-caused environmental perturbations. For example, studies on thermal pollution and global warming may require an understanding of how organisms cope with high temperatures (Sand-Jensen *et al.*, 2007), and organisms from saline lakes may provide clues to species' responses to salinization by agricultural and urban runoff. Furthermore, aquatic microbial ecologists and biotechnologists have isolated useful microbes from extreme habitats, such as those that produce the enzymes essential for the polymerase chain reaction (PCR), an essential tool in modern molecular biology. In this chapter, we discuss how organisms adapt to different extremes and the environments in which the extremes occur. The extremes considered here include high and low temperatures, periodic drying, high salinity, surface layers of water (experiencing damagingly high light), and ultraoligotrophic waters. In the next chapter we will discuss how human stressors, such as pollution, can have detrimental effects on organisms, and adaptations discussed in this chapter will be very relevant to those discussed in the next.

ADAPTATIONS TO EXTREMES

Shelford's law of tolerance suggests that there is an optimal range in which fitness is greatest; as conditions deviate from the optimum, fitness declines and beyond some limit the organism cannot survive and reproduce. With many potential environmental extremes, this is true. However, with some extremes it is not a "law." For example, ionizing radiation has no intermediate level of radiation exposure where fitness is optimal. Still, for many environmental drivers that influence organisms, such as temperature and pH, species adapt to optima and are less fit outside the optimum range.

Many of the adaptations that will be discussed in this section are cellular or molecular, and most of the organisms that inhabit the most extreme environments are microorganisms. Microbes probably dominate because higher plants and animals have complex multicellular systems that cannot evolve to compensate for extremes, such as particularly high temperatures, salinity, and variations in pH, although notable exceptions occur and these are discussed where appropriate.

Understanding the influence of extremes in pH, salinity, and temperature requires knowledge of the structure and function of biological molecules. We begin with a discussion of adaptation to temperature. The main influences of temperature have been related to protein structure, DNA and RNA structures, and lipid fluidity. However, metabolism at high temperatures presents particular difficulties when molecules involved are unstable at high temperatures.

Proteins must maintain structure and have ample thermal energy to function. An enzyme not only needs to maintain an active site in a very specific configuration, but also must be able to translate thermal energy into making or breaking chemical bonds. Thus, enzymes have a temperature range in which they are able to maintain structure and activity, and within that range they have an optimum temperature for activity. Enzyme activity increases with temperature up to a point, and then the enzyme starts to break down (denature) and rapidly loses function. Proteins in organisms in high-temperature environments need enhanced thermal stability. Factors that are correlated with thermal stability of proteins include a hydrophobic core; reduced glycine content; more alpha helical conformation, high ionic interactions, and reduced surface area to volume ratio (better core packing); more stabilizing bonds (hydrogen bonds and salt bridges); and numerous related adaptations (Madigan and Oren, 1999; Kumar *et al.*, 2007).

Biological membranes need to maintain an optimum degree of fluidity to function properly. If the membranes are excessively fluid, they will not maintain cell integrity, and if the membranes are solid they lose biological activity. As temperature increases, the melting point of the lipids increases as well. Lipid melting points increase with greater proportions of single bonds (increasing saturation), increased branching, greater length, and, in the Archaea, ether lipids (Russell and Hamamoto, 1998). For lipids to remain fluid at low temperatures, the opposite properties (unsaturated, short, and unbranched) are required.

DNA and RNA molecules also have a specific temperature range in which they function optimally. RNA molecules of organisms adapted to high temperatures may have more C–G bonds than those growing at lower temperatures because C–G pairs are stabilized by three hydrogen bonds, whereas A–T pairs have only two bonds. The C–G pairs, and increased cation concentrations and specialized proteins, can also stabilize DNA (Stetter, 1998; Madigan and Oren, 1999; Grosjean and Hoshima, 2007). DNA and RNA are damaged at high temperatures, and thermophilic cells have greater rates of DNA repair and RNA turnover. RNA can have modified nucleotides that stabilize in high temperatures (Grosjean and Hoshima, 2007). Messenger and ribosomal RNA need portions of the molecule held in specific configurations (secondary and tertiary structures) to function properly. If these secondary (Fig. 8.3) and tertiary structures are not maintained, protein synthesis (translation) will not occur. In organisms adapted to high temperatures, RNA has extended regions of base pairing (C–G or A–U) to stabilize secondary and tertiary structures.

The requirement for these extended regions is less stringent in low-temperature organisms. Some multicellular organisms can withstand freezing, but a variety of molecular adaptations to low temperatures also occur (Rodrigues and Tiedje, 2008). The molecular adaptations are generally opposite to thermophilic adaptations, but involve different genes.

To survive freezing, the ability to avoid the damaging effect of ice crystal formation in cells is required (Sakai and Larcher, 1987). Ice crystals rupture the plasma membrane and destroy the integrity of the cells. Organisms that live in high latitudes or altitudes use a variety of strategies to avoid or survive freezing including *supercooling*, which is the lowering of the freezing temperature of the water in their bodies (Storey and Storey, 1988). For example, some fishes, amphibians, and invertebrates produce glycerol, which is essentially antifreeze, and concentrate it in critical body parts. Some animals freeze nearly completely, but they use ice nucleators, proteins, or other substances in the blood and tissues that facilitate formation of ice crystals, to control freezing. This process slows the rate of freezing, allowing time for physiological and metabolic adjustments as the body freezes. Many invertebrates simply enter a resting stage called a *diapause*, in which they are more tolerant of freezing. Diapauses often occur during life stages of reduced biological activity, such as an egg, pupae, or cyst, and can also be used to survive other adverse conditions including drought, high temperatures, or anoxic conditions.

Different adaptations are required for organisms living under extremes of salinity. In these situations, ions outside the cells (such as magnesium compounds) become highly hydrated, and water becomes limiting. Animals and microbes can survive moderate levels of salinity by excreting excess salt. Osmoregulation of fishes has received considerable study (Eddy, 1981) and the physiological mechanisms to control salt balance are documented. Only organisms that are able to withstand high intracellular concentrations of salt (above about 10% salinity) can survive in high-salinity environments. Hypersaline microbes have multiple sets of genes coding for sodium and potassium transport (Vellieux *et al.*, 2007). Osmotic pressure will collapse cells and drain their water if they do not maintain an internal concentration of dissolved materials approximately equal to the external ion concentration. For example, *Dunaliella*, a green alga that thrives in saline waters, synthesizes high concentrations of glycerol to counteract the effects of increased salinity (Javor, 1989). Many other species of algae and fungi also use glycerol to counteract osmotic pressure in aquatic environments. Some halobacteria can accumulate up to 5 molar KCl in their cells (Grant *et al.*, 1998). Salinity also alters proteins by increasing hydrophobic interactions, and hypersaline organisms have high amounts of negatively charged amino acids on their external surface to counteract ion effects (Vellieux *et al.*, 2007).

A number of habitats have extremely high or low acidity. For example, hot-springs can be quite acidic or basic. Some contaminated habitats, such as acid mine drainage, also can be highly acidic. Cells must maintain their internal pH close to neutral to allow for normal metabolism, but extracellular proteins must function at the pH of the environment. Acidophilic microorganisms have membranes that are impermeable to hydrogen ions (Johnson, 2007).

Lack of water is a particularly severe condition for aquatic organisms. Temporary pools, wetlands, and intermittent streams all have periods of drying. The ability to withstand desiccation is found in many groups of aquatic organisms. An impressive example is the cyanobacterium *Nostoc*, which accumulates sucrose to maintain biological molecules during drying and has been documented to withstand 107 years of desiccation (Dodds *et al.*, 1995). Similarly, crustaceans that inhabit temporary pools in arid regions produce resting eggs that can remain viable without water for decades (Smith, 2001).

Finally, habitats with high light intensity can be detrimental to many organisms because solar radiation harms cells. Some species that live in shallow, clear water bodies and those that live on the surface need some protection from damaging UV radiation. Solar irradiance causes formation of free radicals that can react with biological molecules. High-energy light, particularly UV, causes the most damage. Compounds such as carotenoids (many types of organisms), mycosporine-like amino acids (diatoms), flavenoids (green algae and higher plants), and scytonemin (cyanobacteria) absorb damaging light. These compounds protect organisms from damage by preventing formation of the harmful free radicals (Long *et al.*, 1994). The drought-resistant eggs of some invertebrates that inhabit temporary waters are also protected from damaging UV by protective coatings. Planktonic cladocerans and copepods that live in clear, high-latitude lakes use protective pigments and behavioral responses, which allow them to feed in the water column during the day when UV exposure is high (Hansson *et al.*, 2007).

SALINE LAKES

There is roughly as much water in saline lakes on Earth as is contained in freshwater lakes, but these lakes are generally not important water supplies, so have received substantially less study than freshwater lakes. Saline lakes are often the only surface water in dry regions of the world. Some large permanent saline lakes include the Great Salt Lake, the Dead Sea, Issyk Kul, the Caspian Sea, the Aral Sea, and Lake Balkash. Many saline lakes, such as the Australian Lake Eyre, are intermittent.

Saline lakes and ponds occur in closed basins in which water leaves primarily through evaporation. In these situations, salts are weathered from the watershed and flow to the lowest point. Water then accumulates until the water body is large enough for evaporation to equal inflow. As the water evaporates, it leaves behind the salts. Saline lakes can have concentrations of salts 10 times or more in excess of those found in marine waters (oceans are approximately 3.5% salt by weight). The relative proportion of ions in the inflowing water and varied solubility of different ion pairs as the salts are concentrated determine the chemistry of the lake.

One interesting feature of some salt lakes is the formation of salt or tufa (carbonate mineral) columns. These are formed as ion-rich water moves up through elevated columns of deposited salts by capillary action and evaporates off of the top, leaving the salt behind. Carbonate deposition by photosynthetic organisms can enhance the process. Such a process forms the bizarre landscape with tufa pillars on the shore of parts of Mono Lake (Fig. 15.1) and similar pillars of salt along the Dead Sea may have been mistaken for Lot's wife.

Depth of salt lakes varies considerably over the years because their levels depend on a balance between evaporation and inflow. During wet years, the depth of the lake increases until the surface area is great enough to allow evaporation to equal inflow. In these cases, salinity decreases and this can have ecological effects on the aquatic community. Thus, a series of wet years led to flooding of the Great Salt Lake in Utah in the 1980s and led to shifts in the lake's food web. Decreases in salinity allowed the predaceous insect *Trichocorixa verticalis* to invade the pelagic zone of the lake. This predator caused decreases in the brine shrimp *Artemia franciscana* and subsequent increases in protozoa and three species of microcrustacean zooplankton (Wurtsbaugh, 1992).

There is a general decrease of diversity of animals and plants as salinity increases and the upper tolerance limit of various organisms is exceeded (Table 15.1). As with other extreme environments, the Bacteria and Archaea dominate in the harshest habitats. Some animals such as brine shrimp (the anostracan *Artemia salina*) can withstand more than 30% salt. The upper salinity limit of some animals may actually be a lower limit for O_2; solubility of O_2 decreases with increased salinity. However, the majority of freshwater species disappears under only moderate salinity, presumably because of an inability to osmoregulate (Bayley, 1972).

Salt lakes have scientific, economic, cultural, recreational, and ecological values. These lakes are very sensitive to decreases in flow. Human appropriation of freshwaters in the dry regions where they are located can have devastating effects. For example, prior to 1960, the Aral Sea in Russia was the fourth largest lake in the world, with a surface area of $68,000 \, km^2$ and volume of $1,090 \, km^3$. By 1993 the lake area had decreased almost by half and the volume

Table 15.1 Upper Salinity Tolerances of Some Members of Selected Groups of Organisms

Organism	Upper Salinity Range (%)
Fishes	11
Nematodes	12.5
Ostracods	13
Gastropods	15.9
Rotifers	16
Isopods	16
Copepods	17.6
Diatoms	20.5
Chironomids (Diptera)	28.5
Ephydra cinerea (Diptera)	30
Artemia salina (brine shrimp, Anastroca)	33
Cyanobacteria	35
Ciliate protozoa	35
Green algae (*Dunaliella*)	35
Parartemia salina (Anostraca)	35.3
Phototrophic bacteria	40
Extreme halophilic bacteria	Saturated

(Data from Javor, 1989)

decreased to $340 \, km^3$. Agricultural uses reduced water inflow from 50 to $7 \, km^3 \, year^{-1}$. The former lake bed is a source of salt and dust storms that have negative impacts on human health. The increase in salinity and numerous introductions of animals have caused sharp decreases in biodiversity (Williams, 1993). Other saline lakes, notably Mono Lake in California, are subject to similar pressures. Such lakes are unique and should be conserved.

ADVANCED: CHEMISTRY OF SALINE LAKES AND SALT PRODUCTION

Most modern nonmarine lake brines are dominated by the anion chloride, followed by carbonate and sulfate. Marine-influenced brines are dominated by chloride. Sodium and potassium are the most common cations, but some lakes are dominated by magnesium or calcium instead (Hardie, 1984). As concentration of salt increases, some salts become insoluble and precipitate.

In general for the major anions, chloride salts are more soluble than sulfate salts, which in turn are more soluble than carbonate salts. For the cations, sodium salts can be at higher concentrations than magnesium, and calcium salts are least soluble (Table 15.2). Although the exact concentration at which chemical precipitation occurs depend upon concentration of salts, interactions among the salts (those in this table and others), pH, and temperature. You can see from Table 15.2 that different salts will be deposited as saline water is dried by concentration. The known series of deposition can be used by saltworks that have a series of evaporation ponds and "harvest" different salts as they are concentrated by evaporation.

A series of ponds is created and as evaporation concentrates the water, different salt compounds are deposited on the bottom of the pond, and the remaining concentrated water transferred to the next basin. Salt works are common in near oceans, but also are found near saline lakes. Since lakes have variable chemistries, the salt works near various saline lakes will yield different types of salts than those near the oceans, or next to lakes with relict marine waters. Biology complicates this process as some species of halophytic organisms can interfere with the precipitation of salts or cause them to precipitate in unpredictable orders.

Table 15.2 Saturation Concentrations, Ordered from Least to Most Soluble, of Various Salts Found in Saline Lakes

Ion	Concentration (Molar)
$CaCO_3$	0.00015
$MgCO_3$	0.0019
$CaSO_4$	0.015
Na_2SO_4	0.33
$NaCO_3$	0.67
$MgSO_4$	2.1
$MgCl_2$	5.7
$CaCl_2$	6.7
$NaCl$	16.4

(Data from Weast, 1978)

HOT SPRINGS

Hot springs associated with geothermal activity have piqued the interest of many scientists. Visitors to thermal areas such as Yellowstone National Park might not be aware that many of the beautiful colors they see in the pools and streams formed by the hot springs are actually living microorganisms. These colors on the bottoms and walls of the hot springs are actually highly organized microbial mats (Castenholz, 1984). Many *thermophilic* (heat-loving) organisms not only tolerate but also actually require the presence of elevated temperatures to grow successfully. Habitats above about 55°C are mainly inhabited by Bacteria and Archaea. These habitats have lower biotic diversity and thus form an attractive system for ecological research.

The chemistry of hot springs is variable; they can range from water that is highly acidic (as low as pH 0.2) to very basic (pH 11). The distribution of pH values is generally bimodal, with acidic springs dominated by sulfates, basic springs dominated by carbonates or silicates, and few neutral springs (Brock, 1978). Different organisms dominate at these pH extremes.

Anoxic waters high in sulfide generally feed acidic springs, which form sulfuric acid on exposure to the atmosphere and have high concentrations of sulfates. The sulfide is oxidized biologically by sulfur-oxidizing bacteria (Fig. 14.1) or abiotically, with both processes leading to formation of

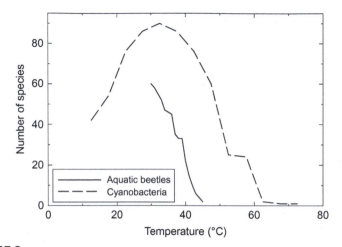

FIGURE 15.2

Number of species of aquatic beetles and cyanobacteria found in springs of different temperatures. *(Data from Brock, 1978).*

sulfuric acid. At the highest temperatures, bacteria that oxidize sulfide are the dominant primary producers.

In general, the biota of hot springs is less diverse as temperature increases, although some groups such as the cyanobacteria apparently prefer warm temperatures (30–40°C) (Fig. 15.2). The thermal tolerance limits of an increasing number of phylogenetic groups are exceeded as temperature increases above 25°C (Table 15.3). Multicellular plants and animals generally cannot withstand temperatures greater than 50°C, but some single-celled Eukarya and the filamentous fungi can withstand temperatures up to 62°C. Photosynthetic bacterial primary producers can be found up to 73°C.

Strains of individual microbial species can also be distributed along a temperature gradient (Fig. 15.3). In this case, unicellular cyanobacteria that appear identical under the microscope have adaptations to different temperature optima and can dominate in a narrowly defined habitat. This range of temperature optima of various strains clearly illustrates that biochemical specialization is necessary for a strain to compete successfully at individual temperatures. The extremely stable nature of hot spring temperatures allows for these strains to dominate in the narrow regions of their temperature optima.

An interesting case of organism distribution related to temperature has been described for Hunter's Hot Spring in Oregon (Wickstrom and Castenholz, 1985). The spring leaves the ground at slightly less than boiling and the water cools as it contacts the atmosphere and ground, creating a gradient

Table 15.3 Upper Temperature Tolerances for Various Groups of Organisms

Group	Approximate Upper Limit (°C)
Fishes	38
Vascular plants	45
Insects	50
Ostracods	50
Mosses	50
Protozoa	56
Algae (eukaryotic)	60
Fungi	62
Cyanobacteria	73
Photosynthetic Bacteria	73
Extreme thermophilic Bacteria and Archaea	110?

(From Brock, 1978)

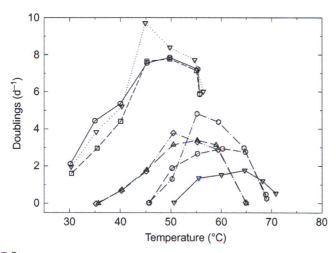

FIGURE 15.3

Growth curves of eight strains of *Senecococcus* isolated from different temperatures in one hot spring. *(Reprinted with permission from Nature. J. A. Peary and R. W. Castenholz. Temperature strains of a thermophilic blue-green alga. Nature 5(64), 720–721. 1964. Macmillan Magazines Limited).*

of decreasing temperature downstream. The cyanobacterium *Synechococcus* dominates from 74 to 54°C because other primary producers are unable to survive (Fig. 15.4). As the stream cools, the motile filamentous cyanobacterium *Oscillatoria terebriformis* dominates, covering the surface of the mat at moderate light levels and contracting to the margins under very high light.

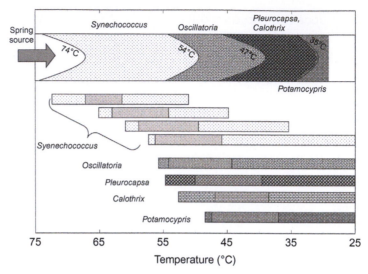

FIGURE 15.4

Distribution of cyanobacterial genera and strains and the grazing ostracod *Potamocypris* in Hunter's Hot Spring, Oregon. Dominant species on top with distribution limits. Bars give temperature ranges of each organism, with gray portion of each bar representing the temperature range for optimum growth. *(Adapted from Wickstrom and Castenholz, 1985).*

As the stream cools further, the herbivorous ostracod *Potamocypris* is able to survive. This algivorous thermophilic ostracod can crop down *Synechococcus* and *Oscillatoria*, allowing for the development of a mixed leathery mat community of two cyanobacteria (*Pleurocapsa* and *Calothrix*), which are poorer competitors but are resistant to grazing (Wickstrom and Castenholz, 1978). Such obvious effects of competition and predation on community structures across a physical gradient are not often observed in nature.

COLD HABITATS

Cold habitats include ice, snow, and polar lakes. These habitats can be present for part of the year in temperate areas or much of the year in polar or high-altitude regions. Organisms that live in these habitats have to be able to function at low temperatures. Data on these organisms may be of applied use in the food industry with respect to organisms that spoil refrigerated food. There are two general groups of organisms. Most commonly, organisms found in low-temperature habitats are also found in more moderate habitats, where they have much higher rates of metabolism. However, some organisms are *psychrophilic*, meaning they require cold temperatures (generally below 5°C) to grow and/or reproduce. The psychrophiles are less common but more interesting physiologically.

Less attention by biotechnology researchers has been focused on psychrophilic organisms. Such organisms produce proteins that are active at low temperatures. These compounds could be useful in cold food preparation and in detergents for washing in cold water (Russell and Hamamoto, 1998).

The lakes in the dry valleys of Antarctica (Fig. 15.1) provide a permanently cold habitat. In general, the same biogeochemical processes occur in these habitats as in temperate systems, but rates are dictated by the extreme environmental conditions (Howard-Williams and Hawes, 2007). These lakes have several meters of ice cover year-round, so they receive very low levels of light. The primary producers (planktonic algae) found in the lakes are adapted to compete for light (steep α and low compensation points for the photosynthesis–irradiance curves; see Chapter 12). Some primary producers are able to consume small particles as well as photosynthesize, and this may allow them to survive the long winter with no light (Roberts and Laybourn-Parry, 1999). The communities are simple, with no fish or large invertebrates. Some of the lakes have warmer regions fed by saline, geothermally heated warm springs (Fig. 15.5). These warmer regions are anoxic, have high nutrients (Green *et al.*, 1993), and have an enhanced population of primary producers located at the chemocline (Fig. 15.5B).

A unique community is associated with liquid inclusion in the ice layers on the surface of the dry-valley lakes (Priscu *et al.*, 1998). Particles from the terrestrial habitat blow onto the ice surface. The particles absorb heat in the summer and melt down into the ice cover. The liquid water surrounding the particles supports a community of algae and bacteria.

Streams feed the dry-valley Antarctic lakes. These streams flow only during a few months of the year when the sun is warm enough to melt the glaciers. The channels dry frequently and generally any remaining water is frozen when the channel is not flowing. Amazingly, the channels have significant biomass of algal primary producers, mostly cyanobacteria. These organisms can be freeze-dried for much of the year, but they are able to actively photosynthesize minutes after being wetted (Vincent, 1988).

Arctic lakes and ponds are also generally very cold. Ponds and lakes can freeze to the bottom; if they freeze completely, they will not contain fishes or many macrophytes. Many fishes can withstand and compete well at temperatures down to 0°C; however, most species have optimum growth above 8°C (Elliott, 1981). Aquatic mosses are often the only macrophytes found in Arctic lakes. These mosses grow slowly and are 7 to 10 years old, which is greater longevity than has been documented previously for any rooted freshwater

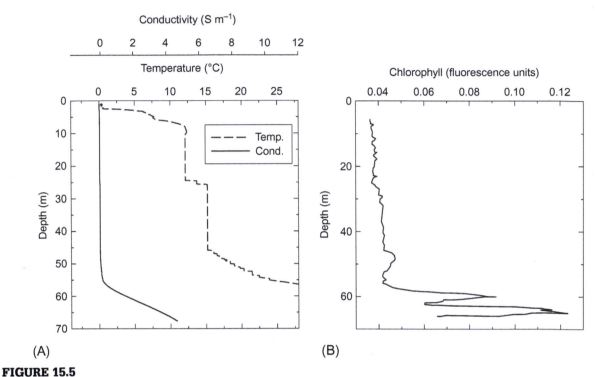

FIGURE 15.5

Vertical profiles of temperature and conductivity (A) and phytoplankton (B) (chlorophyll fluorescence) from Lake Vanda, Antarctica. *(Reproduced with permission from (A) Spigel and Priscu, 1998; and (B) Howard-Williams et al., 1998).*

macrophyte (Sand-Jensen *et al.*, 1999). The cyanobacteria *Nostoc commune* is probably the most common primary producer found in aquatic high arctic habitats (Sheath and Müller, 1997).

High-altitude ponds or lakes are similar to polar habitats because they are ice-free for only a few months a year. The lakes and ponds in high altitudes at temperate or tropical latitudes experience very high levels of light in the summer, and the zooplankton in these habitats are often red or orange because they contain carotenoids that protect against damage by UV-B.

Simple microbial communities can be associated with snowfields that occur in high mountainous or polar regions. Snow algae were noted first by Aristotle, and detailed study began in the early 1800s. The microbial primary producers in snowfields can include chlorophytes, euglenoids, chrysophytes, cyanobacteria, and diatoms. The primary producers can support a community of fungi, bacteria, rotifers, protozoa, and some invertebrates (Hoham, 1980). Increased photosynthesis by the snow algae leads to greater bacterial productivity (Thomas and Duval, 1995). This productivity can be transferred to the terrestrial food web that includes small mammals and birds (Jones, 1999).

The most common algae in snow generally are single-celled green algae (chlorophytes). The most obvious sign of these algae is the pink ("watermelon") snow associated with the psychrophilic *Chlamydamonas nivialis,* a green alga that can acquire a strong reddish color produced by high levels of carotenoids. These pigments are produced by *C. nivialis* and other species of *Chlamydamonas* to protect the cells from ultraviolet irradiance (Bidigare *et al.,* 1993). The irradiance is extremely high at the snow surface because of the high altitude and the reflective properties of snow.

A problem in the life cycle of *C. nivialis* is how to inhabit the upper, lighted portions of snowfields when they are buried each winter. The evolutionary solution to the problem is that the spores of the alga rest in the soil over the winter and hatch and swim to the surface when the snow starts melting. The motile cells then reproduce sexually and produce more resting spores.

Microbial communities have also been described from the slush and snow on the surface of alpine lakes (Felip *et al.,* 1995). These communities include bacteria and autotrophic and heterotrophic ciliates. The production of these communities can be higher than planktonic production in the ice-covered water below. Many of the species present in the slush are either derived from the plankton or from the snow pack above. Apparently, some of the species are adapted to the icy habitats because they are found mainly in the slush and not in the lake or snow nearby.

TEMPORARY WATERS AND SMALL POOLS

Drying is probably the most extreme disturbance that can occur in an aquatic system. However, organisms colonize temporary or ephemeral habitats within days or weeks of rewetting. These habitats include temporary pools (Fig. 15.6), streams, lakes, and wetlands (Williams, 1987). For some organisms, these represent marginal habitats, and for others they are the only habitats that can be exploited successfully. Ephemeral pools can be important sources of mosquitoes in many areas.

Temporary pools can be categorized according to the permanency of their water. One scheme (Williams, 1996) divides pools and streams into those with periods of drying that occur every several years, systems with regular drying that occurs during specific seasons of the year, and systems with very unpredictable drying (few or several times per year). Such classifications have been useful for predicting life histories and invertebrate community structures. Another approach considers streams across gradients of permanence from always flowing to mostly dry (Feminella, 1996). Again, some species are found only in permanent waters and others prefer temporary stream habitats. Obviously, given a gradient of stream and pool types, and cyclic climatic variation, there are not hard and fast definitions of intermittence and no hard

FIGURE 15.6
(Left) A temporary pool formed in granite by freezing and thawing of water and (right) a temporary pool in a tallgrass prairie formed by bison activity. *(Bison pool image courtesy of D. Rintoul).*

line that delineates ephemeral streams. Even a perennial stream or pool could dry during a period of drought.

Most fishes are unable to exploit temporary habitats unless there is a refuge nearby that serves as a source of colonists. Where fish do not occur, large invertebrates can be found that would otherwise be susceptible to predation and can successfully develop and reproduce during the wet period, such as fairy shrimp or tadpole shrimp (Fig. 15.7). Life histories of organisms in temporary pools usually feature resting stages (that are resistant to desiccation) and/or life stages conferring the ability to fly, crawl, or be blown into pools.

Temporary pools serve as an important habitat for amphibians because of the lack of fish predation in such sites. In these systems, amphibians such as ambystomatid salamanders can be top predators that structure prey communities (Wissinger *et al.*, 1999). Temporary pools also allow amphibians with adults that use terrestrial habitats to reproduce in areas without year-round water. The life history of tadpoles is often linked intimately to pond permanence, and species interactions are also related to permanence (Skelly, 1997).

Ephemeral pools or wetlands form an important habitat for other organisms, including waterfowl. Prairie potholes, many of which only hold water for the wet part of the year, provide a key habitat for many of the ducks that live in the central Midwestern United States (Batt *et al.*, 1989; see Chapter 5).

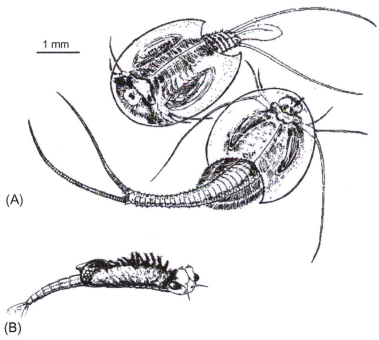

FIGURE 15.7

Tadpole shrimp (*Lepidurus covessi*, A) and a fairy shrimp (*Eubranchipus hundyi*, length 7 mm; B), invertebrates that are typical of temporary pools. *(Reproduced with permission from Dodson and Frey, 1991).*

The waterfowl are mainly migratory and use the pools during wet times of the year. The pools have large populations of invertebrates that serve as a food source for many of the waterfowl.

The vernal pools of California are unique systems in that they contain endemic plant assemblages. Georgia, Texas, Mexico, Chile, South Africa, and Australia also have examples of temporary pools with endemic plant assemblages (Thorne, 1981). Likewise, the fairy shrimps (*Anostraca*) have adapted to vernal pools throughout the world, with some genera that are distributed broadly and others that are endemic to local regions (Belk, 1984). In a good example of endemism in temporary pools, when 58 vernal pools in California were sampled, 67 species of crustaceans were recorded and 30 were probably new species (King *et al.*, 1996).

Impermanent streams have recently received more attention. They had been understudied despite their importance in the many arid regions of the world (Davies *et al.*, 1994). The level of permanence has been clearly related to invertebrate community structure (Miller and Golladay, 1996), and drying probably has stronger effects than flooding (Boulton *et al.*, 1992). Primary production by periphyton in streams is resilient to desiccation (Dodds *et al.*, 1996b) and recovers in days to weeks. Invertebrate grazing pressure on primary producers varies strongly in the weeks following rewetting (Murdock *et al.*, 2010).

Another specialized aquatic habitat that has received attention from ecologists is found in the small pools formed in pitcher plants, tree holes, the leaves of bromeliads, and human refuse such as abandoned car tires. Many different insect larvae can be found in these small pools, and the larvae partition the environment so they will not compete for the same resources. Tadpoles of some amphibians also inhabit these small pools. The pools are attractive study systems because they form a well-defined ecosystem in which all members of the community can be identified and pools can be easily replicated and sampled.

FIGURE 15.8
Darlingtonia, a pitcher plant that contains small pools within its specialized, hollow leaves. The pools are inhabited by some insects, and the plant preys upon others.

Pool size can be very important in these small habitats. Not only does pool size relate to the probability of drying, but also to the ability of adults to find the habitat and lay their eggs. Larger pools are more complex and predators such as damselfly larvae are less effective in larger, more complex habitats (Srivastava, 2006).

Pitcher plants (Fig. 15.8) form small, deep wells with slippery sides, and the pool of water that collects at the bottom serves as an insect trap as well as a habitat for aquatic organisms, including bacteria, protozoa, and aquatic invertebrates. More than 17 invertebrate species are obligate associates of pitcher plants in the southeastern United States (Folkers, 1999). Pitcher plants probably use the trapped insects as a nutrient source. Such carnivorous plants typically grow in nutrient-poor wetlands.

Invertebrates that inhabit the pitcher plant *Sarracenia purpurea* have a positive influence on the plant. The inhabitants include a chironomid larva (*Metriocnemus knabi*) and a culicid (*Wyeomyia smithii*) that inhabit the small pools in the plants and accelerate breakdown of trapped prey and make nutrients and CO_2 more available (i.e., stimulate primary production) than in the absence of the two invertebrates (Bradshaw and

Creelman, 1984). These two invertebrates partition the habitat spatially with a third species, the dipteran *Blaesoxipha fletcheri*, and this allows for their coexistence (Giberson and Hardwick, 1999).

A very large pitcher plant (30 × 16 cm pitchers) has been described from the Philippines (Robinson *et al.*, 2009). The plant, *Nepenthes attenboroughii* (named after the famous naturalist David Attenborough), contains mosquito larvae and probably other species. The pitchers are mostly full of liquid, with the upper layer of clear water and a lower layer of cloudy viscous liquid that contains digestive enzymes. The plant is interesting in that it is large enough to be able to consume small rodents that fall into the pitcher and are subsequently digested.

Small pools with several hundred milliliters of water form in the bracts of the tropical monocot *Heliconia*, supporting a complex community. Studies of insect community interactions have demonstrated that positive and negative interactions in the pools occur among the residents (i.e., competition is not the only structuring force in the community). These studies also provided some of the early direct measurements of interspecific interaction strengths (Seifert and Seifert, 1976).

Mosquito larvae can inhabit tree holes and other small pools. Communities dominated by larvae of *Aedes sierrensis* were investigated for community effects of larval feeding (Eisenberg and Washburn, 2000). The larvae reduced numbers of planktonic protozoa. When biofilms of bacteria and fungi increased, the predation pressure on planktonic protozoa decreased.

ULTRAOLIGOTROPHIC HABITATS

Aquatic systems with very low amounts of available nutrients can be considered extreme environments. Morita (1997) suggested that the normal state of bacteria is one of depletion and starvation with respect to supplies of organic carbon. If this is the case, most bacteria must experience the stress of oligotrophy at least occasionally. Other organisms are subject to the influence of oligotrophy as well when production of photosynthetic organisms and heterotrophs is low, and food webs are severely energy limited. Physiological adaptations to such habitats include slow growth and resting or static stages. Some lakes (e.g., Lake Tahoe and Crater Lake) and many groundwater habitats are extreme oligotrophic environments.

HYPEREUTROPHIC HABITATS

In sharp contrast to ultraoligotrophic habitats, hypereutrophic habitats have excess nutrients and are thus highly productive. However, they are generally dominated by a few species that are tolerant of the harsh conditions that characterize these systems. Inorganic forms of limiting nutrients (mainly phosphorus and nitrogen) are generally not depleted in hypereutrophic systems, and this facilitates dominance by a few microbial species that are best able to exploit a constant supply of nutrients. Livestock and human sewage lagoons are good examples of hypereutrophic habitats, but some temperate zone and tropical wetlands and shallow lakes are naturally hypereutrophic. Human nutrient pollution will be discussed in detail in Chapter 18.

Hypereutrophic systems are generally light limited because of dense growths of algae and associated high *bioturbidity*. Along with light limitation, oxygen is very limited; oxygen supersaturation is common during daylight hours where light can penetrate and photosynthesis proceeds, but high respiration at night leads to frequent anoxia. Even during daylight, only the uppermost layers may be oxygenated because of limited light penetration beyond the surface. Blooms of filamentous cyanobacteria are common in these systems, and planktonic invertebrate communities are generally limited (Scheffer, 1998). Vascular plants are generally poorly represented.

As a result of severe oxygen limitation, invertebrate communities in hypereutrophic habitats are limited to a characteristic community dominated by small forms with low oxygen demands (e.g., many annelids and flatworms; Figs. 10.3 and 10.5), air-breathing insects (e.g., true bugs (Hemiptera) Fig. 10.10, and dipteran larvae with respiratory siphons such as the rat-tailed maggots (Syrphidae; Fig. 10.11), pulmonate snails (Fig. 10.5), and those with other specializations for low oxygen environments such as the "bloodworms," which are chironomid fly larvae with hemoglobin in their blood. Depending upon the degrees of nutrient enrichment, fishes and amphibians that rely on dissolved oxygen in the water are generally poorly represented or absent.

DEEP SUBSURFACE HABITATS

Scientists viewed deep groundwaters as essentially sterile habitats until recently. Such a view is incorrect because bacteria, fungi, and protozoa can be cultivated from subsurface samples (Fig. 15.9), and bacteria have been found as deep as 400 to 500 m (Balkwill and Boone, 1997). A study in Finland documented bacteria at a depth of 940 m (Haveman and Pedersen, 1999). These deep microbial communities include a moderately diverse group of bacteria capable of many common nutrient transformations (e.g., denitrifiers, sulfate reducers, and nitrogen fixers) and a somewhat diverse assemblage of heterotrophic microorganisms (Sinclair and Ghiorse, 1989). Depth limits of organisms may be set by temperature tolerances because geothermal heating increases temperatures with depth (Ghiorse, 1997).

The following are obvious questions that arise upon finding such organisms: How long have they been there and what are they living on? The answer to

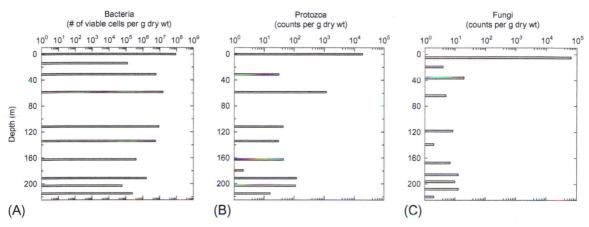

FIGURE 15.9

Distribution of bacteria (A), protozoa (B), and fungi (C) in deep subsurface sediments from near the Savannah River. *(From* Geomicrobiol. J. *Fig. 2, p. 22, and Fig. 3, p. 23, by J. L. Sinclair and W. C. Ghiorse. 1989. Reproduced by permission of Taylor and Francis Inc.).*

the first question is surprising. Many of these sediments were deposited millions of years ago and pore water ages of 1,200 years have been measured where active microbes have been isolated (Kieft *et al.*, 1998). The microbial communities inhabiting at least some groundwaters are likely derived from the microbes present when the sediments were deposited (Amy, 1997). Apparently, some communities in deep groundwaters have been isolated from the surface for more than 10,000 years. Such isolation raises additional questions: What are they eating? Why hasn't it been depleted? How much have the microbes evolved since their isolation?

In some cases, where subsurface hydrocarbon deposits occur, organic material is sufficient to support an active microbial community (Krumholz *et al.*, 1997). In other cases, organic C is limited and the communities must be adapted to a very oligotrophic way of life. Thus, rates of respiration in the deep subsurface are generally extremely slow relative to those in most other aquatic sediments (see Chapter 24).

One interesting study of basalt rocks that formed 6 to 17 million years ago in the Pacific Northwest of the United States suggested that the microbial community present from 200 to 1,000 m deep was supported by chemoautotrophic processes. In this case, Stevens and McKinley (1995) suggested that H_2 gas was used with CO_2 to produce methane and energy. Similar claims have been made by scientists studying deep wells in Sweden (Kotelnikova and Pedersen, 1998), but studies in Finland found no such autotrophic activity (Haveman and Pedersen, 1999). If these studies are correct, this is the only known ecosystem on Earth that is not ultimately dependent on O_2 derived from photosynthesis or photosynthetic products. However, Anderson *et al.* (1998) suggested that the production rates of H_2 are too low in the environment to support microbial growth. More study is necessary to confirm the possibility of chemoautotrophic systems in the deep subsurface maintained by H_2 production.

A provocative paper suggests that one groundwater habitat exists that is dominated by a single microbe (Chivian *et al.*, 2008). These researchers sampled water in fractures 2.8 km deep in an African gold mine. They found that over 99% of the genetic material in the sample was from a single bacterium. The microbe can use organic carbon or live by oxidizing carbon monoxide. It also has the capacity to fix nitrogen. This could be the only environment on Earth where a single organism lives in isolation.

Why is it important to understand the ecology of these deep ecosystems? Subsurface disposal of highly radioactive materials and other waste is common. An active microbial community at these depths could alter transport and containment of such wastes. Subsurface communities can also alter oil deposits and have global geochemical effects (Stevens, 1997). Furthermore, given microbial biomass and the depth at which it has been located, bacteria could have a greater total biomass of active cells than any other type of organism on Earth (Whitman *et al.*, 1998).

THE WATER SURFACE LAYER

The air–water interface is often not studied but represents a distinct habitat that includes organisms with specialized adaptations (Fig. 15.10). Microorganisms living at the surface are called *neustonic* and surface macroorganisms are called *pleustonic*. Those organisms found above the surface are called *epineustonic*, and those below are called *hyponeustonic*. One of the key characteristics of this habitat is the water surface tension. The force at the interface is considerable, and it is quite difficult for a small organism to escape once it has entered (Vogel, 1994). Thus, coming in contact with a lake surface may spell death for some species of *Daphnia* and other zooplankton. Other organisms, such as water striders and whirligig beetles, require the water tension to function. Addition of substances to the water that interrupt the surface tension, such as detergents, renders these insects helpless in the water.

The surface layer of water (within 100 μm) represents a unique chemical environment (Napolitano and Cicerone, 1999). Biogenic surfactants, primarily humic and fulvic acids, accumulate here. Lipids, metals of environmental concern, nutrients, and some microorganisms can accumulate in this layer. Bubbles can interact with the chemicals on the surface, leading to production of foams. The foams are stabilized by lipids and other organic molecules,

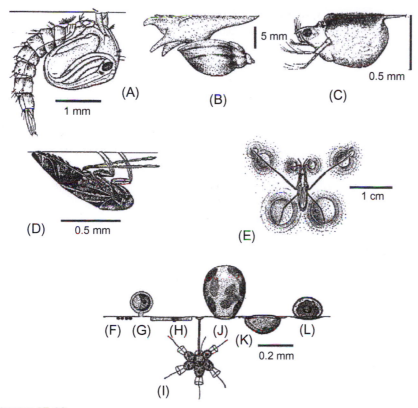

FIGURE 15.10

Some organisms adapted to use the water surface as a habitat. (A) A mosquito pupa, *Anopheles claviger*; (B) a snail, *Lymnaea*; (C) the cladoceran *Scapholeberis mucronata*; (D) *Notonecta*, a water boatman; (E) the water strider, *Gerris*; (F) the bacterium, *Lampropedia hyalina*; (G) the chrysophyte *Ochromonas vischerii*; (H) the diatom *Navicula*; (I) the flagellate *Codonosiga botrytis*; (J) the alga *Botrydiopsis*; (K) the testate amoeba *Arcella*; (L) the alga *Nautococcus*. *(Reproduced with permission from (A–E) Guthrie, 1989; and (F–L) Ruttner, 1963).*

both natural and human produced. Thus, even pristine mountain streams can accumulate foam on their surface.

Organisms that specialize in the surface layer must be able to withstand very high levels of light. Such high light must lead to increased energetic costs associated with repair of cellular damage from free radicals formed by high-energy UV irradiance. This disadvantage is offset by the constant influx of nutrients and organic carbon from the air above.

Surface-dwelling organisms can also alter the properties of the habitat. Surface tension can be manipulated by exuding organic compounds that spread across the surface. An interesting form of locomotion occurs this way; the veliid, *Velia caprai*, and beetles in the genus *Stenus* are able to excrete material that lowers the water tension behind them, so the surface tension in front pulls them forward. The beetle *Dianous coerulescens* can move using water tension in this way at speeds up to $70\,cm\,s^{-1}$ (Hynes, 1970).

SUMMARY

1. Organisms have special adaptations to extreme habitats, allowing them to use a tremendous range of extremes. Bacteria and Archaea dominate in the most extreme habitats.
2. Physiological adaptations to high temperature include lipids with higher melting points and stabilizing features of proteins and nucleic acids.
3. Organisms in high-salinity habitats need to regulate osmotic pressure, as do those that can withstand drying.
4. Diversity decreases as habitats become more extreme.
5. Hot springs have served as attractive communities for study because of their stable nature, low diversity, and the adaptations of the organisms that are able to live in near-boiling water.
6. Temporary pools are colonized quickly by organisms that are able to withstand desiccation or those that can move in from nearby sources.
7. Active microbial communities are found in regions of melted water in ice and snow, in habitats ranging from ultraoligotrophic to hypereutrophic, and in groundwater up to 1,000 m below Earth's surface.
8. The air–water interface is an extreme environment. High surface tension and high irradiance are characteristics of this habitat.

QUESTIONS FOR THOUGHT

1. Can extreme habitats serve as models for early life on Earth or possible life on other planets?
2. Should efforts be made to conserve the biodiversity of unusual habitats such as hot springs?
3. Should companies be able to patent and take full profit from gene sequences taken from organisms collected in national parks without remuneration to the government?
4. Are "extreme" habitats really extreme for organisms adapted to live in them?
5. Why can the depth of a saline lake be highly variable from year to year and from decade to decade, and how may global climate change influence such lakes?
6. Why might saltworks that precipitate brines be interested in the microbiology of saline waters?
7. What features allow certain characteristic species to dominate habitats with excess nutrients?
8. How much (%) is the estimated thickness of the biosphere increased by the understanding that organisms can inhabit up to 500 m depth?

CHAPTER 16

Responses to Stress, Toxic Chemicals, and Other Pollutants in Aquatic Ecosystems

© 2010 Elsevier Inc. All rights reserved.
Doi: 10.1016/B978-0-12-374724-2.00016-7

FIGURE 16.1
Organic pollutants burn on the Cuyahoga River in 1952. *(Courtesy of Cleveland State University,* The Cleveland Press *collection).*

The modern aquatic environment has suffered greatly from physical disturbance as well as organic and inorganic toxic pollution. Although the negative effects of pollutants were recognized by scientists in the 1950s, it was not
until Rachel Carson's book *Silent Spring* was published in 1962 that it became
common public knowledge that organic and inorganic pollutants can have
strongly negative, far-reaching, and unpredictable influences on human health
and ecosystems (Biography 16.1). Furthermore, acid precipitation, thermal
pollution, acid mine wastes, and increases in suspended solids all cause environmental damage. The relative importance of various types and causes of
lake and river pollution have been determined in the United States from state

BIOGRAPHY 16.1 RACHEL CARSON

The positive influence of Rachel Carson (Fig. 16.2) may exceed that of any academic aquatic ecologist. In 1962, she published a book titled *Silent Spring* that became a bestseller and had a tremendous impact on public awareness of the pollution caused by pesticides. Her ability to take a technical subject and make it accessible to the general public led to some of the first laws enacted to control the release of pesticides into the natural environment. Lear (1997) chronicles her life in an informative biography.

Carson's undergraduate studies in biology at the Pennsylvania College for Women (now Chatham College) were followed by a master's degree at Johns Hopkins University. Her research on the developmental biology of catfish eventually led to a job writing for the Bureau of Fisheries. She wrote her first book in 1941 (*Under the Sea-Wind*), followed by two critically acclaimed books and numerous popular articles that translated scientific concepts into lay terms. Then she published *Silent Spring*, in which she chronicles the wanton use of pesticides and some of their effects on the environment, including biomagnification, death of wildlife (including the loss of bird life that leads to a silent spring), and potential influence on human health (toxicity and carcinogenic properties of toxic pollutants).

The completion of *Silent Spring* was a tremendous professional and personal accomplishment. While writing the book, Carson tended her dying mother, and after her sister died, she became a single mother to her orphaned nephew. She also began the battle with breast cancer that claimed her life a few years later. Her careful attention to scientific detail was crucial because her book became the focal point of the debate over pesticide use. The exceptional popular response to her book led to strong backlash from many chemical companies, entomologists, and government officials; the detractors generally had a financial or professional stake in maintaining indiscriminant pesticide use. Carson documented her facts so well

FIGURE 16.2

Rachel Carson. *(Image courtesy of the National Oceanic and Atmospheric Administration).*

that her critics turned to personal attacks in their attempts to discredit *Silent Spring.*

The life and work of Rachel Carson prove that aquatic ecologists can make a difference in the world. She demonstrated that traditional academic and management careers are not the only ways to have a positive impact and that combining two disparate strengths (in her case, excellent popular writing and science) can yield impressive results.

reports (Fig. 16.3). These data suggest that 36% of the river and stream miles and 37% of lakes were impaired in the past decade. Impairment is defined as having evidence of damage to aquatic life, unsuitability for drinking, production of fish that are not safe to eat, or being unsafe for swimming (US Environmental Protection Agency, 1997). Large improvements in water quality

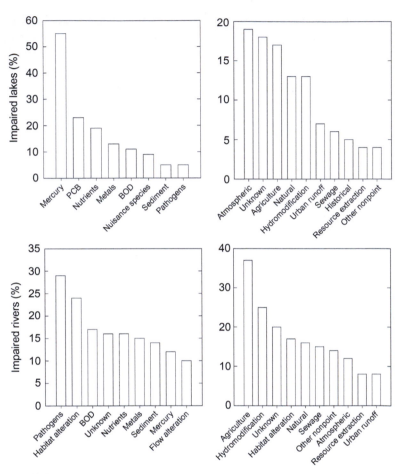

FIGURE 16.3

Percentages of impaired lakes and streams in the United States by type and cause of impairment. Natural impairment is due to natural causes such as floods, droughts or excretion from wildlife. *(Data from the US Environmental Protection Agency for 2004 National Water Quality Inventory).*

have not occurred in the United States over the past decade. Similar patterns are common in developed countries around the world.

Pristine aquatic habitats no longer exist. Pollutants are transported throughout the world in our atmosphere (Ramade, 1989). Climate change is ubiquitous. Most major watersheds are disturbed. The question is no longer if the pollutants are present, but rather in what quantities, and what are their effects?

There is much that is not known about chemical pollution. For example, of the more than 72,000 chemicals in commercial use, only about 10% have been screened for toxicity and only 2% screened as carcinogens. In the United

States, only about 0.5% of these chemicals are regulated by federal and state governments (Miller, 1998). This chapter discusses some general concepts of toxicology, causes and effects of pollution by inorganic and organic contaminants, and thermal pollution. Mitigating solutions are also discussed. Nutrient pollution has had a strong influence on aquatic systems; however, this will be discussed in Chapter 18.

BASIC TOXICOLOGY

Exposure to toxins can either come in large pulses over a short period of time (*acute*) or in low doses over long periods of time (*chronic*). Responses can be *lethal* or *sublethal* (not causing death) and can result from *instantaneous* or *cumulative* (a response to numerous events) reactions to exposure. Studies of toxicity require accounting for variability in responses of organisms. Thus, the *lethal dose*, the amount ingested that causes death, is labeled with a subscript that indicates the percentage of animals killed (e.g., LD_{50} is the lethal dose for 50% of the animals tested). Organisms can also be exposed to toxins through the water, including absorption across cell membranes, gills, and skin. Thus, toxicologists also report lethal concentrations (e.g., LC_{50}). *Effective concentration* (EC) is the concentration that causes some effect other than death (e.g., on reproduction, growth, behavior); a *subscript* is also used to denote the percentage showing the effect (Mason, 1996).

Some toxicants have negative effects on reproduction while having little influence on the general health of the adult organism. These compounds can cause complete extinction of a population, but the effects may be difficult to demonstrate with standard laboratory tests (i.e., the LD_{50} is much higher than environmental concentrations). An example of deleterious effects on reproduction is the response of certain waterbirds to 1,1-dichloro-2,2-bis(p-chlorophenyl)ethylene (DDE), which is a metabolite of dichlorodiphenyltrichloroethane (DDT). DDE causes the birds to lay eggs with thin shells, leading to reproductive failure (Laws, 1993) and extinction of local populations. This effect almost led to the extinction of the bald eagle and still threatens many migratory birds.

The chronic effects of toxins can be delayed. Such delays are particularly the case with mutagenic substances in which prolonged exposure increases the chance of deleterious mutations. If these mutations lead to formation of cancerous cells, a toxin is termed *carcinogenic*.

Several additional issues are important with regard to estimating the influence of pollutants on aquatic organisms and humans. Extrapolating effects to low concentrations of pollutants can be a problem. It has been argued that there is a threshold below which contaminants are not harmful. A minimum threshold of toxicity is expected to be the case if an organism can repair a limited amount of damage caused by a toxicant or can excrete it up to some limited rate, or if the compound does not interact with biological molecules below some concentration. Such a threshold has not been established for most toxic chemicals. A possible threshold is particularly important in regulating human carcinogens in the environment. If no threshold exists, very low concentrations of materials can be predicted to cause a significant number of deaths if a large number of people are exposed. If there is a threshold, then exposure to levels below the threshold is not expected to cause problems.

Low concentrations of toxicants may actually stimulate biological activity (Calabrese and Baldwin, 1999). This further complicates regulation of a toxin and estimation of long-term effects. Such effects mean extensive testing of each suspected toxin is necessary before release into the environment. Nontoxic factors can alter toxicity. For example, benign chemicals and temperature can modify toxic effects. Obviously, it is difficult to predict toxicity of a compound when it is a function of several other variable environmental factors. For example, zinc toxicity is greater for fishes in high temperatures and in low conductivity water (Fig. 16.4). Extrapolating laboratory results such as those from Figure 16.4 to field effects may yield inaccurate results; thus, a combination of field and laboratory approaches may be best for assessments of toxins (Blus and Henny, 1997).

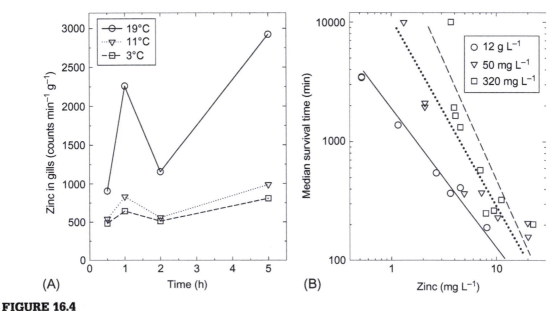

FIGURE 16.4

Effects of temperature on uptake of zinc into salmon gills (A) and influence of calcium carbonate (mg liter^{-1}) on mortality of trout exposed to various concentrations of zinc (B). *(Redrawn from Hodson, 1975, and Lloyd, 1960).*

If two toxicants are present it is difficult to know what their influence will be on one another. In some cases they may alleviate the influence of each chemical alone (*antagonism*), but in others the effects may be strictly *additive*. In the worst-case scenario, the sum of the effects is greater than simply adding the individual toxic effects (*synergistic*). Which of these influences will occur cannot be predicted *a priori*, and direct testing is generally necessary to establish an interactive effect. Additionally, toxic materials can alter sensitivity to other factors. For example, metal toxicity can increase susceptibility to ultraviolet radiation (Kashian *et al.*, 2007).

Toxicants can be concentrated by biota. The first step in this process is *bioconcentration*, or the ability of a compound to move into an organism from the water. *Bioaccumulation* refers to the bioconcentration plus the accumulation of the compound from food. This process can lead to toxicological effects even if environmental concentrations are low. *Biomagnification* refers to the entire increase in concentration from the bottom to the top of the food web. Biomagnification is a particular concern with lipid-soluble organic contaminants and some metals. In general, the less water-soluble the organic compounds, the more they are concentrated by organisms (Fig. 16.5).

Bioconcentration and bioaccumulation factors can be difficult to determine for animals and plants in their natural environment. Factors influencing uptake and retention of a contaminant (such as metabolic rate, rate of assimilation of contaminated food, heterogeneous distribution of the pollutant, and rate of excretion of the contaminant) can all depend on a variable

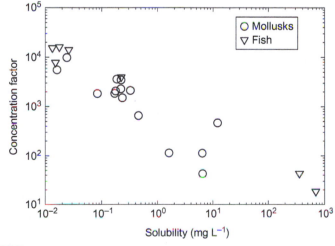

FIGURE 16.5

Relationships between water solubility and bioconcentration factors of various organic compounds in fish and mollusks. *(Adapted from Ernst, 1980).*

Table 16.1 Range of Concentrations (parts per trillion) and Approximate Biomagnification Factors Calculated for DDT and PCBs in Lake Ontario (ranges from Allan, 1989)

Chemical	Water	Benthic Sediments	Suspended Sediments	Plankton	Fish	Herring Gull Eggs
DDT	0.3–57	25,000–18,000	40,000	63,000–72,000	620,000–7,700,000	7,700,000–34,000,000
PCBs	5–60	110,000–1,600,000	600,000–6,000,000	110,000–6,100,000	1,378,000–7,000,000	41,000,000–204,000,000
DDT bioconcentration	1	4,200	1,400	2,300	143,000	719,000
PCB bioconcentration	1	26,000	100,000	94,000	278,000	3,710,000

environment. For example, some compounds that do not bioconcentrate in fully aquatic organisms can bioconcentrate efficiently in air-breathing organisms living in or near the aquatic environment (including humans) because such compounds have volatile phases that can be transmitted in the air but are nonpolar, so they dissolve poorly in water (Kelly *et al.*, 2007). However, despite the uncertainties, biomagnification is a well-documented problem and pollutants can be concentrated many millions of times, even if the range of concentrations and bioaccumulation factors is wide (Table 16.1).

BIOASSESSMENT

Aquatic organisms, particularly invertebrates and fishes, are very useful for assessing the acute and chronic effects of pollutants because the diversity and types of organisms present are related to degree of pollution and environmental gradients (Loeb and Spacie, 1994). For instance, data on many stream invertebrate species (Fig. 16.6) can be used to demonstrate two possible responses to environmental extremes. This evaluation based on diversity of organisms is called *bioassessment*. In the case of O_2 and pH, diversity is maximal at intermediate values (pH about neutral, O_2 about 8 mg liter^{-1}). This is an example of Shelford's "Law of Tolerance" we discussed in the last chapter. In the case of chloride and turbidity, diversity is greatest at the smallest concentrations. These data illustrate that biodiversity can serve as an indicator of environmental conditions.

Specific indices based on more refined taxonomic characteristics are most reliable. Some species or groups are commonly found in eutrophic situations (e.g., cyanobacteria dominate eutrophic lakes, and *Tubifex* worms inhabit sewage outfalls) and others are sensitive to specific environmental factors (e.g., amphibians are susceptible to many types of pollution, and salmonid fishes are limited by water temperature and O_2 concentrations). Bioassessment

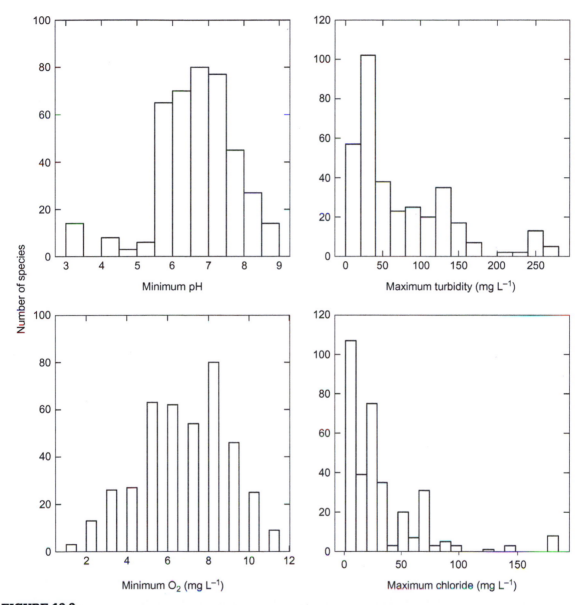

FIGURE 16.6

Number of invertebrate species as a function of pH, chloride, minimum O_2, and turbidity. *(Data from Roback, 1974).*

methods are increasingly used by resource managers for monitoring water quality and overall ecosystem health. Standard methods that use algae, plants, invertebrates, fish, and even riparian birds have been developed for use in streams, lakes, and more recently, wetlands (e.g., Karr, 1981; Barbour *et al.,* 1999; Danielson, 1998; Gerritsen *et al.,* 1998).

A basic community-level indicator of stream health and water quality is the total number of insect taxa in the groups Ephemeroptera, Plecoptera, and Trichoptera (EPT). More species of EPT taxa are usually found in cleaner waters because these groups are generally intolerant of pollution. The EPT index is one of many individual metrics that can be measured and combined into multimetric indices for bioassessment. Many indices include some measure of tolerance, whereby individual taxa are assigned tolerance values based on their distributions across pollution gradients (Hilsenhoff, 1987). For example, a species that is only found in pristine waters would be assigned a score of 0–1 (low tolerance for pollution) and a species found in highly polluted waters would be assigned a tolerance value in the range of 8–10. These values are applied to abundance and diversity data from field samples of fishes, invertebrates, or other groups to calculate an average tolerance value for the community in a habitat.

Individual metrics that assess diversity, community structure, and tolerance are generally scored and combined into an overall rank or score of biotic integrity that can be compared to other regional water bodies, historical data, or modeled predictions of biotic integrity for a given region. For any biological assessment, habitat quality data are usually collected to aid in interpretation; if the physical habitat of a stream is degraded (e.g., a channelized stream with heavy sedimentation), poor biotic index scores might not be related to water quality *per se*. The Index of Biotic Integrity provides a detailed rating for freshwaters using fishes as indicator species. It is a measure of stream quality composed of 12 indicators, including total number of fish species, pollution-tolerant species, food web structure, and fish condition (Karr, 1991). Such indices are useful for determining the suitability of habitats for supporting aquatic life and discerning chronic effects of pollutants and become more useful when calibrated for specific regions.

There is increasing interest in using measures of ecological function for assessing the ecological health of freshwater systems, particularly litter decomposition (Gessner and Chauvet, 2002) and ecosystem metabolism (Young *et al.*, 2008). While most methods of biological assessment focus on elements of ecosystem structure (e.g., community composition and habitat quality), decomposition and metabolism are functions governed by many physical and biological features and processes in a system. As standard, efficient methods are developed, future assessments of freshwater habitat health may include these and other measures of function along with structural metrics.

ORGANIC POLLUTANTS

There are millions of known organic compounds; more than 10,000 have been created and used by humans. Several hundred new chemicals are created each year. The large number of compounds makes regulation difficult. Modern society has a consistent record of releasing toxic organic compounds into the environment, only to determine afterward that they have negative effects on ecosystem and human health. Almost nothing is known about how complex mixtures of these compounds at low concentrations will influence human health (Schwarzenback *et al.*, 2006). The effects of unregulated release of pollutants into a large ecosystem are exemplified by the experiences in the Great Lakes of North America. Problems associated with pollution of these lakes peaked in the 1960s, and the slogan "Lake Erie Is Dying" served as a rallying point for concerned citizens (Sidebar 16.1). Fortunately, the problems have now been mitigated to some degree although organic contaminants linger in the system.

SIDEBAR 16.1
"Lake Erie Is Dying"

Until the 1960s, most sewage and industrial wastes were being released directly into Lake Erie without treatment. The lake seemed so large as to be unaffected by such releases. As the population grew, the problems associated with the releases, such as organic chemical contamination, pathogenic bacteria, and eutrophication, worsened. Such problems led to public pressure to clean up the lake (hence the slogan "Lake Erie Is Dying") and confrontation between citizens, state and federal government officials, and entities causing the pollution (Kehoe, 1997).

Total loads of phosphorus increased five-fold from 1900 to 1970, leading to eutrophication problems. In this sense, the lake was not dying but actually becoming more productive as the phosphorus and nitrogen inputs stimulated algae. This stimulation of algae led to undesirable accumulations of the filamentous green alga *Cladophora* that fouled beaches (Burns, 1985). Some areas of the lake became anoxic and taste and odor problems developed because of algal biomass.

Loading of mercury, lead, cadmium, copper, and zinc increased greatly, with sediment contents 12, 4, 4, 3, and 3 times greater, respectively, than in presettlement times (Burns, 1985). Mercury contents of fishes became so high that they were not healthy for human consumption. Inputs of toxic metals from industry have decreased, but contaminated sediments continue to cause problems.

The Great Lakes Water Quality Agreement of 1978 listed 22 organic compounds that were dumped into the lake as hazardous or potentially hazardous. Of these, polychlorinated biphenyls (PCBs), DDT, and dieldrin caused the greatest concern. DDT use was restricted in 1970, and the concentrations in the smelt taken from the lake decreased from 1.59 to $0.04\,\mu g\,g^{-1}$ between 1967 and the late 1970s. Low levels of DDT contamination continue because DDT is sequestered in the sediments and slowly reenters the food webs. Manufacture and use of PCBs has been banned in the United States since 1976; in 1978, PCBs were entering Lake Erie at about 0.9 metric tons per year, with the majority coming from atmospheric deposition. A decade later, fishes in Lake Erie had enough PCB content that consumption of more than 5 kg of fish per year was deemed unsafe (Burns, 1985). The recently introduced zebra mussel now bioconcentrates PCBs and passes them on to the waterfowl that consume them (Mazak *et al.*, 1997).

Human activities did not kill Lake Erie. However, the system is a good example of how multiple human impacts on a lake can decrease the value for recreation, fisheries, and drinking water. With careful stewardship, the water quality in the lake will continue to improve, and the lake will continue to survive.

The use of organic compounds in agriculture is widespread (Nowell *et al.*, 1999). Worldwide, about 2.3 million metric tons of pesticides are used yearly, and in the United States about 630 different active compounds are employed. Corn, cotton, wheat, and soybean crop management accounts for about 70% of the insecticide use and 80% of the herbicide use in the United States. About 25% of the pesticides are used in urban settings, such as on lawns and golf courses (Miller, 1998). Use of pesticides by individuals in suburban areas is generally unregulated beyond recommendations on product packages, which can lead to overuse. Annual costs associated with the use of pesticides include $1.8 billion for cleaning groundwater, $24 million in fishery losses, and $2.1 billion in losses of terrestrial and aquatic birds (Pimentel *et al.*, 1992).

There is an increasing array of chemical types used as pesticides. In particular, widespread use of organophosphates, carbamates, pyrethrins, and organochlorines, which are all neurotoxins used to control a variety of pest species, has impacted many freshwater habitats and species, even in relatively undisturbed regions. Amphibian population declines in the Sierra Nevada Mountains have been linked to pesticides used in the agricultural Central Valley of California. Pesticides used in the valley, are transported by wind and precipitation into freshwater habitats in the mountains (Sparling and Fellers, 2009).

Eliminating use of pesticides in the near future is unrealistic, and thus careful regulation of their use and disposal is critical. Along with the negative consequences, pesticides and herbicides have had tremendous positive effects on humanity. Increased agricultural production to feed humanity is possible, in part, because of these compounds. Pesticides have also been used successfully to control a variety of insect vectors of important diseases. Control of *Anopheles* mosquitoes has led to eradication of malaria in many parts of the world. Pesticides used on aquatic snails have been effective at controlling river blindness, with only moderate effects on nontarget aquatic species (Resh *et al.*, 2004).

Although biomagnification of toxic organic compounds is a serious problem, compounds that do not biomagnify as readily can still be of concern. Atrazine is a commonly used chemical for control of weeds in croplands. It is fairly water soluble (Nowell *et al.*, 1999), persists 6 to 9 months, and bioconcentrates much less than many other pesticides, although bivalves used in a laboratory study did bioaccumulate atrazine (Jacomini *et al.*, 2006). Atrazine has seen widespread use in the midwestern United States, with 32 million kg applied annually, with less use in Europe where its use has been controlled since the late 1980s. The chemical properties of atrazine lead to efficient transfer through the environment (Pang and Close, 1999). Unfortunately, it is carcinogenic and harms aquatic life (particularly photosynthetic organisms) at levels of $2\,\mu g$ liter^{-1} (Carder and Hoagland, 1998). If the use of atrazine is discontinued, the compounds that are substituted may be even worse

(Vighi and Zanin, 1994). However, given its persistence and water solubility, better management practices are necessary to keep atrazine from entering the surface waters in many agricultural regions.

More recently the glycophosphate herbicide Roundup® has seen significant use, and has been the number one selling herbicide in the world since 1980. The use of this herbicide has increased where crops have been genetically modified to be resistant to the chemical (Roundup-ready). Roundup contains a surfactant that is toxic to wildlife in addition to the glycophosphate. Roundup is toxic to many aquatic wildlife species, and is particularly toxic to amphibians (Relyea, 2005).

Genetically modified crops have also been engineered to contain toxins to their predators. A particularly widespread crop using this practice is *Bt*-modified maize (corn). The potential negative effects of these modified crops on aquatic organisms that come in contact with the pollen and residues are controversial (Sidebar 16.2).

SIDEBAR 16.2
Transgenic Crops and Freshwater Habitats

Considerable controversy has occurred regarding transgenic crops and potential adverse effects on the environment and human health. Although humans have been genetically modifying plants and animals for agriculture and other purposes through selective breeding for centuries, now we have the technology to make more radical changes over short periods of time, including transfer of genes across broad taxonomic lines.

Corn has been genetically modified to express crystalline protein toxins initially derived from the genes in the bacterium *Bacillus thuringiensis* (*Bt*). The proteins expressed by transgenic *Bt* corn plants are endotoxins, which are toxic to many common agricultural pests because they bind to receptors in the gut and cause lethal septicemia. These genetically engineered crops have been at the center of controversy for decades.

The bacterium *Bacillus thuringiensis* and its associated endotoxins have been used since the 1930s in the form of dried spores and crystal toxins to control agricultural pests. Its use increased in the 1980s as resistance to synthetic insecticides increased among agricultural pests. Ironically, *Bt* was developed primarily through organic farming because it is naturally occurring and different forms of the toxin affect specific insects. It was not until *Bt* endotoxin genes were integrated into plant genomes to create genetically modified *Bt* corn and other crops that controversy began. Genetically modified *Bt* crops generally express the endotoxins in all plant tissues and can be used to control pests ranging from the European corn borer to corn rootworm, depending on variety.

In the late 1990s, a study by a group of scientists at Cornell University sparked heated debate over the potential adverse environmental effects of *Bt* crops. Losey *et al.* (1999) found that monarch butterfly larvae that consumed milkweed leaves with *Bt* corn pollen on them had much higher rates of mortality and lower growth than those fed milkweed leaves with non-*Bt* corn pollen. This study was highly publicized and subsequently heavily criticized for being somewhat preliminary and unrealistic in terms of the amounts of pollen to which the caterpillars were exposed. Debate over potential environmental impacts has continued since, with some subsequent studies showing adverse effects, and others showing no effects.

In 2007, a group of researchers at midwestern universities began looking at whether *Bt* crops could influence food webs and ecosystem processes in streams draining midwestern agricultural fields with ever-increasing proportions of *Bt* crops. Rosi-Marshall *et al.* (2007) examined detritus and pollen from *Bt* corn with toxins that target the European corn borer, which represented 57% of the corn planted in the United States in 2008. The endotoxin protein is detectable in crop detritus from crop fields for at least 240 days (Zwahlen *et al.*, 2003), and some portion of this material makes its way into streams via wind and water movements. Wind transport of pollen is another pathway for movement of *Bt* materials into nearby freshwater habitats.

Rosi-Marshall *et al.* (2007) found that significant amounts of crop residues, including corn leaf detritus and pollen, entered streams bordering agricultural fields. They also found that these materials still had active endotoxins in them, that they could be transported significant distances downstream, and that caddisfly larvae, which are closely related to lepidopterans, were adversely affected in laboratory feeding studies with pollen and leaf material. Once again, this study was criticized for being somewhat preliminary and the authors were criticized for overstating the results. However, much of this criticism came from individuals whose research was funded by the industry producing and marketing *Bt* crops.

The possible effects of *Bt* toxins in aquatic ecosystems remain controversial and relatively poorly studied. *Bacillus thuringiensis* var. *israelensis* (B.t.i.) has been used to control black fly larvae in streams for years and studies indicate few effects on nontarget aquatic insects (Jackson *et al.*, 1994). The USEPA has asserted that not enough *Bt* toxin could enter the water to cause harmful effects on aquatic invertebrates (USEPA, 2005), but results of Rosi-Marshall *et al.* (2007) seem to contradict this. Further, the USEPA's stance that there should be no adverse affects in freshwater habitats is based primarily on 48-hour toxicity tests on *Daphnia* performed by scientists employed by the company developing *Bt* crops.

As research on possible negative impacts of *Bt* crops on freshwater habitats progresses, the costs and benefits need careful consideration. Recent field examinations of stream invertebrate communities draining *Bt* and non-*Bt* cornfields showed no patterns; agricultural streams and the communities that inhabit them are subjected to myriad stressors including nutrients, sediments, hydrologic alterations, and channelization, and thus *Bt* toxins may be relatively inconsequential (Chambers *et al.*, in press). Further, a comprehensive assessment of the environmental impacts of transgenic *Bt* crops should consider the environmental and economic consequences of the alternatives ranging from traditional pesticide applications to yield reductions that would occur with no pesticide use.

Some of the toxic organic compounds found in aquatic systems readily move through the atmosphere. Volatile persistent organochlorine compounds are found worldwide (Simonich and Hites, 1995). The compounds condense from the atmosphere depending on temperature, with the most volatile organics condensing in the polar regions (Wania and Mackay, 1993) or at higher elevations. Atmospheric transport, in combination with biomagnification of a long food chain, can account for the unusually high concentrations of the toxic organic compound toxaphene in fish collected from a remote subarctic lake (Kidd *et al.*, 1995). One would assume this lake is a pristine habitat because it is far from civilization. The fact that a toxic organic compound contaminates fish in the lake illustrates the pervasive impacts of humans on aquatic environments.

Petroleum products are another common source of aquatic contamination in many parts of the world. Urban runoff is a significant source of oil contamination, with about 1 g per person per day (Laws, 1993). Multiplying this by the US urban population of 200 million yields 7.3×10^{10} g (about 14 million gallons) of oil entering aquatic habitats per year. Much urban runoff now must be treated in developed countries, so this source of contamination is decreasing. In countries where more automobiles are being used, the source is increasing. Much of the oil that enters freshwaters is likely consumed by microbes or flows to the ocean; the absolute damage to freshwater aquatic habitats is not known. Another common source of contamination is leakage from underground gasoline storage tanks into groundwater. Cleaning spills from such leaks has cost billions of dollars.

Oil and gas also leak into aquatic ecosystems from outboard engines used on watercraft. Visible slicks of oil and gas are commonly observed around busy marinas. Engine exhaust also pollutes water. Two-stroke engines release more pollution than four-stroke engines. The organic compounds in the exhaust of both engine types can kill zooplankton and bacteria. A 15 kW (20 hp), two-stroke engine that operates for 1 h makes 11,000 m^3 of water undrinkable by causing bad taste and odor. Expensive treatment is required to reverse these effects (Jüttner *et al.*, 1995).

Chlorinated hydrocarbons such as polychlorinated biphenyls (PCBs) are of concern in aquatic systems because of their persistence in the environment and their toxicity. Production of PCBs, which were used in a variety of industrial applications as coolants, lubricants, and liquid insulators, was banned for most purposes in the United States in 1979. However, PCB residues remain in sediments of freshwater habitats and organisms, particularly near urban areas. Before they were banned, large-scale production and industrial use of PCBs resulted in severe pollution of some water bodies. The Hudson River in New York was severely contaminated with PCBs and citizens and local governments are now paying a high price because of closure of commercial and recreational fishing and massive, ongoing cleanup operations. Many regions of the world have shut down fishing and issued fish consumption warnings because of the presence of PCBs in fish tissues. Because they bioaccumulate readily, PCBs make their way into riparian predators such as birds, spiders, and amphibians, which feed on insects emerging from contaminated water bodies (Maul *et al.*, 2006a; Walters *et al.*, 2008). In addition to PCBs, many municipal sewage plants treat their final effluent with chlorine to kill pathogens and this treatment forms chlorinated hydrocarbons. Many municipalities are switching to ultraviolet radiation treatment schemes instead.

An increasingly common way to clean up spills of organic materials in the environment is *bioremediation* (Anderson and Lovley, 1997). Bioremediation uses organisms that can break down or inactivate the pollutants. In some cases, organisms are introduced to do the job, and in other cases, native bacteria have the ability to degrade the organic pollutant. Some bacteria are able to metabolize novel organic compounds. Ability to metabolize new organic compounds arises because evolution favors microbes able to use unique carbon sources that are released into the environment by other organisms. The number of individual bacteria is high, and their generation times are short. The probability that an individual microbe will have a mutation that allows use of a unique source of organic carbon is low. However, the probability that one of the millions of bacteria found in each milliliter of water will have a beneficial mutation that helps metabolize the compound is substantial. These features of bacteria lead to the rapid establishment of new genotypes capable of using pollutants.

Bioremediation is probably most important in cases of contaminated groundwater because spills of any size are extremely difficult to remove from underground, particularly if the compounds are not water soluble and are associated with sediments. Several strategies for bioremediation can be used, including pumping the water and treating it at the surface, using plants that bioconcentrate the compound taken into their roots, addition of engineered microbes to the aquifer to consume the pollutants, and use of *in situ* microbial activity to eradicate pollution. In most cases, surfactants (compounds that decrease the ability of organic compounds to associate with solid surfaces) are used to dissociate the compounds from the sediments. Nutrients and oxygen are often added to groundwaters to stimulate microbial activity. Microbes that have the ability to degrade the pollutant may be released into the groundwater. An understanding of the ecology of groundwaters is useful in optimizing rates of bioremediation. For example, protozoan populations can decrease rates of bioremediation (Kota *et al.*, 1999).

The ability to metabolize or inactivate toxins is often coded upon plasmids (small circular pieces of DNA that are free in the cytoplast), which can move within and among microbial species and allow transfer of genetic information. This lateral transfer is of concern in relation to genetically engineered microbes but also may be helpful in bioremediation efforts. In many cases, bacteria capable of metabolizing an organic compound disappear quickly upon release into a contaminated site, but the indigenous bacteria acquire the plasmid that codes for proteins that can degrade the pollutant. Movement of plasmids among natural populations of bacteria is well established.

PHARMACEUTICALS AND ENDOCRINE DISRUPTORS

Humans produce and use numerous chemicals in their daily lives that are specifically produced because of their biological activity, many of which eventually end up in freshwater habitats. Approximately 80,000 chemicals are used by humans today (Pimentel *et al.*, 1992) and there is increasing concern over the widespread occurrence and potential environmental effects of these emerging contaminants in freshwater habitats in Europe (Ternes, 1998) and the United States (Kolpin *et al.*, 2002). Significant amounts of antibiotics, hormones, disinfectants, fragrances, caffeine, and other substances are excreted or dumped by humans and then enter wastewater treatment facilities that are not designed to remove them (Daughton and Ternes, 1999). The average residence time for a given compound in a wastewater treatment facility ranges from less than 1 hour to a few days, which is shorter than the degradation half-lives of many of them (Halling-Sørenson, 1998; Xia *et al.*, 2005). These substances can also be released directly into the environment during storms when wastewater systems are overwhelmed and from farms where livestock are treated with antibiotics and other drugs.

A survey of US surface waters found numerous pharmaceuticals and personal care products, including hormones, caffeine, antacids, and painkillers, at detectable levels (Kolpin *et al.*, 2002; Buxton and Kolpin, 2002). While some of the increased attention to these substances has likely resulted from improved analytical procedures for detecting their presence, there are likely ecological and human health consequences, particularly in urban streams where sewage effluent can dominate discharge. Although most measured concentrations of pharmaceuticals and related contaminants are relatively low (e.g., less than 1 part per billion). However, chronic exposure to low concentrations may result in sublethal effects including changes in behavior, growth, or reproductive capacity (De Lange *et al.*, 2006). While the ultimate consequences of pharmaceuticals and other emerging contaminants in freshwaters remain to be seen, there is evidence that they are assimilated by organisms and can have deleterious effects (Kinney *et al.*, 2008; Vajda *et al.*, 2008). Some emerging contaminants are *endocrine-disrupting compounds* in that they act as biological signals. The presence of endocrine disruptors in the environment is of great concern, as these substances can seriously impact the development and reproduction of organisms (Sidebar 16.3).

The presence of antibiotics in freshwater is a concern because of their potential role in development of antibiotic-resistant bacteria, particularly pathogenic forms, and possible negative impacts on microbially mediated ecological processes (Halling-Sorenson *et al.*, 1998; Maul *et al.*, 2006b). Numerous antibiotics have been detected in streams receiving sewage effluent and below confined animal operations such as cattle feedlots. Bacteria resistant to multiple antibiotics

SIDEBAR 16.3

Ecoestrogens: Compounds That Mimic Natural Hormone Activities

Numerous organic compounds can mimic natural metabolic compounds, leading to endocrine disruption (Stahlschmidt-Allner *et al.*, 1997; Sonnenschein and Soto, 1997). An example of this form of pollution is the release of compounds that mimic estrogen (variously called *oestrogens*, *ecoestrogens*, or *environmental estrogens*). These compounds include pesticides and even ingredients in sunscreens (Schlumpf *et al.*, 2001; Klann *et al.*, 2005).

Exposure to the pesticide DDT has recently been linked to nonfunctional testes in male alligators, and other reports of feminized wildlife have begun to surface. In this case, DDT behaves like estrogen; this adds a new dimension to the documented effects of organic compounds intentionally released into the environment (McLachlan and Arnold, 1996). Endocrine-disrupting compounds influence reproduction of fishes, birds, mollusks, mammals (Colborn *et al.*, 1993), and reptiles (Crain *et al.*, 1998). Other possible cases of influence of environmental estrogens include male fish in polluted waters that produce abnormal amounts of the egg yolk protein normally produced by female fish and sex reversals of turtles when exposed to estrogenic chemicals. Ecoestrogens can bioaccumulate and be passed to offspring (Crew *et al.*, 2000).

An examination from 1995 to 2004 of fishes in 111 US water bodies by the US Geological Survey found that 33% of smallmouth bass and 18% of largemouth bass were intersex, in that they had both male and female reproductive structures (Hinck *et al.*, 2009). Although the exact cause for the intersex condition of so many fishes has yet to be determined, estrogens from human birth control pills and other sources are the most likely cause because rates were highest in densely populated regions and examinations of museum specimens collected decades ago show no intersex individuals.

Some researchers attribute the highly controversial reports of reduced human sperm counts to environmental chemicals, and others report no effects. Meta-analysis is inconclusive on the effects of ecoestrogens on human sperm counts and more research is needed before conclusive results are available (Perry, 2008). Endocrine-disrupting compounds have also been linked to formation of human cancers (Gillesby and Zacharewski, 1998). Apparently, combinations of organic chemicals can also activate the estrogen receptor (Arnold *et al.*, 1996). Such inadvertent biological signaling may have far-reaching and unpredictable effects in aquatic habitats. Fortunately, standard water purification techniques can remove ecoestrogens (Fawell *et al.*, 2001). Regardless of the strength of an individual claim, the topic of ecoestrogens illustrates that wholesale release of organic contaminants into the environment can have unintended effects.

are now common in aquatic environments (Leff *et al.*, 1993; McKeon *et al.*, 1995). Bacteria resistant to human-synthesized antibiotics have been isolated from many rivers and billabongs in remote rural areas of Australia (Boon, 1992); both are environments with low human densities.

A recent study found significantly higher numbers of tetracycline-resistant bacteria gene types in water from wastewater treatment plants compared to nearby lakes (Auerbach *et al.*, 2007), indicating that human antibiotic use can stimulate the prevalence of antibiotic resistance. Antibiotics are used as a routine addition to livestock feed in the United States because they increase growth in healthy animals. In 2006, the European Union banned the feeding of all antibiotics and related drugs to livestock for growth promotion (but not to treat sick animals). Increasing reliance on aquaculture for fish production around the world is further contributing to the problem; fish farms can use significant quantities of antibiotics and other drugs that are applied directly to water. The increasing prevalence of antibiotics or microbes exposed to antibiotics (e.g., animal feedlot runoff) that enter freshwater environments increases the probability that microbes that cause human disease will possess plasmids coding for resistance to antibiotics used to treat those diseases.

It is becoming increasingly difficult to assess and predict the effects of any single substance in freshwater habitats because pollutants rarely occur singly and may interact in antagonistic, additive, or synergistic ways with myriad other contaminants or natural environmental gradients. In many habitats, aquatic species are exposed to *multiple stressors*. For example, fish and invertebrate communities inhabiting streams draining agricultural fields are affected by altered hydrology, degraded physical habitat structure, pesticide runoff, excess nutrients, and numerous other stressors, which may interact in a mind-boggling array of ways. Toxicologists generally avoid this problem by performing laboratory studies on one or a few pollutants at a time. While critical for understanding potential toxic effects of substances, the environmental relevance of these approaches is decreasing as freshwater habitats are subjected to increasing numbers of pollutants and other stressors.

ACID PRECIPITATION

Contamination by acid precipitation has had enormous environmental and economic impacts on aquatic systems. Loss of fisheries and concomitant loss of many tourist dollars are common in affected areas. Here, we discuss sources of acid, distribution of the problem, biological effects, and potential solutions to problems associated with acid precipitation. There are additional sources of acid contamination that are not specifically covered, such as mine drainage (Gray, 1998) and natural acidic systems, but the generalities of the following discussion apply to pH effects regardless of source.

Sources and Geography of Acid Precipitation

Acid precipitation has vast effects on aquatic ecosystems in the vicinity of dense human activity associated with burning hydrocarbons. Acid rain has acidified lakes and streams in all industrialized regions of the world. The US Environmental Protection Agency surveyed more than 1,000 lakes and 211,000 km of streams during the 1980s. About 75% of the lakes and 50% of the streams surveyed were influenced by acid precipitation. The areas most impacted were the Adirondacks, the mid-Appalachian highlands, the upper Midwest, and the high-elevation West. In the worst case, 90% of the streams in the New Jersey Pine Barrens were acidic. In mid-Appalachia, there were 1,350 acidic streams. Furthermore, the Canadian government estimates that 14,000 lakes in eastern Canada are acidic. The Norwegian government sampled 1,000 lakes and found that 52% of the lakes were endangered. In the southern part of Norway, 60% to 70% of the lakes had lost their fishes (Henriksen *et al.*, 1990).

The proximate cause of acid precipitation is sulfuric and nitric acids in rain, snow, and fog. Combustion of coal and oil leads to formation of sulfuric and nitric acids in clouds, ultimately reaching the ground. Acid precipitation is concentrated downwind from industrial and urban areas because they have maximal emissions from factories and automobile exhaust. Acid deposition in the United States is greater in the heavily populated and industrialized northeast. Acid deposition is correlated most closely with sulfate deposition (color plate Fig. 9) and has decreased somewhat with mitigation activities.

When acid precipitation reaches the ground, it can react with the terrestrial ecosystem. If sufficient base is present, the acid will be neutralized. The ability of a soil or water body to absorb acidity without a change in pH is called *buffering capacity*. The most common material that confers the ability to resist changes in pH is the bicarbonate in limestone. The bicarbonate equilibrium (discussed in Chapter 13) leads to neutralization of the acid and release of CO_2. Watersheds and aquatic systems that have a significant amount of limestone have a high buffering capacity and are able to resist the effects of acid precipitation.

Biological Effects of Acidification

Acid rain has major effects on biological systems ranging from altered microbial activity to the ability of fishes to survive and reproduce (Table 16.2). Habitats that are naturally acidic include acid peat bogs (*Sphagnum* bogs) and blackwater swamps (Benner *et al.*, 1989). Lakes and streams in watersheds dominated by such bogs or swamps can be relatively acidic. Some geothermal springs are very acidic and have very distinct microbial communities associated with them. Much of our understanding of the effects of long-term acidification on aquatic ecosystems derives from the study of naturally acidic habitats. Amazingly, an iron-oxidizing archaebacterium isolated from acid mine drainage has been demonstrated to have the ability to grow at pH 0 (Edwards *et al.*, 2000).

One of the basic ecosystem influences of acidification is the lowered rate of decomposition mediated by microbes. Microbes from the naturally acidic Okefenokee Swamp are able to metabolize low-molecular-weight carbon

Table 16.2 Influences of Decreasing pH on Several Groups of Aquatic Organisms

Organism or Process	Approximate pH Value
Bacterial decomposition slows/fungal decomposition takes over	5
Phytoplankton species decline/green filamentous periphyton dominate	6
Most mollusks disappear	5.5–6
Most mayflies disappear	6.5
Beetles, bugs, dragonflies, damselflies disappear	4.5
Caddis flies, stoneflies, *Megaloptera* disappear	4.5–5
Salmonid reproduction fails, aluminum toxicity increases	5
Most adult fishes harmed	4.5
Most amphibians disappear	5
Waterfowl breeding declines	5.5

(After Jeffries and Mills, 1990)

compounds at rates comparable to those of nearby neutral wetlands (Benner *et al.*, 1989). However, the microbes in acidic habitats are less able to metabolize recalcitrant cellulose and lignin, although some degree of adaptation to the acids does occur (Fig. 16.7). Inhibition of microbial activity by low pH leads to greater rates of deposition of organic material and may partially explain the stable existence of acidic depositional wetlands (i.e., once a wetland sediment becomes acidic, microbial activity keeps it so and carbon continues to accumulate). Rates of microbial decomposition of leaves are also lower in acidified streams (Fig. 16.8), which may increase carbon accumulation and alter the associated food webs.

Algal populations are influenced by acidification. Filamentous green algae characteristically bloom in the littoral zones of acidified lakes. When these blooms collapse, the resulting O_2 depletion can have negative impacts on animals (Turner *et al.*, 1995). Diversity of planktonic and benthic algae decreases with lower pH (Dickman and Rao, 1989). Similar decreases in algal diversity and replacement with filamentous green algae have been documented for acidified streams (Meegan and Perry, 1996).

Shifts in algal communities in lake sediment cores resulting from pH changes have been used to verify historical trends in acidification (Mallory *et al.*, 1998). Such verification is required before politicians are willing to enact stringent and potentially costly emission controls. In this technique, existing lakes are used to create an index that correlates algal communities with pH. This index is then used to infer pH from species with parts that are well

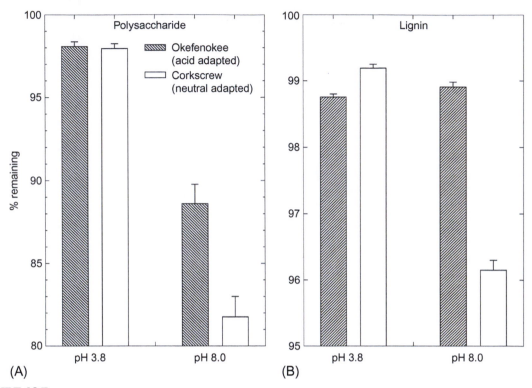

FIGURE 16.7

Percentage of remaining polysaccharide (A) and lignin (B) compounds after degradation by microbes from a naturally acidic swamp (Okefenokee Swamp, pH 3.4–4.2) and a neutral swamp (Corkscrew Swamp, pH 6–8) after incubation at different pH levels. *(Modified from Benner et al., 1989).*

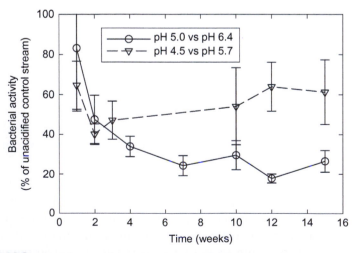

FIGURE 16.8

Microbial activity on leaves placed in acid and neutral streams as measured by the rate of thymidine incorporation into nucleic acids in two acidified streams compared to two nearby neutral streams. *(Modified from Palumbo et al., 1989).*

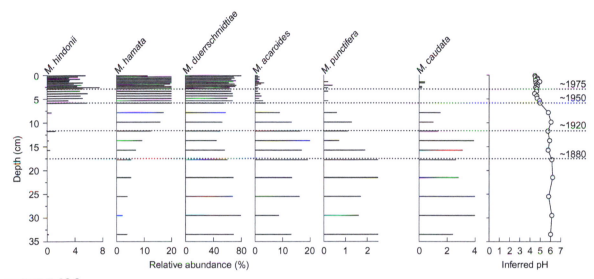

FIGURE 16.9

Distribution of *Mallomonas* spp. scales (a chrysophyte) with depth and reconstructed pH from Big Moose Lake (New York). Sediments were dated by ^{210}Pb content. *(From Majewski and Cumming, 1999, with kind permission from Kluwer Academic Publishers).*

preserved in the sediment, such as diatom frustules or chrysophyte scales. Sediment cores can be used to establish changes in the community over time. The deeper in the sediments, the longer ago the algae were deposited. When the index is coupled with isotope analyses to date specific depths of sediments, it yields a record of pH in a lake over time (Fig. 16.9). In the case of Big Moose Lake, New York, some chrysophyte species are dominant in low pH, whereas others are found only at the higher pH values associated with preindustrial conditions.

Diversity of plants and animals also decreases as aquatic systems become more acidic. Macrophyte diversity is lower in low pH lakes and fungal diversity is less in acidic streams (Fig. 16.10). Macrophyte communities become less diverse as streams acidify (Thiébaut and Muller, 1999). Invertebrates exhibit a wide range of acid sensitivities. Perhaps the most sensitive invertebrates are those that require calcium bicarbonate for shells (e.g., *Mollusca*). These shells dissolve or are unable to form when pH decreases. However, a few species from these groups are adapted to survive in waters with pH < 5 (Freyer, 1993). As aquatic systems become acidified, biomass and diversity of crustaceans (Fig. 16.11) and other invertebrates decrease. Among aquatic insects, mayflies are particularly sensitive to low pH (Herrmann *et al.*, 1993). In lakes, not only does the diversity of zooplankton decrease with increased acidification but also the efficiency of energy transfer up the food web is lowered (Havens, 1992a).

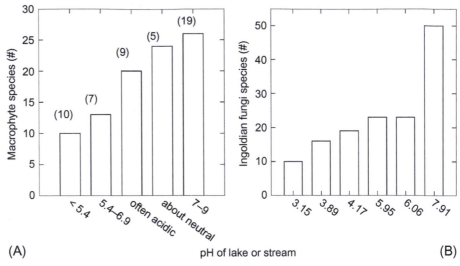

(A)

(B)

pH of lake or stream

FIGURE 16.10

Number of macrophyte species as a function of lake water pH (A) and number of species of Ingoldian fungi as a function of stream water pH (B). The numbers of lakes sampled are shown in parentheses. ((A) Data from Hutchinson, 1975; (B) Dubey et al., 1994).

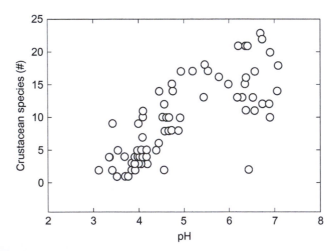

FIGURE 16.11

Crustacean species diversity as a function of pH. (Reproduced with permission from Freyer, 1980).

Fishes are susceptible to acidification, and salmonids have been studied the most because they are of the greatest economic importance in the areas that have been heavily influenced by acid precipitation. Acidification increases the concentration of aluminum (Fig. 16.12), which causes damage to fish gills.

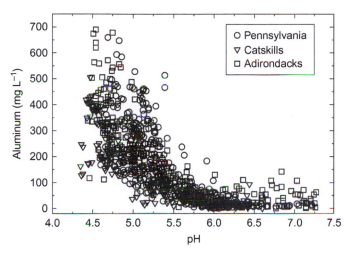

FIGURE 16.12

Relationship of aluminum concentrations to pH. *(Reproduced with permission from Wigington et al., 1996).*

The low pH increases the toxicity of the aluminum (Gensemer and Playle, 1999). Subsequently, the number of sensitive fishes decreases in acidified waters (Fig. 16.13) and the most acidified waters have no fish.

Several treatments are available to counter the effects of acid precipitation. The most obvious is stopping the source by burning low-sulfur fuels for industry and power production and decreasing the emission of nitrogen compounds in automobile exhaust. The most common local treatment is to add lime (calcium carbonate) to lakes and watersheds to neutralize the effects of the acid (Fairchild and Sherman, 1990). Directly adding calcium carbonate to the lake causes short-term (on the order of years or less) increases of pH and recovery of some biota (Hörnström, 1999). Adding calcium carbonate to the entire watershed may have longer lasting effects but is more costly (National Research Council, 1992). Unfortunately, even if acidification is reversed, losses of calcium and magnesium from soils may lead to long-term changes in water chemistry (Likens *et al.,* 1996). Research by Gene Likens has illustrated that knowledge of biogeochemical cycling is important in understanding causes and effects of acid precipitation (Biography 16.2). Another possible solution is to fertilize the lake and allow the biota to reverse the problem (Davison *et al.,* 1995), but as discussed in Chapter 18, eutrophication has its own problems.

Emission controls have led to recent decreases in acid deposition in North America and Europe and associated reversals in surface water acidification (Stoddard *et al.,* 1999). Some areas have not recovered in North America.

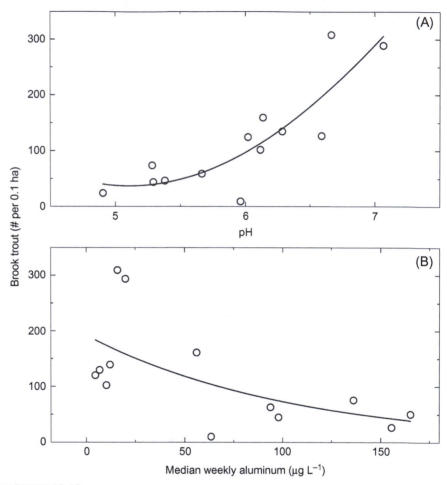

FIGURE 16.13
Biomass of trout in Adirondack streams as a function of pH (A) and aluminum (B). *(Reproduced with permission from Baker et al., 1996).*

Those watersheds were so heavily impacted by acid deposition that they were not able to respond to decreased sulfate loading.

Another cause of acidification of surface waters is mine drainage. Acid mine drainage and metal contamination are related in many instances. Metal pyrites weather when exposed to oxygenated surface waters and metals dissociate with concurrent formation of sulfuric acid. This acid mine drainage results from coal mining and metal mining operations. Treatment options include neutralization with limestone (Hedin *et al.*, 1994), oxidation in wetlands, and various combinations of these treatments (Robb and Robinson, 1995).

BIOGRAPHY 16.2 GENE E. LIKENS

The study of biogeochemistry may not be the most glamorous subject in aquatic ecology, but it is arguably the most related to water quality, the links between aquatic and terrestrial habitats, and the influence of aquatic pollutants. Dr. Likens (Fig. 16.14) is one of the foremost contemporary scientists specializing in the biogeochemistry of ecosystems. He has received numerous honorary degrees and awards, including the Tyler prize (a World Prize for Environmental Achievement), election to the US National Academy of Sciences, and top awards from the American Society of Limnology and Oceanography, the Ecological Society of America, and many other international societies. He has more than 330 publications, including 12 books.

Likens grew up on a farm in northern Indiana, where he fished, collected aquatic organisms, and generally enjoyed exploring aquatic habitats. Likens maintains that a love of natural history is the single best predictor of success for an aquatic ecologist. He attended a small liberal arts college (Manchester) and obtained a PhD from the University of Wisconsin. Following a lecture on the conservation of aquatic resources, he told the professor he was interested in the subject and wondered if he could get paid for that type of work. Obviously, the answer was yes.

Fortunately for the aquatic sciences, Likens did not follow his other career goal. He also wanted to be a professional baseball player and played for two years in the rookie league in Kansas, a league that also gave rise to baseball great Mickey Mantle. Likens was a most valuable player, but he decided that the life of an academician was preferable to that of a professional athlete.

Likens says he feels lucky to have been able to travel to and study some of the most beautiful places in the world, including Hubbard Brook, where he conducted important research on the influence of logging on nutrient transport by streams

FIGURE 16.14
Gene Likens.

(Likens *et al.*, 1978). Hubbard Brook is also the site of much of his research on acid precipitation effects; it has been the site of experimental watershed acidification and produced crucial insights into the long-term impacts of acid leaching of soils (Likens *et al.*, 1996). Likens predicts that a major future challenge in aquatic ecology will be to understand the implications of complexity. He suggests we currently do a good job at assessing the influence of one or two factors, but that to truly understand ecosystems we need to account for the simultaneous influence of multiple biotic and abiotic factors.

METALS AND OTHER INORGANIC POLLUTANTS

A wide variety of metals and some other inorganic materials act as toxic pollutants in aquatic ecosystems (Table 16.3). Arsenic, chromium, lead, zinc, mercury, cadmium, and other metals are naturally occurring, but human activities create situations where concentrations are toxic to aquatic life. Metals can bioaccumulate in many organisms and can be bioconcentrated in food chains. Bioconcentration has led to problems such as excessive lead contamination of

Table 16.3 Maximum Allowable Concentrations of Some Toxic Metals in Natural Waters Used by Humans for the United States and Potential Human Health Problems Associated with Each

Metal	Chemical Symbol	Maximum Conc. ($\mu g\,L^{-1}$)	Some Responses to Acute Poisoning	Some Chronic Effects
Mercury	Hg	0.144	Death within 10 days, severe nausea, abdominal pain, bloody diarrhea, kidney damage	Loss of teeth, kidney damage, muscle tremors, spasms, depression, irritability, birth defects
Lead	Pb	5	Anorexia, vomiting, malaise, convulsions, brain damage	Weight loss, weakness, anemia
Cadmium	Cd	10		Cancer, throat dryness, headache, vomiting
Selenium	Se	10		Nervousness, depression, liver injury (this is an essential element in small amounts, but toxic at higher concentrations)
Thallium	Tl	13	Nausea, vomiting, diarrhea, tingling pain in extremities, weakness, coma, convulsions, death	Weakness and pain in extremities
Nickel	Ni	13.4		Cancer, dermatitis, nausea, vomiting, diarrhea
Silver	Ag	50		Bluish color of skin, skin and mucous membrane irritation
Manganese	Mn	50		Languor, sleepiness, weakness, emotional disturbances, paralysis
Chromium	Cr	50		Cancer, skin, and respiratory irritation, renal damage (chromium III is not toxic; chromium II is)
Iron	Fe	300		
Barium	Ba	1000	Excessive vomiting, violent diarrhea, tremors, death	

(Laws, 1993; Budavari et al., 1989)

fishes. Complex pelagic food webs with many lateral links transfer fewer metals up the food chain (Stemberger and Chen, 1998), an additional argument for maintenance of biodiversity. Atmospheric deposition and industrial waste releases, particularly mining (Table 16.4), are common sources of metal contamination. Such mining activities have had extensive negative impacts in some aquatic habitats (Sidebar 16.4).

Chemical conditions can alter the bioconcentration and toxicity of metals. For example, cadmium, silver, nickel, and zinc uptake by invertebrates is highly

Table 16.4 Some Effects on Aquatic Environments Related to Various Mining Activities

Type of Mining	Potential Impacts
Sulfide ores (copper, nickel, lead, and zinc)	Acidification by sulfuric acid, possible arsenic contamination, sediments
Gold and silver	Same as sulfur ores, possible mercury, cyanide, or arsenic contamination
Uranium	Acid tailing drainage; runoff of radioactive materials, toxic metals, sediments, and organic compounds
Iron	Heavy water demand; runoff of sediments, toxic metals
Coal	High water demand, high sediment load in runoff, acid runoff from high sulfur deposits
Salt	Salinization of waste water

(After Ripley et al., *1996)*

SIDEBAR 16.4

Massive Contamination of the Clark Fork River by Mining Waste

Over 100 years of mining (primarily copper) in the region of Butte, Montana, has resulted in numerous contamination problems in the Clark Fork River. Waste from the mines was discharged directly or washed into tributaries of the river. An estimated 99.8 billion kg of waste was discharged into the system prior to 1959, and 2 or 3 million m^3 of contaminated sediments is present in the floodplain. Contamination has affected the upper 200 km of the river.

Treatment ponds for wastes were installed over the years, and liming was initiated to precipitate metals in the waste in 1959. However, cadmium, copper, lead, and zinc in the water column continue to exceed criteria for protection of aquatic life. Microbial communities are influenced by metal waste that drains into the groundwater (Feris *et al.*, 2004). This extensive contamination has led to designation of the upper Clark Fork River as a Superfund site. Cleanup of the site started in the late 1980s and will continue for at least 20 years, costing millions of dollars.

Historic fish surveys in 1950 showed no fishes in regions of the upper Clark Fork River. With improved water quality, trout have been reintroduced into the upper river. However, thunderstorms cause episodic contamination events, and significant fish kills were recorded in 1983–1985 and 1988–1991. Physiological abnormalities of fish have also been noted, and concentration of the contaminants in the food chain has been observed. The data suggest that metal contamination problems are likely to defy attempts at remediation for significant periods of time after contamination (Phillips and Lipton, 1995). The Milltown dam, upstream of Missoula, Montana, is filled with contaminated mine waste that accumulated in 1908, shortly after it was built. The dam is currently being removed and sediment is being removed from the river.

influenced by reactive sulfides in sediments (Lee *et al.*, 2000). High-sulfide sediments bind the metals and render them less toxic. Also, the redox state of metals can influence toxicity; hexavalent chromium is much more toxic than trivalent chromium.

Lead toxicity for waterfowl has been a particular concern in freshwater systems because of the historical use of lead shot pellets for hunting. Waterfowl such as ducks, geese, and coots ingest the pellets as grit for their crops. Fewer than 10 lead pellets will kill a bird, but marshes frequented by hunters may have 6 or 7 pellets m^{-2} in the sediments. For this reason, the US Fish and Wildlife Service has phased out use of lead shot in favor of steel shot (Laws, 1993). Lead fishing weights are still in use and have been implicated in wildlife deaths. In addition, atmospheric lead deposition increases lead concentration in lakes throughout the world. Analysis of peat bog sediments in Switzerland indicated that anthropogenic inputs increased lead contamination starting 3,000 years ago, and that in 1979 deposition rates were 1,570 times the natural background values found prior to 1,000 BC (Shotyk et al., 1998).

Mercury contamination of fish is a problem that has beset many areas. About half the streams with large piscivorous fishes in the western United States have fish with mercury concentrations exceeding safe limits (Peterson et al., 2007). Methylmercury can be assimilated and concentrated by organisms in aquatic food webs. Mercury enters aquatic systems mostly from atmospheric fallout from coal burning, trash incineration, and industrial emissions; once methylated under anaerobic conditions in saturated sediments, it readily moves into the food web. Mercury accumulation in a Spanish peat bog increased about 2,500 years ago, at a time when mercury mining began in the region (Martínez-Cortizas et al., 1999). In some countries, mercury is used indiscriminately to extract gold in mining operations and can heavily contaminate freshwater systems (Cursino et al., 1999). Periphyton mats appear to be an important site of mercury methylation and its entry into the food web (Cleckner et al., 1999). Biomagnification has resulted in levels of mercury in fish high enough to warrant consumption advisories. Such restriction on consumption may be problematic for people who use fishes as a large component of their diet (Egeland and Middaugh, 1997). Eutrophic systems can have less severe problems with production and concentration of methylmercury in the food web (Gilmour et al., 1998), but generalizations may be difficult since the relationships between organic C and methylmercury concentrations are complex (Hurley et al., 1998). The problem has been studied in the Everglades, where fish consumption advisories have been issued because methylmercury concentrations in fish tissues exceed $30\,\mathrm{ng\,g}^{-1}$ (Cleckner et al., 1998). A landscape-level study found that concentrations of mercury in tissues of most species of stream invertebrates increased significantly with stream size, and this pattern was attributed to in-stream production of methylmercury along the stream network (Tsui et al., 2009).

Selenium has caused severe problems in some wetlands. Irrigation mobilizes selenium naturally found in soils and concentrates the selenium as the water evaporates. In Kesterson Reservoir, a National Wildlife Refuge in central California, selenium contamination caused congenital deformities and mass mortality of waterfowl. Although selenium is a required nutrient in trace levels, it bioaccumulates and becomes toxic at higher concentrations. The US Geological Survey has identified about $500,000 \, km^2$ in the western United States that are susceptible to similar problems. The worst cases occur where irrigation runoff is reused for irrigation, and water ends up in terminal wetlands or lakes. Such lakes and wetlands have no outlets and concentrate selenium by evaporation.

Arsenic can cause problems because it can be present in high concentrations naturally or as runoff from industrial uses. Historically, arsenic was also used as a pesticide and subsequently contaminated aquatic systems. In a particularly terrible case, thousands of drinking water wells in West Bengal, India, are contaminated by naturally occurring arsenic (Bagla, 1996). An estimated 200,000 people in this area have arsenic-induced skin lesions and hardened patches of skin that may become cancerous. More than 1 million Indians may be drinking this contaminated water. The West Bengal problem could be related to recent large-scale withdrawal of groundwater for agriculture, leading to rapid fluctuations in groundwater level and input of O_2, which allows for release of the arsenic from sulfides in the pyrite-rich rocks of the area. Phosphorus from fertilizers also increases arsenic release rates. Microbes can oxidize arsenic (Oremland and Stolz, 2003) and they are implicated in arsenic release, which starts after iron is oxidized from the system (Islam *et al.*, 2004). The microbial effect provides a potential partial explanation for the phosphorus effect since phosphate tends to bind with iron. More research by groundwater geochemists and hydrologists is needed to study this problem and find solutions.

Radioactive compounds can be contaminants of water. These usually occur naturally. The primary contaminants are isotopes of radium, radon, and uranium. Approximately 1% of drinking water supplies are contaminated with radium above acceptable levels, and radium and uranium are found in significant concentrations in many surface and groundwaters. Calculations indicate that numerous human deaths are caused each year by radium (6–120), uranium (2–20), and radon (80–800) in the United States (Milvy and Cothern, 1990).

The effects of natural radioactive materials on aquatic habitats are difficult to gauge. As Laws (1993) states, "While the deaths of 250,000 Americans out of a population of 250 million (i.e., 0.1%) might seem an alarming statistic to many persons, the loss of 0.1% of a population of crabs or tunicates would not be likely to cause much public alarm." Thus, we do not know the effects of many contaminants on aquatic organisms, given the limited information on the effects of most chemicals on humans. Most of the research on the influence of radioactive compounds on aquatic habitats has occurred downstream from nuclear power generation plants. Perhaps the greatest concern is with biomagnification; many radioactive isotopes are retained in body tissues, and concentrations increase with each increase in trophic level.

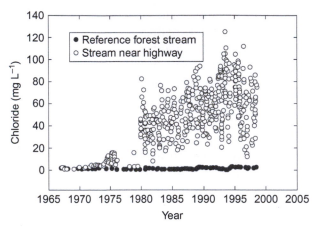

FIGURE 16.15

Long-term data showing significant increases in baseline chloride concentration in a stream following highway construction and a nearby control stream with no roads and no change in chloride concentrations over the same time period. *(Modified and reproduced with permission from* Proceedings of the National Academy of Sciences of the United States of America. *Kaushal* et al., *Increased salinization of fresh water in the northeastern United States.* Proceedings of the National Academy of Sciences of the United States of America *102, 13517–13520).*

SALT POLLUTION

A variety of human activities can increase the salinity of freshwater habitats. Applications of salt to roadways and sidewalks can directly impact nearby freshwater habitats when snow and ice melt (Kaushal *et al.*, 2005). A recent study in Ontario, Canada, indicated that approximately 50% of the total salt applied to a major highway was eventually washed away with overland flow and the remainder entered shallow groundwater habitats, resulting in degradation of groundwater resources (Meriano *et al.*, 2009). Studies in the northeastern United States indicate that salinity of many urban and suburban streams has been increasing for decades and already exceeds lethal limits for many freshwater organisms (Fig. 16.15) (Kaushal *et al.*, 2005). If these trends continue, many surface waters in the northeastern United States will not be fit for human consumption within the next century.

Direct applications of salt are not the only cause of increasing salinity of freshwaters. Agricultural and urban runoff, as well as point sources such as sewage effluent, can increase salinity of receiving waterways by increasing concentrations of chloride and other ions. Dewatering of rivers, groundwaters, and coastal wetlands facilitates salt intrusion into formerly freshwater habitats in coastal regions (Fig. 4.6). Industrial activities, such as production of soda ash, produce large quantities of waste products that can greatly increase salinity in regional freshwater habitats if not carefully contained. Groundwater extraction activities, including coalbed methane extraction, pump deep groundwaters to the surface that are often naturally saline, which can contaminate surface and shallow groundwater habitats.

Agricultural irrigation can increase salinity of downstream waters. Continued application of water will cause concentration of salts as evaporation leaves salt behind. If not enough water is added, then the soil can become too salty to use, so some water is flushed through the system. Runoff from such practices leads to higher concentrations of dissolved salts. Such problems affect many drier irrigated areas; most of the irrigated lands in California's Imperial and San Joaquin Valleys have salinity problems and salts and agricultural chemicals contaminate runoff.

Climate change may also increase salinity of freshwater habitats in some regions as increased frequency and magnitude of droughts and warmer temperatures enhance evaporation of surface waters, concentrating solutes. In some regions, such as the Murray-Darling River Basin of Australia, which has been experiencing increasing salinity for decades, identification of specific causes has proven difficult because of complex interactions among weather, geology, and human activities, but natural weathering processes are contributors to some degree (Morton and Cunningham, 1985; White *et al.*, 2009).

Regardless of the ultimate cause, increased salinity has detrimental effects on aquatic life for reasons discussed earlier; osmotic stress can result in slower growth, reduced reproduction, and ultimately death, depending on the degree of pollution and tolerance of individual species. Unlike most marine habitats, salinity of freshwater habitats can be quite variable in time and space, and thus tolerances of individual species vary greatly. Above certain levels (e.g., above $2.0 \, g \, NaCl \, L^{-1}$), few species can persist in salt-polluted habitats. Chronically salt-polluted habitats have reduced biological diversity and ecosystem functions such as primary production and decomposition rates.

SUSPENDED SOLIDS

Turbidity and suspended solids are natural parts of all freshwater environments. Some habitats are naturally highly turbid, but human activities have increased levels of suspended solids in many habitats (Fig. 16.3). Agricultural and urban runoff, watershed disturbance (e.g., logging, construction, and roads; Forman and Alexander, 1998), removal of riparian vegetation, and alteration of hydrodynamic regimes all can lead to anthropogenic increases in total suspended solids.

Sediments can have different biological and physical effects depending on the types of suspended solids (Table 16.5). High values of suspended solids can lower primary productivity of systems by shading algae and macrophytes, at times leading to almost complete removal. Suspended solids can also have negative effects on aquatic animals by interfering with reproduction, respiratory O_2 transport, filter feeding, and habitat availability.

The negative impact of excessive sediments on stream biota has been known for some time (Hynes, 1970; Waters, 1995). Such sediments lower incoming light and primary production, increase scour, harm sensitive invertebrate species, reduce biodiversity, and lower the aesthetic value of streams. Probably the strongest negative effect is filling gravel and cobble habitat through deposition, leading to increased anoxia. Sediment clogging can harm interstitial invertebrates and eggs of many spawning fishes. Lowering flow through gravels can cause O_2 levels to decrease below what is necessary for eggs to develop successfully. Information is available on the susceptibility of salmonids to sediments (Wilber, 1983). Other fish species vary in their tolerance to suspended solids.

Table 16.5 Classification of Suspended Solids and Their Possible Impacts on Freshwater Systems

Type of Solid	Physical and Chemical Effects	Biological Effects
Clays, silts, sands	Sedimentation, erosion, light attenuation, habitat alteration	Interference with respiration, restriction of habitat, burial, light limitation
Fine particulate organic matter (natural)	Sedimentation, BOD	A food source, anoxia
Fine particulate organic matter (sewage)	Sedimentation, BOD, nutrient enrichment	A food source, anoxia, eutrophication
Toxicants on particles	All of the above	Toxicity

(Wilber, 1983)

Light attenuation in lakes may comprise a large part of the influence of suspended solids on the biota. A highly turbid lake or reservoir may have limited rates of primary production. However, if there is sufficient organic material in the suspended particles, a productive food web based on microbial use of the suspended particulates can occur.

THERMAL POLLUTION

Human alterations to natural temperature regimes of freshwater habitats cause thermal pollution. Thermal pollution is any deviation from the natural temperature in a habitat and can range from elevated temperatures associated with industrial cooling activities to discharges of cold water into streams below large impoundments. Given that the metabolic rates of ectotherms are directly related to temperature and that the vast majority of freshwater organisms are ectothermic, thermal pollution can strongly affect freshwater communities. Alterations to normal water temperature regimes have myriad biological effects, including interfering with temperature cues for spawning fishes, facilitating establishment of exotic species, and altering growth and development of aquatic organisms (Langford, 1990). Further, aquatic organisms evolved in relatively thermally buffered environments, and thus they are generally more sensitive to temperature fluctuations than their terrestrial counterparts.

Most forms of thermal pollution involve temperature increases, and while the effects of extreme temperature increases are obvious, relatively small changes can also be biologically significant. Molecular adaptations to temperature were discussed in Chapter 15. Temperature increases as little as 1 to 2°C can alter communities because they are lethal to some species and can affect growth and reproduction of others. Raising water temperatures just 2 to 3°C above the optimal for some aquatic insects can greatly reduce the number of eggs produced by females because more energy is used to support higher metabolic rates and less is available for egg production (Vannote and Sweeney, 1980; Firth and Fisher, 1992).

Temperature tolerances among species of freshwater organisms are highly variable, but all have an optimal range and low and high limits within which they can survive. Increases in temperature cause an increase in growth rate up to a point. Above some threshold, damage occurs. Because temperature governs rates at all levels of biological organization (e.g., from enzymatic reactions to metabolism of whole organisms), changes in temperature associated with thermal pollution ultimately influence rates of ecosystem processes and functions such as nutrient cycling and decomposition. Thermal pollution associated with sewage effluent has been linked to differences in leaf litter decomposition rates in Lake Titicaca in South America (Costantini *et al.*, 2004). When thermal pollution is released into wetlands, trees can be killed. As temperatures increase, green algae and diatoms are replaced by cyanobacteria. One of the key issues in thermal pollution is the replacement of cold-water fishes with warm-water fishes.

Power plants and industrial factories are the major point source contributors to thermal pollution. In this case, cool water is withdrawn from streams, used for cooling of generators and other machinery, and then returned to the stream at elevated temperatures. Rapid changes in temperature associated with power plant operations can kill fishes by thermal shock (Ottinger *et al.*, 1990). Mitigating the thermal effects of power plant effluent obviously has a significant financial cost. Temperature regimes of small lakes or near-shore portions of lakes are also altered by human activities including effluents from municipal facilities and industry.

Along with industrial sources, urban and suburban runoff can contribute to thermal pollution, particularly during short, intense thunderstorms in watersheds with high amounts of impervious surfaces such as asphalt (Herb *et al.*, 2008). Depending on local groundwater inputs, discharge, and other factors that influence thermal regimes, even small municipal discharges can alter stream temperatures for considerable distances downstream.

Reductions in stream flows or lake volumes alter temperature regimes by reducing the thermal buffering capacity of the water body. Water withdrawals for irrigation, hydroelectric, and other human uses reduce stream flows and lead to significantly increased water temperatures during warm periods and can ultimately result in increased fish kills associated with high temperatures (Caissie, 2006).

Impoundments that release water from the surface can result in higher stream temperatures during warm periods because water velocity is decreased and solar penetration enhanced in the impounded water. Along with the direct effects of warmer temperatures on aquatic life, the solubility of O_2 in water decreases with increasing temperature, and thus O_2 stress increases as temperatures rise. During cold periods when stream water temperatures are normally near freezing, hypolimnetic releases can artificially warm streams.

Deforestation is also a major contributor to thermal pollution, as removal of riparian vegetation greatly increases solar penetration and temperature (Beschta et al., 1987). Small streams and ponds in forested regions are particularly vulnerable because they are normally shaded during warm months and have less thermal buffering capacity. Studies in forested headwater streams show increases in summer maximum temperatures of 5 to 8°C after logging, and recovery periods to normal thermal regimes can take 5 to 15 years (Caissie, 2006).

Anthropogenic cooling of freshwaters can also have strong effects on aquatic life. Releases of cold water from the hypolimnia of large reservoirs alter stream thermal regimes for long distances. Hypolimnetic releases can greatly alter seasonal temperature patterns, lower maximum temperatures by more than 10°C, and dampen normal diurnal temperature fluctuations (Baxter, 1977; Ward and Stanford, 1979; Ward, 1985). The overall effect is a dampening of normal diel and seasonal temperature fluctuations. Water temperatures below large impoundments like the Glen Canyon Dam on the Colorado River in the southwestern United States are now much cooler in the summer than they were historically. Temperate zone streams below large impoundments in regions where natural summer stream temperatures exceed the upper limits for most cold-water species are often stocked with nonnative cold water fishes such as various species of salmonids. Although important for recreation, these tailwater fisheries (fisheries downstream from dams) differ substantially from natural river systems and are often detrimental to native species.

While many forms of thermal pollution originate at point sources, climate change (discussed in Chapter 1) represents a nonpoint source of thermal pollution that is already influencing a wide range of freshwater habitats. The long-term impacts of climate change on freshwater habitats remain to be seen, but predictions indicate significant shifts in structure and function of streams, lakes, and wetlands, particularly in higher latitudes.

HUMAN-INDUCED INCREASES IN UV RADIATION

Although freshwater organisms that live in high light environments have adapted defenses against damaging radiation, human activities are increasing UV-B radiation in freshwater habitats through a variety of mechanisms, including reducing atmospheric ozone, which absorbs incoming UV radiation; reducing water depths and thus attenuation of UV radiation; and altering amounts of suspended particles and dissolved organic matter in water columns, which absorb UV radiation (see sidebar 12.1). Freshwater habitats in polar regions and at high latitudes are most susceptible, as they have higher exposure.

Increased UV-B radiation has been shown to adversely affect oceanic bacterioplankton activity (Herndl *et al.*, 1993), lake phytoplankton productivity (Harrison and Smith, 2009), and lake zooplankton communities (Williamson *et al.*, 1994). Increased UV-B exposure has also been linked to abnormalities and death of amphibian larvae (Bancroft *et al.*, 2008). Exclusion of UV radiation from stream mesocosms in New Zealand increased the abundance of aquatic insects by 54% (Clements *et al.*, 2008). Ultraviolet radiation can interact with other toxic substances such as metals and pesticides, and nutrient levels can determine the level of damage from UV radiation (Carrillo *et al.*, 2008). These interactions can be *synergistic*, in that effects are more than simply additive.

SUMMARY

1. All surface waters on Earth are influenced by human activities.
2. Toxicologists are concerned with the acute and chronic effects of pollutants. Natural variations in uptake, sensitivities, concentration, additive effects of different pollutants, and effects of other environmental factors all complicate predictions of how strongly a particular toxicant will influence a specific organism.
3. Biomagnification of pollutants can cause high concentrations of toxic compounds in the tissues of organisms at the top of food webs. The most lipid-soluble compounds usually are magnified to the greatest degree. Metals, organochlorine pesticides, and PCBs are examples of highly persistent pollutants that readily bioaccumulate.
4. Bioassessment protocols can be used to assess the impacts of pollutants and other habitat alterations on aquatic communities. Indices for bioassessment are generally constructed using data on invertebrates or fish that are present and the state of their habitat, but methods have been developed for all types of aquatic organisms and habitats.
5. Acid precipitation is mainly caused by humans burning fossil fuels, leading to increased sulfuric and nitric acid in the atmosphere. Acid mine drainage is also of concern in many regions of the world.
6. Acidification of aquatic ecosystems impacts all aquatic organisms.
7. Under acidic conditions, microbes degrade complex organic compounds more slowly, cyanobacteria and diatoms are selected against, filamentous green algae are selected for, invertebrates that use calcium carbonate become rare, aluminum has a greater impact on fish gills, and reproduction of many animals is impacted negatively.

8. Metals can have a broad array of negative impacts on aquatic ecosystems. Mining wastes often cause contamination, but runoff from industrial applications and naturally occurring sources can cause problems as well. Mercury contamination of freshwater habitats has resulted in fish consumption warnings in many regions of the world.

9. More than 10,000 organic compounds are discharged by humans into aquatic habitats, including pesticides, oil, and materials in urban runoff. Only a small percentage of these compounds have been tested for toxicity. In some cases, microbes can break down these compounds (bioremediation) given enough time.

10. Endocrine disruptors act as biological triggers and have been linked to reproductive deformities in a variety of freshwater species.

11. Increasing prevalence of antibiotics from human and livestock waste in freshwater habitats has been linked to development of antibiotic-resistant bacteria.

12. Freshwater organisms are often subjected to multiple stressors, which may interact in antagonistic, additive, or synergistic ways. These shifts makes identification of the effects of one particular type of pollutant difficult.

13. Suspended solids can cause harm to aquatic organisms. Generally, interferences with photosynthesis from increased light attenuation and with respiration by clogging flow of water are the main negative impacts. In streams, fine solids fill up and destroy gravel and cobble habitats, can increase scour associated with high flow, and reduce movement of water into subsurface habitats.

14. Thermal pollution can cause shifts in community structure. These shifts may allow for establishment of exotic species and local extinction of native species.

QUESTIONS FOR THOUGHT

1. Why is biomagnification worse with lipid-soluble compounds that are resistant to abiotic and biotic deactivation?

2. Are there any habitats on Earth that have not been influenced by human activities?

3. What political conditions have led to a world in which toxicants are released routinely into the aquatic environment before even cursory testing of their effects on organisms, including humans, has been conducted?

4. Why can controls on emissions of greenhouse gases ultimately decrease acid precipitation?

5. Why can acid precipitation lead to lower iron availability and greater phosphorus in some lakes and wetlands with anoxic sediments?

6. Why is bioremediation of metal contamination more difficult than that of contamination by organic compounds?

7. Why are microbes able to more rapidly evolve ways to inactivate toxicants and develop resistance than fishes?

8. Under what conditions may suspended solids have positive influences on aquatic ecosystems?

9. Should aquatic scientists assume a role of advocacy with regard to issues of aquatic pollution, or should their role be primarily to provide data for informed decisions to be made by managers and policymakers?

Nutrient Use and Remineralization

FIGURE 17.1

Carboys used for an *in situ* bioassay of nutrient limitation at Milford Reservoir, Kansas. The first four experimental additions (from right to left) are control, N, P, and N + P. One week after addition, the nitrogen + phosphorus treatment had the most chlorophyll.

Nutrients can be critical basal resources in aquatic systems because they often limit primary production or heterotrophic activity. Understanding ecosystem production, nutrient pollution, and interactions among heterotrophs, autotrophs, and their environment requires an elucidation of nutrient dynamics. In this chapter, we discuss how nutrients are acquired and assimilated, the relative amounts needed in different systems, availability of nutrients, and the crucial concept of nutrient limitation. How nutrients are made available (recycled) by heterotrophs as the nutrients cycle through the food web of aquatic systems is also discussed.

USE OF NUTRIENTS

In the broadest sense, a *nutrient* is any element required by organisms for growth, but many ecologists use this term primarily in reference to elements that are limiting. For example, even though organisms are composed of large amounts of oxygen and hydrogen, rarely are these elements considered nutrients because they seldom limit primary production. When oxygen limits ecosystem activity, this is due to lack of availability for use as an oxidant of organic carbon, not a lack of availability as material to build cells. Likewise, primary producers can use CO_2 directly, and many researchers think carbon supply ultimately does not limit ecosystems driven by photoautotrophs and should not be considered a "nutrient." Stated differently, while carbon is energy currency for ecosystems, the availability of carbon to build cells is rarely considered limiting for autotrophs. Nitrogen, phosphorus, silicon, and iron are the nutrients most often studied by aquatic ecologists.

The major classes of biological molecules require different amounts of nutrients. Lipids are phosphorus and carbon rich. Amino acids and proteins require relatively more nitrogen than most molecules. Nucleotides and nucleic acids require significant amounts of nitrogen and phosphorus relative to carbohydrates, such as starch and sugars, which are composed entirely of carbon, oxygen, and hydrogen. Other elements such as iron are required in very small amounts as cofactors in enzymes. Specialized requirements include calcium for mollusks and silicon for diatoms. Each organism must acquire nutrients to survive and grow.

Nutrients are acquired in a variety of forms by organisms. Most primary producers and many bacteria have the ability to use inorganic nutrients, such as nitrate or phosphate. In a case that stretches the definition "autotroph," algae such as the dinoflagellates can ingest particulate material for its nitrogen and phosphorus content and to meet part of their carbon requirements. Most animals and heterotrophic eukaryotic unicellular organisms acquire their nutrients in organic form (e.g., nitrogen must be assimilated as proteins or amino acids, and carbon must be assimilated as carbohydrates).

Nutrients in organic form can be dissolved in the surrounding water or contained in particulate material (including organisms). Algae can use some of these dissolved organic forms directly (Berman and Chava, 1999). Individual bacteria have different degrees of ability to use inorganic or organic compounds, but the bacterial assemblage found in natural waters is generally able to use a wide variety of organic and inorganic nutrient sources. Here, we consider the general principles of assimilation and uptake of dissolved nutrients. Uptake of materials in the form of particles (e.g., living and dead organisms) will be considered in later chapters.

Nutrients need to be taken into cells from the water surrounding them (*uptake*) and then incorporated into organic molecules used for growth (*assimilation*). Generally, each of these steps requires energy. Three equations are commonly used to describe the functional relationships among nutrient concentrations, uptake, and assimilation.

The *Michaelis–Menten* relationship is used to describe the influence of nutrient concentration on uptake rate:

$$V = V_{max} \frac{[S]}{K_s + [S]}$$

where V is the uptake rate of substrate, V_{max} is the maximum uptake rate, $[S]$ is the concentration of substrate (nutrient),[1] and K_s is the concentration of S where $V = 1/2 \ V_{max}$. The shape of the curve (Fig. 17.2A) reveals that as concentration increases, uptake increases rapidly only at low concentrations. At high substrate concentrations, the maximal rate of uptake is approached. Values for K_s vary widely among and within algal taxa (Table 17.1) and may even vary among individual cells as they become adapted to surrounding nutrient concentrations. Values of V_{max} can be a useful physiological indicator (Zevenboom *et al.*, 1982). For example, organisms existing in low-nutrient environments tend to have low values of K_s and those in areas with high nutrient availability have high values of V_{max}. Larger cells have greater values for V_{max} and K_s than small cells (Suttle *et al.*, 1988). A calculation with this equation is presented in Example 17.1.

When an organism takes up nutrients, it cannot always immediately use them for growth. Organisms that are under nutrient stress are able to take up nutrients at very high rates upon transient exposure to high nutrient concentrations but are not able to sustain growth rates proportional to this uptake. This high uptake is a selective advantage when nutrients are scarce because pulses of nutrients can be stored. Such consumption in excess of growth requirements is called *luxury consumption*. Enough phosphorus can be stored in the form of the polymer polyphosphate through luxury consumption for several cell divisions (Healey and Stewart, 1973).

Maximum growth rates (μ_{max}) are controlled by nutrient supply, temperature (Fig. 15.3; Eppley, 1972), light (see discussion on photosynthesis–irradiance relationships in Chapter 12), pH, and other factors. Logically, growth rate is related more directly to nutrient concentration inside

---

[1] The term *substrate*, in most of biology with biochemical basis, refers to materials that organisms assimilate for growth (e.g., nutrients and organic carbon). In limnology, this creates confusion because substrate is often used to denote solid surfaces that organisms are associated with. For this reason we follow Wetzel and refer to nutrients as substrate or substrates and solid habitat as substratum or substrata. Also, the square brackets around [S] indicate the concentration of the substrate S, generally in moles per liter. Moles per liter can be converted to grams per liter by multiplying by molecular weight. Units of molecular weight are in grams per mole.

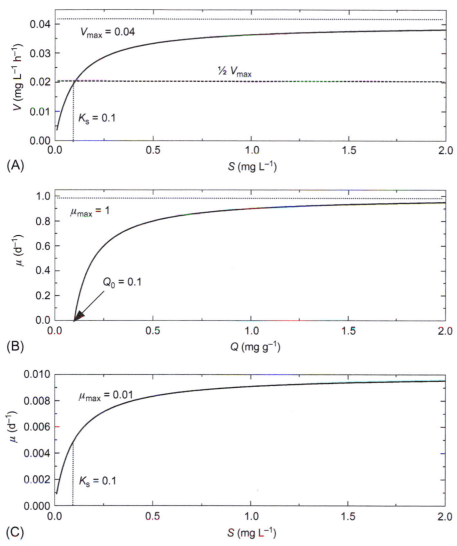

FIGURE 17.2

Graphical representation of equations used to describe nutrient uptake and assimilation: (A) Michaelis–Menten, (B) Droop, and (C) Monod relationships.

([Q]) than outside the cells. The relationship between internal nutrient concentrations and growth is called the *Droop equation*:

$$\mu = \mu_{max} \left(1 - \frac{Q_0}{Q}\right)$$

where μ is the growth rate, μ_{max} is the maximum growth rate, Q is the cell quota (concentration inside the cell), and Q_0 is the minimum cell quota (the

Table 17.1
Some Values for Monod Nutrient Uptake and Droop Equations for Phosphate (After USEPA 1985)

Algal Type	K_s (mg L^{-1})	μ_{max} (d^{-1})	Q_0 (See Title)
NO$_3^-$-N or NH$_4^+$-N			
Phytoplankton	0.114 (0.0014–0.2)	0.039 (0.0024–0.15)	0.024 (0.015–0.04)
Diatoms	0.051 (0.001–0.13)	0.07 (0.015–0.125)	3×10^{-7} (0.5–6×10^{-7})
Green algae	0.016 (0.0014–0.030)	0.0925 (0.06–0.125)	1.7×10^{-6} (0.5–34.2×10^{-7})
Cyanobacteria	0.50 (0.03–0.98)	0.0825 (0.04–0.125)	1.93×10^{-7} (0.52–4.3×10^{-7})
PO$_4^{3-}$-P			
Phytoplankton	0.0357 (0.0028–0.07)	0.222 (0.0014–2.95)	0.0014 (0.001–0.003)
Benthic algae	0.125	0.045	0.0005
Diatoms	0.065 (0.0002–0.06)	0.262 (0.024–0.5)	5.06×10^{-7} (0.01–7×10^{-7})
Green algae	0.28 (0.001–1.5)	0.317 (0.133–0.5)	1.7×10^{-9} (1.7–4.5×10^{-9})
Cyanobacteria	0.5 (0.007–0.98)	0.585 (0.042–0.5)	2.91×10^{-9} (0.58–5.66×10^{-9})

Means are followed by ranges in parentheses. Units of Q_0 are in mg nutrient/mg cell for phytoplankton and in μ moles/cell for benthic algae. Values from literature review of 1 to 5 studies for each parameter

■ Example 17.1

Uptake and Growth Calculations

Two species of phytoplankton have the following characteristics:

Phytoplankton Species	K_s (mg liter^{-1})	V_{max} (μg liter^{-1} h^{-1})	μ_{max} (day^{-1})	Q_0 (gg^{-1})
Anabaena	0.5	3	0.5	0.05
Scenedesmus	0.1	1	0.1	0.01

Problem 1:   Which species will be the best competitor for nutrients (i.e., have the highest nutrient uptake rates) when nutrient concentrations are 0.01 and 0.5 mg liter^{-1}?

The Michaelis–Menten equation yields the following results:

Phytoplankton Species	V (at $S = 0.01$ mg liter^{-1})	V (at $S = 0.5$ mg liter^{-1})
Anabaena	0.059	1.5
Scenedesmus	0.091	0.83

The *Anabaena* will have a competitive advantage at the higher nutrient concentration. The *Scenedesmus* will have the advantage at the lower nutrient concentration. The curves for this and the following two problems are presented in Figure 17.3.

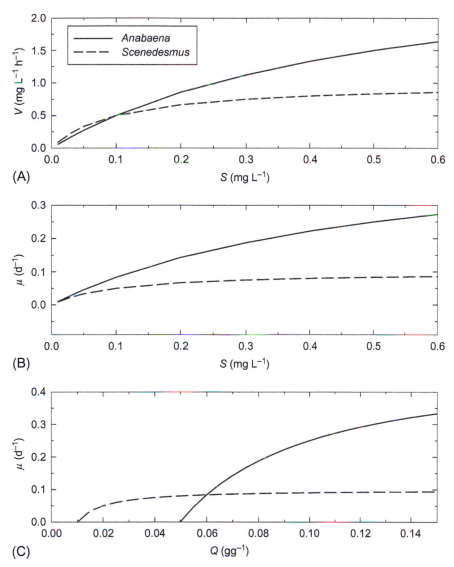

FIGURE 17.3

Graphical representations of the equations used in Example 17.1 for uptake and growth as a function of nutrients for two algae: (A) Michaelis–Menten uptake, problem 1; (B) Monod growth, problem 2; and (C) Droop growth, problem 3.

Problem 2: Which species will have the highest growth rate at $[S] = 0.01$ and $0.5\,\text{mg liter}^{-1}$?

Phytoplankton Species	μ (at $S = 0.01\,\text{mg liter}^{-1}$)	μ (at $S = 0.5\,\text{mg liter}^{-1}$)
Anabaena	0.0098	0.250
Scenedesmus	0.0091	0.0833

The *Scenedesmus* has the disadvantage in growth at both substrate concentrations.

Problem 3: Which species will have the highest growth at a Q of 0.04 and $0.1\,gg^{-1}$?

Phytoplankton Species	μ (at $Q = 0.04\,gg^{-1}$)	μ (at $Q = 0.1\,gg^{-1}$)
Anabaena	0.0	0.25
Scenedesmus	0.075	0.09

The *Scenedesmus* will have a competitive advantage at the lower cell quota because the *Anabaena* cannot grow. The *Anabaena* will have the advantage at the higher cell quota. ■

concentration in the cell below which no growth occurs; Fig. 17.2B). Typical values for the constants in the Droop equation can be found in Table 17.1. This equation is useful if internal nutrient concentrations can be determined. The final step is to link growth to external nutrient concentration.

Growth can be related to concentration of nutrients outside the organism by the *Monod equation*:

$$\mu = \mu_{max} \frac{[S]}{K_s + S}$$

where μ is growth rate, μ_{max} is the maximum growth rate, $[S]$ is the substrate concentration, and K_s is the concentration at which $\mu = 1/2\,\mu_{max}$ (Fig. 17.2C). The equation is in the same form as the Michaelis–Menten equation. This equation works well for single species in culture and moderately well for phytoplankton assemblages. The actual physiological bases for K_s and μ_{max} are complex and can vary over time and with nutrients, even within single cells (Ferenci, 1999). Typical values for these two constants for nitrate, ammonium, and phosphate are presented in Table 17.1.

Nutrient concentration is a prime determinant of uptake, but other factors, such as light, temperature, and metabolic characteristics, control uptake rate as well. For example, low temperature decreases affinity for limiting substrates (Nedwell, 1999). The greater energy requirement for use of nitrate than that of ammonium under oxic conditions provides another example and was discussed in Chapter 14. The energy requirement translates into a greater effect of light on nitrate uptake rates by phytoplankton than ammonium (Fig. 17.4).

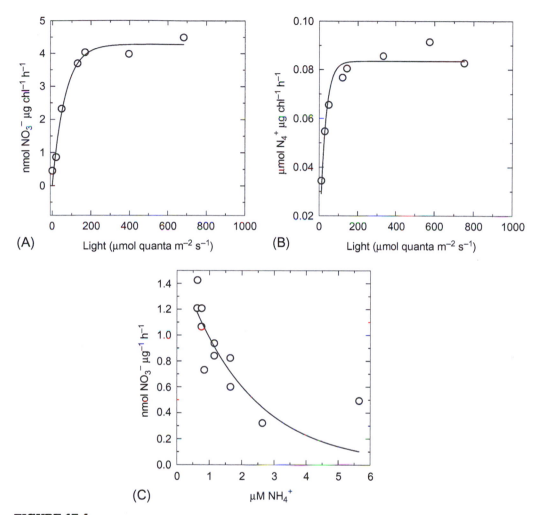

FIGURE 17.4

Uptakes of nitrate (A) and ammonium (B) as a function of light (reproduced with permission from Dodds and Priscu, 1989) and nitrate (C) as a function of ammonium *(reproduced from Dodds et al., 1991, by permission of Oxford University Press)* for epilimnetic plankton in Flathead Lake Montana, July 1987.

Nutrients can also interact. For example, ammonium inhibits nitrate uptake (Fig. 17.4). This adaptation is advantageous because using nitrate takes more energy, so using ammonium when it is available is an efficient strategy. Similarly, energetically expensive nitrogen fixation will not occur at high rates when nitrate or ammonium is plentiful.

Macrophytes have the ability to obtain dissolved nutrients from the sediments through their roots in addition to using nutrients from the water column

(White and Hendricks, 2000). Uptake kinetics are less clear for macrophytes because nutrient uptake from the water column can be important for some submerged macrophytes, and partitioning water column uptake from that of the sediments can be difficult. The relative importance of nutrient supply from sediments compared to supply from the water column has been debated for many years (Sculthorpe, 1967). Stream macrophytes may rely more heavily on nutrients in the water column than lake macrophytes (Pelton *et al.*, 1998). Emergent vegetation in wetlands probably acquires the majority of its nutrients from the sediments.

NUTRIENT LIMITATION AND RELATIVE AVAILABILITY

Relative Availability of Nutrients

The next step after understanding how nutrients are taken into cells is to explore nutrient requirements of cells and how these requirements translate into limitation of growth by nutrients. *Nutrient limitation* is the control of growth or production by a nutrient or nutrients (in contrast to limitation by other factors, such as light, predation, or temperature). The concept of nutrient limitation is central to aquatic ecology because it allows determination of which nutrient or nutrients control primary production or heterotrophic activity of the ecosystem.

All organisms have approximately the same nutrient requirements because they are all built of the same major types of molecules. The typical composition of algal cells with balanced growth (Table 17.2) reveals that various nutrients are required with respect to relative proportions. The relative ratio of these nutrients to each other is called the *stoichiometry*. The stoichiometry of carbon, nitrogen, and phosphorus at balanced growth is generally taken as 106:16:1 (C:N:P by atoms or moles) and is referred to as the *Redfield ratio* (Redfield, 1958). Variations in stoichiometry can be related to physiological state or taxonomy. For example, diatoms require much more silicon than other organisms. The plasticity of nutrient contents varies among individual organisms. Many primary producers can build up significant amounts of starch, lipids, or cellulose to store carbon and alter their C:N:P accordingly. Thus, stoichiometry can be very useful for determining and understanding nutrient limitation.

Some nutrients can be acquired relatively easily by most cells. Oxygen and hydrogen are easily available to photosynthetic organisms able to split water. Carbon is generally available to autotrophs in the form of CO_2. Ultimately, most of the other nutrients must come from weathering of Earth's crust. The relative abundance of materials dissolved in rivers provides an

Table 17.2

Elemental Composition of Algae and Plants Compared to Availability in Freshwater (World Rivers)

Element	Plants and Algae	World Rivers	Average Demand/Supply
H	13,400,000	3,520,000,000	≪1
O	5,880,000	1,780,000,000	≪1
C	2,750,000	31,900	86
N	689,000	525	1,312
Si	163,000	7,390	22
K	34,900	1,880	18
P	24,400	10	2,440
Ca	23,300	12,000	2
Na	17,900	8,340	2
Mg	17,100	5,260	3
S	13,700	3,980	3
Fe	7,240	401	18
Zn	314	5	63
B	264	296	1
Cu	102	5	20
Mn	82	9	9
Mo	1	1	1

All composition data are in moles or atoms relative to molybdenum. Data for algae from Healy and Stewart (1973) combined with data for plants and algae. Data for rivers from Vallentyne (1974). Average demand/supply is algae and plants divided by rivers

approximate guide to the comparative supplies of nutrients available to aquatic organisms (Table 17.2), but it should be kept in mind that standing stock and supply rates are not exactly the same thing. For example, carbon, nitrogen, and phosphorus are difficult to obtain relative to other elements, given relative concentrations in the world's rivers. However, CO_2 readily enters water from Earth's atmosphere (has a high supply rate) and thus can be replenished quickly in most surface waters. Since inorganic carbon is more readily available (and becoming increasingly so from human activities), nitrogen and phosphorus are most likely to limit growth of algae and aquatic macrophytes.

Factors such as chemical characteristics, geology, and human land use alter the availability of nutrients in aquatic systems. For instance, phosphate tends to bind to clays, so it is transported slowly into aquatic systems, and tends to be transported in rivers at times of flooding (Banner *et al.*, 2009). Nitrogen can be lost from anoxic systems by denitrification but can be gained by nitrogen fixation, so the relative supply of nitrogen in individual systems may be difficult to predict. Geology and land use in particular areas can alter relative nutrient availability. For example, phosphorus can be amply available in watersheds with phosphorus-rich volcanic ashes or sedimentary deposits of apatite (mainly calcium phosphate deposits associated with limestone). Variation in surface reference concentration of total phosphorus and nitrogen across US rivers shows areas that are naturally higher in phosphorus than other places in the United States (Fig. 17.5). Current concentrations of both

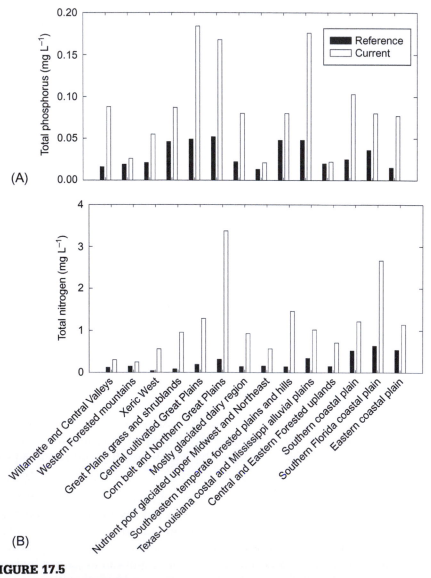

(A)

(B)

FIGURE 17.5

Reference and current median concentrations of (A) total phosphorus and (B) total nitrogen for rivers in the major ecoregions of the United States. *(Data from Dodds et al., 2009).*

nitrogen and phosphorus are at least 10 times greater than reference conditions, mostly related to high densities of human population or intensive agricultural activity. Thus, knowledge of nutrient supply in a specific groundwater aquifer, wetland, lake, or stream requires detailed knowledge of the geological and land-use characteristics of the specific watershed, including point and nonpoint sources of nutrients.

Nutrient Limitation

Growth is limited by the factor or factors present in the lowest relative supply that is required for synthesizing the cellular constituents. The initial application of this idea to aquatic sciences used the concept of *Leibig's law of the minimum*: The rate of a process will be limited by the rate of its slowest subprocess. The original statement of the law was based on crop production. The law presented the idea that a single constituent would eventually disappear and leave the land barren if successive crops were planted in one area with no nutrient amendments.

Consider an automobile construction plant as an analogy for Leibig's law. The plant uses parts shipped from suppliers all over the world. Different amounts of various parts are required for each car (e.g., 4 tires per car, 1 transmission, and 6 engine cylinders). If 400 tires, 100 transmissions, and 540 cylinders are supplied each month, only 90 cars (540/6) can be constructed because the supply rate of cylinders is limiting, even though sufficient tires and transmissions are available for 100 cars per month. Similarly, a living cell requires that elements be supplied at specific ratios for growth. Liebig's law of the minimum has been applied to nutrient limitation of primary producer assemblages in aquatic systems with the assumption that all producers have equal nutrient requirements and nutrients are evenly distributed in space and time in the environment (i.e., a homogeneous equilibrium condition of nutrients exists). The law, combined with the idea of equilibrium nutrient availability, predicts that only one nutrient will limit primary production of a system.

How closely does this prediction match empirical data on nutrient limitation? Several techniques have been used to assay nutrient limitation (Method 17.1), and there is controversy regarding the use of such methods, but the results of the assays indicate some interesting patterns. Results of bioassays on primary producers in lakes, wetlands, and streams are considered here. Little is known about nutrient limitation of microbial activities in groundwaters.

Broad analyses of nutrient limitation of primary producers indicate that nitrogen, phosphorus, or both commonly limit terrestrial, freshwater, and marine waters (Elser *et al.*, 2007). Given that nutrients mostly originate from terrestrial habitats and are transported through freshwaters to marine, the idea that limiting nutrients are similar across habitats is not surprising. Surveys of tests of nutrient limitation in lakes indicate that either nitrogen or phosphorus

METHOD 17.1

Bioassay Tests for Determination of Nutrient Limitation

The most realistic way to determine nutrient limitation is to test the entire system. However, this approach is not practical in many situations because control, multiple treatments, and replication are difficult, if not impossible, and nutrient pollution of an entire system is undesirable. The most commonly used alternative is *in situ* treatments in enclosures or small-scale treatments.

In lakes, mesocosms (limnocorrals) or containers (Goldman, 1962) from 1 to 1,000 liters have been used. Nutrients are added, and the response of the algal biomass is usually measured after about one week (Fig. 17.1, color plate Fig. 10). Enclosing the water may lead to artifacts, such as attached algae proliferating on the walls and lack of external nutrient inputs, but these "container effects" may be minimal in short-term studies in large enclosures.

Benthic systems are often tested with nutrient-diffusing substrata (color plate Fig. 11). In these tests, nutrients are sealed inside a container, out of which they slowly diffuse across a permeable surface that can be colonized by benthic algae. Unglazed clay pots filled with nutrient-enriched agar are commonly used for such tests, but use of other permeable surfaces is increasing (Winterbourn, 1990; Johnson *et al.*, 2009). Factors that can interfere with these tests include grazers that crop algae as it grows and the inability to duplicate natural surfaces (e.g., nutrients rarely diffuse out of solid surfaces in nature).

Another alternative is to take a water sample, return it to the laboratory, and test how well it stimulates production of laboratory cultures of algae (Eaton *et al.*, 1995). In this case, different nutrients can be added to incubations to assess which nutrient is limiting algal growth. The drawbacks to this method are that it does not simulate natural conditions, and species of phytoplankton used in the test may not be found in the system of interest. Duckweed (*Lemna minor*) also has been suggested as a test organism in the laboratory (Eaton *et al.*, 1995) because it is widely distributed in nature, its small size makes it easy to collect, and it is simple to grow in specific media.

Finally, a variety of short-term physiological bioassays have been proposed (Beardall *et al.*, 2001). Of these, the most successful has been use of Redfield ratios for indication of nutrient limitation (Example 17.2). Other physiological methods are generally less reliable although quicker than growth-based bioassays (Dodds and Priscu, 1990).

most commonly limits primary production (Elser *et al.*, 1990a) and that many lakes are *colimited* by both nitrogen and phosphorus (Fig. 17.6A).

Nutrients other than nitrogen and phosphorus can limit some lakes. Studies of Lake Erie suggest that phytoplankton are, at times, colimited by nitrogen, phosphorus, or iron (North *et al.*, 2007), and cross-system analyses suggest that iron might more commonly be limiting in oligotrophic lakes than previously thought (Vrede and Tranvik, 2006). CO_2 can limit phytoplankton in some cases. Floating cyanobacteria can maintain dominance in some eutrophic lakes by intercepting CO_2 so that it is not available to other primary producers (Shapiro, 1997), and phytoplankton in lakes with low CO_2 can exhibit lower photosynthetic rates (Hein, 1997). In the benthic zone of lakes, benthic algae

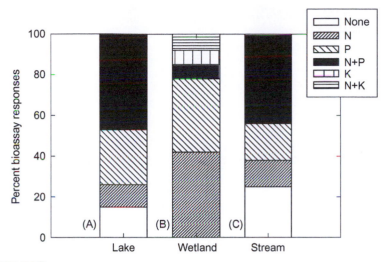

FIGURE 17.6

Summary of nutrient bioassays reported in the literature indicating stimulation of biomass of phytoplankton (A), wetland plants (B), and stream periphyton (C). *(Data in A from Elser* et al., *1990a; data in B from Verhoeven* et al., *1996; and data in C compiled from various sources).*

can be limited by CO_2, particularly in acidified waters with low total inorganic carbon concentrations (Fairchild and Sherman, 1990). Silicon can limit many diatom populations (Schelske and Stoermer, 1972). In some cases, micronutrients such as molybdenum may limit algal growth (Goldman, 1960, 1972; Howarth and Cole, 1985). Early recognition of variable nutrient limitation and use of bioassays to indicate limitation is one of the many contributions by Dr. Charles Goldman (Biography 17.1). Not all lakes are nutrient limited (Tilzer *et al.*, 1991); light and toxins can also limit primary productivity.

Assays on wetlands indicate that nitrogen or phosphorus limit plant production in many peat mires, but potassium may also limit plant growth (Verhoeven, 1986). A literature search on the nitrogen and phosphorus stoichiometry of wetland plants and soils suggests that phosphorus limitation, or nitrogen and phosphorus colimitation, is common in most wetland types (Bedford *et al.*, 1999). Wetlands with high hydraulic throughput (drained wetlands) may be more prone to potassium limitation (Van Duren *et al.*, 1997). Wetland rice production may also be limited by zinc concentrations (Neue *et al.*, 1998). As with lakes and streams, some cases of multiple nutrient limitations occur in wetlands (Fig. 17.6B).

Autotrophs in streams can also be limited by nitrogen, phosphorus, nitrogen and phosphorus, or neither (Fig. 17.6C). It is unwise to assume that any particular nutrient limits primary production in streams or that nutrients cannot be colimiting. CO_2 might also limit some primary producers in streams

BIOGRAPHY 17.1 CHARLES GOLDMAN

Professor Charles Goldman (Fig. 17.7) has dedicated his career to researching factors controlling production in lakes. Early contributions include recognition of the potential importance of trace nutrients (see Chapter 14). Goldman has supervised more than 100 graduate students and 30 postdoctoral researchers. He has published four books and more than 400 scientific articles, and he has produced four documentary films. He has won many prestigious national and international awards for his scientific contributions, including the Albert Einstein World Award of Science.

In 1967, "Goldman Glacier" was named in Antarctica, reflecting his early involvement in limnological research in polar regions; Goldman is an adventurer who has traveled to and studied remote lakes throughout the world. Once, while taking primary production measurements on a very hot day at Lake Victoria, Goldman decided to take a swim and impressed the African crew with back flips off the bridge. After his sixth flip he noticed that a 6-meter-long Nile crocodile had eased up to the opposite side of the boat. He did not try a seventh flip.

Goldman credits his father (an amateur ichthyologist) with getting him started in aquatic sciences and for the idea that "reading is the only way to compensate for the shortness of life." Goldman started as a geologist but changed his focus after taking a limnology course from David Chandler and realizing that "limnology is the queen of the ecological sciences." He suggests that students study as broad of a base in sciences as possible and include humanities because being a successful aquatic ecologist requires a multidisciplinary approach.

Goldman and his students have studied Lake Tahoe since 1958. Research on Lake Tahoe and Castle Lake in California has convinced him of the value of long-term data sets; data sets on

FIGURE 17.7

Charles Goldman showing President Clinton and Vice President Gore a plankton sample from Lake Tahoe in 1998. *(Photograph courtesy of the* Sacramento Bee*).*

these lakes contain trends over a few years that are the opposite of long-term trends in the same data. He has also used his research to support social action to protect Lake Tahoe. His experience with mixing social action and research has also led him to become involved in efforts to protect Lake Baikal. Goldman is an example of an exceptional scientist applying basic scientific knowledge to environmental problems.

(Dodds, 1989; Raven, 1992) and silicon can limit epiphytic diatoms at times (Zimba, 1998). Cases of no nutrient limitation are expected to be more common in streams than lakes or nonforested wetlands because of the greater influence of riparian canopy cover on light (light limitation) in streams and scouring floods that remove algal biomass. Bioassays of heterotrophic metabolism upon nutrient enrichment versus stimulation of primary producers conducted in streams across the United States indicate that heterotrophs in streams can also be limited by nitrogen, phosphorus, both, or neither. These bioassays also indicate that autotrophic and heterotrophic responses often differ within the same stream (Johnson *et al.*, 2009).

Limitation of primary producers by more than one nutrient in wetlands, streams, and lakes raises the question of explaining colimitation that should not occur if Leibig's law is really a law of nutrient limitation (Dodds *et al.*, 1989). Given that streams are commonly far from equilibrium, the existence of colimitation in at least some streams is not surprising. Likewise, wetlands are structured by benthic processes, and it is easy to imagine that spatial variations in sediments and gradients created by different points of nutrient inflow and outflow lead to significant spatial heterogeneity in nutrient limitation. The concept of several nutrients limiting primary production in lakes is more difficult because lakes are more homogeneous than streams and wetlands.

Several of the assumptions used to apply Leibig's law may not hold, even in lakes, because all producers do not have equivalent requirements (e.g., diatoms need more silicon than other algae), nutrients may not be equally available in environments (e.g., pulses of nutrients occur), and not all organisms have the same competitive abilities. Models have been proposed for multiple limiting factors (Verduin, 1988), but these have not been well investigated. In contrast, the links between Leibig's law and phytoplankton diversity have been well investigated and lead to a famous paradox.

The Paradox of the Plankton and Nutrient Limitation

The *paradox of the plankton* was proposed by Hutchinson (1961). Although it may be viewed as a "straw man" given what is currently known about aquatic ecology, the proposed paradox forms a useful starting point for discussion and perhaps forms the basis for the most common question asked in aquatic ecology graduate qualifying exams in the past 40 years.

The paradox is based on applying Leibig's law and the *competitive exclusion principle* (Hardin, 1960) to phytoplankton communities. Competitive exclusion occurs because only one species can be the superior competitor for a single limiting resource. In an environment in which several species are competing for a single resource, the superior competitor eventually will extirpate the others.

Hutchinson argued that because lakes are very well-mixed environments, limiting nutrients are well mixed and equally available to the phytoplankton. Given that most cells have similar requirements and are competing for the same nutrients, the competitive exclusion principle should limit the number of species that are present at any time. The paradox he noted is that a typical mesotrophic or oligotrophic lake has many species (typically 10–100) of phytoplankton present at any one time.

Explanations for plankton diversity include the following: (1) Predation by zooplankton and viruses removes or suppresses dominant competitors (Suttle *et al.*, 1990), (2) pulses and micropatches of nutrients from uneven mixing and excretion lead to nonequilibrium conditions (discussed later in this chapter), (3) mutualistic or beneficial interactions promote otherwise inferior competitors, (4) many lakes are not at equilibrium conditions over timescales greater than 1 month, and the time required for dominant phytoplankton species to outcompete inferior competitors is more than one month (Harris, 1986), (5) different competitive abilities lead to different nutrients limiting different species, and (6) chaos (in the mathematical sense) arises when species compete for three or more resources (Huisman and Weissing, 1999). The idea of different competitive abilities led to the resource ratio theory and the determination of how the Redfield ratio is linked to nutrient limitation.

RESOURCE RATIOS AND STOICHIOMETRY OF PRIMARY PRODUCERS

Primary producers can alter the relative proportions of elements that comprise their cells (their stoichiometry). This adaptation allows producers to acquire and store cellular components when resources are not limiting and use them during times when they are limiting. Luxury consumption (discussed previously) alters the stoichiometry of the primary producers. When nutrients are limiting, photosynthesis still occurs and leads to accumulation of carbon in lipids or starch (i.e., cellular stoichiometry is shifted toward relatively more carbon). This relationship between nutrient supply and cell stoichiometry has led to the extensive use of deviations from the Redfield ratio to indicate nutrient limitations.

The Redfield ratio is derived from nutrient contents of phytoplankton grown with excess concentrations of all nutrients at conditions optimal for maximum growth. Deviations from these ratios indicate nutrient limitation (Example 17.2). A similar approach may be useful in determining nutrient limitation of wetland plants (Boeye *et al.*, 1997; Bedford *et al.*, 1999). For example, nitrogen content of pitcher plant tissues increases when they are placed in higher nitrogen environments (Bott *et al.*, 2008). Interestingly, when nitrogen increases, they change their morphology to one that favors more light capture and less carnivory (a potential nitrogen source).

NUTRIENT REMINERALIZATION

Uptake rates of limiting nutrients at natural concentrations in most aquatic habitats are great enough that the dissolved pools of nutrients will be depleted rapidly if they are not replenished (Axler *et al.*, 1981; Kilham and Kilham,

■ **Example 17.2**

Using Redfield Ratios of Primary Producers to Indicate Nutrient Deficiency

Molecular ratios of C, N, and P are listed for phytoplankton assemblages from three separate lakes. Use the ratios to predict nutrient limitations.

Phytoplankton Stoichiometry	C:N:P (Molar)	C:N (Molar)
Balanced growth	106:16:1	6.6:1
Mirror Lake	212:32:1	6.6:1
Deep Lake	106:3:1	35:1
Clear Lake	400:16:1	25:1

The comparison must be made to balanced growth (the Redfield ratio). The C:N ratio is presented to clarify the example. In Mirror Lake, the N:P ratio is greater than 16, so phosphorus is limiting relative to N. The C:N ratio in Mirror Lake is the same as balanced growth, so N likely is not limiting but phosphorus is limited. In Deep Lake, the N:P ratio is lower than balanced growth, the C:P ratio is the same as balanced growth, and the C:N ratio is greater than balanced growth; thus, nitrogen is limiting in this lake. Finally, in Clear Lake the N:P ratio is the same as balanced growth, but the C:P and C:N ratios exceed those at balanced growth so both nitrogen and phosphorus are limiting. ■

1990). Turnover may take hours or days for the nitrate pool or only seconds for the phosphate pool, but without supply of nutrients from some source, uptake cannot continue at rates measured in the environment. In general, external sources of nutrients (*new nutrients*) such as river and groundwater inflow or atmospheric deposition cannot supply nutrients to lakes and wetlands at measured rates of uptake. Even in unpolluted streams and rivers, nutrients in the water column are depleted in a few hundred meters of flow (see discussion on nutrient spiraling in Chapter 24). Thus, the predominant short-term source of nutrients is from *remineralization* (also known as *regeneration* or *mineralization*), which is the release of inorganic nutrients by organisms. Remineralization is an important ecosystem driver, particularly in the epilimnia of large lakes (Fee *et al.*, 1994). Thus, the productivity of a system is a function of both regeneration and supplies of new nutrients. We describe how the remineralization occurs, what organisms are responsible, and the dynamics of nutrient remineralization.

What Short-Term Processes Control the Levels of Dissolved Inorganic Nutrients Such as Ammonium and Phosphate?

If the water chemistry of a lake, a stream at base flow, wetland pool, or groundwater source is sampled each day for several days, the concentrations of ammonium and phosphate generally vary little. This lack of variability over short periods of time occurs even though uptake rates are often sufficient to completely remove dissolved inorganic nutrients in considerably less than one day.

The balance between uptake and remineralization is the reason that nutrient concentrations are moderately stable in the short term (Dodds, 1993). Uptake removes nutrients at a variable rate, and remineralization replenishes them at an approximately constant rate (Fig. 17.8). This replenishment results in a dissolved nutrient concentration that is moderately resistant to perturbation over hours to days. This balance of uptake and remineralization is common in many planktonic systems (Fig. 17.9). Remineralization is important in wetlands, particularly those with limited hydrologic nutrient inputs (Bridgham *et al.*, 1998), and a balance of concentrations should also occur in them.

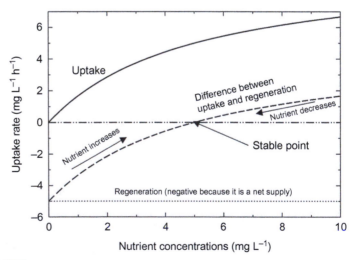

FIGURE 17.8

Graphical representation of the concept that gross nutrient uptake and regeneration can stabilize dissolved nutrient concentrations. This occurs when net uptake = 0 and gross uptake = regeneration. The graph illustrates a net increase in nutrients (excess regeneration) when nutrient concentrations are low and a net decrease (excess uptake) when nutrient concentrations are high. *(Redrawn from Dodds, 1993).*

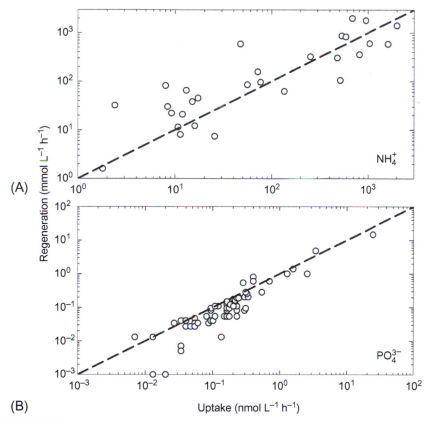

FIGURE 17.9

Uptake and remineralization of ammonium (A) and phosphate (B) from a variety of surface waters. Note the approximate 1:1 correspondence between the rates. *(Reproduced with permission from Dodds, 1993).*

The concept of nutrient balance is illustrated with the relationship between net uptake (V_{net}), gross uptake (V_{gross}), and remineralization (M), where:

$$V_{net} = V_{gross} - M$$

If concentration is constant, then net uptake must equal zero. This equation also implies that gross uptake cannot be measured by change in nutrient concentrations unless nutrient concentrations are far from equilibrium. This is the reason that tracer studies are useful in determining gross uptake rates at ambient nutrient concentrations (Mulholland *et al.*, 2002).

Meaningful discussion about system dynamics is difficult if only the dissolved inorganic nutrient concentrations are known because rates of uptake and remineralization, not standing stocks, of nutrients are the important parameters (Dodds, 2003b). For example, the algae in a highly eutrophic lake can have a very high nutrient demand and thus keep phosphate and ammonium levels near or below detection. Conversely, if the algal bloom in the same lake is declining and dead cells are decomposing, very high levels of ammonium and phosphate can occur. High values for dissolved inorganic nutrients are not confined to eutrophic systems; in a very carbon-limited aquifer, dissolved inorganic nutrients may be high even though the system has very low levels of microbial activity (e.g., it is relatively "oligotrophic" for a groundwater aquifer). Consequently, understanding the processes leading to supply and consumption of nutrients may be more important than just knowing the concentrations of the inorganic nutrients, and nutrient concentrations alone cannot necessarily be used to determine which nutrients are limiting. This is a common problem of assuming that pool (compartment) size is reflective of flux into or out of that pool. Just as lake size does not necessarily indicate flux of water through it (the Great Salt Lake is very large and has no outflow; the Great Lakes have a large river that empties them), nutrient concentration is not the same as nutrient uptake or remineralization.

Processes Leading to Remineralization

Several processes can lead to nutrient remineralization; some remineralized nutrients originate from decomposition of dissolved and particulate organic material and others from living organisms.

Dissolved nutrients in organic form are common in aquatic environments because organisms release them with normal metabolic activity, cells break and release their contents, and organic nutrients enter aquatic habitats from land. Up to half of photosynthetic carbon fixation by algae can be released directly into the dissolved form, even by healthy cells (Zlotnik and Dubinsky, 1989). These dissolved organic molecules often contain nitrogen and phosphorus. Part of this leakage may be associated with release of extracellular enzymes such as phosphatase, and part may be unavoidable losses. In addition, a proportion of cells break from viral infections, damage from feeding activities of algivores, or other causes. These cells release dissolved organic nutrients into solution. In forested streams and wetlands, senescent autumn leaves and other riparian inputs can contribute large quantities of dissolved nutrients when they enter the aquatic environment and leach soluble materials.

Additional organic molecules containing nutrients are released as part of the activities of heterotrophic organisms. Consumers release organic molecules as excreta and through sloppy feeding (e.g., damaging tissues of other organisms but not ingesting them). As these organic molecules are released, they become available to other heterotrophic organisms for consumption. In the process of being broken down, they release their associated nutrients in inorganic form.

Heterotrophic organisms that engulf living and nonliving organic material often acquire nutrients such as nitrogen and phosphorus in excess of their requirements. The excess nutrients are excreted into the aquatic environment. Processes controlling excretion rates are discussed later under the framework of ecological stoichiometry.

Given the variety of processes that can lead to remineralization, which are the most important? Few studies have partitioned out the relative contributions of different organisms to remineralization. The most common way to partition remineralization into functional groups is to *size fractionate* (separate into different size classes by filtration). For the most part, these experiments indicate that very small organisms dominate nutrient remineralization in many planktonic systems (Table 17.3). Similar patterns are observed for size fractionation of uptake (data not shown). This route for nutrient remineralization is often called the *microbial loop*, where small, unicellular algae, bacteria, viruses, protozoa (particularly very small flagellates), and rotifers rapidly recycle carbon and nutrients. The microbial loop dominates nutrient cycling in many groundwater aquifers and sediments in which larger organisms do not occur. The microbial loop will be discussed in greater detail in Chapter 19.

Cases exist in which nutrient supply associated with larger organisms is very important. These cases often involve making nutrients available from outside the system. Examples of external sources in streams include nutrient input from rotting carcasses of salmon after they have spawned and died (Kline *et al.*, 1990; Bilby *et al.*, 1996); nutrient input to streams from vegetation that has fallen from riparian plants and leaches and is processed by invertebrates; and activities of beavers bringing terrestrial nutrients into streams (Fig. 24.4; Naiman *et al.*, 1988, 1994). In lakes, vertical migration of zooplankton from the hypolimnion to the epilimnion and subsequent movement of nutrients from the hypolimnion can occur. Also, movement of benthic organisms, such as the amphipod *Gammarus*, from sediments into the water column can bring as much as 33% of the phosphorus into the water column (Wilhelm *et al.*, 1999).

Fishes in lakes can contain a substantial portion of the phosphorus in the lake, making fish remineralization important (Sereda *et al.*, 2008). One study suggested that as lake productivity increased, the abundant detritivorous fish, *Dorosoma cepedianum*, was responsible for increasing total available nutrients from 18% in reservoirs with less human influence in their watersheds to 50% in more heavily influenced watersheds (Vanni *et al.*, 2006). Different species of fish excrete phosphorus at different rates (Gido, 2002).

Salmon carcasses left after reproduction also bring considerable amounts of marine-derived nitrogen into lakes and streams. In coastal Alaskan lakes

Table 17.3
Size Fractionation of Regeneration of Ammonium and Phosphate in the Epilimnion of Pelagic Freshwater Systems

Lake	Nutrient	High Cutoff (μm)	Low Cutoff (μm)	% High	% Medium	% Small	Reference
Lake Calado (Amazon floodplain)	N	20	3	1	39	60	Fisher et al., 1988
Lake Calado (Amazon floodplain)	P	20	3	0	0–45	55–100	Fisher et al., 1988
Flathead Lake (Montana)	N	280	3	0–10	10–25	75–100	Dodds et al., 1991
Flathead Lake (Montana)	P	280	3	0–30	0–45	50–100	Dodds et al., 1991
Lake Biwa (Japan)	N	98	–	50	–	50	Urabe et al., 1995
Lake Biwa (Japan)	P	98	–	15	–	85	Urabe et al., 1995
Lake Biwa (Japan)	N	100	20	3–16	7–18	63–98	Haga et al., 1995
Lake Kizaki (Japan)	N	100	20	1–62	27–68	40–70	Haga et al., 1995
Ranger Lake (Ontario)	P	40	0.8	52	30	18	Hudson and Taylor, 1996
Mouse Lake (Ontario)	P	40	0.8	15	65	20	Hudson and Taylor, 1996
Lake Herrensee (Germany)	P	150	–	18	–	82	Hantke et al., 1996
Lake Bräuhaussee (Germany)	P	150	–	11	–	89	Hantke et al., 1996
Lake Thaler See (Germany)	P	150	–	2	–	98	Hantke et al., 1996

The % categories are the % regeneration in the size fraction above the high cutoff (% high), between high and low (% medium), and below low (% small). Where several seasons were studied, ranges are presented for the percentages this input is enough to cause substantial effects on lake ecosystem production and phytoplankton and zooplankton community structure (Finney *et al.*, 2000). Salmon moving into small streams and then removed by bears provide about one fourth of the riparian nutrients in some Alaskan streams (Helfield and Naiman, 2006). Excretion into ponds and wetlands by flocks of ducks, egg deposition in wetlands by amphibians, and movement of hippopotami out to graze terrestrial vegetation and excreting the material into rivers or wetlands are other examples of such nutrient movements by larger organisms. Movements of nutrients and energy from one habitat to another (e.g., from forest to stream) are often referred to as *ecological subsidies* (Polis *et al.*, 1997). As large-scale, landscape-level approaches to freshwater ecology become more common, there is increasing interest in ecological subsidies among aquatic and terrestrial habitats (Polis *et al.*, 2004).

Advanced: Remineralization as a Source of Nutrient Pulses in Lentic Systems

Nutrient pulses can be important because it is possible that pulses can persist for some minutes before diffusing to background concentrations at small scales, low Reynolds numbers, and high viscosity. Pulses could allow microscopic organisms lucky enough to encounter the patch the ability to take up large amounts of a limiting nutrient. Pulses might be ecologically important as a partial explanation for large phytoplankton cells doing well in lakes. Large algal cells are poor competitors for nutrients at low concentrations relative to smaller cells (Suttle *et al.*, 1988). Larger cells have a low ratio of surface area to volume and cannot assimilate nutrients as well. These cells may be able to use high concentrations of nutrients associated with pulses and, thus, maintain competitive ability by storing nutrients for use between pulses.

Zooplankton excretion has been suggested as a process that creates patches of elevated nutrients in planktonic environments. In an ingenious study, Lehman and Scavia (1982) proved that zooplankton could excrete pulses that remained stable for a long enough time to give phytoplankton cells in their vicinity a possible competitive advantage. In their study, a culture of the cladoceran *Daphnia* was fed with algal cells that had been labeled with radioactive phosphorus (^{32}P). These radioactive *Daphnia* were transferred into bottles containing unlabeled planktonic algae. After a short time, the algal cells were harvested and some were placed on microscope slides. The slides were coated with a photographic emulsion sensitive to the radioactive ^{32}P. After the slides were exposed and developed, microscopic examination for tracks of beta radiation emission was done. These tracks could be used to determine if each cell had taken up radioactive phosphorus and how much. Analysis of the distribution of the label in the algal cells revealed that some cells had significantly more label in them than would be expected if the phosphorus excreted by the

Daphnia was dispersed completely randomly into the bottle (i.e., the pulses were mixed quickly). Thus, the experiment demonstrated that nutrient pulses could exist in planktonic communities and favor specific cells.

The importance of pulses from zooplankton excretion is uncertain. Artificial nutrient pulses change phytoplankton community composition in algal cultures (Scavia and Fahnenstiel, 1984). But, if it is true that most nutrient remineralization in planktonic communities is dominated by microorganisms smaller than copepods and cladocerans, then such pulses are less likely to have overall importance. Very small cells can only create small pulses of remineralized nutrients. Given Fick's law (see Chapter 3 on properties of water), we know that diffusion rates are greater at smaller scales. Small pulses probably disperse quickly and are unlikely to be present for long enough to stimulate large planktonic algae. Other factors such as grazer resistance may select for larger cells, and nutrient pulses may not be necessary to explain their presence (see Sidebar 3.1 on factors selecting for morphology of plankton).

STOICHIOMETRY OF HETEROTROPHS, THEIR FOOD, AND NUTRIENT REMINERALIZATION

Ecological stoichiometry is a powerful and rapidly expanding field in aquatic and terrestrial ecology (Sterner and Elser, 2002) and has a large influence on how primary producers interact with primary consumers. Heterotrophs rely on organic material and remineralize nutrients such as nitrogen and phosphorus when they are in excess of requirements. Primary producers are very flexible in their stoichiometry and can build cells that are very rich in carbon relative to nitrogen and phosphorus under conditions of nutrient limitation. In contrast to primary producers, the stoichiometry of many heterotrophs is similar to that of the Redfield ratio, and they are generally much less flexible in their ability to alter these ratios. Because heterotrophic organisms need to meet both their energy and carbon demands for growth from the organic material they consume, the nutrients in the food they eat can frequently exceed the amount needed.

As an example of the stoichiometric effects of the carbon requirement for both growth and respiration, consider a fish that is able to convert only 10% of the carbon it consumes into biomass. The remaining 90% of the carbon must be used to create energy for metabolism required for survival and reproduction. In this case, the required molar nutrient ratio is 1060:16:1. If food is consumed that has the Redfield ratio of 106:16:1 mol of C:N:P, only 1/10 of the carbon, nitrogen, and phosphorus can be used for growth. The excess nitrogen and phosphorus will be excreted.

Food for heterotrophs is not always at the Redfield ratio, and requirements of all heterotrophs are not the same as the Redfield ratio. Consideration of stoichiometry has led to much study of the requirements for ratios of nutrients, the stoichiometry of heterotrophs, and the composition of their food.

Most bacterial heterotrophs rely on dissolved organic material for carbon, nitrogen, and phosphorus requirements. This material ultimately comes from primary producers (either phytoplankton in lakes or benthic algae and terrestrial vegetation in wetlands and streams) and can vary considerably in stoichiometry, as discussed previously. Bacteria can retain nitrogen increasingly as the C:N ratio of the dissolved, organic material consumed decreases; thus, net remineralization is high at low C:N ratios (Fig. 17.10).

The dissolved organic carbon available to bacteria may be low in nitrogen and phosphorus, and bacteria may need to meet their requirements for these materials by incorporating (also referred to as immobilizing or assimilating) inorganic forms, such as nitrate, ammonium, and phosphate (Tezuka, 1990). Thus, a significant portion of inorganic nutrient uptake in some lakes (and presumably other aquatic habitats) can be attributed to bacteria (Currie and Kalff, 1984).

Ecosystem processes (e.g., remineralization) can be tied to stoichiometry of organisms (Elser *et al.*, 1996). For example, copepods have a higher N:P ratio than the cladoceran *Daphnia* (Fig. 17.11). The low N:P ratio of *Daphnia* means that it has a relatively high phosphorus requirement for growth (Sterner, 1993).

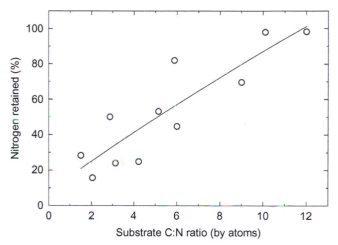

FIGURE 17.10

Nitrogen retention efficiency as a function of C:N ratio of food source for bacteria. Note that when food is relatively nitrogen rich (i.e., C:N is low), a low percentage of the nitrogen is used and most of the nitrogen ingested is remineralized. *(Redrawn from Goldman et al., 1987).*

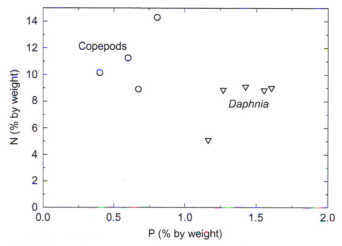

FIGURE 17.11

Data showing N:P ratio of *Daphnia* is lower than that of copepods, indicating different nutrient requirements for both types of grazers. *(Reproduced with permission from Elser et al., 1996. © American Institute of Biological Science).*

This requirement can lead ultimately to more intense phosphorus limitation in lakes (Elser and Hassett, 1994). The high requirement of *Daphnia* for phosphorus can lead to a shift to stronger phosphorus limitation in phytoplankton (Sterner, 1990; Sterner *et al.*, 1992) because of preferential assimilation of phosphorus relative to nitrogen and relatively high ratios of N:P in nutrients remineralized by *Daphnia* (Sterner and Hessen, 1994). Similarly, the high nitrogen requirement of copepods relative to cladocerans such as *Daphnia* can intensify nitrogen limitation and select for nitrogen fixing cyanobacteria (Hambright *et al.*, 2007).

Although less research has been done in streams and wetlands, the constraints of stoichiometry will hold in all environments. For example consumers can alter nutrient content of periphyton. In an experiment with snails (molar N:P = 28) and crayfish (molar N:P = 18), grazer identity had various effects on nutrient content of their food (Evans-White and Lamberti, 2006). The crayfish, with greater relative phosphorus demand, depressed the phosphorus content of periphyton to a greater degree than snails when external supplies

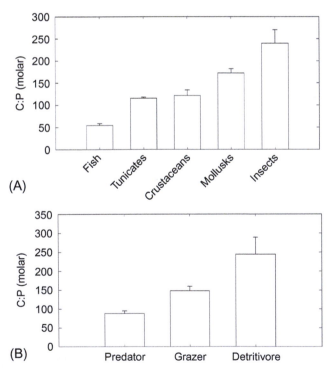

(A)

(B)

FIGURE 17.12

Variation in carbon to phosphorus ratios of aquatic consumers by taxonomic group (A) and functional group (B). Fishes have significantly lower C:P than all other taxonomic groups, and detritivores have significantly greater C:P than predators or grazers. *(Data replotted from Frost et al., 2006).*

of dissolved inorganic phosphorus were low. Periphyton stoichiometry was altered because the crayfish held more phosphorus and excreted less relative to the snails. When the supply rate of dissolved inorganic phosphorus was high, consumers had no influence on phosphorus content. Examination of stoichiometric relationships in forested headwater streams revealed larger elemental imbalances between primary consumers and their detrital resources than has been observed in many other systems (Cross *et al.*, 2003).

There are some broad relationships between taxonomic groups and phosphorus content (Fig. 17.12); fishes have the lowest C:P ratios, and invertebrates the greatest and most variable ratios. Detritivores, which generally subsist on poor quality food, generally have greater body C:P. The general hypothesis is that animals with high growth rates depend upon high phosphorus food and that animals tend to match the food source they have adapted to consuming (Frost *et al.*, 2006).

In a related study, Evans-White *et al.* (2009) found that diversity of macroinvertebrates decreased as total phosphorus in the water column increased, up to a point, and then leveled out at low diversity (Fig. 17.13). The authors hypothesized that fast-growing organisms could compete well at high phosphorus

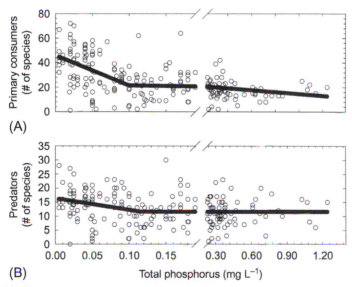

(A)

(B)

FIGURE 17.13

Variation in diversity of stream invertebrate primary consumers (A) and predators (B) from rivers and streams in Kansas, Missouri, and Nebraska in spring and autumn samples as a function of water phosphorus concentration. There were significant breaks in the relationships as denoted by the two lines, with the predator break occurring at greater total phosphorus than the consumers. *(Data from Evans-White et al., 2009).*

concentrations and would dominate under polluted conditions. This hypothesis assumed that primary consumers evolved under a natural range of phosphorus concentrations in their food and that, above this range, only some species would be effective competitors. The effect is mediated by the fact that stoichiometry of food for primary consumers (periphyton and detritus) increases in phosphorus content as stream water becomes more phosphorus-rich. The hypothesis suggests that primary consumers should be more highly influenced by stream nutrient concentrations than predators because predators rely on animals that have relatively constant stoichiometry. Analysis of the data supported the hypothesis because primary consumer diversity was more closely related to total phosphorus concentrations than was predator diversity. Thus, the concepts of stoichiometry and nutrient limitation have implications for food webs and ecosystem function because remineralization feeds back to primary producers and heterotrophic organisms at the base of the food web.

SUMMARY

1. Nutrient uptake can be described by the Michaelis–Menten uptake equation. After nutrients are taken up, they must be assimilated or converted to the chemical compounds that make up cells. The equations that describe this process are the Droop equation, which links intracellular nutrient concentrations with growth, and the Monod equation, which describes the relationship of external dissolved nutrients to growth rate.

2. Other factors that influence uptake and assimilation of nutrients include the ability to acquire and store nutrients for later use (luxury consumption), temperature, and light.

3. Nutrients are required in known ratios (stoichiometry) for growth. The ratio of C:N:P required for algal growth is approximately 106:16:1 by moles and is known as the Redfield ratio. If the relative availability of a nutrient is lower than its requirement relative to availability of others, it can be limiting.

4. Some aquatic scientists maintain that only one nutrient can be limiting at a time, but others argue that more than one nutrient can limit primary producer assemblages at a time. Empirical evidence suggests that N, P, or both are usually the limiting nutrients in lakes, streams, and wetlands.

5. Nutrients can be supplied from outside (new nutrients) or inside the system by the process of remineralization (regeneration). Nutrient remineralization provides the primary source of the nutrients available to primary producers in aquatic ecosystems. Sources of these remineralized nutrients include organic material excreted or lost by producers and processed by heterotrophs and excretion by predators or consumers.

6. The balance of uptake and remineralization often controls dissolved inorganic nutrient concentrations in aquatic ecosystems.
7. Microbes are responsible for the bulk of remineralization in most ecosystems, but in certain cases larger organisms can be important.
8. Nutrient pulses created by larger organisms can persist in the environment, but those produced by smaller microbes likely disperse quickly by diffusion.
9. Stoichiometry of heterotrophs feeds back to alter nutrient limitation of primary producers as well as determining some aspects of community structure.

QUESTIONS FOR THOUGHT

1. Why might dissolved nutrient levels be more variable over days to weeks in streams than in large lakes?
2. Nuisance filamentous benthic algae in the Clark Fork River, Montana, are limited by nitrogen in the summer, despite the fact that phosphorus concentrations dissolved in the water at that time are extremely low. Given that dissolved phosphorus concentrations are very high in the spring, what is a potential reason for the lack of phosphorus limitation in the summer?
3. Why isn't nutrient limitation necessarily additive (e.g., why are there generally small responses to additions of nonlimiting nutrients)?
4. What is an evolutionary argument for why nutrient competition should lead to limitation by multiple nutrients?
5. Why may nutrient pulses be more likely to form and persist in groundwater and wetland sediments than in planktonic habitats?
6. Why are large cells more likely to have high maximum rates of nutrient uptake, high half-saturation constants, and the ability for greater luxury consumption relative to small cells?
7. Why do many scientists think that total phosphorus concentrations are more useful indicators of nutrient supply than dissolved phosphate concentrations?
8. Why can it be misleading to use the ratio of dissolved inorganic nitrogen:dissolved phosphate, rather than the Redfield ratio of organisms, to indicate nutrient limitation?

Trophic State and Eutrophication

FIGURE 18.1
Whole-lake nutrient additions at Lake 226 in northwestern Ontario. The far lake received nitrogen, phosphorus, and carbon, and the near lake received only nitrogen and carbon. The algal bloom in the far lake gives the lake a light color. (Copyright 1974 by the American Association for the Advancement of Science. Photograph courtesy of D. W. Schindler.) *(From Schindler, D. W. Eutrophication and recovery in experimental lakes: Implications for lake management.* Science, *184: May 24, 897–899, 1974).*

Nutrient pollution can alter ecosystem structure and function in all aquatic habitats. This alteration by unnaturally high nutrients is termed *eutrophication*. Nutrient input can be increased by humans (*cultural eutrophication*) or can occur naturally. In this chapter, we describe how *trophic state*, the level of ecosystem productivity, is defined by comparisons across aquatic systems and consider problems that may be associated with eutrophication. Next, the linkages among nutrient loading, nutrients, algal biomass, water clarity, and fish production are examined. Finally, methods are described for controlling eutrophication and several case studies are presented. Given the large economic costs associated with improvement of water quality, and the large costs associated with negative effects of nutrient enrichment (Dodds *et al.*, 2009), eutrophication continues to be a very relevant issue in lakes, streams, and wetlands.

DEFINITION OF TROPHIC STATE

Classifications of the trophic state of aquatic ecosystems are useful because they allow people to compare productivity of ecosystems within and among ecoregions and provide an initial approach for determining the extent of cultural eutrophication. Trophic state is historically signified by the terms *oligotrophic, mesotrophic,* and *eutrophic. Oligo* means "few," *trophic* means "foods," *eutrophic* means "many foods," and *mesotrophic* falls between these two categories. The three categories are only one way to characterize a continuum of ecosystem productivity. Over the years, several systems have been employed to describe the trophic state of lakes; trophic state classifications are not as highly developed for streams, groundwaters, and wetlands.

The recognition of the effect of nutrients on heterotrophic activity, with potential influences up to the ecosystem level, has led to a broader view of eutrophication. This view considers *heterotrophic state* as represented by the whole-system CO_2 production per unit area and *autotrophic state* the whole-system CO_2 fixation per unit area (Dodds and Cole, 2007). Eutrophication can be defined as an increase in the nutritive factor or factors that lead to greater rates of whole-system heterotrophic or autotrophic metabolism (Dodds, 2006).

Early limnologists noticed that certain types of phytoplankton and zooplankton were typically found in high-nutrient lakes and others in nutrient-poor lakes. This observation led to extensive efforts to characterize the trophic state of lakes with regard to their phytoplankton communities (Hutchinson, 1967). Limnologists thus recognized the links among nutrients, phytoplankton biomass and productivity, and water quality. These links will be described quantitatively in this and subsequent sections.

Current classifications of autotrophic trophic state of lakes are generally based on water clarity, phytoplankton biomass, and nutrient concentrations (productivity is not as easy to measure, so it is used less in trophic classification). In general, oligotrophic lakes have low algal biomass, low algal productivity, low nutrients, high clarity, and deep photic zones, and they may support coldwater fisheries. Eutrophic lakes are characterized by cyanobacterial blooms, high total nutrients, and large variation in O_2 concentrations (including potential anoxia in the hypolimnion), and they may have frequent fish kills. The trophic state of lakes is usually based on phytoplankton concentrations, but shallow eutrophic lakes can have extensive macrophyte populations.

One of the commonly used classifications for lakes was constructed by a large group of limnologists interested in eutrophication (OECD, 1982). This classification system was constructed by combining data from many lakes. For this analysis, the lakes were classified by scientists as eutrophic, mesotrophic, or oligotrophic. The lakes in these three categories were subjected to two classification

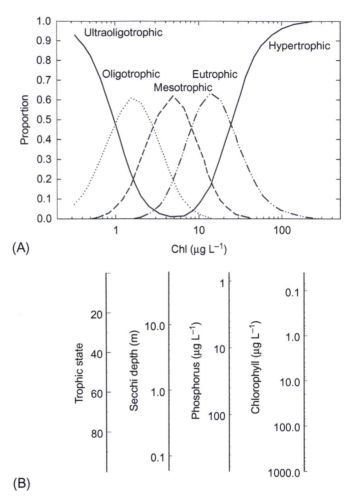

FIGURE 18.2

Two trophic classification systems for lakes. (A) Probability distribution for chlorophyll related to trophic state. The *y* axis is the probability that a lake will have a specific trophic state given a set value of chlorophyll. For example, at 10 μg chl liter⁻¹, the chances are approximately 0% that the lake is ultraoligotrophic, 5% that the lake is oligotrophic, 50% that it is mesotrophic, 45% that it is eutrophic, and 5% that it is hypertrophic (hypereutrophic). *(Adapted from Eutrophication of Waters. Monitoring and Assessment and Control © OECD, 1982).* (B) A logarithmic scale that allows a continuous index to be derived from Secchi depth, total phosphorus, or chlorophyll *a. (Plotted from data of Carlson, 1977).*

approaches: a probability distribution (Fig. 18.2A) and a fixed boundary classification (Table 18.1). There are several fixed boundary classification systems; for the most part, the boundary levels are consistent (Nürnberg, 1996). Another commonly used method of classification involves calculating a trophic index that places trophic state on an exponential scale of Secchi depth, chlorophyll,

Table 18.1 Two Fixed Boundary Trophic Classification Systems for Lakes

Parameter	Ultra-oligotrophic	Oligotrophic	Mesotrophic	Eutrophic	Hypertrophic
OECD (1982)					
Total P (μg L^{-1})	<4	4–10	10–35	35–100	>100
Mean chl (μg L^{-1})	<1	1–2.5	2.5–8	8–25	>25
Maximum chl (μg L^{-1})	<2.5	2.5–8	8–25	25–75	>75
Mean Secchi (m)	>12	12–6	6–3	3–1.5	<1.5
Nürmberg (1996)					
Total P (μg L^{-1})		<10	10–30	30–100	>100
Total N (μg L^{-1})		<350	350–650	650–1,200	>1,200
Mean chl (μg L^{-1})		<3.5	3.5–9	9–25	>25
Mean Secchi (m)		>4	4–2	2–1	<1
O$_2$ depletion rate (mg m^{-2} d^{-1})		<250	250–400	400–550	>550

(*OECD, 1982; **Nürnberg, 1996)

and total phosphorus (Fig. 18.2B), where 10 scale units represent a doubling of algal biomass (Carlson, 1977). This index is useful because it more realistically represents the continuous distribution of trophic state and it offers quick relative comparison among the various factors related to autotrophic state.

Lakes may not clearly fall into an individual category in any of the classification systems. For example, phosphorus could be high enough for a lake to be classified as eutrophic, but light attenuation by suspended sediments could keep chlorophyll levels in the mesotrophic range. Also, total phosphorus and phytoplankton concentrations could be low in a lake with extensive macrophyte biomass and production (Brenner *et al.*, 1999).

Trophic classification of streams can be based on suspended algae, attached algal biomass, or nutrients. Suspended algal mass in streams is usually a function of how much phytoplankton has entered the water column from the stream bottom, except where water flow is slow enough to allow development of a truly planktonic algal assemblage. Trophic classification is difficult because hydrological variation (flooding) and light limitation by riparian canopies translate into high variability in benthic algal biomass over time (weeks or months) and space. Furthermore, many streams have food webs dominated by input from terrestrial organic material, so biomass of primary producers can be a poor indicator of whole-system productivity, and the distinction between autotrophic and heterotrophic state becomes very important.

A trophic classification for temperate streams has been proposed that uses the data available for reference conditions (Table 18.2). This system assigns the

Table 18.2 Autotrophic State Boundaries for Streams

Nutrient	Autotrophic State Boundary	Nutrient Concentration $(mg\,m^{-3})$	% Cases of Mean chl $<100\,mg\,m^{-2}$	% Cases of Maximum chl $<100\,mg\,m^{-2}$
Total N	lower 1/3	330	93	73
Total N	upper 1/3	685	90	71
Total P	lower 1/3	26	95	83
Total P	upper 1/3	60	87	75

(After Dodds, 2006)
The lower third and upper third of the distribution of stream total nitrogen and total phosphorus are pooled across 14 ecoregions. In each of these ecoregions reference values were determined by Smith et al. (2003) and/or Dodds and Oakes (2004). The relationship of the boundary numbers are translated to the percentage of mean or maximum benthic chlorophyll values expected to be less than $100\,mg\,m^{-2}$ when nutrient values were less than the boundary

bottom third of the streams to the oligotrophic category, the middle third to the mesotrophic group, and the streams with the highest chlorophyll or nutrients in the top third to the eutrophic group (Dodds *et al.*, 1998). Again, this is a method to describe a continuum, but it satisfies the convention of assigning systems to one of three trophic categories. The method has not been applied to macrophyte-dominated streams.

Trophic categorizations for groundwater and wetlands have not been developed extensively to our knowledge. Trophic state may correlate to biological characteristics of wetlands. For instance, some oligotrophic, nitrogen-limited wetlands contain carnivorous plants that use captured insects as a nitrogen source. Phosphorus fertilization of the Everglades leads to shifts in plant and periphyton communities. Analysis of published data suggest that N:P ratios alter plant diversity, with greatest wetland plant diversity at intermediate ratios. The same analysis also suggests that at high absolute nutrient concentrations, common weedy species tend to dominate (Güsewell *et al.*, 2005). Thus, it should be possible to create biologically meaningful classifications of wetland trophic state, and nutrients clearly influence biotic integrity of wetlands.

Productivity of groundwaters is generally based on the influx of organic carbon and O_2, so more eutrophic groundwaters could be characterized by high supply rates of organic C, low O_2, and potentially high microbial productivity. The rates of microbial activity may actually decrease in aquifers that are anoxic because anoxic carbon cycling is less efficient than respiration using O_2. Aquifers that are eutrophic because of high organic carbon input could lack a complex invertebrate community due to anoxic conditions, so trophic state may correlate to biological characteristics.

ADVANCED: DETERMINING REFERENCE NUTRIENT CONDITIONS IN FRESHWATER ENVIRONMENTS

The understanding of the natural distribution of nutrient content is of interest for several reasons: (1) These are the conditions under which organisms evolved and could be part of the conditions required for species conservation, (2) regulating nutrients requires understanding what the best possible condition to be expected for an area is, and (3) understanding contributions to downstream loadings requires knowledge of the nonanthropogenic nutrient inputs to a system.

The best way to identify reference conditions is to identify minimally impacted sites (lakes, wetlands, or streams) and measure the nutrient contents in these sites. The sites should be representative of the region and the sampling should be stratified based on the relative occurrence of geology and soil types because nutrient yields from reference watersheds can vary over 100-fold (Smith *et al.*, 2003). An example of this type of stratification of regions is the nutrient ecoregions used by the US Environmental Protection Agency (Omernik, 1995). Unfortunately, there are many areas where all sites are polluted with nutrients. For example, the Corn Belt ecoregion in the midwestern United States and many agricultural areas in Europe have fertilized cropland in almost every watershed. Atmospheric nitrogen deposition influences aquatic ecosystems in all regions inhabited by humans at moderate to high densities.

An alternative method for determining reference nutrient conditions is to use a few small reference watersheds and then modeling methods to scale up to entire watersheds (Smith *et al.*, 2003). These nutrient yield models can account for atmospheric nutrient deposition. These models indicate that current total nitrogen concentrations are more than six times greater than reference conditions, while total phosphorus concentrations have doubled.

Yet another method is to plot human influences on the *x* axis and nutrient concentration on the *y* axis, and use the intercept to extrapolate to reference conditions. In this method, for example, an equation could be obtained such as the following:

$$\log_{10} TP = -0.727 + 0.00668 * \% \text{ cropland} + 0.01465 * \% \text{ urban land}$$

where $\log_{10} TP$ is the log to the base 10 of total phosphorus concentration in mg L^{-1} and % cropland is the percentage of the watershed in croplands. If the % urban land and % cropland is zero, then this should be reflective of lack of human inputs and the reference concentration is simply the constant in the regression ($10^{-0.727}$ results in a total phosphorus concentration of $187 \, \mu\text{g L}^{-1}$

$(0.187 \, \mathrm{mg} \, \mathrm{L}^{-1})$). This method does not account for atmospheric deposition. The method does generally agree with the modeling method (Dodds and Oakes, 2004), and appears to apply well to reservoirs also (Dodds *et al.*, 2006a).

WHY DO ALGAL BLOOMS CAUSED BY NUTRIENT POLLUTION MATTER IN LAKES?

The stimulation of algal blooms and creation of anoxic hypolimnia in lakes leads to many problems. As mentioned in Chapter 1, the monetary value of property on a lake can decrease with eutrophication. Algal blooms are not aesthetically pleasing; they look bad and smell worse. The probability of objectionable algal blooms increases with greater cultural eutrophication (Hart *et al.*, 1999). Water taste and odor problems become more acute as lakes become more eutrophic (Fig. 18.3). Both planktonic and attached cyanobacteria contribute to taste and odor problems (Sugiura *et al.*, 1998). These problems related to eutrophication are difficult to solve with standard water purification methods, leading to greatly increased costs for supplying potable water (Wnorowski, 1992). To make matters worse, algal blooms may be toxic; cyanobacteria and dinoflagellates produce neurotoxins and hepatotoxins (see Sidebar 9.2). Toxin production may be stimulated by phosphorus pollution (Jacoby *et al.*, 2000).

Fish kills related to anoxic events are common symptoms of eutrophication (see Sidebar 12.2). With a series of cloudy days or under an ice cover in a eutrophic lake, fish may die. Cold-water fisheries can be established in deep lakes with cool hypolimnia. If the hypolimnion is anoxic, heat-intolerant fish have no refuge from high temperatures in the epilimnion. High pH associated with algal blooms can also cause fish kills (Kann and Smith, 1999).

Eutrophication can alter biotic integrity. For example, eutrophication can lead to decreases in species richness and diversity of algae, which may have negative consequences for the food web (Proulx *et al.*, 1996). A long-term analysis of Clear Lake, in Iowa, demonstrated that the number of macrophyte species declined from 30 typically found in the 1950s to 12 currently (Egerston *et al.*, 2004). During this time, the lake became more eutrophic going from a Secchi depth of approximately 0.8 to 0.4 m. Even naturally eutrophic lakes have shifts in algal communities associated with agriculture (Räsänen *et al.*, 2006).

Not all the results of eutrophication are bad. Fish biomass may be greatest in eutrophic lakes (Fig. 18.4). Increased productivity can lead to increases in fish productivity, although the fish that dominate in eutrophic lakes may not be desirable species (Lee and Jones, 1991). Some lakes are fertilized artificially to increase fish production (Sidebar 18.1). Eutrophic lakes are also more resistant to the effects of acid precipitation because they are buffered by metabolic activities (Davison *et al.*, 1995) and photosynthesis tends to increase the pH.

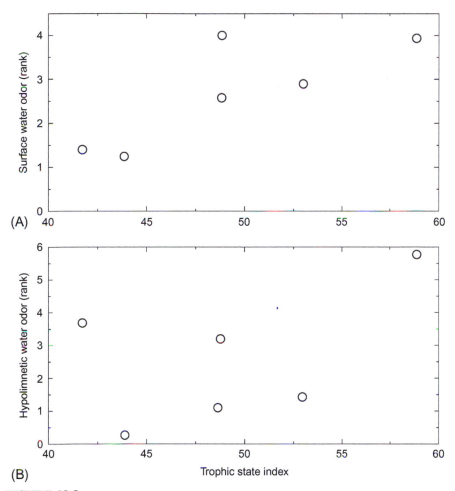

FIGURE 18.3

Relationship of trophic state index (determined by the method of Carlson, 1977; Fig. 18.2B) and water odor of surface (A) and hypolimnetic (B) samples from six Kansas reservoirs. Water odor was ranked by human testers, with a higher rank indicating lower drinking water quality. *(Reproduced with permission from Arruda and Fromm, 1989).*

NATURAL AND CULTURAL PROCESSES OF EUTROPHICATION

The idea that over thousands of years a natural developmental ontogeny of lakes occurs from deep and oligotrophic to shallow and eutrophic, then to a wetland, and then a terrestrial meadow has been present in the ecological literature for decades. The filling of lakes with sediments is a natural process because lakes are depressions in the watershed that collect sediments over time. This idea of a

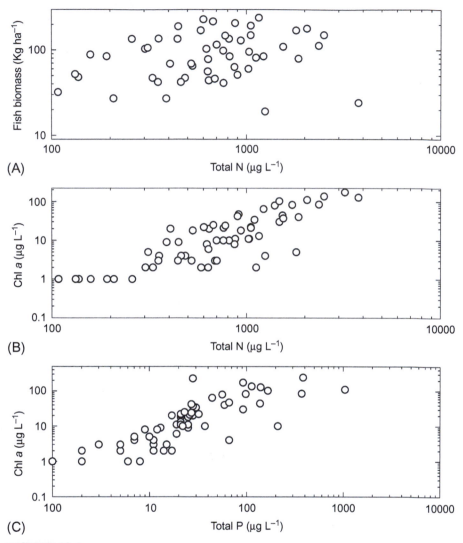

FIGURE 18.4

Relationships of total nitrogen to fish biomass (A) and chlorophyll to total N (B) and total P (C) in 67 Florida lakes. Note that the relationship between fish and nutrients is weaker than that between chlorophyll and nutrients. *(Data from Bachman* et al., *1996).*

succession of lake types is applicable to many small to medium lakes but must be viewed with caution, especially in regard to very deep lakes.

Large tectonic lake basins are generally oligotrophic and likely will remain so for the majority of their histories in the absence of human intervention. For example, Lake Baikal is millions of years old and about 1.5 km deep with up to 7 km of sediment (Fig. 7.5). The time period, if and when the Baikal

SIDEBAR 18.1

Two Examples of Fertilizing Lakes to Increase Fish Production

The following are two examples of situations in which alterations in nutrient regimes led to attempts to fertilize ecosystems to increase fish production. In the first case, lakes on the coast of British Columbia (Vancouver Island) were fertilized to increase survival of young salmon. As human activity decimated natural salmon runs, the numbers of adult salmon returning to spawn in small streams decreased substantially. The adult salmon die after spawning, and as they decay nutrients are remineralized and wash from the streams into the lakes. Historically, coastal lakes in the Pacific Northwest were likely more productive as a result of this fertilization, leading to increased survival and growth of young salmon. Concern over survival of juvenile salmon led to a project of artificial fertilization of 20 lakes over 20 years to determine if increases in survival and growth of juvenile salmon would result (Stockner and MacIsaac, 1996).

The fertilization led to approximate doublings in bacterial abundance, phytoplankton biomass and productivity, and zooplankton biomass. Growth and survival of juvenile sockeye salmon (*Oncorhynchus nerka*) increased more than 60%. The fertilization and associated costs were about $1 million per year, and calculated benefits in increased returns of adult sockeye were about $12 million per year (Stockner and MacIsaac, 1996). These estimates suggest that fertilizing nutrient-poor coastal lakes to levels similar to those thought to occur historically, or to those in pristine ecosystems, is economically feasible. If salmon runs ever recover to near historic levels, no fertilization will be required because the spent adult spawning salmon that acquired their nutrients in the ocean would die and release the nutrients into the system.

A fertilization project in the southwest United States was less successful. Lake Mead is a large reservoir in Nevada and Arizona that supports a sports fishery valued at approximately $7 million per year. Fish production in Lake Mead has decreased; largemouth bass (*Micropterus salmoides*) harvest has declined more than 90% since the 1960s. It was hypothesized that the closure of Glen Canyon Dam in 1963 lowered nutrient input into Lake Mead and led to declines in fish production. A large-scale fertilization experiment was initiated over a four-year period to assess the effect of increased nutrient input on fish production and water quality. This experiment resulted in a moderate decrease in water quality (increased taste and odor problems and chlorophyll *a*). However, increases in zooplankton and forage fish were not significant (Vaux *et al.*, 1995). In this case, more fertilization might be necessary to stimulate fish production, but could cause degradation of the quality of the drinking water from Lake Mead. Furthermore, there is no guarantee that more fertilization would lead to increased fish production.

---

basin fills and becomes a shallow productive lake, would likely be very short relative to the entire geological life span. In long-lived lakes, long-term changes related to geological processes (e.g., deforestation related to glaciation) may lead to periods when lakes are mesotrophic or eutrophic and others when they are oligotrophic.

Paleolimnological methods using isotopic dating and preserved remains of algae in the sediments (primarily diatoms) can be useful for estimating the history of a lake's trophic state (Anderson, 1993). Such methods often reveal that lakes thought to be naturally eutrophic were more oligotrophic thousands of years ago (Anderson, 1995). For example, analyses of the sediments of Lake Okeechobee, Florida, used lead isotopes to establish age of sediments back to the late 1800s. The dated sediments indicated that the phosphorus increased in the lake along with algal pigments (trophic state), with particular increases from the 1950s that correspond to the expansion of crop fertilization in the watershed (Engstrom *et al.*, 2006).

Natural eutrophication can occur with watershed disturbances. In an interesting case, Spirit Lake was altered greatly following the volcanic eruption of Mount St. Helens. The eruption occurred on May 18, 1980, and was the equivalent of a 10-megaton nuclear explosion leading to massive input of downed timber, volcanic ash, and an abrupt temperature increase from 10 to 30°C. Spirit Lake was deep and oligotrophic before the blast. The eruption altered the lake to a shallower basin with a large surface area, ultimately leading to increased macrophyte growth and production (Larson, 1993). Such rapid and drastic change is rare in most natural lakes on human timescales of observation.

Cultural eutrophication is common in the United States and other countries in which there are moderate to high densities of human activity. Cultural eutrophication occurs rapidly (relative to most geological processes) and can be difficult to reverse. Human activities that lead to cultural eutrophication include use of agricultural fertilizers, livestock practices, watershed disturbance such as deforestation, and release of nutrient-rich sewage into surface waters (Loehr, 1974). Road building also leads to increased erosion and infilling of lakes. Historical examples of eutrophication caused by watershed disturbance include road construction of the Via Cassia by the Romans (Sidebar 18.2) and eutrophication caused by agriculture in early Mexico (O'Hara *et al.* 1993).

SIDEBAR 18.2
Lago Di Monterosi: Anthropogenic Eutrophication, BC

In the 1960s, G. E. Hutchinson assembled a group of scientists to study the history of the Italian Lago Di Monterosi (Hutchinson, 1970; Hutchinson and Cowgill, 1970). The scientists included paleontologists who worked on sediment cores from the lake, historians, a geologist, and limnologists. The group members were from Italy, the United States, and Britain.

Lago Di Monterosi was formed by a volcanic blast about 35,000 years ago. It has remained shallower than 10 m since that time, and it approached a depth of 1 m during a very dry period about 10,000 years ago. Analyses of pollen and preserved remains of aquatic plants, animals, and microalgae suggest that the lake remained moderately productive until about 2,000 years ago. During approximately the first 30,300 years, the lake slowly evolved into a shallow oligotrophic lake, became slightly acidic, and contained some *Sphagnum*, indicating peat bog formation, at least along parts of the shores.

Historical records and archeological remains suggest that the Roman Empire built the Via Cassia by 171 BC. The paved road was probably built to improve speed of transit from Rome to and from the strategically important Tuscany. The road passed through the edge of the lake watershed (Ward-Perkins, 1970), which resulted in settlement and deforestation in the watershed.

Analyses of lake sediments dated with ^{14}C to the time when Via Cassia was built reveal a marked increase in the rate of sedimentation, a decrease in the amount of tree pollen, a decrease in the amount of aquatic plant pollen, and increased carbon and nutrient content of the sediments. These and other characteristics are consistent with deforestation of the watershed, increased sedimentation, more nutrient input associated with increased runoff, and greater productivity of the lake. This eutrophic state abated somewhat after the fall of the Roman Empire, and the lake has maintained a moderately eutrophic state since that time.

This study has several important implications. It demonstrates that humans have a long history of causing eutrophication and impacting habitats on a watershed scale. It also demonstrates that lakes do not necessarily undergo a constant succession from oligotrophy to eutrophy over geological time. Finally, this is an early example of study of a limnological problem that was best accomplished by assembling a team of specialists. It illustrates that limnology is a holistic subject, and that observations from both "hard" and "social" sciences can be used to study ecologically and environmentally relevant questions. This groundbreaking approach to ecological issues has now been adopted many places to allow more effective study of environmental issues than would be possible with just one investigator.

Eutrophication control can be costly; thus, political battles over the relative importance of phosphorus control to solve eutrophication problems caused by humans can be intense (Edmundson, 1991). Perhaps the most important scientific verification of the role of phosphorus in eutrophication was the work headed by David Schindler (Biography 18.1) at the Experimental Lakes Area in Canada. These whole-lake experiments demonstrated that phosphorus additions, and not organic carbon additions, are clearly responsible for nuisance algal blooms in lakes (Fig. 18.1).

RELATIONSHIPS AMONG NUTRIENTS, WATER CLARITY, AND PHYTOPLANKTON: MANAGING EUTROPHICATION IN LAKES

The relationships among nutrient loading, algal biomass, and lake clarity were documented clearly by Vollenweider (1976). This correlation of nutrients with lake autotrophic state represented a milestone in lake management because it allowed managers to predict the outcome of nutrient control strategies. The watershed forms the natural unit for this approach to nutrient management (Likens, 2001). The models are based on empirical data relating watershed loading to in-lake nutrients and nutrients to algal biomass (Figs. 18.4 and 18.6) and provide a conceptual framework that links nutrient supply to lakes with phytoplankton biomass and water clarity (Fig. 18.7).

BIOGRAPHY 18.1 DAVID SCHINDLER

David Schindler (Fig. 18.5) is one of the most influential scientists who have studied human-caused pollution in aquatic systems. Although he is best known for his eutrophication work at the Experimental Lakes Area in Canada, he has conducted important research on basic ecosystem processes, acid precipitation, organic carbon contamination, and the influence of global change on aquatic ecosystems. Currently, he is concerned about the cumulative effects of anthropogenic inputs (Schindler, 2001).

Dr. Schindler always loved lakes and ponds, but he entered the aquatic sciences by accident. He began college as a physics major but was hired as a technician in a limnological laboratory. After reading some books off the shelf there, he was hooked. For him, the three most influential books were Hutchinson's *Treatise of Limnology*, Vol. 1 (1957), Elton's book on animal invasions (1958), and Tinbergen's book on animal behavior (1951). Schindler has more than 200 publications, many in the top scientific journals. He has honorary PhD degrees from several universities and has won major awards, including the Stockholm Water prize, the G. E. Hutchinson Award of the American Society of Limnology and Oceanography, the Naumann-Thienemann Medal of the International Limnological Society, the Tyler Prize, and the Volvo Environment prize. Schindler is involved in many national and international committees and panels related to human impacts on aquatic systems.

Schindler is proud that all of his three children chose careers involving aquatic systems. His family also joins him in competitive dogsled racing, his favorite hobby and sport.

FIGURE 18.5
David Schindler.

Schindler suggests that all undergraduates work on writing skills because communicating and publishing scientific discoveries are crucial to a successful scientific career. He sees resurgence in research on eutrophication, particularly on problems related to nonpoint sources of nitrogen and phosphorus. In 1998, for the first edition of this book, Schindler predicted there would be a realization of the problems associated with mercury and organic contaminants. Since that time, mercury contamination has been widely recognized, and problems associated with organic pollutants, including ecoestrogens, have become a matter of increased concern.

A simplified view of the sequence of events that can occur to mitigate eutrophication includes (1) identifying a lake with problems, including determination of uses that interfere with the desirable condition of the lake; (2) characterization of the system, including lake morphology, land use in the watershed, nutrient loading into the lake, lake water retention, and sedimentation rates; (3) identification of feasible strategies for nutrient control considering both point and nonpoint sources; (4) projecting the influence of management actions on nutrient concentrations in the lake; (5) predicting the response of chlorophyll to lower nutrient concentrations in the lake; (6) assessing the potential effect of

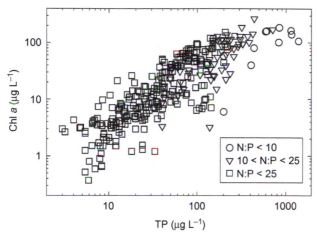

FIGURE 18.6

Relationship between mean growing season concentrations of total phosphorus (*TP*) and chlorophyll in 228 temperate lakes coded by N:P ratios. (*Corrected data plotted following Smith, 1982*).

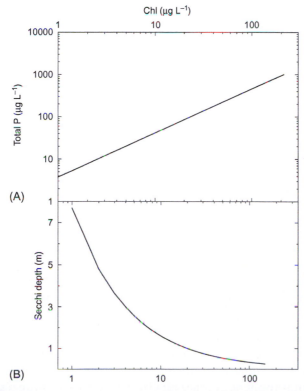

FIGURE 18.7

A nomogram relating epilimnetic chlorophyll concentration to total phosphorus (A) and Secchi depth (B). (*Based on equations in OECD, 1982*). This graph can be used to estimate changes in clarity related to a known change in total phosphorus.

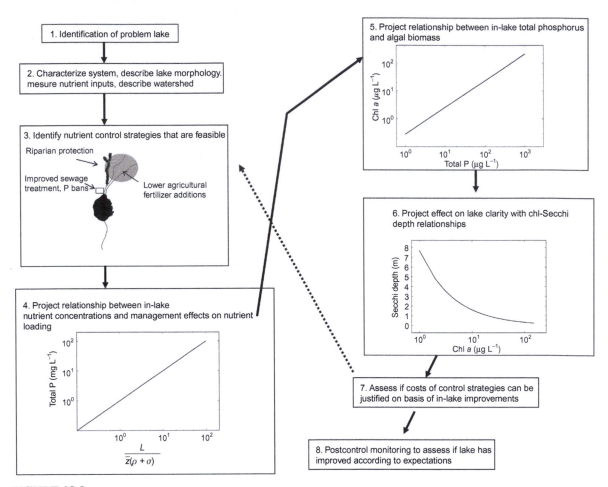

FIGURE 18.8
A simplified diagram of the steps that can be used to modify eutrophication in a lake.

decreased chlorophyll on lake clarity; (7) assessing if the projected costs of the nutrient control strategies justify the predicted benefits to the lake (O'Riordan, 1999); and (8) if nutrient control or mitigation strategies are instigated, monitoring the system to determine if the sought-after improvements have actually occurred (Fig. 18.8).

Lowering *external loading* (supply of nutrients from outside the system) is generally necessary to control eutrophication. The lowering of external loading usually incurs some cost; thus, lake managers may need to estimate the amount of improvement in water quality that will result from a set amount of nutrient control. This estimation involves calculation of in-lake phosphorus concentrations and subsequent algal biomass.

ADVANCED: EMPIRICAL RELATIONSHIPS USED TO PREDICT CONTROL OF EUTROPHICATION

Equations are available that can be used to determine the influence of altered nutrient regimes on productivity for a variety of lake types. We provide a very general description of such equations as an introduction to the method. A more detailed explanation can be found in Cooke *et al.* (1993) and Ryding and Rast (1989). Equations for phosphorus are most commonly used because it is generally assumed that phosphorus limits primary production in lakes (Schindler, 1974; Correll, 1999). However, some have suggested that nitrogen is more likely limiting in tropical lakes (Golterman and de Oude, 1991), and bioassays (see Chapter 17) indicate that colimitation by nitrogen and phosphorus is likely in many lakes.

The following is a general steady-state equation to calculate in-lake total phosphorus concentration:

$$TP = \frac{L}{\overline{z}(\rho + \sigma)}$$

where TP is the total phosphorus in mg m^{-3} (μg L^{-1}), L is the phosphorus loading in mg m^{-2}year^{-1}, z is the mean depth in meters, ρ is the flushing rate in year^{-1}, and σ is the sedimentation rate in year^{-1}, approximately equal to $10/z$.

The general steady-state equation represents one of the simplest cases. It accounts for sources and losses of phosphorus in the lake. A similar equation can also be used to estimate total nitrogen (TN). The source is loading from rivers, groundwater, and atmosphere. Losses are from washout (flushing and sedimentation). Assumptions include steady-state phosphorus concentration, complete mixing of inputs, constant sedimentation, little fluctuation of loading over time, and limited phosphorus input from sediments (*internal loading*). More complex relationships are available to deal with exceptions to most of these assumptions (Cooke *et al.*, 1993).

In practice, L is determined by measurements of total phosphorus in inflowing streams; atmospheric deposition and groundwater inputs are generally ignored. Nutrient input into streams is often heavily dependent on land-use patterns, which will be discussed in the next section. Groundwater input may be difficult to determine, particularly in heterogeneous geological substrata or where septic inflows create areas with exceptionally high phosphorus influx. Determination of mean depth and calculation of flushing rate require morphological mapping of the lake basin and hydrological measurements.

Sedimentation rate can be highly variable among and within lakes, depending on characteristics such as fetch, epilimnion depth, and type and phosphorus content of sediment (i.e., considerable variance occurs in the relationship

$\sigma = 10/\bar{z}$). Direct determination of sedimentation rates may provide more accurate estimates of phosphorus loss from the epilimnion.

Once the TP in the lake is calculated, the next step is to calculate the chlorophyll that can be supported by this amount of nutrient. A clear relationship exists between total phosphorus and chlorophyll (Figs. 18.6 and 18.7) when values for many lakes are plotted. Equations can be derived from such data sets; the following has been proposed by Jones and Bachmann (1976) using data from 143 lakes:

$$\log_{10} \text{chla} = 1.46 \log_{10} TP - 1.09, \ r^2 = 0.90$$

where chla is the summer mean chlorophyll in mg m^3 ($\mu\text{g L}^{-1}$), TP is the summer mean total phosphorus in mg m^{-3} ($\mu\text{g L}^{-1}$), and r^2 is the proportion of the variance that can be described by the relationship. Use of this and the preceding equation is demonstrated in Example 18.1.

■ Example 18.1

Using Loading Equations to Predict Response to Nutrient Control

If a lake manager is able to lower mean total summer phosphorus inputs to a lake by 50% from an initial loading value of $1 \text{g m}^{-2}\text{y}^{-1}$ total P, what will be the expected decrease in chlorophyll given a mean depth of 10 m, a flushing rate (ρ) of 2y^{-1}, and a sedimentation (σ) rate of 1y^{-1}? How does this translate into increased Secchi depth?

First, we need to solve for initial chlorophyll (in practice, this will probably be a measured value). To do so, we solve for TP concentration first, then use the equation relating TP to chlorophyll. To calculate TP, do not forget to convert phosphorus loading (L) into $\text{mg m}^{-2}\text{y}^{-1}$ and that $\text{mg m}^{-3} = \mu\text{g L}^{-1}$:

$$TP = \frac{1000}{10(2 + 1)} = 33.3 \mu\text{g L}^{-1} \text{P}$$

If we rearrange the equation relating chlorophyll to TP, we get:

$$\text{chl} = 10^{1.46 \log_{10} TP - 1.09} = 13.6 \mu\text{g chl L}^{-1}$$

A 50% decrease in loading will reduce the TP by half to $16.7 \mu\text{g L}^{-1}$. Using the second equation, this concentration will yield $4.9 \mu\text{g L}^{-1}$ chl. Inspection of Figure 18.7, which was constructed using slightly different equations, allows the calculated relationships to be checked. If we use the nomogram on Figure 18.7, we can also see that the Secchi depth is expected to increase from approximately 1.5 m to 2.7 m. An additional issue is variance; it is beyond the scope of this discussion, but a significant amount of variance occurs in the empirical relationships and this source of uncertainty must be considered in an actual management situation.   ∎

Smith (1982) used a larger and more variable data set and demonstrated that more variance can be accounted for in his data set if total nitrogen is considered in addition to total phosphorus. If a plot of these chlorophyll values versus total phosphorus is divided into categories of *TN:TP* ratios (Fig. 18.6), it shows that chlorophyll per unit phosphorus is lower when the relative amount of nitrogen is low. An equation relating algal biomass to chlorophyll using nitrogen and phosphorus has been proposed by Smith (1982; corrected equation, $n = 311$ lakes):

$$\log_{10} \text{chla} = 0.640 \log_{10} TP + 0.587 \log_{10} TN - 0.753, \ r^2 = 0.75$$

Units and variables are the same as in the previous equation. This equation is probably most useful in high phosphorus waters (Cooke *et al.*, 1993) and may not apply to tropical lakes (Sarnelle *et al.*, 1998). Tropical lakes also generally have greater variance in the chlorophyll-total phosphorus relationships than do temperate lakes where most research has been done (Huszar *et al.*, 2006).

The probability that an algal bloom will occur, particularly a bloom of cyanobacteria, may be more important than average chlorophyll values. A 21-year data set on phosphorus loading to Lake Mendota, Wisconsin, was used to evaluate the probability of algal blooms (Lathrop *et al.*, 1998). In this analysis, with no change in current levels of loading, there was a 60% chance of a cyanobacterial bloom on any given summer day. When loading was decreased by half, there was only a 20% chance of a bloom. This study illustrates that managers deal with variable and unpredictable systems, and that nutrient control methods may decrease the probability of a noxious bloom but not preclude the possibility. The probability of a cyanobacterial bloom in a lake increases sharply above about $30 \ \mu\text{g} \ \text{L}^{-1}$ total phosphorus (Downing *et al.*, 2001).

MITIGATING LAKE EUTROPHICATION

Eutrophication management can begin with control at the nutrient source (treating the cause) or with in-lake treatment (treating the symptom; Table 18.3). Treating causes is almost always more effective in stopping negative effects of pollutants than treatment after contamination. Just as with other

Table 18.3 Methods for Controlling Causes and Symptoms of Eutrophication Associated with Excessive Phytoplankton

Method	Explanation	Positive Aspects	Negative Aspects
Control of cause			
Control of point sources	Bans on phosphorus in detergents, tertiary sewage treatment.	Clean up at source.	Tertiary treatment can be expensive. Generally ineffective in lakes where hypolimnion goes anoxic.
Control of non-point sources	Control of watershed disturbance, feedlot effluent, intact riparian zones.	Clean up at source, potential for long-term improvement.	Can be politically unpopular. Generally ineffective in lakes where hypolimnion goes anoxic.
In-lake control of symptoms			
Dilution and flushing	Use of low nutrient water to dilute nutrients and phytoplankton.	Where practical can be an easy, inexpensive solution.	Requires large supply of low-nutrient water. Usually only practical in smaller lakes.
Destratification, mixing	Keeping O_2 in the hypolimnion keeps phosphorus in sediments. Deeper mixing increases light limitation of phytoplankton.	Rapid results.	Energy required to mix and destratify lakes. Not practical on large lakes. Can select against desirable cold water fishes.
Hypolimnetic release	Release nutrient-rich water from hypolimnion.	Easy to implement in reservoirs with possibility of hypolimnetic release.	Costly to pump water, greater nutrient input to downstream systems.
Biomanipulation	Manipulate foodweb by increasing piscivores or decreasing planktivores to increase numbers of zooplankton that graze phytoplankton.	Can lead to rapid increases in water quality with minimal costs.	Unpredictable results, does not work on extremely eutrophic systems. May lead to excessive macrophyte growth in shallow systems.
Alum	Alum seals phosphate in hypolimnion, flocculates and settles phytoplankton.	Rapid response.	May need repeated application, may be cost-prohibitive in large lakes.
Copper treatment	Copper kills phytoplankton.	Acts within days.	Repeated treatment necessary, can lead to sediment contamination and negative effects on nontarget species.

(After Cooke et al., 1993)

pollutants, treating the cause of eutrophication by controlling nutrient sources is generally most cost-effective over the long term. Nutrients can come from *point sources*, such as sewage outfalls, factory effluents, septic tanks, and waste flowing from the surface of intensive livestock operations.

Nutrients can also come from *nonpoint sources*, such as agricultural fields, urban storm runoff systems (lawn fertilizers), disturbance of watersheds, addition of fertilizers to golf courses and pastures, and atmospheric deposition. Even heavy use by waterfowl that feed outside the aquatic habitat and return to the water and excrete can lead to substantial nutrient loading (Olson *et al.*, 2005). Nutrient input from point sources is relatively easy to determine and well characterized because it is concentrated and sampling is easy.

Establishing rates of input from nonpoint sources is more difficult, but broad ranges of nutrient loss rates from different types of land uses are known (Fig. 18.9). Agricultural and urban uses lead to the greatest degree of runoff, with human population density in a watershed demonstrating a significant positive correlation to nitrogen and phosphorus runoff (Caraco and Cole, 1999). A 50% increase in agricultural and urban land use can result in a doubling of total nitrogen runoff (Fig. 18.10). Thus, the human uses in the watershed above the water body of interest need to be characterized to estimate the approximate impact of different land uses on nutrients flowing into the system.

Identification of land uses that can increase nutrient inputs has become substantially easier in the past few decades because of advances in information availability and mapping capabilities. A *geographic information system* (GIS) is a powerful tool for mapping land-use patterns and effects of changes in those patterns as related to eutrophication (Hunsaker and Levine, 1995). GIS systems consist of a series of map layers, in which different attributes can be assigned to a spatial grid. The layers can represent different attributes or the changes in an individual attribute over time. Data layers that are currently available include land use, drainage networks, soil types, and elevations. These data layers can often be combined with the extensive data sets on nutrients that many US states and national governments now make available online.

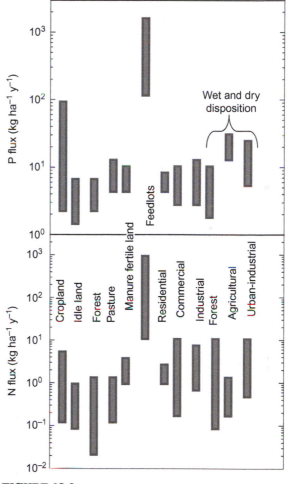

FIGURE 18.9

Ranges of phosphorus and nitrogen fluxes from different land-use categories and the rates of nitrogen and phosphorus loading. The last three bars on the right are rates of loading from the sum of wet atmospheric deposition (in precipitation) and dry deposition (in dust) in areas that are forested, agricultural, or urban. *(Adapted from Loehr et al., 1989).*

Control of Nutrient Sources

Control of nonpoint sources is often difficult because it requires coordination across watersheds and cooperation of many different types of landowners. Patience is also required when enacting steps to control nutrient loading

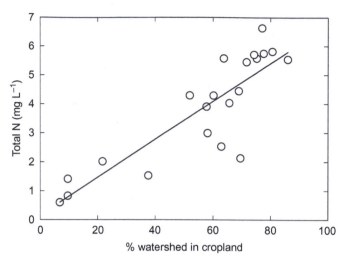

FIGURE 18.10

Relationship between total nitrogen concentration in streams and the percentage of land in agricultural and urban use from a variety of watersheds in Kansas. (*Data courtesy of the Kansas Department of Health and Environment*).

because recovery generally takes a decade or longer to achieve after nutrients are controlled (Bennett *et al.*, 1999; Jeppesen *et al.*, 2005). Generally, agricultural land use is the most important source of nonpoint source pollution (Kronvang *et al.*, 1995). Typical ways to control nonpoint nutrient input include lowering fertilizer applications (to lawns in urban regions and to crops in agricultural regions), proper timing of application, establishing erosion-control strategies (e.g., maintaining riparian vegetation and minimizing exposed soil), keeping livestock out of streams and ponds with fences and by providing stock tanks, and restoration of natural vegetation. Regulation and education are necessary components to nutrient control programs. Given the variety of sources for nonpoint source pollution and the complexity of ecosystem valuation, economic analysis of the costs and benefits of nutrient control can be difficult (Carpenter *et al.*, 1998).

Control of nonpoint sources can have benefits beyond the local effects of lowering lake, stream, and groundwater eutrophication. Since the 1990s, a large anoxic zone has increased in size in the Gulf of Mexico, and the reduced O_2 has damaged fisheries in the region. This anoxic zone likely is caused by riverborne nutrients (Turner and Rabalais, 1994) and continues to increase in size. Similar zones have developed in more than 400 near coastal marine areas (Diaz and Rosenberg, 2008).

An additional effect of eutrophication in streams can be related to relatively lower silicon availability. There is more nitrogen and phosphorus in rivers worldwide but less increase in silicon. The natural weathering rates of silicon remain fairly constant and humans have had little influence on these rates. Nitrogen and phosphorus can stimulate diatom growth leading to uptake and burial of silicon. This shift in stoichiometry and nutrient amount has led to increases in algal productivity and shifts in composition of near-shore marine plankton communities (Justic et al., 1995a). Thus, increased nonpoint source nutrient pollution has increased eutrophication in marine coastal regions throughout the world (Justic et al., 1995b).

One of the first steps toward lowering phosphorus input into watersheds from point sources is generally a ban on phosphate-containing detergents. This restriction can cut in half the phosphate entering sewage works. In these situations, detergents for automatic dishwashers and automatic car washes are generally exempt because of the reduced efficacy of low-phosphate alternatives.

Control of point sources generally puts the majority of the financial burden on fewer institutions (e.g., a municipal sewage treatment plant or a specific factory) than does the control of nonpoint sources. However, regulation of point sources is easier from the regulator's point of view and often laws are written specifically to control point sources. Removing phosphorus from waste streams can be costly. One method of removal involves chemical treatment with alum or Fe^{3+} to precipitate the phosphate. The precipitate is then allowed to settle and the low-phosphorus water is released. This method generally can bring effluent concentrations down to 0.2 to 1 mg phosphorus liter^{-1} (Clasen et al., 1989).

Nitrogen can be removed by converting ammonium to ammonia gas by raising the pH. The solution is stripped of ammonia by bubbling with air, and then the water is neutralized and released. Alternatively, waste can go through an aerobic treatment to convert the nitrogen to nitrate, followed by an anoxic phase in which nitrate is used in denitrification. The resulting N_2 gas enters the atmosphere. Finally, wetlands can be used to remove nitrogen, as discussed later in this chapter.

Removing nitrogen and leaving phosphorus in a system may not solve eutrophication problems because many species of cyanobacteria that form undesirable blooms can use N_2 gas via fixation and do not need nitrate or ammonium to bloom. In the Experimental Lakes Area in Canada phosphorus addition without concurrent nitrogen addition led to large blooms of the nitrogen-fixing cyanobacterium *Aphanizomonon*. This effect continued for several years until nitrogen fixation brought the lake's nitrogen content to match the phosphorus addition rates (Hecky et al., 1994). Even if total biomass is reduced somewhat, undesirable species may be selected for when only nitrogen is removed. Because these cyanobacteria have gas vesicles and are buoyant, they concentrate on the surface. They are more apparent as "scum" and visually indicate the high level of nutrients in the lake, especially phosphorus.

This example illustrates the importance of limiting factors and indicates how phosphorus limitation can be important in regulating algal communities.

Treatment in the Lake

When a lake becomes eutrophic, several methods can be used to treat the symptoms. Treatment becomes more difficult when O_2 has disappeared from the hypolimnion because phosphate that would bind with Fe^{3+} in an oxic hypolimnion is released from the sediments, drastically increasing rates of internal loading. The $FePO_4$ locked in sediments dissociates into Fe^{2+} and PO_4^{3-} in the anoxic hypolimnion and diffuses into the water column. The phosphate released from the sediments becomes available to phytoplankton when the lake mixes. The problem is most severe in medium-depth lakes and more severe under warmer conditions (Genkai-Kato and Carpenter, 2005). Phytoplankton use luxury uptake to acquire and store this phosphate, which provides nutrients for future blooms. Several strategies have been devised to combat this resuspension of phosphate. These strategies are discussed in detail elsewhere (Cooke *et al.*, 1993) and summarized here and in Table 18.3.

One method to counteract the symptoms of eutrophication is to provide O_2 to the hypolimnion so phosphate remains in the sediments. This *hypolimnetic aeration* requires large amounts of energy; thus, it can be prohibitively expensive in any but the smallest of lakes. If the main goal is to protect a cold-water fishery, then only a small part of the hypolimnion needs to be oxygenated and care must be taken to not break stratification. This approach provides low-temperature, oxygenated water as a refuge for salmonid species.

Aeration does not always lower algal biomass (Soltero *et al.*, 1994). However, aeration for many consecutive years has been used successfully to mitigate water quality problems in shallow urban lakes (Lindenschmidt and Hamblin, 1997). Destratification can also keep phosphate in the sediments and can have an inhibitory effect on nuisance cyanobacteria by increasing mixing depth. This mixing can select for green algae and diatoms instead of the toxic cyanobacterium *Microcystis* (Visser *et al.*, 1996).

The use of copper as a method to control algae has been widespread. Copper is particularly toxic to cyanobacteria and thus removes the objectionable algae. In hard waters copper can precipitate as copper carbonate so that repeated applications are necessary to achieve results. The copper contaminates the sediments and can eventually poison other aquatic life, such as crustaceans, if pH is acidic (pH is commonly acidic in anaerobic sediments). Furthermore, the copper may break the cells of cyanobacteria and release toxins into the water (Lam *et al.*, 1995). Extended treatment with copper may become more problematic than the condition it was supposed to cure (Cooke *et al.*, 1993).

Another method of control is to release water from the hypolimnion. This is most easily accomplished in reservoirs that have hypolimnetic release tubes, as lakes and reservoirs with surface release require pumping to remove the water from depths. The nutrient-rich water that builds up in the hypolimnion is removed in this method. This method, if applied correctly (not leading to destratification of lakes), can improve water quality (Nürnberg *et al.*, 2007).

Addition of barley straw has been proposed as a way to control phytoplankton blooms. Some studies of this method have shown measurable lowering of algal biomass (Barrett *et al.*, 1996; Everall and Lees, 1996; Ridge *et al.*, 1999), and dissolved phosphorus (Garbet, 2005). Repeated treatment with barley straw was demonstrated to be effective in lowering cyanobacterial populations and decreasing taste and odor problems in one drinking water supply reservoir (Barrett *et al.*, 1999). The mechanism for this control is not well established but uptake and storage of nutrients by microbes growing on the straw are possible explanations. Finally, alterations of the food web (top-down control) have been advocated to control algal blooms. This method will be discussed thoroughly in Chapter 20.

Macrophyte Removal

One symptom of eutrophication in shallow lakes is excessive growth of macrophyte vegetation (Chambers *et al.*, 1999). Some macrophyte growth is a healthy part of aquatic ecosystems; the plants provide habitat for other desirable species (e.g., fishes) and stands of macrophytes can prevent unwanted sediment suspension (Bachmann *et al.*, 1999). However, macrophytes can interfere with recreation, clog water flow structures, lead to low O_2 conditions, and cause taste and odor problems. Thus, removal of macrophytes is desirable at times. Physical, chemical, or biological controls can remove macrophytes (Table 18.4).

Physical control methods include direct harvesting, sealing aquatic sediments with plastic to prevent establishment of rooted macrophytes, shading, and alternation of water level (dry down; Wade, 1990). The lack of ability to withstand freezing is used to control macrophytes in some temperate zone reservoirs. Water levels can be drawn down during freezing weather, killing some species (Murphy and Pieterse, 1990).

Chemical control methods require application of herbicide. Preferable herbicide properties include a limited lifetime in the water, toxicity primarily to target plants, and no bioconcentration in the food web. In general, physical methods are more expensive than chemical methods (Murphy and Barrett, 1990). Both methods often require repeated use because the macrophytes can recolonize.

Biological control methods include use of fungi, insects, and herbivorous fish. Selective organisms are the most desirable agents of biological control. Nonselective control agents can harm beneficial species and can be extremely difficult to eradicate after they are introduced. Some problems associated with unwanted species introductions are described in Chapter 11.

A common biological control method is the introduction of the herbivorous grass carp (*Ctenopharyngodon idella*). These fish are nonselective and can remove a large amount of beneficial and unwanted aquatic vegetation. The grass carp are used in several areas of the world as a source of protein. A concern regarding grass carp is that they will escape the region where they are introduced and remove desirable macrophytes elsewhere. Triploid grass carp that are not capable of reproduction are available for use in control programs. In addition, spawning requires a temperature of at least 17°C, so reproduction is unlikely in some cool-temperate habitats (van der Zweerde, 1990). Use of grass carp is a relatively inexpensive control method.

In general, there is a balance between too many macrophytes and complete removal. Macrophytes form a refuge for larval fish and can stimulate a productive fishery. The macrophytes also tend to lower algal blooms and lead to clearer water; removal of macrophytes can result in undesirable algal blooms (Genkai-Kato and Carpenter, 2005). For example, decomposing stalks of the reed *Phragmites* can inhibit cyanobacterial growth (Nakai *et al.*, 2006). Excess addition of grass carp commonly results in complete removal of macrophytes, increases in suspended sediment, and greater probability of phytoplankton blooms.

Table 18.4 Methods for Controlling Macrophytes

Method	Explanation	Positive Aspects	Negative Aspects
Physical	Dredge or cut out macrophytes.	Removal rapid; high control on area treated	Costly; ineffective in large systems; needs to be repeated.
Chemical	Apply herbicides.	Removal rapid; moderate control on area treated	May affect nontarget species; needs repeated application; biomagnification of toxins.
Sealing sediment	Seal sediments with plastic film to stop macrophyte establishment.	Also keeps phosphate from reentering lake, thus inhibiting phytoplankton blooms	Expensive, not effective in large systems; plastic lake bottom may be aesthetically unpleasant.
Biological control	Find organisms that specifically graze macrophytes.	Inexpensive, lasting control	May eat desirable species; may become pest species; may be impossible to eradicate after introduction.

(After Cooke et al., 1993)
Whole-lake control of nutrients (Table 18.3) can be useful in controlling excessive macrophytes

MANAGING EUTROPHICATION IN STREAMS AND RIVERS

The relationship in streams between benthic algal biomass and water column nutrients is not as strong as that between nutrients and phytoplankton in lakes (Fig. 18.11). This lack of predictability occurs because floods, grazing, and light limitation can lower algal biomass even when nutrients are high. Eutrophication in streams has become a serious issue, and control of nutrients is likely to be the best solution to the problem (Dodds *et al.*, 1997). Problems with eutrophication in streams include the negative aesthetic impact of excessive algal growth, alteration of food webs, taste and odor problems, low O_2, high pH (Dodds and Welch, 2000), impediment to channel flows (Ferreira *et al.*, 1999), and shifts in macrophyte community structure (Bowden *et al.*, 1994). In addition, there has been increased concern about eutrophication in estuaries and near coastal marine environments caused by nutrients delivered by streams (Turner and Rabalais, 1994; Conley, 2000).

Streams, even more than lakes, can be driven by heterotrophic processes and these processes can be altered by nutrient addition. Thus, control of nutrients in streams is important regardless of the natural condition of light and periphyton dominance (Dodds, 2007).

Historically, one of the worst problems with sewage input to streams and rivers is the loss of O_2 and the introduction of pathogenic bacteria (Fig. 18.12A). This sewage input represented an artificial increase in heterotrophic state.

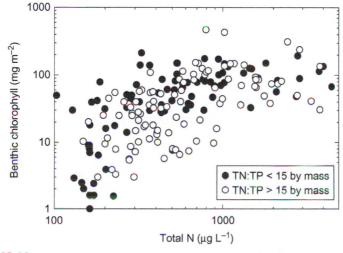

FIGURE 18.11

Relationship between total nitrogen concentration in the water column and mean benthic chlorophyll from about 200 temperate streams coded by N:P ratios. *(Data from Dodds et al., 2006b).*

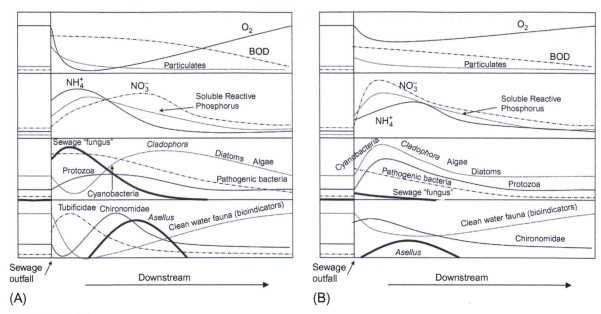

FIGURE 18.12

Chemical and biological parameters as a function of distance downstream from untreated sewage effluent. (*A, Redrawn from Hynes, 1960*, and from a modern sewage treatment plant (B)).

Current sewage treatment does not lower total nutrient input much but does lower loads of biochemical oxygen demand and pathogenic bacteria (primary and secondary treatment are common, but tertiary treatment to remove nitrogen and phosphorus are rarer). Older sewage treatment plants released significant amounts of ammonium, which is toxic to aquatic life. More advanced sewage plants oxygenate treated water, and most ammonium is converted to nitrate by nitrification, leading to different chemical and biological patterns downstream from the sewage outfall (Fig. 18.12B). The most advanced sewage plants remove nitrogen by stripping ammonium or denitrification.

CASE STUDIES OF EUTROPHICATION IN LAKES AND LOTIC SYSTEMS

Examples of pollution control can aid understanding of general ecological concepts and their application to solving eutrophication problems. This section outlines some successes and failures of lake managers that illustrate the complexity of the issues involved and the variety of problems that have arisen.

Lake Washington

The case of Lake Washington is one of the great triumphs for limnologists, lake managers, and environmentalists. This lake has experienced a strong recovery based on scientific understanding of limnological properties and particularly the efforts of the late limnologist Prof. W. Thomas Edmondson (Biography 18.2). A fascinating account has been written of the political and scientific aspects of cleaning up this lake (Edmondson, 1991).

BIOGRAPHY 18.2 W. THOMAS EDMONDSON

The career of Dr. W. Thomas Edmondson (Fig. 18.13) is an excellent example of linking basic scientific principles with successful environmental management. Due to his work, the case of Lake Washington is one of the most visible achievements of modern limnology. Edmondson had a distinguished and scholarly career with numerous publications and awards. He was a member of the National Academy of Sciences of the United States, was academic advisor to some of the top limnologists in the world, and published in the top journals on many aspects of aquatic ecology.

Edmondson got his start in limnology as a high school student studying rotifers in the laboratory of G. E. Hutchinson. Edmondson attended Yale, and by the time he graduated with a bachelor's degree he had eight publications on rotifers. During World War II, he served as an oceanographer making measurements on ocean waves to determine if they conformed to theoretical predictions. He also participated in studies of sound refraction in the deep sea, a dangerous project that required deploying many depth charges by hand at sea.

In 1955, Edmondson was a faculty member at the University of Washington. One of his students discovered that the cyanobacterium, *Oscillatoria rubescens*, had appeared in Lake Washington. The appearance of this alga signaled that eutrophication was beginning, as had been documented in Lake Zurich half a century before. After studying the problem, Edmondson began to correspond with the chairman of the committee appointed to study the pollution of Lake Washington. Ultimately, there was a public vote to determine if all sewage effluent should be diverted from Lake Washington. During the time before the vote, Edmondson was careful not to endorse a specific position but to provide accurate scientific information. He was verbally attacked for his scientific positions and complaints were made to the university president, but he held fast to the scientific facts as he saw them. In 1958, the sewage diversion was approved, leading one prominent political figure to state, "If you explain it well enough, people will do the right thing." The lake made a strong recovery; evidently, Edmondson explained it well enough.

FIGURE 18.13
W. Thomas Edmondson.

The story of Lake Washington illustrates, in part, the difficult position of an environmental scientist when having to balance advocacy and scientific information. If Edmondson had not brought his scientific findings to the attention of the public, the eutrophication of the lake may have been ignored or viewed as unavoidable, the sewage input may have continued, and the lake may have been irreversibly degraded. However, had Edmondson let politics dictate his actions, and strayed too far from his role as a source of scientific information, he may have been discredited. Edmondson's course of action demonstrates one way to be an effective scientist and confront environmental problems.

Table 18.5 Sewage Inputs into Lake Washington

Years	Sewage Input	Sewage Phosphorus Input (1,000 kg y^{-1})
1891–1936	Sewage from Seattle dumped directly to Lake Washington	?
1936–1958	Suburbs dump sewage directly into the lake	20–40
1958–1963	Population of suburbs grows as individual communities work to divert sewage outfalls from lake	100, then decreasing
1963	All major sewage inputs into lake halted—sewage diverted to Puget Sound with large mixing zone	0

(After Edmondson, 1991)

Lake Washington is a large (28 km long and 65 m deep) lake that forms the eastern border of the city of Seattle, Washington, and its suburbs. Lake Washington is a monomictic lake (summer stratification) that was historically oligotrophic. As human population in the lake's watershed grew, pollution of the lake increased (Table 18.5). In 1936, the city of Seattle diverted its sewage from the lake, and by 1963 all major sewage inputs to the lake were halted.

The phosphorus input from sewage dumped into the lake caused a decrease in lake clarity. Following the halt of sewage input into the lake, phosphorus levels in the upper waters of the lake decreased significantly (Edmondson and Lehman, 1981), populations of the eutrophic cyanobacterium *Oscillatoria rubescens* decreased (Fig. 18.14), and a species of *Daphnia* typical of oligotrophic waters became abundant again. Thus, removal of the nutrient input from sewage allowed the lake to return to an oligotrophic state.

In Lake Washington, O_2 never completely disappeared from the hypolimnion of the lake, and nutrient control brought about rapid reversal of eutrophication without release of excessive phosphorus from the sedimentary $FePO_4$. The excess phosphorus added to the lake in the past remains buried in the sediments and some has been washed out of the lake. The lake now receives heavy recreational use and maintains a reasonable clarity and absence of algal blooms.

Lake Trummen

This is the best documented case of the long-term effects of sediment dredging to mitigate internal loading problems. The 100 ha Swedish lake began receiving sewage in the late 1800s and this continued until 1959, leading to poor lake quality and frequent winter fish kills. Even after sewage input was stopped, the lake quality was so bad that citizens considered filling the lake.

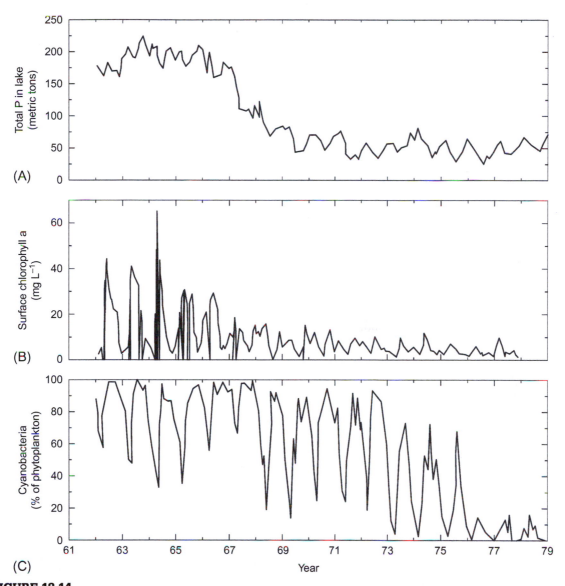

FIGURE 18.14

Changes over time in phosphorus loading (A), epilimnetic chlorophyll (B), and proportion of cyanobacteria (bluegreen algae; C) in Lake Washington. *(Redrawn from Edmondson and Lehman, 1981).*

Rich phosphorus deposits in the top layers of the sediments of the shallow lake (mean depth 1.1 m) caused high rates of internal loading. In 1970 and 1971, dredging was used to remove sediments, increasing the mean depth from 1.1 to 1.75 m. The dredged sediments were drained and sold as topsoil, partially offsetting the dredging costs (Cooke *et al.*, 1993).

Dredging lowered phosphorus concentrations in the sediments from 2.4 to $0.1\,mg\,L^{-1}$, leading to a 90% decrease of phosphorus in the water column, an increase in Secchi depth from 0.23 to 0.75 m, and decreases in nuisance cyanobacterial blooms. The decreased phosphorus and increased mixing depth were responsible for lower phytoplankton biomass. The dredging and minimization of nutrient inputs to the lake resulted in better water quality during the past 25 years, and fishing, swimming, and wind surfing continue to be popular activities on the lake (Cooke *et al.*, 1993).

Lake Tahoe

Lake Tahoe is one of the most visited large oligotrophic lakes in the United States. It is naturally oligotrophic, with a small watershed ($812\,km^2$) to surface area ($501\,km^2$) ratio and great depth (505 m maximum). As development threatens the lake because of associated nutrient input, considerable study on the primary production has occurred. Separating natural variation in processes controlling production of phytoplankton from the effects of nutrient pollution has been important. Dr. Charles Goldman (Biography 17.1) and his research group documented a threefold increase in primary production from 1968 to 1987 (Goldman *et al.*, 1989).

Tahoe is a large tectonic lake on the border between California and Nevada. This lake is ultraoligotrophic, with Secchi depths historically reaching about 40 m and a retention time of 700 years. Tahoe has gone from a regionally popular vacation spot with numerous homes and cabins in the watershed to an international tourist destination. There are gambling casinos on the Nevada side of the lake and several ski areas around the lake.

Part of the tremendous attraction of Lake Tahoe is the steely blue color associated with its ultraoligotrophic nature. Over time, a decrease in clarity occurred that was linked directly to increased nutrient input into the lake. The initial problem was that septic system drain fields were leaking nutrient-rich waters into the lake. The solution to this problem involved installing sewage systems and pumping the treated sewage out of the lake basin. The installation of sewage systems had the unforeseen negative impact of encouraging further construction and development. The associated removal of trees, increases in area of paved surfaces, and road building instigated more nutrient runoff into the lake (nonpoint source runoff).

The nutrient limitation in the lake has switched with pollution input. When septic systems were polluting the lake, an excess of phosphorus was present (primarily from detergents containing phosphate), and nitrogen was limiting. As watershed disturbance and atmospheric deposition became the dominant sources of nutrient pollution, phosphorus pollution became less important, nitrogen additions increased, and the lake passed through a stage of colimitation by nitrogen and phosphorus to a state of phosphorus limitation (Goldman *et al.*, 1993; Jassby *et al.*, 1995).

The future of Lake Tahoe is uncertain. The economic pressures for real estate development along the shore are immense. If development continues, the biomass of the phytoplankton will continue to increase. This increased amount of algae will lead to decreased clarity, and the lake will become a more typical oligotrophic lake instead of one of the purest lakes in the world. The Lake Tahoe Research Group announced in 2009 that clarity had not lessened over the past 8 years, so control measures may be working to at least halt the degradation of the lake clarity.

Lake Okeechobee

Lake Okeechobee in southern Florida is one of the largest lakes in the United States in terms of surface area ($1,840 \, km^2$). However, the mean depth is 3 m, which is less than the maximum wave height possible given the fetch. The shallow lake is characterized by turbid conditions from wind-induced sediment resuspension. The lake has become more eutrophic because agricultural runoff in the watershed has led to a doubling of total phosphorus (Havens *et al.*, 1996a). Paleolimnological evidence supports the idea that development and increases in agriculture have accelerated eutrophication (Engstrom *et al.*, 2006). This eutrophication threatens a recreational fishery (valued at more than $1 million per year) and the lake's value as a domestic and agricultural water supply.

Nutrient enrichment of the lake has caused increases in cyanobacteria known to produce toxins and increases in algal biomass. Phosphorus inputs have apparently stimulated N_2 fixing species. Predicting when and where cyanobacteria are going to bloom is difficult. Wind induces mixing of the lake and leads to lower probabilities of bloom formation by increasing light limitation (Bierman and James, 1995; James and Havens, 1996). Nutrient concentrations are related more directly to bloom formation in the shallower western regions (James and Havens, 1996).

Historically, Lake Okeechobee was probably phosphorus limited, but increases in phosphorus loading related to agriculture increased the degree of nitrogen limitation (Havens, 1995). Improvements in agricultural management practices have lowered phosphorus inputs into the lake by 40% compared to those in the 1980s. Nitrogen inputs from water pumped from nearby agricultural areas have also been decreased, lowering nitrogen inputs by about 50%. However, these reductions have not led to improvements in water quality parameters, probably because of the large amounts of phosphorus stored in the well-mixed sediments of the lake (Havens *et al.*, 1996b).

To further complicate the management scenario, the Everglades receive water from Lake Okeechobee. The complex hydrological management of the Everglades system was discussed in Sidebar 5.2. Eutrophication problems in the Everglades are discussed later.

The Clark Fork River

We discussed the problems associated with metal contamination from mine runoff and this river in Sidebar 16.4. An additional problem is dense algal growth related to point source inputs of nutrients during times of low discharge in the summer (mainly municipal sewage outfall). Benthic chlorophyll values frequently exceeded $100\,\text{mg}\,\text{m}^{-2}$ (Watson, 1989) and this high algal biomass was perceived as a nuisance.

Two approaches were used to calculate total nitrogen and phosphorus in the water column that would lead to acceptable periphyton biomass (Dodds *et al.*, 1997). The first was to identify reaches where chlorophyll levels were generally acceptable and analyze the total nitrogen and phosphorus. The second was to use equations generated from the general relationship between *TN*, *TP*, and benthic chlorophyll similar to those used in lakes to predict what level of nutrient should lead to generally acceptable values of benthic chlorophyll. Both methods suggested about 350 and $30\,\mu\text{g}\,\text{L}^{-1}$ total nitrogen and total phosphorus, respectively. A simple model of nitrogen and phosphorus inputs was then used to estimate the influence of sewage effluent controls on water column total nitrogen and phosphorus. A voluntary nutrient reduction program was put in place, and by 2008 nutrients had been controlled in parts of the river, but the upper part of the river still had extremely high algal biomass.

The Murray-Darling River

The Murray-Darling basin drains a seventh of Australia and is over 3,000 km long. Total water discharge is low, and the river dries in many years leading to a series of isolated pools in the larger portions of the river. Extractive uses have increased the probability of drying of the river. The region produces a third of Australia's agricultural output, and demand for the water is intense.

The relatively high nutrients in the water and the stagnant warm conditions are conducive to algal blooms. The toxic cyanobacterium *Anabaena circinalis* creates water quality problems. Toxic algae commonly occur in samples from the river (Baker and Humpage, 1994). Unlike many other systems, management of hydrology is the primary way to control the problem, including lowering water levels behind some weirs. Water level also influences macrophyte assemblages (Walker *et al.*, 1994) and invertebrate and attached algal populations (Burns and Walker, 2000).

Control of water level is difficult because the climate is arid, and the demand for water is high. In dry years, the problem is worse. Although precipitation could increase with climate change, so will temperature and evaporative demand, potentially putting further demands on the water in the Murray-Darling basin.

MANAGING EUTROPHICATION IN WETLANDS

Wetlands are parts of the landscape that have been largely ignored when water quality problems are considered, and thus only modest attention has been paid to eutrophication in wetlands. Wetlands can have tremendous benefits in nutrient immobilization and sediment trapping. In addition, use of wetlands is increasingly viewed as a way to remove nutrient loads from sewage effluent. Finally, cases now exist in which eutrophication of wetlands is causing problems. For example, atmospheric deposition of nitrogen in wetlands can cause eutrophication (Morris, 1991). Eutrophication in the Everglades is now a serious concern (Sidebar 18.3). Eutrophication control strategies in a European wetland were discussed in Sidebar 14.3.

SIDEBAR 18.3
Eutrophication and the Everglades

The Everglades are a naturally oligotrophic extensive complex of wetlands on the southern tip of Florida that have been impacted by agriculture and urbanization. Among the negative impacts of the activities upstream of the Everglades are those on native plant species assemblages by increased nutrient input and decreased hydrologic flushing. Preservation of the native wetland plants is necessary for animal conservation efforts; approximately one-third of the wading bird populations have decreased, and several other species that rely on the wetlands are nearly extinct, including the Florida panther and snail kites (Davis and Ogden, 1994). This Everglade food web is driven by a natural flow regime that varies seasonally.

The predominant plant cover in the Everglades was historically sawgrass (*Cladium jamaicense*), but it is being replaced by cattail (*Typha*). A shift in the cyanobacterial community is associated with phosphorus enrichment and hydrodynamic alteration (Browder *et al.*, 1994), as well as shifts from *Utricularia*- to *Chara*-dominated communities (Craft *et al.*, 1995). Such community changes have occurred in areas where phosphorus concentrations have increased, mainly from alterations in hydrology (Newman *et al.*, 1998) and increases in agricultural runoff (Doren *et al.*, 1996). Increases in agricultural activities have resulted in almost a threefold increase in input rates of phosphorus compared to historical levels (Davis, 1994). A major increase in sugarcane production in the area was fueled by the crisis in relations between the United States and Cuba in the 1950s and 1960s (Harwell, 1998).

Controlling phosphorus inputs from upstream agriculture will be difficult. Control will require advanced water treatment of agricultural runoff, purchase of sugar farming operations, curtailing fertilization, or some combination of these measures. To further complicate eutrophication management, conversion of mercury deposited from the atmosphere to more readily bioconcentrated toxic forms (methylmercury) by microbes occurs at lower rates under more eutrophic conditions, leading to less contamination of fishes (Gilmour *et al.*, 1998). Thus, eutrophication may lead to fewer fish consumption advisories.

The South Florida Water Management District constructed 1,821 ha of treatment wetlands between 1993 and 2007. These wetlands are designed to reduce phosphorus content of storm water before it flows into the Everglades. Despite the best efforts to preserve the Everglades, they are still severely threatened. The Everglades may retain long-term biotic integrity only if a series of steps are taken to control nutrient inputs, provide a natural hydrologic regime, and control atmospheric inputs of mercury.

Nitrogen pollution of wetlands can alter plant communities and lead to invasion by undesirable species of plants (Tomassen *et al.*, 2004). Experimental addition of nitrogen and nitrogen + phosphorus to phosphorus-limited wetlands in Central America led to a dramatic shift in primary producer communities from mainly cyanobacterial mats to macrophyte beds (*Eleocharis* and *Typha*) and overall shifted the system toward nitrogen limitation (Rejmankova *et al.*, 2008).

Consumer responses to wetland eutrophication can be complex because they are linked to plant responses and associated habitat structure. In a study in the Florida Everglades, invertebrate density increased with phosphorus additions to experimental plots up to the point where periphyton mats were no longer present, and then decreased; periphyton mats were reduced or absent in medium (0.8 g P added per m^2 per year) and high (3.2 g P $m^{-2}y^{-1}$) phosphorus treatments (Liston *et al.*, 2008). Fertilization of cattle pastures in Florida increased water column nutrient concentrations in adjacent wetlands, altered wetland plant communities, and overall reduced aquatic invertebrate richness and diversity (Steinman *et al.*, 2003).

Protecting wetlands from nutrient inputs is accomplished in much the same manner as for other freshwater habitats. Management of riparian zone vegetation can reduce nutrient inputs associated with overland flow and shallow groundwater transport, but atmospheric deposition represents a bigger challenge. Materials such as limestone, loess, and cinder, which adsorb phosphorus, can be used as substrata in constructed wetlands to enhance phosphorus inhibition (Guan *et al.*, 2009), and may be useful for remediation of polluted natural wetlands. Oyster shells, which like limestone are rich in calcium, have also been used in constructed wetlands and associated filtration systems for nutrient removal from wastewaters.

Constructed wetlands are increasingly used for effective wastewater treatment, but in some cases natural wetlands may also be used. Given increasing evidence for adverse affects of nutrient enrichment on wetland communities, managers should give careful consideration before subjecting natural wetlands to high nutrient inputs.

Wetlands as Nutrient Sinks

Wetlands can have major impacts on flows of nutrients, sediments, and water through watersheds. As floodwater moves through a wetland area, it spreads and slows, dropping sediments and surrendering nutrients to the plants growing in the wetlands. Riparian wetlands may be particularly important in this regard.

Wetlands have been used successfully for some time for nutrient removal and general sewage treatment in both North America and Europe (Sloey *et al.*, 1978; Brix, 1994). They can be used for nitrogen and phosphorus removal. When the wetlands are installed, they can have high initial rates of nutrient removal related to nutrient uptake by growing plants and algae (Richardson and Schwegler, 1986). After the first few months of heavy nutrient loads into the wetland, phosphorus removal can decrease significantly, but nitrogen removal can remain at moderate levels. The reason for the difference in nitrogen and phosphorus removal is that denitrification can remove nitrogen in the form of N_2 gas, but when the system becomes saturated with phosphorus little additional removal occurs. However, some phosphorus removal may continue because sedimentation in wetlands can account for a significant loss of phosphorus from the incoming water (Mitsch *et al.*, 1995). If plant biomass is removed continuously from the wetland, plants will add new growth and assimilate additional nutrients. This process is necessary if nutrient removal is to continue. However, finding a use for removed plants, such as mulch or paper making, may be difficult because such uses are not always economically profitable.

Wetlands that are used for nitrate removal can also release organic carbon (Ingersoll and Baker, 1998). Unfortunately, increased organic carbon export increases the biochemical oxygen demand draining from the wetland. Such demand can lead to water quality problems downstream.

The retention efficiencies of wetlands for various materials vary greatly. Such variation is not surprising given the wide assortment of wetland types that occur naturally. Wetlands can retain from 23 to 91% of sediments coming in, from 12 to 1,370 mg N m^{-2} year^{-1}, and from 1.2 to 110 mg P m^{-2} year^{-1} (Johnston, 1991). Essentially all studied wetlands retain nitrogen (from 21 to 95% in those in which inputs and outputs have been monitored) and therefore function as *nutrient sinks*. However, 9 of 24 wetland studies reviewed by Johnston (1991) and 1 of 19 studies reviewed by Kadlec (1994) showed that the wetlands actually serve as a net source of phosphorus. Thus, scientific management of wetlands for nutrient removal is required to ensure that the wetlands serve as nutrient sinks. Wetlands may assist in removal of nutrients from agricultural waters (Woltermade, 2000). Large areas of wetlands are required for effective nutrient removal. These wetlands are particularly effective if incorporated into riparian buffer strips.

SUMMARY

1. Aquatic systems can be classified by trophic state. Trophic state can be based on autotrophic or heterotrophic activity in the system.

2. Eutrophic lakes are characterized by wide swings in O_2 concentration and pH, anoxia in the hypolimnion, fish kills, and algal blooms including increased abundance of cyanobacteria. Eutrophic streams and wetlands can be characterized by high biomass of primary producers and decreased invertebrate diversity. Eutrophic groundwaters can have high inputs of organic carbon and be anoxic.

3. Lakes can naturally become more eutrophic over thousands of years. However, cultural (human-caused) eutrophication is currently far more common.

4. Eutrophication of lakes can lead to taste and odor problems, toxic algal blooms, fish kills, lowered water clarity, and decreased property values.

5. Quantitative equations are available to calculate expected relationships among nutrient loading, nutrient concentration, algal biomass, and Secchi depth in lakes. These equations are used by lake managers to make decisions on efforts to alter lake productivity. Similar equations are also available for streams.

6. Solving eutrophication problems generally requires control of point sources and nonpoint sources of nutrients. Control of nonpoint sources includes limiting excessive application of fertilizers, terracing fields, maintaining riparian and near-shore vegetation, and keeping livestock out of water bodies with fences and provision of stock tanks. Control of point sources includes bans on phosphorus-containing detergents and treatment methods such as chemical precipitation and denitrification to remove nutrients from wastewater.

7. Management of eutrophication in lakes can include oxygenation of the hypolimnion, addition of chemicals to precipitate phosphorus, and chemicals that kill algae.

8. Case studies of eutrophication demonstrate that controlling sources of nutrients and improving water management, rather than treating the symptoms of eutrophication, are the best way to avoid problems associated with excessive nutrients. In lakes, correcting eutrophication problems is much more difficult after the lake has become excessively productive and O_2 disappears from the hypolimnion.

9. Wetlands can also become eutrophic. Eutrophication of wetlands results in changes in microbe, plant, and animal communities, and can facilitate establishment of undesirable exotic species.

10. Wetlands have been used for tertiary treatment of sewage because of their ability to retain nutrients. Natural and constructed wetlands can be very effective for nutrient removal, but the use of natural systems for wastewater treatment can adversely affect wetland communities. The Florida Everglades have been harmed by eutrophication through mismanagement of the drainage basin.

QUESTIONS FOR THOUGHT

1. Why is a common classification system for trophic state useful for aquatic scientists, even if it is mainly a way to classify a continuous gradient of habitat types?

2. Why is the notion of a slow, constant movement toward a more eutrophic state in natural systems probably naive?

3. Why does a lake manager need to be aware of the variance associated with loading equations when making management recommendations?

4. Why might some ecoregions have lakes that naturally have blooms of heterocystous cyanobacteria?

5. Why would addition of Fe^{3+} to remove PO_4^{3-} from the epilimnion cause only temporary relief from eutrophication when the hypolimnion is anoxic?

6. Why is the relationship between total phosphorus and planktonic chlorophyll in lakes stronger (less variable) than the relationship between total phosphorus in streams and benthic chlorophyll?

7. Why might eutrophication of wetlands make insectivorous plants, such as sundews and Venus flytraps, less competitive?

8. Some people have treated eutrophication problems in lakes by diluting them with river flow. What conditions are necessary for this solution to work?

CHAPTER 19

Behavior and Interactions among Microorganisms and Invertebrates

FIGURE 19.1
A composite scanning electron micrograph of a cross section of a periphyton community. The area is approximately 1mm^2. Diatom species present include *Melosira varians* (long cylinders), *Gyrosigma attenuatum* (sygmoid shaped), and various smaller species of *Navicula* and *Nitzschia*. A layer of debris and extracellular exudates is at the bottom of the mat. *(Photo courtesy of Jennifer Greenwood; reproduced with permission from Greenwood et al., 1999).*

Examination of interactions within and among species is central to the study of aquatic ecology, as well as ecology in general. Such interactions are mediated by behavior and metabolism and shaped by evolution. The behavior and interactions of microbes are somewhat simple and provide a basic model for the more complex behaviors of more complex organisms. Many interactions among microorganisms are based on behavioral responses that are mediated by motility and responses to chemicals. Methods of determining bacterial species in their native environment have been developed only recently, so we are just beginning to study microbial communities from classical ecological viewpoints (White, 1995). In this chapter, we describe aspects of the ecology of microbes; Chapters 20 through 23 explore the ecology of macroscale plants and animals. This chapter begins with a discussion of motility, provides a general classification of interaction types, and then discusses species interactions in microbial communities and how macroscopic organisms interact with microscopic organisms.

BEHAVIOR OF MICROORGANISMS

The behavioral ecology of microbes is relatively simple, but important to the survival of many species nonetheless. The ability to control position in the environment is essential to the survival of many microbial species. The cues microorganisms use to move and the modes of motility are discussed in the following sections.

Motility

Several modes of motility occur among microbes. In the Bacteria and Archaea, simple flagella are used for locomotion. The flagella of Bacteria and Archaea are similar, but probably arose independently (they are analogs, not homologs). The molecular biology of flagellar motility has been well characterized for *Escherichia coli* (Glagolev, 1984; Koshland, 1980), but most other mechanisms of motility are less understood. Cyanobacteria have no flagella, but they are capable of gliding on solid substrata. In addition, gas vacuoles allow bacteria and cyanobacteria to float or sink. Behavioral aspects of this motility are discussed later.

The amoeboid protozoa and *Euglena* are capable of moving across solid surfaces by changing the shape of their cells. Other eukaryotic microbes, including diatoms and desmids (a green alga), can glide across solid surfaces; the exact mechanisms for this gliding are poorly understood. Eukaryotic microorganisms also use a variety of strategies to swim through open water, including flagella (protozoa and algae), paddles, or other swimming appendages found in larger multicellular organisms such as Cladocera, or by undulating their body (Copepoda).

In general, the copepods are the aquatic organisms capable of the greatest relative speeds (200 body lengths per second), with the flagellated Bacteria coming in a distant second (Table 19.1). Although fishes are capable of the

Table 19.1 Approximate Velocities of Various Organisms in Water

Organism	Velocity (ms^{-1})	Relative Velocity (body lengths s^{-1})
Desmids	1×10^{-6}	0.01
Amoebae	6×10^{-6}	0.03
Bacteria (gliding)	2.5×10^{-6}	0.1
Diatoms	2.5×10^{-5}	0.1
Cyanobacteria (gliding)	1×10^{-5}	0.1
Cyanobacteria (floating)	2×10^{-5}	0.2
Carp	0.4	1
Eels	0.5	1
Trout	4	8
Salmon	6	10
Ciliate protozoan	5×10^{-4}	10
Bacteria (flagellar movement)	2.0×10^{-5}	20
Copepod	0.1	200

greatest absolute speeds, they are less impressive when velocity is scaled to body size. It is even more amazing that microbial species are so fast given that viscosity is greater at low Reynolds numbers. Microbes live in a distinctly different physical world. Being very small means, among other things, that (1) paddles and fins do not work any better than cylinders as propulsive appendages, (2) a relatively large amount of power is required to move, (3) molecular diffusion predominates unless you can swim fast enough, and (4) sensing changes in the environment requires integration of signals over time rather than instantaneous detection of gradients (Dusenberry, 2009).

Taxis

Moving toward or away from objects or other environmental stimuli is important for many organisms' survival. Many strategies are used, with more elaborate strategies common in more complex animals. Any strategy requires determining the relative change in signal and coupling this to passage of time or change in position. Movement of organisms in response to stimuli is called *taxis* (alternatively they are referred to as *tropisms*). *Chemotaxis, phototaxis,* and *magnetotaxis* are movements of organisms stimulated by chemicals, light, and magnetic fields, respectively. A negative tactic response is called a *phobic* response; for example, negative phototaxis is also a photophobic response.

Perhaps the simplest behavioral taxis involves moving when the environment is not suitable and remaining in place when it is. An example of this is control of the rate of synthesis of gas vesicles used for flotation in plank-tonic cyanobacteria. Under low light and high nutrients, vesicles are synthe-sized; when light is high, synthesis stops (Walsby, 1994). This adaptation for movement allows the cyanobacteria to dominate in eutrophic waters. At the surface of a lake, in a dense cyanobacterial bloom, ample light may be avail-able from above. However, a large, actively growing population of algal cells will locally deplete nutrient concentrations. In this case, the synthesis of gas vesicles stops, and the cells sink. At greater depth, nutrients are high and light is low. Nutrients are assimilated and the rate of gas vesicle synthesis increases again. Concentration of CO_2 may also be involved in buoyancy regulation, but the specific mechanism is less clear (Klemer et al., 1996). This "behavior" may allow the cyanobacteria to compete well in eutrophic waters by moving up and down to avoid light and nutrient limitation, respectively, but field vali-dation of the adaptive value of this behavior is lacking (Bormans et al., 1999).

Geotaxis, movement with respect to gravity, may be useful for benthic organisms. These organisms need to remain in the benthic zone to sur-vive (Hemmersbach et al., 1999). They may seek the bottom by avoiding light, sensing gravity, or using Earth's magnetic field. Magnetotactic bacteria (Fig. 19.2) are examples of organisms that use magnetic fields to move down-ward (Blakemore, 1982). In the Northern Hemisphere, the electromagnetic field toward magnetic north also has a downward component. Thus, bacteria that move toward the north in the Northern Hemisphere tend to move down into sediments. In the Southern Hemisphere, benthic magnetotactic bacteria must move toward the magnetic south to move down into the sediments.

Negative phototaxis can serve as a form of geotaxis. This active movement is a strategy used by some protozoa that prefer anoxic habitats (Doughty, 1991). Photophobic responses allow them to avoid areas where photosynthetic organisms are producing O_2.

Chemotaxis is the movement toward or away from chemicals, and is the most common tactic response of motile microorganisms. Microorganisms are gen-erally small enough that they cannot sense a chemical gradient across the cell (molecular diffusion is rapid enough over several micrometers to disal-low such steep gradients to be maintained). Only the largest multicellular microbes may be able to sense differences from one side of the organism to the other. However, the ability to find and exploit chemical patches offers a selective advantage even to pelagic bacteria (Blackburn and Fenchel, 1999), so the ability to integrate information on chemical concentrations across time is important to some bacteria. Several search strategies can be used to move into regions where the desirable signal originates. Microbes most commonly use the random walk strategy (Fig. 19.3). An example of this is the swim and tumble strategy used by Escherichia coli in which the bacterium swims for

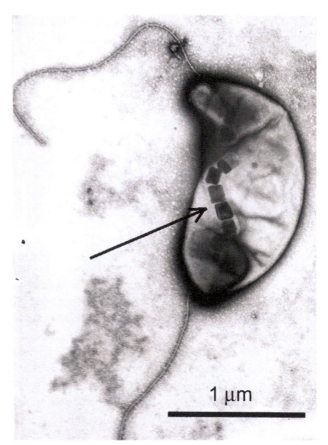

FIGURE 19.2
A bacterium containing magnetosomes, indicated by the arrow. *(Image courtesy of Richard Blakemore).*

a short time and then stops and tumbles rapidly. If conditions continue to improve (e.g., higher sugar concentrations), longer runs are taken; with worse conditions while moving, shorter runs are taken before each tumble. This searching mode allows the bacterium to move toward a chemical signal. The search strategy is not limited to microorganisms; salmon trying to find their home streams exhibit similar behavior (Hasler and Scholz, 1983).

A moderately complex behavioral response is exhibited by the hot spring–inhabiting cyanobacterium, *Oscillatoria terebriformis* (Richardson and Castenholz, 1987a, 1987b). The community setting of this organism was discussed in Chapter 15. This cyanobacterium is positively phototactic under moderate irradiance. This phototaxis causes it to move to the surface of the microbial mat as light increases. Under high irradiance (full sunlight at midday), the cyanobacterium is photophobic and forms self-shading clumps

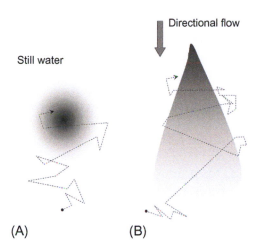

FIGURE 19.3
Random walk strategy for positive chemotaxis in still (A) and flowing (B) water. The organism tumbles and moves a short distance if the concentration of the attractant is not increasing. If the concentration is increasing, the same direction is maintained. Darker regions represent higher concentrations of a diffusing attractant.

to avoid photoinhibition or moves down into the mat. At night, a secondary behavioral response toward sulfide comes into play. The *Oscillatoria* is attracted to moderate concentrations of sulfide but avoids higher concentrations. Sulfide is produced from deeper in the mat and the cyanobacterium moves down several millimeters into the dark mat. These deeper portions of the mat are anoxic and rich in organic carbon, and the *O. terebriformis* is able to use the organic carbon as an anoxic heterotroph. The next morning, phototaxis overrides chemotaxis, and the cyanobacterium moves up into the mat again. The depth of the sulfide concentration also cycles diurnally; oxygenic photosynthesis allows oxidation of the sulfide during the day, but the sulfide gradient is closer to the upper mat surface at night. These behaviors give the cyanobacterium a competitive advantage over some of the other photosynthetic organisms that occur in the mat but do not move up and down in response to light gradients.

INTERACTION TYPES IN COMMUNITIES

We discuss interspecific (not intraspecific) interactions in the following sections. First, we need to introduce the general framework of describing interactions among organisms, be they microbes or macroscopic plants or animals. When interactions among species are examined, it is useful to make the distinction between trait-mediated and density-mediated interactions. *Density-mediated* interactions are those that are defined based on changes of populations of one species in response to another. *Trait-mediated* interactions are those that are evidenced by evolved traits in response to selective pressure associated with the interaction. Most cases of species interactions have been explored with experiments that focus on density-mediated responses; this approach typifies most examples in this chapter and the following two chapters. In general, density-mediated species interactions are important to those interested in questions of how to manage organisms and how to predict the immediate effects of disturbances on organisms. Trait-mediated interactions are important in determining why specific interactions occur and making generalizations about interaction types across communities. The reason that these two interaction types are separate is that population regulation does not necessarily define what is causing natural selection for specific traits.

As discussed in Chapter 8, the interactions that can occur among organisms are exploitation (mainly parasitism and predation), mutualism, competition, commensalism, amensalism, and neutralism. The relative occurrence of each of these interactions in microbial communities is generally not well known. In a study of interactions among the cyanobacterium *Nostoc* and bacteria (Gantar, 1985), positive interactions were about as likely as negative interactions. However, studies of phytoplankton summarized in Hutchinson (1967) show an excess of negative interactions. Twenty-seven species of tropical and subtropical fungi were mostly inhibitory toward each other when grown together in culture (Yuen *et al.*, 1999). If an excessive number of negative interactions occur, amensalism and competition will be more important than commensalism or mutualism. Some possible microbial interactions are discussed in the following sections.

PREDATION AND PARASITISM INCLUDING THE MICROBIAL LOOP

Microbial food webs are central to nutrient cycling and energy transfer in most aquatic systems. Transfer of energy, carbon, and other nutrients through the microbial food web is referred to as the *microbial loop* (Fig. 19.4), which occurs in streams, groundwater, wetlands, and lakes. The idea that microbial assemblages have an active role in transfer of energy in aquatic systems has changed the way that food webs and energy transfer are viewed. The microbial loop is essential to scavenging dissolved organic compounds in water and returning this organic material to the food web. Bacteria release a large variety of enzymes that allow use of organic carbon (Münster and DeHaan, 1998). The majority of organic material originating in terrestrial environments that enters aquatic environments must be processed by microbes. Organisms that eat these bacteria and fungi, or use them in their guts, return the organic carbon (or a portion of it) back into the food web. Without the microbial

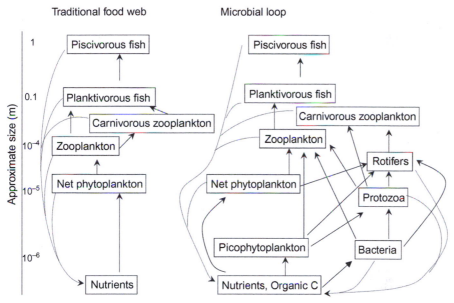

FIGURE 19.4

Representation of traditional views of a pelagic food web and the microbial loop.

loop, the bulk of the organic carbon in aquatic ecosystems would be tied up in nonliving dissolved and particulate organic material. The microbial loop can also be important to understanding how pollutants move into food webs (Wallberg *et al.*, 1997); bactivory can increase rates of bacterial activity (Ribblett *et al.*, 2005), increasing rates that pollutants are processed.

Microbial food webs in lakes are dominated by phytoplankton, bacteria, protozoa, viruses, and rotifers. Phytoplankton can be divided into very small cells (e.g., <2 or 3 μm) and larger cells. The very small cells are referred to as *picophytoplankton*, which are present in a wide variety of lake types (Søndergaard, 1991). Likewise, very small flagellated protozoa have recently been recognized as important consumers of bacteria.

Microbial food webs are a key supplier of carbon in stream food webs (Bott, 1995; Sigee 2005); Judy Meyer (Biography 19.1) demonstrated that bacterial consumption of dissolved organic carbon in streams is significant. Bacteria and fungi break down wood and leaves that enter streams, making the carbon available to primary consumers. Energy transfer via the microbial loop in streams could even be more efficient than in planktonic systems (Meyer, 1994).

The microbial loop is also responsible for a significant amount of energy flow in wetlands, with amoebae rather than ciliates and heterotrophic flagellates predominating (Gilbert *et al.*, 1998). The microbial loop is the primary

BIOGRAPHY 19.1 JUDY MEYER

Dr. Judy Meyer (Fig. 19.5) is a leading investigator in the field of stream ecology. She began to study for a master's degree in aquatic biology at the University of Hawaii because of an interest in dolphins but wound up studying the mathematics of nutrient limitation of phytoplankton. When she realized that a PhD would give her control over her career, she switched to studying freshwater because she did not enjoy organizing and participating in large oceanographic cruises. She decided to specialize in streams because she liked collaborating with the stream ecologists she met.

Meyer has more than 100 publications, many in the top scientific journals. She has been president of the Ecological Society of America, and the North American Benthological Society and has served on numerous scientific boards and committees. She has studied a broad variety of topics from coral reefs to stream food webs. Much of her research is on organic carbon as a food source for stream bacteria and the subsequent use of the microbes as a food source for higher trophic levels. She also has emphasized human impacts on aquatic environments both in recent research and through her involvement in policy issues.

Her suggestions for students are to seek multidisciplinary training to gain a holistic perspective and to remember that everyone starts as a novice. As an undergraduate, Meyer went to work at a premier stream research site, Hubbard Brook, and did not know what a weir was when she started. Also, she thinks students should concentrate on humans as part of the ecosystem. In stream ecology she considers linkages between streams and riparian zones as a crucial component of future research.

Meyer is an excellent example of a top-level scientist who works collaboratively. She has had an outstanding career and is proud of successfully raising a family. She recently retired and now devotes all her working time to environmental issues. She is an original thinker, a hard worker, and a high achiever.

FIGURE 19.5
Judy Meyer.

energy pathway in many groundwater habitats (Gibert *et al.*, 1994) including hyporheic sediments (Findlay and Sobczak, 2000).

Given their small size and difficulty of identification of individual species, microbial communities in general and food webs in specific, are difficult to study. Subsequently, there is not the long history of research that is associated with macroscale aquatic food webs. In the following sections, we discuss important aspects of microbial food webs, including viruses, how organisms can prey upon very small particles, and other forms of parasitism.

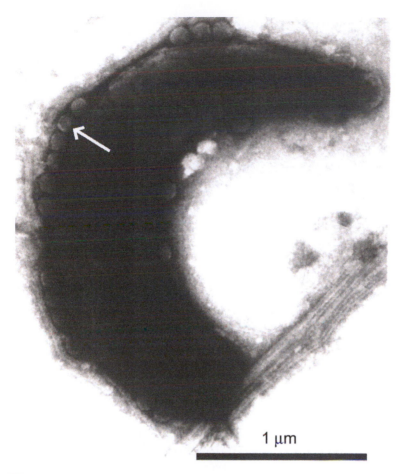

FIGURE 19.6

Virus-like particles attached to a bacterium from a freshwater lake. The arrow points to one of the virus-like particles. *(Reproduced with permission from Pina et al., 1998).*

Viruses

Viral infections occur in all known species and types of organisms where they have been looked for. Microbes are no exception, with *bacteriophages* infecting bacteria, *cyanophages* infecting cyanobacteria, and so on. Virus-like particles are common in freshwaters (Figs. 9.2, 19.6)—about 23 per each bacterial cell (Maranger and Bird, 1995). Viruses are also found in extreme environments such as hot springs (Rice *et al.*, 2001). Viruses are common in an Antarctic lake, but unlike temperate lakes, the viral assemblages appear to be dominated by viruses that infect eukaryotic organisms rather than bacteriophages (López-Bueno *et al.*, 2009). These viruses are indicated by electron microscopy and genomic sequencing methods, but some of these viruses are active and broadly infective. The reason they are referred to as virus-like particles is that they look like viruses when viewed with an electron microscope, but how infective they are and what they infect is not known simply by looking at water samples.

Viral infection of phytoplankton can be on par with zooplankton grazing in a shallow eutrophic lake (Tijdens *et al.*, 2008). Bactivory can be relatively inefficient (see next section), so viral predation on bacteria could be a major route of control of bacterial populations. Population controls by viral infection have been postulated to have several community effects. Many of these mechanisms are based on the epidemiology of infections. The host cell density needs to be great enough that the virus is not inactivated before it reaches a viable host cell.

In addition to controlling populations of organisms in the environment, viruses can be important vectors of gene transfer among organisms (Sigee, 2005). *Transduction* (gene transfer among bacteria by viruses) is one of the three routes of gene transfer. Occasionally both host and viral DNA are included in viral genomes, and this DNA can be subsequently inserted into that of other organisms. In general the amount of DNA transferred is small, but the process is still of interest with respect to evolution and spread of genetically engineered characteristics of organisms. Genes from widely disparate organisms are found in some genomes, and part of this lateral transfer of genetic material may be attributed to actions of viruses over evolutionary time.

Viruses must remain viable between hosts to propagate infection. Factors that inactivate viruses include UV radiation and sunlight (Wommack *et al.*, 1996), absorption to organic (Murray, 1995) and inorganic particles, absorption to nonhost cells, and predation by microflagellates (Gonzalez and Suttle, 1993). Below a certain density, the infection will not spread. As hosts develop immunity to viral attack, the effective density of hosts decreases. These interactions can be modeled with standard ecological predator–prey approaches. The models apply to simple laboratory systems with bacteria and viruses (Bohannan and Lenski, 1997).

Viral infections keeping cell densities below a threshold level may prevent competitive dominants from overrunning less competitive cells. Dramatic collapses of cyanobacterial blooms in lakes have been attributed to viral infection spreading rapidly through the population (Simis *et al.*, 2005). This relationship between host cell density and viral epidemics was invoked by Suttle *et al.* (1990) to help explain the presence of phytoplankton species that are not competitively dominant (i.e., the "paradox of the plankton"; see Chapter 17). Interestingly, viruses transported by rivers into coastal areas can infect the bacteria there and alter bacterial community structure (Auguet *et al.*,

2009), so connection of aquatic habitats across watersheds could influence bacterial communities of lakes, wetlands, and downstream lotic habitats by transport of viruses as well.

Viral infection leads to lysis or breaking of cells and the cell contents may be an important source of dissolved organic compounds that fuel the microbial heterotrophic community. When radioactive carbon dioxide tracers are used to label photosynthetic organisms, they prove that a portion of organic carbon fixed by photosynthetic activity is lost to solution. Viral infections may explain some of this loss. Interestingly, release of dissolved organic matter by healthy planktonic cells may lower rates of viral infection (Murray, 1995). The released substances bind the viruses so they are no longer able to attach to cells. Probably proteins and mucopolysaccharides similar to those found on the cell membrane surface are released into the water and viruses that use these as cues to infect cells attach to them and are not able to unattach. Thus, directly or indirectly, viral activity may be a key feature in providing energy and nutrients to the microbial loop by stimulating release of organic carbon (Bratbak *et al.*, 1994).

Consumption of Small Planktonic Cells

The consumption of small cells (Bacteria, Archaea, and fungi) is necessary if production from the microbial loop is to move into the macroscopic food web. The primary difficulty in consuming very small particles from planktonic systems is the energy required for extracting them from a dilute and viscous aquatic environment. Very small particles are physically difficult to remove via filtration because the Reynolds number is very low and the viscosity is very high. A very fine-retention filter will not allow significant fluid flow between the meshes without a tremendous input of energy. Thus, protozoa engulf the smallest particles individually or rely upon diffusive encounters with bacteria for capture (Sigee, 2005). These strategies allow for effective removal of particles as small as viruses from solution (Gonzalez and Suttle, 1993). Marine planktonic ciliates are major consumers of very small bacteria and phytoplankton (Christaki *et al.*, 1998). Although less research has been done on them, ciliates can also be important consumers of very small plankton in lakes (Porter *et al.*, 1985). Protozoa could be the primary consumers of bacteria in pelagic systems and the only consumers of any importance in anoxic environments (Finlay and Esteban, 1998).

The rate of particle capture is often described as a *clearance rate*, which is essentially the volume of water that can be cleared of particles per unit time. Maximal capture efficiencies for different species of protozoa occur at different particle sizes when plotted as a function of particle size (Fig. 19.7). Specialization for particle size allows for coexistence of protozoa by partitioning the resource base. The upper size limit of particles that can be consumed is probably set by

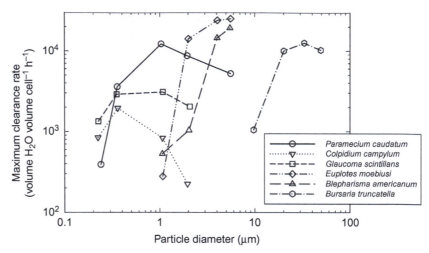

FIGURE 19.7

Consumption of particles by several species of ciliated protozoa as a function of size. Note that different species have different maximum clearance rates and that clearance rates vary with particle concentration within a species. *(Reproduced with permission from Fenchel, 1980).*

the physical constraints of ingesting and handling large particles. The lower limits are set by the ability to locate and ingest large numbers of smaller particles and hydrodynamic constraints related to high viscosity at low Reynolds numbers.

The idea that increased viscosity makes feeding more costly is supported by the observation that rotifer growth decreases with increasing viscosity because ingestion rates decrease (Hagiwara *et al.*, 1998). Temperature has a strong influence on viscosity, so filter feeding becomes more energetically difficult in colder, more viscous water. Coupled with a decrease in metabolic rate at lower temperature, the relative rates of feeding on planktonic bacterial-sized particles should be low in cold water.

The concentration of particles of a specific size also influences ingestion efficiency. Much energy is required to capture and ingest each particle consumed at low concentrations. It is necessary to clear greater volumes for the same amount of food as food particles become more dilute. The predator will not be able to ingest enough food to stay alive if particles are sufficiently diluted. The ingestion rate will increase with concentration up to a point, but then the organism becomes satiated and cannot or does not need to process any more particles per unit time (Fig. 19.8).

Large zooplankton, such as *Daphnia*, may be able to consume bacteria directly at low rates, but they may release organic compounds from sloppy

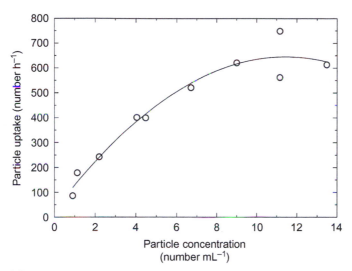

FIGURE 19.8

Consumption of particles by the ciliated protozoan *Glaucoma scintillans* as a function of particle concentration. *(Reproduced with permission from Fenchel, 1980).*

feeding on phytoplankton that stimulate bacteria (Kamjunke and Zehrer, 1999). However, grazing of phytoplankton by zooplankton may ultimately lower bacterial production by depressing production of organic compounds from a decreased algal biomass. Flagellates less than 20 μm in size may be the primary consumers of bacteria in many pelagic environments (Porter *et al.*, 1985; Sherr and Sherr, 1994).

There are many small particles in aquatic ecosystems, ranging from inorganic particles such as clays to living organisms such as phytoplankton and bacteria. Particle feeders may seem to be filtering or consuming particles indiscriminately, but studies using a variety of methods to trace particle consumption have shown that this is not necessarily the case. Inert particles such as fluorescent plastic beads were ingested, but at reduced rates relative to live bacteria of the same size (Sherr *et al.*, 1987), suggesting that organisms that feed on microbes are at least partially selective. Particle feeders will not grow or reproduce as well when ingesting particles that are of poor nutritional quality (Sterner, 1993; Tessier and Consolatti, 1991), and they prefer "soft" algae (DeMott, 1995). Thus, there is an evolutionary advantage for selecting high-quality food, even for organisms that eat microbes.

In general, a review of basic energetics provides a good framework for understanding controls on rates of consumption of very small particles in planktonic systems. More energy is required for survival and reproduction of an

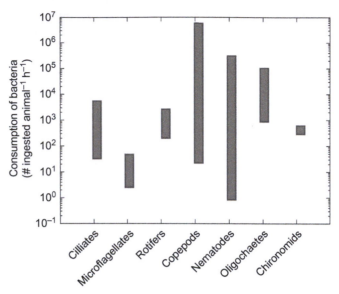

FIGURE 19.9

Ranges of observed consumption rates of bacteria by several groups of benthic organisms from marine and freshwaters. *(Data from 12 laboratory and field studies compiled by Bott, 1995; reproduced with permission).*

organism that consumes small particles when (1) particles are dilute, (2) particles are low quality, (3) particles are small, (4) temperatures are low, and (5) particles protect themselves (e.g., are toxic). These factors all interact and understanding their interrelationships is necessary to characterize planktonic food webs.

In contrast to pelagic habitats, benthic habitats contain larger organisms that are able to process considerable amounts of sediment and thus capture a significant number of bacteria per unit time (Fig. 19.9). In this case, invertebrate animals may equal or exceed the ability of protozoa to ingest bacteria, even if rates are normalized per unit biomass to account for the much larger size of the invertebrates. Several strategies are used by macroorganisms to ingest microorganisms.

Scrapers

Organisms that scrape biofilms can consume a wide variety of microbial species. Scrapers, which are often also referred to as grazers, include snails, some tadpoles, and many types of immature aquatic insects. Scrapers feed on substrata surfaces, consuming attached algae, heterotrophic components of biofilms, and associated deposited organic sediments. Scrapers can also remove heterotrophic biofilms found in groundwaters, so not all scrapers are grazers.

In general, scrapers do not discriminate among individual cells because biofilms are generally complex mixtures of many microbial species. However, studies using ^{15}N tracer additions to streams indicate that some scrapers selectively feed on the more biologically active components of biofilms (Dodds et al., 2000). The total nutritional content of the biofilm could alter scraper feeding and movement rate on the basis of the dominant particles in the biofilm (i.e., some biofilms may be more attractive than others) or the efficiency of digestion of various types of particles.

In general, particulate materials derived from the algal components of biofilms are more easily assimilated than detrital material derived from terrestrial plants. The nutritional ecology of scrapers can be complex because of differential assimilation of ingested materials. Some fishes and tadpoles that have traditionally been considered algivorous scrapers actually derive considerable nutrition from microbes associated with biofilms and detritus, as well as animal material consumed while grazing (Evans-White et al., 2003; Whiles et al., 2010).

A variety of organisms have converged on the scraper lifestyle. Some of the most common scrapers in aquatic environments are snails that possess a radula, a characteristic structure of the gastropods composed of bands of rows of tiny teeth). Grazing fishes (e.g., *Campostoma anomalum* and numerous tropical species) also have subterminal mouthparts that have evolved to optimize scraping. Scrapers that feed on periphyton can greatly reduce algal biomass in freshwater habitats. In some cases, scrapers can enhance biomass-specific production of algae through removal of senescent tissues and reducing light limitation, excretion of nutrients during feeding, and altering algal community structure (Lamberti and Moore, 1984; Steinman, 1996). Scrapers can also be important in *bioturbation*, as their feeding activities produce suspended materials and resuspend deposited particles.

Shredders

Organisms that eat detritus, either in the form of leaves and wood or as finer benthic organic material, are essentially microbial predators, as much of their nutrition is ultimately derived from microbes associated with detritus (Cummins and Klug, 1979). Shredders, including many crustacean and insect species in freshwater habitats, feed on coarse detritus such as wood and leaves. In doing so, they contribute significantly to a key ecosystem function, the decomposition of organic materials. For example, experimental removal of invertebrates, including shredders, from a headwater stream in the Appalachian Mountains resulted in significantly slower leaf litter decomposition and a six-fold decrease in fine particle concentrations in the stream water (Cuffney et al., 1990).

Although they feed on coarse detritus, the nutritional ecology of shredders and other detritivores is linked to microbes. Some animals (insects, mollusks, and crustaceans) have the ability to produce cellulases (Watanabe and Tokuda, 2001; Davison and Blaxter, 2005), but it is thought that most cellulose degradation is accomplished by bacteria and fungi, which are subsequently digested by animals. Archeae also produce cellulases, but their role in detritus decomposition is not well understood as of yet.

Microbial colonization of detritus is termed *conditioning*; when given a choice, shredders preferentially feed on more highly conditioned materials (Golladay *et al.*, 1983; Arsuffi and Suberkropp, 1984). Microbes associated with plant detritus have been termed the "peanut butter on the cracker" because shredders feeding on this material gain more protein from the microbes than the detritus "cracker" (Cummins, 1973). Although wood is generally considered to have relatively low nutritional value among detritus types, wood biofilms, or epixylon, can be more nutritious than leaf detritus to some shredders (Eggert and Wallace, 2007).

Shredders tear leaf and wood material up before ingesting it. This increases rates of microbial degradation of cellulose in their guts and also stimulates breakdown of the remaining material by increasing surface area for further microbial colonization. Because they feed on relatively low quality material, shredders also have high ingestion rates, which facilitate breakdown of coarse organic materials and generation of fine particulates (Wallace and Webster, 1996). Feeding activities of shredders generate small particles that are fed on by a variety of collectors such as midge larvae and filter-feeding caddisflies and mussels. Shredders that burrow enhance organic matter and associated microbial activity in sediments in much the same fashion as earthworms in terrestrial systems (Wagner, 1991). In lotic systems, shredders enhance the entrainment and downstream movement of materials by reducing particle sizes.

The vast majority of research on shredders, microbes, and leaf breakdown has been done in lotic systems, where allochthonous litter inputs and subsequent decomposition are often primary energy flow pathways (Vannote *et al.*, 1980; Webster *et al.*, 1999). Breakdown of leaves is clearly important in wetlands, and probably most shallow lakes surrounded by trees, and even groundwaters that receive leaves as well.

Collector–Gatherers

Freshwater organisms that consume relatively small (<1mm diameter) deposited organic particles from the surface of substrata are termed *collector–gatherers*. Many of these can eat entire microbial assemblages or pick out particles that are rich in cellular material and more biologically active. Oligochaetes and chironomids process large quantities of sediments and assimilate the microbial component and any other digestible organic materials in the sediments. As with shredders, this group receives much of its nutrition from microbes

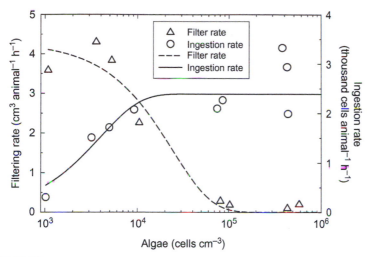

FIGURE 19.10

Filtering and ingestion rates of algal cells (*Chlamydomonas reinhardtii*) by *Daphnia magna* as a function of algal concentration. Although the filtering rate falls off sharply with increased concentration, the ingestion rate increases up to a point and then becomes constant. *(Reproduced with permission from Porter et al., 1982).*

associated with the detritus they consume. For example, stable isotope analyses suggest that chironomid larvae can receive a substantial portion of their diet from methane oxidizing bacteria (Eller *et al.*, 2005, 2007). Use of a ^{13}C sodium acetate tracer indicated that chironomids, copepods, and other gatherers in a headwater mountain stream preferentially assimilated bacterial carbon (Hall, 1995).

Where they have been studied intensively, collector–gatherers contribute significantly to turnover rates of organic particles. In a study of a Sonoran Desert stream, collector–gatherers were estimated to ingest the equivalent of their body weights in 4 to 6 hours, and overall ingestion of particles by collector–gatherers exceeded rates of primary production (Fisher and Gray, 1983).

Collector–gatherers such as oligochaetes and chironomids are often the most abundant macroscopic organisms in many shallow water habitats (Wallace and Webster, 1996), and thus their conversion of detritus and microbial biomass to invertebrate biomass is significant for larger consumers such as fishes and waterfowl, which rely heavily on these groups for food.

Filter Feeders

Organisms such as *Daphnia* are able to filter particles from the medium and maintain a continuous input of food (Fig. 19.10). As with protozoa, the

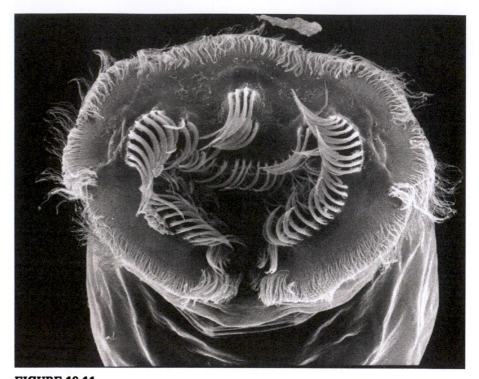

FIGURE 19.11

The filtering apparatus of a rotifer, *Epiphanes senta*. *(Reproduced from G. Melone, 1998,* Hydrobiologia *387/388, 131–134, Fig. 9, with kind permission from Kluwer Academic Publishers).*

ingestion rate is a function of particle size and concentration. Rotifers (Fig. 19.11) and *Daphnia* have many feeding appendages that circulate and filter water. These appendages capture particles and move them toward the mouthparts. Some of the particles are actually filtered, but many others may stick to individual feeding appendages rather than become lodged between them (i.e., impaction rather than filtration is used to capture some particles).

Larger *Daphnia* are more efficient feeders on larger particles than are smaller *Daphnia* (Burns, 1969; Hall and Threlkeld, 1976). Thus, in the absence of fish, they can dominate planktonic communities. This feature is essential to the trophic cascade in lakes described in Chapter 20. The clearance rates of filter feeders can be impressive: The entire volume of a pond can be filtered by zooplankton up to 4.7 times per day. One estimate suggested that more than 100% of the primary production by phytoplankton could be cropped each day (Porter, 1977).

Filter feeders associated with benthic habitats, such as blackfly larvae, clams, mussels, and net spinning caddisfly larvae, can also feed on microbial particles and strongly influence energy flow and nutrient cycling in freshwater habitats. In flowing waters, filter feeding assemblages are often dominated by *passive filter feeders*, those that let the moving water do the work for them. Net spinning caddisflies in the family hydropsychidae are abundant filter feeders found on stable substrata in streams and rivers. Different species of hydropsychids have different mesh sizes of nets, and this alters the optimal particle size that can be captured.

Two species of caddisfly larvae, *Ceratopsyche morosa* and *C. sparna*, were investigated for capture of various sizes (12, 84, and 528 µm diameter) of particles (Edler and Georgian, 2004). *Ceratopsyche morose* has an average mesh size about 20% greater than that of *C. sparna*. The larger particles were captured most efficiently overall, but some of the smaller particles were also captured. *C. sparna*, with smaller mesh size, captured the smaller particles more efficiently, but because Reynolds number dictates less flow through the smaller mesh size, the larger particles were captured less efficiently. The natural seston in this system was dominated by particles in the smaller range of those tested, but large particles, though rare, would have far greater energy content. These two cooccurring species could presumably specialize on different sized particles based on hydrodynamics of their nets.

Blackfly larvae (Simuliidae) can also be very abundant passive filter feeders in streams. Blackflies use specialized cephalic fan structures to filter very fine particles from the water column, including individual bacteria cells. In turn, blackflies produce relatively large fecal pellets that are consumed by larger filter feeders and settle out of the water to fuel benthic food webs. Production of fecal pellets by abundant filterers is not necessarily trivial; estimates from large rivers in Sweden indicate that peak loads of blackfly fecal pellets in transport in the water column can reach over 400 metric tons per day (Malmqvist *et al.*, 2001).

Freshwater bivalves are *active filter feeders* (e.g., they pump water through filtering structures) that can filter large quantities of water and remove significant amounts of phytoplankton and suspended organic materials when they are abundant. Depending on their local densities, freshwater mussels and other bivalves can filter 10% to 100% of the water column per day. In shallow water systems bivalves can be the most significant filter feeders, and they often consume particles at rates higher than downstream transport and settling (Strayer *et al.*, 1999). The pearly mussels, one of the formerly dominant groups of freshwater bivalves, have declined precipitously with increasing human impacts on streams and rivers, with ecosystem-level consequences (Strayer *et al.*, 2004). On the other hand, exotic species such as the zebra mussel (*Dreissena polymorpha*) have invaded many freshwater habitats, with measurable ecological changes (see Sidebar 10.1).

Microbial Adaptations to Avoid Predation

We have already described several adaptations for avoiding predation, including indigestibility, poor nutritional value, and chemical, behavioral, and mechanical defenses. All these adaptations have evolved to be useful in certain circumstances; we briefly discuss each of them.

Some unicellular algae are indigestible. They are able to pass through the guts of grazers because of thick mucilage that protects them from digestion. In addition to protection, these cells actually gain an advantage when they are consumed. The zooplankton gut is a high-nutrient environment, and luxury consumption of phosphorus by the algae in the process of passage through the gut translates into an advantage after expulsion (Porter, 1976). The overall importance of this adaptation is not well quantified for most microbes consumed by animals.

If cells have poor nutritional quality or manufacture toxic substances, they may not be ingested or may cause the death of their predators. Poor nutritional quality or toxin production works only for cells that represent a relatively high proportion of the diet of a predator. Toxins are diluted if few cells are consumed and higher quality food cells make up for low-quality cells. Cells found in mixed assemblages, such as periphyton, a biofilm, or a diverse phytoplankton assemblage, may waste their competitive edge by putting energy into chemical defense. If predators do not eat selectively (as is likely with a more diverse prey assemblage and if predators are very large and individual prey very small), the defended cells will be eaten anyway and will grow more slowly than those that put energy into growth and reproduction rather than defense.

A unicellular algal bloom provides a situation in which production of toxins to deter grazers may be advantageous for planktonic organisms eaten by filter feeders. The fact that blooms of one or a few species can form can explain why it is evolutionarily advantageous for planktonic cyanobacteria to produce broad-spectrum toxins that can harm or even kill grazing zooplankton (Hietala and Walls, 1995; Ward and Codd, 1999). Given the complete dominance of cyanobacteria in many eutrophic habitats, the energy used on chemical defense from predation is not wasted.

The toxic cells leave larger grazers at a disadvantage, and select for smaller grazers such as rotifers that specialize on smaller algae and bacteria that are present in lower concentrations (Hannsen et al., 2007). Some species of zooplankton are adapted to feed on cyanobacteria, but many species are strongly affected by toxins (Tillmanns et al., 2008). Presumably, an evolutionary "arms race" has occurred where algae develop toxins, and their grazers then evolve resistance to the toxins.

Chemical defenses may be useful deterrents to planktonic cells that are eaten by protozoa. Investments in chemical defenses confer an evolutionary advantage in this case because predators select individual cells. For example, one unicellular marine alga has a concentrated compound (dimethyl sulphoniopropionate) that is converted to dimethyl sulfide and toxic acrylate when the cell is grazed (Wolfe *et al.*, 1997). Presumably, such defenses also occur in freshwater algae, as dinoflagellates are known to contain significant quantities of dimethyl sulphoniopropionate (Ginzburg *et al.*, 1998) and are well represented in freshwater habitats.

Spines and large size are defenses because predators cannot ingest the cells efficiently. If the epilimnetic waters of a mesotrophic lake are examined in midsummer, the majority of the phytoplankton is composed of large cells or aggregates that are difficult for zooplankton grazers to ingest. Such mechanical defenses are less useful in benthic habitats, where grazers such as snails, insect larvae, crayfish, and fishes are large enough to be unaffected by microscopic spines or large colony size. The cyanobacterium, *Microcystis*, forms large colonies in the presence of flagellate grazers, rendering them less vulnerable to predation (Yang *et al.*, 2006).

Chemical cues can lead to development of mechanical defense in the green alga *Scenedesmus* (Lürling, 1998; von Elert and Franck, 1999). When *Scenedesmus actus* was grown in water filtered from a *Daphnia* culture, colonies of four to eight cells formed rapidly. Solitary algal cells were produced when the alga was grown without exposure to *Daphnia* water. This defense against *Daphnia* grazing by formation of large colonies occurred only when the predator was present. Larger colonies settle more quickly, so there is an evolutionary cost associated with larger colonies that is offset by resistance to predation (Lürling and Van Donk, 2000).

Parasitism

Parasites or diseases afflict all organisms. Naturalists often treat these opportunistic diseases and infections as oddities. It may be difficult to study parasitism because infections can be sporadic and unpredictable. Diseases of fishes have attracted attention (Table 19.2), particularly those associated with aquaria and fish culture (Untergasser, 1989). Human parasites with intermediate aquatic hosts have also been studied extensively.

Schistosomiasis is one of the most common debilitating human diseases on Earth with an aquatic organism serving as an intermediate host. About 200 million people worldwide have this infection, with about 10% having severe clinical disease (World Health Organization, 1993). Five species of *Schistosoma* trematodes can cause human disease, and each requires a different species of snail as an intermediate host. One option for control of this disease is to

Table 19.2 Some Parasitic Diseases of Freshwater Fishes

Taxonomic Group of Fish Parasite	Example of Diseases (Causative Agent)/ Fish Affected
Virus	Lymphocystis/aquarium fishes
Bacteria	Furunculosis (*Aeromonas salmonicida*)/salmonids
	Fin rot (several species of bacteria)/aquarium fishes
Fungi	*Ichthyophonus hoferi*/all known aquarium fishes
Protozoa	Whirling disease (*Myxospora*)/cold water fishes
	Ich or whitespot disease (*Ichthyophthirius multifiliis*)/many fishes
Trematoda	Hookworms/freshwater and marine fishes
Hirudinea	Leeches/freshwater species
Copepoda	Fish lice/can attack many species
Vertebrata	Lamprey (*Petromyzon*)/salmonids

control the snail species that spread the infection. Molluscicides are generally used to do this, but resistant snails can occur. Understanding the ecology of the snails may increase the effectiveness of control strategies. Proper design of reservoirs, irrigation canals, and drainage ditches can decrease the snail populations and lower disease rates. Control efforts include attempts to develop a human vaccine (Butterworth, 1988) to be used in combination with biological control of the snails, although several decades of research have yet to produce an effective vaccine.

Parasites don't cause only human disease. Blooms of phytoplankton may be heavily infected with fungal pathogens (Van Donk, 1989). Most studied infections are caused by chytrid fungi, and in particular infections of the diatom *Asterionella* (Sigee, 2005). Fungal infection is a potentially important controlling factor in successional sequences. Such parasitic infections are more likely to spread when algal populations are growing more rapidly and population densities are high.

Actinomycetes (bacteria) prey on cyanobacterial cells. About half of the strains of actinomycetes isolated from a lake sediment lysed *Microcystis* cells (Yamamoto *et al.*, 1998). Such predation may offer methods for controlling algal blooms.

Microbial parasites commonly infect frogs (Smyth and Smyth, 1980). Chytrid infection of frogs was discussed in Chapter 11. Frogs are also hosts for all major groups of animal parasites: Protozoa, Trematoda, Cestoda, Acanthocephala, and Nematoda. Some of these life cycles are fairly complex. For example, the trematode *Gorgoderina vitelliloba* infects the frog *Rana temporaria*. Adult insects or tadpoles ingest the parasitic trematode initially and it enters the kidney. After 21 days, the trematode flukes enter the bladder and

FIGURE 19.12

A deformed adult mink frog, *Lithobates septentrionalis*, from central Minnesota. This animal shows multiple hind limbs bilaterally, as well as duplicated and fused pelvic elements. *(Used with permission of the Regents of the University of California and the University of California Press).*

deposit eggs, which are excreted into the water. The trematode eggs hatch, and a small swimming form enters the gills of the freshwater fingernail clam *Pisidium*. From here, they emerge as a cercaria form (a small worm-like form) that is eaten by tadpoles or aquatic insect larvae. The cercariae encyst in the body cavities of their hosts. Adult frogs then eat the infected tadpoles or adult aquatic insects, and the cycle of infection begins again (Smyth and Smyth, 1980).

The trematode *Ribeiroia* has been linked to deformities in frogs because the cercariae larvae of this parasite penetrate and encyst in developing limb buds of larval amphibians (Blaustein and Johnson, 2003). Along with the negative consequences for the infected frog, causing deformities in the host (Fig. 19.12) may be beneficial to the trematode, as deformed frogs are more easily captured by the definitive bird host. There is evidence that infection rates by *Ribeiroia* are enhanced with eutrophication, which leads to higher densities of the intermediate snail host (Szuroczki and Richardson, 2009). Negative impacts of *Ribeiroia* can also be synergistic with pesticides (Keisecker, 2002). While some well-publicized cases of malformed frogs have been linked to trematode parasites, this is by no means the only cause; a variety of pollutants including pesticides, metals, and road de-icing salt can also cause amphibian malformations.

Other Exploitative Interactions

Other exploitative interactions rarely are appreciated as general processes in aquatic communities, but they undoubtedly are important. These interactions are incredibly varied, so we provide only a few examples to illustrate their potential importance.

Macrophytes are often covered by epiphytes, which benefit from living on the macrophytes. The epiphytes compete for light and nutrients with the macrophytes and may increase drag and associated detachment and thus have a negative influence on macrophytes. Such interactions are widespread in aquatic communities (Gross, 2003). Chemicals released by macrophytes can inhibit epiphytes (Dodds, 1991; Burkholder, 1996; Gross, 1999; Nakai *et al.*, 1999; Hilt and Gross, 2008). Most evidence for these interactions is not from the field, but derived from laboratory studies because controlled field experiments on the topic are so difficult (Gross, 2003).

Vibrio cholerae associates with crustaceans and can break down their chitinous exoskeletons. The bacterium causes the deadly human disease cholera. Thus, an exploitive interaction between zooplankton and a bacterium has significant human health implications. This relationship was discovered by one of the most accomplished microbial ecologists of our time, Rita Colwell (Biography 19.2).

When new habitat is exposed in aquatic systems (e.g., floods expose new rock surfaces), a sequence of colonization occurs. Often, the earlier colonists must condition the habitat before the later organisms in the sequence can become established. This conditioning may inhibit or ultimately exclude the initial colonists, but it facilitates the later organisms that can colonize the site. Succession will be discussed more completely in the next section and in Chapter 22.

COMPETITION

Competition can account for many species interactions in aquatic communities and is viewed by many as a key driving force in evolution. In practice,

BIOGRAPHY 19.2 RITA COLWELL

Rita Colwell (Fig. 19.13) is a specialist on *Vibrio cholerae*, the causative agent of cholera. The disease causes thousands of deaths each year and infects many more people, mainly in developing countries without access to clean drinking water. Dr. Colwell established that viable but noncultivable cells of *V. cholerae* can be found in most estuaries and even rivers. She showed that cholera bacteria are often associated with zooplankton and other crustaceans with chitin exoskeletons. She discovered that simple filtration of drinking water through several layers of tightly woven cloth can dramatically reduce occurrence rates of the disease by removing zooplankton. The concept of viable but noncultivable bacteria has become central in the field of microbial ecology and Dr. Colwell's work brought the idea to the forefront, as well as demonstrating that it had implications for public health.

Dr. Colwell is one of the top scientists in the world and has been incredibly successful. She is a member of the US National Academy of Sciences, has received the National Medal for Science, and was the director of the US National Science Foundation for 6 years. Rita has been a coauthor on over 700 publications including numerous books, and has served as president of the American Association for the Advancement of Science. Colwell has numerous other awards and honors, and continues, as an active researcher, to produce top-level scientific work.

Rita says that she gained an appreciation for marine science initially because she grew up near the ocean and spent much time playing on the shore as a child. As an undergraduate she became fascinated by microbiology, and eventually realized that she could spend her life researching aquatic bacteria, and obtained her doctorate degree working on marine *Vibrios*. Her early work demonstrating *V. cholerae* is common in estuaries was initially met with much skepticism; she had substantial difficulty publishing the research and convincing the public health community of the truth of her results.

Dr. Colwell notes that most of her important research results initially were viewed skeptically, but that she trusted the data and ultimately was proven right by others. She learned the hard way to trust the data, when an undergraduate working in her lab showed her a species of *Vibrio* had lateral flagella

FIGURE 19.13
Rita Colwell.

rather than polar flagella as all were thought to have at the time. She ignored the data and a colleague published the same result shortly thereafter.

She claims that growing up in a large family with little money made her competitive and stubborn. So, when her work is criticized, yet she knows it is correct, she works harder to get it published. She suggests that students should trust the data they generate and if at all possible publish their results. She thinks the process of writing up research helps crystallize ideas and separates the "wheat from the chaff," as well as moves the field of science forward.

Now Dr. Colwell, in addition to her research, is working on efforts to improve education for girls and women. She feels that many of the world's problems arise because women are not empowered to control their lives and that education changes this.

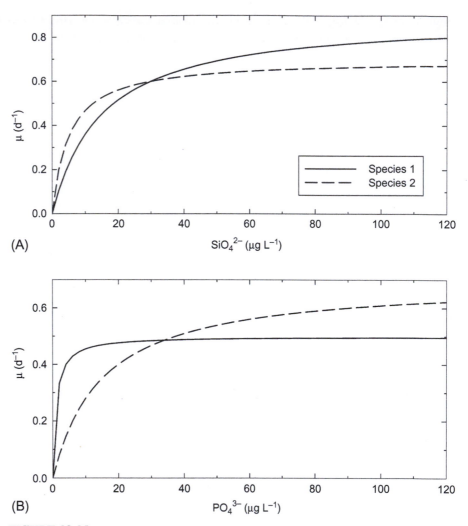

FIGURE 19.14

Competition of two diatom species for silicon (A) and phosphorus (B). Species 1 has a relatively higher affinity for low concentrations of silicate but is outcompeted at higher concentrations. The situation is reversed at higher concentrations.

demonstrating competition directly can be difficult (Connell, 1983; Schoener, 1983). Some competition must occur because many aquatic organisms have similar requirements. Thus, we might expect to find competition for scarce resources if physical conditions remain uniform for a sufficiently long time.

The actual time required for competition to become important is a function of relative growth rates and reproductive rates of the potential competitors. These, in turn, can be related to the relative supply rate of the resources and other environmental factors influencing growth and reproduction. As with all ecological processes, spatial and temporal scales are important considerations.

There are two forms of competition. *Exploitative competition* is competition by organisms that are both exploiting the same resource. Direct negative effects on other organisms are termed *interference competition*. Note that exploitative competition is two organisms using the same resource, whereas the interaction directly between organisms where one benefits and the other is harmed (e.g., predation) is referred to generally as *exploitation*. The consequences of exploitative competition related to simultaneous consumption of several resources have been explored by Tilman (1982). This theory helps us consider the possible ramifications of unequal abilities among organisms to use nutrients. Often, a tradeoff exists between the ability to grow well with low nutrients and the ability to grow well with high nutrients. If one organism can grow well under low nutrient concentrations, and the second can grow well under high nutrient concentrations (Fig. 19.14), they will be competitively dominant at different nutrient concentrations. If their competitive abilities are reversed for a second nutrient, then they are able to coexist under certain ratios of nutrients because they are both limited by different nutrients. In the example presented in Figure 19.14, high Si:P ratio favors species 1, low Si:P ratio favors species 2, and at intermediate ratios both species may coexist (Fig. 19.15). Data suggest that such mechanisms of competition can be important in phytoplankton communities (Tilman *et al.*, 1982; Kilham *et al.*, 1996), but care must be used when extrapolating from results of laboratory experiments to the effect of nutrient ratios on natural populations (Sommer, 1999).

Interference competition is more difficult to establish for microbes. Interference competition can occur through production of toxic compounds that inhibit a potential competitor. Organic compounds that alter growth rates of other organisms are called *allelochemicals* and they are produced in planktonic (Keating, 1977, 1978) and benthic communities. Production of chemicals by microscopic algae that inhibit other primary producers is common and occurs across many taxa (Gross, 2003).

Chemically mediated interactions (allelopathy) can drive successional sequences in phytoplankton (Rice, 1984). Given evolutionary and ecological constraints, such chemicals would be excreted expressly to act as allelochemicals only under certain conditions. The benefits to the organism must exceed the cost of synthesizing the chemicals. In phytoplankton communities, allelochemicals that are synthesized specifically to lower competition are excreted only when cell densities are sufficient to bring total concentration in the water up to effective levels (Lewis, 1986). Such may be the case during algal blooms. However, some chemicals are released into the water for other reasons (e.g., cell lysis and excretion of exoenzymes) and such chemicals may

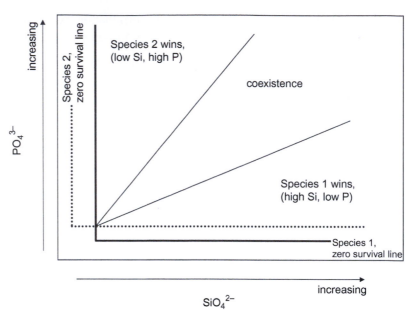

FIGURE 19.15
Scheme showing how the data in Figure 19.14 translate into relative success of two diatom species based on nutrient ratios.

serve as environmental cues for planktonic species, causing them to increase or decrease growth rates (Keating, 1977, 1978).

In biofilms, in which cells are always in proximity to the same competing cells and molecular diffusion dominates (e.g., diffusion is relatively slow), allelochemicals may be more common. Few data are available on chemical interactions among individual organisms living in biofilms (Characklis et al., 1990). There is limited evidence of allelopathy in some species of cyanobacteria inhabiting biofilms (Leao et al., 2009). Still, given the close proximity of cells in biofilms (Fig. 19.1), it is fairly certain that there are numerous and complex chemical inhibitions occurring in biofilms.

Successional sequences are commonly observed in microbial communities, and the sequences may be driven in part by competitive ability. Bacteria and protozoa undergo a successional sequence in sewage treatment (Cairns, 1982). In periphyton communities, a successional sequence may occur as the community grows thicker and competition becomes more intense (Hoagland et al., 1982; Fig. 19.16). The community is reset by disturbances such as flooding and grazing. Research on competition in these communities is sparse. The picture of microbial succession is more complex than simple competition; autotrophic organisms are facilitated in colonizing new habitats by heterotrophic bacteria (Roeselers et al., 2007). Furthermore, as the community changes,

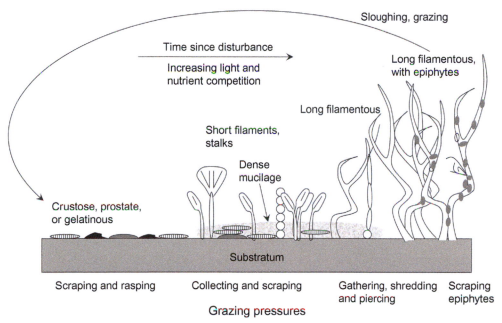

FIGURE 19.16

Conceptual diagram of successional sequence of an algal biofilm and the associated increases in nutrient and light competition. The type of grazer is indicated at each stage. *(After Steinman, 1996).*

the types of grazing activities on the community also change. Given that animals can excrete nutrients, there may be complex feedbacks between microbial producer communities and larger consumers (Bertrand *et al.*, 2009).

MUTUALISM: FACILITATION AND SYNTROPHY

Mutualism is rarely studied in aquatic ecology, possibly because it is rare in aquatic communities relative to other interactions or because researchers have not widely recognized the need to study it. The most common occurrence of two or more species benefiting each other is probably nutrient cycling. Some argue that the example of nutrient cycling stretches the definition of mutualism, but both organisms benefit (albeit indirectly in many cases), thus fitting the definition used in this book. Interactions involving microorganisms in which both species benefit are highly diverse and will be discussed by example.

Even single strains of a single species of bacteria can be selected to cooperate (Rainey and Rainey, 2003). *Pseudomonas fluorescens* in static culture can evolve types that produce polysaccharide that causes them to float and avoid anoxic conditions in the liquid culture. The bacteria "cooperate" to make a floating mat. Thus, microbes can be selected for cooperative characteristics.

Most animals have microorganisms in their guts, and aquatic organisms are no exception. The most commonly reported genera of bacteria found in guts of aquatic invertebrates include *Vibrio*, *Pseudomonas*, *Flavobacterium*, *Micrococcus*, and *Aeromonas* (Harris, 1993). These microbes can benefit organisms by decomposing food that would otherwise be unavailable and by outcompeting pathogenic organisms. Interactions with gut microorganisms have not been well studied, but there are documented cases of aquatic invertebrate digestion being aided by gut microbes and increases in resistance to toxins associated with gut fauna (Harris, 1993).

Association of nitrogen-fixing bacteria or cyanobacteria with plants is probably the most likely interaction involving nutrient cycling to be accepted as mutualistic. Examples from aquatic habitats include the interaction between the water fern *Azolla* (Fig. 19.17A) and the nitrogen-fixing heterocystous cyanobacterium *Anabaena azollae* (Fig. 19.17B). Many wetland or riparian plants may also associate with nitrogen-fixing microbes, including the flowering plant *Gunnera* and the cyanobacterium *Nostoc* (Meeks, 1998). Alder is a common tree in boreal riparian zones that has nitrogen-fixing bacteria associated with its roots and can be a significant source of fixed N to nutrient-limited systems (Rytter *et al.*, 1991). The diatom *Epithemia* contains nitrogen-fixing cyanobacteria (De Yoe *et al.*, 1992), and both organisms may benefit from the interaction (Fig. 19.17E).

Mycorrhizal interactions that can be mutualistic occur in some wetland plants (Søndergaard and Laegaard, 1977; Rickerl *et al.*, 1994; Daleo *et al.*, 2007). Such relationships have been demonstrated to increase competitive ability for nutrients in some lake macrophytes (Wigand *et al.*, 1998). Mycorrhizae can alter wetland plant assemblage composition (Wolfe *et al.*, 2006).

Syntrophy is complementary metabolism. Anaerobic microorganisms in anoxic habitats commonly exhibit syntrophy. In these interactions, each organism provides the other with an organic carbon source or uses a carbon source that would become toxic and limit the other. These interactions are well described by Fenchel and Finlay (1995). A specific example of syntrophy is interspecies H_2 transfer in which hydrogen-generating microbes facilitate activity of methanogens. The removal of H_2 is beneficial to the microbes that produce it because their activity is inhibited by high concentrations of H_2. Understanding this and other syntrophic interactions is central to describing anaerobic sewage digestion. These interactions often involve dense aggregates of organisms in very close proximity to each other.

A general type of mutualistic interaction that is moderately common in aquatic microbial communities involves animals that ingest algal cells and obtain fixed carbon from them. Presumably the alga receives protection from

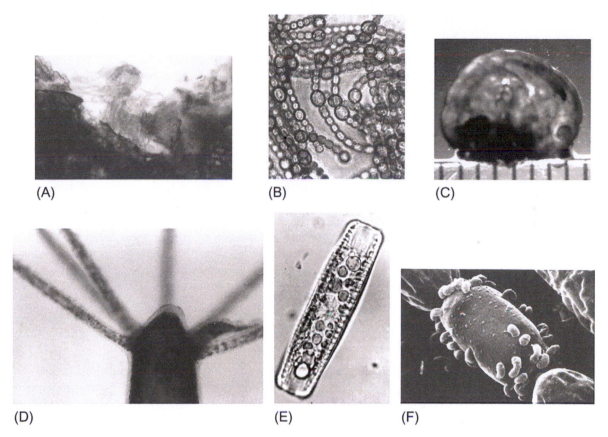

(A)   (B)   (C)

(D)   (E)   (F)

FIGURE 19.17   Some aquatic organisms involved in mutualistic interactions: the pouch of the water fern *Azolla* (A) that contains endosymbiotic *Nostoc* (B), (C) *Nostoc parmelioides* containing the midge *Cricotopus nostocicola*, (D) *Hydra* with endosymbiotic *Chlorella*, (E) the diatom *Epithemia* with cyanobacterial endosymbionts (courtesy of Rex Lowe), and (F) bacteria attached to a heterocyst of *Anabaena flos-aquae* (courtesy of Hans Paerl).

predation and inorganic nutrients from the animal host. Organisms with this type of interaction include *Hydra* (Fig. 19.17D) and *Paramecium bursuri* with the green alga *Chlorella*. Dinoflagellates have ingested several different types of algae in this fashion, including cyanobacteria and green algae.

Similarly, chemoautotrophic bacteria are ingested by protozoa. The most studied of these interactions is the sequestering of methanogenic bacteria by protozoa (Fenchel and Finlay, 1995). This is important because the protozoa and their bacterial endosymbionts can encyst and be transported through oxic habitats to colonize other anoxic habitats. Such an association of organisms may be responsible for much of the unwanted methane production that occurs in landfills and it is likely common in anoxic sediments in all aquatic habitats.

Bacteria are often associated with cyanobacterial heterocysts (Fig. 19.17F). In this case, the photosynthetic cyanobacterium provides fixed carbon and nitrogen. The bacteria respire and remove O_2. Lowering the O_2 tension around heterocysts promotes N_2 fixation because nitrogenase is damaged by O_2. Thus, both microorganisms benefit from the interaction (Paerl, 1990).

Macrophytes that are grazer-resistant may benefit from organisms that remove algae and bacteria from their surface, and the grazers may benefit from the macrophyte that provides growth substrata for their food and perhaps protection from predation. Examples of this type of interaction include snails that remove epiphytic bacteria and algae from *Nostoc* and possibly epiphyte grazers that remove epiphytes from the filamentous green alga *Cladophora* (Dodds, 1991). Grazing snails are commonly seen removing the biofilm from macrophytes, and this may be a mutualistic interaction.

CHEMICAL MEDIATION OF MICROBIAL INTERACTIONS

Given that most microorganisms evaluate the environment around them by sensing chemicals, it is not surprising that many of the interactions among microorganisms (and among macroorganisms; Dodson *et al.*, 1994) are mediated by chemicals. Many studies exist on chemically mediated interactions, including aspects such as attachment cues, toxic chemicals excreted by bacteria, chemicals that alter morphology, and chemicals that attract and repel other microbes (Aaronson, 1981). Chemically mediated interactions are documented for all major groups of microorganisms, including the bacteria, protozoa, fungi, algae (Aaronson, 1981), and rotifers (Snell, 1998).

Examples of chemically mediated interactions among phytoplankton species are accumulating slowly in the literature, but such interactions are likely important (Wetzel, 2001). The work of Keating (1977, 1978), discussed previously, shows how specific ecological predictions can be made if the details of chemically mediated interactions are documented.

An interesting example of chemically mediated interactions involving microbes has been described for *Daphnia* diving in response to fish. A compound that causes *Daphnia* to move to deeper water is excreted by bacteria growing on the fish that prey on *Daphnia* (Ringelberg and Van Gool, 1998). This positive effect on *Daphnia* probably has no effect on the bacteria, but the bacteria have an indirect negative effect on the fish.

The ramifications of directional water flow on chemically mediated interactions are obvious. Generally, such interactions are more likely to be reciprocal among species when they occur within the diffusion boundary layer than when they occur outside it (Dodds, 1990). Thus, Reynolds number, flow dynamics, and Fick's law can be used to make predictions about the type of chemically meditated interactions that will evolve in microorganisms (Dusenbery, 2009).

SUMMARY

1. The behavior of microorganisms is based primarily on factors that control motility. The ability to sense changes in the environment, coupled with control of movement, allows microorganisms to move to more favorable habitats.

2. Chemotaxis is movement toward high or low chemical concentrations, phototaxis is movement toward light, and geotaxis is movement in response to gravity. Different microbes have evolved the ability to use some or several of these taxes. Phobic responses are movements away from stimuli.

3. All the basic interaction types among macroscopic species are also found in microbial species (exploitation, competition, mutualism, commensalism, amensalism, and neutralism).

4. Viruses are important parasites of microbes.

5. The rate of feeding on small cells can depend on their concentration and food quality. Functional feeding descriptions have been developed to describe these relationships.

6. Invertebrates and other macroconsumers in freshwater habitats interact with microbes in a variety of ways. For those that consume detritus, fungi and bacteria can represent the major source of nutrition.

7. Through their feeding activities, invertebrate functional groups can influence rates of critical ecosystem processes such as primary production and decomposition.

8. Microbial adaptations to avoid predation include indigestibility; low nutrient content; and chemical, behavioral, and mechanical defenses.

9. Parasitism by microbial species is an important ecological interaction for aquatic organisms.

10. Competition can influence many microbial species in various ways, including changes in successional sequences and determination of what species will dominate under specific nutrient regimes.

11. Mutualism occurs in aquatic habitats; syntrophic assemblages of anoxic microbes, gut microbes and nutrient remineralization are likely the most common mutualistic interactions.

12. Chemicals excreted into the water mediate many microbial interactions. Examples include successional sequences of phytoplankton and excretion of inhibitory compounds to decrease competition.

QUESTIONS FOR THOUGHT

1. Why do obligate mutualisms appear to be a more important type of interaction in coral reefs than in the benthic zones of lakes, wetlands, and streams?

2. How might shredders and scrapers alter microbial activity on leaves in streams; specifically how could they stimulate production of the microbes they consume on the litter?

3. How can what an organism ingests and what it assimilates vary, and how might this influence its ecological roles?

4. How can interactions among organisms change over time or with changes in biotic conditions?

5. Why should individual diatom species found in periphyton assemblages be less likely to produce chemicals to deter scrapers than would macrophytes?

6. Planktonic bacterial populations may respond in a complex manner to temperature. Can you predict if lower growth or decreases in predation rates related to increased viscosity should be more important controls of biomass?

7. What kind of predator was the precursor to eukaryotic cells (before chloroplasts and mitochondria)?

8. Why is chemical sensing of microbial prey so important but visual identification more likely with larger scale prey?

9. Under what conditions might an algal species exhibit photophobic response? Why may an obligate anaerobe be photophobic?

10. Adhesion to collecting appendages may be an important mode of collection of bacterial prey. Why are viscous forces and Reynolds numbers important to consider in such cases?

11. Why are filter feeders found in the water column of lakes but mainly in the benthic zone (not as much in the water column) of rivers?

Predation and Food Webs

FIGURE 20.1

A walleye (*Stizostedion vitreum*) consuming shiners (cyprinidae). *(Photograph courtesy of Bill Lindauer Photography).*

Doi: 10.1016/B978-0-12-374724-2.00020-9

In this chapter we first consider herbivory, detritivory, omnivory, and predation on animals. Second, adaptations of macroscopic organisms in response to being prey or predator are discussed. Third, food webs and their dynamics are approached. A *food web* is the network of predator–prey interactions that occurs in an ecological community. *Food chains* are the most simplistic view of food webs, in which only trophic levels (e.g., producers and consumers) are considered. Food webs and predation in lakes and streams have received a tremendous amount of attention, less so in wetlands and groundwaters. Some claim that "food webs are a central, if not the central, idea in ecology" (Wilbur, 1997). Though not all ecologists agree with the statement that food webs are "the" central idea in ecology, most recognize that food webs are an essential aspect of ecological interactions and must be considered with other factors, such as abiotic effects and competition (Wilbur, 1987).

HERBIVORY

Herbivory can be divided into consumption of macrophytes or microscopic algae. Microscopic algae can be consumed as phytoplankton or as periphyton. The adaptations of organisms for consuming small cells were discussed in Chapter 19.

Although invertebrates are the primary consumers of phytoplankton, planktivorous and omnivorous fishes may ingest suspended algae (Matthews, 1998). Gizzard shad (*Dorosoma cepedianum*) can effectively use the mucus on their gill rakers to trap cells as small as 20 μm in diameter. Several species of fish use similar strategies to filter and retain small particles (Sanderson *et al.*, 1991). Some small *Tilapia* (including species used in aquaculture) are also able to capture and ingest phytoplankton.

Numerous large organisms consume periphyton. The effects of grazers on benthic algae have been extensively studied and can be divided into functional and structural responses. The structural responses of periphyton to grazers include (1) a general decrease in biomass (but not always); (2) changes in taxonomic composition (but prediction of specific general taxonomic shifts is difficult); (3) changes in the form and structure (physiognomy) of communities, with a general decrease in large erect forms; and (4) alteration of species richness and diversity (perhaps with intermediate levels of grazing leading to maximum diversity).

Functional responses of periphyton to grazing include (1) a general decrease in primary production per unit area, (2) constant or increasing production per unit biomass, (3) changes in nutrient content, (4) increased rate of nutrient cycling, (5) increased rates of export of cells from the assemblage, and (6) alteration of successional trajectories (Steinman, 1996).

SIDEBAR 20.1

Using Grass Carp to Remove Aquatic Vegetation

The grass carp (*Ctenopharyngodon idella*) is a cyprinid that consumes aquatic vegetation as an adult. This herbivorous fish has been introduced into many areas to assist in removal of unwanted macrophyte growth. It was first released in the United States in Arkansas in the early 1960s and has since become widespread.

This species is a very effective herbivore. For example, in Texas, grass carp were stocked at 74 fish per hectare in a reservoir with 40% macrophyte cover. All macrophytes were consumed within one year (Maceina *et al.*, 1992). There was a concurrent increase in cyanobacterial plankton and a decrease in water clarity.

There is concern regarding how much this species can spread, and some states have completely restricted its use for macrophyte control. The temperature range for successful reproduction is 19 to 30°C (Stanley *et al.*, 1978), and because reproduction occurs mainly in larger rivers, it was thought that the grass carp was unlikely to spread when added to ponds. However, ponds have breached and reproductive fish have escaped. As a result, grass carp larvae have been found in the lower Missouri River, and the species has likely become established there (Brown and Coon, 1991).

One approach to keep stocked fish from reproducing is to use triploid grass carp produced with thermal or temperature shocks of newly fertilized eggs. The triploids are unable to reproduce and are the only grass carp allowed in some states (Allen and Wattendorf, 1987). Scientific equipment and training are needed to determine if grass carp are actually triploid.

A problem with use of grass carp to control macrophytes, in a management sense, is inappropriate overstocking. Some people perceive macrophytes as a nuisance in lakes and ponds. If many grass carp are added, all vegetation is removed. This clears the way for large algal blooms to occur in the absence of suppression by macrophytes (except see Lodge *et al.*, 1987, for an alternative outcome). A rational approach is to accept a moderate amount of macrophytes as a healthy component of natural ponds and wetlands and control them only if they become so thick as to completely preclude desired uses such as swimming and fishing. In this case, judicious use of grass carp may be warranted. Sufficiently low densities should be used so that not all macrophytes are removed. Nonfertile triploids should be stocked so that fish will not multiply and completely remove the macrophytes and, worse, spread to other habitats and become a nuisance. Those who use grass carp should be aware that reduced macrophyte densities in lakes can limit recruitment of desired species such as largemouth bass, and the carp can cause a decrease in water clarity by increasing bioturbation and stimulating algal blooms.

Many fishes consume periphyton and small macrophytes (Matthews *et al.*, 1987; Matthews, 1998). Herbivorous fishes that consume periphyton are common in tropical waters. The minnow *Campostoma anomalum* (the central stoneroller) is an important herbivorous fish in small streams in the United States. This minnow can consume significant amounts of periphyton and will be discussed later with respect to its role in food webs. Many organisms consume macrophytes in aquatic habitats, including crayfish, common carp, grass carp (Sidebar 20.1), lepidopteran and trichopteran larvae, moose

(*Alces alces*), wild Asiatic water buffalo (*Bubalis bubalis*), muskrat (*Ondatra zibethicus*), manatees (*Trichechus*), ducks (Anatidae), beavers (*Castor canadensis*), swans (*Cygnus*), and snails. Many tropical fishes also consume macrophytes (Matthews, 1998).

Some macrophyte species are unpalatable to herbivorous animals because they are chemically defended against consumption. Macrophytes can either synthesize or accumulate toxins (Hutchinson, 1975; Porter, 1977; Kerfoot *et al.*, 1998) and may be too tough for some herbivores to process (Brönmark, 1985). Some macrophytes, such as the filamentous green alga *Cladophora*, can be a poor food source for herbivores because they have low nitrogen and phosphorus content (Dodds and Gudder, 1992).

DETRITIVORY

Inputs of terrestrial organic material (allochthonous materials such as autumn shed leaves and wood from riparian trees as well as dissolved carbon) into rivers, lakes, streams, and wetlands can be significant. In some cases, groundwaters can also receive coarse organic materials from terrestrial vegetation (Eichem *et al.*, 1993). In Chapter 19, we discussed detritivory with respect to microbial communities and feeding strategies used; here we discuss the topic from the point of view of food webs and the consumers. This material is consumed by many organisms and is the major source of energy in cases with high input rates or where primary production is limited.

Analysis of benthic and pelagic food webs of a subtropical lake suggests that omnivory and detritus feeding are a general feature of aquatic food webs (Havens *et al.*, 1996b). Most orders of aquatic insects (Cummins and Klug, 1979) and other groups of invertebrates contain omnivorous organisms that consume detritus. For example, in the detailed food web documented for a riffle in an Ontario creek, the majority of the primary consumers were omnivorous (Fig. 20.2). Any benthic organism that does not exclusively specialize on macrophytes, periphyton, or predation on animals probably consumes detritus as a significant portion of its diet. Many species of tadpoles feed on detritus and some fishes are detritivores as well, notably the carp species used in Asian polyculture discussed in Chapter 23.

Detritus itself is not particularly nutritious and it often contains significant quantities of cellulose and lignins, which most consumers cannot digest. The microbial community generally must first condition allochthonous materials before they can be profitably consumed (Allan, 1995). As mentioned in the last chapter, detritivores derive much of their nutrition from the microbes that eat the detritus; microbial colonization of detritus is called *conditioning*, and the "peanut butter on a cracker" analogy has been used in reference to

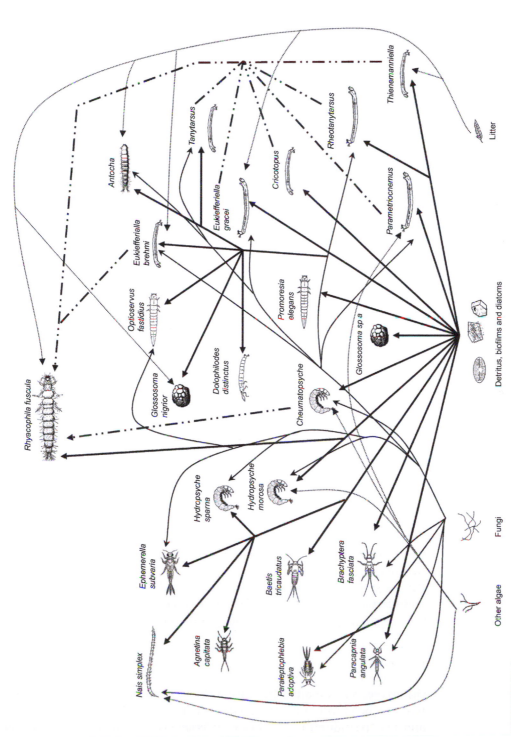

FIGURE 20.2

A food web of Duffin Creek, Ontario, Canada, in December illustrating the high degree of omnivory of the aquatic invertebrates. *(Data from Tavares–Cromar and Williams, 1996).*

the more nutritious microbial "peanut butter" on the detritus "cracker." In streams, fungi are a very important component of the microbial community that conditions leaves and wood entering from the surrounding riparian vegetation. Detritivorous fishes, such as larvae of common carp (*Cyprinus carpio*) and tilapia (*Oreochromis niloticus*), ingest significant quantities of bacteria in the detritus they consume (Matena *et al.*, 1995).

Detritivores can enhance their nutritional intake by selectively feeding on more nutritious components of available detritus. In laboratory studies, leaf-shredding caddisfly larvae offered leaf disks conditioned for different amounts of time selectively fed on the more conditioned disks (Arsuffi and Suberkropp, 1985). Similarly, wood-feeding detritivores apparently derive much of their nutrition from epixylic biofilms composed of microbes, extracellular polysaccharides, and organic particulates (Couch and Meyer, 1992; Eggert and Wallace, 2007). There is also evidence that different species of detritus-colonizing fungi vary in their nutritional quality, and that detritivores select materials colonized by more nutritious types of fungi (Arsuffi and Suberkropp, 1989). Consumption of detritus can create positive feedbacks; feeding activities of many detritivores rupture leaf cuticles, reduce particle sizes and increase surface area for microbial colonization. Egested materials are also readily colonized by microbes, and coprophagy by detritivores is an important process in some freshwater food webs (Wallace and Webster, 1996).

The identity of the food sources also influences its nutritional quality. Several species of riparian tree leaves were fed to the stonefly *Tallaperla maria*. Some leaf species resulted in negative growth rates while others resulted in positive growth. Effects of additions of multiple species of leaves were not additive, suggesting complex nutritional interactions among food-source types, and that identity of riparian vegetation can alter the detritivore community in streams (Swan and Palmer, 2006).

Quality of detritus may be altered by global increases in CO_2. Tuchman *et al.* (2003) grew *Populus tremuloides* trees under doubled current atmospheric CO_2 concentrations and ambient concentrations. The leaves from trees grown under elevated CO_2 were more carbon rich and had more recalcitrant compounds. The leaves were conditioned by microbial communities and then fed to mosquito larvae (*Aedes albopictus*). Mortality of larvae fed leaves from the elevated CO_2 treatments was significantly greater than those fed control leaves.

OMNIVORY

Omnivores consume materials from different trophic levels of the food web. Many, if not most, aquatic animals eat more than one type of food during their life span. Some organisms, such as crayfish, can be predators, herbivores, and detritivores within a life stage. Others, such as many predatory fishes, can be zooplanktivores as fry and piscivores as adults. The prevalence of omnivory greatly complicates food web and functional analyses. It can be difficult to determine the trophic position of organisms given omnivory and technical difficulties associated with determination of ingestion and assimilation. In general, what an organism ingests is more related to its function (e.g., leaf shredding insects facilitating decomposition) and what it actually assimilates is more related to its trophic status.

Several methods have become available to help identify diets and trophic status of omnivores (Method 20.1). Use of these techniques is revealing that omnivory is more the rule than the exception in most ecosystems.

METHOD 20.1

Tracers in Food Web and Nutrition Studies

Gut contents are often used to determine what organisms eat, but this method can be misleading because it does not necessarily describe what organisms assimilate. For example, trout in some streams can have considerable amounts of filamentous algae (e.g., *Cladophora*) in their guts. They are not able to digest this material, but simply ingest it when feeding on invertebrates that live in the algae. The trout ingest but do not assimilate the algae. Feeding and growth experiments in the laboratory are one way to assess the use of various food categories. Analysis of natural abundance of *stable isotopes*, primarily the stable isotopes (not radioactive) of carbon and nitrogen, is becoming an increasingly common method for assessing trophic position of animals and what food sources primarily supply food webs. Fatty acids can also be used to trace food sources; both fatty acid and stable isotope analyses reflect what a consumer has assimilated.

Stable carbon and nitrogen isotopes occur naturally and are useful in the determination of food sources in food webs (Peterson and Fry, 1987; Peterson, 1999). Nitrogen and carbon isotopes ^{15}N and ^{13}C are not radioactive but are heavier than their more abundant counterparts (^{14}N and ^{12}C) in the natural environment. These isotopes are useful because they are fractionated (selected for or against) by physical and biological processes to some degree, causing slight but often consistent variations in natural abundance. *Natural abundance* is the ratio of the trace isotope to the more abundant isotope. Because these ratios are often very small, natural abundance is commonly expressed as a δ value relative to some known standard. The equation used for ^{15}N is

$$\delta^{15}N = \left[\left(\frac{^{15}N_{sample} / ^{14}N_{sample}}{^{15}N_{standard} / ^{14}N_{standard}} \right) - 1 \right] * 1000$$

which is simply the 15/14 N ratio in a sample to that of a standard (in this case atmospheric N_2 gas). Units are parts per thousand (‰). An equation of the same form is used for carbon. A mass spectrometer usually is used to determine the isotopic ratios.

Typically, ^{15}N is the first choice for determining trophic status. There is generally a 3 to 5‰ increase in $\delta^{15}N$ for each trophic transfer, allowing resolution of the feeding levels (Fig. 20.3). In addition, different food sources (e.g., terrestrial vegetation versus algal material) often differ significantly in their natural abundance signatures.

The use of ^{13}C ratios is more difficult. There is not consistent fractionation of ^{13}C across food webs and the fractionation is not as great (Fig. 20.3). However, ^{13}C ratios can often be used to resolve food web differences where there is overlap in the ^{15}N data. Use of natural abundance of both isotopes simultaneously can provide information that laboratory experiments cannot. Additional isotopes (^{18}O, ^{34}S deuterium) can further resolve food sources in some cases.

There are numerous examples of the utility of isotope analyses; a few are noted here. Littoral fishes and crayfish in Canadian Shield lakes are dependent on terrestrial vegetation (France, 1996, 1997b). Habitat usage varies and food preferences at different life history stages changed in fishes in a subtropical lake (Fry *et al.*, 1999). It was demonstrated

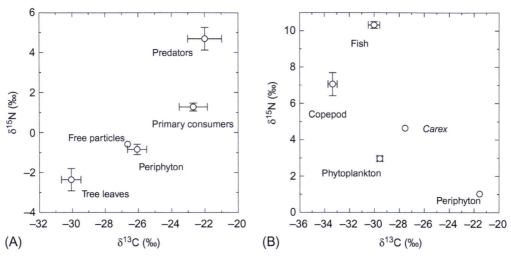

FIGURE 20.3

Stable isotope signatures of components of the food web in Lookout Creek, Oregon (A), and Toolik Lake, Alaska (B). Errors plotted as standard error where data were available. Note that the use of both isotopes allows for clearer separation of the food web components. The primary consumers in the stream probably rely on periphyton, but those in the lake likely rely on phytoplankton. *(Data from Fry, 1991).*

that ocean-derived nitrogen is transported into streams and lakes by spawning salmon (Kline *et al.*, 1990; Finney *et al.*, 2000). Nitrogen fluxes in stream food webs have been quantified (Hall *et al.*, 1998; Peterson *et al.*, 1997; Mulholland *et al.*, 2008). Isotopes have also been used to quantify how important bacteria are as a carbon source in stream food webs (Hall and Meyer, 1998).

Other tracers can be used to examine food web structure in aquatic ecosystems in addition to stable isotopes. Fatty acid profiles can also be used to assess diets and trophic relations and are gaining increasing attention among aquatic ecologists. Fatty acids are essential foundations of energy storage lipids in consumers, but they are not readily synthesized by animals. Thus, they must be obtained through feeding and they are conserved. Various food resources such as algae, bacteria, and detritus have unique combinations of fatty acids, which are reflected in the fatty acid profiles of consumer tissues. To assess diets, samples of consumers and possible food resources are collected and lipids are extracted with organic solvents. Analyses involve separation of polar and nonpolar lipids with a solid-phase extraction system, conversion into fatty acid methyl esters, and separation using a gas chromatograph with a flame ionization or mass spectrometer detector.

Fatty acid analyses have been used to assess diets and food web structure in marine and freshwater systems (Muller-Navarra *et al.*, 2000; Stübing *et al.*, 2003; Sushchnik *et al.*, 2003), as well as aquatic-terrestrial food web linkages (Koussoroplis *et al.*, 2008). In some cases, fatty acid analyses can provide detailed taxonomic information, including specific types of algae assimilated (e.g., Napolitano *et al.*, 1997). In a recent study of the food web of a forested headwater stream, fatty acid analyses revealed that the mayfly *Ephemerella* and the caddisfly *Hydropsyche* assimilated primarily autochthonous materials, even during the summer when shading by the riparian canopy would be expected to limit primary production in the stream (Torres-Ruiz *et al.*, 2007).

Isotopic N data can be used to establish length of food chains. The degree of enrichment of ^{15}N between primary food sources and top consumers can be used as an index of food chain length. Interestingly, omnivory does not alter the ability to estimate chain lengths by using this method, but the method does require assuming a standard amount of enrichment at each level of the food chain. A study involving cross-system analyses using these methods (Vander Zanden and Fetzer, 2007) assumed 3.4 δ^{15}N fractionation per trophic level, and suggested that food chain lengths in streams are significantly less (3.5 levels) than in lakes (4 levels).

ADAPTATION TO PREDATION PRESSURE

Defenses against predation include mechanical, chemical, life history, and behavioral protections. Chemical protection includes both toxic or unpalatable chemicals and low nutritional quality. Mechanical protection includes size (either too large or too small to be eaten) and protective spines or projections. Behavioral responses to predation include avoidance and escape behaviors.

Evolved defenses can be permanent or inducible. Often times it is energetically efficient to defend against predation only when there is a threat. Thus, inducible defenses (chemical, mechanical, or behavioral) can be selected for. We have already discussed some inducible defenses in the last chapter. In other cases, there is always a threat of predation, so permanent defenses are more common.

One common mechanical defense against predation is for the prey to be too large for the predator to effectively consume. In Chapter 19, we discussed how large colonial algae dominate the plankton of lakes with significant zooplankton populations. In a benthic example involving both size and alteration of life history, a snail, *Physella virgata*, increases its growth rate and delays reproduction in the presence of chemical cues from the predatory crayfish *Orconectes virilis*. The more rapid growth increases the probability the snail will reach a size at which the crayfish can no longer effectively prey upon it. In the absence of this predator, the snail reproduces at a smaller size (Crowl and Covich, 1990). Thus, evolution favors growth and deferred reproduction in the presence of the predator. Fishes also increase body size to limit predation, as demonstrated by the crucian carp (*Carassius carassius*), which increases the height of its body in response to the presence of predators (Brönmark *et al.*, 1999). This adaptation requires predators on the crucian carp to have a larger gape to consume them.

Spiny fin rays of fish, hard shells of mollusks, gelatinous sheaths of algae, and cases of caddisfly larvae and other aquatic insects are examples of mechanical defense. Growth of spines in *Daphnia* can serve as a mechanical defense, increasing their effective size. Several invasive species of zooplankton in the United States, such as *Bythotrephes cederstroemi* (the spiny water flea) and

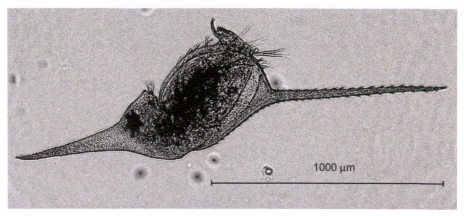

FIGURE 20.4
Daphnia lumholtzi, an invasive species in the central United States with large spines that protect it from predation. *(Photograph courtesy of K. D. Hambright).*

Daphnia lumholtzi, have very long spines (East *et al.*, 1999; Fig. 20.4). This may allow them to resist predation and could be partially responsible for their successful invasion of lakes that do not contain zooplankton species with such long spines.

Chemical protection from predation includes toxic or bad-tasting chemicals (*allomones*, chemicals that affect another species to the benefit of the producer) and poor nutritional quality. Invertebrates may use visual, hydromechanical, or chemical cues to avoid predators (Dodson *et al.*, 1994). Chemical cues produced by predators include *kairomones*, compounds produced by predators that positively affect the behavior, morphology, or life history characteristics of prey species in ways that allow avoidance of predation, and *alarm chemicals*, chemicals that are produced by damaged or stressed organisms. Alteration of behavior by alarm chemicals is common in fishes.

Chemical defenses are probably common in freshwater organisms, particularly those that are prey to fishes. Where such defenses have been sought, they have been well documented; there are numerous examples of toxin production in *Coleoptera* and *Hemiptera* (Scrimshaw and Kerfoot, 1987). Whirligig beetles of the genus *Dineutus* produce a volatile chemical that smells like sour apples, which is used for protection from vertebrate predators. In parts of Africa, some young women use adult gyrinid and dytiscid beetles to stimulate breast development; the beetles are allowed to bite their breasts and, in the process, expose the women to defensive chemicals containing hormone-like steroids that enhance breast development (Kutalek and Kassa, 2005). Some stream-dwelling stoneflies use autohemorrhaging, or reflex bleeding, when threatened, whereby they increase their body fluid pressure until hemolymph (blood) exudes from some joints of the exoskeleton (Williams and Moore, 1990). Insect hemolymph can serve as a chemical defense when it is sticky and coagulates quickly, or it can contain foul tasting chemicals.

Poor food quality can also provide some protection against predation. However, the first response to low-quality food is to consume more, so this adaptation is only effective if quality is so poor that so much energy is expended to consume enough nutritious food that reproduction and survival are compromised. An example of poor food quality leading to decreased success of a predator is the relationship between highly unsaturated fatty acid content in phytoplankton and growth of their zooplankton predators. These fatty acids are almost exclusively synthesized by photosynthetic organisms but are required for animal growth. Low-phosphorus algae that are high in these fatty acids are a better food source for *Daphnia* than high-phosphorus, low-fatty-acid phytoplankton (Brett and Müller-Navarra, 1997), potentially leading to more effective grazer control of trophic state at lower phosphorus concentrations (Müller-Navarra *et al.*, 2004). Single species of diatoms can be nutritionally unsuitable prey for copepods (Jones and Flynn, 2005). Therefore, simple measures of food quality, such as nitrogen and phosphorus content, may not adequately represent nutritional quality.

Animals exhibit numerous behavioral responses to predation. Behavioral defenses may be quite complex and require sensing of predators and complex behaviors to avoid predation. Among invertebrates, behavioral responses range from nocturnal foraging to avoid visual predators, thanatosis (feigning death), and release from the substrata and downstream drifting in stream insects. Behavioral defenses can be combined with chemical defenses, as with reflex bleeding described earlier.

Probably one of the most studied behavioral responses to predation in aquatic systems is the diel vertical migration of *Daphnia* and other zooplankton in lakes (Fig. 20.5). These zooplankton enter surface waters to consume phytoplankton at night when the darkness lowers predation by zooplanktivores that rely on vision to find prey. At daybreak they swim down to darker parts of the lake to avoid the visual feeders. In some cases, larger *Daphnia* that would be more susceptible to predation move deeper than smaller individuals (Hutchinson, 1967; De Meester *et al.*, 1995). Chemical cues excreted by predators or released by predators during feeding may be required to trigger the migration response (Folt and Burns, 1999; Tollrian and Dodson, 1999). Similar migrations can be seen in larger rivers (Jack *et al.*, 2006), probably to avoid sight-feeding predatory fishes.

A chemically mediated behavioral defense of *Daphnia* to predation by fish involves predator-avoidance behavior triggered by alarm chemicals released when other *Daphnia* are eaten. In this case, *Daphnia* swim downward, form

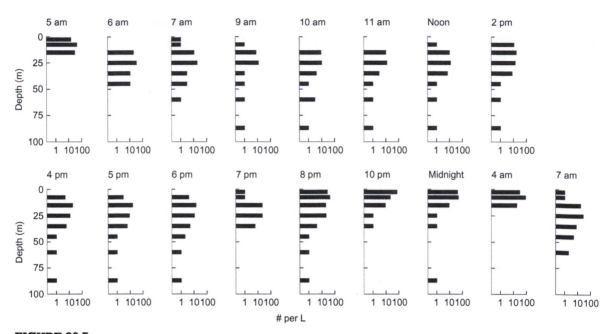

FIGURE 20.5

Diel vertical migration of young *Daphnia longispina* in Lake Lucerne. *(Data from Worthington, 1931).*

aggregate swarms, or increase avoidance behavior after exposure to water with crushed *Daphnia* (Pijanowska, 1997). These behaviors lower predation rates on *Daphnia* from bream (*Abramis brama*). Similar defenses have been documented for the related cladoceran *Ceriodaphnia reticulata* in response to chemicals produced by green sunfish, *Lepomis cyanellus* (Seely and Lutnesky, 1998). Chemicals released by fish can also increase sensitivity of *Daphnia* to mechanical signals arising from movement of predators (Brewer *et al.*, 1999).

Inducible responses to predation have evolved in many species. In Midwest US reservoirs, cooccurring *Dahnia pulicaria* and *Daphnia mendotae* exhibit different evolved responses to predation (Bernot *et al.*, 2006). These species come under increased predation pressure as larval fishes increase in number in the late spring. Exposure to chemicals released by fish into waters (kairomones) caused *Daphnia mendotae* to reproduce at smaller size. Smaller size at first reproduction aids in avoidance of predation because smaller adults are less susceptible to predation. In contrast, *Daphnia pulicaria* produced resting eggs (ephippia) as temperature increased, simply disappearing from the water column as predation pressure increased.

Hydromechanical cues can be an important component of predator avoidance (Peckarsky and Penton, 1989). These pressure waves are transmitted rapidly at the small spatial scales at which insect larvae and smaller organisms operate. Although use of hydromechanical cues may be widespread, much less is known about these than chemical cues (Dodson *et al.*, 1994).

Predator avoidance in time and space is a crucial adaptation to predation in many species. Many species simply hide or swim away when they sense a predator in the vicinity. Timing of reproduction is another approach to avoiding predation. This is the case for some species of zooplankton that produce diapausing eggs in the presence of increased predation pressure (Hairston, 1987). Massive synchronous hatches of aquatic insects not only increase the chance that adults will be able to find a mate but also saturate predators, allowing a greater probability for some adults to survive and mate.

ADAPTATIONS OF PREDATORS

Predation can be characterized logically by a sequence of events. Prey must be encountered and detected, then attacked and captured, and finally ingested (Brönmark and Hansson, 1998). Adaptations are evident at all stages; this section is organized by the natural sequence of events.

The behavioral strategy used by organisms to obtain prey can vary from remaining immobile and allowing prey to approach to active foraging. The specific strategy that has evolved presumably allows for the most efficient harvesting of food given morphological, abiotic, and community constraints. The simplest encounter strategy is to "sit and wait" for prey. For example, pike (*Esox*) lay hidden, waiting for prey to come close. Pike have large tail (caudal) fins and pull their bodies into "S" shapes from which they can accelerate explosively to catch prey. Odonate dragonfly nymphs also wait for prey. They have a hinged labial jaw that ejects very rapidly and snatches prey. *Hydra* capture larval bluegill (*Lepomis macrochirus*) that contact its stinging nematocysts, a form of sit-and-wait predation that can have considerable impact on the populations of the larval fish (Elliot *et al.*, 1997).

Carnivorous plants in wetlands are notable sit-and-wait predators on insects. Most species tolerate or require saturated soils and are wetland species (Juniper *et al.*, 1989). Passive trapping strategies are used, which include the use of adhesive traps such as in sun dew (*Drosera*), chambers that can be entered but not left as in the pitcher plants (e.g., *Sarracenia*), snap traps such as the Venus flytrap (*Dionaea*), and triggered chambers as in the bladderworts (*Utricularia*). These plants also need to attract insects, and adaptations include visual stimuli such as UV patterns visible to insects. Olfactory substances can be produced, including those with nectar scent or the smell of putrefaction, and nectar rewards may be used to lure insects. Tactile stimuli are also important in some cases, such as that of the bladderwort, *Utricularia*, which has filamentous extensions on submerged gas-filled bladders that mimic filamentous algae and attract epiphyte feeders; when the invertebrate contacts the extensions, the bladder fills rapidly with water and sucks in the prey (Juniper *et al.*, 1989). Carnivorous plants are generally found in low-nutrient environments and are photosynthetic, so they use their prey as a source of nitrogen and phosphorus. Carnivorous plants are an excellent example of convergent evolution leading to solution of a problem (nutrient limitation) from divergent plant lineages using a wide variety of capture and attraction mechanisms.

Sensing prey can be accomplished by a variety of adaptations, depending on the organisms and their prey. Invertebrates use visual (in a few organisms with well-developed eyes), mechanical, tactile, and chemical cues (Peckarsky, 1982). In an ingenious demonstration of the importance of the use of mechanical cues, Peckarsky and Wilcox (1989) recorded hydrodynamic pressure wave patterns associated with escaping *Baetis* nymphs. Predatory stonefly nymphs (*Kogotus modestus*) attacked *Baetis* models in greater frequency when the wave patterns were played back than when they were not.

Fishes can sense prey visually, chemically, electrically, or hydrodynamically. Electrical sensory systems in fishes are highly developed; the paddlefish (*Polyodon spathula*) can sense the electrical activity of a swarm of *Daphnia* 5 cm away (Russell *et al.*, 1999).

The idea that evolution through selection leads to maximization of net energy gained per unit time feeding, led to the concept of *optimal foraging* (Pyke *et al.*, 1977; Schoener, 1987). This simple concept can be applied to all phases of predation (encounter, detection, attack, capture, and ingestion). We have already discussed the relative merits of sitting and waiting for prey as opposed to active searching. Food quality, quantity, and spatial distribution are often additional considerations for optimal foraging. Optimal foraging has been well established for several fish species (Mittelbach and Osenberg, 1994). A food item may not be preferred if it is high quality but very rare relative to a lower quality food source. All items, even remotely suitable items, may be taken when food is limiting, but only the most profitable may be taken when more is available. For example, bluegill will capture all sizes of zooplankton in equal amounts when the food is at low density but will take large *Daphnia* preferentially at higher zooplankton concentrations (Fig. 20.6).

Predictions can also be made regarding how long a predator will remain in a patch of food. If a patch is of low quality, then moving to another patch may be more beneficial. However, if the cost of moving to another patch is high, it can be beneficial to extract more food from the current patch. Obviously, temporal and spatial scale are important considerations when making predictions using optimal foraging models. Similar considerations form the basis of the field of landscape ecology, discussed in Chapter 22.

A problem with using optimal foraging theory to make predictions about behavior of predators is that an investigator may not be able to identify the

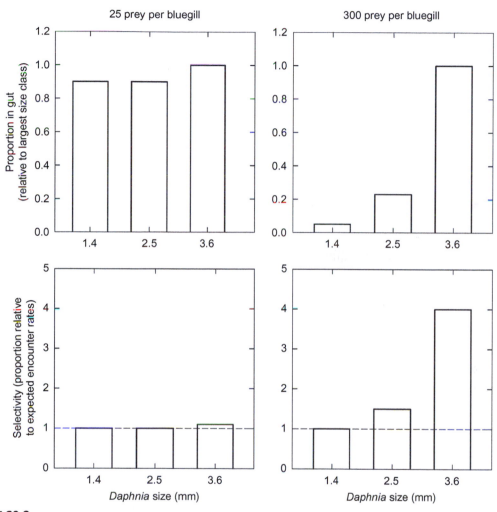

FIGURE 20.6

Selectivity of bluegill (*Lepomis macrochirus*) on size of prey (*Daphnia magna*) at two prey densities. At low densities, all sizes of *Daphnia* are equally represented in the bluegill guts. At high densities of prey, the larger zooplankton are preferred (selectivity of 1 means consumption is the same as expected relative to random encounter rates). *(Data from Werner and Hall, 1974).*

most important selective forces (Gatz, 1983). For example, we could predict that large prey items are preferable to smaller items because they contain more energy. However, large prey may be encountered infrequently and may take more energy to capture. Taking many small prey items may be more energy efficient and seems to be the strategy used by many predatory freshwater fishes (Juanes, 1994).

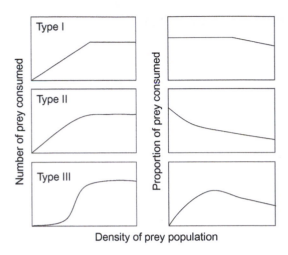

FIGURE 20.7

Holling's three types of functional response curves and the proportion of prey consumed assuming constant predator numbers.

The number of prey eaten per unit prey density is referred to as the *functional response*. The number of predators per unit prey density is referred to as the *numerical response*. A functional response curve can take one of three general forms (Fig. 20.7).

In a type I functional response, predation is linearly related to prey density until some saturation is reached; this is the simplest predation model. In a type II response, there is a hyperbolic response to prey density. The form of this response is similar to that seen in the Michaelis–Menten uptake kinetics described in Chapter 17. This form of response is typical of predators that do not have a complex behavioral response to acquiring prey. Just as an enzyme can react with only so many molecules per unit time, a predator can only take a maximum number of prey items per unit time.

The type III response is seen in more advanced predators. At very low prey densities, few or no prey items are taken (optimal foraging theory predicts that there is not enough energy gain to profitably take prey). At intermediate prey densities the rate of capture increases, and eventually the predator is saturated, as in types I and II. Numerical response generally occurs over longer time periods because it requires changes in the predator's populations (from birth, death, immigration, and emigration), unlike functional responses that are primarily limited by behavioral and physiological aspects of the predator and distribution of prey in the environment.

Several strategies are used to consume prey. Many predators need to consume prey whole. These are referred to as *gape-limited* predators, and the size of prey they can consume is limited by the size of their mouth or gape (Zaret, 1980). The dominant vertebrate predators in freshwaters are mostly of this type. Generally, gape-limited predators are highly dependent on the size of their prey, whereas others may be dependent to various degrees.

NONLETHAL EFFECTS OF PREDATION

The previous discussion assumes that the predator will kill its prey; however, other effects can occur, including injury, restriction in habitat use or foraging behavior, and changes in life history (Allan, 1995). Strategies of nonlethal predators include piercing or sucking prey and removing bites or chunks of prey. Such predators include copepods, predatory cladocera, some midge larvae, fishes that are scale predators, and other insect larvae. Parasites are a special class of predators that have a wide variety of ways to attack prey, but often do not kill their hosts.

Examples of nonlethal predation have been described elsewhere in the book and include herbivores that consume part of a macrophyte, the scale-eating Cichlids of Africa, restricted foraging time in the presence of a predator, and most cases of parasitism. Also, earlier in this chapter we described how the presence of crayfish could lead to delayed reproduction of snails so they more quickly attain a size too large to be consumed. We also described a variety of behavioral responses that alter behavior of prey. Most cases of predator avoidance will be classified as nonlethal effects of predation because of the associated energetic cost.

An additional example of sublethal effects and how they can alter a lake food web was documented by Hill and Lodge (1995). Omnivorous crayfish (*Orconectes*) were exposed to largemouth bass (*Micropterus salmoides*) that were too small to eat them. These bass were apparently perceived as a predation risk because crayfish survival and their feeding on macrophytes and invertebrates decreased. Thus, even though the predator did not directly kill the crayfish, the nonlethal effects on the crayfish and its food species were significant.

TROPHIC LEVELS, FOOD WEBS, AND FOOD CHAINS

Food webs can be simplified into trophic levels for ecological analyses. These levels were described briefly in Chapter 8. Traditionally, levels include *primary producers* (photosynthetic organisms), *decomposers* or detritivores (consume dead organic material), *primary consumers* (herbivores or grazers), and *secondary consumers* (eat primary consumers). This simplification has been criticized because numerous organisms feed on several trophic levels. Thus, some have suggested that species can be assigned fractional trophic levels (e.g., primary producers are assigned level 1, herbivores level 2, predators of the herbivores level 3, and animals that eat half herbivores and half primary producers level 2.5). The idea of trophic levels, regardless of its weaknesses, has proven particularly important in analysis of the effects of top consumers on organisms lower in the food web. These cascading trophic interactions will be discussed next.

THE TROPHIC CASCADE

In 1880, Lorenzo Camerano postulated that animals can control the biomass of lower trophic levels and diagrammed how a carnivore can control herbivore populations and this can have a positive influence on plants (reprinted in Camerano, 1994). The idea that predation at the upper level of food chains can have a cascading effect down through the food chain is called the *trophic*

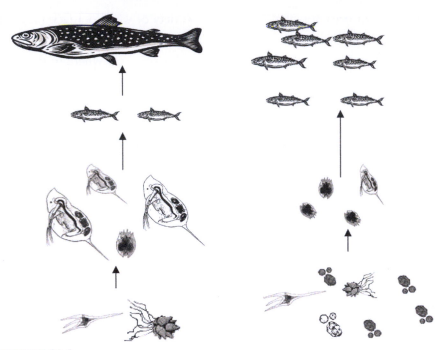

FIGURE 20.8

A conceptual diagram of the trophic cascade in the pelagic zone of lakes. When large piscivores are present, smaller zooplanktivores are uncommon, body size and numbers of zooplankton increase, and phytoplankton decrease.

cascade. Control of primary production by abiotic factors such as nutrients or light is called *bottom-up control*. Control of primary producers from the upper levels of the food chain is referred to as *top-down control*, which is sometimes used interchangeably with trophic cascade. These ideas entered modern ecology in a key paper by Hairston *et al.* (1960), which argued that plants dominate terrestrial systems because predators keep herbivores in check. These arguments were extended to the idea that even numbers of links in food chains (Fig. 20.8) will lead to higher biomass of primary producers (Fretwell, 1977). In this section, we discuss how the trophic cascade could apply in lakes, streams, wetlands, and groundwaters.

In a highly cited paper, Brooks and Dodson (1965) documented that increased predation pressure by a planktivorous fish led to much smaller zooplankton species. Smaller zooplankton are less efficient grazers and the population size shift resulted in increases in chlorophyll. The idea that the trophic cascade was operating in lakes received some subsequent attention (Arruda, 1979; Shapiro, 1979), but a series of papers by Carpenter, Kitchell, and coworkers (e.g., Carpenter and Kitchell, 1987; Carpenter *et al.*, 1987) stimulated abundant research on the trophic cascade in lakes (Power, 1992). This research was stimulated because managers may be able to use *biomanipulation* of lakes to improve water quality by controlling the fish community. The general approach is to encourage the populations of larger fishes that consume zooplanktivores. The trophic manipulation leads to increases in large grazing zooplankton (i.e., average body size of zooplankton increases). The large grazing zooplankton can consume more phytoplankton. Consequently, there is an increase in water clarity related to a decrease in suspended algae. Food web manipulation in concert with nutrient control can lead to significant increases in water quality over decades, as has been demonstrated in Lake Mendota, Wisconsin (Lathrop *et al.*, 1996). However, the approximately 10 decades of research on Lake Mendota illustrate that bottom-up and top-down conditions must both be considered as factors influencing water quality (Kitchell and Carpenter, 1992).

The key issue in propagation of bottom-up or top-down effects in lakes seems to be how well the effects are transmitted between the zooplankton–phytoplankton link (Carney, 1990; Elser *et al.*, 1990; Elser and Goldman, 1991), although the trophic cascade may also break down at the link between zooplankton and fish (Currie *et al.*, 1999). Empirical analysis suggests the grazer link is where top-down and bottom-up effects are decoupled (Brett and Goldman, 1997). A study comparing 11 experiments in shallow lakes found that nutrient effects were three times stronger than top-down effects (Moss *et al.*, 2004). The same analysis indicated that the zooplankton–grazer link transmitted top-down effects. In oligotrophic lakes, nutrients may be so limiting that top-down effects cannot occur (Spencer and Ellis, 1998).

In highly eutrophic lakes, large, grazer-resistant cyanobacterial colonies can dominate, leading to lack of grazer control on phytoplankton and decoupling of top-down effects from the bulk of primary producers. Anoxic hypolimnia of lakes may allow some species to avoid control by trophic manipulation of fish species in the epilimnion. The zooplanktivorous predator *Chaoborous* can find refuge in the anoxic hypolimnion of a eutrophic lake, and alteration of the fish component of the zooplantivorous food web in this case can have little influence (Dawidowicz *et al.*, 2002).

The trophic cascade can extend to physical and chemical aspects of the lake. Low zooplankton grazing rates can lead to greater influx of atmospheric CO_2 related to increased algal biomass with high photosynthetic demand (Schindler *et al.*, 1997). Predation by zooplanktivorous fish can increase nutrient supply by excretion, enhancing algal production both by indirectly lowering grazing pressure and by directly providing nutrients (Persson, 1997; Vanni *et al.*, 1997; Vanni and Layne, 1997). Mazumder *et al.* (1990) demonstrated that lakes with higher abundance of planktivorous fishes had lower zooplankton, higher phytoplankton, and lower light penetration. The lower light penetration led to less heating of the deeper water in the spring, which led to a shallower epilimnion depth. Similar results occurred both in 15-m-deep, 8-m-diameter enclosures in which the plankton was manipulated, and in 27 small Ontario lakes with varied levels of planktivorous fish. These results demonstrated a clear local effect of organisms on heat content and physical structure of the lake as mediated by the food chain.

The trophic cascade concept in lakes requires simplification of food webs into food chains with discrete trophic levels. Several examples illustrate how community complexity can alter the intended effects of food chain manipulation. In shallow eutrophic ponds, very low rates of zooplankton grazing resulting from intense predation can lead to blooms of unwanted cyanobacteria (Spencer and King, 1984). In the same ponds without fishes, zooplankton flourish, phytoplankton decrease drastically, and macrophytes and periphyton growths can reach nuisance levels. Similar results were seen in shallow eutrophic lakes in The Netherlands. In this case, fish biomass was lowered, zooplankton initially flourished, phytoplankton decreased, and then macrophytes filled the lakes, leading to suppression of both zooplankton and phytoplankton numbers (Fig. 20.9). Another example of how complexity of food webs alters efficacy of lake biomanipulation is the fact that large fishes reproduce well when their numbers are increased for trophic control of zooplankton. The young of the year fishes that are produced are zooplanktivorous and can reverse the effects of the biomanipulation, at least temporarily (Hansson *et al.*, 1998).

In warm reservoirs in the United States, gizzard shad (*Dorosoma cepedianum*) can be the most abundant fish. As young of the year, these fish consume individual zooplankton. After growing to several centimeters in length, the fish develop a long gut that includes a grinding chamber and also a filtering structure on their gills. This allows gizzard shad to consume and thrive on phytoplankton, zooplankton, and detritus. Gizzard shad feed on several trophic levels at once and this habit confounds the use of a food chain model to describe trophic dynamics in these reservoirs (Stein *et al.*, 1995). In other shallow lakes, the greatest effect of fishes on algae may not be through the food web but through increased sediment suspension (Havens, 1991) and nutrient mineralization by benthic fishes (Gido, 2002).

Another example in which the idea of a food web rather than a food chain more accurately characterizes aquatic trophic relations is the link between the trophic cascade and the microbial loop. This link is equivocal; some investigators have established the presence of a top-down effect on the microbial loop, and others have not. For example, Pace *et al.* (1998) demonstrated variable effects on heterotrophic flagellates but increased populations of ciliates and rotifers in lakes with low *Daphnia* populations. Gasol *et al.* (1995) showed

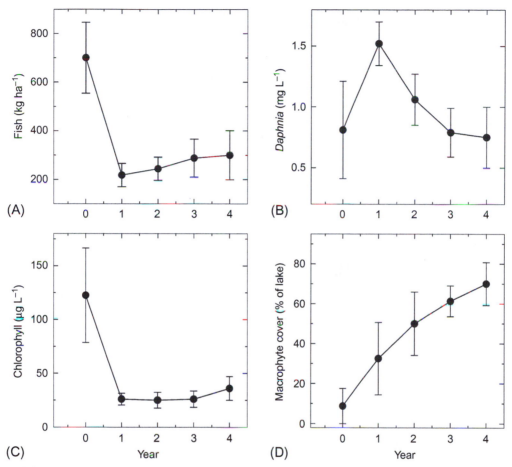

FIGURE 20.9

Effects of biomanipulation on fish, chlorophyll, *Daphnia*, and macrophyte cover in four shallow eutrophic lakes in The Netherlands. Year 0 is before biomanipulation. Points are means from four lakes and error bars equal 1 standard deviation. *(Data from Meijer et al., 1994).*

that cladocerans have the greatest effects on heterotrophic nanoflagellates (very small protozoa) during the summer. Pace and Cole (1994) suggested that there was little influence of *Daphnia* populations on bacteria. Simek *et al.* (1998) described a lake in which a filtering cladoceran controlled bacterial populations in one basin and a ciliate was the primary bactivore in another basin. Jeppesen *et al.* (1998) found a weak link between the microbial loop and zooplankton abundance in an 18 year time series from a hypereutrophic lake.

Consideration of temporal and spatial scale can also alter the response of food webs to manipulation. For example, pulses of nutrients can have different effects depending on food web structure and their timing (Cottingham and

Schindler, 2000). If zooplankton have time to respond to nutrient enhancement of phytoplankton growth, further pulses of nutrients will have little influence. If predation on zooplankton is high, they may not be able to suppress ephemeral phytoplankton blooms in response to nutrient pulses (Strauss *et al.*, 1994).

A strong caution of the trophic cascade concept has been put forward by Wetzel (2001). He mentions the numerous compensatory mechanisms that emerge quickly after biomanipulation. These could include predator protection and avoidance by zooplankton, development of an inedible phytoplankton assemblage, and eventual high macrophyte dominance (in shallow lakes). Eventually, evolution will counter changes in predation pressure; this concept brings up the argument over biotic versus abiotic factors ultimately controlling evolution.

Top-down effects can also occur in benthic habitats of lakes. Snails remove littoral periphyton and are susceptible to predation by sunfish. When sunfish are excluded, algal biomass decreases significantly (Brönmark *et al.*, 1992). Since sunfish are prey for large piscivores, high biomass of large piscivores could lead to a decreased biomass of periphyton through the tropic cascade. Fish were removed from eutrophic Lake Ringsjön, Sweden, including the littoral benthic feeding bream *Abramis brama*, in an attempt to use biomanipulation to improve water quality. Water clarity improved in the pelagic zone and there was an unintended increase in benthic invertebrate populations and concurrent increases in staging waterfowl abundance (Bergman *et al.*, 1999).

Both top-down and bottom-up control of primary production can occur in streams, and in many cases both operate simultaneously (Rosemond *et al.*, 1993). Clear examples of the trophic cascade have been documented for streams. In some California streams (Fig. 20.10), exclosure of fishes resulted in decreases in *Cladophora* (a filamentous green alga) biomass because predation on midge larvae was decreased, and the midge larvae suppressed *Cladophora* (Power, 1990a). These results are consistent with the view that odd numbers of trophic levels lead to high producer biomass.

Research on the herbivorous stoneroller (*Campostoma anomalum*) and piscivorous bass (*Micropterus* spp.) suggests that bass have a top-down effect on primary producers in small prairie streams. Pools with bass have few stonerollers and high algal biomass, whereas pools without bass have low algal biomass and high numbers of stonerollers. Finally, when bass were tethered in a pool, algae proliferated only in the areas around the bass, but not outside their reach (Power and Matthews, 1983). The tethering effect may have occured because bass also can control crayfish in this system, but the effect is still a top-down effect. Grazing by the stonerollers led to lower algal biomass, more cyanobacteria, less hetertrophic bacteria, higher invertebrate density, and less particulate organic material, so the effects of this trophic cascade extended to basic stream ecosystem properties (Gelwick and Matthews, 1992).

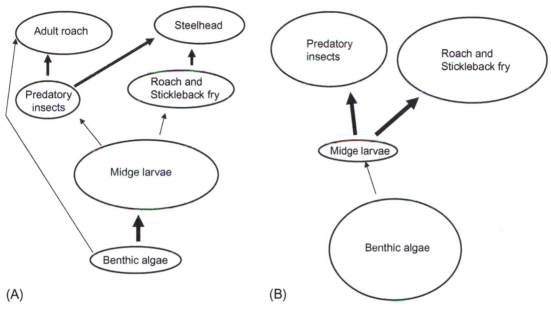

FIGURE 20.10

Summary of effects of enclosure (A) and exclosure (B) on predatory fishes (roach (*Hesperoleucas symmetricus*) and steelhead (*Oncorhynchus mykiss*)), roach and stickleback (*Gasterosteus aculeatus*) fry, invertebrate predators (lestid damselflies), midge larvae (*Pseudochironomus richardsoni*), and benthic algae (*Nostoc* and *Cladophora*). Size of ovals represents amount of biomass at each trophic level. *(Data from Power, 1990a).*

Control of grazers does not always have the anticipated effects in streams. Armored catfish (Loricariidae) are common grazers in the Rio Frijoles in Panama. Exclosure of the catfish led to an initial increase in benthic algal biomass, but after a few weeks, enclosures were smothered with sediments (Power, 1990b). In this case, a secondary effect (sediment resuspension or decrease in sediment trapping) led to results that would not be predicted by a straight trophic view of streams.

Less is known about trophic cascades in wetlands, though the food webs should not differ drastically in structure from other benthic food webs; therefore, there is no reason to believe that the interplay between top-down and bottom-up controls is not important. For example, predaceous hydrophilid beetle larvae can control algae grazing chironomid midge larvae in a seasonal wetland. When predation was not present, the midge larvae outstripped their food supply (Batzer and Resh, 1991). An interesting indirect cascade occurred in Florida wetlands where ponds containing fishes were compared to those that did not. Fish predation decreased dragonfly larvae and subsequently adult populations decreased. The decrease in adult dragonflies led to an increase in pollinators, increasing nearby plant production (Knight *et al.*, 2005). Removal of wolves (*Canis lupus*) in Montana has apparently led to increases in grazing by elk (*Cervis elaphus*) on willows (*Salix*), and decreased riparian cover (Ripple and Beschta, 2004), ultimately influencing the aquatic community by altering shading and litter input. The golden apple snail (*Pomaceacanaliculata*) has invaded numerous wetlands in Southeast Asia, leading to almost complete removal of macrophytes and concurrent increases in phytoplankton (Carlsson *et al.*, 2004).

As with wetlands, the trophic cascade in groundwaters is poorly documented. Grazer control of bacterial production has been demonstrated in several groundwater habitats. The isopod *Caecidotea tridentata* stimulated bacterial production in a limestone aquifer in which carbon additions did not (Edler and Dodds, 1996). A case of bottom-up control was demonstrated in which organic contamination of an aquifer stimulated bacterial and protozoan production relative to nearby uncontaminated groundwater (Madsen *et al.*, 1991). Protozoa depressed bacterial abundance, but increased nitrification rates in a laboratory study using groundwater and associated sediments (Strauss and Dodds, 1997). As more groundwater food webs are described, numerous cases of top-down and bottom-up control are likely to be documented.

SUMMARY

1. Some species from most major groups of aquatic animals can eat phytoplankton, periphyton, macrophytes, detritus, or other animals. Omnivory and detritivory are common in freshwater invertebrates.
2. Herbivores influence plant and algae in a variety of ways, ranging from reducing biomass to increasing biomass-specific production.
3. Detritivores often feed selectively on microbially conditioned materials and gain much of their nutrition from fungi associated with detritus.
4. Stable isotopes and fatty acid profiles are now commonly used to assess trophic interactions.
5. Prey can respond to chemical, visual, and hydromechanical cues.
6. Protective adaptations to predation include mechanical (size and spines), chemical, and behavioral.
7. Predators forage optimally to maximize their efficiency; this may lead to one of several functional responses to prey numbers.
8. Predators can sense their prey by using chemical, visual, tactile, and hydromechanical cues.
9. Trophic cascades are an important part of aquatic food webs and may occur in all freshwater habitats. Effects can be transmitted through food webs from the top (predators) or the bottom (nutrients and light). The grazer–producer link appears to be crucial in transmission of these effects.

QUESTIONS FOR THOUGHT

1. If a predator is consuming prey at a rate lower than the rate at which the prey is able to replace itself, can the effect of predation be considered significant even if the prey population is increasing?

2. How might the increasing use of fungicides in agricultural systems influence adjacent streams, lakes, and wetlands?

3. Why can increasing turbidity of large rivers cause shifts in types of predators that are successful?

4. Why might it be more effective to control algal blooms by biomanipulation with removal of all fish than by imposing fishing regulations to increase the number of piscivorous fish?

5. Why are brightly colored organisms less common in freshwaters than in benthic marine systems?

6. What single cosmopolitan species is the top predator in more freshwater systems throughout the world than any other species?

7. Why are chemical cues to prey easier to follow in streams and benthic habitats than in pelagic habitats?

8. How do selective pressures for streamlining of fishes conflict with limitations on gape width?

9. Are trophic levels a valid concept in most freshwater habitats?

CHAPTER 21

Nonpredatory Interspecific Interactions among Plants and Animals in Freshwater Communities

FIGURE 21.1

An experimental stream mesocosm facility used to study interactions among organisms. Each cylinder represents a pool and each rectangular portion a riffle, with water recirculating in each channel. In the panel below, two herbivorous fish species (*Campostoma anomalum* and *Phoxinus erythrogaster*) are being used in a behavioral experiment in the same mesocosm.

Competition is central to understanding many aspects of aquatic communities and, thus, has been the focus of much research. Other ways that species interact (including indirect interactions, succession, mutualism, and the effects of keystone species) also have consequences in determining community structure. Some of these concepts were discussed in previous chapters including basic definitions of types of interactions (Chapter 8), classification of interaction types (e.g., trait-mediated versus density-mediated interactions; Chapter 19), and interactions involving microorganisms (Chapter 19). This chapter focuses on interactions among larger organisms.

COMPETITION

Competition (both species have a negative effect on each other) has been described as a dominant community interaction. It can occur between organisms from any taxon, but it is more likely to occur among organisms in the same functional group or guild. We discuss competition among species of plants and animals in this section. Competition occurs within a species (*intraspecific*) or between species (*interspecific*).

Intraspecifc competition is *density-dependent* in that competition for a given resource such as food becomes more intense as the relative availability decreases (*density-independent* factors are not a function of population size). At some level of competition, mortality increases or natality decreases, which results in reduced or negative population growth. This type of interaction is important in natural systems, as well as aquaculture and fisheries management, where the goal is to maximize growth of desirable species.

Intraspecific competition can also reduce individual growth rates. For example, tadpoles can reach very high densities in breeding ponds. In experiments where different total numbers of tadpoles are placed in the same sized rearing tanks, reduced growth occurs with increasing densities (Fig. 21.2). In natural ponds in Sweden, experimentally increased densities of *Rana temporaria* tadpoles also resulted in significantly higher mortality rates (Loman, 2004). Density manipulations of other amphibian species have found that tadpoles reared at high densities metamorphose earlier, and at a smaller size, than those reared at lower densities, which has implications for individual fitness (Richter *et al.*, 2009).

Intraspecific competition can be a cause of disruptive selective pressure as different morphologies or strategies of resource use are selected for and the intermediate form is at a disadvantage. For example the three-spine stickleback, *Gasterosteus aculeatus*, lives in lakes and individuals within the same species and the same lake specialize on pelagic or benthic food sources. The morphology of the gill rakers reflects this specialization. In experimental manipulations, when both morphologies were present the difference between the morphologies was more pronounced. The effect was accentuated at higher

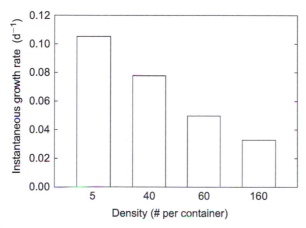

FIGURE 21.2

Daily instantaneous growth rates of *Rana tigrina* tadpoles reared at different densities. Numbers below each bar represent densities of individuals that were reared in containers of the same size. *(Data replotted from Dash and Hota, 1980).*

densities as would be expected if it was a result of intraspecific competition for resources (Bolnick, 2004).

Both intraspecific and interspecific competition can be divided into two general types. *Scramble competition* (*exploitative competition*) occurs when all individuals are "scrambling" to acquire as much as they can of the limiting resource; as availability of the resource decreases, each individual gets less, and all are equally affected. This type of competition does not involve direct interactions among members of the population and is quite common when population densities are high. In *contest competition* (*interference competition*), some individuals get more than others by directly interfering with the ability of others to acquire the resource. Examples of intraspecific contest competition include competition among male dragonflies for territories and mates and competition among male tree frogs for calling perches. Contest competition involves direct interactions among members of the population (e.g., individuals interfering with another individual's ability to acquire the resource). Allelopathy among plant species and territoriality in animals are common forms of interspecific interference competition.

Competition is often difficult to establish directly in the field, in part because evolution leads to decreased overlap in resource use. Intraspecific competition might not be apparent in cases where organisms inhabit defined territories. Interspecific competitors that are usually found in the same habitat at the same time rarely specialize on the same resource. For example, three-spined sticklebacks (*Gasterosteus* spp.) in three lakes of British Columbia have recently evolved from a single marine species. In all these cases, the new pairs of species found in each lake partition the habitat by feeding on benthic or limnetic invertebrates. They have evolved mechanisms of reproductive isolation to ensure that their offspring will inherit their parents' behavioral traits. Limnetic adults prefer to mate with limnetic adults and benthic adults prefer to mate with benthic adults (Rundle *et al.*, 2000).

The concept of competition is often tightly interwoven with that of the *niche*, the ecological role of an organism, particularly as defined by its resource consumption. The idea of niche is somewhat controversial with respect to variability in types of resources used over time, and how much overlap occurs in diets of various species over time. Regardless, species often specialize on resources and partition those resources among species based on varied attributes of those resources.

The idea that competition can be difficult to find in extant communities because evolution has led to species partitioning niches and not competing now is called the "ghost of competition past." It is very difficult to prove that this effect is real (Connel, 1980). Still, the idea provides one good explanation of why it could be difficult to document competition in field experiments.

An additional reason that competition may be difficult to document is that competitive exclusion may lead to conditions where potential competitors do not cooccur. We discussed competitive exclusion in Chapter 17, in the section "The Paradox of the Plankton and Nutrient Limitation." The concept behind the competitive exclusion principle is that under equilibrium conditions, the superior competitor is expected to completely eliminate the inferior competitor. The principle also assumes that both competitors specialize on the same resources. The exclusion may take many generations, but eventually the superior competitor will come to dominate.

The discussion on the paradox of the plankton explored how nonequilibrium conditions, predation, mutualism, disease, and multiple limiting factors could explain coexistence of potentially competitive species of phytoplankton. Many of these concepts can apply equally well to interspecific and intraspecific competitors in any aquatic habitat. For example, if only one genotype was the superior intraspecific competitor, then natural selection should lead to little genetic variation within a species. However, the factors that cause the competitive exclusion principle to fail also select for increased variability within species. Likewise, the types of factors that cause the competitive exclusion principle not to apply to phytoplankton also can increase diversity of potential competitors in other habitats. Nonequilibrium conditions are expected to be more prevalent in benthic habitats, streams, and wetlands than in the pelagic zones of large lakes. Predation and disease are common features of all habitats (see prior chapter). The assumptions of the competitive exclusion principle thus may not always apply and coexistence of interspecific competitors could be a potential feature of many habitats.

Intraspecific competition can lead to strong selective pressures on organisms. These interactions can be avoided by organisms in some instances. For example, intraspecific competition can lead to "Alee effects," where organisms self-limit their population growth so as not to crash the entire population. The interplay between success of individuals and success of the group has led to substantial controversy in the field of evolution.

Sexual selection can result from intraspecific competition. For example, male fishes can have broadly different morphology from females ultimately related to competition for mates (Fig. 21.3). In a more complex example, water striders (*Aquarius remigis*) have intense competition for mates. The water striders congregate in pools and males compete to mate with females. The intense competition to mate actually can harm the females. In conditions where there is only

FIGURE 21.3
Breeding-age adult central stonerollers (*Campostoma anomalum*). The male, in the top panel, has obvious tubercles and fin markings. These features could play a part in aggressive interactions among males competing for mates as well as female mate choice. The female is plain. *(Photograph courtesy of Gerold Sneegas).*

one habitat, the most aggressive males win because they are the most successful at mating. However, when there are several habitats available (e.g., a stream with connected pools, some with many water striders and others with few), females will leave pools with many males to avoid being overwhelmed by mating attempts. If less aggressive males inhabit the pools with few water striders, then they will also mate successfully when females enter their pools to avoid the higher density pools. This is a case where population structure can alter the selective pressures of intraspecific competition (Eldakar *et al.*, 2009).

We now document several types of interspecific competition that commonly occur in freshwater environments. Primary producers compete for nutrients and light. Competition between macrophyte species has been well documented (Gopal and Goel, 1993). Macrophytes can form dense stands that shade all primary producers below them (Haslam, 1978). Floating macrophytes, such as water hyacinth (*Eichhornia*) or duckweed (*Lemna*), can blanket the surface of a lentic habitat and essentially remove all light. Macrophytes can intercept 90% of the incident light before it reaches a lake bottom, leading to a 65% decrease in benthic algal production (Lassen *et al.*, 1997). If systems are sufficiently shallow, emergent primary producers intercept light before it reaches the water's surface.

Competition for light has been documented among macrophytes inhabiting a stream in North Carolina (Everitt and Burkholder, 1991). In this case, riparian vegetation and macrophyte assemblages were characterized in 10 stream segments. The red alga *Lemanea australis* or the aquatic moss *Fontanalis* dominated low-light assemblages in the winter. High-light sites were dominated by *L. australis* and the angiosperm *Podostemum ceratophyllum*. These sites were on the same stream and had similar water velocity and depth, indicating that light availability as influenced by riparian shading was the major abiotic difference between them. Apparently, competition for light structured the macrophyte assemblages.

Macrophytes and phytoplankton compete for light and nutrients in shallow lakes. Several species of macrophytes are capable of interference competition; they release allelopathic chemicals that inhibit phytoplankton species (Gross *et al.*, 2007). In shallow lakes dominated by macrophytes, the deeper portions are often characterized by relatively clear water (low phytoplankton concentrations). Thus, allelopathy can stabilize macrophyte dominance in shallow lakes (Hilt and Gross, 2008).

Wetland plant assemblages can be shaped by competition. Competition between two species of wetland plants may occur aboveground (for light) or belowground (for nutrients), and both may be important simultaneously (Twolan-Strutt and Keddy, 1996). Competitive rankings of individual plants can remain stable against changes in nutrients and flooding in some instances (Keddy *et al.*, 1994). However, wetland plants that occur in disturbed habitats may be released from competition (Keddy, 1989). These studies suggest that competition can determine which plant species dominate in a particular wetland.

An example of competition determining spatial patterns in macrophytes is found in two species of the cattail, *Typha latifolia* and *T. angustifolia*. *Typha latifolia* has broad leaves and can outcompete *T. angustifolia* in shallow water, but not in deeper water (Grace and Wetzel, 1998). As the season progresses, the two species are almost completely segregated (Fig. 21.4). This segregation

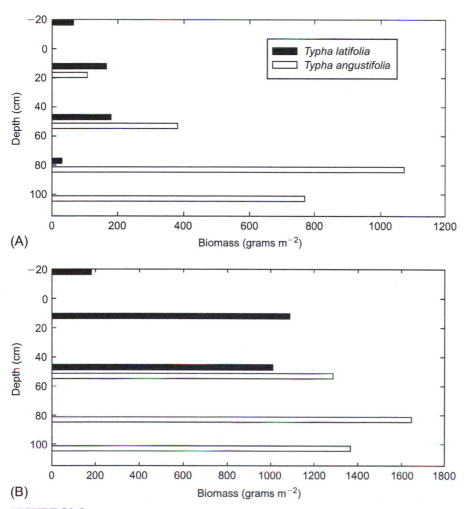

FIGURE 21.4

Competitive exclusion of two species of *Typha*. Distributions at the same locations in the same wetland in May (A) and September (B). *(Data plotted from Grace and Wetzel, 1998).*

occurs even though both species have optimum growth rates at 50cm depth when occurring in isolation.

Zooplankton species often compete for food resources. In Chapter 19, we discussed the idea that ratios of nutrients may alter competitive ability or lead to coexistence of phytoplankton species. Similar resource partitioning could lead to coexistence of potentially competing zooplankton species (Lampert, 1997; DeMott, 1995). Such application of *resource ratio theory* has vastly expanded in application (Sterner and Elser, 2002). Resource ratio theory predicts how trade-offs associated with using two or more limited resources influence how species coexist (Tilman, 1982). For example, two species that are differentially limited by two resources will coexist at intermediate ratios of resource supply. At more skewed ratios of supply, the best competitor for the most limiting resource can exclude other species.

In some cases, a trade-off may occur in competitive ability. For example, one species of rotifer (*Keratella cochlearis*) may compete better at a lower food concentration and the other (*Keratella earlinae*) at higher concentrations (Fig. 21.5). Both species could coexist in a spatially or temporally variable environment. Competitive interactions among zooplankton species probably vary over space and time. In Chapter 20, we discussed the idea that large zooplankton are superior competitors for phytoplankton cells. However, this competitive interaction may not be so simplistic. Large zooplankton may slow or cease

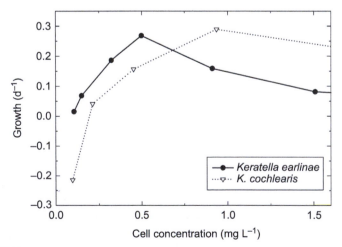

FIGURE 21.5

Competitive ability of two species of rotifers (*Keratella cochlearis* and *K. earlinae*) that feed on the cryptomonad *Rhodomonas*. Note that *K. cochlearis* is able to grow more rapidly at lower concentrations of food, but *K. earlinae* grows better at higher concentrations of food. *(Redrawn from Stemberger and Gilbert, 1985)*.

feeding in blooms of large inedible algae while smaller zooplankton continue to feed unhindered on subdominant small phytoplankton and bacteria (Gliwicz, 1980). Similarly, cladocerans may be superior competitors for food compared to rotifers, but suspended clay can decrease the growth of cladocerans while having little effect on rotifers. Thus, the clay allows rotifers to be released from competition with cladocerans (Kirk and Gilbert, 1990).

There are situations in which competitive exclusion should operate but may not be strong enough to drive out weaker competitors. An example occurred with five species of herbivorous zooplankton in a small humic lake (Hessen, 1990). These species coexisted, but bottle experiments showed no evidence of significant predation or a single dominant competitor. Weak competition for abundant nutrient-poor food in this lake was hypothesized to allow coexistence of the species. This study and those noted previously are a few of many that suggest that competition is an important but context-dependent feature of zooplankton assemblages.

Invasive species are often very strong competitors. For example, purple loosestrife (*Lythrum salicaria*) has expanded rapidly in the United States, displacing other wetland species (Blossey *et al.*, 2001). The reed canary grass (*Phalaris arundinacea*) is also a recent invasive species. In the wetlands where they have invaded, they often become dominant; purple loosestrife makes up about 30% of the total abundance, and canary grass about 8% overall (Fig. 21.6). In most cases where both were present (6 of 7), purple loosestrife was dominant. In 6 of 24 wetlands, reed canary grass was the most abundant species, and in 7 of 24 cases

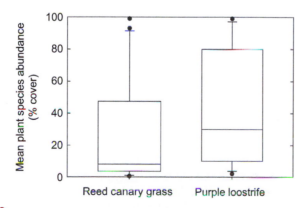

FIGURE 21.6

Relative abundance of two invasive species (purple loosestrife *Lythrum salicaria* and reed canary grass *Phalaris arundinacea*) in Pacific Northwest US wetlands. Box plots show median, the box the 25th and 75th percentiles, and the whiskers the 90th and 10th percentiles. While overall neither species dominates in all wetlands, there is a high degree of dominance in some wetlands. *(Data replotted from Schooler et al., 2006).*

purple loosestrife was dominant (Schooler et al., 2006). These data indicate that the invaders are strong competitors and that in situations where they attain high abundance they likely decrease plant diversity of the wetlands they have invaded.

Competition for food has led to clear specialization of benthic freshwater macroinvertebrates over evolutionary time. Thus, invertebrates can be classified into functional feeding groups based on mode of food acquisition, such as collectors, filterers, shredders, and scrapers. Within these groups, competition can still occur. For example, net-spinning hydropsychid caddisflies collect particles from the water column with silk nets of different mesh sizes, with each mesh size characteristic of a species. Smaller mesh nets are more efficient at collecting small particles, whereas large meshes capture large particles more effectively (Loudon and Alstad, 1990). Thus, species with fine nets will likely compete more effectively for fine particles. Such competitive specialization may determine which species dominate.

Although competition among coexisting species using similar resources, such as assemblages of net-spinning caddisflies, might be expected, interactions can be complex and sometimes counterintuitive. Laboratory experiments using different numbers of species of net-spinning caddisflies found that the particle capture efficiency of individual species was actually enhanced by the presence of other species because of changes in near bed flow dynamics (Cardinale *et al.*, 2002).

Competition among pond-breeding amphibian species has been studied. In some ponds, predators remove the competitive dominant frogs, and predation thus allows less competitive species to coexist with stronger competitors (Morin, 1983). The chytrid fungus *Batrachochytrium dendrobatidis*, which has been linked to amphibian declines around the world (Sidebar 11.2), can alter competitive interactions among amphibian species that do not die from infection. In the presence of the fungus, both fowler's toad (*Bufo fowleri*) and gray tree frog (*Hyla versicolor*) tadpoles metamorphosed at smaller body masses when reared together in tanks compared to when they were reared separately (Parris and Cornelius, 2004). The toad tadpoles also had strong negative effects on tree frog tadpole development, but only in the presence of the fungus.

Competition among fishes may be an important consideration for fisheries managers. For instance, Hodgson *et al.* (1991) demonstrated that introduction of rainbow trout (*Oncorhynchus mykiss*) into lakes with 2- and 3-year-old largemouth bass (*Micropterus salmoides*) can lower the condition (weight to length ratio) of bass. The diet of the bass shifted from *Daphnia* to odonate naiads, and bass condition was lower after the introduction of trout into one lake, particularly when compared to a nearby lake with no trout. Cage experiments suggest that sunfishes (Centrarchidae) compete when predators drive them to take refuge in macrophyte beds (Mittelbach, 1988), and competition may occur with juveniles and translate into lower production of adult fish (Osenberg *et al.*, 1992).

MUTUALISM AND FACILITATION

Mutualisms (both species have a positive effect on each other) are less conspicuous in freshwater than in marine systems, possibly because the continuous time for evolution of mutualisms has been less in freshwaters than in marine systems (i.e., many freshwater habitats have a shorter continuous history than marine or terrestrial habitats). However, it seems that some of the conditions for mutualism occur in freshwater. For example, fish that clean other fishes are common on marine reefs, and this mutualism involves many species of fishes from diverse taxonomic groups, but the same interaction has not been identified in freshwaters despite comparable benefits to freshwater fishes and the long geological age of some river systems. Many of the mutualisms that occur in freshwaters involve microorganisms and were discussed in Chapter 19.

Mutualisms based on behavior require coevolved systems and organisms capable of complex behavioral patterns, such as fishes. Cichlids from Lake Tanganyika demonstrate parental care, including guarding eggs and fry from predators, and two species can brood in the same region and mutually defend their broods (Keenleyside, 1991). Fishes in the same lake have evolved cooperation in which predators hunt in mixed groups and this cooperation increases success (Nakai, 1993). Mixed-feeding schools may occur in other fish assemblages but have not been well studied. This is not a completely unique adaptation; birds have demonstrated a similar cooperative strategy of mixed-feeding flocks.

Facilitation is any unidirectional positive effect of one species on another. Facilitation can include mutualism and commensalism, as well as some exploitation relationships. Facilitation may be an important and overlooked aspect of community interactions (Bertness and Callaway, 1994). Plants in stressful environments can facilitate each other (Callaway, 1995; Callaway and Walker, 1997). Few macrophyte and wetland plant assemblages have been studied with regard to facilitation, but it could be important in stressful freshwater habitats, as has been demonstrated for estuarine marshes (Bertness and Hacker, 1994). For instance, emergent freshwater marsh plants that are aerenchymous (transport oxygen to their roots) can facilitate other emergent plants living nearby by aerating the sediments (Callaway and King, 1996). As more research is done on aquatic plant assemblages, more examples of facilitation will likely be documented, given the importance of facilitation among terrestrial plants (Brooker *et al.*, 2008).

Indirect facilitation may occur in streams. Crayfish exclude the green alga *Cladophora* from pools. The exclusion of *Cladophora* facilitates the growth of

Table 21.1 Influence of Mutualistic Midge Larvae on *Nostoc* Nitrogen Fixation Rates, O_2 Concentrations, and Photosynthetic Rates

Measurement	Units	Condition	Mean (95% confidence band)
$^{15}N_2$ incorporation	Del ^{15}N of *Nostoc*	With midge	0.188 (0.005)
		Without midge	0.153 (0.009)
Microscale O_2 concentration with microelectrodes	mmol L^{-1}	Over midge	0.67 (0.09)
		Away from midge	0.81 (0.06)
Photosynthetic rate with microelectrodes	μmol O_2 L^{-1} s^{-1}	Over midge	1.4 (0.3)
		Away from midge	1.1 (0.2)

(Data from Dodds, 1989)

epilithic diatoms. The diatoms in turn support an increased biomass of grazing insect larvae (Hart, 1992; Creed, 1994). Such complex interactions may be a common feature of communities.

An interesting mutualism occurs between the chironomid midge larva *Cricotopus nostocicola* and the cyanobacterium *Nostoc parmeliodes* (Fig. 19.15C). The midge receives sustenance from the *Nostoc* and lives inside it until pupation and emergence as an adult (Brock, 1960). The midge is a poor swimmer and highly susceptible to predation without the shelter in the tough leathery *Nostoc* colony. In turn, the midge increases the photosynthetic rate of the *Nostoc* (Ward *et al.*, 1985) by altering its morphology and by attaching it more firmly onto rocks so it can extend into flow and have a smaller diffusive boundary layer (Dodds, 1989). Respiration by the midge lowers O_2 concentration in the vicinity of the larva and enhances nitrogen fixation rates as well as photosynthetic rates (Table 21.1). This example is particularly curious because the eukaryotic organism is endosymbiotic to the bacterial organism.

Snails could facilitate other grazing snails of different species by providing substrata for algae to grow on their shells (Abbot and Bergey, 2007). In this case the nutrients excreted by grazing snails can stimulate growth of diatoms that attach to their shells. Other snails then graze the diatoms off of the snail shells. Grazing tadpoles in tropical streams facilitate small grazing mayflies by removing sediments from substrata and exposing underlying periphyton; grazing mayflies respond by increasing their densities in tadpole-grazed patches (Ranvestel *et al.*, 2004; Whiles *et al.*, 2006).

Facilitation was demonstrated in a group of three oligochaete species (Fig. 21.7), but the mechanisms were not clear (Brinkhurst *et al.*, 1972). In this case, two of three species tested had greater weight gains when grown with the others than when grown in a single-species culture. Facilitation may also be a

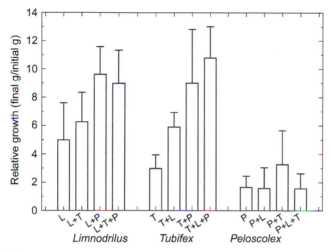

FIGURE 21.7

Growth of three tubificid oligochaetes alone and in culture with the other species. L, *Limnodrilus hoffmeisteri*, T, *Tubifex tubifex*, and P, *Peloscolex multisetosus*. Note that in two of three experiments, growth was greater in the presence of the other two species. *(Data from Brinkhurst et al., 1972).*

common feature of stream invertebrates that process litter. Shredders excrete fine particulate organic material that is ingested by collectors. Collectors may remove fine material that interferes with shredders or excrete nutrients that stimulate the microbes and make litter usable for shredders. Such facultative links merit additional study.

Facilitation could be more important for some predators in streams than thought previously. For example, it has been assumed that wading birds compete with bass for prey, but a cage experiment challenged this assumption. In a prairie stream in Illinois where the dominant predatory wading birds are great blue heron (*Ardea herodias*), green heron (*Butorides virescens*), and great egret (*Ardea alba*), cages were used to manipulate access by birds and smallmouth bass (*Micropterus dolomieu*) (Steinmetz *et al.*, 2008). The birds are effective predators only in shallow waters, whereas bass can consume prey in deeper water. In this case, more small prey fishes were consumed with both types of predators (wading birds and fish) present than either alone, and the effect was greater than additive. This experiment suggests facilitation occurs between the predators because prey fishes could not find refuge in deep or shallow waters.

OTHER SPECIES INTERACTIONS

In addition to predation (+/−), competition (−/−), and mutualism (+/+), other species interactions (neutralism (0/0), amensalism (−/0), and commensalism (+/0)) may be important but are rarely studied. Several examples are given here to explain potentially important interactions in which one species has an influence on a species that does not have an influence in return.

Tadpoles of a common frog (*Rana temporaria*) have a negative effect on the snail, *Lymnaea stagnalis*, by competing for microalgae (Brönmark *et al.*, 1991). The snail then consumes lower quality *Cladophora* and excretes nutrients that stimulate microalgal growth. The tadpole has a strong negative effect on the snail, but the snail has a weak positive effect on the tadpole.

Macrophytes in lakes and ponds may provide a vital habitat for survival of small fishes. The macrophytes apparently receive little direct benefit from the fishes that live in them. However, the macrophytes could receive some mineralized nutrients from fish excretion. Also, indirect relationships could occur if the small fishes eat invertebrates that remove competitive epiphytes or prey upon herbivorous insects that eat the macrophytes.

Macrophytes can also provide nutrients to epiphytes growing on them. The macrophyte *Najas flexilis* was planted in sediments labeled with radioactive phosphorus. The amount of the radioisotope taken into epiphytes was quantified using track autoradiography. A substantial portion of epiphyte phosphorus was derived from the macrophytes. Furthermore, different species obtained their phosphorus in varied proportions with the filamentous and more erect forms relying more heavily on the water column for nutrients, which indicated niche separation (Moeller *et al.*, 1988). Similarly, nitrogen-fixing microbes could leak nitrogen and stimulate nearby organisms that are unable to use N_2. This interaction is manipulated in rice paddies under traditional cultivation as discussed in Chapter 19.

Most interactions of aquatic organisms with humans are amensal. Humans have negative effects on many aquatic species, but the effects of most of these species on humans are negligible. With some thought, the reader could identify more examples.

SUMMARY

1. Competition has been well documented for species from a variety of taxonomic groups found in freshwaters.
2. Competition can occur between members of the same species (intraspecific) and members of different species (interspecific).
3. Competition can be divided into two general types; scramble competition, where individuals exploit the same resource, and inference competition, where individuals interfere with the ability of others to use a resource.
4. It can be difficult to establish competition in field studies.

5. Mutualism, amensalism, and commensalism have been studied much less than predation or competition, but they may be important at times.
6. Facilitation among plant species in harsh environments is an example of positive interactions that may be important in wetland communities.
7. Facilitation may be much more common than previously thought.

QUESTIONS FOR THOUGHT

1. How can disturbance in a habitat act as an agent of natural selection?
2. Are indirect interactions so strong and numerous that they complicate predicting the effects of interactions within an ecological community?
3. Is there a maximum diversity based on competition?
4. How might temporal patterns of disturbance act so that competition becomes unimportant?
5. Are positive or negative interactions more important in natural communities?
6. Why are both per capita interaction strength and the number of species interacting important?
7. Can you think of additional cases of facilitative interactions that were not included earlier?

Complex Community Interactions

FIGURE 22.1

Mt. St. Helens in Washington State erupted and drastically altered the community of Spirit Lake (foreground). This led to documentation of an unusual case of primary succession. *(Photograph courtesy of the US Geological Survey, Lynn Topinka).*

Doi: 10.1016/B978-0-12-374724-2.00022-2

Communities are characterized by a complex interplay among abiotic factors and interactions among many species. Community ecology is beset with difficulties in making predictions and is prone to numerous "hot topics" that have ended up being false leads in the quest to understand pattern and process in complex groups of interacting organisms (Dodds, 2009). Still, there are some strong generalizations that can be made about aquatic communities. Generalizations can be made about temporal (e.g., time of recovery since disturbance) and spatial (e.g., landscape ecology) patterns in aquatic ecosystems. We attempt to cover some of the major areas and delineate where predictions can be made and describe patterns that may not hold in all systems, but are important to consider because they can occur in some systems.

DISTURBANCE

Defining *disturbance* is difficult. White and Pickett (1985) suggest that a disturbance is "any relatively discrete event in time that disrupts ecosystem, community, or population structure and changes resources, substrate availability, or the physical environment." This definition is very broad and allows many important aspects of disturbance to be included: spatial distribution, frequency, return interval, predictability, area, and intensity can all be aspects of disturbance. Community properties and the associated disturbance must be characterized at appropriate spatial and temporal scales relative to the life histories of the organisms of interest.

One thorny issue in disturbance ecology is how to classify natural events. If a species has evolved to take advantage of what, for others, is a disturbance, how do we classify an event? For example, a flood in a stream may decimate the diversity of insect larvae, but some benthic stream algae may grow better after a flood because of reduced grazing pressure; thus, is a flood a disturbance for the algae? Likewise, dry periods in intermittent streams and wetlands may seem like disturbances to some species, but others may depend on these events to complete their life cycles. One key issue regarding disturbance is predictability; predictable events such as summer drying in intermittent streams and spring floods in large rivers may not constitute disturbances because they fall within the normal range of conditions for those systems (Resh et al., 1988). Human alteration of natural disturbance regimes (e.g., regulation of rivers to reduce high and low flows) can adversely affect aquatic communities and favor exotic species (Lytle and Poff, 2004; Poff et al., 2007). The idea that flooding is a natural part of flowing water habitats is now being extended to habitat management, including operation of dams and reservoirs.

Natural disturbance has variable influence across freshwater habitats. In general, groundwaters are least influenced by disturbance. However, karst aquifers may respond rapidly to surface precipitation events and cave streams can flood. Deeper groundwaters in unconsolidated sediments are expected to have very slow rates of change. Lakes are only moderately affected by disturbances in most cases; severe weather can cause mixing, and shallow or small lakes could be greatly influenced by floods. Streams, rivers, and some wetlands are more prone to disturbance. Flooding is an integral part of many stream ecosystems and riparian wetlands; recovery from flooding and adaptations to such disturbances are important to many aquatic organisms. Along with hydrologic events, wetlands with emergent vegetation can be subject to fires. A nonequilibrium view of streams and wetlands has become essential to understanding their ecology (Palmer and Poff, 1997).

The nonequilibrium view of disturbance and its relationship to competition as a determinant of species diversity has been termed the dynamic equilibrium model (Huston, 1994). This model explains some responses to disturbance by stream invertebrate communities. In a study by McCabe and Gotelli (2000), frequency of disturbance did not decrease invertebrate species richness, but intensity and area of disturbance did alter richness. Similarly, the dynamic equilibrium model was consistent with riparian plant richness in riverine wetlands in Alaska (Pollock *et al.*, 1998). In the study by Pollock *et al.*, plant productivity (a gauge of competition) altered the effect of disturbance on species richness.

The overall impact of a disturbance depends on the *resistance* (the ability to withstand the disturbance) and *resilience* (the ability to recover) of the system. Resistance is often related to the magnitude and intensity of the disturbance, as well as features of the system such as the types and amounts of refugia (Lake, 2000). Resilience depends on the ability of colonists to reach an area (Cushing and Gaines, 1989), their ability to reproduce, the severity of the disturbance, and the existence of refugia (Townsend, 1989; Lancaster and Hildrew, 1993). The response to disturbance can also depend upon resource availability; disturbance of rocks in an alpine river had substantial effects on the macroinvertebrate community under natural nutrient concentrations, but there was much less response to disturbance with nutrient fertilization (Gafner and Robinson, 2007).

The harshness of disturbance in an aquatic environment is exemplified by research on recovery of macroinvertebrate species found in intermittent prairie streams following flooding and drought. Fritz and Dodds (2005) found that species richness could be predicted as a function of length of drought, severity of flood, time since flooding or drought, and distance from spring-fed pools that served as refuges from disturbance (Fig. 22.2). They created an index of harshness that incorporated both spatial and temporal aspects related to recovery from disturbance.

Consideration of natural disturbance dynamics is crucial in wetland restoration (Middleton, 1999). Such disturbance includes flood pulses, fires

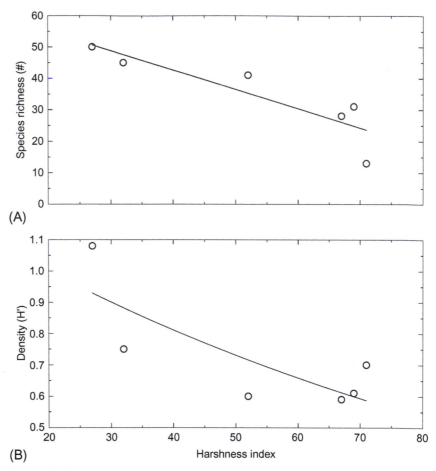

FIGURE 22.2

Relationship between macroinvertebrate species richness (A), Shannon diversity (B), and harshness index in Kings Creek, Kansas. The harshness index is a score based on time since flood or drought, intensity of last flood or drought, and the distance from sources of new colonists. A higher index indicates a harsher habitat. The two points on the left are from permanent sites; the rest are from intermittent sites. Both lines are statistically significant ($p < 0.05$). *(Data from Fritz and Dodds, 2005).*

(de Szalay and Resh, 1997), hurricanes, beaver activity, and herbivory. A large-scale example of managing disturbance in wetland restoration is the Kissimmee River project in Florida (Middleton, 1999). This project has continued through multiple phases and includes backfilling a drainage channel with tons of soil and levee removal. It will also include creation of flood pulses from upstream control structures to mimic natural patterns of discharge. This project attempted to reverse the effects of channelization and flood control structures put in place between 1964 and 1971 by returning the system's natural hydrology (Toth, 1996). Disruption of the natural floods led to more dryland plant species in the floodplains around the river, more lentic species of macrophytes (including the invading water hyacinth), and more animals in the pools adjacent to and within the river adapted to lentic habitats (Toth, 1996; Harris *et al.*, 1995). Several species of wading birds and waterfowl declined with channelization (Weller, 1995). Restoration to reintroduce flooding and reverse channelization started in 1984 and will ultimately involve 70 km of river channel. These efforts have already increased flooding, increased the numbers of wetland plants and waterfowl, and restored lotic species to the river. The newly restored habitats are being colonized rapidly by invertebrate communities (Merritt *et al.*, 1999).

Disruption of flooding can cause major changes in a river ecosystem, from geomorphological to biotic aspects. Such is the case with the Glen Canyon Dam on the Colorado River. Lack of flooding and continuous release of clear, cold water have drastically changed the communities in the river, leading to a depauperate river dominated by fewer algal and invertebrate species than occur naturally (Stevens *et al.*, 1997). The cold-water releases have allowed establishment of introduced trout and have had a negative impact on native warm-water fishes. However, the lack of flooding has also led to increases in fluvial marshes bordering the Colorado River (Stevens, 1995). These marshes increase wildlife habitat and lead to an increased diversity of terrestrial and wetland species. In March 1996, a limited artificial flood was instigated. The scouring reestablished sandbars but did not completely remove invasive species, such as carp, catfish, and tamarisk (Middleton, 1999). There was a short-term reduction in primary producer and invertebrate biomass followed by recovery (McKinney *et al.*, 1999; Shannon *et al.*, 2001). The cost of the release was $1.8 million of power-generating capacity to the dam.

Since the initial experimental flood in 1996, experimental high flows were conducted in 2004 and 2008. Although these floods were of lesser magnitude than peak pre-dam high flows (i.e., 1,133 versus 5,947 m^3 s^{-1}; Topping *et al.*, 2003) they did reestablish lateral sandbars. Sandbar habitats throughout the Grand Canyon benefit various aspects of the system, from native fishes to recreational users to preservation of archeological sites. For these reasons, applications of high flow releases from Glen Canyon Dam that are specifically designed to promote sandbar habitats will likely continue in the Grand Canyon. Restoration of flood regimes is becoming a more common management approach worldwide (e.g., Robinson and Uehlinger, 2008).

Disturbance of terrestrial habitats can cascade to aquatic habitats. For example, riparian disturbance can increase sediment inputs, alter inputs of detritus, decrease the amount of pebble and cobble substrata, and influence invertebrate and algal communities (Stevens and Cummins, 1999). Forest fires can result in large inputs of ash and nutrients into streams, increase sedimentation, and alter light regimes (Minshall *et al.*, 1989). Many pollutants enter aquatic habitats from terrestrial sources and influence communities. Watershed disturbance is a prime source of nutrients that often cause eutrophication.

SUCCESSION

Succession has long been hypothesized to drive the seasonal cycle of planktonic organisms in lakes. Long-lived fishes and macrophytes respond less strongly to seasonal cycles than organisms with shorter life spans. However, most lake species exhibit predictable seasonal cycles in their life histories. The ecological literature differentiates between *primary* and *secondary succession*. In primary succession, all the organisms have been removed from a habitat or a habitat is newly created. In secondary succession, disturbance removes some of the species or substantially reduces populations. In contrast, *seasonal succession* occurs on an annual basis and is the predictable change in species composition and biomass as abiotic conditions change seasonally.

The Phytoplankton Ecology Group created the most complete model of seasonal succession of plankton for temperate lakes with summer stratification. This group of 30 limnologists created a model consisting of 24 steps (Sommer, 1989). We present a simplified version here:

1. Nutrients build in the euphotic zone of lakes during the winter because nutrients mix from the hypolimnion during fall mixing, light is limiting, and there is little demand for nutrients by phytoplankton. Light becomes limiting because of less solar irradiance and phytoplankton may be mixed deeply below the compensation point, or ice cover severely limits light input.
2. During the spring, populations of small, rapidly growing phytoplankton species peak because nutrients and light are high and zooplankton grazing is low.
3. In late spring, reproduction allows zooplankton populations to increase, and their grazing causes a clear water phase that lasts until grazer-resistant species of phytoplankton develop during the summer.
4. Zooplankton populations decrease because of declining food and increased fish predation (particularly by newly hatched fish larvae).
5. A later bloom of cyanobacteria may occur with higher lake temperatures and lower nitrogen availability.
6. Fall mixing can stimulate a second bloom of edible phytoplankton and larger zooplankton.

This hypothetical successional cycle is diagramed in Figure 22.3.

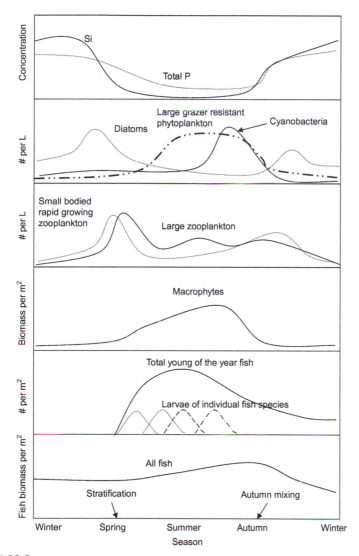

FIGURE 22.3

A hypothetical successional sequence in the epilimnion of a temperate lake with summer stratification.

Fishes can have life histories that allow different coexisting species to hatch at distinct and predictable times in a seasonal sequence. Amundrud *et al.* (1974) demonstrated an order of appearance of larvae in a small eutrophic lake (Fig. 22.4). In this lake, yellow perch (*Perca flavescens*) larvae appeared first, followed by log perch (*Percina caprodes*), black crappies (*Pomoxis nigromaculatus*), and finally pumpkinseeds and bluegills (*Lepomis* spp.). The sequence suggests that competition has led these fishes to habitat partitioning, or that they specialize on prey that occur at different times during the seasonal successional cycle.

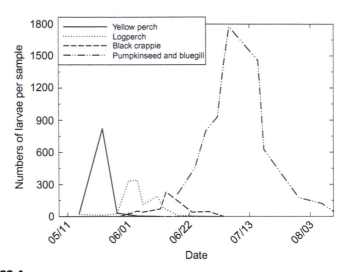

FIGURE 22.4

Numbers of larval fishes in a small eutrophic lake during a summer successional sequence. *(Redrawn from Amundrud et al., 1974).*

On a longer time frame, succession can occur in newly created lakes or existing lakes that are subject to very strong disturbance (i.e., primary succession). There have been geological time periods when new freshwater habitats were created relatively rapidly; following periods of glaciation, numerous lakes, streams, and wetlands are created. As the glaciers retreat, new streams are created, leading to primary succession. In alpine regions around the world, global warming is now causing retreats of glaciers. The stream invertebrate communities will reflect this succession (Burgherr and Ward, 2001). Presumably the rate of creation of new habitat is slow enough relative to dispersal capabilities of many stream organisms that limitation by colonization rates is not an important factor. Creation of new lentic habitat is also a common occurrence in our modern world due to the establishment of new reservoirs.

In reservoirs, the general view is that an early eutrophication phase is common (Donar *et al.*, 1996). Sequential colonization of the reservoir by invertebrates also occurs, related to formation of a sedimented bottom and establishment of macrophytes (Voshell and Simmons, 1984; Bass, 1992). The sequence of events that occur following installation of a new reservoir can vary depending on the area being dammed, the morphology of the reservoir, and a variety of other factors. For example, community formation was contrasted in two reservoirs in south Saskatchewan (Hall *et al.*, 1999). Paleolimnological techniques were used to analyze algal and chironomid midge assemblages in the two reservoirs over time. One reservoir was formed by damming a river (Lake Diefenbaker) and the other by raising the level of an existing lake (Buffalo Pound Lake). The river reservoir was characterized by fluctuations in water level of 6 m per year and the flooded lake by fluctuations of 1 to 3 m per year. Lake Diefenbaker (500 km²) exhibited a typical sequence of succession found in newly formed reservoirs; an initial period (4 years) of eutrophic conditions occurred, followed by a decade of mesotrophy and a recent shift back to eutrophic conditions. Buffalo Pound Lake (50 km²) had lower phytoplankton biomass after flooding, but midge larvae and macrophyte populations expanded. The different routes of formation, lake level fluctuations, and sizes of reservoirs formed are hypothesized to have led to different successional trajectories.

One of the more detailed studies of succession in a reservoir was done on the Gocsalkowice reservoir in Poland. This reservoir was filled in 1955 and observations were summarized in 1986. The Secchi depth increased from 1 to 2 m during the first 20 years but then decreased to less than 1m because of anthropogenic nutrient inputs in the watershed (Kasza and Winohradnik, 1986). Diatoms and green algae dominated in high numbers in the 1950s and 1960s but were replaced by cyanobacteria in the 1980s (Pajak, 1986). Macrophyte growth was extensive within a year after the reservoir was filled, but few species were present. Emergent macrophytes dominated the reservoir margins in later years, and macrophyte community diversity increased and stabilized after two decades (Kuflikowksi, 1986). Riverine fish and invertebrate species were replaced by species more common in lakes within the first few years (Starmach, 1986; Krzyzanek, 1986).

Succession in lakes can also be driven by external storms. The planktonic bacterial assemblages of a subalpine freshwater humic lake in Taiwan changed predictably following typhoons that destabilized the water column of the lake (Jones et al., 2008). Following typhoons, the bacterial communities were generaly similar to those immediately following other typhoons. As time since disturbance increased, the epilimnetic and hypolimnetic assemblages diverged, and successional trajectories varied following each disturbance. In contrast, the phytoplankton communities did not respond predictably to this disturbance.

An unusual but interesting case of primary succession occurred when Mt. St. Helens in Washington State (Fig. 22.1) erupted and drastically altered Spirit Lake (Larson, 1993). The lake was an oligotrophic mountain lake before eruption. The eruption superheated the lake's waters, killed most of the animals and plants, and filled the lake basin with volcanic ash and tree trunks. Soon after the eruption, the water went completely anoxic, leading to a community dominated by heterotrophic and chemoautotrophic bacteria, a few protozoa, and rotifers. By the next spring, O_2 returned to the epilimnion of the lake, and a moderately diverse phytoplankton community developed. Five years later, a diverse phytoplankton assemblage had colonized the lake as well as at least four zooplankton species. Eight years after eruption, macrophytes (milfoil, (*Myriophyllum*) and stonewort (*Chara*)) had become established, as had snails and other macroinvertebrates.

Seasonal wetlands may experience a strong successional sequence. For example, in the Pantanal wetlands in Brazil, a seasonal wet–dry cycle leads to large and predictable changes in the aquatic communities (Heckman, 1994). During the wet season, considerable flow can occur in the main channels, but during the dry season these same channels become lentic. Macrophyte populations develop during the dry seasons and are flushed out during the wet seasons. During the dry periods, the large number of fishes trapped in drying ponds attract many waterbirds and caimans (a crocodilian) that consume them. These isolated ponds become hypereutrophic from the nutrients released from the dying fishes and excretion from the waterbirds and caimans, and they exhibit algal blooms. Eventually, all but the deepest pools dry and then fill in the following wet season, when the cycle begins again.

Successional patterns of macrophytes in riparian wetlands are commonly associated with any river that exhibits seasonal variability. In the Rhone River in France, successional sequence depends on the degree of connectivity to the main channel. The diversity is highest in frequently flooded habitats because more propagules are available to establish plants (Bornette *et al.*, 1998b). In a more spatially restricted study, four species of macrophytes were found to coexist in a side channel: *Sparganium emersum* was found in the least disturbed areas of the channel, *Hippuris vulgaris* and *Groenlandia densa* were found in moderately disturbed areas, and *Luronium natans* was found in the most disturbed areas (Greulich and Bornette, 1999). This pattern illustrates the relationship between competitive ability and successional processes. Without disturbance and succession, *S. emersum* would probably be the only species present because it is the competitive dominant.

Understanding successional patterns may be crucial in wetland restoration schemes (Middleton, 1999). Two general lines of thought are taken: The first is that succession depends on the initial inhabitants, and the second is that a natural successional series will restore a wetland if the abiotic features are reproduced. Thus, some people assume that the manager must plant the species desired, and others assume that species will naturally populate a restored wetland (Middleton, 1999). The reality probably lies in-between these two points on a continuum. If a severely disturbed wetland is to recover, reintroduction of lost or rare species may be necessary. Alternatively, in a wetland of any size, it is difficult to control exactly which species become established over the years and where.

Rivers and streams may also undergo succession on seasonal timescales. In small streams in deciduous forests (Fig. 22.5), the light regime and leaf litter inputs vary over the seasons. Increased light and limited litter inputs favor periphyton and species that consume it (scrapers). Increased litter inputs favor species that depend on leaves for nutrition (shredders). Seasonal flooding may also alter the diversity and biomass of species in such a stream. Invertebrates in temperate streams can be categorized into three groups: slow-seasonal, fast-seasonal, and nonseasonal life cycles (Anderson and Wallace, 1984). Many slow-seasonal insects reproduce in the fall and grow through the winter (e.g., winter stoneflies and some Trichoptera). These species can specialize

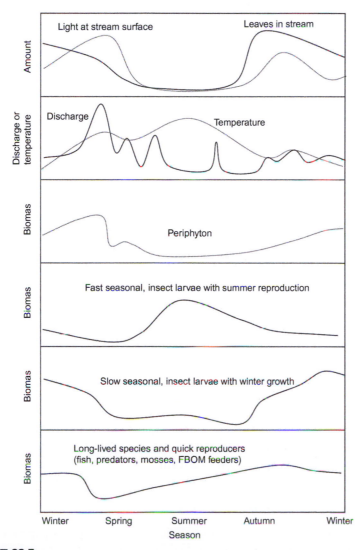

FIGURE 22.5

A hypothetical seasonal successional sequence in a small temperate stream in a deciduous forest.

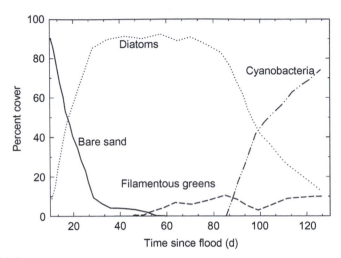

FIGURE 22.6

A successional sequence of algal groups in a desert stream following a flood event. *(Reproduced with permission from Fisher and Grimm, 1991).*

on the large amount of leaf litter entering the stream in fall. Fast-seasonal insects have a prolonged diapause and a short period of rapid growth followed by reproduction. These species may reproduce in spring, summer, or fall. Nonseasonal species may have very long life spans or very short and overlapping generations. Data collected from a small Michigan stream suggest that total invertebrate biomass remains approximately constant throughout the year despite the dominance of different species in specific seasons (Cummins and Klug, 1979). For instance, insect larvae that are shredders attain maximum biomass in spring after a winter of feeding on detritus. Seasonal cycles can also influence invertebrate responses to flooding (Robinson and Minshall, 1986).

Seasonal flooding is also an important feature of desert streams. A predictable sequence of colonization by algae occurs in Sycamore Creek, Arizona (Fig. 22.6). Quick colonizers include diatoms. Filamentous green algae take more time to establish. Finally, cyanobacteria dominate as nitrogen becomes limiting. The successional patterns discussed here provide generalized and simplified models, to which exceptions clearly exist, mainly to illustrate the potential for seasonal succession in streams.

INDIRECT INTERACTIONS

Indirect interactions are interactions between two species mediated by one or more other species (Wootton, 1994). The number of indirect interactions in a community is tremendous because of all the linkages among species in each ecosystem. The complexity of understanding indirect interactions can be daunting, so we first start with some specific examples rather than a general theoretical treatment.

One of the strongest examples of indirect interactions is the trophic cascade, which we discussed in Chapter 20, in which changes in the abundance of a top predator can alter the abundance of primary producers as mediated by interactions among several other species. While the trophic cascade is a relatively clean example of indirect interactions, a simple food chain is not a completely accurate characterization of trophic interactions with most food webs characterized by omnivory and feeding on more than one trophic level. Even more indirect interactions can occur in food webs than food chains. For example, removal of grazers may lead to a decrease in algal production because the grazers remineralize nutrients and increase the growth efficiency of the algae (Darcy-Hall, 2006; Bertrand *et al.*, 2009).

How indirect interactions alter predictions made by considering only direct interactions can be illustrated by the case of one predator with two competing prey species, and how the predator can favor continued existence of an inferior competitor. A predator can extirpate its prey if the predator and prey are present only in one habitat. Likewise, a competitor can extirpate another species if the second species is an inferior competitor (competitive exclusion principle discussed in the last chapter). If two prey species are present, the predator may switch to prefer the more abundant prey species. This switching can occur because it can be more energy efficient for a predator to forage preferentially for the more abundant prey (see Chapter 20). Because the preferred prey species is not necessarily the competitive dominant, the predator may promote the coexistence of both species.

Indirect interactions can give rise to mutualistic relationships in lake communities (Lane, 1985). Mathematical analyses of lake food webs confirm this for a variety of pelagic ecosystems. The example from the previous paragraph of a predator allowing an inferior competitor to remain in a community is but one of the possible ways such mutualistic interactions can arise. Although the details of such analyses are beyond the scope of this book, this analysis does suggest that trophic cascades in food webs are not the only important indirect interactions.

A specific example of indirect and complex interactions is illustrated by the common algal macrophyte *Cladophora*, its epiphytes, and grazers of the epiphytes (Dodds, 1991; Dodds and Gudder, 1992). In this case, the epiphytes generally compete for nutrients and light with the *Cladophora*. However, the *Cladophora* is N limited, and some of the epiphytes fix N and so may ultimately provide N to the *Cladophora*.

Contrary to initial expectations, epiphytes reduce drag on *Cladophora*. This occurs because the epiphytes decrease the amount of space between the filaments, lowering Reynolds number, and decreasing flow through the algal mass. Forcing the water to flow around the algal mass decreases friction and thus decreases drag. At the same time, forcing water around the mass decreases advective transport through the *Cladophora* filaments. Thus, the epiphytes may harm or help the *Cladophora*, depending on the environmental conditions. Invertebrate insect larvae remove sediments and epiphytes from the surface of the *Cladophora* and remineralize nutrients; this may facilitate *Cladophora* growth. At high densities, when other food sources are unavailable to invertebrates, they will eat the *Cladophora*. The *Cladophora* provides habitat and protection from predation for the invertebrates. Thus, the three groups of organisms have complex interactions that vary over space and time in the environment, can be mediated by other organisms, and fluctuate between being positive and negative.

An interesting indirect interaction occurs when the snail, *Physa acuta*, suffers from trematode parasites. The parasites often reach biomass of more than 30% of the snail animal tissue. The infection stresses the snails and causes them to graze more rapidly than uninfected snails. Snail populations with 50% infection rates grazed 20% more than uninfected populations. The increased grazing also caused a shift in algal community structure with fewer diatoms and cyanobacteria and increased dominance by the grazing-resistant filamentous alga *Cladophora* (Bernot and Lamberti, 2008).

Another example of complex, and sometimes indirect, interactions was shown in an experiment that was performed on competition between tufted ducks (*Aythya fuligula*) and fishes in a small pond formed by quarry activities in the south of England (Giles, 1994). This shallow pond had a fish community including bream, roach, perch, and pike and was turbid with few macrophytes. Most of the fishes were removed during a two-year period, after which midge larvae, macrophytes, and snails increased (Fig. 22.7). Fishes were reintroduced to enclosures inside the lake, and midge larvae and snail numbers returned to levels found in the entire lake before fish removal. Predation on ducklings by pike is a direct interaction that occurred before the fishes were removed. But fish removal had indirect effects as well; feeding by ducks, numbers of brooding ducks, and brooding success increased after fish removal because of their dependence on snails and midge larvae. Indirect effects of fishes on the ducks include competition for midge larvae. Indirect effects of fishes also included the removal of macrophytes, which provided habitat for snails, an additional food source for the tufted ducks. These data are instructive because they compare competition between birds and fishes and document an indirect effect mediated by macrophytes.

Linkages can occur among ecosystems, and indirect interactions can be propagated across ecosystems. For example, predation by fish can reduce dragonfly

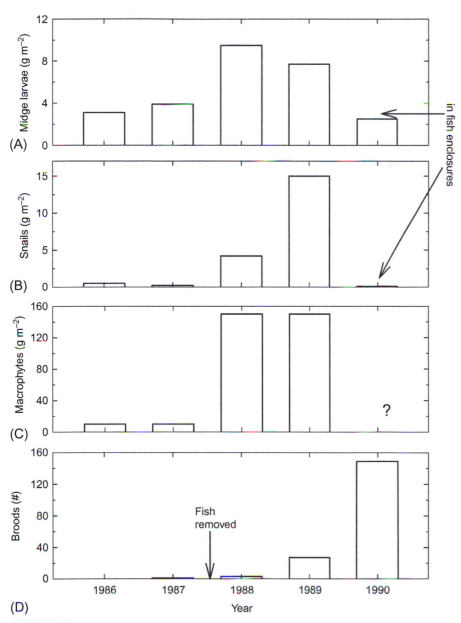

FIGURE 22.7

Interactions of fish and tufted ducks in a small English lake. Fish were removed in 1987 and 1988. Peak densities of chironomid midge larvae (A), snails (B), and macrophytes (C), and number of duck broods (D) all changed. Fish were added to the enclosures in 1990, and midge and snail densities are reported from inside the enclosures, demonstrating that fish were able to lower both populations. The ? in panel C means not reported. *(Data from Giles, 1994).*

larvae abundance in ponds. The terrestrial areas around ponds with fewer adult dragonflies preying upon aerial insects have more pollinators. More pollinators can lead to changes in terrestrial plant communities (Knight *et al.*, 2005). This chain of interactions indicates how aquatic and terrestrial habitats cannot be considered in isolation from each other, and how knowledge of the life history of organisms is essential in understanding complex interactions among them.

The difficulty with characterizing the influence of indirect interactions is the vast number of interactions that potentially occur. Theoretically, in the case where no species can be involved in one indirect interaction chain more than once, given a group of S species, there are S*(S-1) possible direct interactions, or interactions with one link. However, the number of possible indirect interaction chains with n links is S*(S-1)*(S-2)*...*(S-n). Consider a group of three species; there are six potential direct interactions and six potential indirect interactions chains. When there are seven species there are 42 potential direct interactions, 210 with two links, 840 with three links, 2,520 with four links and 5,040 potential indirect interaction chains with five and six links. How can an experimental ecologist deal with this complexity? Most all communities have far more than seven members. Luckily, there are weak and zero interactions that can attenuate or truncate indirect interactions, and a community may have a defined size (Dodds and Nelson, 2006). The issue still remains to be resolved.

STRONG INTERACTORS

Keystone species are those that have a disproportionately large impact on their community or ecosystem relative to their abundance (Power *et al.*, 1996). An alternative form of the same idea is that some species are strong interactors. The term keystone species has been used loosely in the ecological literature, so claims that a species is a keystone species should be viewed with caution. Identifying keystone species requires knowledge of the particular system of study and the organisms found in it. Stated differently, there is not always a keystone species, and which species will be a strong interactor is difficult to predict in advance, but the potential existence of strongly interacting species can have major consequences for understanding the community structure and, by extension, conservation efforts as well as dynamics of introduced species.

An example of a potential keystone species is the detritivorous fish *Prochilodus mariae* found in the Orinoco river basin in South America. The fish removes sediments from the stream bottom through its normal feeding activities (Flecker, 1996). These activities lower algal biomass and invertebrate mass, though some species of invertebrates and algae are stimulated. With experimental removal of this detritivore, large amounts of organic detritus built up, and the carbon flux into the food web and the patterns of carbon flux were substantially altered (Taylor *et al.*, 2006). These results were surprising because of the high diversity in the system (80 species of fishes) and the expectation that another species would take over the role of the large detritivore. This detritivore is an example of a strong interactor that has direct and indirect influences at several levels of the food web.

Bluegill sunfish (*Lepomis macrochirus*) may structure communities in small lakes and ponds throughout the eastern United States (Smith *et al.*, 1999). Frog tadpoles (except the bullfrog, *Lithobates catesbeianus*), newts, and predatory aquatic insects are much less abundant when bluegill are present. Since bluegills are common in many small ponds, their presence could alter community structure. Power *et al.* (1996) provided additional examples of possible keystone species in freshwater systems that included predatory fishes in river and stream food webs, piscivorous fishes in the trophic cascade of lakes, predatory salamanders in temporary ponds that prey on anuran tadpoles, and planktivorous fishes that have strong impacts on zooplankton communities. Beaver (*Castor canadensis*) have substantial effects on streams, from altering hydrology, to importing nutrients, and causing strong changes in the aquatic plant community by herbivory (Parker *et al.*, 2007). Finally, a weevil that attacks Eurasian water milfoil (*Myriophyllum spicatum*) could be a keystone species because it stops the milfoil from dominating (Creed, 2000).

THEORETICAL COMMUNITY ECOLOGY AND AQUATIC FOOD WEBS

Lake food webs are well characterized and have been analyzed by theoretical community ecologists in attempts to describe general patterns (Pimm, 1982). General questions that have been asked include the following: What limits the length of food chains? Are complex systems more or less stable? How interconnected are large food webs? Does aggregation of species into trophic groups alter the results obtained from analyses of food webs? How variable are food webs over space and time? Can state changes (regime shifts) be predicted?

It is not clear what limits the lengths of aquatic food chains. The length of the food chain is of particular importance when determining biomagnification of lipid-soluble pollutants. General analyses suggest that the amount of primary production at the base of the food chain is not a good indicator of food chain length (Briand and Cohen, 1990); but these analyses are not necessarily consistent with the idea that energy is lost at each tropic level. However, at least one study suggests that trophic transfer of energy in the Okefenokee Swamp may be very efficient (Patten, 1993), so our concepts of food web efficiency and how it should link to the number of trophic levels may need to accommodate situations with high energy transfer efficiencies. Groundwaters in which the number of large animals is physically limited by their ability to move through aquifers typically have short food chains. Marine pelagic food chains are considerably longer than most pelagic freshwater chains. It has been suggested that stream food webs tend to be short but wider relative to those of lakes (i.e., with more species as primary consumers). Large rivers and lakes tend to have similar food web structure (Briand, 1985). Movement of energy in freshwater ecosystems will be discussed in more detail in Chapter 24.

The relationship between food web stability and complexity is also unclear. Early ecologists viewed simple systems as less stable (MacArthur, 1955; Elton, 1958). Mathematical analyses of simple linear models of randomly assembled "communities" then suggested decreased stability with more interacting species (May, 1972). Since then, empirical (Frank and McNaughton, 1991) and modeling (Dodds and Henebry, 1996) approaches have suggested that increased diversity begets greater stability. We are not aware of any data from aquatic systems that can resolve the controversy regarding food web complexity and stability.

The degree of connectivity (the number of links per species) and how it changes with community size are also controversial. It has been suggested that connectivity increases as food webs become more complex (Havens, 1992), but this analysis is controversial (Martinez, 1993; Havens, 1993). Interesting results have been obtained that indicate that the proportion of species in each trophic category does not change with food web size (Havens, 1992). However, even this idea of constant proportions of species at each trophic level independent of food web size has been demonstrated to be false in one stream community (Tavares-Cromar and Williams, 1996).

As a convenience, scientists conducting community analyses of food webs lump organisms into trophic categories. For example, in benthic food webs, one of the primary food sources is often detritus. In reality, detritus is a complex microbial and invertebrate community associated with decaying organic material. Analysis of a highly resolved food web from Little Rock Lake, Wisconsin (Fig 22.8), suggests that lumping or aggregation of food webs into functional groups strongly influences properties such as the proportion of species at each trophic level and the connectivity of the food webs (Martinez, 1991).

Food webs can have variable structure over space and time. Descriptive characteristics of a detritus-based food web varied over time where the food web was more or less complex across seasons depending on the life cycle of the benthic invertebrates (Tavares-Cromar and Williams, 1996). Temporal variation in stream food webs has been linked to seasonal hydraulic cycles (Power *et al.*, 1995). Spatial variation also has clear effects, and benthic and pelagic food webs in lakes are distinct (Havens *et al.*, 1996a), pools and riffles in streams contain different organisms, and there are strong effects of water depth on species composition in wetlands.

FIGURE 22.8

The food web of Little Rock Lake as described by Martinez (1991). Producers are on the bottom ring, consumers above. Species are indicated by the balls, and the sticks connecting the balls indicate which species below are being eaten. Note some of the consumers have loops to themselves representing cannibalism. *(Image produced with FoodWeb3D, written by R.J. Williams and provided by the Pacific Ecoinformatics and Computational Ecology Lab; www.foodwebs.org; Yoon et al., 2004).*

THRESHOLDS AND ALTERNATIVE STABLE STATES

One concern about ecological systems is that they will experience regime changes (cross thresholds that are difficult to return from) as humans influence them. Such thresholds are, by definition, difficult to predict because they are abrupt and completely change system structure. A system that does not return to the initial state after the initial conditions are reestablished is said to exhibit *hysteresis*. Predicting these changes based on mechanistic understanding is becoming increasingly important as humans have global directional influences on the environment (Scheffer *et al.*, 2001). One such threshold we have already discussed occurs with eutrophication, where an anoxic hypolimnion leads to increased internal phosphorus loading and large difficulty in recovering an oligotrophic state. Another obvious state change is that caused by extinction; once a species is extinct, the probability is essentially zero that it will evolve again.

How do we predict when a system is reaching a point where a threshold will occur? Forecasting is difficult because new states are often outside our range of experience (Groffman *et al.*, 2006). Regime changes may be preceded by alterations in system variance (Carpenter and Brock, 2006). Making things more difficult, multiple stressors may be involved in changes. Interaction effects of warming, drought, and acidification, rather than a simple summation of individual stressor effects, best explained responses of consumer and primary producer communities (Christensen *et al.*, 2006). With global climate change and strong local influences, multiple interactive stressors are probably the norm in freshwater ecosystems.

One example of alternative stable states that is well investigated is the state of shallow lakes. There are two stable states, phytoplankton dominated and macrophyte dominated (a clear water phase). Eutrophication is able to drive these shallow lakes from macrophyte dominance to phytoplankton dominance. In Lake Veluwe (located in The Netherlands) increases in phosphorus above $20\,\mu g\,L^{-1}$ led to disappearance of the clear water state. Nutrient control was implemented, and the lake was flushed with low phosphorus water. The macrophyte-dominated clear water state was not reattained until the total phosphorus concentrations fell below $10\,\mu g\,L^{-1}$. This system exhibited strong hysteresis (Ibelings *et al.*, 2007).

The threshold of internal phosphorus loading was discussed in Chapter 18. This is a good example where understanding mechanism is important in understanding how a threshold and an alternative stable state might occur. When increases in nutrients that are great enough to lead to disappearance of dissolved oxygen from the hypolimnion change the state of the lake there is a fundamental change in biogeochemical processes. In this case, internal loading is much greater if the surface of the sediment is not oxidized and $FePO_4$ is not held in the sediments. A similar eutrophication threshold may also occur in the Everglades, where rapid changes in biotic components occur when total phosphorus concentrations exceed 12 to $15\,\mu g\,L^{-1}$ (Richardson *et al.*, 2007).

A problem is that predicting when a system will change state rapidly is difficult. Such prediction is important because once a threshold is crossed it can be difficult or expensive to restore the system. Rising variance may be an indicator that a threshold is being approached (Carpenter and Brock, 2006; Carpenter *et al.*, 2008). Using variance as an indicator needs further study, and the authors stress that once a threshold is detected, it could be too late. Prediction is difficult, but the fact that thresholds can occur when a system is pushed is important to keep in mind. There are several important points to managing ecological systems with potential thresholds. First, a mechanistic understanding of the system is required to understand the specific peculiarities and drivers of a system. Second, the precautionary principle (don't change the system unless you know the change will not cause an undesirable outcome) should be taken.

INVASION AND EXTINCTION REVISITED

Some ecologists have termed this the era of homogenization, or *homogocene*, because increasing globalization is spreading species around the planet (Rosenzweig, 2001). While *exotic species* simply refers to species located outside of their natural range, *invasive exotic species* are those that proliferate and cause damage in new habitats; invasive exotics are often weedy, generalist taxa with high reproductive capacities. Humans are transporting organisms ranging from microbes to large herbivores around the world, some with devastating consequences (see Chapter 11 for further discussion on species invasions). At the same time, most ecologists agree that we are causing the sixth mass extinction event in Earth's history (Wake and Vredenburg, 2008), and exotic species can be a major cause. Certainly, humans have invaded many habitats where they have not been abundant, and have had tremendous negative effect on many species. Nonhuman exotic species can outcompete or increase predation pressure on increasingly stressed native species, and subsequently the biodiversity of most regions of the planet is rapidly declining.

Invasive exotic freshwater species have been linked to declines and extirpations of native species through a variety of direct and indirect mechanisms, which are often difficult to identify. For example, rusty crayfish (*Orconectes rusticus*) and other invasive crayfish in the United States are replacing native species in parts of the northeast and Great Lakes where they have been introduced (Thorp and Covich, 2001). Rusty crayfish are large, aggressive crayfish that can displace smaller species from preferred habitats such as rock shelters, and it is hypothesized that displaced crayfish are rendered more vulnerable to fish predation (an indirect effect of the interaction among crayfish). Other hypothesized mechanisms for crayfish species replacements include reproductive interference, direct predation, and competition for food and habitat. However, definitive experimental evidence for these various mechanisms is generally lacking.

Determining if exotic invasive species are in fact the causes of native species declines is difficult, as the presence of invasive species and habitat degradation are almost inevitably correlated (Didham *et al.*, 2005). As an example, the direct, negative impacts of zebra mussels (*Dreissena polymorpha*) on native mussels and other species seem obvious (see Sidebar 10.1). However, some have questioned whether habitat degradation and loss of native unionid mussels opened the door for zebra mussel invasions (Gurevich and Padilla, 2004), and experimental evidence for either scenario is lacking. Regardless, the end result of the spread of exotic invasive species and loss of native species is a biologically less diverse and homogenized planet. Along with the ecological consequences, invasive species are having major economic impacts. As of 2005, it was estimated that there were 50,000 exotic species in the United States, which cost an estimated $120 billion per year (Pimentel *et al.*, 2005). Major costs associated with exotic species include agricultural losses, recreational and ecotourism losses, and control and eradication efforts.

Although some of the consequences of the loss of so much biodiversity remain to be seen, many adverse ecological and economic effects are already evident, including declines and losses of fisheries, reductions in water quality, increased costs of water treatment, and adverse effects on agriculture and aquaculture (Edwards and Abivardi, 1998). Fully understanding these effects requires understanding the complex interactions among organisms. Studies of relationships between biodiversity and ecosystem function (discussed more fully in Chapter 24) suggest that the structure (e.g., community structure), function (e.g., resource use efficiency and productivity), and stability (e.g., resistance and resilience to disturbance) of ecosystems is being compromised as species are lost (Tilman *et al.*, 1997; Loreau *et al.*, 2001).

The ecological consequences of declining species diversity are less studied in freshwater systems compared to terrestrial (Covich *et al.*, 2004), but there is ample evidence that loss of species will alter freshwater ecosystem function and reduce stability (e.g., Vanni, 2002; Cardinale *et al.*, 2006). As discussed in the earlier section, "Strong Interactors," experimental manipulations with the detritivorous fish, *Prochilodus mariae*, in a Venezuelan river found that their feeding, egestion, and associated bioturbation accounted for over half of downstream particulate carbon export; removal of the fish from a stream reach resulted in an over 400% increase in deposited organic sediments (Taylor *et al.*, 2006). Removal of this fish also resulted in significant increases in respiration and primary production.

Studies of fish assemblages in Lake Tanganyika, Africa, and a South American river provide further evidence that declining fish diversity will affect ecosystem functioning. Modeled excretion rates and extinction scenarios indicate that some species in these assemblages have stronger influences on nutrient cycling rates than others; unfortunately, some of the most important contributors to nutrient recycling are targets of commercial harvest (McIntyre *et al.*, 2007). The ultimate consequences of extinctions depend on which species are lost and if other species compensate (e.g., their populations increase in response to the loss of other species). Either way, the models indicate that declining freshwater fish diversity will negatively influence nutrient recycling. Clearly, declining diversity of fishes and other consumers will have ecological consequences. Given that extinction rates are highest in freshwater habitats, understanding linkages among biodiversity and ecosystem function and stability is an active area of research in freshwater ecology.

SUMMARY

1. Disturbance is an important ecological process. Perhaps the best documented form of disturbance in aquatic ecosystems is river flooding. This is a natural part of the ecosystem, and river restoration projects are beginning to consider this aspect of aquatic ecosystem dynamics.

2. Succession occurs both seasonally and over longer time frames in most aquatic habitats. The best examples of seasonal succession are from the series of events that occur in the epilimnion of stratified temperate lakes. Succession after reservoir construction, when a shift from lotic to lentic communities occurs, is also well documented.

3. Streams may also exhibit seasonal succession related to changes in deciduous vegetation or predictable seasonal flooding. Riparian wetlands also exhibit successional patterns related to seasonal flooding.

4. Indirect interactions are likely important in all aquatic habitats. Such interactions may be difficult to predict without sound knowledge of the life history and ecological roles of organisms in a particular habitat.

5. Some species in aquatic ecosystems have disproportionately strong effects on many other species. These are termed strong interactors or keystone species.

6. Complex interactions are common in aquatic ecosystems.

7. Thresholds and alternative stable states complicate understanding and management of aquatic ecosystems, and the potential difficulty in returning some systems to their original ecological state suggests that the precautionary approach to management is warranted until system dynamics are well understood.

8. Human activities are facilitating biological homogenization through the spread of exotic species and loss of native biodiversity. We are currently experiencing one of the most dramatic extinction events in the history of the planet, and freshwater species are among the hardest hit.

9. The ecological consequences of species losses and introductions are not fully understood, and this is an active area of research in freshwater ecology.

QUESTIONS FOR THOUGHT

1. How predictable are successional trajectories (i.e., can sequences of species colonization be predicted or just general patterns)?

2. How might understanding species interactions be important for predicting the effects of introduced species?

3. How can disturbance make competition less intense?

4. If a species is a keystone species, should greater attention be paid to conservation of that species than others in a habitat?

5. Why would introduction of a keystone species potentially be a threshold in an aquatic environment?

6. How many potential indirect interactions, excluding loops, are there in a community of 10 species?

7. How could you design an experiment to test whether an invasive species or habitat degradation caused the extinction of a native species?

Fish Ecology and Fisheries

FIGURE 23.1
A large pike (*Esox lucius*). (*Photograph courtesy of Chris Guy*).

Doi: 10.1016/B978-0-12-374724-2.00023-4

Much of the economic impetus behind protecting freshwaters is related to maintenance of productive sport and commercial fisheries. Fishes dominate many freshwater food webs as the top predator in streams and lakes. Fishes are good indicators of water quality and model organisms for physiological and behavioral research. Knowledge of fish ecology and traits is essential to fish conservation (Angermeier, 1995). In this chapter, we discuss the biogeography of fish communities and factors that influence growth, survival, and reproduction of fishes. Population dynamics, communities, and production of fishes are briefly described, as are harvesting, management, and aquaculture methods.

BIOGEOGRAPHICAL AND ENVIRONMENTAL DETERMINANTS OF FISH ASSEMBLAGE DIVERSITY

Diversity of organisms was discussed Chapter 11, but this chapter will provide more detail on the ecology of fishes and practical aspects of fisheries at a very basic level. Over the past couple of decades, fisheries biologists in developed countries have become more interested in managing species other than those used for harvest or recreation. Efforts to conserve aquatic species have required agencies to be involved with management of nongame species, including maintenance and even restoration of biodiversity and protection of endangered species. These aspects of fisheries biology require consideration of factors that influence diversity and distribution of fishes. Patterns will be described from the most general to the smallest scale.

The numbers of species are approximately the same for freshwater and marine fishes. This is an interesting observation because marine habitats dominate the planet and marine fishes have had a longer evolutionary history and would be expected to be more diverse. However, continental waters are more geographically isolated, so the possibility for reproductive isolation and subsequent speciation is greater in freshwaters.

Speciation, in large part, occurs within drainage basins, and larger drainage basins tend to have more species (Oberdorff *et al.*, 1995; Guégan *et al.*, 1998). Islands tend to have small watersheds and lower fish diversity. This diversity pattern is consistent with the theory of island biogeography covered in Chapter 11. Also, consistent with many other animal and plant groups, fish species are richer in the tropics than in temperate zones. Data compiled by Matthews (1998) demonstrated that approximately 7% of samples from both tropical and temperate streams had no species in them (completely unsuitable for fishes), but the maximum possible biodiversity is considerably higher in tropical streams and rivers. Several explanations could explain the high diversity in the tropics. Lower diversity in temperate waters could

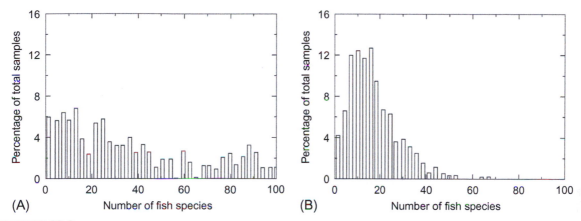

FIGURE 23.2

The proportion of total samples containing various numbers of fish species from tropical (A) and temperate (B) streams. The total numbers of samples were 204 in the tropical streams and 815 in the temperate streams. *(After Matthews, 1998).*

be partially related to glaciation of higher latitudes that led to local extirpations of fish species. Time for evolution (i.e., total generations) of fish species has also been greater in the tropics because more generations per year can be produced.

Tropical areas also have greater terrestrial net productivity, and this has been linked to fish diversity (Oberdorff *et al.*, 1995; Guégan *et al.*, 1998). Thus, the number of species per sample peaks at 10 to 20 in many temperate streams, but many samples in tropical streams have 50 or more species (Fig. 23.2). A striking example of tropical adaptive radiation is the diverse and highly coevolved assemblage of fishes associated with riparian habitats in the Amazon basin (Sidebar 23.1). The spectacular diversity of fishes in some African Rift lakes was described in Chapter 11.

Within continents, different ecoregions have varied diversity. For example, in North America exclusive of Mexico, there are about 740 fish species. About 300 of these can be found in the Mississippi basin. This basin has an area about equal to that of the Hudson Bay drainage, which has about 100 species. Lack of glaciation and a long evolutionary history (over 20 million years since the sea level fell during the Tertiary period) probably led to evolution of more species in the Mississippi basin (but terrestrial net primary productivity is also greater in the Mississippi basin, making strict explanations difficult).

Within a basin, fish diversity in rivers and streams increases at greater distances from the headwater streams (Fig. 23.3) because larger rivers are less likely to dry, small streams and rivers are more subject to debris torrents and catastrophic flooding, and the floodplain of a large river contains a larger

---

SIDEBAR 23.1
Fishes in the Forest

Many fishes associated with the Amazon River drainage are dependent on annual flooding of the riparian forest (Goulding, 1980). The Amazon and its tributaries flowing through tropical forest annually flood the low-lying areas around the river channels, allowing fishes access to the forest floor, and temporarily connecting many small ponds and lakes with the main river channel. The total number of Amazonian fish species may be between 2500 and 3000, and a diverse fish assemblage forages in the floodplains. Some of the species move from the main river into tributaries to spawn, and others move from the flooded forest through the tributaries to the main river channel to spawn. These migrations allow predatory fishes and local fisherman to capture spawning fish as they move predictably in large schools through specific places during specific times.

A most unusual group of fishes has evolved to eat fruits and seeds. These frugivorous fishes are confined almost entirely to South America. In a coevolutionary response, some of the riparian forest species have evolved fruits that are dispersed by the fishes. These trees and shrubs fruit during the flooding season, and their fruits are under or near the water surface. Fishermen take advantage of the feeding habits of the frugivores and bait their hooks with fruits that are preferred by the fishes.

Other fishes are seed predators and crush, consume, and digest seeds. Seeds form a significant proportion of the diet of some species of piranhas (*Serrasalmus* spp.), which are known to most people as aggressive animal predators. These fishes can have impressive jaws and teeth that are adapted to breaking the hard shells protecting nuts and seeds.

The fishes in the forest provide a clear example of the linkage between riparian wetlands and production of fishes. Many of these fish species have an absolute requirement in their life cycle for the seasonal availability of resources in the flooded riparian zone. Some plants benefit from the interaction as well.

---

range of habitat types (e.g., wetlands, oxbows, small side channels, small streams entering the main river, and open channels with high water velocity). While human impacts such as nutrient pollution and channel modifications have greatly decreased fish diversity in many large rivers, the observed maximum number of fish species per sample is generally higher in larger rivers compared to lower order streams.

Within basins, factors such as flooding, pollution, turbidity, reservoirs, channel modifications, and exploitation by humans can alter the numbers and types of fishes that are found. For example, in large rivers in which reservoirs have been constructed without fish ladders, species that move upstream to spawn can be excluded from upriver areas. Reduced diversity and restricted movements of fishes in large rivers has also adversely affected freshwater mussels, whose larval stages attach to fishes for dispersal. Fish species that required flowing waters to keep their eggs in the water column until they hatch are also harmed by reservoirs. In these cases, the riverine fishes are replaced mostly by

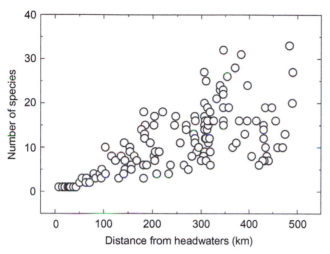

FIGURE 23.3

Number of fish species as a function of distance from headwaters in the Kali Gandaki River, Nepal. Statistical analysis of the data suggests that the number of fish species increases with river distance up to about 300 km. *(Data from David Edds)*.

a lentic fish assemblage in the reservoirs. Heavily polluted rivers, particularly those in which O_2 levels are very low, support few or no fishes.

Within a stream segment or reach or within a lake, the type of habitat present can be very important. Any angler knows about specific areas that fishes are more likely to inhabit. In streams, for example, undercut banks, woody debris, and deep pools all are places where fishes may congregate. Humans tend to remove habitat structure (e.g., snagging operations in large rivers to clear channels for navigation), and much effort has been made in many lakes and streams to restore habitat that is essential to maintaining a diverse fish assemblage and a productive fishery.

Reservoirs generally have a blend of lotic and lentic habitats. At the upstream ends, where rivers enter reservoirs, fish species characteristic of flowing waters are more likely to be found. In the deeper waters pelagic species are common. The numerous embayments found in many reservoirs can be more like shallow lakes and have high densities of submerged and emergent wetlands. The face of the dam may be rock, earth, or concrete. The rocky or concrete dam faces may be unusual benthic habitat for many areas, particularly in reservoirs that have been constructed in large depositional valleys where rocky habits are rare in natural rivers.

Analysis of the fisheries of the Parana River in South America before and after construction of a major impoundment showed that, along with the negative ecological impacts on the river, the overall value of the fisheries declined (Hoeinghaus *et al.*, 2009). The impoundment shifted the local fisheries from one focused on high-value migratory fishes to one based on relatively low-value reservoir-adapted introduced species. As a result, the energetic costs of fisheries production increased, while the market value decreased.

Restoration of streams and lakes requires consideration of habitat characteristics that lead to successful recovery of fish biodiversity (National Research Council, 1992). The ecological basis for river management must consider what constitutes a natural hydrological regime (Petts *et al.*, 1995; Palmer *et al.*, 2005). Restoring physical structure to stream channels may include wing deflectors, low dams, placement of debris such as boulders or logs, and other measures (De Jalón, 1995). Boat traffic may also need to be controlled to decrease turbidity and wakes that can interfere with reproduction (Murphy *et al.*, 1995). Restoration or habitat enhancement is also practiced in lakes and reservoirs. Measures include increasing cover for small prey fishes and juvenile fishes of larger species and altering the hydrological regime to match spawning requirements of desirable fishes. Addition of old Christmas trees or other large structural material is commonly used to increase heterogeneity of reservoir fish habitats.

The diversity of prey species may also alter the diversity of piscivorous species. For example, it is unlikely that the number of predatory species will exceed that of prey species (Fig. 23.4). However, this is a weak relationship and water quality, habitat structure, and large-scale patterns of species dispersal are stronger determinants of fish diversity.

Along with their taxonomic diversity, fishes exhibit a great diversity of functional roles and life histories. Many of the species we associate with recreational and commercial fisheries are invertivores or piscivores, but many are also important grazers of macrophytes or periphyton, some consume detritus, and a large number are omnivorous. While predatory species are often the most desirable species for food and sport, they can be more vulnerable to overexploitation because of their longer life cycles and naturally lower population densities.

Multivariate analyses of life history characteristics of 216 North American fish species showed some consistent patterns among groups of freshwater fishes (Winemiller and Rose, 1992). Larger, later-maturing fishes generally had high fecundity, small eggs, low parental care, and a short spawning season; the opposite was true for smaller fish species. There was a strong effect of phylogeny, whereby higher taxonomic groups often shared suites of life history traits. Fishes that live in similar habitats also shared suites of traits. These patterns allow for grouping fishes according to life history traits and provide a framework for effective management and conservation of various fish species.

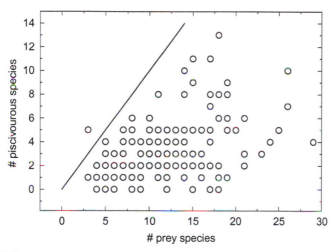

FIGURE 23.4

Number of piscivorous species as a function of number of prey species from 178 fish assemblages in eastern North America. The line denotes a 1:1 relationship. Note that more predatory than prey species occur in only one case. *(After Matthews, 1998).*

PHYSIOLOGICAL ASPECTS INFLUENCING GROWTH, SURVIVAL, AND REPRODUCTION

Energetics can be thought of as the overriding factor controlling growth, survival, and reproduction of fishes (or any other organism for that matter). Environmental extremes can cause excessive energy drains affecting the success of fishes. Survival requires a set amount of energy that varies depending on the external environment. Growth requires additional energy, and successful reproduction requires the most energy (Fig. 23.5).

Osmoregulation requires energy and is necessary for survival. Osmoregulation mechanisms vary among freshwater fishes. The lower limit of a fish's distribution in an estuary may be determined by its salt tolerance, as may the response to runoff from road salt additions and hypersalinity in closed-basin lakes.

All fishes that enter freshwater must osmoregulate because their internal salt concentrations exceed those found in most freshwaters (they are hypertonic relative to the water). In general, excess water is excreted by well-developed kidneys. Because water is not limiting, up to one-third of the fish's body weight may be excreted each day (Moyle and Cech, 1996). Diadromous species must have versatile osmoregulation strategies to deal with the transitions from one level of salinity to another. Fishes living in highly dilute waters may have ion-specific active transport mechanisms to acquire and

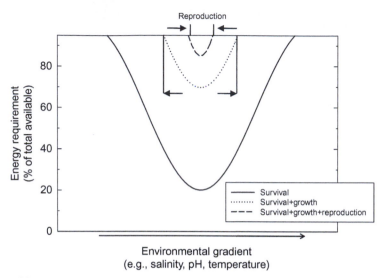

FIGURE 23.5

Conceptual model of energy requirements for survival, growth, and reproduction as a function of an environmental gradient. If 100% or more of the available energy is required for survival, the fish will not survive. If the sum of energy required for survival and growth exceeds 100%, the fish will not grow, and if the sum of energy for survival, growth, and reproduction is less than 100%, the fish can reproduce.

retain useful ions. All fishes must regulate the concentration of ions such as sodium and potassium in order to survive, and this requires energy.

Temperature also controls growth, survival, and reproduction of fishes. At low temperatures, metabolic rates are depressed, and swimming requires more energy because of the increased viscosity of water (see Chapter 2); an increase in temperature from 13 to 23°C is accompanied by a 22% decrease in viscosity but roughly a doubling in metabolic rate. General molecular adaptations to temperature extremes were described in Chapter 15. Physiological characteristics interact to determine the optimal temperature for growth as mediated by energetic constraints. For example, even though salmonid species seem very similar, they have different temperature requirements for growth and survival (Table 23.1). Temperature influences a variety of fish life history characteristics, so fishes actively seek water that has a temperature near the optimal for growth.

Temperature requirements for reproduction may be more stringent than those for survival or growth. Temperature constraints of fishes in freshwaters and interactions with trophic state have led fisheries managers to classify lakes in several general types of thermal regimes with regard to commercial or sport fishes (Fig. 23.6): (1) warm waters dominated by black, white, yellow, and striped basses, sunfish, perch, or catfish; (2) cool waters with some of the previously mentioned species plus walleye and pike; (3) cold waters dominated

Table 23.1 Temperature Requirements for Various Salmonid Species

Species	Maximum Growth (°C)	Maximum Temperature (°C)	Spawning Temperature (°C)
Atlantic salmon (*Salmo salar*)	13–15	16–17	0–8
Brown trout (*Salmo trutta*)	12	19	2–10
Rainbow trout (*Onchorhynchus mykiss*)	14	20–21	4–10
American brook trout (*Salvelinus fontinalis*)	12–14	19	2–10
Grayling (*Thymallus thymallus*)	10–16	20	10

(After Templeton, 1995) Note how maximum growth rate occurs at a temperature only slightly less than that for maximum survival, and the temperature required for reproduction is much less than that for maximum growth

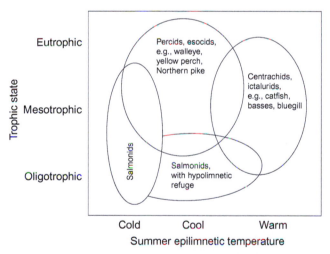

FIGURE 23.6
Conceptual illustration of fishery types in temperate lakes with summer stratification as a function of temperature and trophic status.

by salmonids; and (4) two-story waters in stratified lakes with oxygenated hypolimnia in which cool water and cold water species can find refuge from high water temperatures during the summer. During summer months, cold- or cool-water species are hypolimnetic, and warm water fishes reside in the epilimnion. If thermal refugia are present in streams (e.g., groundwater input), cold water species can survive in fairly warm streams and rivers. In some warm regions, cold water species are stocked and maintained in streams below the cold outflows of hypolimnetic release impoundments.

Temperature optima and considerations are important to predict how fishes will respond to climate change (see Chapter 1). Recent analyses of fish communities in France show shifts toward warmer water species assemblages (Daufresne and Boet, 2007). Furthermore, the authors noted that impoundments fragment populations and stop migrations of species to cooler waters, further altering the ability of fishes to respond to changes in temperature. The systems mentioned earlier, with cool water refugia, may not remain cool enough to maintain cold water fishes.

Oxygen is crucial for fishes. Those that breathe with gills require dissolved O_2. The problem of fish kills related to O_2 depletion was described in Chapter 12. The O_2 requirement is greater as temperature increases, and the concentration of O_2 that can be dissolved in warmer water is lower. Fishes can respond to lower O_2 by increasing ventilation rate. Although increased ventilation allows survival under marginal O_2 concentrations, it uses energy and causes stress. Some species of fishes (e.g., lungfish) can use atmospheric O_2, but most others will die if dissolved O_2 concentration becomes too low. Temporary anoxia can be withstood by many species, but repeated bouts of anoxia can lead to cumulative harmful effects (Hughes, 1981). Early life stages of fishes (e.g., larvae) are often most sensitive; adults may be able to tolerate anoxic events, but they may not successfully reproduce.

Food quantity and quality can alter the survival, reproduction, and growth of fishes. Thus, fishes are selective in what they eat (Hughes, 1997). All fishes require protein, carbohydrates, lipids, vitamins, and minerals. A fish that is piscivorous may be able to obtain reasonable amounts of protein and lipids but can be limited by the energy content of its food. Zooplanktivorous fishes may have diets that are rich in lipids, and many teleost fishes have a high lipid requirement. Thus, copepods that contain high concentrations of wax esters are an excellent food source for fish species with high lipid requirements. Herbivorous fishes can be more limited by proteins and lipids, and detritivorous fishes are more limited by protein than herbivores.

Fishes that live in warmer waters can require lipids with a higher melting point. Herbivorous fishes (such as grass carp) can have special enzymes, extra long digestive tracts, or symbiotic bacteria in their guts that allow them to use low food quality plant materials more efficiently as a source of carbohydrate. If one type of food is of a high quality, it can partially compensate for lower quality food. For example, catfish can use cellulose as a carbon source almost as efficiently as starch, if a high proportion of protein is present in their diet (Fig. 23.7). Less is known about the roles of vitamins and minerals in fish diets (Pitcher and Hart, 1982).

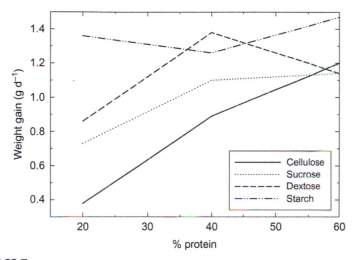

FIGURE 23.7
Growth of channel catfish (*Ictalurus punctatus*) on four carbohydrate sources as a function of protein content of food. *(Data from Simco and Cross, 1966)*.

The rate at which food must be consumed to allow for growth and reproduction depends on both the requirements imposed by physiological demands of the environment and the actual energy investment in obtaining food. Under stressful conditions, a greater portion of the food ingested must be used to maintain basic metabolism, and less must be used for growth. Determining actual energy requirements can be very complex. For example, in a heterogeneous environment, metabolic costs can vary over space and time, as can the amount of energy required to locate and capture prey. Consider, for instance, the idea that foraging is more difficult when a predator is present (see discussion on optimal foraging in Chapter 20). Ample food may be available in the environment, but when a predator is nearby, a fish must use energy to avoid being eaten (Milinski, 1993) and balance that against its need to forage for food (Fig. 23.8).

POPULATION DYNAMICS OF FISHES

Fish population dynamics can be determined by several key characteristics, including the size of the population (*stock*) and the population growth (*production*). The population can be divided into a number of *age classes* or *size classes*. The number of new fish entering each size or age class (*recruitment*) and the number of fish that are lost between each class (*mortality*) are also related to population dynamics. Also, the number of eggs females produce (*fecundity*) and the number of larval fish that successfully hatch from those eggs can influence population size.

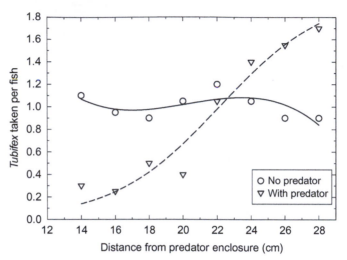

FIGURE 23.8

Effect of predator (*Oreochromis mariae*) on the rate at which a stickleback (*Gasterosteus aculeatus*) consumes prey (*Tubifex*). This experiment included an adjacent fish cage that could hold the predator, and worms were placed in tubes at various distances from the adjacent cage. The stickleback had to enter the tube and lose sight of the predator to take a worm. With no predator, the worms were taken from all tubes equally. With the predator, the sticklebacks preferred to feed as far away from the predator as possible. *(Modified from Milinski, 1993)*.

When sampling a fish population, the first questions that arise are usually how many fish, how big are they, and how old are they? Fishes have indeterminate growth, and keep increasing in size as they get older, if their energetic needs are met. However, there is not a strict relationship between age and size, so additional information such as relationships among mass and length and population structure also need to be determined. This information can indicate several important things about the reproduction, growth, and mortality of fishes and assist with management of populations for recreational or commercial use. An important consideration is how to catch the fish and determine their population size (Method 23.1).

After the fish are captured, the total number and sizes of individual fish must be determined. Length is the easiest to measure but is not a linear function of mass (Fig. 23.9B) because mass is related more closely to volume than length; mass is a logarithmic function of length for most animals, including fishes. The relationship between length and mass can be used to assess the condition of fishes. A fish that has a high mass per unit length often is considered to be healthy and in better condition. Indices have been constructed to characterize mass–length relationships (Anderson and Neumann, 1996). These indices

METHOD 23.1

Sampling Fish Populations

A wide variety of techniques are available for sampling fish populations. Determining the method to use depends on the physical constraints of the habitat being sampled, the species of interest, the information being sought, and the relative effectiveness of the method. The capture methods can be categorized as passive capture techniques, active netting, and electrofishing. All three are used in freshwaters and require determination of catch per unit effort. If all things are equal, twice the effort should lead to capture of twice as many fish. Of course, as population density decreases, additional effort to capture individuals will lead to a lower amount of catch per unit effort.

Passive capture techniques work by entanglement, entrapment, or angling with set lines (Hubert, 1996). Gill nets are commonly used to entangle fish in meshes of specific size. The larger the mesh size, the greater the selectivity for larger fish. Trammel nets also entangle fish. They are fine-mesh nets with larger mesh next to the net. When the fish pushes the smaller mesh net through the larger mesh, it forms a pocket in the smaller mesh net and is caught. Entrapment gear allows the fish to enter, but not leave, and may include bait to lure the fish. One type of entrapment gear has cylindrical hoop nets with a series of funnels in which the fish are able to move into the cylinder but not out. Fyke and other types of trap nets have additional panels of netting at the entrance to a cylindrical trap that guides fishes into the trap. Pot gears also use entrapment. These are rigid traps with funnels or one-way entrances. Minnow traps, crayfish pots, and similar equipment are included in this group. Weirs are barriers built across a stream to divert fish into a trap; these work well on migratory (usually reproductive) fishes.

Active gears require moving nets through the water to capture organisms (Hayes *et al.*, 1996). Nets can be towed (trawled) at the surface, through the water column, or across the bottom. Alternatively, purse seines can be used to encircle the fish and confine them into successively smaller space as the net is drawn closed. Other active gears include nets that are thrown or pulled up rapidly from the bottom, spears, dipnets, dynamite, and rotenone (poison). Obviously, some of these methods are quite destructive.

Electrofishing is a common and effective way to capture fishes. Either AC or DC current is used to alter the behavior of fishes or to stun or kill them (Reynolds, 1996). The most common strategy is to stun fish and then catch them with a net. To stun fish, the current and its pulsing frequency are controlled at a level that causes paralysis but little injury and death. Water conductivity is an important factor in determining effectiveness of the technique. Low-conductivity waters are poor conductors of electricity and decrease the effectiveness of the method; very high-conductivity waters allow the charge to dissipate too rapidly.

Fishes can be counted directly when removal is not necessary. Scuba diving or snorkeling is commonly employed. Habitats that are very turbid, with excessive water velocity, or unsafe for human contact are obviously not suitable for such techniques.

Several methods are used to determine total population size, depending on habitat and method of capture. If sampling is highly effective, such as in a small pond with a clean bottom, the fishes are collected until all are captured. It is rarely possible to capture all the fishes in a habitat, and several other methods are used to compensate. A depletion method repeatedly samples the same area, and the decrease in catch per unit effort is used to estimate the original population. Alternatively, mark and recapture techniques are used. In this method, a sample of fish is taken and marked. Fishes are later recaptured, and the proportion of marked fish is used to estimate the population size. There are several marking techniques, including fin clipping, external and internal tags, and marks (Guy *et al.*, 1996).

---

can be used to assess the relative condition of individual fish within a habitat or to compare the condition of fish populations among habitats.

As with many animal populations, small fish are usually more numerous than large fish (Fig. 23.9A) because mortality occurs over time and ultimately decreases the number of large (older) fish in a population. Figure 23.9 does

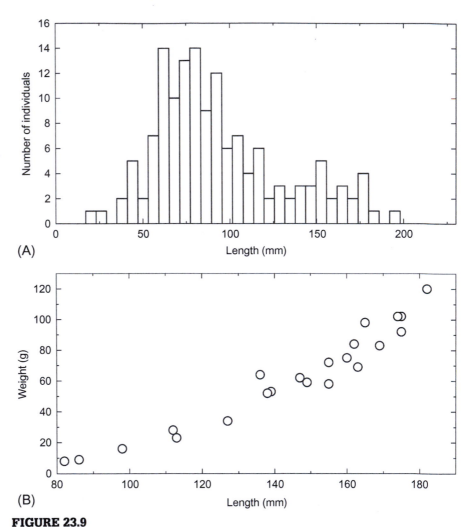

FIGURE 23.9

(A) The number of bluegill (*Lepomis macrochirus*) at each length and (B) the relationship between length and wet weight for fish taken by electrofishing from Pottawatomie State Fishing Lake, Kansas, on September 29, 1999.

not include data on the very youngest fish because they are not sampled effectively by electrofishing. The large numbers of fish of about 80 mm in length are probably 1 year old. An obvious exception to the trend of more small fish than large fish occurs when adult fish are stocked.

Many indices are used by fisheries managers to quantify growth, mortality, recruitment, condition, and size structure. We discuss one of these indices that is based on length data as an example. Relative stock density (RSD) is the percentage of fish in a specific length range relative to the total number of fish above a minimum size estimated to be in the population (Anderson and Neumann, 1996). In other words, it is the proportion of fish in a specific size range relative to all fish that can be sampled effectively. The RSD offers an index that describes the distribution of fish lengths in a population. It can be used in several specific situations. If the index is restricted to just the largest fish in the population, it can be used to calculate the relative number of trophy fish in a population. If the index considers all fish large enough for anglers to keep, it can indicate the quality of the fishery. The RSD can also be used to determine if a fishery is balanced (e.g., that there are not too many small fish or that competition or some other factor is causing poor conditions in a specific size class).

The next aspect of population structure we consider is the size class of each age of fish. If the size class is known as well as the population and biomass, the rate of growth or the production of the fish population can be calculated. Two of the most commonly used approaches to estimate ages of fishes are length frequency analysis and analysis of anatomical hard parts (Devrie and Frie, 1996). Length frequency analysis is used where distinct peaks in the distribution of lengths are observable. Population peaks often occur at specific size classes, each corresponding to a reproductive cohort. Such peaks may not be discernable when growth and time of reproduction are variable among individuals and for older fish in the population.

Scales, otoliths, bones, and spines are hard parts that are used to age fishes. As a fish grows, it adds to these hard structures. Daily marks can sometimes be discerned in young fishes, and where growth has a distinct seasonal pattern as in temperate zones, annual marks can be observed as well and can be used in much the same manner as tree rings (Fig. 23.10). Scales and fin rays are commonly used to determine age because they can be removed without killing the fish. Otoliths (calcified structures from the inner ear) can yield more reliable aging results but require sacrifice and dissection of the fish.

Once age distribution of a population is determined, recruitment and mortality for each size class can be calculated. Reproduction and mortality are influenced by numerous abiotic and biotic factors and can vary tremendously (up to 400-fold) among year classes (Pitcher and Hart, 1982). Explaining this variation is an important part of fish population management and ecology. Life tables such as those used by demographers and population ecologists allow mortality and recruitment rates to be calculated for each age class in a fish population. Age-specific life tables, for which a cohort is followed for years, are the most useful. These tables are constructed by calculating the proportion of fish surviving each time interval, generally a year. The recruitment and mortality at each size class can be correlated with environmental factors

FIGURE 23.10
A fin ray from a river carp sucker (*Carpiodes carpio*) showing growth rings. *(Photograph courtesy of Chris Guy).*

and used to calculate the influences of natural factors and exploitation on fish populations.

If the age distribution is known (age and mass of each year class of fish), then the production of the population can be calculated. Production incorporates both static (biomass, population size) and dynamic (growth, recruitment, mortality) measures. The mass produced per age class can be of interest to fish managers wanting to know how productive a water body will be of catchable fish. The total production could be of interest if the fish is a food source for sports fish, or if an ecosystem approach is being taken. More detailed treatment of calculation of animal production in the ecosystem context is covered in Chapter 24.

Generally, a negative correlation exists between recruitment and growth. If a large number of fish occur in a size class, then significant, intraspecific competition can occur. Intense intraspecific competition leads to poor fish yields in the size class. If food is not limiting (rarely the case in natural situations but possible in aquaculture), recruitment and growth may both be high.

REGULATING EXPLOITATION OF FISH STOCKS

Any fishery containing desirable fishes and with unlimited human access will be overexploited without management (tragedy of the commons). The amount of fish that are taken per unit time is the *yield* of a fishery. The ability to predict potential fish yields is central to attempts to manage exploited fish stocks. For many years, it was thought that the *maximum sustainable yield* (MSY) could be calculated and used to estimate the maximum potential harvest of any fish population that can be sustained in a particular water body. In theory, the MSY is the point when the yield or harvest in a given period is equal to the production over the same time period, a situation that is sometimes referred to as full exploitation because all "excess" production is removed. The MSY was calculated by William Ricker (Biography 23.1) using a curve that plotted the number of reproductive adult fish against the number of reproductive adults they could produce. The Ricker curve has seen much use in fisheries, in which the numbers of reproductive adults and their offspring can be determined.

While still informative, the MSY concept has several problems, including that it estimates yield without regard to size, the number of fish that are produced is often not related to the number of reproductive adults because many offspring can be produced per adult, and survival of eggs and fry is highly variable. Accurate estimation of the MSY requires a level of precision in quantifying population abundance or stock biomass that is often not realistic. Actual yield may be highly dependent on unpredictable biotic and abiotic factors, and most freshwater fisheries managers do not calculate MSY. Rather, managers rely on data on growth and recruitment of specific size classes of harvested fish for recommending management options. Finally, the only way to know the MSY for a population for sure is to find the inflection point by increasing harvest until production begins to drop; at this point, of course, the resource has been overexploited.

Management of fish stocks usually involves setting regulations on the numbers and sizes of fish that are removed. Such regulations are set depending on the type of fishery that is desired. Management techniques include closed seasons, limited access, limitation on the number of fishers, regulations on the numbers and sizes of fish taken by anglers, and regulations on methods used to take fishes (including types of gears, baits, and net mesh size).

As discussed previously, a negative correlation generally exists between recruitment and growth. Thus, a fishery can be managed either for a large number of small fish or for fewer large fish, but generally not both. Sports fisheries are

BIOGRAPHY 23.1 WILLIAM RICKER

Dr. William Ricker (Fig. 23.11) is one of the most renowned fish biologists in the world. He described the stock recruitment curve that is known as the Ricker curve. He was also one of the first scientists to introduce the concept of chaos into population dynamics. He had more than 100 scientific publications and an equal number of popular articles. Ricker was the recipient of many prestigious awards, including the Eminent Ecologist Award of the Ecological Society of America. He was a fellow of the Royal Society of Canada and of the American Association for Advancement of Science. He was chief scientist of the Fisheries Research Board of Canada.

Ricker was a man of many talents. He identified 80 new species and 46 new genera of stoneflies, wrote a Russian/ English book of fisheries terms, and he played musical instruments. Ricker got started in fisheries in 1938 when he was offered a job for the Fisheries Research Board of Canada. At that time, any job was welcome. It was a job that excited him, and he turned it into a career.

His views were not always popular. In a 1984 article in *Sports Illustrated* he was quoted as saying, "Practically everyone who has ever gone fishing considers himself an expert in fish management and doesn't hesitate to say so. Also, the man who uses any particular type of fishing gear invariably regards all other types as pernicious and destructive; but he can insist, with a straight face, that his kind of fishing couldn't possibly do the stock any harm."

When interviewed for this biography, Ricker suggested that it is important to arrange a research career around problems

FIGURE 23.11
William Ricker.

in which you are interested or you will not be successful. His advice was to get hands-on experience in a career before you commit to it so that you will be sure that you will like it. He thought an important part of fisheries management in the future will be to keep people from extirpating fish. Ricker said that if there are no more fish, there will be nothing more for fisheries biologists to research.

often managed for larger fish, and commercial fisheries are managed for a high yield of usable fish (e.g., just above some minimal size). The uncertainties in predicting recruitment and growth make absolute prediction of the effects of exploitation difficult. This uncertainty has led to numerous cases of overexploitation of fish populations (Hilborn, 1996). Thus, assessments of major fisheries will increasingly include risk analyses. If risk assessment is used, a manager will set limits based in part on the potential risk of long-term harm to production of the fishery. Risk management must be considered in fisheries, and the precautionary principle, or not taking actions until it is certain they will not cause harm, is often necessary (Hilborn *et al.*, 2001).

One of the most common regulations in sports fisheries is creel limits, or the numbers of fish that can be taken per day. Such limits rarely are effective in managing fish populations because most anglers do not catch their limit. However, reaching the limit does provide a degree of satisfaction for the angler.

Setting size limits is another method that is used to ensure that fish reach a size that is acceptable to anglers. Generally, these size limits are set so that several year classes of reproductive fish are less than the catchable size. Thus, in a poor year for recruitment or reproductive failure in a specific age class, the fishery will still have some recruits entering the catchable size range. Given a single length limit and heavy fishing pressure, the size distribution can consist of many fish just below the allowable size. This skewed size distribution can lead to intraspecific competition and low recruitment to fish of allowable size (Noble and Jones, 1999).

Slot limits that do not allow harvest of fishes of intermediate size are also used. In this case, surplus young fish can be taken. If the small fish are taken, intraspecific competition and recruitment into intermediate size lengths are lower, and growth rates of these fish are high. Slot limits could potentially create high production of large fish that are desirable for anglers.

Seasonal limits are often imposed. The most common of these is not allowing any taking of fishes while they are spawning. Fishes can be particularly vulnerable to angling pressure at this point in their life cycle. They often feed aggressively to assimilate energy needed for reproduction. Fishes that aggregate to spawn in highly predictable parts of the environment are most susceptible.

STOCKING FISH FOR FISHERIES

Many fisheries need additions of human-produced stock to allow exploitation to occur at desirable rates. Stocking is done for many reasons, including (1) introduction of new exploitable species, (2) introduction of new prey species, (3) introduction of biological control agents, (4) provision of fish to be caught immediately or after they grow, (5) satisfying public pressure to enhance fisheries, (6) reestablishing species where they have been extirpated, and (7) manipulating the size or age class structure of existing fish populations (Heidinger, 1999). These programs are expensive. For example, approximately 2.5 billion sports fishes are stocked each year in United States and Canada (Heidinger, 1999). Survey methods and accounting for purchases of fish licenses and fishing gear can be used to estimate if the cost of stocking programs is offset by revenues and economic stimulation.

The stocking of fish can be a useful management tool. In cases in which reservoirs and ponds are constructed, native stream and river fishes may not survive. Thus, introduction of sport fishes could be desirable (assuming they will not move into rivers and streams and cause problems there or exacerbate the disconnection of streams by the reservoir). In this case, native species are the safest species to stock.

Stocking can also be crucial to the recovery of some endangered fishes. Many fishes have been stocked in habitats in which they are not native under the impression that they would improve the fishery. Large-scale introductions can also include inadvertent introductions of other unwanted species or fish diseases. Unfortunately, the ultimate biological effects of such introductions usually have not been carefully considered or researched in advance. Precautionary approaches should be taken for new species introductions (Bartly, 1996) to avoid negative impacts in the future. These approaches include assuming that a species will have a negative impact until proven otherwise and putting the burden of proof of impact on the agency, company, or individual that releases the fish.

Several problems can arise when fishes are stocked, including competition with or predation on native fish species, hybridization with native fish stocks, lack of genetic diversity in fish populations that may lead to poor fitness, and transfers of parasites and disease (Moyle *et al.*, 1986). All of these factors should be considered when designing a stocking program.

Whirling disease is caused by infection by the microscopic cnidarian *Myxobolus cerebralis*. This parasite entered salmonid assemblages of the United States through stocked trout from Europe in 1958. By 2009, the disease had spread through hatcheries and river networks to 25 states. Initially, the disease was thought to be isolated to just hatcheries. Since then, however, the disease has spread to native trout and salmon populations and led to drastic declines in some populations. This disease can have negative economic impacts and may endanger native fishes. Introduction of whirling disease might have been avoided by careful assessment of stocking programs and more cautious aquaculture procedures.

The size and type of fishes to be stocked are often of concern. Large fish survive better, but small fish are less expensive to rear. Eggs are stocked only rarely because of poor survival. Fishes should also be stocked when and where they are most likely to survive and grow. If too many fish are stocked, interspecific and intraspecific competition may lead to poor growth. Thus, it is crucial to be aware of growth, mortality, and recruitment characteristics associated with the environment to be stocked, in addition to the ecology and the life history of the species used.

AQUACULTURE

The culture of aquatic organisms is called *aquaculture*. Fishes, amphibians, crustaceans, mollusks, and some algae are grown for consumption, stocking,

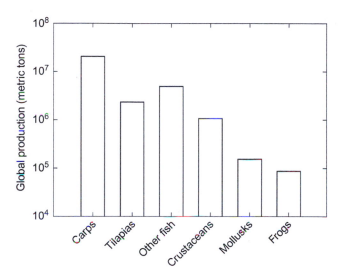

FIGURE 23.12

Global production of selected freshwater organisms by aquaculture in 1989; note the log scale on the *y* axis. *(Data from Stickney, 1994).*

and the aquarium trade. Fish culture may be an important source of protein in developing countries. Fish comprised 15.3% of the animal protein consumed by humans in 2000 (Food and Agriculture Organization, 2007). A billion people rely on fish as their primary source of protein (Laurenti, 2002, as described in Allan *et al.*, 2005). World aquaculture production is dominated by production of carp and the amount of carp produced is about 10-fold greater than that of any other fish (Fig. 23.12). Most fish production occurs in Southeast Asia.

Culture of fishes and crustaceans is generally more profitable if techniques are based on an understanding of the ecology of aquatic ecosystems. The trick in aquaculture is to maximize productivity but avoid the negative effects of eutrophication. For example, to grow at maximum rates, fishes need a significant amount of food. Excessive food additions can lead to increased biochemical oxygen demand, anoxia, and death of fish. In addition, aquaculturists grow organisms at high densities, which provides ideal conditions for propagation of diseases and parasites, and attracts predators (Meade, 1989).

The diseases of cultured aquatic organisms include protozoa, fungi, bacteria, and viruses. Stressed and crowded fish are more prone to infections. Bacteria are more commonly problematic in cold water species of fish, but antibiotics are generally effective in controlling these infections. Several species of viruses also cause problems for fishes. These diseases cannot be treated once the stock is infected. Rather, strict handling practices including use of disease-free stock must be followed. Some invertebrate parasites are transmitted via intermediate invertebrate hosts, such as snails and polycheate worms. In this case, disease prevention may include use of pesticides and pond draining to kill the intermediate hosts.

Perhaps the most ecologically advanced form of aquaculture is the polyculture of carp in Southeast Asia. This polyculture system was developed over centuries by fish farmers using close observation of the ponds and trial and error. The first book on aquaculture was written in Chinese in 460 BC (Fichter, 1988). The systems were developed to use several species with varied food requirements, and the excretion of one species fertilizes the growth of food for the other (Zweig, 1985). The dominant fishes in the Chinese systems are common carp (*Cyprinus carpio*, omnivore, benthic), bighead carp (*Hypophthalmichthys nobilis*, zooplanktivore), grass carp (*Ctenopharyngodon idella*, herbivore), and silver carp (*Hypophthalmichthys molitrix*, phytoplanktivore). In India, the catla (*Catla catla*), rohu (*Labeo rohita*), and mrigal (*Cirrhinus mrigala*) are used. There are about 2.5 million ha of carp ponds in India and China. In most cases, these ponds are fertilized with manure, and fish are fed with vegetation or invertebrates. It takes 20 years of training to become adept at all the techniques of disease control, fish feeding, manipulation of reproduction, and fertilization associated with these traditional forms of fish culture (Zweig, 1985).

SUMMARY

1. Biodiversity of fish is related to factors that operate at a variety of temporal and spatial scales. In general, fish communities are more diverse in tropical areas, in large drainage basins, in areas with high terrestrial productivity, where greater habitat diversity exists, and where more prey species exist.
2. Physiological ecology of fishes is described successfully by the energetic requirements of various processes, including osmoregulation, O_2 requirements, responses to temperature, food quality and quantity, and behavioral considerations.
3. Stock size, production, recruitment, and mortality are central aspects of fish populations that are used in their management.
4. A variety of methods are used to capture fishes depending on their size, the habitat being sampled, and reasons for obtaining data on the fishes.
5. Many indices are used by managers to assess fish populations and potential yields, including the relative stock density, which is the percentage of fish in a specified length or age class relative to the total number of fish estimated by population sampling.
6. A variety of regulations are used to control overexploitation of fish populations. These regulations require consideration of human dimensions but also generally involve a trade-off between high numbers of small fish and fewer large fish.

7. Aquaculture of fishes requires understanding of aquatic ecology of fishes. Problems commonly encountered with aquaculture include hypereutrophy of fish culture facilities and diseases.

QUESTIONS FOR THOUGHT

1. Do lakes and streams commonly contain unused niches that can be exploited by fisheries managers wishing to improve sport fisheries?
2. Why are unregulated fisheries overexploited?
3. How might fisheries management for sport fisheries lower diversity of native fish populations?
4. Do the indices used by fisheries managers, such as RSD, have any ecological relevance?
5. Why are herbivorous and detritivorous fishes more often used for food in Asia than in North America?
6. Why should genetic diversity of fish stocks be an important aspect of aquaculture?

Freshwater Ecosystems

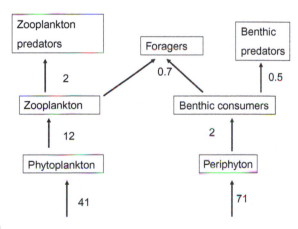

FIGURE 24.1

Graphic representation of one of the first accountings of energy flux through an ecosystem. Flux rates in g-cal cm^{-2} y^{-1}. The data are for Cedar Bog Lake. *(Lindeman, 1942)*.

635

The ecosystem viewpoint of ecology was initiated during the first half of the twentieth century. Raymond Lindeman (1942) described how fluxes of energy could be used to characterize ecosystems in an extremely influential paper published posthumously and summarized in Figure 24.1. Explicitly accounting for energy fluxes arose from the new ecological idea that an *ecosystem* is the sum of the biotic and abiotic parts of an environment. A more comprehensive, contemporary definition is provided by Covich (2001):

> Ecosystems are thermodynamically open, hierarchically organized communities of producers, consumers, and decomposers together with the abiotic factors that influence species growth, reproduction, and dispersal. These abiotic factors include the flow of energy and the circulation of materials together with the geological, hydrological, and atmospheric forces that influence habitat quality, species distributions, and species abundances. Energy flows through many species, and the way in which this flow affects the persistence of ecosystems is influenced by land-use changes, precipitation, soil erosion, and other physical constraints such as geomorphology.

This holistic view of ecology is powerful because it allows for analyses of entire systems rather than subsets of systems. However, ecosystems are so complex that they still must be reduced conceptually to units manageable for study. In this chapter, we discuss general methods of approaching ecosystems (including trophic energy transfers, nutrient budgets, and the link between biodiversity and ecosystem function) and then focus on ecosystem properties in groundwaters, rivers, streams, lakes, and wetlands.

GENERAL APPROACHES TO ECOSYSTEMS

Some of the earliest attempts to reduce ecosystems to manageable units revolved around assigning organisms to *trophic levels*. The levels are decomposers, primary producers, primary consumers (herbivores), and higher levels of consumers (secondary, tertiary, etc.). For example, in Figure 24.1, phytoplankton are primary producers, zooplankton are primary consumers, and zooplanktivores are secondary consumers. Movement of energy through these trophic levels has been a focal point of ecosystem research and can be used to illustrate basic ecosystem concepts.

We discuss ecosystem concepts of energy flow using data from an early ecosystem study on Silver Springs, Florida (Odum, 1957; Odum and Odum, 1959). Biomass of the various trophic levels is considered first (Fig. 24.2A). The *biomass* of the primary producers is 10 times greater than that of any other trophic level. This situation, where biomass of each tropic level is less than the level below it, forms what is traditionally called a *biomass pyramid* because the

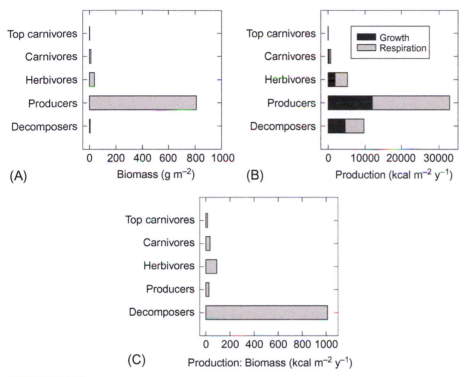

FIGURE 24.2
Biomass (A), production (B), and production per unit biomass (C) of Silver Springs, Florida. Note that production is divided into respiration and growth. *(Data from Odum and Odum, 1959).*

base of producers has much more biomass than the top-level carnivores. A biomass pyramid with a broad base does not always occur, because production rate per unit biomass also plays a role in ecosystem structure.

The second community characteristic is *production,* or flux of carbon or energy through an ecosystem compartment. It is essential to remember that production is not equivalent to biomass. Biomass is an amount, and production is a rate. A common way to contrast these two values is to use the ratio of production to biomass (P:B). The problem with assuming that high biomass is equivalent to high production is well illustrated by the decomposers in Figure 24.2. The biomass of the decomposers is relatively low, but their production is second only to that of the primary producers. This is because metabolic activity per unit biomass (P:B is a ratio that reflects this) is much greater for decomposers than for any of the other trophic levels for Silver Springs. Thus, P:B can be thought of as an index of relative efficiency.

Net ecosystem production is a gauge of how much carbon is fixed by all primary producers in an ecosystem. As discussed in Chapter 12, gross ecosystem production – ecosystem respiration = net ecosystem production. Ecosystems can have a positive or negative net production based upon the balance between external carbon imports (*allochthonous*) and primary production within the ecosystem (*autochthonous*), and the relationship between the two is often expressed as the ratio of production to respiration (P:R).

SECONDARY PRODUCTION

As carbon moves up the food web, production is defined somewhat differently. Secondary production is tissue elaboration by heterotrophs regardless of its fate (e.g., whether it is consumed by predators, dies and decomposes, or is molted off). Like primary production, it is a rate that is usually expressed as mass per unit area per unit time (e.g., grams $m^{-2} y^{-1}$), and it represents actual energy available to higher trophic levels.

Estimates of secondary production can also be used to link animal populations directly to other rates such as nutrient cycling. For example, if a given animal species' tissues are 11% nitrogen, and its production is $10 g\ m^{-2}y^{-1}$, we know that $1.1\ g\ Nm^{-2} y^{-1}$ is cycling through that species $(0.11*10 = 1.1)$. Production estimates are also relevant to the study of populations because they incorporate densities, biomass, individual growth rates, survivorship, and reproduction.

Secondary production is the end product of a series of ecological efficiencies. When an animal ingests food, a certain portion is assimilated and the rest is egested as feces. The difference between what is ingested and what is egested is *assimilation* (ingestion – egestion = assimilation). Assimilation is often expressed as *assimilation efficiency*, which is assimilation/ingestion. Of the portion assimilated, some is used for respiration and some is lost through excretion of metabolic wastes. That which is left represents production, which can be applied to growth, reproduction, or storage. The efficiency at which assimilated materials are converted to production is the *net production efficiency*; the overall efficiency at which ingested materials are converted to production (e.g., assimilation efficiency * net production efficiency) is *gross production efficiency.*

Secondary production estimates have been used to assess food availability for managed species such as game fish and waterfowl, examine ecosystem responses to natural and anthropogenic disturbances, and quantify roles of consumers in energy flow and nutrient cycling. Production estimates vary considerably as a function of biomass and growth rates. Estimates for single species of freshwater invertebrates range from $<1 mg\ m^{-2} y^{-1}$ up to extremes of $\sim3000\ g\ m^{-2} y^{-1}$ for species with very high biomass and rapid growth rates (Benke, 1993; Huryn and Wallace, 2000). Likewise, estimates for fishes and amphibians are highly variable depending on biomass and growth rates.

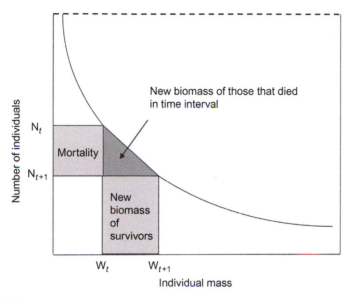

FIGURE 24.3

An Allen curve showing changes in density and individual weights of a cohort of animals over time. N is number and W is individual weight; subscript t and $t + 1$ indicate points in time. Over time, the number of individuals declines (moving down the y axis; difference between N_t and N_{t+1} = mortality) and individuals grow (moving left to right on the x axis). The sum of the new biomass created by survivors (represented by the light gray rectangle) and the biomass produced in the interval but lost to mortality (darker grey triangle) represent production for the time interval depicted. Total production for the cohort is the area under the curve; note that the area under the dashed line at the top would represent total production for the cohort if no mortality occurred.

Secondary production can be estimated by following changes in density and individual weights of a *cohort*, which is a group of individuals that start life at the same time. The relationship between density and individual weight over time can be plotted as an *Allen curve* (Allen, 1951; Benke and Huryn, 2006), which can be used to actually estimate production and is also useful for conceptualizing production (Fig. 24.3). Production for the cohort is the total area under the curve; there are numerous ways to estimate this, some of which are described as follows.

One of the simplest ways to estimate production is the *instantaneous growth method*. This method only requires estimates of average biomass and individual growth rates over a time interval, and thus can be used in situations where a clear cohort cannot be followed, or as a simplified approach when a cohort can be followed. Biomass estimates are generated from quantitative field sampling, but estimating growth rates can be more difficult, particularly in cases where rapid growth or overlapping cohorts occur. When growth rates cannot

■ Example 24.1

Calculating Secondary Production with the Instantaneous Growth Method

Using the data in Table 24.1, calculate the secondary production of the caddisfly population. Because a clear cohort structure is present, G can be calculated from the change in average individual weights between sampling intervals. For example, for the first interval of 18 May to 1 June:

$$G = \ln\left(\frac{0.057}{0.021}\right) = \ln(2.714) = 0.998 \text{ interval}^{-1}$$

Using biomass estimates for the same interval (5.94 mg on 18 May and 12.94 mg on 1 June), average biomass is 9.44 mg m^{-2} and thus production for the interval is:

$$P = 0.988 * 9.44 = 9.42 \text{ mg m}^{-2}$$

Although the instantaneous growth method requires less detailed information and is easier to calculate, production estimates using this approach are generally considered less precise than those obtained with cohort procedures such as the method described in Method 24.1. However, in some cases, the instantaneous growth method may be the best option because of unclear cohort structure. ■

be determined through field sampling, they may be derived through laboratory or field rearing studies, or mark-recapture procedures. Changes in individual weight over time are then used to estimate instantaneous growth as follows:

$$G = \ln\left(\frac{W_{t+1}}{W_t}\right)$$

where G = instantaneous growth rate, W_t = weight at the beginning of the interval, and W_{t+1} = weight at the end of the interval. The growth rate can be standardized to a daily rate by dividing by the number of days in the interval. Once G and average biomass for the interval are obtained, interval production is estimated as the product of the two. The instantaneous growth method can be used to estimate production using data on a population of stream-dwelling caddisflies (Example 24.1). The increment summation method can give better results when cohorts are clearly identifiable (Method 24.1). Also a third method, known as the size-frequency method, is now commonly used. This

METHOD 24.1

Advanced: Calculating Secondary Production Using the Increment Summation Method

Figure 24.3 shows how production of a cohort for a given time interval is a trapezoid underneath the Allen curve, which is composed of the biomass produced by those that survived and those that died during the time interval. The trapezoid for a given time interval can be calculated from densities and individual weights by multiplying the average number of individuals present during the time interval by the change in individual weight over the interval:

$$\frac{N_t + N_{t+1}}{2}(W_{t+1} - W_t)$$

where N_t is the number of individuals at the beginning of the time interval and N_{t+1} is the number at the end of the interval; W_t is average individual weight at the beginning of the interval and W_{t+1} is average individual weight at the end. Using data for the first interval in Table 24.1 yields:

$$\frac{283 + 227}{2}(0.057 - 0.021) = 255*0.036 = 9.18 \text{ mg m}^{-2} \text{ interval}^{-1}$$

This is production for a given time interval; total production is the sum of the interval values; in this case, 256 mg m^{-2} y^{-1}. This is the *increment summation method* (Waters, 1977; Benke and Huryn, 2006), which is considered fairly accurate, but can only be used when a cohort can be followed through time. In this case, the estimate from the instantaneous growth method (method 24.1, 9.42 mg m^{-2} interval^{-1}) and the increment summation method are similar; this is because a clear cohort structure is present, and thus accuracy of both methods is high. A quick test for method suitability is to plot an Allen curve (density vs. individual mass) with the data; if density does not decrease and individuals do not get larger through time, the increment summation method cannot be used.

method involves constructing an "average cohort," somewhat like a life table, and thus can be used in situations where a cohort can or cannot be followed. Details of this method are beyond the scope of this text, but are presented along with other secondary production methods by Benke and Huryn (2006).

Annual *production-to-biomass ratios* (P:B) are useful for assessing growth and turnover and explain why biomass is not always directly related to production (e.g., high biomass does not necessarily mean high production). For the example in Table 24.1, P:B is 256.1/41 = 6.2, which is fairly typical for an invertebrate with a one-year life cycle. Actual *turnover time* (actual biomass turnover rate or replacement rate) can be calculated by dividing the number of days in a year by the annual P:B; 365/6.2 = 59.0 days. Annual P:B is a function of growth; faster growing species have high P:B values and annual P:Bs can range from 1 to 2 for very slow growing taxa to over 100 for small, rapidly growing species such as some midges and small crustaceans (Benke, 1993). For obvious reasons, biomass can be a poor proxy for secondary production.

Table 24.1 Data and Production Calculations for *Brachycentrus Spinae*, a Stream-Dwelling Caddisfly with a One-Year Life Cycle

Date	Days in Interval	Density (no. m^{-2}) N	Avg. Individual Weight (mg) W	Biomass (mg m^{-2}) N*W	Weight Change ($W_{t+1} - W_t$) Δ Weight	Avg. N	Production (mg m^{-2}) Avg. N*Δ Weight
18 May	14	283	0.021	5.94	0.036	255.0	9.2
1 Jun	12	227	0.057	12.94	0.031	204.5	6.3
13 Jun	16	182	0.088	16.02	0.085	160.5	13.6
29 Jun	14	139	0.173	24.05	0.179	124.0	22.2
13 Jul	13	109	0.352	38.37	0.588	98.5	57.9
26 Jul	22	88	0.940	82.72	0.266	74.0	19.7
17 Aug	13	60	1.206	72.36	0.590	54.0	31.9
30 Aug	16	48	1.796	86.21	0.025	42.5	1.1
15 Sep	18	37	1.821	67.38	1.378	32.0	44.1
3 Oct	42	27	3.199	86.37	0.358	20.0	7.2
14 Nov	23	13	3.557	46.24	1.074	11.0	11.8
7 Dec	51	9	4.631	41.68	2.222	6.5	14.4
27 Jan	21	4	6.853	27.41	1.624	3.5	5.7
17 Feb	24	3	8.477	25.43	3.071	2.5	7.7
13 Mar	45	2	11.548	23.10	3.252	1.0	3.3
27 Apr		0	14.800	0.00			

Annual average biomass = 41.0   Total production = 256.1 mg m^{-2} y^{-1}

(Data are from Ross and Wallace, 1981)
Dates are Days When Samples Were Collected From the Stream; May is Used as a Start Date Because This is When Newly Hatched Larvae First Appear. Production is Calculated Using The Increment Summation Method

ENERGY FLUXES AND NUTRIENT CYCLING

A central concept of thermodynamics that can be related to efficiency is the Second Law, which roughly states that all processes must lose some energy. However, all processes could be 99.999% efficient, or any other value less than 100%, according to the Second Law. In nature ecological transformation is considerably less efficient than 100%; organisms may be able to turn as much as 75% of the mass of food consumed into biomass or less than 10%, with the remainder lost to respiration. Aquatic herbivorous insects assimilate approximately 30 to 60% of the materials they ingest, whereas aquatic predatory insects assimilate anywhere from 15 to 77% (Wiegert and Peterson, 1983).

For each individual organism, the amount of energy required to obtain, ingest, and assimilate food; the sum of efficiencies of all the metabolic pathways; and the energy required for reproduction and survival determine how efficient the organism is in obtaining and using energy. Assimilation efficiency is influenced by a variety of factors including food quality (e.g., C:N, % refractory materials such as lignins in plant materials), digestive system morphology and physiology, and temperature. In general, endothermic organisms are more efficient at assimilating materials than ectotherms, but less efficient at converting materials to production (e.g., production efficiency) because of higher metabolic demands. All of these efficiencies can also vary with abiotic factors over space and time, such as changes in temperature.

The data in Figure 24.2B from Silver Springs, Florida, suggest that about half of all the carbon taken up by producers is translated into biomass available for the next trophic level, 10% of the decomposer's production becomes biomass, and about 30% of the herbivore's biomass becomes available to predators (Odum and Odum, 1959). Interestingly, this early energetic study on Silver Springs continues to be relevant as it has been used to document a shift in the basic ecosystem energetics related to human influence in the area. The primary production has shifted from dominance by macrophytes to benthic algae as nitrate levels have doubled in the water (Quinlan *et al.*, 2008).

NUTRIENT BUDGETS

Although energy flux certainly makes sense as a primary unit when discussing production of food for secondary consumers and above, it may not be as important for primary producers, herbivores, and microbes, which can be limited by nutrients other than carbon. Stoichiometric analysis indicates that many primary food sources (detritus and primary producers) are so poor in nitrogen and phosphorus that primary consumers are not carbon limited (Sterner and Elser, 2002). For example, rates of fungal and bacterial activity and growth in streams can be limited by nitrogen or phosphorus (Suberkropp, 1995; Tank and Webster, 1998). Also, pelagic bacteria are important consumers of phosphorus and compete well for it with phytoplankton (Currie and Kalff, 1984). Thus, consideration of energy limitation alone does not always provide an accurate description of functional ecosystem relationships. Therefore, another branch of ecosystem science is concerned with *nutrient budgets*, which quantify fluxes in nutrient cycles.

We present two examples of nutrient budgets, a flux diagram and a table that tallies total inputs and outputs. A flux diagram allows representation of fluxes that occur within a system. Often, input and output budgets treat the ecosystem as a black box and account for only materials entering and leaving the system.

The nutrient flux diagram is exemplified by the description of a single nutrient budget that was undertaken to describe the importance of nitrogen fixation by a dominant cyanobacterium (*Nostoc*) in a cold water pond (Fig. 24.4). Nitrogen fixation was responsible for only about 5% of the nitrogen input to the pond. However, this nitrogen input was directly into biota in the pond,

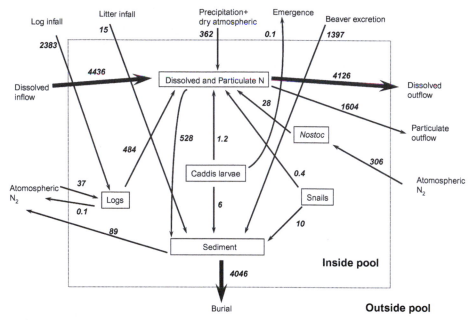

FIGURE 24.4

Diagram of nitrogen fluxes in a cold water spring (Mare's Egg Spring) dominated by *Nostoc*, a nitrogen-fixing cyanobacterium. Fluxes are given in moles N year^{-1}. Note: Fluxes are measured so where the budget does not balance there are measurement errors. *(Redrawn from Dodds and Castenholz, 1988).*

whereas much of the nitrogen input that entered in the spring inflow rapidly flowed out of the system. A similar approach was used to demonstrate that nitrogen limitation in wetlands can be relieved by activity of nitrogen-fixing organisms (Scott *et al.*, 2007).

Nutrient budgets and accounting for internal nutrient cycling fluxes can be accomplished using isotopic tracers. Hutchinson and Bowen (1947) pioneered this approach when they added radioactive phosphate to a lake and traced the fate of the materials. They found that the macrophytes in a pond were very good competitors for the added phosphate relative to the phytoplankton. More recently, a large number of whole-stream stable nitrogen isotope releases have been used to trace movement of nitrogen through streams and model budgets of streams (Peterson *et al.*, 2001; Mulholland *et al.*, 2008). These approaches are powerful because they allow for gross fluxes (internal dynamics) to be determined in addition to the net fluxes that can be determined by a budgeting approach.

The method of accounting for influx (inputs) and outflux (outputs) of nutrients is demonstrated with a general analysis of nutrient budgets to assess the importance of various sources and losses of nitrogen across ecosystem types (Table 24.2). Such budgets are central to understanding the effects of nitrogen

Table 24.2 Nitrogen Budgets of Some Aquatic Systems Showing the Importance of Nitrogen Fixation and Denitrification

Ecosystem	Inputs				Outputs			Reference
	Inflow	Litter Deposition	Atmospheric	N₂ Fixation	Outflow	Accumulation	Denitrification	
Sphagnum bog	0	0	1.4 (95)	0.05 (5)	0.2 (15)	0.9 (71)	0.2 (13)	Urban and Eisenreich, 1988
Riparian zone in agricultural area	2.9 (56)	?	1.2 (24)	1.1 (20)	1.3 (33)	5.2 (54)	3.2 (33)	Lowrance et al., 1984
Groundwater-fed fen in agricultural area	2.1 (28)	?	4.2 (56)	1.3 (17)	21 (24)	6.6 (75)	0.1 (1)	Koerselman et al., 1990
River-fed fen in agricultural area	0.7 (14)	?	4.4 (82)	0.2 (4)	1.0 (20)	3.8 (76)	0.1 (3)	Koerselman et al., 1990
Eutrophic lake (Clear Lake, California)	0.3 (45)	0	0.06 (9)	0.3 (45)	0.4 (55)	0.3 (40)	0.03 (5)	Horne and Goldman, 1972, 1994
Eutrophic lake (Lake Okeechobee, Florida)	3.2 (58)	0	1.7 (31)	0.6 (11)	1.6 (29)	3.0 (53)	1 (18)	Messer and Brezonik, 1983
Spring pool with dominant Nostoc	27 (50)	24 (42)	2 (4)	2 (4)	36 (59)	25 (41)	0.6 (1)	Dodds and Castenholz, 1988
Desert stream	0.47 (92)	?	?	0.04 (8)	0.49 (89)	0.05 (11)	?	Grimm and Petrone, 1997
Forest stream	11 (73)	3 (22)	?	1 (5)	11 (75)	4 (25)	?	Triska et al., 1984
Alpine streams/lakes	0.52 (57)	0	0.39 (43)	?	3 (99)	?	0.025 (1)	Baron and Campbell, 1997

Values in g N m⁻² y⁻¹. % of total input or output in parentheses. *Accumulation includes burial and storage in plant parts. When only accumulation is listed, the value includes burial + denitrification contamination and related ecosystem processes. For example, budgets are necessary to determine the ability to use wetlands to remove nutrient pollution from wastewaters. The data in Table 24.2 suggest that some wetland types are more retentive of nitrogen than others and that burial or assimilation in biomass is a significant component of many nitrogen budgets. Such approaches demonstrate that nitrogen fixation rarely accounts for more than 5% of nitrogen inputs in streams (Marcarelli *et al.*, 2008). Small habitats have substantial input of nitrogen from litter, but larger habitats are dominated by water inflows or dry and wet depositions. In some systems, nitrogen fixation (generally by cyanobacteria) can be a major nitrogen input, and denitrification can be a notable loss. A more in-depth analysis of such budgets for streams and lakes for a variety of elements is synopsized in Wetzel (2001).

BIODIVERSITY AND ECOSYSTEM FUNCTION

The biodiversity of ecosystems may be related to rates of ecosystem processes and biological structure of ecosystems in a variety of ways. For example, diverse assemblages of net-spinning caddisflies filter higher portions of suspended particles from the water column than monocultures because of facilitation among different species (Cardinale *et al.*, 2002). However, relationships between biodiversity and ecosystem function can be variable and complex. A study of functionally and competitively dominant leaf-shredding *Pycnopsyche* caddisflies found a negative relationship between species diversity and leaf litter decomposition because *Pycnopsyche* reduced the presence of other shredder species (strong competitor) and decomposed large amounts of litter and thus was functionally dominant (Creed *et al.*, 2009).

Given the current biodiversity crisis, particularly in freshwater habitats, efforts to establish relationships between *biodiversity and ecosystem function* have been a central issue in freshwater ecology (Schulze and Mooney, 1994). The first problem is how to define ecosystem function. Function can refer to rates of basic processes, such as photosynthesis, respiration, denitrification, or phosphorus retention. It can also refer to more specific things, such as production of plant biomass for herbivores. The crux of this issue is if species are *functionally redundant* with regard to their role in the ecosystem (i.e., can a species be removed from a community without a change in a specific ecosystem process?). Functional redundancy can vary with the ecosystem process that is being considered, the specific habitat, and the time. Many continental habitats have diverse assemblages with a considerable redundancy.

The links between diversity and ecosystem processes driving nutrient cycling are unclear because little is known about microbial diversity and the degree of redundancy of specific functional groups (Meyer, 1994). Although Hutchinson made early arguments about the functional redundancy of phytoplankton

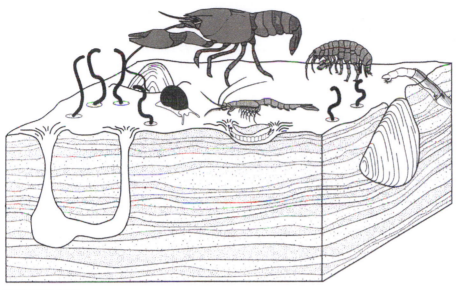

FIGURE 24.5

Benthic macroinvertebrates that burrow into layered sediments and accelerate nutrient cycling and movement of materials into the water column. Burrowing bivalve, crayfish, tubificid worms, and aquatic insect larvae mix O_2 into the sediments through their activities. Surface-dwelling invertebrates increase turnover of microbial communities and increase sediment suspension. *(Reproduced with permission from Covich et al., 1999).*

species with regard to the paradox of the plankton (see Chapter 17), the degree of functional separation of different phytoplankton species has not been established (Steinberg and Geller, 1994). Finlay *et al.* (1997) argue that microbial diversity is never so impoverished in natural communities that biogeochemical cycling is seriously altered.

Benthic animal diversity can have strong influences on processes of material exchange between the water column and the benthic zone (Covich *et al.*, 1999). Benthic animals can be important in energy flows and nutrient cycling (Fig. 24.5) and different species alter nutrient flux in different ways (e.g., burrowing, digging tunnels, stirring up sediments, actively pumping oxygenated water into the sediments, and processing different types of benthic materials). A study of a tropical river with a diverse fish assemblage discussed previously indicated that loss of just one fish species, a large migratory detritivore, significantly decreased downstream transport of carbon (Taylor *et al.*, 2006). This ingenious study used a fence along the middle of the stream and excluded or included the fish of interest. The data suggest that traits of a single species can alter ecosystem function, even in a diverse system where functional redundancy is expected to be high.

Some of the best data for functional separation of ecosystem processes related to biodiversity derive from studies of two shrimp species that break down leaf litter in Puerto Rican streams (Covich, 1997). The two species (*Xiphocaris elongata* and *Atya lanipes*) can both degrade leaf litter, but breakdown is significantly more efficient and the streams are more retentive of organic particles when both species are present. *Atya* does not break down intact leaves as rapidly as does *Xiphocaris*, but it scrapes microbes from the leaves and filters fine particles from the water column more efficiently. Particulate transport is highest in streams in which both species of shrimp are rare because predatory fishes are present (Pringle *et al.*, 1999). Such a relationship between biodiversity and ecosystem function may occur in mainland streams as well (Jonsson *et al.*, 2001). The Puerto Rican example is but one instance of how facilitation among diverse species can mediate species effects on ecosystem processes, a finding that has been replicated in marsh plants (Bertness and Hacker, 1994; Bertness and Leonard, 1997).

Variance in ecosystem structure or function can also be tied to diversity (France and Duffy, 2006). Experiments with multiple microbial communities assembled differently suggest that variation in respiration rate is tied to biodiversity. More diverse communities were more similar in respiration rates, but less diverse communities varied more widely (Morin and McGrady-Steed, 2004).

As numerous studies have been published, a more in-depth meta-analysis of the relationships between diversity and ecosystem function has become possible. A report commissioned by the Ecological Society of America (Hooper *et al.*, 2005) expressed certainty that (1) functional characteristics of species influence ecosystems, (2) human-caused species extinctions and introductions have influenced ecosystems in ways that are costly to humans, (3) effects of species loss or shifts in community structure on ecosystems are contingent upon local conditions, (4) some ecosystem properties are decoupled from diversity because of functional redundancy, and (5) as spatial and temporal variability increases, more species are required to generate a stable supply of ecosystem goods and services. Analysis of 446 terrestrial and aquatic studies found the strongest effects on primary producer and primary consumer abundances, as well as decomposer activity (Balvanera *et al.*, 2006). Analysis of 111 studies indicated that species identity was the single most important determinant of ecosystem effects related to species loss (Cardinale *et al.*, 2006).

GROUNDWATER ECOSYSTEMS

Groundwater ecosystems often rely on organic material derived from surface habitats (Gibert *et al.*, 1994). Alternatively, chemoautotrophic processes, such as sulfide oxidation (e.g., Sidebar 14.2), ammonium oxidation, or use of iron or manganese as electron donors, can form the basis of autotrophic

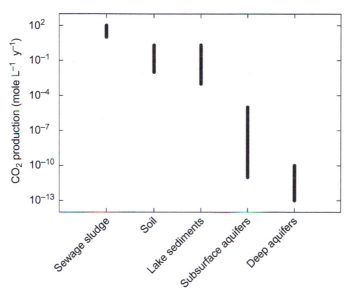

FIGURE 24.6

Ranges of documented respiration rates from various types of sediments. *(Data from Kieft and Phelps, 1997).*

production in some systems. Investigation of respiration rates of groundwater sediments suggests that respiration rates decrease with depth (Fig. 24.6). Deep groundwater sediments have the slowest rates of biological activity of any known habitats.

Groundwater ecosystems can be classified on a continuum of permeability and average interstitial space. The size and connectivity of the pores or channels through an aquifer can control the transfer of materials through the aquifer and limit the size of the organisms that inhabit the aquifer. Movement of water and organisms has direct effects on energy flux through ecosystems. In aquifers with very fine pore sizes, only dissolved materials, fine particles, bacteria, and very small protozoa can move through the sediments. Thus, only bacterial "producers" and primary consumers are present. Karst aquifers have the largest channels, but there are still few trophic levels relative to streams and lakes; however, there are more trophic levels in karst aquifers that those with fine pore sizes. Predatory fishes, amphibians, crayfish, or planarians can be the top carnivores; four trophic levels may be the highest number found in karst groundwaters (Culver, 1994), but usually there are less.

Other ways to classify groundwater ecosystems are by their depth or by their degree of connectivity with surface waters and terrestrial habitats. Connection of groundwaters and streams has been well described in some cases and is dependent on the type of substrata and spatial and temporal scales being considered (Harvey and Wagner, 2000). Over the short term (mainly less than a few months), in aquifers with low hydraulic conductivity, only shallow groundwater interacts with lakes, streams, and wetlands. Over longer time periods, deep groundwater can have considerable interaction with surface waters.

Groundwater can link lakes across the landscape over years. In northern Wisconsin, lakes occur in sandy glacial outwash, and groundwater links many of them (Kratz *et al.*, 1997). Drought leads to changes in groundwater flow and increases in ion concentrations in some lakes. It can take up to 5 years for drought effects on ion concentrations to move through lakes that are substantially affected by groundwater dynamics (Webster *et al.*, 1996, 2000). Groundwater effects in these lakes can sometimes override the effects of climate depending on lake position (Baines *et al.*, 2000).

Given the difficulty of sampling groundwater habitats, detailed trophic analyses and energetic or nutrient budgets are not as well documented as they are for surface waters. However, knowledge of groundwater habitats increased tremendously in the past two decades. What was a virtually undescribed group of biomes is beginning to be understood.

One of the most interesting developments in groundwater ecology is the documentation of very deep, very simple microbe-dominated habitats. Geological processes give rise to hydrogen gas and this can combine with carbon dioxide to generate methane and energy. These ecosystems were discussed in the chapter on extreme habitats (Chapter 15). These habitats are the only ones that do not depend upon photosynthesis or the redox gradient ultimately generated when the part of the biosphere that is oxidized by photosynthesis is in interface with the lower redox ions generated in Earth's crust.

STREAMS AND RIVERS

The major concepts associated with streams considered here are (1) the flood pulse concept, (2) autochthonous versus allochthonous production, (3) inverted biomass pyramids (and the related Allen paradox), (4) nutrient spiraling, and, (5) the river continuum concept. Although other important concepts have been explored for stream ecosystems, these exemplify some of the key features of stream ecosystem science.

The flood pulse concept is a change in paradigm from viewing floods as disturbances that alter an ecosystem that is otherwise at equilibrium to viewing flooding as a characteristic property of river and stream ecosystem function. A central idea of the flood pulse concept is that lateral habitats, particularly in large rivers, are connected in times of floods and disconnected during lower flow periods. The lateral habitats are much more strongly influenced by local than upstream processes and can strongly influence ecosystems and communities of the main river (Junk *et al.*, 1989). For example, the length of time of inundation substantially influences sediment respiration (Valett *et al.*, 2005) and can significantly influence biogeochemical cycles. Flood pulses can be important in smaller braided rivers as well. These rivers have multiple side channels that are variously connected depending upon discharge, and their degree of connection alters macroinvertebrate diversity in both backwater and main channel habitats (Whiles and Goldowitz, 2005).

It is sometimes difficult for people to view flooding as a natural process in rivers that does not need to be controlled. For example, debris flows in small streams can exceed $10,000\,m^3$ of mud, logs, rocks, and sediment that sweep through portions of steep watersheds in a matter of minutes. Such flows may appear disastrous, but recovery does occur (Lamberti *et al.*, 1991). Flood disturbances in small streams can control primary producers (Biggs, 1995, 2000) and, thus, ecosystem function. Flooding and flow are becoming important to ecosystem and river management, such as in efforts to protect the endangered whooping crane (Sidebar 24.1). Stream ecologists and resource managers must consider the ramifications of flooding in management plans.

The concept of allochthonous (organic material provided from outside the system) versus *autochthonous* (organic material from photosynthetic organisms within the system) production has been stressed in streams because of the potentially strong influence of terrestrially derived organic material and a substantial standing stock and production of periphyton (Minshall, 1978) and macrophytes (Hill and Webster, 1983) in some systems. The source of organic material is important because different invertebrates specialize in different types of carbon (Cummins, 1973), and varied sources of carbon can alter pathways of carbon transfer through the food web. For example, invertebrates that process leaf litter would be expected to provide important routes for energy flux into the food web in a small forested stream (allochthonous input). Exclusion of litter from forest streams has a profound effect on stream invertebrate communities (Wallace *et al.*, 1999; color plate Fig. 12). Wood inputs can be tremendous, but not all wood is available to consumers because few are equipped to digest it. In the Queets River, Washington, most large wood is less than 50 years old, but some in the channels is up to 1,400 years old (Hyatt and Naiman, 2001).

The relative contributions of primary production and external sources of organic carbon can be difficult to identify. In small streams that are heavily wooded, the input of leaves and wood is high, and shading limits primary production; some slow-growing mosses may be abundant. In larger streams, algal biomass is high when sufficient light can reach the bottom. However, dissolved and particulate organic carbon enters the stream from the surrounding terrestrial

---

SIDEBAR 24.1

Management of the Platte River for Water Quality, Water Quantity, and Species Preservation

The Platte River and its tributaries drain a large portion of Nebraska and about one-third of Wyoming and Colorado. The endangered pallid sturgeon (*Scaphirhynchus albus*) is found in the river. The Platte serves as a major stopover for migrant waterfowl and provides vital habitat for endangered whooping cranes (*Grus americana*), piping plovers (*Charadrius melodus*), and least terns (*Sterna antillarum*). About 80% of all sandhill cranes (*Grus canadensis*) stop there during their migrations. Approximately 70% of the Platte's discharge is diverted by consumptive uses in Colorado, Wyoming, and western Nebraska. There is concern about how this diversion will influence the Platte River ecosystem (US Fish and Wildlife Service, 1981).

A major result of flow reduction and control has been an alteration of channel morphology since the early 1900s. Historically, the river was sandy and braided with a wide, shallow channel. Flow modification has resulted in the invasion of woody species (*Populus* and *Salix*) and loss of much of the initial channel width (Johnson, 1994). Human modifications to the river and its floodplain have also facilitated invasions by saltcedar, purple loosestrife, and other exotic plants. Flow modification has also decreased water in the channel during winter, when seedlings are vulnerable to damage by ice. Lack of spring flooding has further encouraged establishment of vegetation on sandbars. Since the 1960s, the width of the channel has stabilized, vegetation has trapped sediments, and riparian forests have developed. This has necessitated active management efforts such as mechanical and chemical vegetation removal, controlled burns, and sandbar disking to remove vegetation.

Channel modification and flow alteration influence the extent and ecosystem characteristics of riparian wetlands. Cranes obtain much of their nutrition in wet riparian meadows, acquiring needed fat for continued migration. These wet meadows have a unique assemblage of organisms associated with them and they are rapidly disappearing (Whiles *et al.*, 1999; Whiles and Goldowitz, 2001). Areas with narrow channels have fewer associated wetlands and lower than historical usage rates by whooping cranes and other waterfowl (US Fish and Wildlife Service, 1981). Increasing recognition of the importance of the floodplain wetlands in the Platte valley has led to extensive restoration efforts, with some degree of success, but the restoration of natural hydrologic regimes is difficult (Meyer *et al.*, 2008; Meyer and Whiles, 2008).

The management of this ecosystem requires knowledge of how hydrology relates to habitat and organisms over long time scales. Vegetation removal to widen the channel may not be advisable because it causes sudden large sediment releases (Johnson, 1997). Mimicking the natural discharge regime to discourage establishment of riparian vegetation may be the preferred alternative. Given the tremendous demand for water from the Platte River basin, it may be difficult to obtain a discharge regime similar to that occurring historically to maintain the desirable biotic features of the Platte.

---

areas and from upstream. One way to establish the relative importance of internal versus external supplies of carbon is to compare the respiration and photosynthesis occurring in the stream (stream metabolism). These metabolic estimates can be used as indicators of a trophic state (Dodds, 2006).

Estimates of stream metabolism can be used to determine the ratio of photosynthesis to respiration (P:R), which serves as an index to the degree of autotrophy (relative autochthonous production) in the system. Two methods have been used to make such estimates based mainly on rates of O_2 production and consumption: isolation of shallow benthic substrata in sealed recirculating chambers and measurements of whole-stream diurnal O_2 flux (see Chapter 12). In general, chamber methods have indicated that primary production exceeds respiration in well-lighted streams (Minshall *et al.*, 1983; Naiman, 1983; Bott *et al.*, 1985). Whole-stream estimates suggest that production over a 24-hour period rarely exceeds respiration, and that P:R is usually less than 1, even in streams that receive substantial sunlight (Young and Huryn, 1999; Mulholland *et al.*, 2001). The discrepancy between these two methods occurs because the chambers include only the top layer of benthic organisms, whereas the whole-stream methods include significantly more subsurface respiration (the influence of the hyporheic component). Because the hyporheic metabolic activity is linked to instream O_2 dynamics, it makes sense to include it in estimates of whole-system metabolic activity.

The trophic dynamics of stream invertebrates and the relationship between standing stocks (biomass) and production have received attention with regard to stream ecosystems. Observations of invertebrate biomass have yielded examples of greater biomass of secondary and higher consumers than of primary consumers. This inverted biomass pyramid has been termed the *Allen Paradox* following the observation that fish in a stream required 100 times more benthic prey than was available at any one time (Allen, 1951; Hynes, 1970). The Allen Paradox is an example of problems that can arise when biomass is assumed to be directly proportional to production. The production of primary consumers per unit biomass can be very high because of high turnover (growth rates) and can support the secondary and tertiary consumer biomass, even when the biomass of primary consumers is relatively low (Allan, 1983; Benke, 1984). Such analyses have become very detailed, including description of carbon flux associated with each species in invertebrate communities. This detailed description involves determination of the *trophic basis of production*, or the relative contributions of individual food sources to the production of each species (Benke and Wallace, 1980).

Materials cycle as they move downstream. Cycling of materials in unidirectional flow environments is termed *nutrient spiraling* (Webster, 1975). Molecules are in the water column for an average amount of time while they move downstream. They are then taken up or adsorbed and moved downstream more slowly. Thus, the nutrient cycle that is typically conceptualized as a wheel in lakes becomes a spiral in streams. The spiral length (S) for a nutrient is the sum of the distance that an average molecule travels in the water column in the dissolved form (S_w) and how far it is transported in the

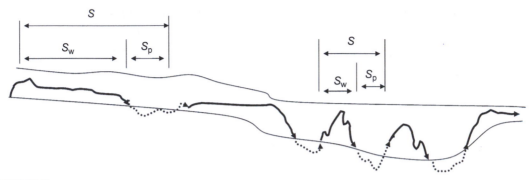

FIGURE 24.7

A diagram of nutrient spiraling in streams. S is the total spiral length, S_p is the time spent in particulate form in water column or the benthic zone and S_w is the average time spent in the water. Average velocity is greater in the riffle on the left, so spiral length is greater than in the pool at the right.

primary particulate compartments (algae, microbes, suspended particles, and animals; S_p):

$$S = S_w + S_p$$

The relationship can be described graphically (Fig. 24.7) or in more mathematical detail (Newbold *et al.*, 1981). In practice, modeling spiraling length can be difficult (Stream Solute Workshop, 1990) because understanding the movement of nutrients and environmental contaminants through streams requires detailed descriptions of the processes of dilution, uptake, and remineralization. Description of the influence of the hyporheic zone may be particularly important (Mulholland and DeAngelis, 2000).

Some generalizations are possible with regard to spiral length: Length should be greater with greater stream discharge, increased average water velocity, decreased uptake rates per unit area, increased disturbance of the benthic zone, and increased insect drift. Spiral lengths of inorganic nutrients (S_w) that are in high demand are generally short; uptake lengths of phosphate and ammonium are often less than 100 m (Mulholland *et al.*, 1990; Hart *et al.*, 1992; Butturini and Sabater, 1998). A variety of metrics is related to spiral lengths and can be used to parse out various factors influencing spiral length. These additional metrics include uptake rate per unit area of stream bottom and average movement of nutrients toward the benthic zone (Stream Solute Workshop, 1990). The concept of resource spiraling is powerful because it allows comparison of how nutrients are retained by a variety of streams (Elwood *et al.*, 1983), a process that is particularly important in small streams (Peterson *et al.*, 2001; Mulholland *et al.*, 2008).

Understanding how streams move nutrients downstream has become particularly important because of the global increase in episodes of near-shore hypoxia fueled by nutrient enrichment from streams (Diaz and Rosenberg, 2008). The nutrients stimulate phytoplankton blooms along the coast that sink and then decompose, causing low O_2 in the deeper marine waters. Nutrients that most commonly come from agriculture and river networks retain some of the nutrients that enter them (Alexander et al., 2008), albeit with less efficiency as loading increases (O'Brien et al., 2007; Mulholland et al., 2008), and eventually deposit them in coastal habitats. Spatial and temporal variability occurs in nutrient retention and there is some indication that instream biological processes are more important in nitrogen than phosphorus retention (Martí and Sabater, 1996).

The *river continuum concept* (Vannote et al., 1980) has been one of the most influential ideas in stream ecosystem theory. This concept views flowing waters as a connected continuum from small forested headwater streams to large rivers. It uses the associated gradient in abiotic and riparian characteristics to make specific predictions about the biological community. The concept posits that the dense canopy cover and low light found in small forested streams supply leaf material as the primary carbon source, and the invertebrate community is dominated by shredders. As the stream increases in width downstream, light increases and leaf input becomes less important. Benthic algal productivity and fine organic material washed from upstream contribute most heavily to production of available carbon, and grazing and collecting invertebrates dominate. In the largest rivers, benthic production is low, suspended particulate material is high, zooplankton and phytoplankton can become established in the water column, and collectors dominate the invertebrate community (Fig. 24.8).

This original model is a simple description of the possible ecosystem parameters that can vary from headwaters to large rivers (Sedell et al., 1989). An expanded view (Table 24.3) considers other abiotic (e.g., temperature and inorganic substrata) and biotic (e.g., woody debris) factors. This model is powerful in part because it encourages consideration of stream ecosystems across landscapes and as influenced by the watershed (including terrestrial and in-stream processes) above each point.

There are clear exceptions to the generalizations of the river continuum concept (as there are to any general ecological model). The original paper acknowledges the specificity of their model for forested streams. For example, streams that are frequently disturbed do not exhibit spatial trends in functional feeding groups of insects (Winterbourn et al., 1981). Large rivers may be as influenced by carbon input from their side channels as upstream (Sedell et al., 1989). Also, small streams can flow directly into oceans without ever

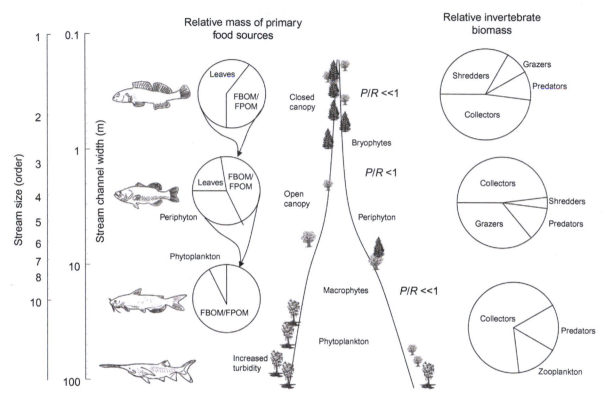

FIGURE 24.8
Diagram of the river continuum concept. See text and Table 24.3 for explanation. *(Modified from Vannote et al., 1980).*

moving into higher order rivers. Further, grassland and tundra streams and rivers should have limited leaf input in the smallest streams (Gurtz *et al.,* 1988). Finally, the idea of serial discontinuity (Ward and Stanford, 1983) suggests that dams disrupt the expected natural river continuum. Dams can cause settling of organic particles and boost populations of zooplankton and phytoplankton downstream. Exceptions not withstanding, the conceptual model provides one way for stream ecologists to think critically about streams in their ecosystem context (Cummins, 1977; Cummins *et al.,* 1984).

In contrast to the river continuum and flood pulse concepts, the *riverine productivity model* suggests that in-stream primary production (autochthonous energy) and direct riparian litter inputs (not linked to the flood pulse) can be the main source of energy in some large rivers (Thorp and Delong, 1994). This conceptual model is most appropriate for large rivers with constricted channels and limited floodplains, such as the upper Ohio River, where it was originally developed. The riverine productivity model also requires that stable substrata are present within the photic zone to support local primary production.

Table 24.3 Summary of an Expanded View of the River Continuum Concept

Feature	Headwaters	Middle Reaches	Large Rivers
Physical			
Stream order	1–3	4–7	>7
Discharge	Low	Medium	High
Flooding	Flashy, short, unpredictable	Medium	Regular, predictable
Gradient	High	Medium	Low
Temperature	Cool, constant when shaded	Moderate, variable	Warm, constant
Substrata	Rocky, large wood	Intermediate	Silt, sand
Riparian canopy	Dense, covering stream channel	Above stream channel open	Important only in flood zone
Turbidity	Low	Low	High
Light	Low	High	Low
Metabolic			
Photosynthesis (P)	Low	High	Moderate-low
Respiration (R)[*]	?	?	?
P/R	<<1	<1	<<1
Organic carbon	Coarse	Intermediate	Fine
CPOM/FPOM ratio[**]	>1	<1	<<<1
Woody debris	Large wood, debris dams	Along margins	Relatively rare, but an important substrate in sandy or silty rivers (Haden *et al.*, 1999)
Producers			
Periphyton	Moderate	High	Low
Phytoplankton	Low	Low	Relatively high
Macrophytes	Low, but mosses may predominate	Moderate	Low except in side pools
Consumer invertebrates			
Shredders	High	Moderate	Low
Filter feeders	Low	Moderate	High
Scrapers/ grazers	Moderate	High	Low
Collector gatherers	Moderate	Moderate	High
Predators	Moderate	Moderate	Moderate
Fishes			
Diversity	Low-cool water	Medium	High-warm water
Sight feeders	High	High	Low
Prey	Invertebrates	Invertebrates/fishes	Invertebrates/plankton/fishes

[*]*Relative patterns not established*
[**]*CPOM = Coarse particulate organic matter, FPOM = Fine particulate organic matter*

The river continuum concept, flood-pulse concept, and riverine productivity model all have obvious merits as conceptual foundations for studying rivers, and the appropriateness of the various components of each varies among and within different types of river systems. The recently developed *riverine ecosystem synthesis model* draws on elements of all of these and incorporates patterns in time and space in river networks at multiple scales. For example, due to changes in geomorphology along a river system, there may be reaches where the channel is constricted and there is limited floodplain in contrast to reaches with extensive floodplains and backwater habitats. These different reaches would obviously differ in a variety of physical and biological ways. These, and other sections of the network with different geomorphology and hydrology, can be viewed as *hydrogeomorphic patches*. In turn, these hydrogeomorphic patches give rise to different *functional process zones* because ecological communities, and thus ecosystem structure and function, differ among patches (Benda *et al.*, 2004; Thorp *et al.*, 2006). In some cases, patches and zones may be discrete, whereas in other cases they may be represented by subtle changes along the river network.

LAKES AND RESERVOIRS

One general way to classify lake ecosystems is based on lake autotrophic state (Table 24.4). Oligotrophic lakes are not very productive and are limited by nutrients, and oxic processes predominate. On the opposite side of the spectrum, eutrophic lakes are prone to cyanobacterial blooms, have anoxic hypolimnia, and have high rates of production in the water column, and production tends to be limited by nitrogen (because nitrogen is lost to denitrification) or light. Important exceptions to this classification scheme include dystrophic lakes (with high concentrations of humic compounds) that have low planktonic production but high macrophyte production, limitation by light for the phytoplankton, and heavily anoxic sediments with high rates of denitrification.

Table 24.4 Some Generalized Ecosystem Characteristics of Temperate Lakes of Different Trophic Levels

Type	Productivity ($mg\ C\ m^{-2}\ d^{-1}$)	Anoxic Hypolimnion (in Deep Enough Lakes)	Hypolimnetic O_2 Depletion Rate ($mg\ m^{-2}\ d^{-1}$)	Factors Limiting Production	Relative Rates of Denitrification and Nitrogen Fixation
Oligotrophic	<300	No	<250	N and P	Low
Mesotrophic	300–600	Maybe	250–400	N, P, and grazing	Medium
Eutrophic	>600	Yes	>400	N or light	High

The classical view is that a lake ecosystem has cleanly defined boundaries and river inflow and outflow. The view assumes that carbon dynamics are dominated by biomass produced by phytoplankton photosynthesis that is consumed by zooplankton, and zooplankton are consumed by fishes. This view is useful because it allows simple models of lake ecosystems to be constructed. A model of this type is presented in Fig. 24.9. The idea that all fluxes can be accounted for in the closed basin is of particular predictive value; models of lake eutrophication (see Chapter 18) can represent material balances and planktonic algal biomass of lakes reasonably well. Of course, as with all ecological constructs, there are exceptions to the simplification.

Many ecological studies have assumed that benthic primary production is not important. In most large, deep lakes, this approximation is probably reasonable. However, there are more small lakes than large lakes, and reservoirs tend to be shallow (see Chapter 7). Thus, benthic primary production may play a significant role in lake ecosystems (i.e., half or more of the production may be attributed to littoral algae or macrophytes in shallow lakes; Wetzel, 1983; Figs. 24.1 and 24.9). Now, models even suggest that benthic algal production is important in large lakes as well (Vadeboncoeur et al., 2008). Wetzel (2001)

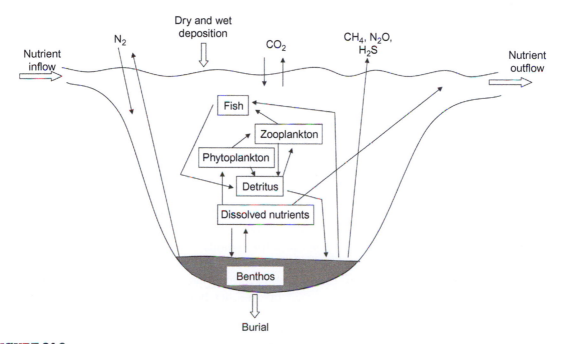

FIGURE 24.9

A simple diagram of nutrient flux through a lake ecosystem. The system is represented as a two-compartment bioreactor with a pelagic zone and the benthic zone. *(Modified from Covich et al., 1999).*

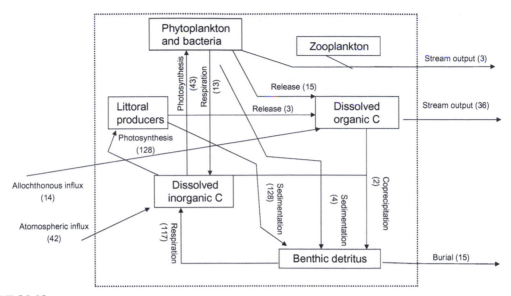

FIGURE 24.10

Diagram of carbon flux in Lawrence Lake, Michigan. *(Data from Wetzel, 1983).*

has long defended the importance of benthic processes in small lakes, and data continue to provide increased support for his views.

A carbon budget for Lawrence Lake, Michigan, illustrates some of the primary carbon flux pathways in lakes (Fig. 24.10). This lake has significantly greater rates of autochthonous production than allochthonous inputs. Thus, primary producers in the system dominate carbon cycling. Macrophytes and associated algae were responsible for about two-thirds of the primary production. A large portion of the macrophyte production ended up as benthic detrital carbon, whereas less than one-third of the phytoplankton production ended up in the sediments. Rates of carbon burial were about half of export via streams, and the lake was a net source of organic carbon to the watershed.

More recent analyses of whole-ecosystem metabolism indicates that many lakes may be net heterotrophic (Dodds and Cole, 2007), giving credence to the concept of separating autotrophic and heterotrophic states of lakes. Rates of heterotrophy exceed photosynthetic rates when a broad number of lakes are considered. The evidence for net heterotrophy is based on analysis of the degree of saturation of CO_2 in 4,665 lakes throughout the world (Cole *et al.*, 1994). The data showed that 87% of the lakes were supersaturated with CO_2, indicating that respiration rates exceed carbon sequestration and export. These results can be explained most easily if externally derived carbon (allochthonous sources) exceeds washout plus burial in the sediments.

Thus, the CO_2 data indicate that lakes are generally net heterotrophic. Of course, this generalization covers a range of lake types. High rates of respiration relative to photosynthesis may be common in more oligotrophic aquatic ecosystems (Duarte and Agustí, 1998). Trophic cascades may alter the relative importance of heterotrophy in lakes (Carpenter *et al.*, 2001).

In some large lakes, such as the Great Lakes of North America, photosynthesis is likely high relative to allochthonous organic carbon input. Other systems may be driven by external carbon inputs (in fine and dissolved organic material) that are consumed by bacteria, and bacteria are consumed by zooplankton. Different sources of terrestrial carbon influenced different parts of the food web (Cole *et al.*, 2006). In general, watersheds influence metabolic characteristics of lakes more than was previously thought (Cole and Caraco, 2001).

Whole ecosystem additions of stable isotope tracer ^{13}C indicated about half the energy for the food web was derived from allochthonous sources in three Wisconsin lakes, two oligotrophic and one dystrophic (Pace *et al.*, 2004; Carpenter *et al.*, 2005). Some more directed inputs, such as terrestrial invertebrates for surface feeding fishes, can have unexpectedly high inputs in lakes as well (Cole *et al.*, 2006). Lakes with high concentrations of nonliving suspended particles can support a productive fish community despite very low algal biomass and productivity, and small humic lakes may have high production of bacteria that consume humic substances (Münster *et al.*, 1999).

The idea that benthic primary production and allochthonous carbon provide considerable energy input into the food web complicates the view of energetics of lakes compared to that of a simple model considering phytoplankton–zooplankton–fish linkages. A model incorporating allochthonous inputs and the role of the microbial loop may more accurately characterize lake and reservoir ecosystems. The actual importance of each path of energy flux is context-dependent. If a lake is shallow and clear, macrophytes may dominate, whereas a large, deep, clear lake will be dominated by phytoplankton. A lake with high throughput and an extensive littoral zone may function more similarly to a stream and be dominated by allochthonous carbon sources.

In Chapter 20, we discussed food webs in lakes and the trophic cascade systems of interacting populations of organisms, but not from the perspective of ecosystem energy flux. An interesting aspect of ecosystem energy flux is related to the fact that primary producers are usually limited by nutrients, but consumers are limited by energy. Where the switch from nutrient to energy limitation occurs depends on the stoichiometry of the system. The stoichiometry of grazers can feed back and intensify or relieve nutrient limitation (Elser and Urabe, 1999). Thus, predicting ecosystem energy flux may require knowledge of community structure. For example, large *Daphnia* lower phytoplankton by grazing and intensify phosphorus limitation because of their high phosphorus demand

(Elser and Hassett, 1994). Changes in trophic structure that alter *Daphnia* populations can thus affect factors that limit primary production.

Viewing lakes from a regional or landscape perspective can yield important information (Magnuson and Kratz, 1999; Kratz and Frost, 2000). One of the major aspects of groups of lakes is the coherence of lake properties with time (Magnuson *et al.*, 1990). Documenting this coherence allows estimation of how well research results from one lake in an area can be extrapolated to another. For example, lakes tend to have more similar chemical and biological properties across a landscape when hydrological throughput is high (Soranno *et al.*, 1999). Lakes have also been classified by how well they are linked to other lakes by hydrology and by how far down in the drainage they are (similar to stream ordering). This classification correlates with patterns of species richness, chlorophyll concentrations, and major ion concentration (Riera *et al.*, 2000).

ADVANCED: RESERVOIRS AS UNIQUE ECOSYSTEMS

Throughout the chapters we have described a number of features that tend to distinguish reservoirs from other lentic habitats. In general, these are (1) dendritic shape, (2) river influence in upper areas with shallower habitats common in the upstream portions, (3) deepest portion of the water body right next to the dam, (4) possible water releases from deep in the pool, and (5) unnatural flow regimes, with potentially large swings in depth. The fact that reservoirs generally are operated for power generation, flood protection, and water storage means that the trends of water removal and depth are highly unnatural.

Wetzel (1990) provided a more comprehensive list of generalizations about reservoirs as compared to lakes including (1) more reservoirs are found outside of areas of glaciations; (2) greater relative number of reservoirs with drier climate; (3) reservoirs tend to have longer narrower watersheds; (4) reservoirs have greater watershed development indices and are shallower and thus less likely to stratify; (5) reservoirs tend to be dominated by larger inflows, irregular outflows, and higher flushing rates; (6) reservoirs experience high sediment loading and more sediment deposition in riverine zones; (7) reservoirs are warmer; (8) variable sediment loading and flushing in reservoirs lead to more variable light extinction; (9) reservoirs display low diversity of benthic fauna and are characterized by warm water fisheries. Of course these are generalizations, but they illustrate why reservoirs deserve special consideration.

Most reservoirs are relatively shallow, particularly nearer to the river or stream inputs. Shallower benthic habitat suggests that benthic primary production could be quite important in many reservoirs. The fluctuations in water level may make it difficult for macrophytes to establish; desiccation harms many macrophyte species. Pelagic production should be more important closer to the dam where water is deeper and the ecosystem is characterized by deep open water.

Nutrient limitation gradients can occur with more nutrient limitation closer to the dam, and storms that create pulses of nutrients influence productivity, particularly close to river inputs (Vanni *et al.*, 2006). However, nutrient pulses with riverine influence can be offset by turbidity gradients with more light limitation near river inputs. In this case there is a trade-off between light and nutrient limitation as water progresses from the river into the main reservoir and maximum productivity will occur somewhat downstream from the input.

Reservoirs have large watersheds and are substantially influenced by terrestrial processes. This terrestrial influence can have strong effects on production of food webs, particularly if omnivorous organisms that can eat detritus or algae are abundant (Vanni *et al.*, 2005). As with nutrients, the terrestrial materials entering reservoirs are expected to have their greatest influence near the river and stream inputs with attenuation of availability of terrestrial carbon with distance from inflows. Reservoirs are expected to be more likely net heterotrophic because they interact more with their watershed than do many lakes.

Reservoirs can be important zones of carbon and nitrogen retention in river networks. Reservoirs can act to "starve" rivers of organic carbon. By slowing movement of water, large organic particles will sink to the sediments where they are processed. The water leaving reservoirs can be impoverished in large particulate material, including organic carbon that would have subsidized downstream food webs. The larger area of fine anoxic sediments can also stimulate denitrification, leading to significant losses of total nitrogen as water passes through a reservoir.

WETLANDS

Wetland ecosystems can be classified along a hydrologic continuum from those that have very little hydrologic throughput to those that are closely coupled to rivers, estuaries, or lakes. In Chapter 5, we introduced the idea that some wetlands can have high hydrologic throughput (minerotrophic), whereas others are fed mainly by precipitation and have low hydrologic throughput (ombrotrophic). This variation in hydrology has implications for ecosystem function. Minerotrophic wetlands use nutrients from outside and nutrients can be washed from them easily. Ombrotrophic wetlands must rely more heavily on nutrient input from precipitation and internal nutrient cycling.

Autochthonous production in wetlands is usually extremely high; wetlands are characterized by some of the greatest rates of primary productivity of any

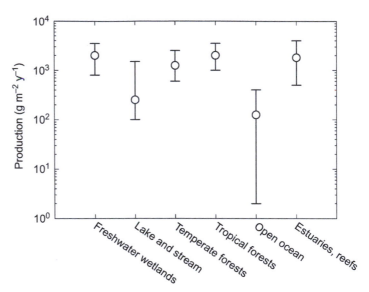

FIGURE 24.11
Ranges and means of production by various ecosystems. *(Data from Whittaker, 1975).*

habitats on Earth (Fig. 24.11). Production of even temporary wetlands may be important across dry landscapes because the majority of dry terrestrial habitats have very low production (Robarts *et al.*, 1995). Freshwater marshes have very high rates of production; peatlands and deepwater swamps have lower rates of production (Table 24.5). Production of methane appears to follow primary production trends, with more productive wetlands producing greater amounts of methane. Greater production presumably increases the extent of anoxia and leads to greater methane production (Table 24.5).

The majority of primary production by macrophytes is not grazed directly by herbivores in wetlands (Mitsch and Gosselink, 1993); rather, it is deposited as detritus, and much of this organic production can be stored in the sediment (Fig. 24.12) or consumed by invertebrates. Although less than 4% of Earth's surface is wetlands, wet soils contain about one-third of all organic matter stored in the world's soils. The vast deposits of coal are remnants of such organic storage from the swamps of the Carboniferous period (because the ability to degrade lignin was not as widespread). Given this potentially great production of carbon, variable rates of storage, respiration, and hydrological throughput, wetlands can serve as either sinks or sources of organic matter in the landscape. Natural wetlands with high hydrological throughput can be significant sources of organic C in the watershed (Mulholland and Kuenzler, 1979). However, artificial wetlands are used as sewage treatment systems; in this case, the wetland has a net consumption of organic carbon.

Table 24.5 Ecosystem Function in Some Wetland Types

Type	Distribution	Production (g C m^{-2} y^{-1})	Methane Production (mg C m^{-2} d^{-1})	Nutrient Retention
Freshwater marsh	Worldwide	1,000–6,000	45–285	Sometimes N and P sink
Tidal freshwater marsh	Mid- to high latitude, in regions with a broad coastal plain	1,000–3,000	440	N and P sink
Riparian wetland	Worldwide	600–1,300	?	Sometimes N and P sink
Northern wetland	Cold temperate climates of high humidity, generally in Northern Hemisphere	240–1,500	0.1–90	Usually N and P sink, may be an N source
Deepwater swamp	Southeast United States	200–1,700	1–15	

(After Mitsch and Gosselink, 1993) See Table 5.2 for Description of Wetland Types

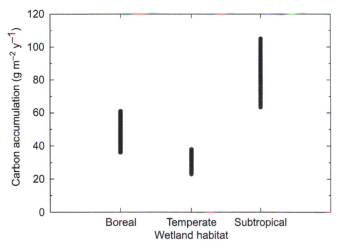

FIGURE 24.12

Sediment deposition rates for wetlands. *(Data from Schlesinger, 1997).*

Wetlands represent a hybrid between terrestrial systems and aquatic systems. The carbon flux diagram of Creeping Swamp, North Carolina, illustrates some unique features of wetland ecosystems (Fig. 24.13). Carbon production was dominated by trees, followed by algae and small plants. The production of carbon in coarse particulate organic material fuels the food web of the wetland. Most of the carbon in ecosystem compartments is in trees and

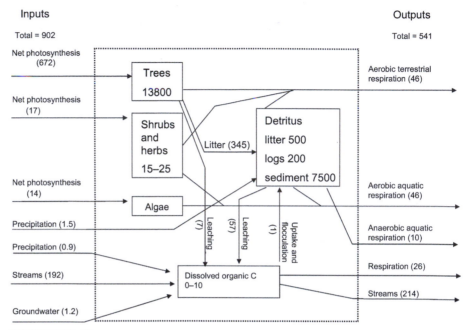

FIGURE 24.13

Carbon biomass and flux rates for the Creeping Swamp ecosystem. Biomass values in g cm^2 are listed in the compartment boxes, and flux rates in g C m^{-2} y^{-1} are in parentheses. *(Data from Mullholland, 1981).*

sediments. The majority of the carbon flux into the detrital pool is derived from trees in the wetland. The majority of the carbon fixed by photosynthesis is released by respiration and burial, but the swamp serves as a net source of organic carbon to the stream water.

Invertebrate communities in wetlands rely heavily on detritus from emergent plants and algae, with macrophyte tissue being less important (Batzer and Wissinger, 1996). In contrast to streams, shredders that are specifically adapted to directly process plant litter are rare, with some notable exceptions (e.g., Whiles *et al.*, 1999). Numerous invertebrates, particularly midge larvae, specialize on algal producers. Availability of algae can limit production of these invertebrates (Batzer and Wissinger, 1996). Insect herbivory on the areal portions of emergent plants can be important (Wissinger, 1999). Top predators in the water are generally insects, crayfish, small fishes, and amphibians (Porter *et al.*, 1999). Vertebrate predators, such as raccoon and birds, rely on the larger animals in the water.

Wetlands associated with rivers and streams are strongly affected by floods. The flood pulse concept provides a view of the river as a dynamic system that is connected with its flood plain, and it may provide valuable ways to describe the riparian ecosystem (Lewis *et al.*, 2000). In large rivers, this pulsing is particularly important because such systems are characterized by seasonal flooding and associated connections with riparian wetlands. As discussed previously, flooding can be important in the biology of large river fishes (see Sidebar 23.1). The flood–riparian connection provides riverine species with food resources and spawning habitat and riparian lake and wetland species with an avenue for dispersal. The connection is also important in material transport.

Floods inundate riparian areas, which slows water velocity and allows for settling of sediments. Dry, coarse organic debris initially floats and can be moved from the riparian zones into the large channels or moved within the wetland (Molles *et al.*, 1998). Likewise, nutrients move from the river into the side pools (Knowlton and Jones, 1997). This process can be important in tropical floodplain rivers, in which lakes and wetlands in the riparian zone become progressively less productive during the dry season as nutrients are taken up by organisms and are deposited into the sediments. Flooding then provides a new pulse of nutrients that boosts productivity (Hamilton and Lewis, 1990). Flooding in arid-zone rivers also provides nutrients to the riparian zone. This subsidy by flooding can occur even in the absence of overland flow because the flooding can pulse water and nutrients into the hyporheic zone (Martí *et al.*, 2000).

Flood pulsing is an essential characteristic of riparian wetland ecosystems. These wetlands are highly endangered by river channelization and modification. Addition of dams and regulation of extreme flow can severely alter their natural cycles. Restoration of riparian wetlands requires management of floods and connectivity to the main river channel (Middleton, 1999). The Pantanal is the largest wetland in the world and is threatened by changes in the hydrologic regime (Sidebar 24.2) as are many others globally.

SIDEBAR 24.2

The Pantanal, the World's Largest Wetland Ecosystem Complex

The Pantanal (color plate Fig. 13) is a vast complex of seasonally flooded wetlands, lakes, and streams along the Paraguay River in Brazil, Bolivia, and Paraguay. The total area of wetlands and savannah includes 140,000 km^2 in Brazil and 100,000 km^2 in Bolivia and Paraguay; the portion in Brazil is larger than the state of New York. During the wet season, about 80% of the area is flooded. During the dry season, the grasses emerge, and the area is used heavily for cattle grazing. This wetland is highly dependent upon a wet-season flood pulse and vulnerable to hydrologic alteration (Junk *et al.*, 2006).

The wetland undergoes a strong seasonal succession. During the wet season, several meters of water cover all but the highest tree islands. Massive growths of macrophytes occur and aquatic species disperse. During the dry season, the ponds become isolated from the main channel. Fishes are trapped in these ponds and waterbirds and caimans congregate to feed on the trapped fishes (color plate Fig. 13). The ponds become eutrophic from nutrients released from the fish carcasses and excretia of the predators (Heckman, 1994).

The Pantanal is a wetland of international importance for ecological conservation. It is habitat for the endangered spotted jaguar (*Panthera onca*), giant anteater (*Myrmecophaga tridactyla*), and giant river otters (*Pteronura brasiliensis*). Numerous waterbirds use the wetland, as does a diverse assemblage of parrots. The Pantanal also has more than 400 species of fish and attracts large numbers of sport fisherman to the mostly unregulated fishery.

The greatest threats from development are channelization and wetland modification that could lower water levels. A major project to increase the navigability of the Paraguay River is being considered. The dredging and channelization project will cost approximately $1 billion, and economic assessments have not carefully considered ecological impacts and the altered hydrology with associated flooding down river (Gottgens *et al.*, 1998). With only a 0.25 m decrease in water level, the inundated area of the wetland will be decreased by more than half in the upper regions of the wetland and 5,790 km^2 overall (Hamilton, 1999). The project will negatively impact the livelihoods of thousands of indigenous people that inhabit the region. Political pressure has caused a decrease in the scale of the plans, but a comprehensive development plan is lacking, and dredging and channelization plans continue. Expansion of agriculture, industry and, population all influence the wetland (de Olivera Neves, 2009). As of 2005, only moderate conservation progress had been made (Harris *et al.*, 2005). If past cases of exploitation of wetlands are any indication (e.g., the Everglades), the Pantanal ecosystem will suffer greatly as humans develop the area (Gottgens *et al.*, 1998).

WHOLE ECOSYSTEM EXPERIMENTS

Although sometimes plagued with statistical issues such as low replication, whole ecosystem studies have provided important insight into the structure and function, and relationships between the two, of freshwater systems. A classic series of studies in Wisconsin involving whole lake consumer and nutrient manipulations revealed trophic cascades that ultimately determined if lakes were net sources or sinks of CO_2 (Carpenter *et al.*, 2001). Similarly, a long-term phosphorus addition to an Alaskan tundra stream showed responses at multiple trophic levels, from periphyton to predatory fish (Peterson *et al.*, 1993) and the system continued to shift over more than a decade of phosphorus addition; after 8 years bryophytes replaced diatoms as the dominant primary producers, which in turn altered nutrient uptake rates, primary production, and invertebrate communities (Slavik *et al.*, 2004).

Development of new techniques, such as using stable isotopes to trace fluxes through ecosystems, has facilitated whole system studies. For example, ^{15}N tracers have been used to examine nutrient cycling at the reach scale in streams and has greatly improved our understanding of nutrient uptake, cycling, and export from a variety of stream types (Peterson *et al.*, 2001; Mulholland *et al.*, 2008). Use of a ^{13}C tracer revealed the importance of bacteria to higher trophic levels such as macroinvertebrates in headwater streams (Hall and Meyer, 1998).

Numerous whole system experimental manipulations have taken place at the Coweeta Hydrologic Laboratory in the Appalachian Mountains, and these studies have greatly enhanced our understanding of the ecology of detritus-based headwater streams. Experimental removal of insects from headwater streams at Coweeta resulted in significantly reduced leaf litter decomposition and organic seston export, underscoring the importance of invertebrates in detritus processing and energy flow (Cuffney et al., 1990; Wallace et al., 1991; Whiles et al., 1993). Exclusion of leaf litter inputs to Coweeta streams resulted in greatly reduced invertebrate abundance, biomass, and production, illustrating the tight linkage between headwater streams and riparian forests (Wallace et al., 1997). Multiple years of nitrogen additions to a headwater stream at Coweeta greatly enhanced invertebrate production, but severely reduced amounts of leaf litter in the stream channel, which may have negative long-term consequences for instream energy cycling and food web structure (Cross et al., 2006).

Whole ecosystem studies (*holistic approaches*) have the advantage of taking into account the entire intact system and thus results do not need to be extrapolated. However, difficulties with statistical replication at this scale and the complexity of natural systems sometimes necessitates a *reductionist approach*, whereby individual components or subsets of systems are examined. Reductionist approaches range in scale from microcosms in controlled environments, to mesocosms (e.g., replicated ponds, pools, or experimental streams), to field manipulations such as experimental exclusions. Reductionist approaches have the distinct advantage of being more statistically robust and they allow for isolation of specific variables, but results may not always reflect processes and patterns in whole systems. For example, recent analyses of biodiversity and ecosystem function studies indicate that small-scale manipulations likely underestimate the importance of biodiversity to ecosystem functioning (Duffy, 2009). Both holistic and reductionist approaches have their merits and have greatly contributed to our understanding of freshwater systems; combined approaches have proved particularly valuable.

COMPARISON OF FRESHWATER ECOSYSTEMS

All the generalizations presented in previous chapters can be used to classify ecosystems. The difficult issue is how to weight the importance of the various factors. Keep in mind that classification is mainly a tool for scientists to deal with a very complex world. Primary abiotic differences are associated with hydrologic throughput, the availability of light, the amount of allochthonous input, and the extent of benthic habitat (depth). If we view each of these

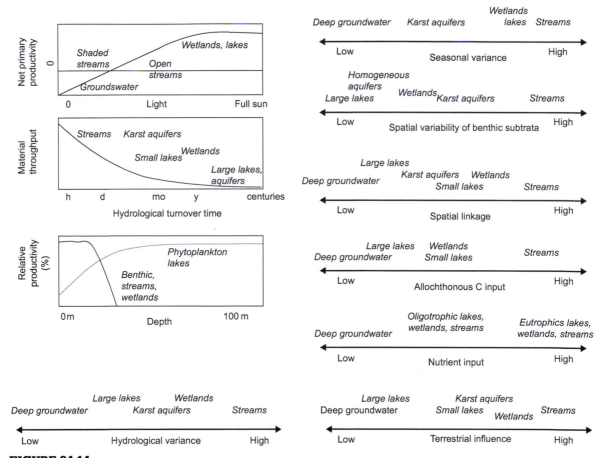

FIGURE 24.14

Freshwater habitats aligned across gradients that drive ecosystem properties.

factors as gradients, where most combinations are possible, in addition to variance associated with each of the factors over spatial and temporal scales, we can describe many essential characteristics of aquatic habitats (Fig. 24.14). Classification of abiotic parameters can provide information about constraints on evolutionary and physiological processes. For example, a forested temperate headwater stream can be characterized as being highly variable (hydrologically and with seasons), receiving low light, and having a high degree of terrestrial influence (including allochthonous carbon input). In contrast, deep groundwaters are highly stable, receive no light, receive low amounts of allochthonous input and even less autochthonous input, and have very low hydrologic throughput. The river continuum concept is an excellent example of a conceptual model that uses classifications of gradients and spatial linkages to make specific predictions about ecosystem function and community structure.

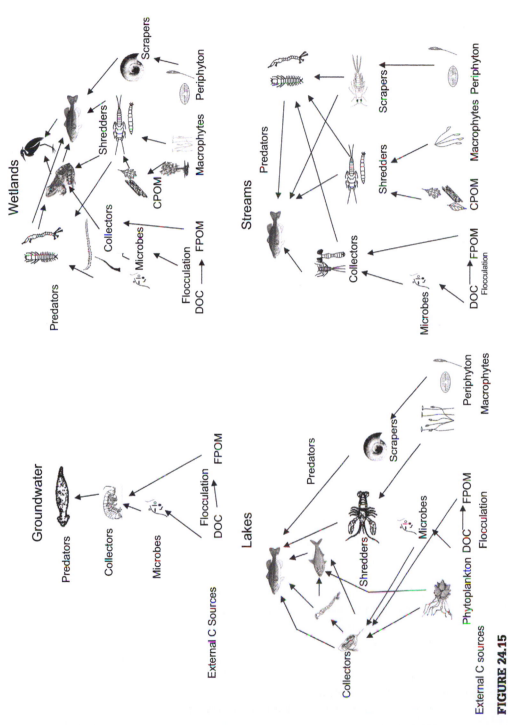

FIGURE 24.15

Comparison of carbon flow pathways and trophic complexity of groundwater, wetland, lake, and stream habitats.

Of course, problems arise when few abiotic gradients are used to classify ecosystems. Very shallow lakes may have very little emergent vegetation or macrophyte production but high cyanobacterial biomass. Wind, unstable benthic substrata, high grazing, and high nutrients all lead to these blooms. Once the bloom is established, it is difficult for the macrophyte assemblage to become reestablished. The complexity of natural systems precludes our ability to completely classify and make predictions about many of their processes (Dodds, 2009). Perhaps the most interesting ecosystems are those that are the most difficult to classify.

Abiotic gradients lead to biotic differences in the food webs of various freshwater habitats (Fig. 24.15). Simple planktonic food webs characterize the pelagic zone of lakes and simple food webs based on consumption of biofilms characterize groundwaters. Shallow benthic habitats with heterogeneous substrata (wetlands, streams, and lake littoral zones) can contain diverse assemblages of organisms that process large amounts of detritus in addition to grazers and predators. The vagaries of chance, dispersal, and time have acted in concert with these abiotic gradients to allow evolution and community assembly to produce the diversity of organisms associated with each of the habitats.

Many of the human-caused disturbances of freshwater habitats can be characterized by shifts of systems on these axes of classification, particularly those related to habitat modification. Description and understanding of how these complex ecosystems are constrained are essential to describing and mitigating the effects that people have on freshwater, our most valuable resource.

SUMMARY

1. Ecosystems can be represented by trophic levels and the fluxes of carbon, energy, or nutrients between the trophic levels.

2. Biomass is an amount; production is a rate. The two should not be confused. In a more general sense, standing stocks are not the same as fluxes.

3. Secondary production is tissue elaboration by heterotrophs, regardless of its fate. It reflects energy available to higher trophic levels and because it is a rate it can be used to quantitatively link consumers to processes such as nutrient cycling.

4. Biodiversity may have an influence on ecosystem function, particularly in low-diversity systems.

5. Groundwater ecosystems are driven mostly by external inputs of organic carbon. Little is known about energetic fluxes in most groundwater ecosystems and nutrient budgets are not well characterized for groundwaters.

6. Streams are essentially nonequilibrium ecosystems in which flooding and unidirectional downstream flow provide strong abiotic influences on the ecosystem.

7. Nutrients spiral in streams as they cycle and are moved downstream.

8. The river continuum concept describes a series of physical and biological changes that are expected when moving from small headwater streams to large lowland streams. Alternative models of stream ecosystem structure and function have been developed, mostly focused on the complexity of large rivers.

9. Planktonic production and consumption can be very important in deep lakes. Allochthonous inputs are minimal, but benthic production can dominate in shallow lakes and reservoirs.

10. Reservoirs have unique ecosystem properties compared to lakes such as being shallower, more terrestrially dominated, and more influenced by riverine inputs.

11. Wetland ecosystems are highly dependent on autochthonous production. This production is deposited into wetlands as detritus from emergent plants or produced by algae.

12. Aquatic habitats can be aligned on a continuum of abiotic axes that describe many of the essential parameters that constrain ecosystem function and biotic capacities.

13. Whole system studies have greatly enhanced our understanding of freshwater ecosystems, but statistical limitations and the complexity of systems sometimes dictate a reductionist approach; both are important in freshwater ecology.

QUESTIONS FOR THOUGHT

1. At what point on an Allen curve would you expect secondary production to be highest?

2. Why are wetland ecosystems generally more productive than streams?

3. Why do wetland ecosystems store a greater amount of carbon than lakes or streams?

4. Can a lake ecosystem be described adequately with a two-compartment model, one for the pelagic zone and another for the limnetic zone?

5. Do whole communities evolve over time to optimally exploit ecosystems?

6. If global warming increased the number of freshwater marshes by converting northern peatlands into marshes, what would happen to methane production?

7. How well does an equilibrium model represent stream ecosystems relative to lake and groundwater ecosystems?

8. Should ecosystems be preserved in addition to endangered species?

9. Some people refer to "biotic integrity" and "ecosystem health" in the context of conservation of the environment. What do you think these terms mean and how should they be defined?

10. How can some ecosystems have a higher biomass of predators than primary consumers?

11. How would you design a study to look at predator–prey interactions in a pond using a reductionist approach and holistic approach; what about a combined approach?

CHAPTER 25

Conclusions

We have passed the point at which we can continue to take unabatedly without casting back some comprehensive understanding and wise use in return. Nature is remarkably resilient to human insults. Yet, humans must learn what are nature's dynamic capacities because excessive violation without harmony will only unleash her intolerable vengeance. The very survival of humankind depends on our understanding of our finite freshwater resources.

—Robert Wetzel (2001)

FIGURE 25.1

A stream on Konza Prairie in northeastern Kansas (top), a nearby suburban stream (middle), and an agricultural stream in Indiana (bottom). *(Agricultural stream photo by N. A. Griffiths).*

675

Doi: 10.1016/B978-0-12-374724-2.00025-8

Aquatic ecology is a tremendously rich and detailed field of study. We have covered only the most basic aspects of it in this text. Our hope is that students will have gained some useful insight that can be applied in the making of informed decisions regarding how our aquatic resources are used or studied. Hopefully, some of these students will be inspired to further academic study and careers in aquatic ecology. Regardless, we have tried to make the concepts and applications understandable and interesting to a wide variety of students. Details were presented in some cases to illustrate complexity of real systems. Comments from students and instructors that would improve the text are welcome. Those of us in developed countries who are able to read (and write) this book live in a golden age. It is an incredible luxury and privilege to pursue academic interests. As the world's population increases and our appetite for resources grows, the existence of unspoiled aquatic habitats will become increasingly rare. An ever-increasing number and proportion of the world's human population is malnourished and impoverished. The minority of the people who hold the majority of the economic and political power do so, in part, at the expense of the environment. Neither the rich nor the poor have shown much inclination to conserve our aquatic resources in the past, and it is unlikely that they will do so in the future. The depressing consequence across Earth is that biodiversity in, and quality and quantity of, freshwaters will decrease drastically in our lifetimes; these realities are already well underway. As competition for limited aquatic resources increases, aquatic ecologists will have fewer "natural" habitats to study and to use as baselines against which to compare impacted ecosystems. Thus, we need to study and appreciate what we have now because there may not be an opportunity to do so later.

There is reason for hope. Many rivers and streams in developed areas are cleaner than they were in the middle of the twentieth century. Knowledge of methods for lowering eutrophication, ways to avoid contamination by sewage, techniques for mitigation of sources of acid precipitation, and other technological advances have led to some improvements. These improvements are also being adopted in developing countries, and the importance of aquatic resources is being increasingly recognized in the scientific literature and by the public.

Unfortunately, some of the problems associated with human activities are not reversible over a human life span. Extinction is clearly not reversible. Groundwater pollution can be mitigated, but there is no known case of complete removal of a pollutant from groundwater. Unwanted species introductions are the same; once they are established, eradication is generally impossible. The effects of pollution are now global. There are no completely pristine habitats left.

Much of the future job of aquatic ecologists may be in "damage control" or restoration. We will be asked what is necessary to maintain ecosystem function and preserve desirable species. A more detailed knowledge of aquatic ecology than we currently possess is necessary to provide this information. For example, the link between diversity and ecosystem function is not well understood. The redundancy of ecosystem services by species (i.e., what is the minimum assemblage of species necessary to maintain productivity and the ability to neutralize pollutants in aquatic ecosystems) is not well documented. We simply cannot predict any but the most extreme effects of our impacts on aquatic habitats. Likewise, detailed knowledge of the biology of species is often required before they can be preserved. Such knowledge is sorely lacking for all but the most popular game fishes. Natural history is not currently fashionable, but specific details of life history are needed for most species.

Many aquatic ecologists enter the field because of a love of water and the organisms living in it. This leads to high levels of satisfaction derived from studying the aquatic environment. Many scientists with such motivation never directly study environmental problems. However, their contributions may increase our ability to understand aquatic habitats. The graduate study of the first author centered on a spring-fed pool in which the cyanobacterium *Nostoc* grows unusually large. This study was motivated by nothing beyond scientific curiosity. However, the research was eventually used to develop a management plan to protect the *Nostoc* in this pool. It can be difficult to predict what scientific information will ultimately be useful.

Some information presented in this book may seem marginally useful to students taking a class as a requirement for a program specializing in other aspects of aquatic sciences. For example, a fisheries student may have difficulty motivating interest for nutrient cycling. However, experience has shown that managers who do not take a holistic approach are doomed to make mistakes. Such mistakes may lead to financial loss and permanent damage to an ecosystem. Many aquatic species have been intentionally introduced to provide benefits, only to later cause unintended problems that were not anticipated. More knowledge of how aquatic ecosystems work may decrease the chances of making such mistakes.

If you are reading this, you are a persistent reader. Why read a part of the text that is unlikely to be on a test? This conclusion is for you. Please take time to reflect on what you have learned from this text, and take with you the valuable parts. Get your feet wet, enjoy the water.

Experimental Design in Aquatic Ecology

FIGURE A.1

A replicated field experiment using flow-through enclosures to estimate *in situ* growth rates of different species of tadpoles in a Panamanian stream. *(Photograph courtesy of Roberto Brenes).*

679

Experimental results supplied in a text or course setting are often presented as facts with little information on the scientific process that was used to decide if an effect was "real" or not. Obtaining much of the following information from a statistics course or text would be preferable (e.g., Sokal and Rohlf, 1981); at a minimum, some background in statistical methods is recommended. Because much ecological knowledge is gained from formal statistical treatment, the following brief introduction to experimental design is provided for those having little or no statistical experience. In our years of teaching experience, we have found that it is common for students to reach the advanced undergraduate level without much understanding about how science is really done with respect to design and analysis of experiments. Although this treatment is not specific to freshwater ecology, we hope this will convince even the math-phobic students of the necessity of statistical courses for the study of aquatic ecology.

Generally, when a scientist approaches a problem, she or he has an idea of what causes a particular phenomenon to occur. This initial expectation can be stated formally as a hypothesis (though such formal statements may not always occur, and initial hypotheses are often discarded as naive). The hypothesis states that some factor (or factors) has an effect. The hypothesis can also be stated in the form of a question. The job of the scientist then is to test the hypothesis. The construction of hypotheses and their testing and usefulness to ecologists have been discussed in detail by numerous authors (Pickett *et al.*, 1994; Quinn and Dunham, 1981). It will be helpful to know some basic statistical terms before discussion on the methods (Table A.1).

Table A.1 Some Common Statistical Terms

Term	Description
Mean	Average
Variance	The amount of variation about the mean, often denoted by standard error or standard deviation
N	The size of sample (total replicates)
Normal distribution	The expected distribution when a factor is sampled randomly; most statistical tests rely upon the assumption that the data are sampled from a normal distribution
Nonparametric test	Statistical tests that can be used when data distribution is not normal
p value	The probability that a statistical test shows that the hypothesis is correct
Replication	More than one sample from an independent treatment
Independent variable	A variable that influences the dependent variable
Dependent variable	A variable that changes as a function of the independent variable or variables

Several methods may be used to test a hypothesis; some rely on formal experimental methods, and others are based on observation of natural history or simulation. Although formal experimental methods are thought to provide unequivocal results, they are not always possible and can be limited in scope. All these methods have a place in the aquatic ecologist's toolbox.

NATURAL OBSERVATIONS AND EXPERIMENTS

Historically, much of biology has been based on observation of patterns called natural history and the use of these patterns to support explanations of the way the natural world works. Many of these approaches can be called natural experiments. An excellent example is the theory of evolution. Prior to advances in genetics and ultimately molecular biology, the precise mechanisms by which evolution could occur were not understood. This notwithstanding, Darwin set forth an impressive array of observations that strongly supported his theory. Biologists continued to amass support for evolution over the years, and today it is the central pillar of biological science. The original theory was proposed and accepted because of the vast weight of natural history observations that supported it. Only relatively recently have experiments been devised to directly test predictions related to evolutionary processes. Currently, much of molecular biology takes a similar approach. Molecular biologists describe genes and molecular interactions (pathways) in much the same way that natural historians describe communities. Other fields, such as astronomy, rely heavily upon observation of pattern (Dodds, 2009).

Likewise, certain facets of ecology and environmental sciences rest mainly on correlation. Correlation is a way to measure the cooccurrence of two variables, and it may provide the only available method to study some systems. For example, hypotheses regarding the causes and effects of global change cannot be tested experimentally because there is only one Earth. If a polluter contaminates a stream with an unusual organic compound of unknown toxicity and all the fish and aquatic invertebrates die, the most prudent option is to assume that the chemical caused the deaths. A similar release is not likely to be done on a nearby stream to replicate the event. Likewise, when large-scale changes are observed in large lakes, river systems, or aquifers there is no feasible way to perform replicated experiments on these systems, so other techniques are required. Correlation is one of these techniques.

The primary problem with correlative approaches lies in the inability to separate cause from effect. A tongue-in-cheek example of this is a plot of the number of churches per town against the number of saloons per town for a variety of towns with different populations. The plot will show a positive correlation. The correlation occurs not because saloons cause more churches to be built but because larger towns have more churches and more saloons.

Correlation can support causation but does not unequivocally prove ecological hypotheses.

A classic example of the power of correlation from limnology is the controversy regarding the causes of eutrophication. More phosphorus certainly was correlated with greater algal biomass in lakes. However, not until whole lake phosphorus addition experiments were conducted (Fig. 18.1) did polluters' denial that increased phosphorus led to eutrophication become very ineffective. After the causation was determined, the correlation was more likely to be used to predict responses of lakes to alterations in phosphorus supply.

Natural experiments can be strengthened in several ways (Carpenter, 1989). Time series can be used to provide replication, as described later for whole-system studies. For example, if a lake exhibits one level of fish production for several years, then nitrogen is added and fish production increases, we can be fairly certain that the nitrogen addition caused the increase in production. The certainty is greater if the nitrogen addition is discontinued and the fish production decreases to its original level. Another way to strengthen such observations is to compare two similar, if not identical, ecosystems.

Even given the potential problems, natural experiments have undeniable benefits. Perhaps the strongest of these benefits is that the observations and correlations occur in entire systems, not in a beaker or a bottle. Natural experiments may be most likely to have relevance to community- or ecosystem-level processes that operate in the real world (Carpenter *et al.*, 1995), and the results may be contrary to those extrapolated from small-scale replicated experiments (Schindler, 1998).

MULTIVARIATE METHODS

Examination of complex natural observational data sets is now assisted by multivariate methods. Many of these are becoming more common in the ecological literature, so at least a passing familiarity with the methods is desirable. Multivariate techniques are commonly used for community or assemblage level analyses and can be powerful tools for finding patterns in large data sets. An overview of some common multivariate methods is presented in Table A.2.

Multiple regression is one of the more commonly used multivariate techniques. Whereas regression examines the relationship between one dependent variable and one independent variable, multiple regression allows for examination of multiple independent variables and a dependent variable. For example, an ecologist may be interested in how nutrient concentrations, turbidity, and substrata composition (independent variables) influence the biotic integrity scores of streams (dependent variable). Multiple regression can be used to assess the relative influence of each independent variable, and

Table A.2 Description of Some Common Multivariate Statistical Methods

Method	Description	Comments
Multiple regression	Examines influence of multiple independent variables on a dependent variable	Sensitive to correlation among independent variables
Multiple analysis of variance	Examines influence of independent variables on multiple dependent variables	Accounts for correlation among dependent variables
Ordination	Finds patterns in complex data sets	Often used for data exploration and reduction of variables
Cluster analysis	Groups samples based on similarity	Creates groups based on distance measures
Discriminant analysis and classification	Identifies variables that define predesignated groups	Classification used to assign samples or sites to groups

to predict biotic integrity scores based on measured independent variables. An important limitation of multiple regression, which is often overlooked, is that correlation among the independent variables is problematic and limits the reliability of the results. As such, careful consideration of which independent variables to include in the analysis is important. In cases where there are numerous potentially correlated variables, ordination techniques described next can be used to reduce the number of variables used.

Multivariate analysis of variance (MANOVA) is used to examine responses of multiple dependent variables to independent variables. In analysis of variance (ANOVA), multiple independent variables may be included, but only one dependent variable is considered. For example, an ecologist may want to examine how stream temperature, conductivity, and pH differ among defined geographic regions, a question best addressed using MANOVA because three dependent variables are being examined. An alternative approach would be to run separate ANOVAs for each dependent variable, but this increases the probability of committing a type I error, whereby the chance of finding a significant difference increases with the number of tests that are run, particularly if any of the variables are correlated. MANOVA accounts for potential correlation among the multiple dependent variables.

Ordination and cluster analysis techniques often are used to examine patterns, including similarities and differences, among communities and assemblages and are thus a powerful tool of community ecologists. For example, Figure A.2, which was generated using nonmetric multidimensional scaling analysis, a common ordination technique, depicts clear differences in macroinvertebrate assemblages associated with different substrata in a river. Patterns

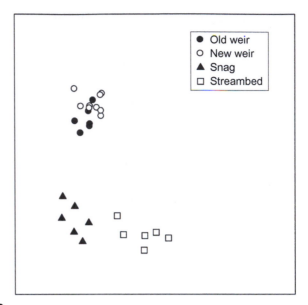

FIGURE A.2

An ordination plot (generated using nonmetric multidimensional scaling) showing differences in macroinvertebrates associated with all dominant substrata types in the Cache River, Illinois, except for those on old and new rock weirs. The other substrata are submerged wood (snags) and the ambient scoured clay streambed. Analysis of similarity indicated that all assemblages except for those on old and new weirs were significantly different. *(Data from Walther and Whiles, 2008).*

like these can be difficult to identify by just looking at data or using standard univariate techniques. Various statistical procedures such as analysis of similarity (ANOSIM) can be used to test for statistical significance of patterns (Clarke, 1993). There are also statistical methods for identifying taxa or other variables that drive observed patterns.

There are numerous different ordination techniques, including principal components analysis and canonical correspondence analysis, each with its own set of assumptions and advantages and limitations; use of a particular method depends on the specifics of the data set and the research questions being asked (Ludwig and Reynolds, 1988). Cluster analysis is similar in that it is often used to group samples and is based on similarities. Cluster analysis uses measures of distance among samples to construct groups and does not rely on the multivariate normal distribution.

Discriminant analysis and classification techniques are used to separate samples or sites into groups or allocate samples or sites to groups based on multivariate data. Discriminant function analysis often is used in an exploratory manner to identify variables that define predesignated groups. For example, an ecologist might designate ponds in a region as "impaired" or "unimpaired" based on field observations. Discriminant function analysis could then be performed on water quality and biological data collected from the ponds to identify the variables that separate impaired and unimpaired ponds. Similarly, classification techniques are used to assign a new pond to a category based on measured variables.

SIMULATION MODELING

An additional method that can be used to explore possible hypotheses is simulation modeling. Perhaps these models could be called virtual experiments. In this case, computer models of a system with the desired level of detail and representation of processes are built, and the system can be perturbed as desired. This is the main approach used to investigate global climate change, atmospheric dynamics, and large-scale physical oceanography. Benefits to modeling are that it is cheap, easily replicated, and may indicate critical factors in complex systems that control the observed behavior. Once the critical factors are determined by modeling, more detailed studies targeting these critical factors can occur. Thus, modeling can improve efficiency of environmental research.

The results of such models can be strengthened in several ways. *Sensitivity analysis* is used to systematically test the sensitivity of the model to the assumptions used to construct it. For this type of analysis, input parameters are varied and outputs are compared under different model scenarios. If much uncertainty exists over the value of a parameter that has a great influence over the behavior of a model, the predictions of the model are suspect. The scientist may then design the research program to more thoroughly characterize the uncertain parameter. Another method for validating a model is called *hind-casting*. In this method, the model is used to predict a system response that is already known. This is possible in systems in which sufficient data are available from the past but may be difficult where conditions are changing over long periods of time (e.g., under global change).

MANIPULATIVE EXPERIMENTS

To provide more formal proof of a hypothesis, manipulative experiments are often necessary. To test a hypothesis experimentally, all other factors must be held constant, the factor of interest must be varied, and the effect must be noted. If a change is related to variation in several factors, then the hypothesis is not proved because the change may have occurred without changing the factor of interest (correlation, not causation). Thus, a control, in which no factors are varied, is necessary for comparison within the treatment. In the example given in the previous section, the polluter that released the organic chemical into the stream could argue that the chemical was not necessarily responsible for the fish and invertebrate deaths and that they would have died anyway. If the fish and invertebrates in nearby, similar streams did not die, such an argument would hold less water.

Deciding which is the control and which is the experimental treatment can be difficult because this designation depends on how the hypothesis is stated. For example, if we form the hypothesis that ultraviolet (UV) light lowers fertility of frog eggs, then no ultraviolet light is the control, and ultraviolet light exposure is the experimental treatment. Conversely, if the hypothesis is that lack of UV increases fertility of frog eggs, then UV light can be the control and removal of UV the experimental treatment. Neither way is right or wrong, but often it makes ecological sense to frame a hypothesis in one particular way. In the previous example, ambient UV exposure from sunlight might be a control, and increased UV equal to that expected to result from decreased ozone in the upper atmosphere could be the treatment.

The next problem that arises is the certainty of any particular result. Say that we treated one frog egg with increased UV and it did not hatch, and the control egg treatment with ambient UV did hatch. Have we proven the hypothesis that increased UV will lower frog egg fertility? Perhaps the UV-treated egg was not fertilized properly in the first place. Replicating or repeating the experiment could increase the certainty of the result. If we did the experiment five more times and always had the same result, this would increase the certainty that we had proved the hypothesis. This replication is another key point in experimental design.

Statistics are involved when an experiment is replicated. Statistics allow comparisons between experimental and control treatments and formal expression of the degree of certainty that the hypothesis is true (Tables A.1 and A.3). Many tests can be applied to different experimental designs. These tests result in a probability value that can be used to express certainty. The convention is that a result is not significant until the investigator is 95% certain. However, this is purely an arbitrary value. When human lives are at stake, 95% is probably not good enough. When publication of an ecological paper is at stake, 95% is probably sufficient.

Experimental design is presented here in a simplistic fashion. Replication is often a problem for the environmental scientist. On the one hand, many do not understand replication; on the other hand, some experiments cannot be replicated as discussed previously.

When scientists do not understand replication correctly, pseudoreplication can be a problem because it can lead to undue faith in results. Pseudoreplication occurs when "replicate" samples are taken from one treatment (Hurlbert, 1984). For example, if we are testing the effects of nitrogen

Table A.3 Description of Some Common Statistical Methods

Method	Description	Comments
T-test	Comparison of one mean and a number or of two means	Common, simple procedure for assessing differences between means
Regression	Fits the best line to the data (minimizes variance)	Can be linear or nonlinear, with one or more independent variables
Correlation	Fit to a simple linear relationship	Does not denote causation; normally used for data exploration
Analysis of variance (ANOVA)	Tests the effects of multiple treatments on one dependent variable	Commonly used in replicated manipulative experiments
Chi square	Compares expected versus observed occurrences in categories	Commonly used to assess whether observed patterns are random or not
Meta-analysis	Compares results from many studies	Arnqvist and Wooster (1995); Gurevitch and Hedges (1993)

on water quality in two watersheds, and fertilize one watershed with nitrogen and leave the second untreated, it is not appropriate to take "replicate" water samples from each watershed and apply experimental statistics to them. We may be able to strongly infer results, but assigning statistical significance to the results is incorrect.

WHOLE SYSTEM MANIPULATIONS

Whole system manipulations can be very informative, but are often plagued with statistical problems because of low, or no, replication. Freshwater systems can be hard to replicate (e.g., no two streams are exactly alike), and funding and logistical constraints often preclude replicating ecosystem-scale manipulations even if suitable replicates were available. Nonetheless, some landmark studies in freshwater ecology have involved system-level manipulations; some of these studies were discussed in Chapter 24. Ultimately, combined approaches that use large-scale manipulations to examine "real world" responses, coupled with smaller-scale experiments to isolate variables, will be most powerful.

To compensate for statistical limitations, large scale manipulations generally involve collection of robust, long-term data sets. Ideally, the system to be manipulated and a designated control system are sampled for a long period of time to develop a statistical relationship between them prior to the manipulation. The two systems do not need to be identical, but there needs to be a consistent statistical relationship between them and they should respond similarly to regional climate, etc. The frequency and magnitude of sampling necessary will depend on the variability of the systems and measured variables; highly variable systems or variables might require multiple years of monthly sampling to develop statistical relationships. Once a robust premanipulation data set has been developed, the manipulation is applied and sampling continues. Again, the frequency and duration of sampling following the manipulation will depend on the variability of systems, along with the magnitude of the response to the manipulation.

Large-scale manipulations of this type are often analyzed using a before-after, control-impact design (BACI; Snedcor and Cochran, 1980; Stewart-Oaten *et al.*, 1986, 1992). BACI analyses are based on statistical relationships of response variables between control and manipulated systems before and after a manipulation, and whether those relationships change after the manipulation is applied. Randomized intervention analysis (RIA) is a nonparametric alternative to BACI; that is, RIA and other nonparametric tests do not require that data are normally distributed, as is the case for parametric techniques. RIA has been used to effectively show changes in system-level parameters associated with manipulations in a variety of whole-system lake and stream studies (e.g., Carpenter *et al.*, 1989; Wallace *et al.*, 1997; Connelly *et al.*, 2008). Like BACI, RIA is based on relationships between systems before and after a manipulation, and whether those relationships change after manipulation.

Ultimately, the results of any experimental endeavor are a function of the quality of the experimental design and data. Careful consideration of the questions addressed and associated hypotheses should guide experimental design and selection of specific statistical analyses. Freshwater systems present some significant challenges, and consultation with a statistician during the design and analysis phases of a project is recommended.

SUMMARY

1. There are several general types of experiments: natural experiments, simulation models, and manipulative experiments. Each has its own strengths and weaknesses.
2. Natural observation and experiments may be the most realistic but often are not replicated, and assigning causation to statistical results is difficult.
3. Multivariate techniques can be powerful tools for identifying patterns in large, complex data sets. Ordination techniques can be used to graphically convey patterns in data sets.
4. Simulation models may offer a way to approach intractably large or complex systems and provide insights into key factors in these systems.

5. Manipulative experiments can be replicated and the results subject to statistical determination of certainty of outcome. However, such experiments require replication, and this may lead to small-scale treatments or artificial conditions that make relevance of the results to the system of interest questionable.

6. Whole-system experiments can be powerful for understanding ecosystem responses to disturbances, but they require careful consideration because of limited replication and associated statistical issues.

7. All the methods are useful to aquatic ecologists. The trick is in asking the important questions and determining the best method to use to obtain satisfactory answers. Consulting with a statistician while planning an experiment and analyzing data is good practice.

Glossary

This glossary explains words used in the text, and also words common in the freshwater literature that do not appear in this text. We hope that this glossary will be useful for students reading the scientific literature in aquatic ecology.

A area

A_d drainage basin area

A_0 surface area of a lake

A_z area at a specific lake depth (z)

Abiotic oxidation a chemical oxidation in the absence of organisms

Absorb soak up (compare to adsorb)

Acidity the proton ion concentration (pH); also the ability of a solution to react with a base

Acute toxicity poisoning with large pulses over short periods of time

Adaptive radiation relatively rapid evolution of a single species into many species

Adhesion sticking to other things

Adsorb to take onto a surface

Advection movement by flow

Advective transport movement of materials by movement of parcels of water (as opposed to molecular diffusion)

Aeolian process formed by wind action

Aerobic oxic, with oxygen

Age class organisms within a specific age range

Aggradation deposition of excess stream load (sediment) to its channel

AHOD (oxygen deficit per unit area) decrease in O_2 in the hypolimnion per unit area surface; $mg\ O_2\ cm^{-2}\ day^{-1}$ or $mg\ O_2\ m^{-2}\ day^{-1}$

Airlift pump pump that uses compressed gas to move water up a tube; used for groundwater sampling and hypolimnetic aeration

Akinete resistant resting cell of cyanobacteria and some green algae

Alarm chemical chemical produced by a damaged or stressed prey that alarms other prey

Algae nonvascular organisms capable of oxygenic photosynthesis, without sterile cells covering gametangia

Alkaline phosphatase enzyme that hydrolyzes organic phosphorus compounds under basic conditions, rendering some P obtainable as soluble inorganic phosphate; used in some P deficiency assays and cyanobacterial toxin bioassays

Alkalinity acid-neutralizing capacity; the sum of all titratable bases; usually a function of carbonate, bicarbonate, and hydroxide contents

Allelochemical a chemical produced by one species that alters the behavior or growth of another species

Allen curve a plot of individual size vs. number of individuals in a cohort through time

Allen's paradox based on studies in New Zealand streams; invertebrate prey production appears lower than predatory fish production than the system would demand

Allochthonous originating from outside the system; often refers to organic carbon

Allomone chemical released by one species that affects the behavior of an individual of another species, and benefits the species that produces and releases it

Alpha diversity within-habitat diversity (α diversity)

Alternative stable states the idea that ecosystems can exist under different biotic and abiotic conditions that persist in the face of normal environmental variation

Amensalism interspecific interaction in which one species is harmed and the other is not influenced

Amictic when a lake almost never mixes

Ammonia gas NH_3, ammonium ion is converted to its toxic form, ammonia gas, at basic pH

Ammonification ammonium (NH_4^+) produced from organic nitrogenous compounds by metabolism of living organisms and decomposition of organic matter

Ammonium an inorganic N compound; the ion NH_4^+

Amphidromous regular migration from fresh- to salt water (see *diadromous*); not always for breeding purposes

Anadromous moving from marine habitats to freshwater for spawning

Anaerobic without oxygen; anoxic

Anaerobic ammonium oxidation conversion of nitrate and ammonium directly to dinitrogen gas under anaerobic conditions

Anammox see *anaerobic ammonium oxidation*

Anchor ice ice that forms on the bottom of open, flowing streams when temperatures are below 0 °C

Angiosperms true flowering plants

Anoxic without oxygen

Anoxygenic photosynthesis photosynthesis that does not evolve oxygen, either as cyclic photophosphorilization or using sulfide as an electron donor

Anthropogenic originating from human activity

Apatite calcium phosphate

Aquaculture the farming of aquatic organisms

Aquiclude a layer that groundwater flows through slowly (unsuitable for a well)

Aquifer permeable deposit that can yield water by a well

Aquifuge an impermeable layer through which groundwater cannot flow

Archaea a domain of organisms, unicellular without organelles; one of three domains (super kingdoms) of organisms; distinguished on the basis of biochemical and genetic differences from the Bacteria and Eukarya

Arenchymous containing spongy, air-filled tissues; occurs in some wetland plants and allows for movement of oxygen among parts of the plant, as well as release of methane to the atmosphere

Argillotrophy a mode of obtaining nutrition in turbid systems in which the primary source is organic material associated with clay particles

Artesian well a well with water naturally under pressure so it continuously flows from the wellhead

Ash-free dry mass (AFDM) the mass of material that will combust from a dry sample; used as an estimate of biomass

Assimilation the process of using nutrients to synthesize components of a cell; with regard to consumers, ingested material that is actually digested and absorbed (difference between ingestion and egestion)

Astatic unstable; refers to water level

Astrobleme a meteor crater; lakes can form in these

Atelomixis unfinished vertical blending of stratified water, combining layers of varied chemical properties without affecting the hypolimnion

Athalassohaline saline water having relative ionic proportions very different from those of seawater

Atmospheric loading nutrient input from dry and wet atmospheric depositions

Attenuation absorption of light as it passes through a medium

Attenuation coefficient a value that indicates how rapidly light is absorbed

Aufwuchs see *periphyton*

Autecology study of the ecology of a species; population, not community or ecosystem ecology

Autochthonous originating within the system (e.g., organic carbon supplied by primary producers in the system)

Autofluorescence natural fluorescence in pigments such as chlorophyll

Autotrophic the capacity to perform primary production; self-feeding; able to use CO_2 as a source of carbon using chemical (chemoautotrophy) or light (photoautotrophy) energy

Autotrophic index chlorophyll:ATP ratio

Bacteria a domain of organisms, unicellular without organelles; one of three domains (super kingdoms) of organisms; distinguished on the basis of biochemical differences from the Archaea

Bacterioplankton suspended bacteria

Bailer a tube with a one-way valve on the bottom (allowing only inflow of water) used to sample well water

Baseflow the level of stream discharge in the absence of recent storms

Bathylimnion the deepest part of a stratified lake

Bathymetric map a topographical map of a lake that indicates the distribution of depths and shape of the bottom

Bed load sediment transported by rolling, sliding, or saltation on or close to the stream or river bed

Beer's law law and associated equations relating light absorption to properties of the medium that light is traveling through; the law states that light absorption is linearly proportional to concentration

Benthic associated with the bottom

Benthos benthic animals

Beta diversity (β diversity) diversity between habitats; sum of species unique to each habitat when examining more than one habitat

Bicarbonate HCO_3^- an ionic form of inorganic carbon that dissolves in water

Bicarbonate equilibrium the chemical equilibrium involving the dissolved inorganic forms of carbon dioxide, carbonic acid, bicarbonate, and carbonate; the relative proportions of these ions depends upon pH

Bioaccumulation bioconcentration plus the accumulation of a compound in an organism from food

Bioassessment use of organisms to evaluate environmental quality

Biochemical oxygen demand (BOD) the demand for O_2 created by compounds that can be respired by organisms plus the chemicals that will react with O_2

Bioconcentration ability of a compound to move into an organism from the water

Biodiversity the number of different species, organisms, genotypes, or genes of ecological functional groups in a region (also referred to as biocomplexity)

Biofilm a film of organisms and associated materials attached to a solid surface (substratum)

Biogenic meromixis an increase in the density of the hypolimnion caused by dissolved materials released from sinking organic matter so that mixing cannot occur

Biological assessment see *bioassessment*

Biologically available phosphorus (BAP) the amount of phosphorus that can be used by organisms; can refer to the instantaneously available phosphate or the P that will become available with long-term decomposition

Biomagnification increase in concentration of a chemical at higher trophic levels of a food web

Biomanipulation to improve water quality by controlling the fish community

Biomass mass of organisms

Bioremediation cleanup of pollution using organisms

Biotic oxidation oxidation of compounds by organisms

Bioturbation stirring of sediments by movement and activity of sediment-dwelling organisms

Bittern bromide, magnesium, and chloride blend left when seawater almost completely evaporates

Bivoltine producing two generations annually

Blackfish species that are resistant to deoxygenated conditions and survive in standing floodplain water during dry periods

Blackwater river water colored by dissolved organic matter (humic substances) low in dissolved inorganic and suspended solids and usually with a low pH

Bloom a large dense population of algae

BOD see *biochemical oxygen demand*

BOD$_5$ oxygen consumed in 5 days by biological and chemical processes

Bog a wetland in which peat accumulates; with minimal inflow or outflow; supports acid-loving mosses such as *Sphagnum*

Borehole a hole drilled to sample geological material or for well installation

Bottomland lowland along a stream or river that is flooded periodically

Bottom-up control control of system productivity by nutrients or light

BPOM benthic particulate organic matter

Brackish saline water with salinity less than that of seawater, as low as 100 ppm

Braided stream a stream having elaborate, multiple channels rather than one large channel; also referred to as an anastomosing stream

Brine water more saline than seawater

Brook small stream

Brownian motion molecules or particles moving independently at microscopic scales

Brownwater New Zealand term for "blackwater"

Bryophyte mosses, liverworts and other small, non-vascular plants

Budget an accounting of the relative magnitude of the fluxes between compartments in an environment

Buffering the ability of a solution to resist changes in pH

Byssal threads (byssus) strong, proteinaceous fibers used by some bivalves to attach to substrata

^{14}C a radioactive isotope of carbon often used as a tracer in ecological studies, particularly to measure photosynthetic rates

^{13}C a stable isotope of carbon used in some ecological studies; naturally present in the environment at trace levels relative to the more abundant ^{12}C

Caldera a collapsed volcanic crater; some calderas contain lakes

Capacity the amount that a stream can transport when full to the banks

Capillary involving or resulting from surface tension

Capillary action movement of water by capillary forces

Capillary fringe belt of soil above groundwater that contains some water drawn up by capillary action; immediately above the water table

Carbonate, CO_3^{2-} an inorganic ion with carbon

Carbon dioxide, CO_2 a gas

Carbon sequestration atmospheric carbon capture and storage

Carbonic acid, H_2CO_3 the form first assumed by CO_2 when it dissolves in water

Carnivore animal that eats other animals

Catadromous moving from freshwater to saltwater to reproduce

Catchment (watershed) surface area drained by a network of stream channels; although "watershed" is used synonymously, watershed has been defined in European literature as a line that joins the highest points of the perimeter of a catchment

Cellulose complex carbohydrate synthesized by plants

Certainty something that rarely happens in ecology

Chemical diffusion movement of dissolved materials in water

Chemoautotrophic obtains energy from chemicals other than organic C

Chemocline a steep chemical gradient or pycnocline in a lake; often found at the metalimnion

Chemolithotrophy autotrophy with the energy sources of inorganic chemical bonds and inorganic substances as electron donors (chemosynthesis); also called chemoautotrophy

Chemophobic repelled by a chemical

Chemotactic attracted to a chemical

Chlorophyll a the primary pigment of photosynthesis in cyanobacteria and eukaryotic autotrophs; often used to indicate biomass; absorbs red and blue light

Chronic over long periods of time

Chronic toxicity toxicity with long-term exposure

Cirque a bowl formed by glacial action at the head of a valley; can contain a lake

Clear water stream water with high transparency, lacking visible suspended material and brown color; ranging from acidic to slightly alkaline in pH

Cline any continuum or gradient

Clinograde distribution showing a gradient; often used to describe temperature or oxygen curves

Clinolimnion part of a lake in which temperature distribution declines exponentially from the eplimnion to the hypolimnion due to turbulence

Cnidocyte (nematocyst) specialized venomous cells of cnidarians used to secure food

Coarse particulate organic matter (CPOM) any materials greater than an arbitrary size of 0.5 mm; includes leaves, wood, and aquatic organisms

COD$_5$ oxygen removed by chemical process from water in 5 days

Cohesion sticking together

Cohort a group of individuals that started life at the same time

Cold monomixis one annual total circulation without ice cover; includes cold thereimictic and warm thereimictic lakes

Cold thereimixis circulation in summer at approximately 4 °C; cold monomixis that occurs in a polar lake

Collector an organism that makes its living collecting fine particles, either by filtering from the water column (collector–filterer) or feeding on deposited particles (collector–gatherer)

Colloidal particles particles not settled by gravity

Combined nitrogen organic N, nitrate, nitrite, or ammonium (not N_2 gas)

Commensalism interspecific interaction in which one species is influenced positively and the other is not influenced

Community all organisms in an area or a group of species in an area

Community assembly theory the concept that environmentally similar sites with different species assemblages arise from random fluctuations of a common pool of species, different species traits, or the order that they come into the environment

Compartments parts of a budget into which materials are divided; for example, in the global hydrologic cycle the ocean is a compartment

Compensation point light level at which O_2 production by photosynthesis equals consumption by respiration, or CO_2 assimilation is equal to production by respiration

Competition an interspecific interaction in which both species harm each other

Competitive exclusion principle the idea that only the competitively dominant organism will survive in an equilibrium environment

Conductivity the ability of water to conduct electricity, a function of the number of dissolved ions in the water; measured in units of mhos or Seimens per unit distance

Confined aquifer an aquifer between two impermeable layers (aquifuges)

Consumptive use a use of water that does not return it to the stream channel (causes loss by evaporation or to groundwater)

Contest competition a form of intraspecific competition; some individuals of a species get more of a resource to the detriment of other individuals of the same species

Control an experimental condition to which the treatments are compared

Coprophagy ingestion of feces

Core of depression an area around a well where the pumping lowers the level of the groundwater (drawdown)

Coriolis force a force that induces circulation in very large lakes; caused by rotation of Earth

Correlation a statistical way to measure relatedness of two variables; not necessarily causation

Cosine collectors sensors used to measure radiation from above

Cosmopolitan distribution having a very large range; found almost everywhere on the planet

CPOM see *coarse particulate organic matter*

Creek low-order, small stream (crick)

Crenogenic meromixis differences in density between waters of monimolimnion and mixolimnion attributable to below-surface flows of spring or seep saline water

Critical mixing depth mixing depth below which phytoplankton growth does not occur

Cryogenic lake lake located in a thaw basin of permanently frozen ground

Cryoperiphyton (kryoperiphyton) periphyton attached to the bottom of the ice

Cryptic species organisms that are morphologically indistinguishable, but genetically different enough to be considered different species

Cryptorheic concealed drainage with below-surface stream flow, usually in limestone, karstic areas

Cultural eutrophication nutrient enrichment caused by humans

Cumulative a response to numerous events

Cycle all the fluxes of a material that occur in an environment

Cyclic colonization colonization of less permanent habitats by colonists from more permanent habitats

Cyclomorphosis successive emergence of distinctive morphologies in the same species; often observed in microcrustaceans and rotifers

D_L shoreline development index

Dam barrier preventing the flowing of water; impoundment

DAPI a fluorescent dye specific for DNA used to count total bacteria using epifluorescent microscopy or flow cytometry

Darcy's law a relationship used to calculate groundwater flow rates

Decomposition the breakdown of organic materials

Decomposer consumes or breaks down organic materials

Deflation basin a basin formed by action of the wind; can contain a lake or wetland

Degradation erosion of stream channels; opposite of aggradation

Denitrification conversion of nitrate to N_2 gas by microorganisms; a form of respiration that uses nitrate rather than O_2 to oxidize organic carbon

Density current current going along the bottom or through a stratified layer; current of different temperature, ionic strength, or turbidity than the water through which it flows

Density-mediated interactions interactions evidenced by changes in population size

Desalinization removal of salt from water

Detritivores organisms that eat detritus; also called saprophytes

Detritus decaying organic material

Dewatering removing water from a river or stream (also abstraction)

DGGE (denaturing gradient gel electrophoresis) molecular fingerprinting technique that separates polymerase chain reaction (PCR)-generated DNA products

Diadromous migrating between fresh and salt waters

Diagenesis conversion of sediment into rock

Diapause part of the normal life cycle that is stationary, physiologically dormant, and often resistant to environmental extremes such as drying or heat

Diatoms single-celled algae with silica shell (frustule) and golden brown coloring (Bacillariophyceae)

Diel 24 h day with a light–dark cycle (as opposed to a period of light)

Diffusion boundary layer the thin layer near a solid surface where diffusion is dominated by molecular diffusion; its thickness can control metabolic rates of microorganisms

Diffusion coefficient a constant used to describe diffusion of a compound or heat independent of distance and concentration

Diffusion flux the amount of a compound diffusing across an area per unit time

Dimethylsulfide (DMS; C_2H_6S) water-insoluble flammable liquid organosulfur compound

Dimictic a lake that mixes twice each year

Dinoflagellates one-celled algae that move by means of flagella (Pyrrhophyta)

Direct interaction occurs between two species and involves no others

Discharge the volume of a fluid flowing past a point per unit time

Disproportionation breakdown of a compound that has two atoms of the same element in different redox states to yield energy (for example, acetate to methane and CO_2 or thiosulfate to sulfate and sulfide)

Dissimilatory nitrate reduction to ammonium (DNRA), nitrate reduction to ammonium without assimilation involved; an alternative pathway to oxidize organic carbon or sulfide using nitrate with ammonium as the end product

Dissimilatory process a chemical transformation mediated by organisms that does not involve assimilation

Dissimilatory sulfur reduction using oxidized sulfur as an electron receptor to respire organic carbon or react with reduced inorganic compounds to yield energy

Dissolved inorganic nitrogen (DIN) the sum of ammonium, nitrate, and nitrite

Dissolved materials materials smaller than a particular size (e.g., $0.45\,\mu m$) that remain in solution

Dissolved organic carbon (DOC) organic carbon compounds dissolved in solution

Dissolved organic nitrogen (DON) organic nitrogen able to pass through a $0.45\,\mu m$ filter

Dissolved oxygen (DO) O_2 dissolved in water

Disturbance an event that disrupts ecosystem, community, or population structure

DNA barcoding analysis of short genetic markers in DNA to identify species

DNRA see *dissimilatory nitrate reduction to ammonium*

Dolina depression caused by dissolution of limestone substrata; sink or swallow hole

DOM see *dissolved organic carbon, dissolved organic matter*

Drainage area the land area that is drained by all the tributary streams above a chosen point in the channel

Drawdown lowering a water table by pumping

Drift the material that washes downstream, particularly invertebrates

Droop equation relationship between intracellular nutrient concentrations and growth

Dy sediment of dystrophic lakes, generally with a high organic content and allochthonous origin

Dynamic equilibrium model the idea that response of species richness to disturbance is a function of competition intensity in a community and productivity of the system

Dynamic viscosity a constant that describes the intrinsic viscosity of a fluid

Dystrophic a lake that is not productive because it has been influenced by factors that attenuate light or retard photosynthetic organisms; a lake that is high in tannin and lignin is dystrophic

Eckman dredge a dredge that is lowered to the bottom, with a messenger that is sent down its line and triggers jaws on the bottom to shut, taking in sediment

Ebullition release of bubbles

EC$_{50}$ concentration of a toxin that induces a response halfway between baseline and maximum after a given time period

Ecohydrology study of ecology as influenced by hydrology

Ecoregion an area with a geographically distinct assemblage of communities

Ecosystem all living and nonliving community constituents

Ecosystem engineer organisms that strongly alter their environments

Ecosystem goods products of value to humans from ecosystems

Ecosystem function the rate of specific basic ecosystem processes

Ecosystem services services of value to humans associated with ecosystem processes

Ecotone transitional zone between two habitats, e.g., where there is a change between groundwater and surface water organisms

Ectogenic meromixis meromixis caused by inflow of materials from exterior sources (e.g., saline water entering a dilute lake)

Edaphic relating to the ground or soil, particularly matter influenced by them or from them

Eddy current or small whirlpool moving counter to the main flow

Eddy diffusion diffusion by transport or mixing of a diffusing substance or heat; much faster than molecular diffusion; also called transport diffusion or advective transport

Effective population size the number of individuals in a population that actually contribute offspring

Effluent the water released from a sewage plant, factory, or other point source

Emergent growing above the water

Endemic species having a distribution that is restricted to a relatively small region

Endocrine disrupting compounds organic compounds that mimic natural metabolic compounds leading to disruption of endocrine function

Endogenic endogenous, created or produced internally; occasionally used instead of autochthonous

Endorheic stream or basin that does not drain to a larger stream or basin (a closed basin)

Endosymbiont an organism living inside another organism

Entrainment mixing of part of the hypolimnion into the epilimnion with high winds

Ephemeral not holding surface water most of the year

Ephippium resistant egg case produced by a cladoceran

Epifluorescent microscopy microscopy using a light source from above that excites fluorescent molecules and then filters the excitation wavelengths from the observed light, rendering the emitted fluorescence visible

Epigean living in surface water (above the ground or sediment)

Epilimnion the surface layer of a stratified lake, above the metalimnion

Epilithic growing on rocks

Epineuston organisms on the upper part of the water surface film

Epipelion community inhabiting mud surfaces

Epiphytic growing on a plant or macrophyte

Epipsammic growing on sand

Epixylon microbial communities associated with wood

Epizooic growing on an animal

EPT index component of biological assessment that measures the number of Ephemeroptera, Plecoptera, and Trichoptera taxa present; three groups that are generally intolerant of pollution

Erosional processes movement of particles off land by water or wind

Euphotic zone the region where light is above the compensation point, so net photosynthesis is positive

Eutrophic very productive

Eutrophication the process of becoming eutrophic

Evaporation to convert into vapor

Evapotranspiration evaporation plus plant transpiration

Evenness a measure of the degree of equal distribution of numbers of each species in a community

Exploitation interaction that harms one species and helps another

Exploitative competition competition between organisms that are using the same resource; one species reduces the resource or uses it more efficiently, reducing its availability for the other species

External loading supply of nutrients from outside the system

Extinction a condition where no living individuals of a species exist anywhere on Earth

Extinction coefficient same as the absorption coefficient

Extinction vortex the mutual reinforcement of biotic and abiotic factors that drive small declining populations extinct

Extirpated species that no longer occur in a region (though are not necessarily extinct)

Fall mixing the autumnal period in temperate lakes after summer stratification breaks down

Fault a fracture in rock layers where adjacent layers have moved parallel to the fracture

Fecundity the number of offspring that reproductive females produce

Fen like a bog, with peat accumulation but more input of water from outside

Fermentation organisms using organic carbon in the absence of oxygen

Ferric iron ion in oxidized state (Fe^{3+})

Ferric hydroxide flocculent precipitate of ferric ions and hydroxyl ions

Ferrous iron ion in reduced state (Fe^{2+})

Fetch the longest uninterrupted distance on a lake that wind can move across to create waves

Fick's law a mathematical formula describing the relationship between diffusion flux, distance, and concentration gradients

Filterers organisms that sieve small particles from the water column

Fine particulate organic matter (FPOM) organic material between 0.45 and 500 μm in diameter

Fixed groundwater groundwater trapped in rocks (confined)

Fjord lake a steep glacial valley (fiord) containing a lake

Flashy having repeated, rapid discharge or floods

Floating attached macrophytes with floating leaves that are rooted in the sediments

Floating unattached free-floating macrophytes

Flood rising and overflowing of a body of water, generally above the banks

Floodplain a flat region in the bottom of a valley that is, or historically was, influenced by river flooding

Flow movement of a fluid; generally a velocity, but can be used to mean discharge

Flow boundary layer the zone in which turbulent flow is rare and laminar flow dominates; near solid surfaces

Fluvial produced by action of a stream or river

Fluxes movements of materials between pools or compartments in a cycle (e.g., of water through the global hydrologic cycle)

Food chain the most simplistic view of food webs in which only trophic levels are considered

Food web a network of consumer–resource/predator–prey interactions that occur in an ecological community

FPOM fine particulate organic material

Free groundwater groundwater not trapped or confined

Freshwater water with less than 1 part per thousand of dissolved salts

Froude number a dimensionless measure relating inertia forces to gravitation effects; important when gravity is dominant (i.e., flow in open channels)

Frustule hard cell wall of diatoms composed mostly of silica

Fulvic acids the humic fraction that does not precipitate when a solution is acidified

Functional feeding group a subset of organisms from a community that feed using similar strategies (e.g., filterers, scrapers, and shredders)

Functional redundancy the degree to which different species provide the same ecosystem function

Functional response the number of prey eaten per unit prey density; increased consumption rates of predators in response to increasing prey densities

Fungi a kingdom of parasitic and saprophytic organisms

Gaining stream channel is below the water table so it gains water and increases in discharge as it flows down slope

Gape limited size of prey consumed is limited by mouth size

Gas vesicles intracellular protein structures that lend buoyancy to cyanobacteria

Gemmule resistant, asexual buds of sponges

GIS (Geographic Information System) a computer-based mapping system that can be used for complex spatial and temporal geographic analysis

Glacial relating to or produced by glaciers

Glacier a large body of flowing ice

Global water budget estimated water movement (fluxes) between compartments throughout the world

Glochidia specialized larval stage of freshwater mussels that attach to fish for dispersal

Graben a depression between two faults when one block slips down relative to two others and creates a basin (see also *horst*); lakes can form in the depression

Gravitational water water in rocks and soils above groundwater, includes perched aquifers

Gravitoidal particles particles that will settle

Grazer primary consumer; eats algae or plants

Greyfish Australian fish living in backwater, shoreline vegetation, lake edges, or stagnant channels in dry seasons

Gross photosynthesis the total amount of photosynthesis; in bottle experiments, the O_2 production rate in the light bottle minus that in the dark bottle

Groundwater water in or below the water table

Groundwater recharge replenishment of groundwater

Guild organisms that use a similar resource in a similar manner

Gyttja partially reduced, minute-grained, organic, profundal sediments of eutrophic lakes; copropel, mainly of autochthonous origin

[^{3}H]tritium a radioactive isotope of hydrogen used as a tracer

Habitat diversity beta diversity, between-habitat diversity; alpha diversity, within-habitat diversity

Halophilic salt loving; requires saline water for growth or reproduction

Hardness a characteristic of water that does not allow soap to dissolve; primarily caused by Ca^{2+} and Mg^{2+}

Head the potential energy of water from gravity

Heat budget an account of heat incorporated and lost by a water body during a specified time period

Heat capacity amount of energy required to increase the temperature of a specific mass of material

Heat of fusion the number of calories required for a specific mass of solid substance to form from a liquid

Herbivore organism that eats primary producers

Hermaphrodite an organism with both male and female sexual reproductive organs

Heterocyst specialized cyanobacterial cell in which nitrogen fixation occurs at high rates

Heterogeneous consisting of dissimilar ingredients or constituents

Heterogeneous aquifer an aquifer with many impermeable layers or areas where water cannot flow evenly

Heterotrophy metabolic energy and growth from degradation of organic molecules; carnivory, detritivory, herbivory, microbial decay, and omnivory

Hind-casting when a model is used to predict system response when the data from the system are already known

Hochmoor elevated bog with peaty matter higher than the rim of the cavity housing the bog

Holomixis total circulation or blending as in a holomictic lake

Homogeneous of the same or similar kind of nature

Homogeneous aquifer an aquifer with evenly distributed substrata and even groundwater flows

Homeothermy relatively constant temperature; see *isothermy*

Horst a depression formed by a fault when blocks tilt, slide against each other, and create a basin; lakes can form in them

HPLC high-performance liquid chromatography; an analytical system in which a carrier fluid flows through materials that can separate different types of dissolved molecules

Humic acids humic material that is precipitated by acid

Humic compounds from decomposition of plant material, includes humic acids, fulvic acids, and humin

Humin the fraction of humic compounds that cannot be extracted by acid or base

Humus high-molecular-weight organic molecules, polymeric, primarily from decayed plants; humic acids with the –COOH radical; humolimnic acids in lake waters and sediments

Hydraulic conductivity the ability of a material to allow water flow (coefficient of permeability)

Hydraulic gradient in groundwater, a line that will connect the level of water in wells or that follows the top of the water table; in streams, the drop in the channel per unit distance

Hydraulic head difference in water elevation between two connected sites

Hydraulic regimes degree and duration of water inundation and depth

Hydric soils soils with characteristics related to constant water inundation

Hydrochory seed dispersal by water

Hydrodynamics temporal and spatial variations in movement and distribution of water

Hydrogen bonding when molecules are polar (i.e., there is an uneven distribution of charge across the molecule), the positive and negative parts of separate molecules are attracted toward each other; particularly important in water

Hydrograph plot of discharge as a function of time for a stream or river

Hydrologic cycle the fluxes of water across the landscape

Hydroperiod length of time that surface water is present

Hydrophilic water loving

Hydrophobic repelled by water

Hydrophyte an aquatic plant, usually a macroscopic, rooted variety; macrophyte

Hydropsammon interstitial zone in sand below shallow water; sand-dwelling organisms live here

Hydrostatic pressure the pressure that fluids exert (density × depth)

Hygropetric in water on vertical rock surfaces

Hypereutrophic (hypertrophic) extremely productive, with abundant nutrients and very high primary producer biomass

Hypogean existing underground, subterranean, interstitial, or cave dwelling

Hyponeuston organisms in the lower part of the water surface film

Hyporheic zone a region of groundwater influenced by a nearby stream or river

Hyporheos a region of groundwater influenced by nearby surface water

Hypolimnion the bottom layer of a stratified lake; below the metalimnion

Hypothesis initial expectation; statement of cause to be tested by observation or experiment

Hypsography mapping and measuring of elevations and contours

Hysteresis a memory or lag effect

Ichthyochory seed dispersal by fishes

Impervious surface a surface that does not allow infiltration of water

Indirect interaction interaction between two species mediated by one or more other species

Inertia resistance of a body to a change in its state of motion

Infauna animals that dwell in the substrata or sediments

Infiltrate flow into; permeate

Infiltration gradual movement or permeation (e.g., of water through soils)

Interception the process of plants stopping precipitation before it reaches the ground

Interference competition direct negative competitive effects between species

Intermediate disturbance hypothesis the idea that species diversity is maximal at intermediate levels of natural disturbance

Intermittent usually drying at least once per year

Internal loading availability of nutrients from within the system; often associated with lake mixing

Internal seiche rocking of the hypolimnion while the surface of epilimnion stays still

Interspecific between two species or among numerous species

Interstices voids between sediment particles or in rocks

Interstitial between particles

Intraspecific within a species

Inverse thermal stratification warm water under cold water in a vertical temperature profile; can occur below an ice cover during winter stratification or when warm saline water sits below cooler, more dilute water

Irradiance radiance flux density on a given surface

Isobath contour line of lake depth; a bathymetric map of a lake is composed of isobaths

Isothermal with the same temperature; homeothermal

Kairomone chemical produced by a species that benefits another species with adverse consequences for the species that released it

Karst irregular limestone region with sinks, underground streams, and caverns

Kemmerer sampler a tube that closes by gravity when a messenger is sent down a line to it; used to sample water at a specific depth

Kerogen marine and lacustrine residues, not soil humus

Kettles when large blocks of ice melt and leave lakes, ponds, or wetlands

Keystone species species that have a major impact on their community or ecosystem; impact is disproportionately large relative to abundance

Lacustrine shallow lake habitat

Lake very slowly flowing body of water in a depression of ground not in contact with the sea

Laminar flow flow all in one direction, with little lateral mixing (as opposed to turbulent flow)

Langmuir circulation large spiral circulation patterns in lakes induced by wind

Larvae early life form of an animal, usually not capable of reproduction

LC$_{50}$ the concentration of a toxic compound that will kill half of the test organisms

LD$_{50}$ the dose of a toxic compound that will kill half of the test organisms

Leaf pack natural or artificial amassing of leaves in a stream

Leibig's law of the minimum a law that states that the rate of a process is limited by the rate of its slowest subprocess

Lentic still water habitat

Lethal causing death

Lichen symbiotic mutualistic partnership between fungi and alga

Lignin a complex polymer produced by plants that is resistant to microbial degradation

Limnocrene water from a spring or artesian well forming a pool

Limnology study of continental waters

Lithotrophy primary production or autotrophy with inorganic substances providing electrons; chemolithotrophy, photolithotrophy, chemoautotrophy

Littoral zone shallow, shoreline area of a body of water; often considered the portion of benthos from zero depth to the deepest extent of rooted plants

Logarithmic an exponential relationship

Lotic moving water

Losing stream channel is above the water table and the stream loses water to groundwater; discharge decreases downstream

Luxury consumption uptake of a nutrient in excess of needs

Lysimeter a sampler used to sample soil water

Maar volcanic eruption crater; can contain a lake or wetland

Macrobenthos small (about 1 mm) bottom-dwelling organisms

Macrophyte large colony of algae visible without magnification, or aquatic plants

Madicolous thin sheets of water flowing over rock

Magnetotaxis preferential movement toward north or south magnetic lines

Mainstem primary flowing section of a river or stream (as opposed to forks, tributaries, or other divisions)

Marl calcareous sediments, primarily soft

Manipulative experiment all other factors held constant and factors of interest are varied

Marsh continuously or usually inundated wetland with saturated soils and emergent vegetation

Mass spectrometer an instrument used to determine molecular weight of molecules and elements; used in stable isotope studies and to identify organic pollutants

Maximum density of water 1.000 g mL^{-1}; occurs at $3.9\,^\circ\text{C}$

Maximum sustainable yield the largest catch or take that can be harvested from a population over an indefinite period of time

Meander a natural feature of flowing water where S-shaped curves form

MEI morphoedaphic index to fish production; TDS (total dissolved solids) of water divided by the depth of lake; milligrams per liter divided by meters

Meiobenthos bottom-dwelling organisms (about 0.05 mm in size)

Meniscus an upsloping surface where water meets a hydrophilic solid

Meromictic a lake that is permanently stratified or mixes irregularly

Mesolimnion see *metalimnion*

Mesotrophic moderately productive system

Messenger a weight that attaches to a line and can be dropped to trigger sampling devices

Meta-analysis a statistical technique that tests the significance of combined results of many different studies

Metacommunity two or more communities linked by dispersal of multiple, potentially interacting species; a community of communities

Metalimnion the intermediate zone in a stratified lake in which the temperature change with depth is rapid, below the epilimnion and above the hypolimnion; also known as the thermocline or the mesolimnion

Metapopulation two or more geographically separated populations that interact to some degree

Methanogenesis anoxic bacteria producing methane

Methanotrophy oxidizing methane to obtain energy; a type of methylotrophy

Methylotrophy bacteria harvesting energy by oxidizing chemical compounds with methyl groups

Michaelis–Menten a relationship used to describe the influence of nutrient concentration on uptake rate

Microbenthos bottom-dwelling microorganisms, including bacteria, small algae, ciliates, gastrotrichs, and rotifers

Microbial conditioning colonization of organic materials such as dead leaves with microbes, which generally increases nutritional quality

Microbial loop the part of the food web based on consumption of bacteria and bacteria-sized algae, including protozoa, rotifers, and other microbes; often refers to the flow of carbon through the microbial community

Microphytobenthos periphyton

Minerotrophic a wetland with a high hydrologic throughput

Minimum viable population the minimum number of individuals of a species required for survival in the wild

Mire a peat-accumulating wetland (European definition)

Mixolimnion upper layer that sometimes mixes in meromictic lakes

Mixotrophy capacity to employ both organic and inorganic carbon sources for nourishment (using both autotrophy and heterotroph)

Molecular diffusion diffusion that occurs by random movement of molecules (Brownian motion)

Monimolimnion layer in meromictic lakes under the thermocline

Monod equation relationship of growth to concentration of nutrients outside the organism

Monomictic a lake that mixes once a year

Moor a peatland; can be raised or a depression

Moraine a wall of material deposited by a glacier (e.g., when the forward movement of a glacier is approximately equal to its backward melting rate a terminal moraine is deposited); lateral moraines are formed at the margins of a glacier; can dam streams to form lakes

Morphometry shape and size of lakes and their watersheds, or the shape and size of any object

Mortality the number of organisms lost between each size or age class

Moss lower plant; a small filamentous bryophyte

Multivoltine (polyvoltine) producing many generations annually

Muskeg a large peatland or bog, particularly in northern North America

Mutualism an interspecific interaction in which both species benefit

^{15}N a stable isotope of nitrogen used to trace nitrogen flux

^{14}N the most common stable isotopic form of N in the environment

Naiad another term for the larva of an aquatic insect

Nanoplankton plankton particles between approximately 3 and 50 μm

Natural abundance amount of a stable isotope found naturally

Natural experiment observation of natural patterns

Nekton plankton able to control their position in water columns by swimming

Net photosynthetic rate the amount of photosynthesis that is used for growth; in bottle experiments, the O_2 concentration in the light minus the initial concentration

Net plankton plankton particles between approximately 50 and 500 μm

Neuston microorganisms living at the water surface

New nutrients nutrients from outside the system

Niche position of a species in its environment, including all relationships with biotic and abiotic components

Nitrate, NO^{3-} an oxidized ionic form of inorganic N in natural waters

Nitrification microbial processes that convert nitrate to ammonium, yielding energy

Nitrite, NO_2^- an ionic form of dissolved inorganic N occasionally found in significant concentrations in natural waters

Nitrogenase an enzyme with molybdenum as a cofactor that reduces N_2 to NH_4^+ (nitrogen fixation); found only in bacteria

Nitrogen fixation the process of converting N_2 to a form of nitrogen used by organisms (to combined nitrogen)

N:P nitrogen-to-phosphorus ratio; can be expressed per mole or by weight

Nomogram a graphical analog computation device for deriving unknown values based on known values

No net loss regulations that require no net loss of wetland or other habitat area

Nonpoint source a diffuse source of pollution from the landscape, such as that resulting from cultivation, urban lawn spraying, or runoff from parking lots

Numerical response predator densities increasing with prey densities

Nutrient element or chemical compound required by organisms for growth

Nutrient budget a quantification of fluxes in a nutrient cycle

Nutrient cycling the transformation of nutrients from organic to inorganic forms and among different oxidation states

Nutrient limitation control of growth or production by a nutrient or nutrients

Nutrient loading input of nutrients to a system from river or stream inflow, dry or wet deposition, and internal sources

Nutrient sinks compartments that store nutrients over time

Nutrient spiraling alternate uptake and release of a nutrient; another term for nutrient cycling, with "spiraling" referring to downstream motion between uptake and release

Nymph another term for the larva of an aquatic insect

Oligomixis infrequent circulation of water masses

Oligotrophic nutrient-poor system with relatively low primary production

Ombrotrophic system that receives most of its water and minerals from precipitation; ombrotrophic bog; generally nutrient-poor

Optimal foraging evolution through selection leads to maximization of energy gain while foraging

Organic carbon carbon compounds bonded with hydrogen, usually oxygen, and sometimes other elements

Organism a complex living structure made up of a cell or cells

Organotrophy heterotrophic nutrition in which energy is taken from fermenting or oxidizing organic compounds

ORP oxygen reduction potential; also called redox

Orthograde straight distribution as in a vertical oxygen or temperature profile

Osmoregulation regulation of salt and water balance of the body or cell

Otolith small, hard particles in the inner ear associated with maintaining orientation. In fishes, they can be used to estimate age in much the same way that tree rings are used

Oxbow lake section of an old river channel that has been cut off from the current channel to form a lake or wetland

Orthophosphate phosphate free, as a polymer, or bound to organic compounds

Oxic with O_2 (aerobic)

Oxidation–reduction potential see *redox*

Oxidized microzone the small region in sediment between oxic water above and anoxic water below

^{32}**P** a radioisotope of phosphorus used as a tracer to study phosphorus dynamics

Paleolimnology the study of ancient lakes and their sediments

Palynology the study of pollen grains and spores

PAR see *photosynthetically available radiation*

Paradox of the plankton the paradox that multiple competing phytoplankton species are able to coexist without competitive exclusion removing less competitive species

Parasitism predation where the predator is substantially smaller than the prey

Parthenogenesis reproduction by development of an unfertilized gamete

Particulate large enough to be retained on a filter (often $0.45\,\mu m$)

Particulate organic carbon (POC) organic carbon particles larger than $0.45\,\mu m$

Particulate organic nitrogen (PON) organic N retained on a filter

Paternoster lakes formed by glacial scour in mountain valleys

P:B ratio of production to biomass

PCB (polychlorinated biphenyls) persistent organic pollutants that were once widely used in industry as insulators and coolants

PCR (polymerase chain reaction) molecular technique used to amplify small amounts of DNA

Peatland any wetland that accumulates organic matter

Péclet number ratio of molecular to advective diffusion of heat

Pelagic in open water

Perched water table a small pocket of groundwater held above the main water table by an impermeable layer

Percolate when a solvent passes through a permeable substance

Periphyton the mixed assemblage of organisms attached to solid substrates in lighted benthic habitats, including algae, bacteria, protozoa, and invertebrates; a biofilm containing algae; also called aufwuchs and microphytobenthos

Permeability ability of a substance to transmit water

Perennial holding surface water year-round

pH activity of hydrogen ions (which is closely related to concentration); expressed as log10 (moles H^+ liter^{-1})

Phobic a response away from, such as a photophobic response in which light is avoided

Pholeterous living in crayfish burrows

Phosphatase enzyme that cleaves organic phosphorus compounds to liberate phosphate

Phosphate, PO_4^{3-} dominant ionic form of inorganic phosphorus in natural waters

Photoautotrophs photosynthetic organisms that use light

Photoinhibition the deleterious effects of high light

Photophobic repelled by light

Photosynthesis–irradiance (P–I) describes the effect of light on photosynthetic rate

Photosynthetically available radiation (PAR) light from 400 to 700 nm that is generally available to cyanobacteria and eukaryotic photosynthetic organisms; often expressed in μmole photons $m^{-2}s^{-1}$

Photosynthetic photon flux density (PPFD) photosynthetically available radiation, often used interchangeably with PAR

Photosynthetic quotient moles of oxygen produced divided by moles of CO_2 fixed

Phototactic attracted toward light

Phreatic from the zone below the water table

Phycobilins protein pigments that collect light for photosynthesis in the cyanobacteria and red algae absorbing in the 550 to 650 nm range

Physiography the natural geography of a region

Phytoperiphyton the photosynthetic organisms in periphyton, usually algae

Phytoplankton suspended algae

Picophytoplankton algal picoplankton

Picoplankton planktonic organisms smaller than 3 μm in diameter

Piezometer an instrument for measuring pressure or compressibility; used in determining groundwater flow patterns

Plankton suspended organisms

Playa a shallow marsh-like pond, not formed glacially

Pleuston macroorganisms living on the surface of the water

Pneumatophore specialized aerial roots of plants living in saturated soils

Poikilothermy having variable body temperature; also, complicated vertical temperature profile

Point bar a depositional area on the inside of a river bend

Point source a clearly defined source of pollution such as a sewage outfall (as opposed to nonpoint source)

Polar lake pond or lake with a surface temperature of 4 °C or lower in the warm season

Polar molecule a molecule with unequal distribution of charge

Polymictic almost continuous circulation or many flowing or mixing periods annually

Polyphosphate anionic phosphate polymers used in luxury uptake and for phosphorus storage

POM particulate organic matter that is retained by an approximately 0.45 μm membrane filter

Pool a slow-moving portion of a river or stream

Porosity the maximum water that can be stored in hydrated sediment

Potamodromous with migratory pattern entirely in freshwater

Potamology older term for stream biology, river ecology, and lotic limnology

Potamon stream or larger water body environment (rhithron)

Potential energy stored energy that can do work

Pothole a shallow pond or wetland formed in glacial till

ppb parts per billion (μg liter^{-1})

PPFD photosynthetic photon flux density; see *photosynthetically available radiation*

ppm parts per million (mg liter^{-1})

ppt parts per trillion (ng liter^{-1})

P:R ratio of gross primary production to community respiration

Prairie pothole a sizeable, rounded, and often water-filled depression in grassland habitat; formed in glacial drift

Precipitation the deposit of hail, mist, fog, rain, sleet, or snow

Predation one organism eating another

Predator an organism that feeds on other organisms

Predator-permanence gradient increasing prevalence of predators with increasing hydroperiod

Primary consumer herbivore

Primary producer photosynthetic organism

Primary production rate of carbon fixation by autotrophs (=gross primary production); this minus respiration and maintenance is net primary production

Probability value used to express significance

Production efficiency the efficiency at which an animal converts assimilated (net production efficiency) or ingested (gross production efficiency) materials into actual production

Profundal zone benthic zone in lakes deep enough that there is not enough light to support photosynthetic organisms

Psammon a habitat in interstitial water between sand grains

Pseudoreplication repeated samples from one treatment incorrectly treated as true replicates

Psycrophilic cold-loving; requires cold temperatures to grow or reproduce

Putrefaction decomposition of organic materials, primarily proteins, resulting in formation of ammonium (NH_4^+) and sulfide (H_2S)

Pycnocline density gradient

Pyrite a complex of sulfide and metal

Radula hard, rasping mouthpart of gastropods generally used for scraping biofilms

Random by chance

Rating curve a plot of measured stream discharge against depth that is used to predict discharge from depth (stage)

Reach a series of pools and riffles in a stream or river

Recharge water entering an aquifer

Recruitment the number of fish entering each size or age class

Redd area on lake or stream bottom where salmon or trout spawn; often a round, cleared depression in gravel

Redfield ratio algal composition under balanced growth, 106:16:1 C:N:P, by moles

Redox the relative number of free electrons in solution; measured as oxidation–reduction potential

Reedswamp a wetland dominated by reeds (e.g., *Phragmites*)

Reflect to turn, throw, or bend off backward at an angle

Reflex bleeding (autohemorrhaging) exuding hemolymph (blood) containing chemical deterrents to avoid predation

Refraction light bending as it passes through materials of different densities

Regeneration see *remineralization*

Remineralization organisms excreting nutrients

Replication repeating treatments in an experiment

Reset the return of a habitat to preceding (in time) conditions or to conditions upstream (in a continuum); a flood may "reset" a stream to earlier conditions

Residence time amount of time a material spends in an ecosystem compartment

Resilience the ability to bounce back or recover

Resistance the ability to remain unchanged

Resource ratio theory predictive model of species interactions based on their use of shared resources; predicts that species dominance varies with ratios of resource availability

Retention time average time materials spend in an ecosystem compartment

Reynolds number the ratio of inertia to dynamic viscosity; a unitless number that describes the properties of fluids as related to spatial scale and movement

RFLP (restriction fragment length polymorphism) DNA profiling technique that separates restriction fragments of DNA by their length

Rheocrene a spring-fed brook

Rhithron brook environment

Rice paddies wetlands that allow for rice culture

Richness see *species richness*

Riffle a rapidly flowing portion of a river or stream where the influence of the bottom can be seen at the surface

Riparian related to or located near water; often a transition zone between terrestrial and aquatic habitats

Riparian buffer an area adjacent to a stream or other water body where vegetation is left intact to reduce pollution inputs

Rising limb the portion of the hydrograph that is increasing during or immediately after a storm

River continuum concept an ecosystem-based view of streams and rivers as a continuum from small forested headwater streams to large rivers

rRNA (ribosomal RNA) focus of molecular taxonomy methods because it is highly conserved (low variability)

Run a portion of a river or stream where flow is rapid, but the surface is smooth

Runoff coefficient the proportion of precipitation falling on a catchment that enters the river or stream without entering the soil

Salinity amount of inorganic salts dissolved in water, stated as %, ‰, $g\ kg^{-1}$, ppm, $mg\ liter^{-1}$, $mg\ dm^{-3}$; in seawater, closely connected to Cl^- concentration and conductance; salinity of seawater is approximately 3.5% (35,000 ppm)

Saltern saline water with a similar composition to seawater

Sapropel reduced, nondescript sediment of polluted or hypereutrophic lakes; black with FeS, often odorous with H_2S

Saprophytes heterotrophs that decompose organic carbon

Saturating concentration equilibrium concentration when pure water is left in full contact with the atmosphere or in full contact with an undissolved substance

Scalar collector a spherical light collector

Scale to arrange in a graduated series

Scintillation counter an instrument used to quantify radioactive isotopes

Scramble competition a form of intraspecific competition; multiple individuals of the same species compete for a resource and all get equally less than they need

Scraper (grazer) an organism that obtains its food by scraping biofilms and periphyton

Scum a layer of floating algae or organic material

Seasonality index ratio of runoff in wet season to runoff in dry season

Secchi disk a circular disk with black and white quadrants that is lowered into the water to estimate water clarity

Secondary consumer eats primary consumers

Secondary production tissue production by heterotrophic organisms

Sediment the particulate matter that settles to the bottom of a liquid

Segment several reaches of a stream together or the length between major tributaries

Sensitivity analysis to systematically test the sensitivity of a model to the assumptions used to construct it

Seiche rocking of a lake or a layer of a lake

Seine a long net used to trap fish

Sessile attached to bottom

Seston suspended organic particles

Shannon diversity an index of diversity that includes both species richness and evenness (also called Shannon–Wiener or Shannon–Weaver index)

Shear stress tractive force per unit area

Sheet flow a shallow layer of water flowing across the surface of soil

Shelford's law of tolerance concept that species distributions are ultimately limited by their range of tolerance for local environmental factors

Sherwood number ratio of rates of molecular diffusion to transport diffusion

Shoreline development the circumference of a lake divided by the circumference of a circle with the same area; used as an indicator of trophic state

Shredder an organism that makes its living shredding organic material for food

Significance an estimate of the statistical certainty

Silica SiO_2

Silicic acid H_4SiO_4

Simulation modeling nonexperimental method used to explore possible hypotheses

Sinkhole surface depression formed in karst area when a subsurface cavity collapses

Sink population among metapopulations, a population that does not produce enough individuals to be self-sustaining; relies on immigration to persist through time

Sinter sediments from mineral springs, including the siliceous geyserite and the calcareous tufa or travertine

Sinuosity the degree of meandering

Size class discrete size range of an organism

Size fractionate separate into size classes

Slough swamp, shallow lake, or slowly flowing marsh

Solubility the ability to dissolve in a liquid

Soluble reactive phosphorus a chemically determined fraction containing phosphate and other forms of orthophosphate; also called DRP (dissolved reactive phosphorus)

Solvents can dissolve both gases and ions

Source population among metapopulations, a population that produces enough individuals to sustain itself, and often produces excess individuals to contribute to other populations

Species area relationship relationship between number of species present and geographical area

Species diversity generally measured as the number of species in an area and their evenness (relative abundance)

Species richness number of species in an area

Specific runoff precipitation runoff per unit catchment area

Spectrophotometer an instrument that measures the absorption of light as a function of wavelength

Spherical collector having the form of a sphere or part of a sphere; spherical irradiance collectors are used to estimate total light available to photosynthetic plankton

Spicule small, hard spine-like structures; form the skeleton of sponges

Spiraling length horizontal distance on a river or stream between successive uptake or release events (nutrient spiraling)

Spring mixing lack of differences in temperature with depth allowing wind to mix an entire lake during spring

SRP see *soluble reactive phosphorus*

Stability of stratification (S) the energy required to blend a body of water to constant density without adding or subtracting heat; expressed as g-cm cm-2 of lake surface

Stable isotope natural form of an element that does not emit radioactivity

Stenothermial able to survive in only a limited range of water temperatures

Stock size of population

Stokes law law describing sinking rates of small spherical objects

Stoichiometry the ratio of elements to each other

Strahler classification system used to describe locations in dendritic patterns of stream drainage systems; first-order streams are those without tributaries, second-order streams are the result of two first-order streams joining, third-order streams are the result of two second-order streams joining, and so on; a smaller order entering a larger order does not change the numeric designation of the larger

Stratification density differences in water that can maintain stable layers

Streamline the path of a fluid particle relative to a solid body past which the fluid is moving in smooth flow without turbulence

Streamlined shaped to avoid causing turbulence

Stream order classification of streams; the most commonly used system is the Strahler system

Stygobite obligate groundwater dweller; not found in surface waters

Styigophile groundwater organism with adaptations to living in groundwater, but also found in surface waters

Sublethal not causing death

Sublittoral bottom region between littoral and profundal zones

Submersed a macrophyte that grows below the water surface

Succession the sequence of organisms that colonize and inhabit a disturbed or new habitat over time

Sulfate (SO_4^{2-}) anion of sulfur

Sulfide sulfur ion, S^{2-}

Summer stratification in temperate lakes the hot period in which the lake water is in stable layers of discreet temperatures

Supercooling dropping the temperature of a liquid or gas below the freezing point without it becoming a solid

Supersaturation containing dissolved materials in excess of equilibrium concentrations

Surber sampler an apparatus with a square, fixed-area frame (often 1 ft^2 or 0.1 m^2) hinged to another frame that has a net; the open first frame is located on the stream bed, the net is elevated to a vertical position, and rocks within the open frame are rubbed and shaken to allow stream flow to move organisms into the net

Surface area to volume relationships a geometric ratio that can be used to indicate the ability of an organism to exchange materials with its surroundings

Surface tension hydrogen bonding pulls water into a tight surface

Suspended load fine materials suspended and transported in stream water

Suspended particles material retained by a 0.45 μm filter that stays suspended (seston in streams)

Swamp wetland dominated by woody vegetation (United States) or forested fen or reedswamp (Europe)

Symbiosis two organisms that live in close proximity; does not determine interaction effect; incorrectly used by some to refer to mutualism

Synecology older term for study of communities (as opposed to single species)

Syntrophy a group of organisms with complimentary metabolic capabilities; characteristic of anoxic microbial communities

T_{50} the amount of time necessary for 50% of original dry leaf mass to break down in decomposition studies

Tannins derivatives of multimeric gallic acid that leach from bark and leaf litter with properties similar to those of humic substances

Taxis movement toward stimuli

Tarn a small mountain lake (often in a cirque)

Tectonic basin a basin formed by movement of Earth's crust

Thalassohaline having ionic proportions similar to seawater

Thalweg the part of a stream channel through which the main or most rapid flow travels

Thanatosis feigning death to avoid predation

Thermal bar vertical or horizontal mass of water separating two areas of less dense water

Thermocline see *metalimnion*

Thermophilic hot-temperature-loving; requires warm temperatures to grow or reproduce

Thiosulfate ($S_2O_3^{2-}$) an oxyanion of sulfur produced by sulfate ions reacting with elemental sulfur in boiling water; the two sulfur atoms are at different redox states

Throughfall precipitation falling through vegetation

Thymidine uptake a method using [^{3}H] thymidine incorporation into DNA to assess bacterial growth rate

Time series temporal replication in experiments

Top-down control regulation of lower trophic levels or basal resources by higher trophic levels

Torenticole adapted to live in swift water

Tortuosity average length of the flow path between two points

Total dissolved solids (TDS) total mass of material left after a filtered water sample is dried

Total suspended solids (TSS) total weight of particulate material per unit volume of water

Tracer a dye, isotope, or ion used to trace the movement of water or other chemicals through the environment

Trait-mediated interactions interactions evidenced by evolved traits

Transient storage zone regions of slower flow in streams

Transmit to cause or allow spreading (particularly transmission of light in water)

Transport diffusion diffusion with water movement

Treatment method to remove contaminants from water

Troglobitic obligatory cave-dwelling or hypogean organisms

Troglophilic facultative cave-dwelling or hypogean organisms

Trophic basis of production the contribution of individual food sources to production of each species

Trophic cascade influence of consumer organisms on those lower in the food web with alternating effects at each trophic level (top-down control)

Trophic level position in a food web

Trophic state total ecosystem productivity, including heterotrophic and autotrophic pathways

Trophogenic zone area of photosynthetic production

Tropholytic zone area of decomposition

Tropism growth or movement toward a stimulus; similar to taxis (e.g., chemotaxis or chemotropism)

Tufa sedimentary rock formed by precipitation of carbonate minerals from water

Turbidity the amount of suspended particles in water (TSS) or absorption of light by those particles

Turbulent flow flow with swirls and eddies not in the direction of the main flow

Turnover calculation of residence time of water in a lake given volume and amounts of water entering and leaving; can also be the rate of flux of any material through a compartment per unit mass in the compartment (e.g., in nutrient budgets)

Ultraoligotrophic exceptionally unproductive system

Unconfined groundwater water not constrained from above

Unit hydrograph the hydrograph resulting from a single storm event

Univoltine producing one generation annually

Unsaturated zone vadose zone

Uptake nutrients taken into cells from the water surrounding them

Uptake length the average distance a nutrient travels in the water column before being taken up; one component of spiral length

Urea a nitrogen-containing organic compound that can be excreted by organisms

V volume of a body of water

Vadose zone above the water table; below the surface soil (also called unsaturated zone)

Van Dorn bottle a tube that can be lowered to a specific depth on a rope; when a messenger is sent down the rope, it triggers two ends to snap onto the tube, trapping the sampled water

Vaporization evaporation

Veliger a planktonic larval stage of many mollusks

Vernal pool shallow pool or pond that holds water in spring but is usually dry for much of the year

Vicariant event an atmospheric or geophysical event resulting in disturbance or fragmentation of a previously constant distribution and thus leading to allopatric speciation and radiation

Viscosity the resistance of a fluid to change, an internal friction; two types of viscosity can be considered—the dynamic viscosity (intrinsic property of the fluid) and the viscous force (a property of scale)

Volcanic related to action of volcanoes

Voltinism relating to the number of generations produced annually by an organism (e.g., bivoltine)

Vortex circular motion associated with turbulent flow

VPOM very fine (ultrafine) particulate organic matter (0.45–$53\,\mu m$ diameter)

Warm monomixis a period of total circulation during the cold time of year in a lake without ice; temperature stratification occurs in summer

Warm thereimixis summer mixing or cold monomixis with water warmer than $4\,°C$

Water abstraction removal of water to be used for drinking water, irrigation, industry, etc. from rivers, lakes, aquifers, and other sources

Watershed area above a point in a stream that catches the water that flows down to that point (in Europe, called catchment, and the watershed is the high point that divides catchments)

Water table the top of the groundwater or saturated zone

Water velocity speed of water

Water yield the amount of water (depth) per unit time from a specific area

Weathering dissolution of materials from rocks

Weir a device to concentrate all the fluid in a channel into one place and allow for measurement of discharge

Well casing a pipe used to keep a well from collapsing

Wetland areas inundated or saturated by surface or groundwater at a frequency and duration sufficient to support a prevalence of vegetation adapted for life in saturated soil conditions

Wet meadow a meadow without open water but saturated soils

Wet prairie a shallow soil wetland with grass

Whitefish a fish species that migrates to elude severe conditions or, alternately, a type of North American coreogonid fish

Whitewater streams that carry large quantities of suspended solids and appear muddy or silty or large quantities of dissolved inorganic solids and are slightly alkaline or circumneutral; also sections of rivers or streams with extensive rapids and entrained air causing the water to appear white

Wind a natural movement of air of any velocity

Winterbourne upper reach of a chalk stream that channels water in fall after dry summer periods when aquifer levels decrease

Withdrawal taking something out (abstraction)

Yield the amount of fish that are taken per unit time from a fishery

z depth; vertical distance from surface

$\bar{Z}$ mean depth of a body of water; V/A

z_m maximum depth of a body of water

ZSD the maximum depth of Secchi disc visibility; usually expressed in meters

Zoophyte an organism growing on an animal

Zooplankton suspended invertebrates, generally multicellular and generally smaller than 1 mm

References

Aaronson, S. (1981). *Chemical communication at the microbial level*. Boca Raton, FL: CRC Press, Inc.

Abbott, L. L., & Bergey, E. A. (2007). Why are there few algae on snail shells? The effects of grazing, nutrients and shell chemistry on the algae on shells of *Helisoma trivolvis*. *Freshwater Biology, 52*, 2112–2120.

Abe, T., Lawson, T., Weyers, J. D. B., & Codd, G. A. (1996). Microcystin-LR inhibits photosynthesis of *Phaseolus vulgaris* primary leaves: implications for current spray irrigation practice. *New Phytology, 133*, 651–658.

Abebe, E., Decraemer, W., & De Ley, P. (2008). Global diversity of nematodes (Nematoda) in freshwater. *Hydrobiologia, 595*, 67–78.

Abell, R. A., Olson, D. M., Dinerstein, E., Hurley, P. T., Diggs, J. T., Eichbaum, W., et al. (2000). *Freshwater ecoregions of North America, a conservation assessment*. Washington, DC: Island Press.

Alexander, G. G., & Allan, J. D. (2007). Ecological success in stream restoration: case studies from the midwestern United States. *Environmental Management, 40*, 245–255.

Alexander, R. B., Smith, R. A., & Schwarz, G. E. (2000). Effect of stream channel size on the delivery of nitrogen to the Gulf of Mexico. *Nature, 403*, 758–761.

Alexander, R. B., Smith, R. A., Schwarz, G. E., Boyer, E. W., Nolan, J. V., & Brakebill, J. W. (2008). Differences in phosphorus and nitrogen delivery to the Gulf of Mexico from the Mississippi River Basin. *Environmental Science and Technology, 42*, 822–830.

Alexopolus, C. J., Mims, C. W., & Blackwell, M. (1996). *Introductory mycology* (4th ed.). New York: John Wiley & Sons, Inc.

Alford, R. A. (1999). Ecology resource use, competition, and predation. In R. W. McDiarmid & R. Altig (Eds.), *Tadpoles: The biology of anuran larvae* (pp. 240–278). Chicago, IL: The University of Chicago Press.

Alford, R. A., & Richards, S. J. (1999). Global amphibian declines: a problem in applied ecology. *Annual Review of Ecology and Systematics, 30*, 133–165.

Alfreider, A., Krössbacher, M., & Psenner, R. (1997). Groundwater samples do not reflect bacterial densities and activity in subsurface systems. *Water Research, 31*, 832–840.

Allan, J., Abell, R., Hogan, Z., Revenga, C., Taylor, B., Welcomme, R., et al. (2005). Overfishing of inland waters. *BioScience, 55*, 1041–1051.

Allan, J. D. (1983). Food consumption by trout and stoneflies in a Rocky Mountain stream, with comparison to prey standing crop. In T. D. Fontaine, III, & S. M. Bartell (Eds.), *Dynamics of lotic ecosystems* (pp. 371–390). Ann Arbor, MI: Ann Arbor Science Publishers.

Allan, J. D. (1995). *Stream ecology: Structure and function of running waters*. London: Chapman and Hall.

Allan, J. D., & Carillo, M. M. (2007). *Stream ecology: Structure and function of running waters* (2nd ed.). Dordrecht: Springer.

Allan, R. J. (1989). Factors affecting source and fate of persistent toxic organic chemicals: examples from the Laurentian Great Lakes. In A. Boudou & F. Ribeyre (Eds.), *Aquatic ecotoxicology: Fundamental concepts and methodologies* (pp. 219–248). Boca Raton, FL: CRC Press.

Allen, K. R. (1951). The Horokiwi stream: a study of a trout population. *New Zealand Department of Fisheries Bulletin, 10*, 1–238.

Allen, S. K., Jr., & Wattendorf, R. J. (1987). Triploid grass carp: status and management implications. *Fisheries, 12,* 20–24.

Allen, T. F. H., & Hoekstra, T. W. (1992). *Toward a unified ecology.* New York: Columbia University Press.

American Rivers, Friends of the Earth, and Trout Unlimited. (1999). *Dam removal success stories: Restoring rivers through selective removal of dams that don't make sense.* Washington, DC: American Rivers. Also available online at http://www .americanrivers.org/damremovaltoolkit/successstoriesreport.htm. American Rivers; Friends of the Earth; Trout Unlimited, Washington, DC.

Amundrud, J. R., Faber, D. J., & Keast, A. (1974). Seasonal succession of free-swimming perciform larvae in Lake Opinicon, Ontario. *Canadian Journal of Fisheries and Aquatic Sciences, 31,* 1661–1665.

Amy, P. S. (1997). Microbiology of the terrestrial deep subsurface. In P. S. Amy & D. L. Haldeman (Eds.), *Microbial dominance and survival in the subsurface* (pp. 185–203). Boca Raton: Lewis Publishers.

Anderson, A. F., & Banfield, J. F. (2008). Virus population dynamics and acquired virus resistance in natural microbial communities. *Science, 320,* 1047–1050.

Anderson, M. G. (1995). Interactions between *Lythrum salicaria* and native organisms: a critical review. *Environmental Management, 19,* 225–231.

Anderson, N. H., & Wallace, J. B. (1984). Habitat, life history, and behavioral adaptations of aquatic insects. In R. W. Merritt & K. W. Cummins (Eds.), *An introduction to the aquatic insects of North America* (pp. 38–57). Dubuque, IA: Kendall/Hunt Publishing Co.

Anderson, N. J. (1993). Natural versus anthropogenic change in lakes: the role of the sediment record. *Trends in Ecology and Evolution, 8,* 356–361.

Anderson, N. J. (1995). Naturally eutrophic lakes: reality, myth or myopia? *Trends in Ecology and Evolution, 10,* 137–138.

Anderson, R. O., & Neumann, R. M. (1996). Length, weight, and associated structural indices. In B. R. Murphy & D. W. Willis (Eds.), *Fisheries techniques* (pp. 447–482). Bethesda, MD: American Fisheries Society.

Anderson, R. T., Chapelle, F. H., & Lovley, D. R. (1998). Evidence against hydrogen-based microbial ecosystems in basalt aquifers. *Science, 281,* 976–977.

Anderson, R. T., & Lovley, D. R. (1997). Ecology and biogeochemistry of *in situ* groundwater bioremediation. In *Advances in microbial ecology* (pp. 289–350). New York: Plenum Press.

Angermeier, P. L. (1995). Ecological attributes of extinction-prone species: loss of freshwater fishes of Virginia. *Conservation Biology, 9,* 143–158.

Angradi, T. R. (1998). Observations of freshwater jellyfish *Craspedacusta sowerbyi* Lankester (Trachylina: Petasidae) in a West Virginia reservoir. *Brimleyana, 25,* 34–42.

Arnold, S. F., Klotz, D. M., Collins, B. M., Vonier, P. M., Guillette, L. J., Jr., & McLachlan, J. A. (1996). Synergistic activation of estrogen receptor with combinations of environmental chemicals. *Science, 272,* 1489–1492.

Arnqvist, G., & Wooster, D. (1995). Meta-analysis: synthesizing research findings in ecology and evolution. *Trends in Ecology and Evolution, 10,* 236–240.

Arruda, J. A. (1979). A consideration of trophic dynamics in some tallgrass prairie farm ponds. *American Midland Naturalist, 10,* 254–262.

Arruda, J. A., & Fromm, C. H. (1989). The relationship between taste and odor problems and lake enrichment from Kansas lakes in agricultural watersheds. *Lake and Reservoir Management, 5,* 45–52.

Arscott, D. B., Bowden, W. B., & Finlay, J. C. (1998). Comparison of epilithic algal and bryophyte metabolism in an arctic tundra stream, Alaska. *Journal of the North American Benthological Society, 17,* 210–227.

Arsuffi, T. L., & Suberkropp, K. (1984). Leaf processing capabilities of aquatic hyphomycetes: interspecific differences and influence on shredder feeding preferences. *Oikos, 42,* 144–154.

Arsuffi, T. L., & Suberkropp, K. (1985). Selective feeding by stream caddisfly (Trichoptera) detritivores on leaves with fungal-colonized patches. *Oikos, 45,* 50–58.

Arsuffi, T. L., & Suberkropp, K. (1989). Selective feeding by shredders on leaf colonizing stream fungi: comparison of macroinvertebrate taxa. *Oecologia, 79,* 30–37.

Atlas, R. M., & Bartha, R. (1998). *Microbial ecology, fundamentals and applications* (4th ed.). Menlo Park, CA: Addison Wesley Longman, Inc.

Auer, A. (1991). Qualitative diatom analysis as a tool to diagnose drowning. *American Journal of Forensic Medicine and Pathology, 12*, 213–218.

Auer, M. T., & Effler, S. W. (1989). Variability in photosynthesis: impact on DO models. *Journal of Environmental Engineering, 115*, 944–963.

Auerbach, E. A., Seyfried, E. E., & McMahon, K. D. (2007). Tetracycline resistance genes in activated sludge wastewater treatment plants. *Water Research, 41*, 1143–1151.

Auguet, J., Montanié, H., Hartmann, H., Lebaron, P., Casamayor, E., Catala, P., et al. (2009). Potential effect of freshwater virus on the structure and activity of bacterial communities in the Marennes-Oléron Bay (France). *Microbial Ecology, 57*, 295–306.

Axelrod, H. R. (1973). *African cichlids of lakes malawi and tanganyika.* Neptune, NJ: T. F. H. Publications, Inc.

Axler, R. P., Redfield, G. W., & Goldman, C. R. (1981). The importance of regenerated nitrogen to phytoplankton productivity in a subalpine lake. *Ecology, 62*, 345–354.

Baattrup-Pedersen, A., & Riis, T. (1999). Macrophyte diversity and composition in relation to substratum characteristics in regulated and unregulated Danish streams. *Freshwater Biology, 42*, 375–385.

Bachmann, R. W., Hoyer, M. V., & Canfield, D. E., Jr. (1999). The restoration of Lake Apopka in relation to alternative stable states. *Hydrobiologia, 394*, 219–232.

Bachmann, R. W., Jones, B. L., Fox, D. D., Hoyer, M., Bull, L. A., & Canfield, D. E., Jr. (1996). Relations between trophic state indicators and fish in Florida (U.S.A.) lakes. *Canadian Journal of Fisheries and Aquatic Sciences, 53*, 842–855.

Bagla, P., & Kaiser, J. (1996). India's spreading health crisis draws global arsenic experts. *Science, 274*, 174–175.

Bahr, M., Hobbie, J. E., & Sogin, M. L. (1996). Bacterial diversity in an arctic lake: a freshwater SAR1 cluster. *Aquatic Microbial Ecology, 11*, 271–277.

Baines, S. B., Webster, K. E., Kratz, T. K., Carpenter, S. R., & Magnuson, J. J. (2000). Synchronous behavior of temperature, calcium, and chlorophyll in lakes of northern Wisconsin. *Ecology, 81*, 815–825.

Baker, J. P., Van Sickle, J., Gagen, C. J., DeWalle, D. R., Sharpe, W. E., Carline, R. F., et al. (1996). Episodic acidification of small streams in the northeastern United States: effects on fish populations. *Ecological Applications, 6*, 422–437.

Baker, P. D., & Humpage, A. R. (1994). Toxicity associated with commonly occurring cyanobacteria in surface waters of the Murray-Darling Basin, Australia. *Australian Journal of Freshwater Research, 45*, 773–786.

Balkwill, D. L., & Boone, D. R. (1997). Identity and diversity of microorganisms cultured from subsurface environments. In P. S. Amy & D. L. Haldeman (Eds.), *Microbiology of the terrestrial deep subsurface* (pp. 105–117). Boca Raton, FL: Lewis Publishers.

Balvanera, P., Pfisterer, A. B., Buchmann, N., He, J. S., Nakashizuka, T., Raffaelli, D., et al. (2006). Quantifying the evidence for biodiversity effects on ecosystem functioning and services. *Ecology Letters, 9*, 1146–1156.

Bancroft, B. A., Baker, N. J., & Blaustein, A. R. (2007). Effects of UVB radiation on marine and freshwater organisms: a synthesis through meta-analysis. *Ecology Letters, 10*, 332–345.

Bancroft, B. A., Baker, N. J., & Blaustein, A. R. (2008). A meta-analysis of the effects of ultraviolet B radiation and its synergistic interactions with pH, contaminants, and disease on amphibian survival. *Conservation Biology, 22*, 987–996.

Banner, E., Stahl, A., & Dodds, W. (2009). Stream discharge and riparian land use influence in-stream concentrations and loads of phosphorus from central plains watersheds. *Environmental Management, 44*, 552–565.

Barbour, M. T., Gerritsen, J., Snyder, B. D., & Stribling, J. B. (1999). *Rapid bioassessment protocols for use in wadeable streams and rivers: Periphyton, benthic macroinvertebrates, and fish.* EPA 841-B-99-002 (2nd ed.). Washington DC: Office of Water, United States Environmental Protection Agency.

Barel, C. D. N., Dorit, R., Greenwood, P. H., Fryer, G., Hughes, N., Jackson, P. B. N., et al. (1985). Destruction of fisheries in Africa's lakes. *Nature, 315*, 19–20.

Barnett, T. P., Pierce, D. W., Hidalgo, H. G., Bonfils, C., Santer, B. D., Das, T., et al. (2008). Human-induced changes in the hydrology of the western United States. *Science, 319*, 1080–1083.

Baron, J. S., & Campbell, D. H. (1997). Nitrogen fluxes in a high elevation Colorado Rocky Mountain basin. *Hydrological Processes, 11*, 783–799.

Barrett, P. R. F., Curnow, J. C., & Littlejohn, J. W. (1996). The control of diatom and cyanobacterial blooms in reservoirs using barley straw. *Hydrobiologia, 340*, 307–311.

Barrett, P. R. F., Littlejohn, J. W., & Curnow, J. (1999). Long-term algal control in a reservoir using barley straw. *Hydrobiologia, 415*, 309–313.

Bartley, D. M. (1996). Precautionary approach to the introduction and transfer of aquatic species. In *Technical consultation on the precautionary approach to capture fisheries (including species introductions)* (pp. 159–189). Lysekil, Sweden: Food and Agriculture Organization of the United Nations.

Bass, D. (1992). Colonization and succession of benthic macroinvertebrates in Arcadia Lake, a South-Central USA reservoir. *Hydrobiologia, 242*, 123–131.

Batt, D. J., Anderson, M. G., Anderson, C. D., & Caswell, F. D. (1989). The use of prairie potholes by North American ducks. In A. van der Valk (Ed.), *Northern prairie wetlands* (pp. 204–227). Ames, IA: Iowa State University Press.

Battaglia, L. L., Fore, S. A., & Sharitz, R. R. (2000). Seedling emergence, survival and size in relation to light and water availability in two bottomland hardwood species. *Journal of Ecology, 88*, 1041–1050.

Battin, T. J., & Sengschmitt, D. (1999). Linking sediment biofilms, hydrodynamics, and river bed clogging: evidence from a large river. *Microbial Ecology, 37*, 185–196.

Batzer, D. P., & Resh, V. H. (1991). Trophic interactions among a beetle predator, a chironomid grazer, and periphyton in a seasonal wetland. *Oikos, 60*, 251–257.

Batzer, D. P., & Sharitz, R. R. (2006). *Ecology of freshwater and estuarine wetlands*. Berkeley: University of California Press.

Batzer, D. P., & Wissinger, S. A. (1996). Ecology of insect communities in nontidal wetlands. *Annual Review of Entomology, 41*, 75–100.

Baxter, R. M. (1977). Environmental effects of dams and impoundments. *Annual Review of Ecology and Systematics, 8*, 255–283.

Bayley, I. A. E. (1972). Salinity tolerance and osmotic behavior of animals in a thalassic saline and marine hypersaline waters. *Annual Review of Ecology and Systematics, 3*, 233–268.

Bebout, B. M., & Garcia-Pichel, F. (1995). UVB-induced vertical migrations of cyanobacteria in a microbial mat. *Applied and Environmental Microbiology, 61*, 4215–4222.

Bedford, B. L., Walbridge, M. R., & Aldous, A. (1999). Patterns in nutrient availability and plant diversity of temperate North American wetlands. *Ecology, 80*, 2151–2169.

Behrenfeld, M. J., Lean, D. R. S., & Lee, I. H. (1995). Ultraviolet-B radiation effects on inorganic nitrogen uptake by natural assemblages of oceanic plankton. *Journal of Phycology, 31*, 25–36.

Belk, D. (1984). Patterns in Anostracan distribution. In S. Jain & P. Moyle (Eds.), *Vernal pools and intermittent streams* (pp. 168–172). Davis, CA: Institute of Ecology, University of California.

Belt, D. (1992). The world's great lake. *National Geographic, 181*, 2–39.

Bencala, K. E., & Walters, R. A. (1983). Simulation of solute transport in a mountain pool-and-riffle stream—a transient storage model. *Water Resources Research, 19*, 718–724.

Benda, L., Poff, N. L., Miller, D., Dunne, T., Reeves, G., Pess, G., et al. (2004). The network dynamics hypothesis: how channel networks structure riverine habitats. *BioScience, 54*, 413–427.

Benfield, E. F. (2006). Decomposition of leaf material. In F. R. Hauer & G. A. Lamberti (Eds.), *Methods in stream ecology* (pp. 711–720). San Diego: Academic Press.

Benke, A. C. (1984). Secondary production of aquatic insects. In V. H. Resh & D. M. Rosenberg (Eds.), *The ecology of aquatic insects* (pp. 289–322). New York: Praeger.

Benke, A. C. (1993). Concepts and patterns of invertebrate production in running waters. *Internationalen vereinigung für theoretische und angewandte Limnologie. Verhandlungen, 25*, 15–38.

Benke, A. C., & Huryn, A. D. (2006). Secondary production of macroinvertebrates. In F. R. Hauer & G. A. Lamberti (Eds.), *Methods in stream ecology* (pp. 691–710). London, UK: Academic Press.

Benke, A. C., Van Arsdall, T. C, Jr., Gillespie, D. M., & Parrish, F. K. (1984). Invertebrate productivity in a subtropical blackwater river: the importance of habitat and life history. *Ecological Monographs, 54*, 25–63.

Benke, A. C., & Wallace, J. B. (1980). Trophic basis of production among net-spinning caddisflies in a southern Appalachian stream. *Ecology, 61*, 108–118.

Benner, R., Lewis, D. L., & Hodson, R. E. (1989). Biogeochemical cycling of organic matter in acidic environments: are microbial degradative processes adapted to low pH? In S. S. Rao (Ed.), *Acid stress and aquatic microbial interactions* (pp. 34–43). Boca Raton, FL: CRC Press.

Bennett, A. M. R. (2008). Aquatic hymenoptera. In R. W. Merritt, K. W. Cummins, & M. B. Berg (Eds.), *An introduction to the aquatic insects of North America* (pp. 673–686). Dubuque, IA: Kendall-Hunt.

Bennett, E. M., Reed-Andersen, T., Houser, J. N., Gabriel, J. R., & Carpenter, S. R. (1999). A phosphorus budget for the Lake Mendota watershed. *Ecosystems, 2*, 69–75.

Bennett, P. C., Rogers, J. R., & Choi, W. J. (2001). Silicates, silicate weathering, and microbial ecology. *Geomicrobiology Journal, 18*, 3–19.

Benson, A. J. (2000). Documenting over a century of aquatic introductions in the United States. In R. Claudi & J. H. Leach (Eds.), *Nonindigenous freshwater organisms* (pp. 1–31). Boca Raton, FL: Lewis Publishers, CRC Press.

Benstead, J. P., March, J. G., Pringle, C. M., & Scatena, F. N. (1999). Effects of a low-head dam and water abstraction on migratory tropical stream biota. *Ecological Applications, 9*, 656–668.

Bergeron, M., & Vincent, W. F. (1997). Microbial food web responses to phosphorus supply and solar UV radiation in a sub-arctic lake. *Aquatic Microbial Ecology, 12*, 239–249.

Bergman, E., Hansson, L.-A., Persson, A., Strand, J., Romare, P., Enell, M., et al. (1999). Synthesis of theoretical and empirical experiences from nutrient and cyprinid reductions in Lake Ringsjön. *Hydrobiologia, 404*, 145–156.

Berman, T., & Chava, S. (1999). Algal growth on organic compounds as nitrogen sources. *Journal of Plankton Research, 21*, 1423–1437.

Berner, E. K., & Berner, R. A. (1987). *The global water cycle*. Englewood Cliffs, NJ: Prentice Hall.

Bernert, J. A., Eilers, J. M., Elers, B. J., Blok, E., Daggett, S. G., & Bierly, K. F. (1999). Recent wetlands trends (1981/82-1994) in the Willamette Valley, Oregon, USA. *Wetlands, 19*, 545–559.

Bernhardt, E. S., & Palmer, M. A. (2007). Restoring streams in an urbanizing world. *Freshwater Biology, 52*, 738.

Bernhardt, E. S., Palmer, M. A., Allan, J. D., Alexander, G., Barnas, K., Brooks, S., et al. (2005). Synthesizing US river restoration efforts. *Science, 308*, 636–637.

Bernot, R. J., Dodds, W. K., Quist, M. C., & Guy, C. S. (2006). Temperature and kairomone induced life history plasticity in coexisting *Daphnia*. *Aquatic Ecology, 40*, 361–372.

Bernot, R. J., & Lamberti, G. A. (2008). Indirect effects of a parasite on a benthic community: an experiment with trematodes, snails and periphyton. *Freshwater Biology, 53*, 322–329.

Bertness, M. D., & Callaway, R. (1994). Positive interactions in communities. *Trends in Ecology and Evolution, 9*, 191–193.

Bertness, M. D., & Hacker, S. D. (1994). Physical stress and positive associations among marsh plants. *The American Naturalist, 144*, 363–372.

Bertness, M. D., & Leonard, G. H. (1997). The role of positive interactions in communities: lessons from intertidal habitats. *Ecology, 78*, 1976–1989.

Bertrand, K., Gido, K., Dodds, W., Murdock, J., & Whiles, M. (2009). Disturbance frequency and functional identity mediate ecosystem processes in prairie streams. *Oikos, 118*, 917.

Beschta, R. L., Bilby, R. E., Brown, G. W., Holtby, L. B., & Hofstra, T. D. (1987). Stream temperature and aquatic habitat: fisheries and forestry interactions. In E. O. Salo & T. W. Cundy (Eds.), *Streamside management: Forestry and fishery interactions* (pp. 191–232). Seattle: University of Washington, Institute of Forest Resources.

Beveridge, M. C. M., Ross, L. G., & Kelly, L. A. (1994). Aquaculture and biodiversity. *Ambio, 23*, 497–502.

Bidigare, R. R., Ondrusek, M. E., Kennicutt, M. C., II, Iturriaga, R., Harvey, H. R., Hoham, R. W., et al. (1993). Evidence for a photoprotective function for secondary carotenoids of snow algae. *Journal of Phycology, 29*, 427–434.

Bierman, V. J, Jr., & James, R. T. (1995). A preliminary modeling analysis of water quality in Lake Okeechobee, Florida: diagnostic and sensitivity analyses. *Water Research, 29*, 2767–2775.

Biggs, B. J. F. (1995). The contribution of flood disturbance, catchment geology and land use to the habitat template of periphyton in stream ecosystems. *Freshwater Biology, 33*, 419–438.

Biggs, B. J. F. (2000). Eutrophication of streams and rivers: dissolved nutrient-chlorophyll relationships for benthic algae. *Journal of the North American Benthological Society, 19*, 17–31.

Bilby, R. E., Fransen, B. R., & Bisson, P. A. (1996). Incorporation of nitrogen and carbon from spawning Coho salmon into the trophic system of small streams: evidence from stable isotopes. *Canadian Journal of Fisheries and Aquatic Sciences, 53*, 164–173.

Billen, G. (1991). Protein degradation in aquatic environments. In R. J. Chróst (Ed.), *Microbial enzymes in aquatic environments* (pp. 123–143). New York: Springer-Verlag.

Bitton, G. (1994). *Wastewater microbiology*. New York: John Wiley & Sons.

Blackburn, N., & Fenchel, T. (1999). Influence of bacteria, diffusion and shear on micro-scale nutrient patches, and implications for bacterial chemotaxis. *Marine Ecology Progress Series, 189*, 1–7.

Blakemore, R. P. (1982). Magnetotactic bacteria. *Annual Review of Microbiology, 36*, 217–238.

Blaustein, A. R., Edmond, B., Kiesecker, J. M., Beatty, J. J., & Hokit, D. G. (1995). Ambient ultraviolet radiation causes mortality in salamander eggs. *Ecological Applications, 5*, 740–743.

Bloom, A. A., Palmer, P. I., Fraser, A., Reay, D. S., & Frakenberg, C. (2010). Large-scale controls of methanogenesis inferred from methane and gravity spaceborne data. *Science, 327*, 322–325.

Blossey, B., Skinner, L., & Taylor, J. (2001). Impact and management of purple loosestrife (*Lythrum salicaria*) in North America. *Biodiversity and Conservation, 10*, 1787–1807.

Blus, J. B., & Henny, C. J. (1997). Field studies on pesticides and birds: unexpected and unique relations. *Ecological Applications, 7*, 1125–1132.

Bockstael, N. E., Freeman, A. M., III, Kopp, R. J., Portney, P. R., & Smith, V. K. (2000). On measuring economic values for nature. *Environmental Science and Technology, 34*, 1384–1389.

Boeye, D., Verhagen, B., Haesebroeck, V. V., & Verheyen, R. F. (1997). Nutrient limitation in species-rich lowland fens. *Journal of Vegetation Science, 8*, 415–424.

Bohannan, B. J. M., & Lenski, R. E. (1997). Effect of resource enrichment on a chemostat community of bacteria and bacteriophage. *Ecology, 78*, 2303–2315.

Böhme, H. (1998). Regulation of nitrogen fixation in heterocyst-forming cyanobacteria. *Trends in Plant Science, 3*, 346–351.

Bolnick, D. (2004). Can intraspecific competition drive disruptive selection? An experimental test in natural populations of sticklebacks. *Evolution, 58*, 608–618.

Boon, P. I. (1992). Antibiotic resistance of aquatic bacteria and its implications for limnological research. *Australian Journal of Marine and Freshwater Research, 43*, 847–859.

Borman, S., Korth, R., & Temte, J. (1997). *Through the looking glass: A field guide to aquatic plants*. Merrill, WI: Wisconsin Lakes Partnership.

Bormans, M., Sherman, B. S., & Webster, I. T. (1999). Is buoyancy regulation in cyanobacteria an adaptation to exploit separation of light and nutrients? *Marine and Freshwater Research, 50*, 897–906.

Bornette, G., Amoros, C., & Lamouroux, N. (1998a). Aquatic plant diversity in riverine wetlands: the role of connectivity. *Freshwater Biology, 39*, 267–283.

Bornette, G., Amoros, C., Piegay, H., Tachet, J., & Hein, T. (1998b). Ecological complexity of wetlands within a river landscape. *Biological Conservation, 85*, 35–45.

Borror, D., Triplehorn, C., & Johnson, N. (1989). *An introduction to the study of insects*. Philadelphia, PA: Saunders College Publishing.

Bothe, H. (1982). Nitrogen fixation. In N. G. Carr & B. A. Whitton (Eds.), *The biology of cyanobacteria* (pp. 87–104). Berkeley, CA: University of California Press.

Bothwell, M. L., Sherbot, D. M. J., & Pollock, C. M. (1994). Ecosystem response to solar ultraviolet-B radiation: influence of trophic-level interactions. *Science, 265*, 97–100.

Bott, T., Meyer, G. A., & Young, E. B. (2008). Nutrient limitation and morphological plasticity of the carnivorous pitcher plant *Sarracenia purpurea* in contrasting wetland environments. *New Phytologist, 180*, 631–641.

Bott, T. L. (1995). Microbes in food webs. *American Society of Microbiology News, 61*, 580–585.

Bott, T. L. (2006). Primary productivity and community respiration. In F. H. Hauer & G. A. Lamberti (Eds.), *Methods in stream ecology* (pp. 663–690). Boston: Academic Press.

Bott, T. L., Brock, J. T., Baattrup-Pedersen, A., Chambers, P. A., Dodds, W. K., Himbeault, K. T., et al. (1997). An evaluation of techniques for measuring periphyton metabolism in chambers. *Canadian Journal of Fisheries and Aquatic Sciences, 54*, 715–725.

Bott, T. L., Brock, J. T., Dunn, C. S., Naiman, R. J., Ovink, R. W., & Petersen, R. C. (1985). Benthic community metabolism in four temperate stream systems: an inter-biome comparison and evaluation of the river continuum concept. *Hydrobiologia, 123*, 3–45.

Boulton, A. (2000). The subsurface macrofauna. In J. B. Jones & P. J. Mulholland (Eds.), *Streams and ground waters* (pp. 337–361). San Diego, CA: Academic Press.

Boulton, A. J., Peterson, C. G., Grimm, N. B., & Fisher, S. G. (1992). Stability of an aquatic macroinvertebrate community in a multiyear hydrologic disturbance regime. *Ecology, 73*, 2192–2207.

Bowden, W. B. (1999). Roles of bryophytes in stream ecosystems. *Journal of the North American Benthological Society, 18*, 151–184.

Bowden, W. B., Finlay, J. C., & Maloney, P. E. (1994). Long-term effects of PO_4 fertilization on the distribution of bryophytes in an arctic river. *Freshwater Biology, 32*, 445–454.

Bowen, R. (1986). *Groundwater* (2nd ed.). New York: Elsevier Applied Science Publishers.

Bradshaw, W. E., & Creelman, R. A. (1984). Mutualism between the carnivorous purple pitcher plant and its inhabitants. *American Midland Naturalist, 112*, 294–304.

Bratbak, G., Thingstad, F., & Heldal, M. (1994). Viruses and the microbial loop. *Microbial Ecology, 28*, 209–221.

Brem, F. M. R., & Lips, K. R. (2008). *Batrachochytrium dendrobatidis* infection patterns among Panamanian amphibian species, habitats and elevations during epizootic and enzootic. *Diseases of Aquatic Organisms, 81*, 189–202.

Brenner, M., Whitmore, T. J., Lasi, M. A., Cable, J. E., & Cable, P. H. (1999). A multi-proxy trophic state reconstruction for shallow Orange Lake, Florida, USA: possible influence of macrophytes on limnetic nutrient concentrations. *Journal of Paleolimnology, 21*, 215–233.

Brett, M. T., & Goldman, C. R. (1997). Consumer versus resource control in freshwater. *Science, 275*, 384–386.

Brett, M. T., & Müller-Navarra, D. C. (1997). The role of highly unsaturated fatty acids in aquatic foodweb processes. *Freshwater Biology, 38*, 483–499.

Brewer, M. C., Dawidowicz, P., & Dodson, S. I. (1999). Interactive effects of fish kairomone and light on *Daphnia* escape behavior. *Journal of Plankton Research, 21*, 1317–1335.

Brezonik, P. L. (1994). *Chemical kinetics and process dynamics in aquatic systems*. Boca Raton, FL: CRC Press, Inc.

Briand, F. (1985). Structural singularities of freshwater food webs. *Verhandlungen—Internationale Vereinigung fuer Theoretische und Angewandte Limnologie, 22*, 3356–3364.

Briand, F., & Cohen, J. E. (1990). Environmental correlates of food chain length. In J. E. Cohen, F. Briand, & C. M. Newman (Eds.), *Community food webs data and theory* (pp. 55–62). Berlin: Springer-Verlag.

Bridgham, S. D., Updegraff, K., & Pastor, J. (1998). Carbon, nitrogen, and phosphorus mineralization in northern wetlands. *Ecology, 79*, 1545–1561.

Briggs, J. C. (1986). Introduction to the zoogeography of North American fishes. In C. H. Hocutt & E. O. Wiley (Eds.), *The zoogeography of North American freshwater fishes* (pp. 1–16). New York: John Wiley & Sons.

Brinkhurst, R. O., Chua, K. E., & Kaushik, N. K. (1972). Interspecific interactions and selective feeding by tubificid oligo-chaetes. *Limnology and Oceanography, 17*, 122–133.

Brinkhurst, R. O., & Gelder, S. R. (1991). Annelida: Oligochaeta and Branchiobdellida. In J. H. Thorp & A. P. Covich (Eds.), *Ecology and classification of North American freshwater invertebrates* (pp. 401–436). San Diego, CA: Academic Press, Inc.

Brinson, M. M., Kruczynski, W., Lee, L. C., Nutter, W. L., Smith, R. D., & Whigham, D. F. (1994). Developing an approach for assessing the functions of wetlands. In W. J. Mitsch (Ed.), *Global wetlands old world and new* (pp. 615–624). Amsterdam, The Netherlands: Elsevier.

Brix, H. (1994). Constructed wetlands for municipal wastewater treatment in Europe. In W. J. Mitsch (Ed.), *Global wetlands old world and new* (pp. 325–333). Amsterdam, The Netherlands: Elsevier.

Brock, E. M. (1960). Mutualism between the midge *Cricotopus* and the alga *Nostoc. Ecology, 41*, 474–483.

Brock, T. D. (1978). *Thermophilic microorganisms and life at high temperatures*. New York: Springer-Verlag.

Brönmark, C. (1985). Interactions between macrophytes, epiphytes and herbivores: an experimental approach. *Oikos, 45*, 26–30.

Brönmark, C., & Hansson, L.-A. (1998). *The biology of lakes and ponds*. New York: Oxford University Press, Inc.

Brönmark, C., & Hansson, L.-A. (2000). Chemical communication in aquatic systems: an introduction. *Oikos, 88,* 103–109.

Brönmark, C., Pettersson, L. B., & Nilsson, P. A. (1999). Predator-induced defense in Crucian Carp. In E. Tollrian & C. D. Harvell (Eds.), *The ecology and evolution of inducible defenses* (pp. 203–217). Princeton, NJ: Princeton University Press.

Brönmark, C., Rundle, S. D., & Erlandsson, A. (1991). Interactions between freshwater snails and tadpoles: competition and facilitation. *Oecologia, 87,* 8–18.

Brönmark, C. S., Klosiewski, P., & Stein, R. A. (1992). Indirect effects of predation in a freshwater, benthic food chain. *Ecology, 73,* 1662–1674.

Brooker, R., Maestre, F., Callaway, R., Lortie, C., Cavieres, L., Kunstler, G., et al. (2008). Facilitation in plant communities: the past, the present, and the future. *Journal of Ecology, 96,* 18–34.

Brooks, J. L. (1946). Cyclomorphosis in *Daphnia*. *Ecological Monographs, 16,* 409–447.

Brooks, J. L. (1950). Speciation in ancient lakes. *The Quarterly Review of Biology, 25*(30–60), 131–176.

Brooks, J. L., & Dodson, S. I. (1965). Predation, body size, and composition of plankton. *Science, 150,* 28–35.

Brooks, R. P., Wardrop, D. H., Cole, C. A., & Campbell, D. A. (2005). Are we purveyors of wetland homogeneity? A model of degradation and restoration to improve wetland mitigation performance. *Ecological Engineering, 24,* 331–340.

Browder, J. A., Gleason, P. J., & Swift, D. R. (1994). Periphyton in the Everglades: spatial variation, environmental correlates, and ecological implications. In S. M. Davis & J. C. Ogden (Eds.), *Everglades, the ecosystem and its restoration* (pp. 379–418). Delray Beach, FL: St. Lucie Press.

Brown, D. J., & Coon, T. G. (1991). Grass carp larvae in the lower Missouri River and its tributaries. *North American Journal of Fisheries Management, 11,* 62–66.

Brown, K. M. (1991). Mollusca: Gastropoda. In J. H. Thorp & A. P. Covich (Eds.), *Ecology and classification of North American freshwater invertebrates* (pp. 285–314). San Diego, CA: Academic Press, Inc.

Brown, L. R. (1995). Nature's limits. In *State of the world. A Worldwatch Institute Report on Progress toward a Sustainable Society.* New York: W. W. Norton & Co. (pp. 3–20).

Brown, L. R. (1996). The acceleration of history. In: *State of the World. A Worldwatch Institute Report on Progress toward a Sustainable Society* (pp. 3–20). New York: W. W. Norton & Co.

Brown, T. C. (2000). Projecting U.S. freshwater withdrawals. *Journal of Water Resources Research, 36,* 769–780.

Brunke, M., & Gonser, T. (1997). The ecological significance of exchange processes between rivers and groundwater. *Freshwater Biology, 37,* 1–33.

Budavari, S., O'Neil, M. J., Smith, A., & Heckelman, P. E. (1989). *The Merck Index* (11th ed.). Rahway, NJ: Merck and Co.

Burgherr, P., & Ward, J. (2001). Longitudinal and seasonal distribution patterns of the benthic fauna of an alpine glacial stream (Val Roseg, Swiss Alps). *Freshwater Biology, 46,* 1705–1721.

Burgin, A. J., & Hamilton, S. K. (2007). Have we overemphasized the role of denitrification in aquatic ecosystems? A review of nitrate removal pathways. *Frontiers in Ecology and Environment, 5,* 89–96.

Burkholder, J. M. (1996). Interactions of benthic algae with their substrata. In R. J. Stevenson, M. L. Bothwell, & R. L. Lowe (Eds.), *Algal ecology* (pp. 253–297). San Diego, CA: Academic Press.

Burkholder, J. M., & Glasgow, H. B., Jr. (1997). *Pfiesteria piscicida* and other *Pfiesteria*-like dinoflagellates: behavior, impacts, and environmental controls. *Limnology and Oceanography, 42,* 1052–1075.

Burns, A., & Walker, K. F. (2000). Effects of water level regulation on algal biofilms in the River Murray, South Australia. *Regulated Rivers: Research & Management, 16,* 433–444.

Burns, C. W. (1969). Particle size and sedimentation in the feeding behavior of two species of *Daphnia*. *Limnology and Oceanography, 14,* 392–402.

Burns, N. M. (1985). *Erie: The lake that survived.* Totowa, NJ: Rowman & Allanheld.

Bury, N. R., Eddy, F. B., & Codd, G. A. (1995). The effects of the cyanobacterium *Microcystis aeruginosa*, the cyanobacterial toxin microcystin-LR, and ammonia on growth rate and ionic regulation of brown trout. *Journal of Fisheries Biology, 46,* 1042–1054.

Butler, J. N. (1991). *Carbon dioxide equilibria and their applications.* Chelsea, MI: Lewis Publishers Inc.

Butterworth, A. E. (1988). Control of schistosomiasis in man. In P. T. Englund & A. Sher (Eds.), *The biology of parasitism. A molecular and immunological approach* (pp. 43–59). New York: Alan R. Liss, Inc.

Butturini, A., & Sabater, F. (1998). Ammonium and phosphate retention in a Mediterranean stream: hydrological versus temperature control. *Canadian Journal of Fisheries and Aquatic Sciences, 55,* 1938–1945.

Buxton, H. T., & Kolpin, D. W. (2002). *Pharmaceuticals, hormones, and other organic wastewater contaminants in US streams* US Geological Survey Fact Sheet FS-027-02. United States Geological Survey.

Cairns, J., Jr. (1982). Freshwater protozoan communities. In A. T. Bull & A. R. K. Watkinson (Eds.), *Microbial interactions and communities* (pp. 249–285). London: Academic Press.

Cairns, J., Jr. (1993). Can microbial species with a cosmopolitan distribution become extinct? *Speculations in Science and Technology, 16,* 69–73.

Cairns, M. A., & Lackey, R. T. (1992). Biodiversity and management of natural resources: the issues. *Fisheries, 17,* 6–18.

Caissie, D. (2006). The thermal regime of rivers: a review. *Freshwater Biology, 51,* 1389–1406.

Calabrese, E. J., & Baldwin, L. A. (1999). Reevaluation of the fundamental dose-response relationship. *BioScience, 49,* 725–732.

Callander, R. A. (1978). River meandering. *Annual Review of Fluid Mechanics, 10,* 129–158.

Callaway, R. M. (1995). Positive interactions among plants. *The Botanical Review, 61,* 306–348.

Callaway, R. M., & King, L. (1996). Temperature-driven variation in substrate oxygenation and the balance of competition and facilitation. *Ecology, 77,* 1189–1195.

Callaway, R. M., & Walker, L. R. (1997). Competition and facilitation: a synthetic approach to interactions in plant communities. *Ecology, 78,* 1958–1965.

Camerano, L. (1994). On the equilibrium of living beings by means of reciprocal destruction. In S. A. Levin (Ed.), *Frontiers in mathematical biology* (pp. 360–379). Berlin: Springer-Verlag.

Canter-Lund, H., & Lund, J. W. G. (1995). *Freshwater algae: Their microscopic world explored.* Bristol, UK: Biopress Ltd.

Caraco, N., & Cole, J. (1999). Regional-scale export of C, N, P, and sediment: what river data tell us about key controlling variables. In J. D. Tenhunen & P. Kabat (Eds.), *Integrating hydrology, ecosystem dynamics, and biogeochemistry in complex landscapes* (pp. 239–254). New York: John Wiley & Sons Ltd.

Caraco, N. F., Cole, J. J., Findlay, S. E. G., Fischer, D. T., Lampman, G. G., Pace, M. L., et al. (2000). Dissolved oxygen declines in the Hudson River associated with the invasion of the zebra mussel (*Dreissena polymorpha*). *Environmental Science and Technology, 34,* 1204–1210.

Carder, J. P., & Hoagland, K. D. (1998). Combined effects of alachlor and atrazine on benthic algal communities in artificial streams. *Environmental Toxicology and Chemistry, 17,* 1415–1420.

Cardinale, B. J., Palmer, M. A., & Collins, S. L. (2002). Species diversity enhances ecosystem functioning through interspecific facilitation. *Nature, 415,* 426–429.

Cardinale, B. J., Srivastava, D. S., Duffy, J. E., Wright, J. P., Downing, A. L., Sankaran, M., et al. (2006). Effects of biodiversity on the functioning of trophic groups and ecosystems. *Nature, 443,* 989.

Carignan, R., Blais, A.-M., & Vis, C. (1998). Measurement of primary production and community respiration in oligotrophic lakes using the Winkler method. *Canadian Journal of Fisheries and Aquatic Sciences, 55,* 1078–1084.

Carlsson, N., Bronmark, C., & Hansson, L. (2004). Invading herbivory: the golden apple snail alters ecosystem functioning in Asian wetlands. *Ecology, 85,* 1575–1580.

Carlton, J. T. (1993). Dispersal mechanisms of the zebra mussel (*Dreissena polymorpha*). In T. F. Nalepa & D. W. Schloesser (Eds.), *Zebra mussels: Biology, impacts and control* (pp. 677–697). Boca Raton, FL: Lewis Publishers.

Carmichael, W. W. (1994). The toxins of cyanobacteria. *Scientific American, 270,* 78–86.

Carmichael, W. W. (1997). The cyanotoxins. *Advances in Botanical Research, 27,* 211–240.

Carney, H. J. (1990). A general hypothesis for the strength of food web interactions in relation to trophic state. *Verhandlungen— Internationale Vereinigung fuer Theoretische und Angewandte Limnologie, 24,* 487–492.

Carpenter, S., & Brock, W. (2006). Rising variance: a leading indicator of ecological transition. *Ecology Letters, 9,* 311–318.

Carpenter, S. R. (1989). Replication and treatment strength in whole-lake experiments. *Ecology, 70,* 453–462.

Carpenter, S. R., Bolgrien, D., Lathrop, R. C., Stow, C. A., Reed, T., & Wilson, M. A. (1998). Ecological and economic analysis of lake eutrophication by nonpoint pollution. *Australian Journal of Ecology, 23,* 68–79.

Carpenter, S. R., Brock, W. A., Cole, J. J., Kitchell, J. F., & Pace, M. L. (2008). Leading indicators of trophic cascades. *Ecology Letters, 11*, 128–138.

Carpenter, S. R., Chisholm, S. W., Krebs, C. J., Schindler, D. W., & Wright, R. F. (1995). Ecosystem experiments. *Science, 269*, 324–327.

Carpenter, S. R., Cole, J. J., Hodgson, J. R., Kitchell, J. F., Pace, M. L., Bade, D., et al. (2001). Trophic cascades, nutrients, and lake productivity: whole-lake experiments. *Ecological Monographs, 71*, 163–186.

Carpenter, S. R., Cole, J. J., Pace, M. L., Bogert, M. V.d., Bade, D. L., Bastviken, D., et al. (2005). Ecosystem subsidies: terrestrial support of aquatic food webs from ^{13}C addition to contrasting lakes. *Ecology, 86*, 2737–2750.

Carpenter, S. R., Frost, T. M., Heisey, D., & Kratz, K. K. (1989). Randomized intervention analysis and the interpretation of whole-ecosystem experiments. *Ecology, 70*, 1142–1152.

Carpenter, S. R., & Kitchell, J. F. (1987). The temporal scale of variance in limnetic primary production. *American Naturalist, 129*, 417–433.

Carpenter, S. R., Kitchell, J. F., Hodgson, J. R., Cochran, P. A., Elser, J. J., Elser, M. M., et al. (1987). Regulation of lake primary productivity by food web structure. *Ecology, 68*, 1863–1876.

Carrillo, P., Delgado-Molina, J. A., Medina-Sanchez, J. M., Bullejos, F. J., & Villar-Argaiz, M. (2008). Phosphorus inputs unmask negative effects of ultraviolet radiation on algae in a high mountain lake. *Global Change Biology, 14*, 423–439.

Carson, R. (1962). *Silent spring*. New York: Houghton Mifflin Company.

Castenholz, R. W. (1984). Composition of hot springs microbial mats: a summary. In Y. Cohen, R. W. Castenholz, & H. O. Halvorson (Eds.), *Ancient stromatolites and microbial mats* (pp. 101–119). New York: Alan R. Liss, Inc.

Castenholz, R. W. (2004). Phototrophic bacteria under UV stress. In J. Seckbach (Ed.), *Origins: Genesis, evolution and diversity of life*. Kluwer Academic Publ. Dordrecht.

Chambers, C. P., Whiles, M. R., Rosi-Marshall, E. J., Tank, J. L., Royer, T. V., Griffiths, N. A. (In press). Responses of stream macroinvertebrates to Bt maize leaf detritus. *Ecological Applications*.

Chambers, P. A., DeWreede, R. E., Irlandi, E. A., & Vandermeulen, H. (1999). Management issues in aquatic macrophyte ecology: a Canadian perspective. *Canadian Journal of Botany, 77*, 471–487.

Chapnick, S. D., Moore, W. S., & Nealson, K. H. (1982). Microbially mediated manganese oxidation in a freshwater lake. *Limnology and Oceanography, 27*, 1004–1015.

Characklis, W. G., McFeters, G. A., & Marshall, K. C. (1990). Physiological ecology in biofilm systems. In W. G. Characklis & K. C. Marshall (Eds.), *Biofilms* (pp. 341–394). New York: John Wiley & Sons, Inc.

Chen, J., & Xie, P. (2008). Accumulation of hepatotoxic microcystins in freshwater mussels, aquatic insect larvae and oligochaetes in a large, shallow eutrophic lake (Lake Chaohu) of subtropical China. *Fresenius Environmental Bulletin, 17*, 849–854.

Chick, J. H., Cosgriff, R. J., & Gittinger, L. S. (2003). Fish as potential dispersal agents for floodplain plants: first evidence in North America. *Canadian Journal of Fisheries and Aquatic Sciences, 60*, 1437–1439.

Chivian, D., Brodie, E., Alm, E., Culley, D., Dehal, P., DeSantis, T., et al. (2008). Environmental genomics reveals a single-species ecosystem deep within earth. *Science, 322*, 275.

Christaki, U., Dolan, J. R., Pelegri, S., & Rassoulzadegan, F. (1998). Consumption of picoplankton-size particles by marine ciliates: effects of physiological state of the ciliate and particle quality. *Limnology and Oceanography, 43*, 458–464.

Christensen, D. L., Herwig, B. R., Schindler, D. E., & Carpenter, S. R. (1996). Impacts of lakeshore residential development on coarse woody debris in north temperate lakes. *Ecological Applications, 6*, 1143–1149.

Christensen, M. R., Graham, M. D., Vinebrooke, R. D., Findlay, D. L., Paterson, M. J., & Turner, M. A. (2006). Multiple anthropogenic stressors cause ecological surprises in boreal lakes. *Global Change Biology, 12*, 2316–2322.

Christensen, T. H., Bjerb, P. L., Banwart, S. A., Jakobsen, R., Heron, G., & Albrechtsen, H.-J. (2000). Characterization of redox conditions in groundwater contaminant plumes. *Journal of Contaminant Hydrology, 45*, 165–241.

Chróst, R. J. (1991). Environmental control of the synthesis and activity of aquatic microbial ectoenzymes. In R. J. Chróst (Ed.), *Microbial enzymes in aquatic environments* (pp. 29–59). New York: Springer-Verlag.

Clarke, K. R. (1993). Non-parametric multivariate analyses of changes in community structure. *Australian Journal of Ecology, 18*, 117–143.

Clasen, J., Rast, W., & Ryding, S.-O. (1989). Available techniques for treating eutrophication. In S.-O. Ryding & W. Rast (Eds.), *The control of eutrophication of lakes and reservoirs* (pp. 169–212). Paris, France: UNESCO and the Parthenon Publishing Group.

Cleckner, L. B., Garrison, P. J., Hurley, J. P., Olson, M. L., & Krabbenhoft, D. P. (1998). Trophic transfer of methyl mercury in the northern Florida Everglades. *Biogeochemistry, 40*, 347–361.

Cleckner, L. B., Gilmour, C. C., Hurley, J. P., & Drabbenhoft, D. P. (1999). Mercury methylation in periphyton of the Florida Everglades. *Limnology and Oceanography, 44*, 1815–1825.

Clements, W. H., Brooks, M. L., Kashian, D. R., & Zuellig, R. E. (2008). Changes in dissolved organic material determine exposure of stream benthic communities to UV-B radiation and heavy metals: implications for climate change. *Global Change Biology, 14*, 2201–2214.

Clymo, R. S., & Hayward, P. M. (1982). The ecology of *Sphagnum*. In A. J. E. Smith (Ed.), *Bryophyte ecology* (pp. 229–289). London: Chapman and Hall.

Codd, G. A. (1995). Cyanobacterial toxins: occurrence, properties and biological significance. *Water Science and Technology, 32*, 149–156.

Codd, G. A., Bell, S. G., Kaya, K., Ward, C. J., Beattie, K. A., & Metcalf, J. S. (1999). Cyanobacterial toxins, exposure routes and human health. *European Journal of Phycology, 34*, 405–415.

Codd, G. A., Ward, C. J., Beattie, K. A., & Bell, S. G. (1999). Widening perceptions of the occurrence and significance of cyanobacterial toxins. In G. A. Peschek, W. Löffelhardt, & G. Schmetterer (Eds.), *The phototrophic prokaryotes* (pp. 623–632). New York: Kluwer Academic.

Cogley, J. (1991). GGHYDRO-Global Hydrographic data release 2.0. Trent Climate Note **19**, Department of Geography, Trent University.

Cohen, A. S. (1995). Paleoecological approaches to the conservation biology of benthos in ancient lakes: a case study from Lake Tanganyika. *Journal of the North American Benthological Society, 14*, 654–668.

Cohen, J. E. (1995). Population growth and earth's human carrying capacity. *Science, 269*, 341–346.

Colborn, T., Saal, F. S.v., & Soto, A. M. (1993). Developmental effects of endocrine-disrupting chemicals in wildlife and humans. *Environmental Health Perspectives, 101*, 378–384.

Cole, G. A. (1994). *Textbook of limnology* (4th ed.). Prospect Heights, IL: Waveland Press, Inc.

Cole, J. J., & Caraco, N. F. (2001). Carbon in catchments: connecting terrestrial carbon losses with aquatic metabolism. *Marine and Freshwater Research, 52*, 101–110.

Cole, J. J., Caraco, N. F., Kling, G. W., & Kratz, T. K. (1994). Carbon dioxide supersaturation in the surface waters of lakes. *Science, 265*, 1568–1570.

Cole, J. J., Carpenter, S. R., Pace, M. L., Van de Bogert, M. C., Kitchell, J. L., & Hodgson, J. R. (2006). Differential support of lake food webs by three types of terrestrial organic carbon. *Ecology Letters, 9*, 558–568.

Coles, B., & Coles, J. (1989). *People of the wetlands: Bogs, bodies and lake-dwellers. German Democratic Republic:* Thames and Hudson Publishing Co.

Collins, J., & Crump, M. (2008). *Extinction in our times: Global amphibian decline*. Oxford University Press, Oxford.

Committee on Characterization of Wetlands. (1995). *Commission on geosciences, environment, and resources. Wetlands characteristics and boundaries*. Washington, DC: National Research Council.

Conant, R., & Collins, J. T. (1998). *A field guide to reptiles and amphibians: Eastern and central North America* (3rd ed.). Boston: Houghton Mifflin Company.

Conley, D. J. (2000). Biogeochemical nutrient cycles and nutrient management strategies. *Hydrobiologia, 410*, 87–96.

Connell, J. H. (1978). Diversity in tropical rain forests and coral reefs. *Science, 199*, 1302–1310.

Connell, J. H. (1980). Diversity and the coevolution of competitors, or the ghost of competition past. *Oikos, 35*, 131–138.

Connell, J. H. (1983). On the prevalence and relative importance of interspecific competition: evidence from field experiments. *American Naturalist, 122*, 661–696.

Connelly, S., Pringle, C. M., Bixby, R. J., Brenes, R., Whiles, M. R., Lips, K. R., et al. (2008). Changes in stream primary producer communities resulting from large-scale catastrophic amphibian declines: can small-scale experiments predict effects of tadpole loss? *Ecosystems, 11*, 1262–1276.

Conrad, H. S., & Redfearn, P. L., Jr. (1979). *How to know the mosses and liverworts*. Dubuque, IA: William C. Brown Company Publishers.

Cook, C. D. K. (1996). *Aquatic plant book*. Amsterdam, The Netherlands: SPB Academic Publishing.

Cooke, G. D., Welch, E. B., Peterson, S. A., & Newroth, P. R. (1993). *Restoration and management of lakes and reservoirs* (2nd ed.). Boca Raton, FL: Lewis Publishers.

Cooper, G. P., & Washburn, G. N. (1949). Relation of dissolved oxygen to winter mortality of fish in Michigan lakes. *Transactions of American Fisheries Society, 76*, 23–32.

Correa, S. B., Winemiller, K. O., Lopez-Fernandez, H., & Galetti, M. (2007). Evolutionary perspectives on seed consumption and dispersal by fishes. *BioScience, 57*, 748–756.

Correll, D. L. (1999). Phosphorus: a rate limiting nutrient in surface waters. *Poultry Science, 78*, 674–682.

Costantini, M. L., Sabetta, L., Mancinelli, G., & Rossi, L. (2004). Spatial variability of the decomposition rate of *Schoenoplectus tatora* in a polluted area of Lake Titicaca. *Journal of Tropical Ecology, 20*, 325–335.

Costanza, R., D'Arge, R., Groot, R.D., Farber, S., Grasso, M., Hannon, B., et al. (1997). The value of the world's ecosystem services and natural capital. *Nature, 387*, 253–260.

Costanza, R., & Farley, J. (2007). Ecological economics of coastal disasters: introduction to the special issue. *Ecological Economics, 63*, 249–253.

Cotner, J. B., Gardner, W. S., Johnson, J. R., Sada, R. H., Cavaletto, J. F., & Heath, R. T. (1995). Effects of zebra mussels (*Dreissena polymorpha*) on bacterioplankton evidence for both size-selective consumption and growth stimulation. *Journal of Great Lakes Research, 21*, 517–528.

Cottingham, K. L., & Schindler, D. E. (2000). Effects of grazer community structure on phytoplankton response to nutrient pulses. *Ecology, 81*, 183–200.

Couch, C. A., & Meyer, J. L. (1992). Development and composition of the epixylic biofilm in a blackwater river. *Freshwater Biology, 27*, 43–51.

Coulter, G. W. (1991). The benthic fish community. In G. W. Coulter (Ed.), *Lake Tanganyika and its life* (pp. 151–199). New York: Oxford University Press.

Coulter, G. W. (1991). Composition of the flora and fauna. In G. W. Coulter (Ed.), *Lake Tanganyika and its life* (pp. 200–274). New York: Oxford University Press, Oxford.

Covich, A. P. (1993). Water and ecosystems. In P. H. Gleick (Ed.), *Water in crisis: A guide to the world's freshwater resources* (pp. 40–55). New York: Oxford University Press.

Covich, A. P. (1997). Leaf litter processing: the importance of species diversity in stream ecosystems. In: *Biodiversity in benthic ecology, Proceedings from Nordic Benthological Meeting* (pp. 15–20). Silkeborg, Denmark: National Environmental Research Institute.

Covich, A. P. (2001). *Energy flow and ecosystems. Encyclopedia of biodiversity.* San Diego, CA: Academic Press.

Covich, A. P., Austen, M. C., Barlocher, F., Chauvet, E., Cardinale, B. J., Biles, C. L., et al. (2004). The role of biodiversity in the functioning of freshwater and marine benthic ecosystems. *BioScience, 54*, 767–775.

Covich, A. P., Palmer, M. A., & Crowl, T. A. (1999). The role of benthic invertebrate species in freshwater ecosystems. *BioScience, 49*, 119–127.

Cowardin, M. F., Carter, V., Golet, F. C., & LaRoe, E. T. (1979). *Classification of wetlands and deepwater habitats of the United States.* Washington, DC: Office of Biological Services, Fish and Wildlife Service, US Dept. of the Interior.

Cowley, D. E. (2008). Estimating required habitat size for fish conservation in streams. *Aquatic Conservation—Marine and Freshwater Ecosystems, 18*, 418–431.

Craft, C., Clough, J., Ehman, J., Joye, S., Park, R., Pennings, S., et al. (2009). Forecasting the effects of accelerated sea-level rise on tidal marsh ecosystem services. *Frontiers in Ecology and the Environment, 7*, 73–78.

Craft, C. B., Vymazal, J., & Richardson, C. J. (1995). Response of Everglades plant communities to nitrogen and phosphorus additions. *Wetlands, 15*, 258–271.

Crain, D. A., Guillette, L. J., Jr., Pickford, D. B., Percival, H. F., & Woodward, A. R. (1998). Sex-steroid and thyroid hormone concentrations in juvenile alligators (*Alligator mississippiensis*) from contaminated and reference lakes in Florida, USA. *Environmental Toxicology and Chemistry, 17*, 446–452.

Creed, R. P., Jr. (1994). Direct and indirect effects of crayfish grazing in a stream community. *Ecology, 75*, 2091–2103.

Creed, R. P., Jr. (2000). Is there a new keystone species in North American lakes and rivers? *Oikos, 91*, 405–408.

Creed, R. P., Jr., Cherry, R. P., Pflaum, J. R., & Wood, C. J. (2009). Dominant species can produce a negative relationship between species diversity and ecosystem function. *Oikos, 118*, 723–732.

Creuze des Chatelliers, M., Poinsart, D., & Bravard, J. P. (1994). *Geomorphology of alluvial groundwater ecosystems. Groundwater ecology* (pp. 157–185). Aquatic Biology Series. San Diego: Academic Press.

Crew, D., Willingham, E., & Skipper, J. K. (2000). Endocrine disruptors: present issues, future directions. *The Quarterly Review of Biology, 75*, 243–260.

Cross, W. F., Benstead, J. P., Rosemond, A. D., & Wallace, J. B. (2003). Consumer-resource stoichiometry in detritus-based streams. *Ecology Letters, 6*, 721–732.

Cross, W. F., Wallace, J. B., Rosemond, A. D., & Eggert, S. L. (2006). Whole-system nutrient enrichment increases secondary production in a detritus-based ecosystem. *Ecology, 87*, 1556–1565.

Crowl, T. A., & Covich, A. P. (1990). Predator-induced life-history shifts in a freshwater snail. *Science, 247*, 949–952.

Cuffney, T. F., Wallace, J. B., & Lugthart, G. H. (1990). Experimental evidence quantifying the role of benthic invertebrates in organic matter dynamics of headwater streams. *Freshwater Biology, 23*, 281–299.

Culver, D. C. (1994). Species interactions. In J. Gibert, D. L. Danielopol, & J. A. Stanford (Eds.), *Groundwater ecology* (pp. 271–285). San Diego, CA: Academic Press.

Culver, D. C., & Fong, D. W. (1994). Small scale and large scale biogeography of subterranean crustacean faunas of the Virginias. *Hydrobiologia, 287*, 3–9.

Cummins, K. W. (1973). Trophic relations of aquatic insects. *Annual Review of Entomology, 18*, 183–206.

Cummins, K. W. (1974). Structure and function of stream ecosystems. *BioScience, 24*, 631–641.

Cummins, K. W. (1977). From headwater streams to rivers. *The American Biology Teacher, May*, 305–312.

Cummins, K. W., & Klug, M. J. (1979). Feeding ecology of stream invertebrates. *Annual Review of Ecology and Systematics, 10*, 147–172.

Cummins, K. W., Minshall, G. W., Sedell, J. R., Cushing, C. E., Petersen, R. C. (1984). Stream ecosystem theory. *Diversity and the coevolution of competitors, or the ghost of competition past, 22*, 1818–1827.

Currie, D. J., Dilworth-Christie, P., & Chapleau, F. (1999). Assessing the strength of top-down influences on plankton abundance in unmanipulated lakes. *Canadian Journal of Fisheries and Aquatic Sciences, 56*, 427–436.

Currie, D. J., & Kalff, J. (1984). The relative importance of bacterioplankton and phytoplankton in phosphorus uptake in freshwater. *Limnology and Oceanography, 29*, 311–321.

Cursino, L., Oberdá, S. M., Cecílio, R. V., Moreira, R. M., Chartone-Souza, E., & Nascimento, A. M. A. (1999). Mercury concentration in the sediment at different gold prospecting sites along the Carmo stream, Minas Gerais, Brazil, and frequency of resistant bacteria in the respective aquatic communities. *Hydrobiologia, 394*, 5–12.

Cushing, C. E., & Gaines, W. L. (1989). Thoughts on recolonization of endorheic cold desert spring-streams. *Journal of the North American Benthological Society, 8*, 277–287.

Dahl, T. E., Johnson, C. E., & Frayer, W. E. (1991). Status and trends of wetlands in the conterminous United States, mid-1970's to mid-1980's. Washington, DC: US Department of the Interior, Fish and Wildlife Service.

Daleo, P., Panjul, E., Casariego, A. M., Silliman, B. R., Bertness, M. D., & Iribarne, O. (2007). Ecosystem engineers activate mycorrhizal mutualism in salt marshes. *Ecology Letters, 10*, 902–908.

Danielopol, D. L., des Châtelliers, M. C., Moeszlacher, F., Pospisil, P., & Popa, R. (1994). Adaptation of crustacea to interstitial habitats: a practical agenda for ecological studies. In J. Gibert, D. L. Danielopol, & J. A. Stanford (Eds.), *Groundwater ecology* (pp. 217–243). San Diego, CA: Academic Press.

Danielson, T. J. (1998). *Wetland bioassessment fact sheets*. Washington, DC: EPA843-F-98-001, US Environmental Protection Agency, Office of Wetlands, Oceans, and Watersheds, Wetlands Division.

Darcy-Hall, T. (2006). Relative strengths of benthic algal nutrient and grazer limitation along a lake productivity gradient. *Oecologia, 148*, 660–671.

Dash, M. C., & Hota, A. K. (1980). Density effects on the survival, growth rate, and metamorphosis of *Rana tigrina* tadpoles. *Ecology, 61*, 1025–1028.

Daszak, P., Strieby, A., Cunningham, A. A., Longcore, J. E., Brown, C. C., & Porter, D. (2004). Experimental evidence that the bullfrog (*Rana catesbeiana*) is a potential carrier of chytridiomycosis, an emerging fungal disease of amphibians. *Herpetological Journal, 14*, 201–207.

Daufresne, M., & Boet, P. (2007). Climate change impacts on structure and diversity of fish communities in rivers. *Global Change Biology, 13*, 2467–2478.

Daughton, C. G., & Ternes, T. A. (1999). Pharmaceutical and personal care products in the environment: agents of subtle change? *Environmental Health Perspectives, 107,* 907–938.

Davidson, C. (2004). Declining downwind: amphibian population declines in California and historical pesticide use. *Ecological Applications, 14,* 1892–1902.

Davies, B. R., Thoms, M. C., Walker, K. F., O'Keefe, J. H., & Gore, J. A. (1994). Dryland rivers: their ecology, conservation and management. In P. Calow & G. E. Petts (Eds.), *The rivers handbook. Hydrological and ecological principles* (pp. 484–511). London: Blackwell Scientific Publications.

Davies, R. W. (1991). Annelida: Leeches, Polychaetes, and Acanthobdellids. In J. H. Thorp & A. P. Covich (Eds.), *Ecology and classification of North American freshwater invertebrates* (pp. 437–480). San Diego, CA: Academic Press, Inc.

Davies-Colley, R. J. (1997). Stream channels are narrower in pasture than in forest. *New Zealand Journal of Marine and Freshwater Research, 31,* 599–608.

Davis, S. M. (1994). Phosphorus inputs and vegetation sensitivity in the Everglades. In S. M. Davis & J. C. Ogden (Eds.), *Everglades, the ecosystem and its restoration* (pp. 357–378). Delray Beach, FL: St. Lucie Press.

Davis, S. M., & Ogden, J. C. (1994). Introduction. In S. M. Davis & J. C. Ogden (Eds.), *Everglades: The ecosystem and its restoration* (pp. 3–8). Delray Beach, FL: St. Lucie Press.

Davis, S. N., & De Wiest, J. M. (1966). *Hydrogeology.* New York: Wiley.

Davison, A., & Blaxter, M. (2005). Ancient origin of glycosyl hydrolase family 9 cellulase genes. *Molecular Biology and Evolution, 22,* 1273–1284.

Davison, W., George, D. G., & Edwards, N. J. A. (1995). Controlled reversal of lake acidification by treatment with phosphate fertilizer. *Nature, 377,* 504–507.

Dawidowica, P., Prejs, A., Engelmayer, A., Martyniak, A., Kozlowski, J., Kufel, L., et al. (2002). Hypolimnetic anoxia hampers top-down food-web manipulation in a eutrophic lake. *Freshwater Biology, 47,* 2401–2409.

De Jalón, D. G. (1995). Management of physical habitat for fish stocks. In D. M. Harper & A. J. D. Ferguson (Eds.), *The ecological basis for river management* (pp. 363–374). West Sussex, England: John Wiley & Sons, Ltd.

De Lange, H. J., Morris, D. P., & Williamson, C. E. (2003). Solar ultraviolet photodegradation of DOC may stimulate freshwater food webs. *Journal of Plankton Research, 25,* 111–117.

De Lange, H. J., Noordoven, W., Murk, A. J., Lurling, M., & Peeters, E. (2006). Behavioural responses of *Gammarus pulex* (Crustacea, Amphipoda) to low concentrations of pharmaceuticals. *Aquatic Toxicology, 78,* 209–216.

De Meester, L., Weider, L. J., & Tollrian, R. (1995). Alternative antipredator defenses and genetic polymorphism in a pelagic predator-prey system. *Nature, 378,* 483–485.

de Oliveira Neves, A. C. (2009). Conservation of the Pantanal wetlands: the definitive moment for decision making. *AMBIO: A Journal of the Human Environment, 38,* 127–128.

de Szalay, F. A., & Resh, V. H. (1997). Responses of wetland invertebrates and plants important in waterfowl diets to burning and mowing of emergent vegetation. *Wetlands, 17,* 149–156.

de Vaate, A. B. (1991). Distribution and aspects of population dynamics of the zebra mussel, *Dreissena polymorpha* (Pallas, 1771), in the Lake Ijsselmeer area (The Netherlands). *Oecologia, 86,* 40–50.

Dean, J. A. (1985). *Lange's handbook of chemistry* (13th ed.). New York: McGraw-Hill Book Company.

Deborde, D. C., Woessner, W. W., Kiley, Q. T., & Ball, P. (1999). Rapid transport of viruses in a floodplain aquifer. *Water Research, 33,* 2229–2238.

Delorme, L. D. (1991). Ostracoda. In J. H. Thorp & A. P. Covich (Eds.), *Ecology and classification of North American freshwater invertebrates* (pp. 691–722). San Diego, CA: Academic Press, Inc.

DeMott, W. R. (1995). The influence of prey hardness on *Daphnia*'s selectivity for large prey. *Hydrobiologia, 307,* 127–138.

DeNicola, D. M. (1996). Periphyton responses to temperature at different ecological levels. In R. J. Stevenson, M. L. Bothwell, & R. L. Lowe (Eds.), *Algal ecology: Freshwater benthic ecosystems* (pp. 150–183). San Diego, CA: Academic Press.

Denny, M. W. (1993). *Air and water: The biology and physics of lifes media.* Princeton, NJ: Princeton University Press.

des Châtelliers, M. C., Poinsart, D., & Bravard, J.-P. (1994). Geomorphology of alluvial groundwater ecosystems. In J. Gibert, l. L. Danielopo, & J. A. Stanford (Eds.), *Groundwater ecology* (pp. 157–185). San Diego, CA: Academic Press, Inc.

Devrie, D. R., & Frie, R. V. (1996). Determination of age and growth. In B. R. Murphy & D. W. Willis (Eds.), *Fisheries techniques* (pp. 483–512). Bethesda, MD: American Fisheries Society.

DeYoe, H. R., Lowe, R. L., & Marks, J. C. (1992). Effects of nitrogen and phosphorus on the endosymbiont load of *Rhopalodia gibba* and *Epithemia turgida* (Bacillariophyceae). *Journal of Phycology, 28,* 773–777.

Diaz, R. J., & Rosenberg, R. (2008). Spreading dead zones and consequences for marine ecosystems. *Science, 321,* 926–929.

Dickman, M., & Rao, S. S. (1989). Diatom stratigraphy in acid-stressed lakes in The Netherlands, Canada, and China. In S. S. Rao (Ed.), *Acid stress and aquatic microbial interactions* (pp. 116–138). Boca Raton, FL: CRC Press.

Didham, R. K., Tylianakis, J. M., Hutchison, M. A., Ewers, R. M., & Gemmell, N. J. (2005). Are invasive species the drivers of ecological change? *Trends in Ecology and Evolution, 20,* 470–474.

Dillard, G. E. (1999). *Common freshwater algae of the United States, an illustrated key to the genera (excluding the diatoms).* Berlin, Germany: Gebrüder Borntraeger.

Dittmann, E., & Wiegand, W. (2006). Cyanobacterial toxins—occurrence, biosynthesis and impact on human affairs. *Molecular Nutrition and Food Research, 50,* 7–17.

Dodds, M. (2008). Water bomb—how big is the biggest possible raindrop? *New Scientist, 198,* 81.

Dodds, W. K. (1989). Photosynthesis of two morphologies of *Nostoc parmelioides* (*Cyanobacteria*) as related to current velocities and diffusion patterns. *Journal of Phycology, 25,* 258–262.

Dodds, W. K. (1990). Hydrodynamic constraints on evolution of chemically mediated interactions between aquatic organisms in unidirectional flows. *Journal of Chemical Ecology, 16,* 1417–1430.

Dodds, W. K. (1991). Community interactions between the filamentous alga *Cladophora glomerata* (L.) Kuetzing, its epiphytes, and epiphyte grazers. *Oecologia, 85,* 572–580.

Dodds, W. K. (1992). A modified fiber-optic light microprobe to measure spherically integrated photosynthetic photon flux density: characterization of periphyton photosynthesis-irradiance patterns. *Limnology and Oceanography, 37,* 871–878.

Dodds, W. K. (1993). What controls levels of dissolved phosphate and ammonium in surface waters? *Aquatic Sciences, 55,* 132–142.

Dodds, W. K. (1997). Distribution of runoff and rivers related to vegetative characteristics, latitude, and slope: a global perspective. *Journal of the North American Benthological Society, 16,* 162–168.

Dodds, W. K. (1997). Interspecific interactions: constructing a general neutral model for interaction type. *Oikos, 78,* 377–383.

Dodds, W. K. (2003). Misuse of inorganic N and soluble reactive P concentrations to indicate nutrient status of surface waters. *Journal of the North American Benthological Society, 22,* 171–181.

Dodds, W. K. (2003a). The role of periphyton in phosphorus retention in shallow freshwater aquatic systems. *Journal of Phycology, 39,* 840–849.

Dodds, W. K. (2006). Eutrophication and trophic state in rivers and streams. *Limnology and Oceanography, 51,* 671–680.

Dodds, W. K. (2007). Trophic state, eutrophication and nutrient criteria in streams. *Trends in Ecology and Evolution, 22,* 669–676.

Dodds, W. K. (2008). *Humanity's footprint: Momentum, impact, and our global environment.* New York: Columbia University Press.

Dodds, W. K. (2009). *Laws, theories, and patterns in ecology.* Berkeley: University of California Press.

Dodds, W. K., Banks, M. K., Clenan, C. S., Rice, C. W., Sotomayor, D., Strauss, E. A., et al. (1996a). Biological properties of soil and subsurface sediments under abandoned pasture and cropland. *Soil Biology and Biochemistry, 28,* 837–846.

Dodds, W. K., Bouska, W. W., Eitzmann, J. L., Pilger, T. J., Pitts, K. L., Riley, A. J., et al. (2009). Eutrophication of US freshwaters: analysis of potential economic damages. *Environmental Science and Technology, 43,* 12–19.

Dodds, W. K., & Brock, J. (1998). A portable flow chamber for *in situ* determination of benthic metabolism. *Freshwater Biology, 39,* 49–59.

Dodds, W. K., Carney, E., & Angelo, R. T. (2006a). Determining ecoregional reference conditions for nutrients, Secchi depth and chlorophylla in Kansas lakes and reservoirs. *Lake and Reservoir Management, 22,* 151–159.

Dodds, W. K., & Castenholz, R. W. (1988). The nitrogen budget of an oligotrophic cold water pond. *Archive für Hydrobiology Supplement, 79,* 343–362.

Dodds, W. K., & Cole, J. J. (2007). Expanding the concept of trophic state in aquatic ecosystems: it's not just the autotrophs. *Aquatic Sciences, 69,* 427–439.

Dodds, W. K., Evans-White, M. A., Gerlanc, N. M., Gray, L., Gudder, D. A., Kemp, M. J., et al. (2000). Quantification of the nitrogen cycle in a prairie stream. *Ecosystems, 3,* 574–589.

Dodds, W. K., Gido, K., Whiles, M. R., Fritz, K. M., & Matthews, W. J. (2004). Life on the edge: the ecology of great plains prairie streams. *BioScience, 54*, 205–216.

Dodds, W. K., & Gudder, D. A. (1992). The ecology of *Cladophora. Journal of Phycology, 28*, 415–427.

Dodds, W. K., Gudder, D. A., & Mollenhauer, D. (1995). The ecology of *Nostoc. Journal of Phycology, 31*, 2–18.

Dodds, W. K., & Henebry, G. M. (1996). The effect of density dependence on community structure. *Ecological Modeling, 93*, 33–42.

Dodds, W. K., Hutson, R. E., Eichem, A. C., Evans, M. A., Gudder, D. A., Fritz, K. M., et al. (1996b). The relationship of floods, drying, flow and light to primary production and producer biomass in a prairie stream. *Hydrobiologia, 333*, 151–159.

Dodds, W. K., Johnson, K. R., & Priscu, J. C. (1989). Simultaneous nitrogen and phosphorus deficiency in natural phytoplankton assemblages: theory, empirical evidence and implications for lake management. *Lake and Reservoir Management, 5*, 21–26.

Dodds, W. K., Jones, J. R., & Welch, E. B. (1998). Suggested classification of stream trophic state: distributions of temperate stream types by chlorophyll, total nitrogen, and phosphorus. *Water Research, 32*, 1455–1462.

Dodds, W. K., & Nelson, J. A. (2006). Redefining the community: a species-based approach. *Oikos, 112*, 464–472.

Dodds, W. K., & Oakes, R. M. (2004). A technique for establishing reference nutrient concentrations across watersheds affected by humans. *Limnology and Oceanography Methods, 2*, 333–341.

Dodds, W. K., & Priscu, J. C. (1989). Ammonium, nitrate, phosphate, and inorganic carbon uptake in an oligotrophic lake: seasonal variation among light response variables. *Journal of Phycology, 25*, 699–705.

Dodds, W. K., & Priscu, J. C. (1990). A comparison of methods for assessment of nutrient deficiency of phytoplankton in a large oligotrophic lake. *Canadian Journal of Fisheries and Aquatic Sciences, 47*, 2328–2338.

Dodds, W. K., Priscu, J. C., & Ellis, B. K. (1991). Seasonal uptake and regeneration of inorganic nitrogen and phosphorus in a large oligotrophic lake: size-fractionation and antibiotic treatment. *Journal of Plankton Research, 13*, 1339–1358.

Dodds, W. K., Smith, V. H., & Lohman, K. (2006b). Nitrogen and phosphorus relationships to benthic algal biomass in temperate streams (vol 59, pg 865, 2002). *Canadian Journal of Fisheries and Aquatic Sciences, 63*, 1190–1191.

Dodds, W. K., Smith, V. H., & Zander, B. (1997). Developing nutrient targets to control benthic chlorophyll levels in streams: a case study of the Clark Fork River. *Water Research, 31*, 1738–1750.

Dodds, W. K., & Welch, E. B. (2000). Establishing nutrient criteria in streams. *Journal of the North American Benthological Society, 19*, 186–196.

Dodds, W. K., Wilson, K. C., Rehmeier, R. L., Knight, G. L., Wiggam, S., Falke, J. A., et al. (2008). Comparing ecosystem goods and services provided by restored and native lands. *BioScience, 58*, 837–845.

Dodson, S. I., Crowl, T. A., Peckarsky, B. L., Kats, L. B., Covich, A. P., & Culp, J. M. (1994). Non-visual communication in freshwater benthos: an overview. *Journal of the North American Benthological Society, 13*, 268–282.

Dodson, S. I., & Frey, D. G. (1991). Cladocera and other branchiopoda. In J. H. Thorp & A. P. Covich (Eds.), *Ecology and classification of North American freshwater invertebrates* (pp. 725–745). San Diego, CA: Academic Press, Inc.

Dole-Olivier, M.-J., Marmonier, P., Creuzé des Châtelliers, M., & Martin, D. (1994). Interstitial fauna associated with the alluvial floodplains of the Rhône River (France). In J. Gibert, D. L. Danielopol, & J. A. Stanford (Eds.), *Groundwater ecology* (pp. 313–346). San Diego, CA: Academic Press, Inc.

Donar, C. M., Neely, R. K., & Stoermer, E. F. (1996). Diatom succession in an urban reservoir system. *Journal of Paleolimnology, 15*, 237–243.

Doolittle, W. F. (1999). Phylogenetic classification and the universal tree. *Science, 284*, 2124–2128.

Doren, R. F., Armentano, T. V., Whiteaker, L. D., & Jones, R. D. (1996). Marsh vegetation patterns and soil phosphorus gradients in the Everglades ecosystem. *Aquatic Botany, 56*, 145–163.

Doughty, M. J. (1991). Mechanism and strategies of photomovement in protozoa. In F. Lenci, F. Ghetti, G. Colombetti, D.-P. Häder, & P.-S. Song (Eds.), *Biophysics of photoreceptors and photomovements in microorganisms* (pp. 73–102). New York: Plenum Press.

Douglas, M., & Lake, P. S. (1994). Species richness of stream stones: an investigation of the mechanisms generating the species-area relationship. *Oikos, 69*, 387–396.

Downing, J. A., Osenberg, C. W., & Sarnelle, O. (1999). Meta-analysis of marine nutrient-enrichment experiments: variation in the magnitude of nutrient limitation. *Ecology, 80*, 1157–1167.

Downing, J. A., Watson, S. B., & McCauley, E. (2001). Predicting cyanobacteria dominance in lakes. *Canadian Journal of Fisheries and Aquatic Sciences, 58*, 1905–1908.

Duarte, C. M., & Agustí, S. (1998). The CO_2 balance of unproductive aquatic ecosystems. *Science, 281*, 234–236.

Dubey, T., Stephenson, S. L., & Edwards, P. J. (1994). Effect of pH on the distribution and occurrence of aquatic fungi in six West Virginia mountain streams. *Journal of Environmental Quality, 23*, 1271–1279.

Duever, M. J., Meeder, J. F., Meeder, L. C., & McCollom, J. M. (1994). The climate of South Florida and its role in shaping the Everglades ecosystem. In S. M. Davis & J. C. Ogden (Eds.), *Everglades* (pp. 225–248). Delray Beach, FL: St. Lucie Press.

Duff, J. H., & Triska, F. J. (2000). Nitrogen biogeochemistry and surface-subsurface exchange in streams. In J. B. Jones & P. J. Mulholland (Eds.), *Streams and ground waters* (pp. 197–220). San Diego, CA: Academic Press.

Duffy, J. E. (2009). Why biodiversity is important to the functioning of real-world ecosystems. *Frontiers in Ecology and the Environment, 7*, 437–444.

Dugan, P. (1993). *Wetlands in danger: A world conservation atlas*. New York: Oxford University Press.

Dumont, H. J. (1995). The evolution of groundwater Cladocera. *Hydrobiologia, 307*, 69–74.

Dusenbery, D. (2009). *Living at micro scale: The unexpected physics of being small*. Boston: Harvard University Press.

Dynesius, M., & Nilsson, C. (1994). Fragmentation and flow regulation of river systems in the northern third of the world. *Science, 266*, 753–762.

East, T. L., Havens, K. E., Rodusky, A. J., & Brady, M. A. (1999). *Daphnia lumholtzi* and *Daphnia ambigua*: Population comparisons of an exotic and a native cladoceran in Lake Okeechobee, Florida. *Journal of Plankton Research, 21*, 1537–1551.

Eaton, A. D., Clesceri, L. S., Rice, E. W., Greenberg, A. E., & Franson, M. A. H. (2005). *Standard methods for the examination of water and wastewater* 21st edition. American Public Health Association.

Eaton, A. E., Clesceri, L. S., & Greenberg, A. E. (1995). *Standard methods for examination of water and wastewater* (19th ed.). Washington, DC: American Public Health Association.

Eddy, F. B. (1981). Effects of stress on osmotic and ionic regulation in fish. In A. D. Pickering (Ed.), *Stress and fish* (pp. 77–102). London, UK: Academic Press, Inc.

Eddy, S., & Underhill, J. C. (1969). *How to know the freshwater fishes*. Dubuque, IA: Wm. C. Brown Company Publishers.

Edler, C., & Dodds, W. K. (1996). The ecology of a subterranean isopod, *Caecidotea tridentata*. *Freshwater Biology, 35*, 249–259.

Edler, C., & Georgian, T. (2004). Field measurements of particle-capture efficiency and size selection by caddisfly nets and larvae. *Journal of the North American Benthological Society, 23*, 756–770.

Edmondson, W. T. (1991). *The uses of ecology: Lake Washington and beyond*. Seattle, WA: The University of Washington Press.

Edmondson, W. T., & Lehman, J. T. (1981). The effect of changes in the nutrient income on the condition of Lake Washington. *Limnology and Oceanography, 26*, 1–29.

Edwards, K. J., Bond, P. L., Gihring, T. M., & Banfield, J. F. (2000). An archaeal iron-oxidizing extreme acidophile important in acid mine drainage. *Science, 287*, 1796–1799.

Edwards, P. J., & Abivardi, C. (1998). The value of biodiversity: where ecology and economy blend. *Biological Conservation, 83*, 239–246.

Effler, S. W., Boone, S. R., Sigfired, C., & Ashby, S. L. (1998). Dynamics of zebra mussel oxygen demand in Seneca River, New York. *Environmental Science and Technology, 32*, 807–812.

Egeland, G. M., & Middaugh, J. P. (1997). Balancing fish consumption benefits with mercury exposure. *Science, 278*, 1904–1905.

Egertson, C. J., Kopaska, J. A., & Downing, J. A. (2004). A century of change in macrophyte abundance and composition in response to agricultural eutrophication. *Hydrobiologia, 524*, 145–156.

Eggert, S. L., & Wallace, J. B. (2007). Wood biofilm as a food resource for stream detritivores. *Limnology and Oceanography, 52*, 1239–1245.

Eichem, A. C., Dodds, W. K., Tate, C. M., & Edler, C. (1993). Microbial decomposition of elm and oak leaves in a karst aquifer. *Applied and Environmental Biology, 59*, 3592–3596.

Eisenberg, J. N. S., Washburn, J. O., & Schreiber, S. J. (2000). Generalist feeding behaviors of *Aedes sierrensis* larvae and their effects on protozoan populations. *Ecology, 81*, 921–935.

Eldakar, O., Dlugos, M., Pepper, J., & Wilson, D. (2009). Population structure mediates sexual conflict in water striders. *Science, 326*, 816.

Eller, G., Deines, P., Grey, J., Richnow, H.-H., & Krüger, M. (2005). Methane cycling in lake sediments and its influence on chironomid larval ^{13}C. *FEMS Microbiology Ecology, 54*, 339–350.

Eller, G., Deines, P., & Kruger, M. (2007). Possible sources of methane-derived carbon for chironomid larvae. *Aquatic Microbial Ecology, 46*, 283–293.

Elliott, J. K., Elliott, J. M., & Legett, W. C. (1997). Predation by *Hydra* on larval fish: field and laboratory experiments with bluegill (*Lepomis macrochirus*). *Limnology and Oceanography, 42*, 1416–1423.

Elliott, J. M. (1981). Some aspects of thermal stress on freshwater teleosts. In A. D. Pickering (Ed.), *Stress and fish* (pp. 209–245). London, UK: Academic Press, Inc.

Elser, J. J., Bracken, M. E. S., Cleland, E. E., Gruner, D. S., Harpole, W. S., Hillebrand, H., et al. (2007). Global analysis of nitrogen and phosphorus limitation of primary producers in freshwater, marine and terrestrial ecosystems. *Ecology Letters, 10*, 1135–1142.

Elser, J. J., Carney, H. J., & Goldman, C. R. (1990). The zooplankton-phytoplankton interface in lakes of contrasting trophic status: an experimental comparison. *Hydrobiologia, 200/201*, 69–82.

Elser, J. J., Dobberfuhl, D. R., MacKay, N. A., & Schampel, J. H. (1996). Organism size, life history, and N:P stoichiometry, toward a unified view of cellular and ecosystem processes. *BioScience, 46*, 674–684.

Elser, J. J., & Goldman, C. R. (1991). Zooplankton effects on phytoplankton in lakes of contrasting trophic status. *Limnology and Oceanography, 36*, 64–90.

Elser, J. J., & Hassett, R. P. (1994). A stoichiometric analysis of the zooplankton-phytoplankton interaction in marine and freshwater ecosystems. *Nature, 370*, 213–221.

Elser, J. J., Marzolf, E. R., & Goldman, C. R. (1990). Phosphorus and nitrogen limitation of phytoplankton growth in the freshwaters of North America: a review and critique of experimental enrichments. *Canadian Journal of Fisheries and Aquatic Sciences, 47*, 1468–1477.

Elser, J. J., & Urabe, J. (1999). The stoichiometry of consumer-driven nutrient recycling: theory, observations, and consequences. *Ecology, 80*, 735–751.

Elton, C. S. (1958). *The ecology of invasion by animals and plants*. New York: Wiley.

Elwood, J. W., Newbold, J. D., O'Neill, R. V., & Van Winkle, W. (1983). Resource spiraling: an operational paradigm for analyzing lotic ecosystems. In T. D. Fontaine, III, & S. M. Bartell (Eds.), *Dynamics of lotic ecosystems* (pp. 3–27). Ann Arbor, MI: Ann Arbor Science Publishers.

Engstrom, D. R., Schottler, S. P., Leavitt, P. R., & Havens, K. E. (2006). A reevaluation of the cultural eutrophication of Lake Okeechobee using multiproxy sediment records. *Ecological Applications, 16*, 1194–1206.

Eppley, R. W. (1972). Temperature and phytoplankton growth in the sea. *Fishery Bulletin, 70*, 1063–1085.

Ernst, W. (1980). Effects of pesticides and related organic compounds in the sea. *Helgoländer Meeresunters, 33*, 301–312.

Etnier, D. A. (1997). Jeopardized southeastern freshwater fishes: a search for causes. In G. W. Benz & D. E. Collins (Eds.), *Aquatic fauna in peril, the southeastern perspective* (pp. 87–104). Decatur, GA: Southeast Aquatic Research Institute.

Euliss, N. H., Jr., Wrubleski, D. A., & Mushet, D. M. (1999). Wetlands of the prairie pothole region: invertebrate species composition, ecology, and management. In D. P. Batzer, R. B. Rader, & S. A. Wissinger (Eds.), *Invertebrates in freshwater wetlands of North America, ecology and management* (pp. 471–514). New York: John Wiley and Sons.

Evanno, G., Castella, E., Antoine, C., Paillat, G., & Goudet, J. (2009). Parallel changes in genetic diversity and species diversity following a natural disturbance. *Molecular Ecology, 18*, 1137–1144.

Evans, W. C., Kling, G. W., Tuttle, M. L., Tanyileke, G., & White, L. D. (1993). Gas buildup in Lake Nyos, Cameroon: the recharge process and its consequences. *Applied Geochemistry, 8*, 207–221.

Evans, W. C., White, L. D., Tuttle, M. L., Kling, G. W., Tanyileke, G., & Michel, R. L. (1994). Six years of change at Lake Nyos, Cameroon, yield clues to the past and cautions for the future. *Geochemical Journal, 28*, 139–162.

Evans-White, M. A., Dodds, W. K., Huggins, D. G., & Baker, D. S. (2009). Thresholds in macroinvertebrate biodiversity and stoichiometry across water-quality gradients in central plains (USA) streams. *Journal of the North American Benthological Society, 28*, 855–868.

Evans-White, M. A., Dodds, W. K., & Whiles, M. R. (2003). Ecosystem significance of crayfishes and stonerollers in a prairie stream: functional differences between co-occurring omnivores. *Journal of the North American Benthological Society, 22*, 423–441.

Evans-White, M. A., & Lamberti, G. A. (2006). Stoichiometry of consumer-driven nutrient recycling across nutrient regimes in streams. *Ecology Letters, 9*, 1186–1197.

Everall, N. C., & Lees, D. R. (1996). The use of barley-straw to control general and blue-green algal growth in a Derbyshire Reservoir. *Water Research, 30*, 269–276.

Everitt, D. T., & Burkholder, J. M. (1991). Seasonal dynamics of macrophyte communities from a stream flowing over granite flatrock in North Carolina, USA. *Hydrobiologia, 222*, 159–172.

Fagan, W. F., & Holmes, E. E. (2006). Quantifying the extinction vortex. *Ecology Letters, 9*, 51–60.

Fahnenstiel, G. L., Bridgeman, T. B., Lang, G. A., McCormick, M. J., & Nalepa, T. F. (1995). Phytoplankton productivity in Saginaw Bay, Lake Huron: effects of zebra mussel (*Dreissena polymorpha*) colonization. *Journal of Great Lakes Research, 21*, 465–475.

Fahnenstiel, G. L., Lang, G. A., Nalepa, T. F., & Johengen, T. H. (1995). Effects of zebra mussel (*Dreissena polymorpha*) colonization on water quality parameters in Saginaw Bay, Lake Huron. *Journal of Great Lakes Research, 21*, 435–448.

Fairchild, G. W., & Sherman, J. W. (1990). Effects of liming on nutrient limitation of epilithic algae in an acid lake. *Water, Air, and Soil Pollution, 52*, 133–147.

Falconer, I. R. (1999). An overview of problems caused by toxic blue-green algae (*Cyanobacteria*) in drinking and recreational water. *Environmental Toxicology, 14*, 5–12.

Falconer, I. R., & Humpage, A. R. (2006). Cyanobacterial (blue-green algal) toxins in water supplies: cylindrospermopsins. *Environmental Toxicology, 21*, 299–304.

Falke, J. A., & Gido, K. B. (2006). Spatial effects of reservoirs on fish assemblages in Great Plains streams in Kansas, USA. *River Research and Applications, 22*, 55–68.

Falkenmark, M. (1992). Water scarcity generates environmental stress and potential conflicts. In W. James & J. Niemczynowicz (Eds.), *Water, development and the environment* (pp. 279–294). Ann Arbor, MI: Lewis Publishers.

Fawell, J. K., Sheahan, D., James, H. A., Hurst, M., & Scott, S. (2001). Oestrogens and oestrogenic activity in raw and treated water in Severn Trent water. *Water Research, 35*, 1240–1244.

Fee, E. J., Hecky, R. E., Regehr, G. W., Hendzel, L. L., & Wilkinson, P. (1994). Effects of lake size on nutrient availability in the mixed layer during summer stratification. *Canadian Journal of Fisheries and Aquatic Sciences, 51*, 2756–2768.

Feinberg, G. (1969). Light. In *Lasers and light: Readings from* Scientific American (pp. 4–13). San Francisco, CA: W. H. Freeman & Co.

Felip, M., Sattler, B., Psenner, R., & Catalan, J. (1995). Highly active microbial communities in the ice and snow cover of high mountain lakes. *Applied and Environmental Microbiology, 61*, 2394–2401.

Feminella, J. W. (1996). Comparison of benthic macroinvertebrate assemblages in small streams along a gradient of flow permanence. *Journal of the North American Benthological Society, 15*, 651–669.

Fenchel, T. (1980). Suspension feeding in ciliated protozoa: functional response and particle size selection. *Microbial Ecology, 6*, 1–11.

Fenchel, T., Esteban, G. F., & Finlay, B. J. (1997). Local versus global diversity of microorganisms: cryptic diversity of ciliated protozoa. *Oikos, 80*, 220–225.

Fenchel, T., & Finlay, B. J. (1995). *Ecology and evolution in anoxic worlds*. Oxford, UK: Oxford University Press.

Ferenci, T. (1999). Regulation by nutrient limitation. *Current Opinion in Microbiology, 2*, 208–213.

Feris, K., Ramsey, P., Frazar, C., Rillig, M., Moore, J., Gannon, J., et al. (2004). Seasonal dynamics of shallow-hyporheic-zone microbial community structure along a heavy-metal contamination gradient. *Applied and Environmental Microbiology, 70*, 2323.

Ferreira, M. T., Franco, A., Catarino, L., Moreira, I., & Sousa, P. (1999). Environmental factors related to the establishment of algal mats in concrete irrigation channels. *Hydrobiologia, 415*, 163–168.

Fichter, G. S. (1988). *Underwater farming*. Sarasota, FL: Pineapple Press, Inc.

Field, D. W., Reyer, A. J., Genovese, P. V., & Shearer, B. D. (1991). *Coastal wetlands of the United States*. Washington, DC: National Oceanic and Atmospheric Administration and US Fish and Wildlife Service.

Figuerola, J., & Green, A. J. (2002). Dispersal of aquatic organisms by waterbirds: a review of past research and priorities for future studies. *Freshwater Biology, 47*, 483–494.

Findlay, S., Pace, M. L., & Fischer, D. T. (1998). Response of heterotrophic planktonic bacteria to the zebra mussel invasion of the tidal freshwater Hudson River. *Microbial Ecology, 36*, 131–140.

Findlay, S., & Sobczak, W. V. (2000). Microbial communities in hyporheic sediments. In J. B. Jones & P. J. Mulholland (Eds.), *Streams and ground waters* (pp. 287–306). San Diego, CA: Academic Press.

Finlay, B. J. (2002). Global dispersal of free-living microbial eukaryote species. *Science, 296*, 1061–1063.

Finlay, B. J., & Esteban, G. F. (1998). Freshwater protozoa: biodiversity and ecological function. *Biodiversity and Conservation, 7*, 1163–1186.

Finlay, B. J., Maberly, S. C., & Cooper, J. I. (1997). Microbial diversity and ecosystem function. *Oikos, 80*, 209–213.

Finney, B. P., Gregory-Eaves, I., Sweetman, J., Douglas, M. S. V., & Smol, J. P. (2000). Impacts of climatic change and fishing on Pacific salmon abundance over the past 300 years. *Science, 290*, 795–799.

Firth, P., & Fisher, S. G. (Eds.) (1992). *Global climate change and freshwater ecosystems*. New York: Springer-Verlag.

Fisher, S. G., & Gray, L. J. (1983). Secondary production and organic matter processing by collector macroinvertebrates in a desert stream. *Ecology, 64*, 1217–1224.

Fisher, S. G., & Grimm, N. B. (1991). Streams and disturbance: are cross-ecosystem comparisons useful? In J. Cole, G. Lovett, & S. Findlay (Eds.), *Comparative analyses of ecosystems. Patterns, mechanisms, and theories* (pp. 196–221). New York: Springer-Verlag.

Fisher, T. R., Doyle, R. D., & Peele, E. R. (1988). Size-fractionated uptake and regeneration of ammonium and phosphate in a tropical lake. *Verhein Internationale Verein Limnologie, 23*, 637–641.

Fitts, C. (2002). *Groundwater science*. London: Academic Press.

Fitzgerald, D. J., Cunliffe, D. A., & Burch, M. D. (1999). Development of health alerts for cyanobacteria and related toxins in drinking water in South Australia. *Environmental Toxicology, 14*, 203–209.

Fitzhugh, T. W., & Richter, B. D. (2004). Quenching urban thirst: growing cities and their impacts on freshwater ecosystems. *BioScience, 54*, 741–754.

Flecker, A. S. (1996). Ecosystem engineering by a dominant detritivore in a diverse tropical stream. *Ecology, 77*, 1845–1854.

Flecker, A. S., & Townsend, C. T. (1994). Community-wide consequences of trout introduction in New Zealand streams. *Ecological Applications, 4*, 798–807.

Flower, R. J., & Battarbee, R. W. (1983). Diatom evidence for recent acidification of two Scottish lochs. *Nature, 305*, 130–133.

Folkers, D. (1999). Pitcher plant wetlands of the southeastern United States. In D. P. Batzer, R. B. Rader, & S. A. Wissinger (Eds.), *Invertebrates in freshwater wetlands of North America: Ecology and management* (pp. 247–275). New York: John Wiley & Sons, Inc.

Folkerts, G. W. (1997). State and fate of the world's aquatic fauna. In G. W. Benz & D. E. Collins (Eds.), *Aquatic fauna in peril, the southeastern perspective* (pp. 1–16). Decatur, GA: Southeast Aquatic Research Institute.

Folt, C. L., & Burns, C. W. (1999). Biological drivers of zooplankton patchiness. *Trends in Ecology and Evolution, 14*, 300–305.

Fong, D. W., & Culver, D. C. (1994). Fine-scale biogeographic differences in the crustacean fauna of a cave system in West Virginia, USA. *Hydrobiologia, 287*, 29–37.

Food and Agriculture Organization. (2007). Yearbook of fishery statistics *in* Food and Agriculture Organization of the United Nations, editor. Rome, Italy.

Food and Agriculture Organization of the United Nations. (1995). Food and Agriculture Organization of the United Nations 1993 Yearbook. Rome, Italy.

Forman, R. T. T., & Alexander, L. E. (1998). Roads and their major ecological effects. *Annual Review of Ecology and Systematics, 29*, 207–231.

Forney, L., Zhou, X., & Brown, C. (2004). Molecular microbial ecology: land of the one-eyed king. *Current Opinion in Microbiology, 7*, 210–220.

France, K., & Duffy, J. (2006). Diversity and dispersal interactively affect predictability of ecosystem function. *Nature, 441*, 1139–1143.

France, R. (1996). Ontogenetic shift in crayfish ^{13}C as a measure of land-water ecotonal coupling. *Oecologia, 107*, 239–242.

France, R. L. (1997a). Land-water linkages: influences of riparian deforestation on lake thermocline depth and possible consequences for cold stenotherms. *Canadian Journal of Fisheries and Aquatic Sciences, 54*, 1299–1305.

France, R. L. (1997b). Stable carbon and nitrogen isotopic evidence for ecotonal coupling between boreal forests and fishes. *Ecology of Freshwater Fish, 6*, 78–83.

Francoeur, S. N., & Lowe, R. L. (1998). Effects of ambient ultraviolet radiation on littoral periphyton: biomass accrual and taxon-specific responses. *Journal of Freshwater Ecology, 13*, 29–37.

Frank, D. A., & McNaughton, S. J. (1991). Stability increases with diversity in plant communities: empirical evidence from the 1988 Yellowstone drought. *Oikos, 62*, 360–362.

Freckman, D., Blackburn, T. H., Brussaard, L., Hutchings, P., Palmer, M. A., & Snelgrove, P. V. R. (1997). Linking biodiversity and ecosystem functioning of soils and sediments. *Ambio, 26*, 556–562.

French, J. R. P., III (1993). How well can fishes prey on zebra mussels in eastern North America? *Fisheries, 18*, 13–19.

Fretwell, S. D. (1977). The regulation of plant communities by the food chains exploiting them. *Perspectives in Biology and Medicine, Winter*, 169–185.

Freyer, G. (1980). Acidity and species diversity in freshwater crustacean faunas. *Freshwater Biology, 10*, 41–45.

Freyer, G. (1993). Variation in acid tolerance of certain freshwater crustaceans in different natural waters. *Hydrobiologia, 250*, 119–125.

Friedl, G., & Wüest, A. (2002). Disrupting biogeochemical cycles—consequences of damming. *Aquatic Sciences, 64*, 55–65.

Friedman, J. M., Osterkamp, W. R., Scott, M. L., & Auble, G. T. (1998). Downstream effects of dams on channel geometry and bottomland vegetation: regional patterns in the great plains. *Wetlands, 18*, 619–633.

Frissell, C. A., Liss, W. J., Warren, C. E., & Hurley, M. D. (1986). A hierarchical framework for stream habitat classification: viewing streams in a watershed context. *Environmental Management, 10*, 199–214.

Fritz, K. M., & Dodds, W. K. (2005). Harshness: characterisation of intermittent stream habitat over space and time. *Marine and Freshwater Research, 56*, 13–23.

Frost, P. C., Benstead, J. P., Cross, W. F., Hillebrand, H., Larson, J. H., Xenopoulos, M. A., et al. (2006). Threshold elemental ratios of carbon and phosphorus in aquatic consumers. *Ecology Letters, 9*, 774–779.

Frost, T. M. (1991). Porifera. In J. H. Thorp & A. P. Covich (Eds.), *Ecology and classification of North American freshwater invertebrates* (pp. 95–120). San Diego, CA: Academic Press, Inc.

Fry, B. (1991). Stable isotope diagrams of freshwater food webs. *Ecology, 72*, 2293–2297.

Fry, B., Mumford, P. L., Tam, F., Fox, D. D., Warren, G. L., Havens, K. E., et al. (1999). Trophic position and individual feeding histories of fish from lake Okeechobee, Florida. *Canadian Journal of Fisheries and Aquatic Sciences, 56*, 590–600.

Fryer, G. (2006). Evolution in ancient lakes: radiation of Tanganyikan atyid prawns and speciation of pelagic cichlid fishes in Lake Malawi. *Hydrobiologia, 568*, 131–142.

Fryer, G., & Iles, T. D. (1972). *The cichlid fishes of the great lakes of Africa. Their biology and evolution*. Great Britain: Oliver & Boyd.

Fuller, M. M., & Drake, J. A. (2000). Modeling the invasion process. In R. Claudi & J. H. Leach (Eds.), *Nonindigenous freshwater organisms* (pp. 411–414). Boca Raton, FL: Lewis Publishers, CRC Press.

Gafner, K., & Robinson, C. T. (2007). Nutrient enrichment influences the responses of stream macroinvertebrates to disturbance. *Journal of the North American Benthological Society, 26*, 92–102.

Galat, D. L., Fredrickson, L. H., Humburg, D. D., Bataille, K. J., Bodie, J. R., Dohrenwend, J., et al. (1998). Flooding to restore connectivity of regulated, large-river wetlands. *BioScience, 48*, 721–734.

Galatowitsch, S. M., & van der Valk, A. G. (1995). Natural revegetation during restoration of wetlands in the southern prairie pothole region of North America. In B. D. Wheeler, S. C. Shaw, W. J. Fojt, & R. A. Robertson (Eds.), *Restoration of temperate wetlands* (pp. 129–142). Chichester, UK: J. W. Wiley.

Galaziy, G. I. (1980). Lake Baikal's ecosystem and the problem of its preservation. *Marine Technological Society Journal, 14*, 31–38.

Gantar, M. F. (1985). The effect of heterotrophic bacteria on the growth of *Nostoc* sp. (*Cyanobacterium*). *Archives fur Hydrobiologie, 103*, 445–452.

Garbett, P. (2005). An investigation into the application of floating reed bed and barley straw techniques for the remediation of eutrophic waters. *Water and the Environment, 19*, 174–180.

Gardner, W. S., Cavaletto, J. F., Johengen, T. H., Johnson, J. R., Heath, R. T., & Cotner, J. J. B. (1995). Effects of the zebra mussel, *Dreissena polymorpha*, on community nitrogen dynamics in Saginaw Bay, Lake Huron. *Journal of Great Lakes Research, 21*, 529–544.

Gasith, A., & Gafny, S. (1990). Effects of water level on the structure and function of the littoral zone. In M. M. Tilzer & C. Serruya (Eds.), *Large lakes ecological structure and function* (pp. 156–172). New York: Springer-Verlag.

Gasol, J. M., Simons, A. M., & Kalff, J. (1995). Patterns in the top-down versus bottom-up regulation of heterotrophic nano-flagellates in temperate lakes. *Journal of Plankton Research, 17*, 1879–1903.

Gatz, J., Jr. (1983). Do stream fishes forage optimally in nature? In T. D. Fontaine, III, & S. M. Bartell (Eds.), *Dynamics of lotic ecosystems* (pp. 391–403). Ann Arbor, MI: Ann Arbor Science Publishers.

Gavrieli, I. (1997). Halite deposition from the Dead Sea: 1960-1993. In T. M. Niemi, Z. Ben-Avraham, & J. R. Gat (Eds.), *The Dead Sea. The lake and its setting* (pp. 161–183). New York: Oxford University Press.

Gehrke, C. (1998). Effects of enhanced UV-B radiation on production-related properties of a *Sphagnum fuscum* dominated subarctic bog. *Functional Ecology, 12*, 940–947.

Gelwick, F. P., & Matthews, W. J. (1992). Effects of an algivorous minnow on temperate stream ecosystem properties. *Ecology, 73*, 1630–1645.

Genkai-Kato, M., & Carpenter, S. R. (2005). Eutrophication due to phosphorus recycling in relation to lake morphometry, temperature, and macrophytes. *Ecology, 86*, 210–219.

Gensemer, R. W., & Playle, R. C. (1999). The bioavailability and toxicity of aluminum in aquatic environments. *Critical Reviews in Environmental Science and Technology, 29*, 315–450.

Gerba, C. P. (1987). Transport and fate of viruses in soils: field studies. In V. C. Rao & J. L. Melnick (Eds.), *Human viruses in sediments, sludges, and soils* (pp. 141–154). Boca Raton, FL: CRC Press.

Gerlanc, N. M., & Kaufman, G. A. (2003). Use of bison wallows by anurans on Konza Prairie. *American Midland Naturalist, 150*, 158–168.

Gerritsen, J., Carlson, R. E., Dycus, D. L., Faulkner, C., Gibson, G. R., Harcum, J., et al. (1998). *Lake and reservoir bioassessment and biocriteria*. Washington, DC: EPA 841-B-98-007, USEPA Office of Water.

Gessner, M. O., & Chauvet, E. (1994). Importance of stream microfungi in controlling breakdown rates of leaf-litter. *Ecology, 75*, 1807–1817.

Gessner, M. O., & Chauvet, E. (2002). A case for using litter breakdown to assess functional stream integrity. *Ecological Applications, 12*, 498–510.

Ghiorse, W. C. (1997). Subterranean life. *Science, 275*, 789–790.

Giberson, D., & Hardwick, M. L. (1999). Pitcher plants (*Sarracenia purpurea*) in Eastern Canadian peatlands. In D. P. Batzer, R. B. Rader, & S. A. Wissinger (Eds.), *Invertebrates in freshwater wetlands of North America: Ecology and management* (pp. 401–422). New York: John Wiley & Sons, Inc.

Gibert, J., Danielopol, D. L., & Stanford, J. A. (Eds.). (1994). *Groundwater ecology*. San Diego: Academic Press.

Gibert, J., Stanford, J. A., Dole-Olivier, M.-J., & Ward, J. V. (1994). Basic attributes of groundwater ecosystems and prospects for research. In J. Gibert, D. L. Danielopol, & J. A. Stanford (Eds.), *Groundwater ecology* (pp. 7–40). San Diego, CA: Academic Press.

Gido, K. (2002). Interspecific comparisons and the potential importance of nutrient excretion by benthic fishes in a large reservoir. *Transactions of the American Fisheries Society, 131*, 260–270.

Gilbert, D., Amblard, C., Bourdier, G., & Francez, A.-J. (1998). The microbial loop at the surface of a peatland: structure, function, and impact of nutrient input. *Microbial Ecology, 35*, 83–93.

Giles, N. (1994). Tufted duck (*Aythya fuligula*) habitat use and broad survival increases after fish removal from gravel pits. *Hydrobiologia, 279/280*, 387–392.

Gillesby, B. E., & Zacharewski, T. R. (1998). Exoestrogens: mechanisms of action and strategies for identification and assessment. *Environmental Toxicology and Chemistry, 17*, 3–14.

Gillespie, R. G., Howarth, F. G., & Roderick, G. K. (2001). Adaptive radiation. In *Encyclopedia of biodiversity* (pp. 25–27). San Diego, CA: Academic Press.

Gillis, A. M. (1995). What's at stake in the Pacific Northwest salmon debate? *BioScience, 45*, 125–128.

Gilmour, C. C., Riedel, G. S., Ederington, M. C., Bell, J. T., Benoit, J. M., Gill, G. A., et al. (1998). Methylmercury concentrations and production rates across a trophic gradient in the northern everglades. *Biogeochemistry, 40*, 327–345.

Gilpin, M. E., & Hanksi, I. A. (1991). *Metapopulation dynamics: Empirical and theoretical investigations*. London: Academic Press.

Gilpin, M. E., & Soulé, M. E. (1986). Minimum viable populations: the processes of species extinctions. In M. E. Soulé (Ed.), *Conservation biology: The science of scarcity and diversity* (pp. 13–34). Sunderland, MA: Sinauer Associates.

Ginzburg, B., Chalifa, I., Gun, J., Dor, I., Hadas, O., & Lev, O. (1998). DMS formation by dimethylsulfoniopropionate route in freshwater. *Environmental Science and Technology, 32*, 2130–2136.

Giovannoni, S. J., Britschgi, T. B., Moyer, C. L., & Field, K. G. (1990). Genetic diversity in Sargasso Sea bacterioplankton. *Nature, 345*, 60–62.

Giovannoni, S. J., Turner, S., Olsen, G. J., Barns, S., Lane, D. J., & Pace, N. R. (1988). Evolutionary relationships among *Cyanobacteria* and green chloroplasts. *Journal of Bacteriology, 170*, 3584–3592.

Gladyshev, E. A., Meselson, M., & Arkhipova, I. R. (2008). Massive horizontal gene transfer in bdelloid rotifers. *Science, 320*, 1210–1213.

Glagolev, A. N. (1984). *Motility and taxis in prokaryotes*. Cher, Switzerland: Harwood Academic Publishers.

Gleason, P. J., & Stone, P. (1994). Age, origin, and landscape evolution of the everglades peatland. In S. M. Davis & J. C. Ogden (Eds.), *Everglades* (pp. 149–198). Delray Beach, FL: St. Lucie Press.

Gleick, P. H. (1998). Water in crisis: paths to sustainable water use. *Ecological Applications, 8*, 571–579.

Gleick, P. H. (2008). *Water conflict chronology 11/10/08*. Pacific Institute for Studies in Development, Environment, and Security.

Gliwicz, Z. M. (1980). Filtering rates, food size selection, and feeding rates in cladocerans—another aspect of interspecific competition in filter-feeding zooplankton. In W. C. Kerfoot (Ed.), *Evolution and ecology of zooplankton communities* (pp. 282–291). University Press of New England.

Glud, R. N., & Fenchel, T. (1999). The importance of ciliates for interstitial solute transport in benthic communities. *Marine Ecology Progress Series, 186*, 87–93.

Goldman, C. R. (1960). Molybdenum as a factor limiting primary productivity in Castle Lake, California. *Science, 132*, 1016–1017.

Goldman, C. R. (1962). A method of studying nutrient limiting factors *in situ* in water columns isolated by polyethylene film. *Limnology and Oceanography, 7*, 99–101.

Goldman, C. R. (1972). The role of minor nutrients in limiting the productivity of aquatic ecosystems. In G. E. Likens (Ed.), *Symposium on nutrients and eutrophication* (pp. 21–38). American Society of Limnology and Oceanography.

Goldman, C. R., Jassby, A., & Powell, T. (1989). Interannual fluctuations in primary production: meteorological forcing at two subalpine lakes. *Limnology and Oceanography, 34*, 310–323.

Goldman, C. R., Jassby, A. D., & Hackley, S. H. (1993). Decadal, interannual, and seasonal variability in enrichment bioassays at Lake Tahoe, California-Nevada, USA. *Canadian Journal of Fisheries and Aquatic Sciences, 50*, 1489–1496.

Goldman, J. C., Caron, D. A., & Dennett, M. R. (1987). Regulation of gross growth efficiency and ammonium regeneration in bacteria by substrate C:N ratio. *Limnology and Oceanography, 32*, 1239–1252.

Goldschmidt, T., Witte, F., & Wanink, J. (1993). Cascading effects of the introduced Nile perch on the detritivorous/phytoplanktivorous species in the sublittoral areas of Lake Victoria. *Conservation Biology, 7*, 686–700.

Golladay, S. W., Webster, J. R., & Benfield, E. F. (1983). Factors affecting food utilization by a leaf shredding aquatic insect: leaf species and conditioning time. *Holarctic Ecology, 6*, 157–162.

Golterman, H. L., & de Oude, N. T. (1991). Eutrophication of lakes, rivers and coastal seas. In O. Hutzinger (Ed.), *The handbook of environmental chemistry* (pp. 79–124). Berlin: Springer-Verlag.

Gonzalez, J. M., & Suttle, C. A. (1993). Grazing by marine nanoflagellates on viruses and virus-sized particles: ingestion and digestion. *Marine Ecology Progress Series, 94*, 1–10.

Gopal, B., & Goel, U. (1993). Competition and allelopathy in aquatic plant communities. *The Botanical Review, 59*, 155–193.

Gordon, N. D., McMahon, T. A., & Finlayson, B. L. (1992). *Stream hydrology. An introduction for ecologists*. West Sussex, England: John Wiley & Sons Ltd.

Gore, J. A., & Shields, F. D., Jr. (1995). Can large rivers be restored? *BioScience, 45*, 142–152.

Gorham, E. (1991). Northern peatlands: role in the carbon cycle and probable response to climactic warming. *Ecological Applications, 1,* 182–195.

Gotelli, N., & Taylor, C. (1999). Testing metapopulation models with stream-fish assemblages. *Evolutionary Ecology Research, 1,* 835–845.

Gottgens, J. F., Fortney, R. H., Meyer, J., Perry, J. E., & Rood, B. E. (1998). The case of the Paraguay-Paraná waterway ("Hidrovia") and its impact on the Pantanal of Brazil: a summary report to the society of wetland scientists. *Wetlands Bulletin, 1998,* 12–18.

Goulding, M. (1980). *The fishes and the forest.* Berkeley: University of California Press.

Gounot, A. M. (1994). Microbial ecology of groundwaters. In J. Gibert, D. L. Danielopol, & J. A. Stanford (Eds.), *Groundwater ecology* (pp. 189–215). San Diego, CA: Academic Press.

Govedich, F. R., Blinn, D. W., Hevly, R. H., & Keim, P. S. (1999). Cryptic radiation in erpobdellid leeches in xeric landscapes: a molecular analysis of population differentiation. *Canadian Journal of Zoology, 77,* 52–57.

Grace, J., & Wetzel, R. (1981). Habitat partitioning and competitive displacement in cattails *(Typha):* experimental field studies. *American Naturalist,* 463–474.

Graham, J. L., Jones, J. R., Jones, S. B., Downing, J. A., & Clevenger, T. E. (2004). Environmental factors influencing microcystin distribution and concentration in the midwestern United States. *Water Research, 38,* 4395–4404.

Graham, L. E., & Wilcox, L. W. (2000). *Algae.* Upper Saddle River, NJ: Prentice Hall.

Granick, S., & Bae, S. C. (2008). A curios antipathy for water. *Science, 322,* 1477–1478.

Grant, W. D., Gemmell, R. T., & McGenity, T. J. (1998). Halophiles. In K. Horikoshi & W. D. Grant (Eds.), *Extremophiles: Microbial life in extreme environments* (pp. 93–132). New York: Wiley-Liss, Inc.

Gray, L. J. (1989). Emergence production and export of aquatic insects from a tallgrass prairie stream. *Southwestern Naturalist, 34,* 313–318.

Gray, N. F. (1998). Acid mine drainage composition and the implications for its impact on lotic systems. *Water Research, 32,* 2122–2134.

Green, W. J., Canfield, D. E., Shensong, Y., Chave, K. E., Ferdelman, T. G., & DeLanois, G. (1993). Metal transport and release processes in Lake Vanda: the role of oxide phases. In W. J. Green & E. I. Friedmann (Eds.), *Physical and biogeochemical processes in Antarctic lakes* (pp. 145–164). Washington, DC: American Geophysical Union.

Greenwood, J. L., Clason, T. A., Lowe, R. L., & Belanger, S. E. (1999). Examination of endopelic and epilithic algal community structure employing scanning electron microscopy. *Freshwater Biology, 41,* 821–828.

Greenwood, J. L., Rosemond, A. D., Wallace, J. B., Cross, W. F., & Weyers, H. S. (2007). Nutrient stimulate leaf breakdown rates and detritivore biomass: bottom-up effects via heterotrophic pathways. *Oecologia, 151,* 637–649.

Greenwood, P. H. (1974). *The cichlid fishes of Lake Victoria, East Africa: The biology and evolution of a species flock.* London: Bulletin of the British Museum.

Gregory, S. V., Swanson, F. J., McKee, W. A., & Cummins, K. W. (1991). An ecosystem perspective of riparian zones. *BioScience, 41,* 540–551.

Greulich, S., & Bornette, G. (1999). Competitive abilities and related strategies in four aquatic plant species from an intermediately disturbed habitat. *Freshwater Biology, 41,* 493–506.

Grimm, N. B., & Petrone, D. C. (1997). Nitrogen fixation in a desert stream ecosystem. *Biogeochemistry, 37,* 33–61.

Grist, D. H. (1986). *Rice.* New York: Longman.

Groffman, P., Baron, J., Blett, T., Gold, A., Goodman, I., Gunderson, L., et al. (2006). Ecological thresholds: the key to successful environmental management or an important concept with no practical application? *Ecosystems, 9,* 1–13.

Groffman, P. M., Dorsey, A. M., & Mayer, P. M. (2005). N processing within the geomorphic structures in urban streams. *Journal of the North American Benthological Society, 24,* 613–625.

Grosjean, H., & Oshima, T. (2007). How nucleic acids cope with high temperature. In C. Gerday & N. Glandsdorff (Eds.), *Physiology and biochemistry of extremophiles* (pp. 39–56). Washington, DC: ASM Press.

Gross, E. (2003). Allelopathy of aquatic autotrophs. *Critical Reviews in Plant Sciences, 22,* 313–339.

Gross, E. M. (1999). Allelopathy in benthic and littoral areas: case studies on allelochemicals from benthic cyanobacteria and submersed macrophytes. In K. M. M. Inderjit & C. L. Foy (Eds.), *Principles and practices in plant ecology* (pp. 179–199). Boca Raton, FL: CRC Press.

Gross, E. M., Hilt, S., Lombardo, P., & Mulderij, G. (2007). Searching for allelopathic effects of submerged macrophytes on phytoplankton—state of the art and open questions. *Hydrobiologia, 584,* 77–88.

Guan, B. H., Yao, X., Jiang, J. H., Tian, Z. Q., An, S. Q., Gu, B. H., et al. (2009). Phosphorus removal ability of three inexpensive substrates: physicochemical properties and application. *Ecological Engineering, 35,* 576–581.

Guégan, J., Lek, S., & Oberdorff, T. (1998). Energy availability and habitat heterogeneity predict global riverine fish diversity. *Nature, 391,* 382–384.

Gulis, V., Rosemond, A. D., Superkropp, K., Weyers, H. S., & Benstead, J. P. (2004). Effects of nutrient enrichment on the decomposition of wood and associated microbial activity in streams. *Freshwater Biology, 49,* 1437–1447.

Gurevitch, J., & Hedges, L. V. (1993). Meta-analysis: combining the results of independent experiments. In S. M. Scheiner & J. Gurevitch (Eds.), *Design and analysis of ecological experiments* (pp. 378–426). New York: Chapman & Hall.

Gurevitch, J., & Padilla, D. K. (2004). Are invasive species a major cause of extinctions? *Trends in Ecology and Evolution, 19,* 470–474.

Gurtz, M. E., Marzolf, G. R., Killingbeck, K. T., Smith, D. L., & McArthur, J. V. (1988). Hydrologic and riparian influences on the import and storage of coarse particulate organic matter in a prairie stream. *Canadian Journal of Fisheries and Aquatic Sciences, 45,* 655–665.

Gusewell, S., Bailey, K. M., Roem, W. J., & Bedford, B. L. (2005). Nutrient limitation and botanical diversity in wetlands: can fertilization raise species richness? *Oikos, 109,* 71–80.

Gustafsson, Ö., & Gschwend, P. M. (1997). Aquatic colloids: concepts, definitions, and current challenges. *Limnology and Oceanography, 42,* 519–528.

Guthrie, M. (1989). *Animals of the surface film.* Slough, UK: The Richmond Publishing Co. Ltd.

Guy, C. S., Blankenship, H. L., & Nielsen, L. A. (1996). Tagging and marking. In B. R. Murphy & D. W. Willis (Eds.), *Fisheries techniques* (pp. 353–383). Bethesda, MD: American Fisheries Society.

Haag, W. R., & Warren, M. L., Jr. (1999). Mantle displays of freshwater mussels elicit attacks from fish. *Freshwater Biology, 42,* 35–40.

Haden, G. A., Blinn, D. W., Shannon, J. P., & Wilson, K. P. (1999). Driftwood: an alternative habitat for macroinvertebrates in a large desert river. *Hydrobiologia, 397,* 179–186.

Häder, D.-P. (1997). Effects of UV radiation on phytoplankton. In J. G. Jones (Ed.), *Advances in microbial ecology* (pp. 1–26). New York: Plenum Press.

Häder, D.-P., Worrest, R. C., Kumar, H. K., & Smith, R. C. (1995). Effects of increased solar ultraviolet on aquatic ecosystems. *Ambio, 24,* 174–180.

Haga, H., Nagata, T., & Sakamoto, M. (1995). Size-fractionated NH_4^+ regeneration in the pelagic environments of two mesotrophic lakes. *Limnology and Oceanography, 40,* 1091–1099.

Hagerthey, S. E., & Kerfoot, W. C. (1998). Groundwater flow influences on the biomass and nutrient ratios of epibenthic algae in a north temperate seepage lake. *Limnology and Oceanography, 43,* 1227–1242.

Hagiwara, A., Yamamiya, N., & de Araujo, A. B. (1998). Effects of water viscosity on the population growth of the rotifer *Brachionus plicatilis* Müller. *Hydrobiologia, 387/388,* 489–494.

Hairston, N. G., Jr. (1987). Diapause as a predator-avoidance adaptation. In W. C. Kerfoot & A. Sih (Eds.), *Predation. Direct and indirect impacts on aquatic communities* (pp. 281–290). Hanover, NH: University Press of New England.

Hairston, N. G., Jr. (1996). Zooplankton egg banks as biotic reservoirs in changing environments. *Limnology and Oceanography, 41,* 1087–1092.

Hairston, N. G., Sr., Smith, F. E., & Slobodkin, L. B. (1960). Community structure, population control, and competition. *The American Naturalist, 94,* 421–425.

Hairston, N. G., Jr., Van Brunt, R. A., Kearns, C. M., & Engstrom, D. R. (1995). Age and survivorship of diapausing eggs in a sediment egg bank. *Ecology, 76,* 1706–1711.

Hakenkamp, C. C., & Palmer, M. A. (2000). The ecology of hyporheic meiofauna. In J. B. Jones & P. J. Mulholland (Eds.), *Streams and ground waters* (pp. 307–336). San Diego, CA: Academic Press.

Halbwachs, M., Sabroux, J., Grangeon, J., Kayser, G., Tochon-Danguy, J.-C., Felix, A., et al. (2004). Degassing the "Killer Lakes" Nyos and Monoun, Cameroon. EOS Transactions. *American Geophysical Union, 85,* 281–288.

Hall, D. J., & Threlkeld, S. T. (1976). The size-efficiency hypothesis and the size structure of zooplankton communities. *Annual Review of Ecology and Systematics, 7,* 177–208.

Hall, R. I., Leavitt, P. R., Dixit, A. S., Quinlan, R., & Smol, J. P. (1999). Limnological succession in reservoirs: a paleolimnological comparison of two methods of reservoir formation. *Canadian Journal of Fisheries and Aquatic Sciences, 56,* 1109–1121.

Hall, R. O., Jr. (1995). Use of a stable carbon-isotope addition to trace bacterial carbon through a stream food-web. *Journal of the North American Benthological Society, 14,* 269–277.

Hall, R. O., Jr., & Meyer, J. L. (1998). The trophic significance of bacteria in an detritus-based stream food web. *Ecology, 79,* 1995–2012.

Hall, R. O., Jr., Peterson, B. J., & Meyer, J. L. (1998). Testing of a nitrogen-cycling model of a forest stream by using a nitrogen-15 tracer addition. *Ecosystems, 1,* 283–298.

Halling-Sorensen, B., Nielsen, S. N., Lanzky, P. F., Ingerslev, F., Lutzhoft, H. C. H., & Jorgensen, S. E. (1998). Occurrence, fate and effects of pharmaceutical substances in the environment—a review. *Chemosphere, 36,* 357–394.

Hambright, K. D., Hairston, N. G., Schaffner, W. R., & Howarth, R. W. (2007). Grazer control of nitrogen fixation: synergisms in the feeding ecology of two freshwater crustaceans. *Fundamental and Applied Limnology, 170,* 89–101.

Hamilton, S. K. (1999). Potential effects of a major navigation project (Paraguay-Paraná Hiedrovía) on inundation in the Pantanal floodplains. *Regulated Rivers: Resource and Management, 15,* 289–299.

Hamilton, S. K., & Lewis, W. M., Jr. (1990). Basin morphology in relation to chemical and ecological characteristics of lakes on the Orinoco River floodplain, Venezuela. *Archiv für Hydrobiologie, 119,* 393–425.

Hamner, W. M., Gilmer, R. W., & Hamner, P. P. (1982). The physical, chemical, and biological characteristics of a stratified, saline, sulfide lake in Palau. *Limnology and Oceanography, 27,* 896–909.

Hansen, S. K., Rainey, P. B., Haagensen, J. A. J., & Molin, S. (2007). Evolution of species interactions in a biofilm community. *Nature, 445,* 533.

Hansson, L. A., Hylander, S., & Sommaruga, R. (2007). Escape from UV threats in zooplankton: a cocktail of behavior and protective pigmentation. *Ecology, 88,* 1932–1939.

Hansson, L.-A., Annadotter, H., Bergman, E., Hamrin, S. F., Jeppesen, E., Kairesalo, T., et al. (1998). Biomanipulation as an application of food-chain theory: constraints, synthesis, and recommendations for temperate lakes. *Ecosystems, 1,* 558–574.

Hantke, B., Fleischer, P., Domany, I., Koch, M., Pleb, P., Wiendl, M., et al. (1996). P-release from DOP by phosphatase activity in comparison to P excretion by zooplankton. Studies in hardwater lakes of different trophic levels. *Hydrobiologia, 317,* 151–162.

Hardie, L. A. (1984). Evaporites: marine or non-marine? *American Journal of Science, 284,* 193–240.

Hardin, G. (1960). The competitive exclusion principle. *Science, 131,* 1292–1297.

Harris, G. P. (1986). *Phytoplankton ecology, structure, function and fluctuation.* Cambridge, UK: Chapman and Hall.

Harris, J. M. (1993). The presence, nature, and role of gut microflora in aquatic invertebrates: a synthesis. *Microbial Ecology, 25,* 195–231.

Harris, M., Tomas, W., Mourao, G., Da Silva, C., Guimaraes, E., Sonoda, F., et al. (2005). Safeguarding the Pantanal wetlands: threats and conservation initiatives. *Conservation Biology, 19,* 714.

Harris, S. C., Martin, T. H., & Cummins, K. W. (1995). A model for aquatic invertebrate response to Kissimmee River restoration. *Restoration Ecology, 3,* 181–194.

Harrison, J. W., & Smith, R. E. H. (2009). Effects of ultraviolet radiation on the productivity and composition of freshwater phytoplankton communities. *Photochemical & Photobiological Sciences, 8,* 1218–1232.

Harrison, S. (1991). Local extinction in a metapopulation context: an empirical evaluation. *Biological Journal of the Linnean Society, 42,* 73–88.

Hart, B. T., Freeman, P., & McKelvie, I. D. (1992). Whole-stream phosphorus release studies: variation in uptake length with initial phosphorus concentration. *Hydrobiologia, 235/236,* 573–584.

Hart, B. T., Jaher, B., & Lawrence, I. (1999). New generation water quality guidelines for ecosystem protection. *Freshwater Biology, 41,* 347–359.

Hart, D. D. (1992). Community organization in streams: the importance of species interactions, physical factors, and chance. *Oecologia, 91*, 220–228.

Hartman, K. J., Kaller, M. D., Howell, J. W., & Sweka, J. A. (2005). How much do valley fills influence headwater streams? *Hydrobiologia, 532*, 91–102.

Hartman, P. E. (1983). Nitrate/nitrite ingestion and gastric cancer mortality. *Environmental Mutagenesis, 5*, 111–121.

Harvey, J. W., & Wagner, B. J. (2000). Quantifying hydrologic interactions between streams and their subsurface hyporheic zones. In J. B. Jones & P. J. Mulholland (Eds.), *Streams and ground waters* (pp. 3–44). San Diego, CA: Academic Press.

Harwell, M. A. (1998). Science and environmental decision making in south Florida. *Ecological Applications, 8*, 580–590.

Haselkorn, R., & Buikema, W. J. (1992). Nitrogen fixation in cyanobacteria. In G. Stacey, R. H. Burris, & H. J. Evans (Eds.), *Biological nitrogen fixation* (pp. 166–190). New York: Chapman and Hall.

Haslam, S. M. (1978). *River plants: The macrophytic vegetation of water courses.* London, UK: Cambridge University Press.

Hasler, A. D., & Scholz, A. T. (1983). *Olfactory imprinting and homing in salmon.* Berlin, Germany: Springer-Verlag.

Haveman, S. A., & Pedersen, K. (1999). Distribution and metabolic diversity of microorganisms in deep igneous rock aquifers of Finland. *Geomicrobiology Journal, 16*, 277–294.

Havens, K. E. (1991). Fish-induced sediment resuspension: effects on phytoplankton biomass and community structure in a shallow hypereutrophic lake. *Journal of Plankton Research, 13*, 1163–1176.

Havens, K. E. (1992a). Acidification effects on the algal-zooplankton interface. *Canadian Journal of Fisheries and Aquatic Sciences, 49*, 2507–2514.

Havens, K. E. (1992b). Scale and structure in natural food webs. *Science, 257*, 1107–1109.

Havens, K. E. (1993). Effect of scale on food web structure. *Science, 260*, 242–243.

Havens, K. E. (1995). Secondary nitrogen limitation in a subtropical lake impacted by non-point source agricultural pollution. *Environmental Pollution, 89*, 241–246.

Havens, K. E., Aumen, N. G., James, R. T., & Smith, V. H. (1996a). Rapid ecological changes in a large subtropical lake undergoing cultural eutrophication. *Ambio, 25*, 150–155.

Havens, K. E., Bull, L. A., Warren, G. L., Crisman, T. L., Philips, E. J., & Smith, J. P. (1996b). Food web structure in a subtropical lake ecosystem. *Oikos, 75*, 20–32.

Hayes, D. B., Ferreri, C. P., & Taylor, W. W. (1996). Active fish capture methods. In B. R. Murphy & D. W. Willis (Eds.), *Fisheries techniques* (pp. 193–220). Bethesda, MD: American Fisheries Society.

Healey, F. P., & Stewart, W. P. D. (1973). Inorganic nutrient uptake and deficiency in algae. *Critical Reviews of Microbiology, 3*, 69–113.

Heath, R. T., Fahnenstiel, G. L., Gardner, W. S., Cavaletto, J. F., & Hwang, S.-J. (1995). Ecosystem-level effects of zebra mussels (*Dreissena polymorpha*): an enclosure experiment in Saginaw Bay, Lake Huron. *Journal of Great Lakes Research, 21*, 501–516.

Heckman, C. W. (1994). The seasonal succession of biotic communities in wetlands of the tropical wet-and dry-climatic zone: I. Physical and chemical causes and biological effects in the Pantanal of Mato Grosso, Brazil. *International Revue gest Hydrobiologie, 79*, 397–421.

Hecky, R. E., Rosenberg, D. M., & Campbell, P. (1994). The 25th anniversary of the experimental lakes area and the history of lake 227. *Canadian Journal of Fisheries and Aquatic Sciences, 51*, 2243–2246.

Hedin, L. O., Fischer, J. C.v., Ostrom, N. E., Kennedy, B. P., Brown, M. G., & Robertson, G. P. (1998). Thermodynamic constraints on nitrogen transformations and other biogeochemical processes at soil-stream interfaces. *Ecology, 79*, 684–703.

Hedin, R. S., Watzlaf, G. R., & Nairn, R. W. (1994). Passive treatment of acid mine drainage with limestone. *Journal of Environmental Quality, 23*, 1338–1345.

Heidinger, R. C. (1999). Stocking for sport fisheries enhancement. In C. C. Kohler & W. A. Hubert (Eds.), *Inland fisheries management in North America* (pp. 375–401). Bethesda, MD: American Fisheries Society.

Hein, M. (1997). Inorganic carbon limitation of photosynthesis in lake phytoplankton. *Freshwater Biology, 37*, 545–552.

Heino, J., Virkkala, R., & Toivonen, H. (2009). Climate change and freshwater biodiversity: detected patterns, future trends and adaptations in northern regions. *Biological Reviews, 84*, 39–54.

Helfield, J., & Naiman, R. (2006). Keystone interactions. *Salmon and bear in riparian forests of Alaska, 9*, 167–180.

Hemmersbach, R., Volkmann, D., & Häder, D.-P. (1999). Graviorientation in protists and plants. *Journal of Plant Physiology*, *154*, 1–15.

Henriksen, A., Lien, L., Traaen, T. S., Rosseland, B. O., & Sevalrud, I. S. (1990). The 1000-lake survey in Norway 1986. In B. J. Mason (Ed.), *The surface waters acidification programme* (pp. 199–213). Cambridge: Cambridge University Press.

Herb, W. R., Janke, B., Mohseni, O., & Stefan, H. G. (2008). Thermal pollution of streams by runoff from paved surfaces. *Hydrological Processes*, *22*, 987–999.

Herdendorf, C. E. (1990). Distribution of the world's largest lakes. In M. M. Tilzer & C. Serruya (Eds.), *Large lakes: Ecological structure and function* (pp. 3–38). New York: Springer-Verlag.

Herndl, G. J., Mullerniklas, G., & Frick, J. (1993). Major role of ultraviolet-B in controlling bacterioplankton growth in the surface-layer of the ocean. *Nature*, *361*, 717–719.

Herrmann, J., Degerman, E., Gerhardt, A., Johansson, C., Lingdell, P., & Muniz, I. P. (1993). Acid-stress effects on stream biology. *Ambio*, *22*, 298–307.

Hesse, L. W., Chaffin, G. R., & Brabander, J. (1989). Missouri River mitigation: a system approach. *Fisheries*, *14*, 11–15.

Hessen, D. O. (1990). Niche overlap between herbivorous cladocerans; the role of food quality and habitat homogeneity. *Hydrobiologia*, *190*, 61–78.

Hewlett, J. D., & Hibbert, A. R. (1967). Factors affecting the response of small watersheds to precipitation in humid regions. In W. E. Sopper & H. W. Lull (Eds.), *Forest hydrology* (pp. 275–290). Oxford: Pergamon Press.

Hey, D. L., & Philippi, N. S. (1995). Flood reduction through wetland restoration: the Upper Mississippi River basin and a case history. *Restoration Ecology*, *3*, 4–17.

Hickman, C. P., & Roberts, L. S. (1995). *Animal diversity*. Dubuque, IA: Wm. C. Brown Pub.

Hiebert, F. K., & Bennett, P. C. (1992). Microbial control of silicate weathering in organic-rich ground water. *Science*, *258*, 278–281.

Hietala, J., Reinikainen, M., & Walls, M. (1995). Variation in life history responses of *Daphnia* to toxic *Microcystis aeruginosa*. *Journal of Plankton Research*, *17*, 2307–2318.

Hilborn, E. D., Carmichael, W. W., Soares, R. M., Yuan, M., Servaites, J. C., Barton, H. A., et al. (2007). Serologic evaluation of human microcystin exposure. *Environmental Toxicology*, *22*, 459–463.

Hilborn, R. (1996). The development of scientific advice with incomplete information in the context of the precautionary approach. In *Technical consultation on the precautionary approach to capture fisheries (including species introductions)* (pp. 77–101). Lysekil, Sweden: Food and Agriculture Organization of the United Nations.

Hilborn, R., Maguire, J., Parma, A., & Rosenberg, A. (2001). The precautionary approach and risk management: Can they increase the probability of successes in fishery management? *Canadian Journal of Fisheries and Aquatic Sciences*, *58*, 99–107.

Hill, A. M., & Lodge, D. M. (1995). Multi-trophic-level impact of sublethal interactions between bass and omnivorous crayfish. *Journal of the North American Benthological Society*, *14*, 306–314.

Hill, A. R., & Lymburner, D. J. (1998). Hyporheic zone chemistry and stream-subsurface exchange in two groundwater-fed streams. *Canadian Journal of Fisheries and Aquatic Sciences*, *55*, 495–506.

Hill, B. H., & Webster, J. R. (1983). Aquatic macrophyte contribution to the new river organic matter budget. In T. D. Fontaine, III, & S. M. Bartell (Eds.), *Dynamics of lotic ecosystems* (pp. 273–282). Ann Arbor, MI: Ann Arbor Science Publishers.

Hill, W. R., Dimick, S. M., McNamara, A. E., & Branson, C. A. (1997). No effects of ambient UV radiation detected in periphyton and grazers. *Limnology and Oceanography*, *42*, 769–774.

Hilsenhoff, W. L. (1987). An improved biotic index of organic stream pollution. *Great Lakes Entomologist*, *20*, 31–40.

Hilsenhoff, W. L. (1991). Diversity and classification of insects and Collembola. In J. H. Thorp & A. P. Covich (Eds.), *Ecology and classification of North American freshwater invertebrates* (pp. 593–664). San Diego, CA: Academic Press, Inc.

Hilt, S., & Gross, E. M. (2008). Can allelopathically active submerged macrophytes stabilise clear-water states in shallow lakes? *Basic and Applied Ecology*, *9*, 422–432.

Hinck, J., Blazer, V., Schmitt, C., Papoulias, D., & Tillitt, D. (2009). Widespread occurrence of intersex in black basses (*Micropterus* spp.) from US rivers, 1995–2004. *Aquatic Toxicology*.

Hinck, S., Neu, T. R., Lavik, G., Mussmann, M., Beer, D.d., & Jonkers, H. M. (2007). Physiological adaptation of a nitrate-storing *Beggiatoa* sp. to diel cycling in a phototrophic hypersaline mat. *Applied and Environmental Microbiology, 73*, 7013–7022.

Hoagland, K. D., Roemer, S. C., & Rosowski, J. R. (1982). Colonization and community structure of two periphyton assemblages, with emphasis on the diatoms (*Bacillariophyceae*). *American Journal of Botany, 69*, 188–213.

Hobbie, J. E. (1992). Microbial control of dissolved organic carbon in lakes: research for the future. *Hydrobiologia, 229*, 169–180.

Hobbie, J. E., Peterson, B. J., Bettez, N., Deegan, L., O'Brien, W. J., Kling, G. W., et al. (1999). Impact of global change on the biogeochemistry and ecology of an Arctic freshwater system. *Polar Biology, 18*, 207–214.

Hobbs, H. H., III (1991). Decapoda. In J. H. Thorp & A. P. Covich (Eds.), *Ecology and classification of North American freshwater invertebrates* (pp. 823–858). San Diego, CA: Academic Press, Inc.

Hodgson, J. R., Hodgson, C. J., & Brooks, S. M. (1991). Trophic interaction and competition between largemouth bass (*Micropterus salmoides*) and rainbow trout (*Oncorhynchus mykiss*) in a manipulated lake. *Canadian Journal of Fisheries and Aquatic Sciences, 40*, 1704–1712.

Hodoki, Y. (2005). Effects of solar ultraviolet radiation on the periphyton community in lotic systems: comparison of attached algae and bacteria during their development. *Hydrobiologia, 534*, 193–204.

Hodson, P. V. (1975). Zinc uptake by Atlantic salmon (*Salmo salar*) exposed to a lethal concentration of zinc at 3, 11, and 19 C. *Journal of the Fisheries Research Board of Canada, 32*, 2552–2556.

Hoeinghaus, D. J., Agostinho, A. A., Gomes, L. C., Pelicice, F. M., Okada, E. K., Latini, J. D., et al. (2009). Effects of river impoundment on ecosystem services of large tropical rivers: embodied energy and market value of artisanal fisheries. *Conservation Biology, 23*, 1222–1231.

Hoham, R. W. (1980). Unicellular chlorophytes-snow algae. In E. R. Cox (Ed.), *Phytoflagellates: Developments in marine biology* (pp. 61–84). Amsterdam, Netherlands: Elsevier/North Holland.

Holeck, K. T., Mills, E. L., MacIssac, H. J., Dochoda, M. R., Colautii, R. I., & Ricciardi, A. (2004). Bridging troubled waters: Biological invasions, transoceanic shipping, and the Laurentian Great Lakes. *BioScience, 54*, 919–929.

Holland, H. D. (1978). *The chemistry of atmosphere and oceans*. New York: Wiley Interscience.

Holland, R. E., Johengen, T. H., & Beeton, A. M. (1995). Trends in nutrient concentrations in Hatchery Bay, western Lake Erie, before and after *Dreissena polymorpha*. *Canadian Journal of Fisheries and Aquatic Sciences, 52*, 1202–1209.

Holt, J. G., Krieg, N. R., Sneath, P. H., Staley, J. T., & Williams, S. T. (1994). *Bergey's manual of determinative bacteriology* (9th ed.). Baltimore, MD: Williams and Wilkins.

Holyoak, M., & Leibold, M. (2005). *Metacommunities: Spatial dynamics and ecological communities*. Chicago: University of Chicago Press.

Hooper, D. U., Chapin, F. S., III, Ewel, J. J., Hector, A., Inchausti, P., Lavorel, S., et al. (2005). Effects of biodiversity on ecosystem functioning: a consensus of current knowledge. *Ecological Monographs, 75*, 3–35.

Hori, M., Gashagaza, M. M., Nshombo, M., & Kawanabe, H. (1993). Littoral fish communities in Lake Tanganyika: irreplaceable diversity supported by intricate interactions among species. *Conservation Biology, 7*, 657–666.

Horne, A. J., & Goldman, C. R. (1972). Nitrogen fixation in Clear Lake, California. I. Seasonal variation and the role of heterocysts. *Limnology and Oceanography, 17*, 678–692.

Horne, A. J., & Goldman, C. R. (1994). *Limnology* (2nd ed.). New York: McGraw-Hill, Inc.

Hörnström, E. (1999). Long-term phytoplankton changes in acid and limed lakes in SW Sweden. *Hydrobiologia, 394*, 93–102.

Howard-Williams, C., & Hawes, I. (2007). Ecological processes in Antarctic inland waters: interactions between physical processes and the nitrogen cycle. *Antarctic Science, 19*, 205–217.

Howard-Williams, C., Schwarz, A.-M., Hawes, I., & Priscu, J. C. (1998). Optical properties of the McMurdo dry valley lakes, Antarctica. In J. C. Priscu (Ed.), *Ecosystem dynamics in a polar desert* (pp. 189–203). Washington, DC: American Geophysical Union.

Howarth, R. W., & Cole, J. J. (1985). Molybdenum availability, nitrogen limitation, and phytoplankton growth in natural waters. *Science, 229*, 653–655.

Hu, D. L., & Bush, J. W. M. (2005). Meniscus-climbing insects. *Nature, 437*, 733–736.

Huang, C., Wikfeldt, K. T., Tokushima, T., Nordlund, D., Harada, Y., Bergmann, U., et al. (2009). The inhomogeneous structure of water at ambient conditions. *Proceedings of the National Academy of Sciences, 106*, 15214–15218.

Hubert, W. A. (1996). Passive capture techniques. In B. R. Murphy & D. W. Willis (Eds.), *Fisheries techniques* (pp. 157–181). Bethesda, MD: American Fisheries Society.

Hudson, J. J., & Taylor, W. D. (1996). Measuring regeneration of dissolved phosphorus in planktonic communities. *Limnology and Oceanography, 41*, 1560–1565.

Hudson, J. J., & Taylor, W. D. (2005). Rapid estimation of phosphate at picomolar concentrations in freshwater lakes with potential application to P-limited marine systems. *Aquatic Sciences, 67*, 316–325.

Hughes, G. M. (1981). Effects of low oxygen and pollution on the respiratory systems of fish. In A. D. Pickering (Ed.), *Stress and fish* (pp. 212–246). New York: Academic Press.

Hughes, R. N. (1997). Diet selection. In J.-G. J. Godin (Ed.), *Behavioural ecology of teleost fishes* (pp. 134–162). New York: Oxford University Press.

Huisman, J., & Weissing, F. J. (1999). Biodiversity of plankton by species oscillations and chaos. *Nature, 402*, 407–410.

Hunsaker, C. T., & Levine, D. A. (1995). Hierarchical approaches to the study of water quality in rivers. *BioScience, 45*, 193–203.

Hunter, R. G., Faulkner, S. P., & Gibson, K. A. (2008). The importance of hydrology in restoration of bottomland hardwood wetland functions. *Wetlands, 28*, 605–615.

Hunter-Cevera, J. C. (1998). The value of microbial diversity. *Current Opinion in Microbiology, 1*, 278–285.

Hurd, C. L., & Stevens, C. L. (1997). Flow visualization around single- and multiple-bladed seaweeds with various morphologies. *Journal of Phycology, 33*, 360–367.

Hurlbert, S. H. (1984). Pseudoreplication and the design of ecological field experiments. *Ecological Monographs, 54*, 187–211.

Hurley, J. P., Krabbenhoft, D. P., Cleckner, L. B., Olson, M. L., Aiken, G. R., & Rawlik, J. P. S. (1998). System controls on the aqueous distribution of mercury in the northern Florida Everglades. *Biogeochemistry, 40*, 293–311.

Huryn, A. D., & Wallace, J. B. (2000). Life history and production of stream insects. *Annual Review of Entomology, 45*, 83–110.

Huston, M. A. (1994). *Biological diversity, the coexistence of species on changing landscapes.* Cambridge, UK: Cambridge University Press.

Huszar, V., Caraco, N., Roland, F., & Cole, J. (2006). Nutrient–chlorophyll relationships in tropical-subtropical lakes: Do temperate models fit? *Biogeochemistry, 79*, 239–250.

Hutchens, E., Radajewski, S., Dumont, M. G., McDonald, I. R., & Murrell, J. C. (2004). Analysis of methanotrophic bacteria in Movile Cave by stable isotope probing. *Environmental Microbiology, 6*, 111–120.

Hutchin, P. R., Press, M. C., Lee, J. A., & Ashenden, T. W. (1995). Elevated concentrations of CO_2 may double methane emissions from mires. *Global Change Biology, 1*, 125–128.

Hutchinson, G. E. (1957). *A treatise on limnology. Geography, Physics and Chemistry.* New York: John Wiley & Sons, Inc.

Hutchinson, G. E. (1959). Homage to Santa Rosalia or why are there so many kinds of animals? *The American Naturalist, 93*, 145–159.

Hutchinson, G. E. (1961). The paradox of the plankton. *The American Naturalist, 95*, 137–145.

Hutchinson, G. E. (1967). *A treatise on limnology. An introduction to lake biology and limnoplankton.* New York: John Wiley & Sons, Inc.

Hutchinson, G. E. (1975). *A treatise on limnology. Limnological botany.* New York: John Wiley & Sons, Inc.

Hutchinson, G. E. (1993). *A treatise on limnology. The zoobenthos.* New York: John Wiley & Sons, Inc.

Hutchinson, G. E., & Bowen, V. T. (1947). A direct demonstration of the phosphorus cycle in a small lake. *Proceedings of the National Academy of Sciences of USA, 33*, 148–153.

Hutchinson, G. E., & Cowgill, U. (1970). Ianula: an account of the history and development of the Lago di Monterosi, Latium, Italy. The History of the lake: a synthesis. *Transactions of the American Philosophical Society, 60*, 163–170.

Hyatt, T. L., & Naiman, R. J. (2001). The residence time of large woody debris in the Queets River, Washington, USA. *Ecological Applications, 11*, 191–202.

Hynes, H. B. N. (1960). *The biology of polluted waters.* Liverpool, UK: Liverpool University Press.

Hynes, H. B. N. (1970). *The ecology of running waters.* Toronto, Ontario, Canada: University of Toronto Press.

Hynes, H. B. N. (1975). Edgardo Baldi memorial lecture; the stream and its valley. *Verein Internationale Verein Limnologie, 19,* 1–15.

Ibelings, B. W., Portielje, R., Lammens, E. H. R. R., Noordhuis, R., vandenBerg, M. S., Joosse, W., et al. (2007). Resilience of alternative stable states during the recovery of shallow lakes from eutrophication: Lake Veluwe as a case study. *Ecosystems, 10,* 4–16.

Ingersoll, T. L., & Baker, L. A. (1998). Nitrate removal in wetland microcosms. *Water Research, 32,* 677–684.

IPCC. (2007). Climate change 2007: the physical science basis: summary for policymakers. In M. L. Parry, O. F. Canziani, J. P. Palutikof, P. J. van der Linden, & C. E. Hanson (Eds.), *Fourth Assessment Report of the IPCC.* Cambridge, UK: Cambridge University Press.

Islam, F. S., Gault, A. G., Boothman, C., Polya, D. A., Charnock, J. M., Chatterjee, D., et al. (2004). Role of metal-reducing bacteria in arsenic release from Bengal delta sediments. *Nature, 430,* 68.

Jack, J. D., Fang, W., & Thorp, J. H. (2006). Vertical, lateral and longitudinal movement of zooplankton in a large river. *Freshwater Biology, 51,* 1646–1654.

Jacoby, J. M., Collier, D. C., Welch, E. B., Hardy, F. J., & Crayton, M. (2000). Environmental factors associated with a toxic bloom of *Microcystis aeruginosa. Canadian Journal of Fisheries and Aquatic Sciences, 57,* 231–240.

Jacomini, A. E., Avelar, W. E. P., Martinez, A. S., & Bonato, P. S. (2006). Bioaccumulation of atrazine in freshwater bivalves *Anodontites trapesialis* (Lamarck, 1819) and *Corbicula fluminea* (Muller, 1774). *Archives of Environmental Contamination and Toxicology, 51,* 387–391.

Jahn, T. L., Bovee, E. C., & Jahn, F. F. (1979). *How to know the protozoa.* Dubuque, IA: Wm. C. Brown Company Publishers.

James, R. T., & Havens, K. E. (1996). Algal bloom probability in a large subtropical lake. *Water Resources Bulletin, 32,* 995–1006.

Jana, B. B. (1994). Ammonification in aquatic environments: a brief review. *Limnologica, 24,* 389–413.

Jansson, R., Nilsson, C., Dynesius, M., & Andersson, E. (2000). Effects of river regulation on river-margin vegetation: a comparison of eight boreal rivers. *Ecological Applications, 10,* 203–224.

Jassby, A. D., Goldman, C. R., & Reuter, J. E. (1995). Long-term change in Lake Tahoe (California-Nevada, U.S.A.) and its relation to atmospheric deposition of algal nutrients. *Arch. Hydrobiol., 135,* 1–21.

Javor, B. (1989). *Hypersaline environments. Microbiology and biogeochemistry.* Berlin, Germany: Springer-Verlag.

Jeffrey, W. H., Pledger, R. J., Aas, P., Hager, S., Corrin, R. B., Haven, R. V., et al. (1996). Diel and depth profiles of DNA photodamage in bacterioplankton exposed to ambient solar ultraviolet radiation. *Marine Ecology Progress Series, 137,* 283–291.

Jeffries, M., & Mills, D. (1990). *Freshwater ecology: Principles and applications.* London: Belhaven Press.

Jenkins, D. G., & Buikema, A. L., Jr. (1998). Do similar communities develop in similar sites? A test with zooplankton structure and function. *Ecological Monographs, 68,* 421–443.

Jenkins, D. G., & Underwood, M. O. (1998). Zooplankton may not disperse readily in wind, rain, or waterfowl. *Hydrobiologia, 387/388,* 15–21.

Jenkins, M. (2003). Prospects for biodiversity. *Science, 302,* 1175–1177.

Jeppesen, E., Sondergaard, M., Jensen, J. P., Havens, K. E., Anneville, O., Carvalho, L., et al. (2005). Lake responses to reduced nutrient loading—an analysis of contemporary long-term data from 35 case studies. *Freshwater Biology, 50,* 1747–1771.

Jeppesen, E., Søndergaard, M., Jensen, J. P., Mortensen, E., Hansen, A.-M., & Jørgensen, T. (1998). Cascading trophic interactions from fish to bacteria and nutrients after reduced sewage loading: An 18-year study of a shallow hypertrophic lake. *Ecosystems, 1,* 250–267.

Joabsson, A., Christensen, T. R., & Wallén, B. (1999). Vascular plant controls on methane emissions from northern peatforming wetlands. *Trends in Ecology and Evolution, 14,* 385–388.

Jöhnk, K. D., Huisman, J., Sharples, J., Sommeijer, B., Visser, P. M., & Stroom, J. M. (2008). Summer heatwaves promote blooms of harmful cyanobacteria. *Global Change Biology, 14,* 495–512.

Johns, C. (1995). Contamination of riparian wetlands from past copper mining and smelting in the headwaters region of the Clark-Fork River, Montana, US. *Journal of Geochemical Exploration, 52,* 193–203.

Johnson, D. (2007). Physiology and ecology of acidophilic microorganisms. In C. Gerday & N. Glandsdorff (Eds.), *Physiology and biochemistry of extremophiles* (pp. 257–270). Washington, DC: ASM Press.

Johnson, L., Tank, J., & Dodds, W. (2009). The influence of land use on stream biofilm nutrient limitation across eight North American ecoregions. *Canadian Journal of Fisheries and Aquatic Sciences, 66,* 1081–1094.

Johnson, T. C., Scholz, C. A., Talbot, M. R., Kelts, K., Ricketts, R. D., Ngobi, G., et al. (1996). Late pleistocene desiccation of Lake Victoria and rapid evolution of cichlid fishes. *Science, 273,* 1091–1092.

Johnson, W. C. (1994). Woodland expansion in the Platte River, Nebraska: patterns and causes. *Ecological Monographs, 64,* 45–84.

Johnson, W. C. (1997). Equilibrium response of riparian vegetation to flow regulation in the Platte River, Nebraska. *Regulated Rivers Research & Management, 13,* 403–415.

Johnston, C. A. (1991). Sediment and nutrient retention by freshwater wetlands: effects on surface water quality. *Critical Reviews in Environmental Control, 21,* 491–565.

Jones, H. G. (1999). The ecology of snow-covered systems: a brief overview of nutrient cycling and life in the cold. *Hydrological Processes, 13,* 2135–2147.

Jones, J. B., Jr., & Holmes, R. M. (1996). Surface-subsurface interactions in stream ecosystems. *Trends in Ecology and Evolution, 11,* 239–242.

Jones, J. R., & Bachmann, R. W. (1976). Prediction of phosphorus and chlorophyll levels in lakes. *Journal of the Water Pollution and Control Federation, 48,* 2176–2183.

Jones, R., & Flynn, K. (2005). Nutritional status and diet composition affect the value of diatoms as copepod prey. *Science, 307,* 1457.

Jones, S., Chiu, C., Kratz, T., Wu, J., Shade, A., & McMahon, K. (2008). Typhoons initiate predictable change in aquatic bacterial communities. *Limnology and Oceanography, 53,* 1319–1326.

Jonsson, M., Malmqvist, B., & Hoffsten, P.-O. (2001). Leaf litter breakdown in boreal streams: does shredder species richness matter? *Freshwater Biology, 46,* 161–171.

Jørgensen, B. B., & Marais, D. J. D. (1988). Optical properties of benthic photosynthetic communities: fiber-optic studies of cyanobacterial mats. *Limnology and Oceanography, 33,* 99–113.

Jouzel, J., Petit, J. R., Souchez, R., Barkov, N. I., Lipenkov, V. Y., Raynaud, D., et al. (1999). More than 200 meters of lake ice above subglacial Lake Vostok, Antarctica. *Science, 286,* 2138–2141.

Joyce, D. A., Lunt, D. H., Bills, R., Turner, G. F., Katongo, C., Dufner, N., et al. (2005). An extinct cichlid fish radiation emerged in an extinct Pleistocene lake. *Nature, 435,* 90–95.

Juanes, F. (1994). What determines prey size selectivity in piscivorous fishes? In D. J. Stouder, K. L. Fresh, & R. J. Feller (Eds.), *Theory and application in fish feeding ecology* (pp. 80–100). Columbia, SC: University of South Carolina Press.

Juneau, P., & Harrison, P. J. (2005). Comparison by PAM fluorometry of photosynthetic activity of nine marine phytoplankton grown under identical conditions. *Photochemistry and Photobiology, 81,* 649–653.

Juniper, B. E., Robins, R. J., & Joel, D. M. (1989). *The carnivorous plants.* London, UK: Academic Press.

Junk, W. J. (1999). The flood pulse concept of large rivers: learning from the tropics. *Archiv Fur Hydrobiologie Suppl, 115/3,* 261–280.

Junk, W. J., Bayley, P. B., & Sparks, R. E. (1989). The flood pulse concept in river-floodplain systems. In *Proceedings of the International Large River Symposium.* Canadian special publications of fisheries and aquatic sciences (pp. 110–127).

Junk, W. J., de Cunha, C. N., Wantzen, K. M., Petermann, P., Strüssmann, C., Marques, M. I., et al. (2006). Biodiversity and its conservation in the Pantanal of Mato Grosso, Brazil. *Aquatic Sciences, 68,* 278–309.

Justic, D., Rabalais, N. N., & Turner, R. E. (1995). Stoichiometric nutrient balance and origin of coastal eutrophication. *Marine Pollution Bulletin, 30,* 41–46.

Justic, D., Rabalais, N. N., Turner, R. E., & Dortch, Q. (1995). Changes in nutrient structure of river-dominated coastal waters: stoichiometric nutrient balance and its consequences. *Estuarine, Coastal and Shelf Science, 40,* 339–356.

Jüttner, F., Backhaus, D., Matthias, U., Essers, U., Greiner, R., & Mahr, B. (1995). Emissions of two- and four-stroke outboard engines. II. Impact on water quality. *Water Research, 28,* 1983–1987.

Kadlec, R. H. (1994). Wetlands for water polishing: free water surface wetlands. In W. J. Mitsch (Ed.), *Global wetlands old world and new* (pp. 335–349). Amsterdam, The Netherlands: Elsevier.

Kalbus, E., Reinstorf, F., & Schirmer, M. (2006). Measuring methods for groundwater–surface water interactions: a review. *Hydrology and Earth System Sciences, 10,* 873–887.

Kamjunke, N., & Zehrer, R. F. (1999). Direct and indirect effects of strong grazing by *Daphnia galeata* on bacterial production in an enclosure experiment. *Journal of Plankton Research, 21,* 1175–1182.

Kann, J., & Smith, V. H. (1999). Estimating the probability of exceeding elevated pH values critical to fish populations in a hypereutrophic lake. *Canadian Journal of Fisheries and Aquatic Sciences, 56*, 2262–2270.

Kapitsa, A. P., Ridley, J. K., Robin, G. D. Q., Siegert, M. J., & Zotikov, I. A. (1996). A large deep freshwater lake beneath the ice of central East Antarctica. *Nature, 381*, 684–686.

Kareiva, P., Marvier, M., & McClure, M. (2000). Recovery and management options for spring/summer Chinook salmon in the Columbia River Basin. *Science, 290*, 977.

Karentz, D., Bothwell, M. L., Coffin, R. B., Hanson, A., Herndl, G. J., Kilham, S. S., et al. (1994). Impact of UV-B radiation on pelagic freshwater ecosystems: report of working group on bacteria and phytoplankton. *Archive für Hydrobiologie Beih, 43*, 31–69.

Karl, D. M., Bird, D. F., Björkman, K., Houlihan, T., Shackelford, R., & Tupas, L. (1999). Microorganisms in the accreted ice of Lake Vostok, Antarctica. *Science, 286*, 2144–2147.

Karr, J. R. (1981). Assessment of biotic integrity using fish communities. *Fisheries, 6*, 21–27.

Karr, J. R. (1991). Biological integrity: a long-neglected aspect of water resource management. *Ecological Applications, 1*, 66–84.

Kashian, D. R., Zuellig, R. E., Mitchell, K. A., & Clements, W. H. (2007). The cost of tolerance: sensitivity of stream benthic communities to UV-B and metals. *Ecological Applications, 17*, 365–375.

Kasza, H., & Winohradnik, J. (1986). Development and structure of the Gozalkowice reservoir ecosystem VII. Hydrochemistry. *Ekologia Polska, 34*, 365–395.

Kaufman, L. (1992). Catastrophic change in species-rich freshwater ecosystems. *BioScience, 42*, 846–858.

Kaushal, S. S., Groffman, P. M., Likens, G. E., Belt, K. T., Stack, W. P., Kelly, V. R., et al. (2005). Increased salinization of fresh water in the northeastern United States. *Proceedings of the National Academy of Sciences of the United States of America, 102*, 13517–13520.

Keating, K. I. (1977). Allelopathic influence on blue-green bloom sequence in a eutrophic lake. *Science, 196*, 885–887.

Keating, K. I. (1978). Blue-green algal inhibition of diatom growth: transition from mesotrophic to eutrophic community structure. *Science, 199*, 971–973.

Keddy, P. A. (1989). Effects of competition from shrubs on herbaceous wetland plants: a 4-year field experiment. *Canadian Journal of Botany, 67*, 708–716.

Keddy, P. A., Fraser, L. H., Solomeshch, A. I., Junk, W. J., Campbell, D. R., Arroyo, M. T. K., et al. (2009). Wet and wonderful: the world's largest wetlands are conservation priorities. *BioScience, 59*, 39–51.

Keddy, P. A., Twolan-Strutt, L., & Wisheu, I. C. (1994). Competitive effect and response rankings in 20 wetland plants: Are they consistent across three environments? *Journal of Ecology, 82*, 635–643.

Keenan, C. W., & Wood, J. H. (1971). *General college chemistry* (4th ed.). New York: Harper & Row, Publishers.

Keenleyside, M. H. A. (1991). Parental care. In M. H. A. Keenleyside (Ed.), *Cichlid fishes, behaviour, ecology and evolution* (pp. 191–208). Cambridge, UK: Chapman and Hall.

Kehew, A. E., & Lord, M. L. (1987). Glacial-lake outbursts along the mid-continent margins of the Laurentide ice-sheet. In *Catastrophic flooding* (pp. 95–120). Boston, MA: Allen & Unwin.

Kehoe, T. (1997). *Cleaning up the Great Lakes*. Dekalb, IL: Northern Illinois University Press.

Kelly, B. C., Ikonomou, M. G., Blair, J. D., Morin, A. E., & Gobas, F. A. P. C. (2007). Food web-specific biomagnification of persistent organic pollutants. *Science, 317*, 236–239.

Kenefick, S. L., Hrudey, S. E., Peterson, H. G., & Prepas, E. E. (1993). Toxin release from *Microcystis aeruginosa* after chemical treatment. *Water Science and Technology, 27*, 433–440.

Kent, G. (1987). *Fish, food and hunger*. Boulder, CO: Westview Press.

Kerfoot, W. C., Newman, R. M., & Hanscom, Z., III (1998). Snail reaction to watercress leaf tissues: reinterpretation of a mutualistic "alarm" hypothesis. *Freshwater Biology, 40*, 201–213.

Kidd, K., Schindler, A. D. W., Muir, D. C. G., Lockhart, W. L., & Hesslein, R. H. (1995). High concentrations of toxaphene in fishes from a subarctic lake. *Science, 269*, 240–242.

Kieft, T. L., Murphy, E. M., Haldeman, D. L., Amy, P. S., Bjornstad, B. N., McDonald, E. V., et al. (1998). Microbial transport, survival, and succession in a sequence of buried sediments. *Microbial Ecology, 36*, 336–348.

Kieft, T. L., & Phelps, T. J. (1997). Life in the slow lane: activities of microorganisms in the subsurface. In P. S. Amy & D. L. Haldeman (Eds.), *Microbiology of the terrestrial deep subsurface* (pp. 137–163). Boca Raton, FL: Lewis Publishers.

Kiesecker, J. M., Blaustein, A. R., & Belden, L. K. (2001). Complex causes of amphibian population declines. *Nature, 410,* 681–684.

Kilham, P., & Kilham, S. S. (1990). Endless summer: internal loading processes dominate nutrient cycling in tropical lakes. *Freshwater Biology, 23,* 379–389.

Kilham, S. S., Theriot, E. C., & Fritz, S. C. (1996). Linking planktonic diatoms and climate change in the large lakes of the Yellowstone ecosystem using resource theory. *Limnology and Oceanography, 41,* 1052–1062.

Kilroy, C., Biggs, B. J. F., & Vyverman, W. (2007). Rules for macroorganisms applied to microorganisms: patterns of endemism in benthic freshwater diatoms. *Oikos, 116,* 550–564.

King, J. L., Simovich, M. A., & Brusca, R. C. (1996). Species richness, endemism and ecology of crustacean assemblages in northern California vernal pools. *Hydrobiologia, 328,* 85–116.

Kinney, C. A., Furlong, E. T., Kolpin, D. W., Burkhardt, M. R., Zaugg, S. D., Werner, S. L., et al. (2008). Bioaccumulation of pharmaceuticals and other anthropogenic waste indicators in earthworms from agricultural soil amended with biosolid or swine manure. *Environmental Science and Technology, 42,* 1863–1870.

Kinzie, R. A., III, Banaszak, A. T., & Lesser, M. P. (1998). Effects of ultraviolet radiation on primary productivity in a high altitude tropical lake. *Hydrobiologia, 385,* 23–32.

Kiørboe, T., Saiz, E., & Visser, A. (1999). Hydrodynamic signal perception in the copepod *Acartia tonsa. Marine Ecology Progress Series, 179,* 97–111.

Kiørboe, T., & Visser, A. W. (1999). Predator and prey perception in copepods due to hydromechanical signals. *Marine Ecology Progress Series, 179,* 81–95.

Kirk, J. T. O. (1994). *Light and photosynthesis in aquatic ecosystems* (2nd ed.). Cambridge, UK: Cambridge University Press.

Kirk, K. L., & Gilbert, J. J. (1990). Suspended clay and the population dynamics of planktonic rotifers and cladocerans. *Ecology, 71,* 1741–1755.

Kitchell, J. F., & Carpenter, S. R. (1992). Summary: accomplishments and new directions of food web management in Lake Mendota. In J. F. Kitchell (Ed.), *Food web management. A case study of Lake Mendota* (pp. 539–544). New York: Springer-Verlag.

Klann, A., Levy, G., Lutz, I., Muller, C., Kloas, W., & Hildebrandt, J. P. (2005). Estrogen-like effects of ultraviolet screen 3-(4-methylbenzylidene)-camphor (Eusolex 6300) on cell proliferation and gene induction in mammalian and amphibian cells. *Environmental Research, 97,* 274–281.

Klemer, A. R., Cullen, J. J., Mageau, M. T., Hanson, K. M., & Sundell, R. A. (1996). Cyanobacterial buoyancy regulation: the paradoxical roles of carbon. *Journal of Phycology, 32,* 47–53.

Kline, T. C., Jr., Goering, J. J., Mathisen, O. A., & Poe, P. H. (1990). Recycling of elements transported upstream by runs of Pacific salmon I. ^{15}N and ^{13}C evidence in Sashin Creek, Southeastern Alaska. *Canadian Journal of Fisheries and Aquatic Sciences, 47,* 136–144.

Kling, G. W. (1987). Seasonal mixing and catastrophic degassing in tropical lakes, Cameroon, West Africa. *Science, 237,* 1022–1024.

Kling, G. W., Clark, M. A., Compton, H. R., Devine, J. D., Evans, W. C., Humphrey, A. M., et al. (1987). The 1986 Lake Nyos gas disaster in Cameroon, West Africa. *Science, 236,* 175–179.

Knapp, A. K., Blair, J. M., Briggs, J. M., Collins, S. L., Hartnett, D. C., Johnson, L. C., et al. (1999). The keystone role of bison in North American tallgrass prairie—bison increase habitat heterogeneity and alter a broad array of plant, community, and ecosystem processes. *BioScience, 49,* 39–50.

Knight, T. M., McCoy, M. W., Chase, J. M., McCoy, K. A., & Holt, R. D. (2005). Trophic cascades across ecosystems. *Nature, 437,* 880–883.

Knowlton, M. F., & Jones, J. R. (1997). Trophic status of Missouri River floodplain lakes in relation to basin type and connectivity. *Wetlands, 17,* 468–475.

Koerselman, W., Bakker, S. A., & Blom, M. (1990). Nitrogen, phosphorus and potassium budgets for two small fens surrounded by heavily fertilized pastures. *Journal of Ecology, 78,* 428–442.

Kolar, C. S., & Lodge, D. M. (2001). Progress in invasion biology: predicting invaders. *Trends in Ecology and Evolution, 16,* 199–204.

Kolar, C. S., & Wahl, D. H. (1998). Daphnid morphology deters fish predators. *Oecologia, 116,* 556–564.

Kolasa, J. (1991). Flatworms: turbellaria and nemertea. In J. H. Thorp & A. P. Covich (Eds.), *Ecology and classification of North American freshwater invertebrates* (pp. 145–172). San Diego, CA: Academic Press, Inc.

Kolpin, D. W., Furlong, E. T., Meyer, M. T., Thurman, E. M., Zaugg, S. D., Barber, L. B., et al. (2002). Pharmaceuticals, hormones, and other organic wastewater contaminants in US streams, 1999–2000: a national reconnaissance. *Environmental Science and Technology, 36,* 1202–1211.

Kondolf, G. M., Boulton, A. J., O'Daniel, S., Poole, G. C., Rahel, F. J., Stanley, E. H., et al. (2006). Process-based ecological river restoration: visualizing three-dimensional connectivity and dynamic vectors to recover lost linkages. *Ecology and Society, 11,* 5.

Koprivnjak, J.-F., Blanchette, J. G., Bourbonniere, R. A., Clair, T. A., Heyes, A., Lum, K. R., et al. (1995). The underestimation of concentrations of dissolved organic carbon in freshwaters. *Water Research, 29,* 91–94.

Koshland, D. E., Jr. (1980). *Bacterial chemotaxis as a model behavioral system.* New York: Raven Press Books, Ltd.

Kota, S., Borden, R. C., & Barlaz, M. A. (1999). Influence of protozoan grazing on contaminant biodegradation. *FEMS Microbiology Ecology, 29,* 179–189.

Kotak, B. G., Hrudey, S. E., Kenefick, S. L., & Prepas, E. E. (1992). Toxicity of cyanobacterial blooms in Alberta lakes. In E. G. Baddaloo, S. Ramamoorthy, & J. W. Moore (Eds.), *Proceedings of the Nineteenth Annual Aquatic Toxicity Workshop* (pp. 172–179). Edmonton, Alberta, Canada: Alberta Environmental Protection.

Kotelnikova, S., & Pedersen, K. (1998). Distribution and activity of methanogens and homoacetogens in deep granitic aquifers at Äspö Hard Rock Laboratory, Sweden. *FEMS Microbiology Ecology, 26,* 121–134.

Koussoroplis, A. M., Lemarchand, C., Bec, A., Desvilettes, C., Amblard, C., Fournier, C., et al. (2008). From aquatic to terrestrial food webs: decrease of the docosahexaenoic acid/linoleic acid ratio. *Lipids, 43,* 461–466.

Kovalak, W. P., Longton, G. D., & Smithee, R. D. (1993). Infestation of power plant water systems by the zebra mussel (*Dreissena polymorpha* Pallas). In T. F. Nalepa & D. W. Schloesser (Eds.), *Zebra mussels. Biology, impacts, and control* (pp. 359–379). Boca Raton, FL: Lewis Publishers.

Kozhov, M. (1963). *Lake Baikal and its life.* The Hague: Dr. W. Junk, Publishers.

Kratz, T. K., & Frost, T. M. (2000). The ecological organization of lake districts: general introduction. *Freshwater Biology, 43,* 297–299.

Kratz, T. K., Webster, K. E., Bowser, C. J., Magnuson, J. J., & Benson, B. J. (1997). The influence of landscape position on lakes in northern Wisconsin. *Freshwater Biology, 37,* 209–217.

Krienitz, L., Ballot, A., Kotut, K., Wiegand, C., Pütz, S., Metcalf, J. S., et al. (2003). Contribution of hot spring cyanobacteria to the mysterious deaths of lesser flamingos at Lake Bogoria, Kenya. *FEMS Microbiology Ecology, 43,* 141–148.

Kroeze, C., Dumont, E., & Seitzinger, S. P. (2005). New estimates of global emissions of N_2O from rivers and estuaries. *Environmental Sciences, 2,* 159–165.

Kromm, D. E., & White, S. E. (1992). Groundwater problems. In D. E. Kromm & S. E. White (Eds.), *Groundwater exploitation in the high plains* (pp. 44–63). Lawrence, KS: University Press of Kansas.

Kromm, D. E., & White, S. E. (1992). The high plains Ogallala region. In D. E. Kromm & S. E. White (Eds.), *Groundwater exploitation in the high plains* (pp. 1–27). Lawrence, KS: University Press of Kansas.

Kronvang, B., Grant, R., Larsen, S. E., Svendsen, L. M., & Kristensen, P. (1995). Non-point-source nutrient losses to the aquatic environment in Denmark: impact of agriculture. *Marine Freshwater Research, 46,* 167–177.

Krumholz, L. R., McKinley, J. P., Ulrich, G. A., & Suflita, J. M. (1997). Confined subsurface microbial communities in Cretaceous rock. *Nature, 386,* 64–66.

Krzyzanek, E. (1986). Development and structure of the Goczalkowice Reservoir ecosystem XIV. Zoobenthos. *Ekologia Polska, 34,* 491–513.

Kuflikowski, T. (1986). Development and structure of the Goczalkowice Reservoir ecosystem X. Macrophytes. *Ekologia Polska, 34,* 429–445.

Kumar, A., Smith, R. P., & Häder, D.-P. (1996). Effect of UV-B on enzymes of nitrogen metabolism in the cyanobacterium *Nostoc calcicola. Journal of Plant Physiology, 148,* 86–91.

Kumar, S., Arya, S., & Nussinov, R. (2007). Temperature-dependent molecular adaptation features in proteins. In C. Gerday & N. Glansdorff (Eds.), *Physiology and biochemistry of extremophiles* (pp. 75–85). Washington, DC: ASM Press.

Kutalek, R., & Kassa, A. (2005). The use of gyrinids and dyctiscids for stimulating breast growth in East Africa. *Journal of Ethnobiology, 25*, 115–128.

la Rivière, J. W. M. (1989). Threats to the world's water. *Scientific American, 9/89*, 80–94.

Lake, P. S. (2000). Disturbance, patchiness, and diversity in streams. *Journal of the North American Benthological Society, 19*, 573–592.

Lam, A. K.-Y., & Prepas., E. E. (1997). In situ evaluation of options for chemical treatment of hepatotoxic cyanobacterial blooms. *Canadian Journal of Fisheries and Aquatic Sciences, 54*, 1736–1742.

Lam, A. K.-Y., Prepas, E. E., Spink, D., & Hrudey, S. E. (1995). Chemical control of hepatotoxic phytoplankton blooms: implications for human health. *Water Research, 29*, 1845–1854.

Lam, B., Baer, A., Alaee, M., Lefebvre, B., Moser, A., Williams, A., et al. (2007). Major structural components in freshwater dissolved organic matter. *Environmental Science and Technology, 41*, 8240–8247.

Lamberti, G. A., Gregory, S. V., Ashkenas, L. R., Wildman, R. C., & Moore, K. M. S. (1991). Stream ecosystem recovery following a catastrophic debris flow. *Canadian Journal of Fisheries and Aquatic Sciences, 48*, 196–208.

Lamberti, G. A., & Moore, J. W. (1984). Aquatic insects and primary consumers. In V. H. Resh & D. M. Rosenberg (Eds.), *The ecology of aquatic insects* (pp. 164–195). New York: Praeger.

Lampert, W. (1997). Zooplankton research: the contribution of limnology to general ecological paradigms. *Aquatic Ecology, 31*, 19–27.

Lancaster, J., & Hildrew, A. G. (1993). Characterization of in-stream flow refugia. *Canadian Journal of Fisheries and Aquatic Sciences, 50*, 1663–1675.

Lane, P. A. (1985). A food web approach to mutualism in lake communities. In D. H. Boucher (Ed.), *The biology of mutualism* (pp. 344–374). New York: Oxford University Press.

Langford, T. E. (1990). *Ecological effects of thermal discharges*. New York: Elsevier Science Publishing, Inc.

Langhelle, A., Lindell, M. J., & Nyström, P. (1999). Effects of ultraviolet radiation on amphibian embryonic and larval development. *Journal of Herpetology, 33*, 449–456.

Lannoo, M. (2008). *Malformed frogs: the collapse of aquatic ecosystems*. Berkeley: University of California Press.

Larson, D. (1993). The recovery of Spirit Lake. *American Scientist, 81*, 166–177.

Lassen, C., Revsbech, N. P., & Pedersen, O. (1997). Macrophyte development and resuspension regulate the photosynthesis and production of benthic microalgae. *Hydrobiologia, 350*, 1–11.

Lathrop, R. C., Carpenter, S. R., & Rudstam, L. G. (1996). Water clarity in Lake Mendota since 1900: responses to differing levels of nutrients and herbivory. *Canadian Journal of Fisheries and Aquatic Sciences, 53*, 2250–2261.

Lathrop, R. C., Carpenter, S. R., Stow, C. A., Soranno, P. A., & Panuska, J. C. (1998). Phosphorus loading reductions needed to control blue-green algal blooms in Lake Mendota. *Canadian Journal of Fisheries and Aquatic Sciences, 55*, 1169–1178.

Laurion, I., Lean, D. R. S., & Vincent, W. F. (1998). UVB effects on a plankton community: results from a large-scale enclosure assay. *Aquatic Microbial Ecology, 16*, 189–198.

Lavrentyev, P. J., Gardner, W. S., Cavaletto, J. F., & Beaver, J. R. (1995). Effects of the zebra mussel (*Dreissena polymorpha* Pallas) on protozoa and phytoplankton from Saginaw Bay, Lake Huron. *Journal of Great Lakes Research, 21*, 545–557.

Laws, E. A. (1993). *Aquatic pollution*. New York: John Wiley & Sons.

Lawton, J. H. (1991). Are species useful? *Oikos, 62*, 3–4.

Lean, D. R. S., & Pick, F. R. (1981). Photosynthetic response of lake plankton to nutrient enrichment: a test for nutrient limitation. *Limnology and Oceanography, 26*, 1001–1019.

Leao, P., Vasconcelos, M., & Vasconcelos, V. (2009). Allelopathy in freshwater cyanobacteria. *Critical Reviews in Microbiology*, 1–12.

Lear, L. (1997). *Rachel Carson: witness for nature*. New York: H. Holt & Co.

Leavitt, P. R., Vinebrooke, R. D., Donald, D. B., Smol, J. P., & Schindler, D. W. (1997). Past ultraviolet radiation environments in lakes derived from fossil pigments. *Nature, 388*, 457–459.

Lee, B.-G., Griscom, S. B., Lee, J.-S., Choi, H. J., Koh, C.-H., Luoma, S. N., et al. (2000). Influences of dietary uptake and reactive sulfides on metal bioavailability from aquatic sediments. *Science, 287*, 282–284.

Lee, G. F., & Jones, R. A. (1991). Effects of eutrophication on fisheries. *Reviews in Aquatic Sciences, 5*, 3–4.

Lee, J. J., Leedale, G. F., & Bradbury, P. (2000). *An illustrated guide to the Protozoa* (2nd ed.). Lawrence, KS: Society of Protozoologists.

Leff, L. G., McArthur, J. V., & Shimkets, L. J. (1993). Spatial and temporal variability of antibiotic resistance in freshwater bacterial assemblages. *FEMS Microbiology Ecology, 13*, 135–144.

Lehman, J. T., & Scavia, D. (1982). Microscale nutrient patches produced by zooplankton. *Proceedings of the National Academy of Science, 789*, 5001–5005.

Lehmkuhl, D. M. (1979). *How to know the aquatic insects*. Dubuque, IA: Wm. C. Brown Company Publishers.

Leopold, L. B. (1994). *A view of the river*. Cambridge, MA: Harvard University Press.

Leopold, L. B., & Davis, K. S. (1996). *Water*. New York: Time Inc.

Leopold, L. B., Wolman, M. G., & Miller, J. P. (1964). *Fluvial processes in geomorphology*. San Francisco, CA: W. H. Freeman and Company.

Lever, C. (1994). *Naturalized animals: The ecology of successfully introduced species*. London: T & AD Poyser Ltd.

Levine, J. M. (2000). Species diversity and biological invasions: relating local process to community pattern. *Science, 288*, 852–854.

Lewis, W. M., Jr. (1986). Evolutionary interpretations of allelochemical interactions in phytoplankton algae. *The American Naturalist, 127*, 184–194.

Lewis, W. M., Jr., Hamilton, S. K., Lasi, M. A., Rodríguez, M., & Saunders, J. F., III (2000). Ecological determinism on the Orinoco floodplain. *BioScience, 50*, 681–692.

Li, H. W., Rossignol, P. A., & Castillo, G. (2000). Risk analysis of species introductions: insights from qualitative modeling. In R. Claudi & J. H. Leach (Eds.), *Nonindigenous freshwater organisms* (pp. 431–447). Boca Raton, FL: Lewis Publishers, CRC Press.

Liao, J. C., Beal, D. N., Lauder, G. V., & Triantafyllou, M. S. (2003). Fish exploiting vortices decrease muscle activity. *Science, 302*, 1566–1569.

Licht, L. E. (2003). Shedding light on ultraviolet radiation and amphibian embryos. *BioScience, 53*, 551–561.

Lieth, H. F. H. (1975). Primary production of the major vegetation units of the world. In H. F. H. Lieth & R. H. Whittaker (Eds.), *Primary productivity of the biosphere* (pp. 203–215). New York: Springer-Verlag.

Light, S. S., & Dineen, J. W. (1994). Water control in the Everglades: a historical perspective. In S. M. Davis & J. C. Ogden (Eds.), *Everglades* (pp. 47–84). Delray Beach, FL: St. Lucie Press.

Likens, G. E. (2001). Biogeochemistry, the watershed approach: some uses and limitations. *Marine Freshwater, 52*, 5–12.

Likens, G. E., Bormann, F. H., Pierce, R. S., & Reiners, W. A. (1978). Recovery of a deforested ecosystem. *Science, 199*, 492–496.

Likens, G. E., Driscoll, C. T., & Buso, D. C. (1996). Long-term effects of acid rain: response and recovery of a forest ecosystem. *Science, 272*, 244–246.

Lillesand, T. M., Johnson, W. L., Deuell, R. L., Lindstrom, O. M., & Meisner, D. E. (1983). Use of Landsat data to predict the trophic state of Minnesota lakes. *Photogrammetric Engineering and Remote Sensing, 12*, 2045–2063.

Lind, O. T. (1974). *Handbook of common methods in limnology*. St. Louis, MO: C. V. Mosby.

Lindeman, R. L. (1942). The trophic-dynamic aspect of ecology. *Ecology, 23*, 399–418.

Lindenschmidt, K.-E., & Hamblin, P. F. (1997). Hypolimnetic aeration in Lake Tegel, Berlin. *Water Research, 31*, 1619–1628.

Lips, K. R., Brem, F., Brenes, R., Reeve, J. D., Alford, R. A., Voyles, J., et al. (2006). Emerging infectious disease and the loss of biodiversity in a neotropical amphibian community. *Proceedings of the National Academy of Sciences of the United States of America, 103*, 3165–3170.

Lips, K. R., Diffendorfer, J., Mendelson, J. R., & Sears, M. W. (2008). Riding the wave: reconciling the roles of disease and climate change in amphibian declines. *Plos Biology, 6*, 441–454.

Lips, K. R., Reeve, J. D., & Witters, L. R. (2003). Ecological traits predicting amphibian population declines in Central America. *Conservation Biology, 17*, 1078–1088.

Lipson, S. M., & Stotzky, G. (1987). Interactions between clay minerals and viruses. In V. C. Rao & J. L. Melnick (Eds.), *Human viruses in sediments, sludges, and soils* (pp. 197–230). Boca Raton, FL: CRC Press.

Liston, S. E., Newman, S., & Trexler, J. C. (2008). Macroinvertebrate community response to eutrophication in an oligotrophic wetland: an in situ mesocosm experiment. *Wetlands, 28*, 686–694.

Litchman, E. (1998). Population and community responses of phytoplankton to fluctuating light. *Oecologia, 117*, 247–257.

Little, T. J., & Hebert, P. D. N. (1996). Endemism and ecological islands: the ostracods from Jamaican bromeliads. *Freshwater Biology, 36*, 327–338.

Liu, K., Brown, M. G., Carter, C., Saykally, R. J., Gregory, J. K., & Clary, D. C. (1996). Characterization of a cage form of the water hexameter. *Nature, 381*, 501–503.

Lloyd, R. (1960). The toxicity of zinc sulphate to rainbow trout. *Annals of Applied Biology, 48*, 84–94.

Lodge, D. M., Barko, J. W., Strayer, D., Melack, J. M., Mittelbach, G. G., Howarth, R. W., et al. (1987). Spatial heterogeneity and habitat interactions in lake communities. In S. R. Carpenter (Ed.), *Complex interactions in lake communities* (pp. 181–208). Berlin: Springer-Verlag.

Loeb, S. L., & Spacie, A. (Eds.). (1994). *Biological monitoring of aquatic systems*. Ann Arbor, MI: Lewis Publishers.

Loehr, R. C. (1974). Characteristics and comparative magnitude of non-point sources. *Journal of the Water Pollution Control Federation, 46*, 1849–1872.

Loehr, R. C., Ryding, S.-O., & Sonzogni, W. C. (1989). Estimating the nutrient load to a waterbody. In S.-O. Ryding & W. Rast (Eds.), *The control of eutrophication of lakes and reservoirs* (pp. 115–146). Paris, France: The Parthenon Publishing Group.

Loman, J. (2004). Density regulation in tadpoles of *Rana temporaria*: a full pond field experiment. *Ecology, 85*, 1611–1618.

Lomans, B. P., Smolders, A. J. P., Intven, L. M., Pol, A., denCamp, H. J. M. O., & vanderDrift, C. (1997). Formation of dimethyl sulfide and methanethiol in anoxic freshwater sediments. *Applied and Environmental Microbiology, 63*, 4741–4747.

Long, S. P., Humphries, S., & Falkowski, P. G. (1994). Photoinhibition of photosynthesis in nature. *Annual Revue of Plant Physiology and Plant Molecular Biology, 45*, 633–662.

Lopez-Bueno, A., Tamames, J., Velazquez, D., Moya, A., Quesada, A., & Alcami, A. (2009). High diversity of the viral community from an Antarctic Lake. *Science, 326*, 858.

Lorang, M., & Hauer, F. R. (2007). Fluvial geomorphological processes. In F. R. Hauer & G. A. Lamberti (Eds.), *Methods in stream ecology* (pp. 145–168). San Diego, CA: Academic Press.

Loreau, M., Naeem, S., Inchausti, P., Bengtsson, J., Grime, J. P., Hector, A., et al. (2001). Ecology—biodiversity and ecosystem functioning: current knowledge and future challenges. *Science, 294*, 804–808.

Losey, J. E., Raynor, L. S., & Carter, M. E. (1999). Transgenic pollen harms monarch larvae. *Nature, 399*, 214.

Loudon, C., & Alstad, D. N. (1990). Theoretical mechanics of particle capture: predictions for hydropsychid caddisfly distribution ecology. *The American Naturalist, 135*, 360–381.

Lovley, D. R., Coates, J. D., Blunt-Harris, E. L., Phillips, E. J. P., & Woodward, J. C. (1996). Humic substances as electron acceptors for microbial respiration. *Nature, 382*, 445–448.

Lowe, R. L., & Pillsbury, R. W. (1995). Shifts in benthic algal community structure and function following the appearance of zebra mussels (*Dreissena polymorpha*) in Saginaw Bay, Lake Huron. *Journal of Great Lakes Research, 21*, 558–566.

Lowrance, R., Todd, R., Fail, J., Jr., Hendrickson, O., Jr., Leonard, R., & Asmussen, L. (1984). Riparian forests as nutrient filters in agricultural watersheds. *BioScience, 34*, 374–377.

Lucassen, E. C. H. E. T., Smolders, A. P., VanderSalm, A. L., & Roelofs, J. G. M. (2004). High groundwater nitrate concentrations inhibit eutrophication of sulphate-rich freshwater wetlands. *Biogeochemistry, 67*, 249–267.

Ludes, B., Coste, M., Tracqui, A., & Mangin, P. (1996). Continuous river monitoring of the diatoms in diagnosis of drowning. *Journal of Forensic Science, 41*, 425–428.

Ludwig, J. A., & Reynolds, J. F. (1988). *Statistical ecology: a primer on methods and computing*. New York: John Wiley & Sons.

Ludyanskiy, M. L., McDonald, D., & MacNeill, D. (1993). Impact of the zebra mussel, a bivalve invader. *BioScience, 43*, 533–544.

Lund, J. W. G. (1964). Primary production and periodicity of phytoplankton. *Verhein Internationale Verein Limnologie, 15*, 37–56.

Lüring, M. (1998). Effect of grazing-associated infochemicals on growth and morphological development in *Scenedesmus acutus* (Chlorophyceae). *Journal of Phycology, 34*, 578–586.

Lürling, M., & Donk, E. V. (2000). Grazer-induced colony formation in *Scenedesmus*: are there costs to being colonial? *Oikos, 88*, 111–118.

Luzar, A., & Chandler, D. (1996). Hydrogen-bond kinetics in liquid water. *Nature, 379*, 55–57.

Lytle, D. A., & Poff, N. L. (2004). Adaptation to natural flow regimes. *Trends in Ecology and Evolution, 19*, 94–100.

MacArthur, R. H. (1955). Fluctuations of animal populations and a measure of community stability. *Ecology, 36,* 533–536.

MacArthur, R. H., & Wilson, E. O. (1967). *The theory of island biogeography.* Princeton, NJ: Princeton University Press.

Maceina, M. J., Cichra, M. F., Betsill, R. K., & Bettoli, P. W. (1992). Limnological changes in a large reservoir following vegetation removal by grass carp. *Journal of Freshwater Ecology, 7,* 81–95.

Mack, R. N., Simberloff, D., Lonsdale, W. M., Evans, H., Clout, M., & Bazzaz, F. A. (2000). Biotic invasions: causes, epidemiology, global consequences and control. *Ecological Applications, 10,* 689–710.

Madigan, M. T., & Oren, A. (1999). Thermophilic and halophilic extremophiles. *Current Opinion in Microbiology, 2,* 265–269.

Madronich, S., McKenzie, R. L., Caldwell, M. M., & Bjorn, L. O. (1995). Changes in ultraviolet radiation reaching the earth's surface. *Ambio, 24,* 143–152.

Madsen, E. L., & Ghiorse, W. C. (1993). Groundwater microbiology: subsurface ecosystem processes. In T. E. Ford (Ed.), *Aquatic microbiology: An ecological approach* (pp. 167–213). Oxford, UK: Blackwell Scientific Publications.

Madsen, E. L., Sinclair, J. L., & Ghiorse, W. C. (1991). In situ biodegradation: microbiological patterns in a contaminated aquifer. *Science, 252,* 830–833.

Magnuson, J. J., Benson, B. J., & Kratz, T. K. (1990). Temporal coherence in the limnology of a suite of lakes in Wisconsin, U.S.A. *Freshwater Biology, 23,* 145–159.

Magnuson, J. J., & Kratz, T. K. (1999). Lakes in the landscape: approaches to regional limnology. *Verein Internationale Verein Limnologie, 27,* 1–14.

Magnuson, J. J., Robertson, D. M., Benson, B. J., Wynne, R. H., Livingstone, D. M., Arai, T., et al. (2000). Historical trends in lake and river ice cover in the Northern Hemisphere. *Science, 289,* 1743–1746.

Magnuson, J. J., Webster, K. E., Assel, R. A., Bowser, C. J., Dillon, P. J., Eaton, J. G., et al. (1997). Potential effects of climate changes on aquatic systems: Laurential Great Lakes and precambrian shield region. *Proceedings of Hydrology, 11,* 825–871.

Magurran, A. E. (2004). *Measuring biological diversity.* Oxford: Blackwell.

Maire, R., & Pomel, S. (1994). Karst geomorphology and environment. In J. Gibert, D. L. Danielopol, & J. A. Stanford (Eds.), *Groundwater ecology* (pp. 129–155). San Diego, CA: Academic Press, Inc.

Majewski, S. P., & Cumming, B. F. (1999). Paleolimnological investigation of the effects of post-1970 reductions of acidic deposition on an acidified Adirondack lake. *Journal of Paleolimnology, 21,* 207–213.

Makarewics, J. C., Lewis, T. W., & Bertram, P. (1999). Phytoplankton composition and biomass in offshore waters of Lake Erie: Pre- and post-*Dreissena* introduction (1983–1993). *Journal of Great Lakes Research, 25,* 135–148.

Mal, T. K., Lovett-Doust, J., Lovett-Doust, L., & Mulligan, G. A. (1992). The biology of Canadian weeds. 100. *Lythrum salicaria. Canadian Journal of Plant Science, 72,* 1305–1330.

Malard, F., & Hervant, F. (1999). Oxygen supply and the adaptations of animals in groundwater. *Freshwater Biology, 41,* 1–30.

Malard, F., Reygrobellet, J.-L., Mathieu, J., & Lafont, M. (1994). The use of invertebrate communities to describe groundwater flow and contaminant transport in a fractured rock aquifer. *Archiv für Hydrobiologie, 131,* 93–110.

Malecki, R. A., Blossey, B., Hight, S. D., Schroeder, D., Kok, L. T., & Coulson, J. R. (1993). Biological control of purple loosestrife. *BioScience, 43,* 680–686.

Malin, G., & Kirst, G. O. (1997). Algal production of dimethyl sulfide and its atmospheric role. *Journal of Phycology, 33,* 889–896.

Mallory, M. L., McNicol, D. K., Cluis, D. A., & Laberge, C. (1998). Chemical trends and status of small lakes near Sudbury, Ontario, 1983–1995: evidence of continued chemical recovery. *Canadian Journal of Fisheries and Aquatic Sciences, 55,* 63–75.

Malmqvist, B., Wotton, R. S., & Zhang, Y. X. (2001). Suspension feeders transform massive amounts of seston in large northern rivers. *Oikos, 92,* 35–43.

Mangin, A. (1994). Karst hydrogeology. In J. Gibert, D. L. Danielopol, & J. A. Stanford (Eds.), *Groundwater ecology* (pp. 43–67). San Diego, CA: Academic Press, Inc.

Maranger, R., & Bird, D. F. (1995). Viral abundance in aquatic systems: a comparison between marine and fresh waters. *Marine Ecology Progress Series, 121,* 217–226.

Marcarelli, A. M., Baker, M. A., & Wursbaugh, W. A. (2008). Is in-stream N_2 fixation an important N source for benthic communities and stream ecosystems? *Journal of the North American Benthological Society, 27,* 186–211.

March, J. G., Benstead, J. P., Pringle, C. M., & Scatena, F. N. (1998). Migratory drift of larval freshwater shrimps in two tropical streams, Puerto Rico. *Freshwater Biology, 40,* 261–273.

Marmonier, P., Vervier, P., Gibert, J., & Dole-Olivier, M.-J. (1993). Biodiversity in ground waters. *Trends in Ecology and Evolution, 8,* 392–395.

Martí, E., Fisher, S. G., Schade, J. D., & Grimm, N. B. (2000). Flood frequency and stream-riparian linkages in arid lands. In J. B. Jones & P. J. Mulholland (Eds.), *Streams and ground waters* (pp. 111–136). San Diego, CA: Academic Press.

Martí, E., & Sabater, F. (1996). High variability in temporal and spatial nutrient retention in Mediterranean streams. *Ecology, 77,* 854–869.

Martinez, N. D. (1991). Artifacts or attributes? Effects of resolution on the Little Rock Lake food web. *Ecological Monographs, 61,* 367–392.

Martinez, N. D. (1993). Effect of scale on food web structure. *Science, 260,* 242–243.

Martínez-Cortizas, A., Pontevedra-Pombal, X., García-Rodeja, E., Nóvoa-Muñoz, J. C., & Shotyk, W. (1999). Mercury in a Spanish peat bog: archive of climate change and atmospheric metal deposition. *Science, 284,* 939–942.

Marzolf, E. R., Mulholland, P. J., & Steinman, A. D. (1994). Improvement to the diurnal upstream-downstream dissolved oxygen change technique for determining whole-stream metabolism in small streams. *Canadian Journal of Fisheries and Aquatic Sciences, 51,* 1591–1599.

Marzolf, E. R., Mulholland, P. J., & Steinman, A. D. (1998). Reply: improvements to the diurnal upstream-downstream dissolved oxygen change technique for determining whole-stream metabolism in small streams. *Canadian Journal of Fisheries and Aquatic Sciences, 55,* 1786–1787.

Mason, C. F. (1996). *Biology of freshwater pollution* (3rd ed.). Essex, UK: Longman Press.

Matena, J., Simek, K., & Fernando, C. H. (1995). Ingestion of suspended bacteria by fish: a modified approach. *Journal of Fish Biology, 47,* 334–336.

Mats, V. D. (1993). The structure and development of the Baikal rift depression. *Earth-Science Reviews, 34,* 81–118.

Matthews, W. J. (1998). *Patterns in freshwater fish ecology.* Norwell, MA: Kluwer Academic Publishers.

Matthews, W. J., Stewart, A. J., & Power, M. E. (1987). Grazing fishes as components of North American stream ecosystems: effects of *Campostoma anomalum.* In W. J. Matthews & D. C. Heins (Eds.), *Community and evolutionary ecology of North American fishes* (pp. 128–135). Norman, OK: University of Oklahoma Press.

Maul, J. D., Belden, J. B., Schwab, B. A., Whiles, M. R., Spears, B., Farris, J. L., et al. (2006a). Bioaccumulation and trophic transfer of polychlorinated biphenyls by aquatic and terrestrial insects to tree swallows (*Tachycineta bicolor*). *Environmental Toxicology and Chemistry, 25,* 1017–1025.

Maul, J. D., Schuler, L. J., Belden, J. B., Whiles, M. R., & Lydy, M. J. (2006b). Effects of the antibiotic ciprofloxacin on stream microbial communities and detritivorous macroinvertebrates. *Environmental Toxicology and Chemistry, 25,* 1598–1606.

May, R. M. (1972). Will large complex systems be stable? *Nature, 238,* 413–414.

May, R. M. (1988). How many species are there on earth? *Science, 241,* 1441–1448.

Mazak, E. J., MacIsaac, H. J., Servos, M. R., & Hesslein, R. (1997). Influence of feeding habits on organochlorine contaminant accumulation in waterfowl on the great lakes. *Ecological Applications, 7,* 1133–1143.

Mazumder, A., Taylor, W. D., McQueen, D. J., & Lean, D. R. S. (1990). Effects of fish and plankton on lake temperature and mixing depth. *Science, 247,* 312–315.

McAuliffe, J. R. (1983). Competition, colonization patterns, and disturbance in stream benthic communities. In J. R. Barnes & G. W. Minshall (Eds.), *Stream ecology: Application and testing of general ecological theory* (pp. 137–156). New York: Plenum Press.

McCabe, D. J., & Gotelli, N. J. (2000). Effects of disturbance frequency, intensity, and area on assemblages of stream macroinvertebrates. *Oecologia, 124,* 270–279.

McCafferty, W. P. (1988). *Aquatic entomology, the fisherman's and ecologists' illustrated guide to insects and their relatives.* Sudbury, MA: Jones and Bartlett Publishers.

McClain, M. E., Richey, J. E., & Pimentel, T. P. (1994). Groundwater nitrogen dynamics at the terrestrial-lotic interface of a small catchment in the Central Amazon Basin. *Biogeochemistry, 27,* 113–127.

McCormick, P. V., Chimney, M. J., & Swift, D. R. (1997). Diel oxygen profiles and water column community metabolism in the Florida Everglades, U.S.A. *Archiv für Hydrobiologie, 140,* 117–129.

McFadden, C. H., & Keeton, W. T. (1995). *Biology: An exploration of life*. New York: Norton.

McInnes, S. J., & Pugh, P. J. A. (1998). Biogeography of limno-terrestrial Tardigrada, with particular reference to the Antarctic fauna. *Journal of Biogeography, 25*, 31–36.

McIntyre, P. B., Jones, L. E., Flecker, A. S., & Vanni, M. J. (2007). Fish extinctions alter nutrient recycling in tropical freshwaters. *Proceedings of the National Academy of Sciences of the United States of America, 104*, 4461–4466.

McKeon, D. M., Calabrese, J. P., & Bissonnette, G. K. (1995). Antibiotic-resistant gram-negative bacteria in rural groundwater supplies. *Water Research, 29*, 1902–1908.

McKinney, T., Rogers, R. S., Ayers, A. D., & Persons, W. R. (1999). Lotic community responses in the Lees Ferry Reach. In R. H. Webb, J. C. Schmidt, G. R. Marzolf, & R. A. Valdez (Eds.), *The controlled flood in the Grand Canyon* (pp. 249–258). Washington, DC: American Geophysical Union, Geophysical Monograph Series.

McKnight, D. M., Boyer, E. W., Westerhoff, P. K., Doran, P. T., Kulbe, T., & Andersen, D. T. (2001). Spectrofluorometric characterization of dissolved organic matter for indication of precursor organic material and aromaticity. *Limnology and Oceanography, 46*, 38–48.

McLachlan, J. A., & Arnold, S. F. (1996). Environmental estrogens. *American Scientist, 84*, 452–461.

McMahon, R. F. (1991). Mollusca: Bivalvia. In J. H. Thorp & A. P. Covich (Eds.), *Ecology and classification of North American freshwater invertebrates* (pp. 315–400). San Diego, CA: Academic Press.

McMahon, R. F. (2000). Invasive characteristics of the freshwater bivalve *Corbicula fluminea*. In R. Claudi & J. H. Leach (Eds.), *Nonindigenous freshwater organisms* (pp. 315–343). Boca Raton, FL: Lewis Publishers, CRC Press.

Meade, J. W. (1989). *Aquaculture management*. New York: Van Nostrand Reinhold.

Mearns, L. O., Gleick, P. H., & Schneider, S. H. (1990). Climate forecasting. In P. E. Waggoner (Ed.), *Report of the American Association for the Advancement of Science Panel on Climatic Variability, Climate Change and the Planning and Management of U.S. Water Resources* (pp. 87–138). New York: John Wiley & Sons, Inc.

Meegan, S. K., & Perry, S. A. (1996). Periphyton communities in headwater streams of different water chemistry in the central Appalachian Mountains. *Journal of Freshwater Ecology, 11*, 247–256.

Meehl, G. A., Stocker, T. F., Collins, W. D., Friedlingstein, P., Gaye, A. T., Gregory, J. M., et al. (2007). Global climate projections. In S. Solomon, D. Qin, M. Manning, Z. Chen, M. Marquis, & K. B. Averyt (Eds.), *Climate Change 2007: The Physical Science Basis. Contribution of Working Group I to the Fourth Assessment Report of the Intergovernmental Panel on Climate Change*. Cambridge, UK, and New York: Cambridge University Press.

Meeks, J. C. (1998). Symbiosis between nitrogen-fixing cyanobacteria and plants. *BioScience, 48*, 266–276.

Megonigal, J. P., & Schlesinger, W. H. (1997). Enhanced CH_4 emissions from a wetland soil exposed to elevated CO_2. *Biogeochemistry, 37*, 77–88.

Meijer, M.-L., Jeppesen, E., van Donk, E., Moss, B., Scheffer, M., Lammens, E., et al. (1994). Long-term responses to fish-stock reduction in small shallow lakes: interpretation of five-year results of four biomanipulation cases in the Netherlands and Denmark. *Hydrobiologia, 275/276*, 457–466.

Mellina, E., Rasmussen, J. B., & Mills, E. L. (1995). Impact of zebra mussel (*Dreissena polymorpha*) on phosphorus cycling and chlorophyll in lakes. *Canadian Journal of Fisheries and Aquatic Sciences, 52*, 2553–2573.

Melone, G. (1998). The rotifer corona by SEM. *Hydrobiologia, 387/388*, 131–134.

Memmott, J., Craze, P. G., Waser, N. M., & Price, M. V. (2007). Global warming and the disruption of plant–pollinator interactions. *Ecology Letters, 10*, 710–717.

Meriano, M., Eyles, N., & Howard, K. W. F. (2009). Hydrogeological impacts of road salt from Canada's busiest highway on a Lake Ontario watershed (Frenchman's Bay) and lagoon, City of Pickering. *Journal of Contaminant Hydrology, 107*, 66–81.

Merritt, R. W., Cummins, K. W., & Berg, M. B. (Eds.). (2008). *An introduction to the aquatic insects of North America* (4th ed.). Dubuque, IA: Kendall/Hunt Pub. Co.

Merritt, R. W., Higgins, M. J., Cummins, K. W., & Vandeneeden, B. (1999). The Kissimmee River-riparian marsh ecosystem, Florida. Seasonal differences in invertebrate functional feeding group relationships. In D. P. Batzer, R. B. Rader, & S. A. Wissinger (Eds.), *Invertebrates in freshwater wetlands of North America: Ecology and management* (pp. 55–79). New York: John Wiley & Sons, Inc.

Messer, J., & Brezonik, P. L. (1983). Comparison of denitrification rate estimation techniques in a large, shallow lake. *Water Research, 17,* 631–640.

Meybeck, M. (1982). Carbon, nitrogen, and phosphorus transport by world rivers. *American Journal of Science, 282,* 401–450.

Meybeck, M. (1993). Riverine transport of atmospheric carbon: sources, global typology and budget. *Water, Air, and Soil Pollution, 70,* 443–463.

Meybeck, M. (1995). Global distribution of lakes. In A. Lerman, D. M. Imboden, & J. R. Gat (Eds.), *Physics and chemistry of lakes* (pp. 1–36). Berlin, Germany: Springer-Verlag.

Meybeck, M., Chapman, D. V., & Helmer, R. (1989). *Global freshwater quality. A first assessment.* Published on behalf of the World Health Organization and the United Nations Environment Programme by Blackwell Inc., Cambridge, MA.

Meyer, C. K., Baer, S. G., & Whiles, M. R. (2008). Ecosystem recovery across a chronosequence of restored wetlands in the Platte River valley. *Ecosystems, 11,* 193–208.

Meyer, C. K., & Whiles, M. R. (2008). Macroinvertebrate communities in restored and natural Platte River slough wetlands. *Journal of the North American Benthological Society, 27,* 626–639.

Meyer, F. P. (1990). Introduction. In F. P. Meyer & L. A. Barklay (Eds.), *Field manual for the investigation of fish kills* (pp. 1–5). Washington, DC: United States Department of the Interior, Fish and Wildlife Service.

Meyer, J. L. (1994). The microbial loop in flowing waters. *Microbial Ecology, 28,* 195–199.

Meyer, J. L., Paul, M. J., & Taulbee, W. K. (2005). Stream ecosystem function in urbanizing landscapes. *Journal of the North American Benthological Society, 24,* 602–612.

Meyer, O. (1994). Functional groups of microorganisms. In E.-D. Schulze & H. A. Mooney (Eds.), *Biodiversity and ecosystem function* (pp. 67–96). Berlin: Springer-Verlag.

Michael, H. J., Boyle, K. J., Bouchard, R. (1996). Miscellaneous Report 398, Maine Agricultural and Forest Experiment Station, Orono, ME.

Michel, E. (1994). Why snails radiate: a review of gastropod evolution in long-lived lakes, both recent and fossil. *Archiv für Hydrobiologie -Beiheft Ergebnisse der Limnologie, 44,* 285–317.

Middleton, B. (1999). *Wetland restoration, flood pulsing, and disturbance dynamics.* New York: John Wiley & Sons.

Milinski, M. (1993). Predation risk and feeding behavior. In T. J. Pitcher (Ed.), *Behavior of teleost fishes* (pp. 285–306). London, UK: Chapmand and Hall.

Millenium Ecosystem Assessment. (2003). *Ecosystems and human well-being: A framework for assessment.* Washington: Island Press.

Miller, A. M., & Golladay, S. W. (1996). Effects of spates and drying on macroinvertebrate assemblages of an intermittent and a perennial prairie stream. *Journal of the North American Benthological Society, 15,* 670–689.

Miller, G. T., Jr. (1998). *Living in the environment* (10th ed.). Belmont, CA: Wadsworth.

Milliman, J. D., Broadus, J. M., & Gable, F. (1989). Environmental and economic implications of rising sea level and subsiding deltas: the Nile and Bengal examples. *Ambio, 18,* 340–345.

Mills, E. L., Leach, J. H., Carlton, J. T., & Secor, C. L. (1994). Exotic species and the integrity of the Great Lakes. *BioScience, 44,* 666–676.

Milly, P. C. D., Betancourt, J., Falkenmark, M., Hirsch, R. M., Kundzewicz, Z. W., Lettenmaier, D. P., et al. (2008). Climate change—stationarity is dead: whither water management? *Science, 319,* 573–574.

Milvy, P., & Cothern, C. R. (1990). Scientific background for the development of regulations for radionuclides in drinking water. In C. R. Cothern & P. A. Rebers (Eds.), *Radon, radium and uranium in drinking water* (pp. 1–16). Chelsea, MI: Lewis Publishers.

Minshall, G., Brock, J., & Varley, J. (1989). Wildfires and Yellowstone's stream ecosystems. *BioScience,* 707–715.

Minshall, G. W. (1978). Autotrophy in stream ecosystems. *BioScience, 28,* 767–771.

Minshall, G. W., Petersen, R. C., Cummins, K. W., Bott, T. L., Sedell, J. R., Cushing, C. E., et al. (1983). Interbiome comparison of stream ecosystem dynamics. *Ecological Monographs, 53,* 1–25.

Mitsch, W. J., Cronk, J. K., Wu, X., & Nairn, R. W. (1995). Phosphorus retention in constructed freshwater riparian marshes. *Ecological Applications, 5,* 830–845.

Mitsch, W. J., & Gosselink, J. G. (1993). *Wetlands* (2nd ed.). New York: Van Nostrand Reinhold.

Mitsch, W. J., & Gosselink, J. G. (2007). *Wetlands* (4th ed.). Hoboken, NJ: John Wiley & Sons.

Mittelbach, G. (1988). Competition among refuging sunfishes and effects of fish density on littoral zone invertebrates. *Ecology*, 614–623.

Mittelbach, G. G., & Osenberg, C. W. (1994). Using foraging theory to study trophic interactions. In D. J. Stouder, K. L. Fresh, & R. J. Feller (Eds.), *Theory and application in fish feeding ecology* (pp. 45–59). Columbia, SC: University of South Carolina Press.

Moeller, R., Burkholder, J., & Wetzel, R. (1988). Significance of sedimentary phosphorus to a rooted submersed macrophyte (*Najas flexilis* (Willd.) Rostk. and Schmidt) and its algal epiphytes. *Aquatic Botany, 32*, 261–281.

Molles, M. C., Jr., Crawford, C. S., Ellis, L. M., Valett, H. M., & Dahm, C. N. (1998). Managed flooding for riparian ecosystem restoration. *BioScience, 48*, 749–756.

Moore, W. S. (1996). Large groundwater inputs to coastal waters revealed by ^{226}Ra enrichments. *Nature, 380*, 612–614.

Morin, P. J. (1983). Predation, competition, and the composition of larval anuran guilds. *Ecological Monographs, 53*, 119–138.

Morin, P. J., & McGrady-Steed, J. (2004). Biodiversity and ecosystem functioning in aquatic microbial systems: a new analysis of temporal variation and species richness-predictability relations. *Oikos, 104*, 458–466.

Morisawa, M. (1968). *Streams their dynamics and morphology*. New York: McGraw-Hill Book Co.

Morita, R. Y. (1997). *Bacteria in oligotrophic environments*. New York: Chapman and Hall.

Morris, D. P., Zagarese, H., Williamson, C. E., Balseiro, E. G., Harbraves, B. R., Modenutti, B., et al. (1995). The attenuation of solar UV radiation in lakes and the role of dissolved organic carbon. *Limnology and Oceanography, 40*, 1381–1391.

Morris, J. T. (1991). Effects of nitrogen loading on wetland ecosystems with particular reference to atmospheric deposition. *Annual Review of Ecology and systematics, 22*, 257–279.

Mortimer, C. H. (1941). The exchange of dissolved substances between mud and water in lakes. *Journal of Ecology, 29*, 280–329.

Morton, R., & Cunningham, R. B. (1985). Longitudinal profile of salinity in the Murray River. *Australian Journal of Soil Research, 23*, 1–13.

Moss, B., Stephen, D., Balayla, D., Bécares, E., Collings, S., Fernández-Aláez, C., et al. (2004). Continental-scale patterns of nutrient and fish effects on shallow lakes: synthesis of a pan-European mesocosm experiment. *Freshwater Biology, 49*, 1633–1649.

Moyle, P. B., & Cech, J. J., Jr. (1996). *Fishes an introduction to ichthyology*. Upper Saddle River, NJ: Prentice Hall.

Moyle, P. B., Li, H. W., & Barton, B. A. (1986). The Frankenstein effect: impact of introduced fishes on native fishes in North America. In R. H. Stroud (Ed.), *Symposium on the role of fish culture in fisheries management* (pp. 415–426). Lake of the Ozarks, MO: American Fisheries Society.

Moyle, P. B., & Light, T. (1996). Biological invasions of fresh water: empirical rules and assembly theory. *Biological Conservation, 78*, 149–161.

Mulholland, P. J. (1981). Organic carbon flow in a swamp-stream ecosystem. *Ecological Monographs, 51*, 307–322.

Mulholland, P. J., & DeAngelis, D. L. (2000). Surface-subsurface exchange and nutrient spiraling. In J. B. Jones & P. J. Mulholland (Eds.), *Streams and ground waters* (pp. 149–166). San Diego, CA: Academic Press.

Mulholland, P. J., Fellows, C. S., Tank, J. L., Grimm, N. B., Webster, J. R., Hamilton, S. K., et al. (2001). Inter-biome comparison of factors controlling stream metabolism. *Freshwater Biology, 46*, 1503–1517.

Mulholland, P. J., Helton, A. M., Poole, G. C., Hall, R. O., Hamilton, S. K., Peterson, B. J., et al. (2008). Stream denitrification across biomes and its response to anthropogenic nitrate loading. *Nature, 452*, 202–246.

Mulholland, P. J., & Kuenzler, E. J. (1979). Organic carbon export from upland and forested wetland watersheds. *Limnology and Oceanography, 24*, 960–966.

Mulholland, P. J., & Sale, M. J. (1998). Impacts of climate change on water resources: findings of the IPCC regional assessment of vulnerability for North America. *Water Resources, 112*, 10–15.

Mulholland, P. J., Steinman, A. D., & Elwood, J. W. (1990). Measurement of phosphate uptake rate in streams: comparison of radio tracer and stable PO_4 releases. *Canadian Journal of Fisheries and Aquatic Sciences, 47*, 2351–2357.

Mulholland, P. J., Tank, J. L., Webster, J. R., Bowden, W. B., Dodds, W. K., Gregory, S. V., et al. (2002). Can uptake length in streams be determined by nutrient addition experiments? Results from an interbiome comparison study. *Journal of the North American Benthological Society, 21*, 544–560.

Muller-Navarra, D. C., Brett, M. T., Liston, A. M., & Goldman, C. R. (2000). A highly unsaturated fatty acid predicts carbon transfer between primary producers and consumers. *Nature, 403*, 74–77.

Münster, U., & Haan, H. D. (1998). The role of microbial extracellular enzymes in the transformation of dissolved organic matter in humic waters. *Ecological Studies, 133*, 199–257.

Münster, U., Heikkinen, E., Likolammi, M., Järvinen, M., Salonen, K., & Haan, H. D. (1999). Utilisation of polymeric and monomeric aromatic and amino acid carbon in a humic boreal forest lake. *Arch. Hydrobiol. Spec. Issues Advanc. Limnol., 54*, 105–134.

Murdock, J. N., Gido, K. B., Dodds, W. K., Bertrand, K. N., Whiles, M. R. (2010). Consumers alter the recovery trajectory of stream ecosystem structure and function following drought. *Ecology., 91*, 1048–1062

Murphy, K., Willby, N. J., & Eaton, J. W. (1995). Ecological impacts and management of boat traffic on navigable inland waterways. In D. M. Harper & A. J. D. Ferguson (Eds.), *The ecological basis for river management* (pp. 427–442). West Sussex, England: John Wiley & Sons, Ltd.

Murphy, K. J., & Barrett, P. R. F. (1990). Chemical control of aquatic weeds. In A. H. Pieterse & K. J. Murphy (Eds.), *Aquatic weeds: the ecology and management of nuisance aquatic vegetation* (pp. 136–173). New York: Oxford University Press.

Murphy, K. J., & Pieterse, A. H. (1990). Present status and prospects of integrated control of aquatic weeds. In A. H. Pieterse & K. J. Murphy (Eds.), *Aquatic weeds: the ecology and management of nuisance aquatic vegetation* (pp. 222–227). New York: Oxford University Press.

Murray, A. B., & Paola, C. (1994). A cellular model of braided rivers. *Nature, 371*, 54–57.

Murray, A. G. (1995). Phytoplankton exudation: exploitation of the microbial loop as a defense against algal viruses. *Journal of Plankton Research, 17*, 1079–1094.

Naiman, R. J. (1983). The annual pattern and spatial distribution of aquatic oxygen metabolism in boreal forest watersheds. *Ecological Monographs, 53*, 73–94.

Naiman, R. J., & Décamps, H. (1997). The ecology of interfaces: Riparian zones. *Annual Review of Ecology and Systematics, 28*, 621–658.

Naiman, R. J., Décamps, H., & Pollock, M. (1993). The role of riparian corridors in maintaining regional biodiversity. *Ecological Applications, 3*, 209–212.

Naiman, R. J., Johnston, C. A., & Kelley, J. C. (1988). Alteration of North American streams by beaver. *BioScience, 38*, 753–762.

Naiman, R. J., Pinay, G., Johnston, C. A., & Pastor, J. (1994). Beaver influences on the long-term biogeochemical characteristics of boreal forest drainage networks. *Ecology, 75*, 905–921.

Naiman, R. J., & Rogers, K. H. (1997). Large animals and system-level characteristics in river corridors. *BioScience, 47*, 521–528.

Nakai, K. (1993). Foraging of brood predators restricted by territory of substrate-brooders in a cichlid fish assemblage. In H. Kawanabe, J. E. Cohen, & K. Iwasaki (Eds.), *Mutualism and community organization. behavioural, theoretical, and food-web approaches* (pp. 84–108). New York: Oxford University Press.

Nakai, S., Inoue, Y., Hosomi, M., & Murakami, A. (1999). Growth inhibition of blue-green algae by allelopathic effects of macrophytes. *Water Science and Technology, 39/8*, 47–53.

Nakai, S., Zhou, S., Hosomi, M., & Tominaga, M. (2006). Allelopathic growth inhibition of cyanobacteria by reed. *Allelopathy Journal, 18*, 277–285.

Nakano, S., & Murakami, M. (2001). Reciprocal subsidies: Dynamic interdependence between terrestrial and aquatic food webs. *Proceedings of the National Academy of Science, 98*, 166–170.

Nalepa, T. F. (1994). Decline of native unionid bivalves in Lake St. Clair after infestation by the zebra mussel, *Dreissena polymorpha. Canadian Journal of Fisheries and Aquatic Sciences, 51*, 2227–2233.

Napolitano, G. E., & Cicerone, D. S. (1999). Lipids in water-surface microlayers and foams. In M. T. Arts & B. C. Wainman (Eds.), *Lipids in freshwater ecosystems* (pp. 235–262). New York: Springer-Verlag.

Napolitano, G. E., Pollero, R. J., Gayoso, A. M., MacDonald, B. A., & Thompson, R. J. (1997). Fatty acids as trophic markers of phytoplankton blooms in the Bahia Blanca Estuary (Buenos Aires, Argentina) and in Trinity Bay (Newfoundland, Canada). *Biochemistry and Systematic Ecology, 25*, 739–755.

Nasri, H., El Herry, S., & Bouaïcha, N. (2008). First reported case of turtle deaths during a toxic *Microcystis* spp. bloom in Lake Oubeira, Algeria. *Ecotoxicology and Environmental Safety, 71*, 535–544.

National Atmospheric Deposition Program (NRSP-3)/National Trends Network. (1997). *Illinois state water supply*. Champaign, IL: NADP/NTN Coordination Office.

National Research Council. (1992). *Restoration of aquatic ecosystems. Science, technology, and public policy*. Washington, DC: National Academy Press, 552.

National Research Council. (1996). *Stemming the tide. Controlling introductions of nonindigenous species by ships' ballast water*. Washington, DC: National Academy Press.

Nedunchezhian, N., Ravindran, K. C., Abadia, A., Abadia, J., & Kulandaivelu, J. (1996). Damages of photosynthetic apparatus in *Anacystis nidulans* by ultraviolet-B radiation. *Biologia Plantarum*, *38*, 53–59.

Nedwell, D. B. (1999). Effect of low temperature on microbial growth: Lowered affinity for substrates limits growth at low temperature. *FEMS Microbiology Ecology*, *30*, 101–111.

Needham, J. G., & Needham, P. R. (1975). *A guide to the study of freshwater biology*. San Francisco, CA: Holden-Day, Inc.

Nelson, D. R. (1991). Tardigrada. In J. H. Thorp & A. P. Covich (Eds.), *Ecology and classification of North American freshwater invertebrates* (pp. 501–522). San Diego, CA: Academic Press, Inc.

Neue, H. U., Quijano, C., Senadhira, D., & Setter, T. (1998). Strategies for dealing with micronutrient disorders and salinity in lowland rice systems. *Field Crops Research*, *56*, 139–155.

Neves, R. J., Bogan, A. E., Williams, J. D., Ahlstedt, S. A., & Hartfield, P. W. (1997). Status of aquatic mollusks in the southeastern United States: A downward spiral of diversity. In G. W. Benz & D. E. Collins (Eds.), *Aquatic fauna in peril, the southeastern perspective* (pp. 43–86). Decatur, GA: Southeast Aquatic Research Institute.

Newbold, J. D., Elwood, J. W., O'Neill, R. V., & Winkle, W. V. (1981). Measuring nutrient spiraling in streams. *Canadian Journal of Fisheries and Aquatic Sciences*, *38*, 860–863.

Newman, S., Schuette, J., Grace, J. B., Rutchey, K., Fontaine, T., Reddy, K. R., et al. (1998). Factors influencing cattail abundance in the northern Everglades. *Aquatic Botany*, *60*, 265–280.

Nikora, V. I., Goring, D. G., & Biggs, B. J. F. (1997). On stream periphyton-turbulence interactions. *New Zealand Journal of Marine and Freshwater Research*, *31*, 435–448.

Nilsson, C., Jansson, R., & Zinko, U. (1997). Long-term responses of river-margin vegetation to water-level regulation. *Science*, *276*, 798–800.

Noble, R. L., & Jones, T. W. (1999). Managing fisheries with regulations. In C. C. Kohler & W. A. Hubert (Eds.), *Inland fisheries management in North America* (pp. 455–477). Bethesda, MD: American Fisheries Society.

North, R. L., Guildford, S. J., Smith, R. E. H., Havens, S. M., & Twiss, M. R. (2007). Evidence for phosphorus, nitrogen, and iron colimitiano of phytoplankton communities in Lake Erie. *Limnology and Oceanography*, *52*, 315–328.

Nowell, L. H., Capel, P. D., & Dileanis, P. D. (1999). *Pesticides in stream sediment and aquatic biota, distribution, trends, and governing factors*. Boca Raton, FL: Lewis Publishers.

Nürnberg, G. K. (1996). Trophic state of clear and colored, soft- and hardwater lakes with special consideration of nutrients, anoxia, phytoplankton and fish. *Journal of Lake and Reservoir Management*, *12*, 432–447.

Nürnberg, G. K. (2007). Lake responses to long-term hypolimnetic withdrawal treatments. *Lake and Reservoir Management*, *23*, 388–409.

Oberdorff, T., & Hugueny, B. (1995). Global scale patterns of fish species richness in rivers. *Ecography*, *18*, 345–352.

Oberemm, A., Becker, J., Codd, G. A., & Steinberg, C. (1999). Effects of cyanobacterial toxins and aqueous crude extracts of cyanobacteria on the development of fish and amphibians. *Environmental Toxicology*, *14*, 77–88.

Obermeyer, B. K., Edds, D. R., Miller, E. J., Prophet, C. W. (1995). Range reductions of southeast Kansas unionids. In: K. S. Cummings, A. C. Buchanan, C. A. Mayer, T. J. Naimo (Eds.), *Conservation and management of freshwater mussels II. initiatives for the future* (pp. 1–11). Proceedings of a UMCRR symposium. Upper Missouri Conservation Committee, St. Louis, MO.

O'Brien, J. M., Dodds, W. K., Wilson, K. C., Murdock, J. N., & Eichmiller, J. (2007). The saturation of N cycling in Central Plains streams: N-15 experiments across a broad gradient of nitrate concentrations. *Biogeochemistry*, *84*, 31–49.

Ochumba, P. B. O. (1990). Massive fish kills within the Nyanza Gulf of Lake Victoria, Kenya. *Hydrobiologia*, *208*, 93–99.

Odum, H. T. (1956). Primary production in flowing waters. *Limnology and Oceanography*, *1*, 102–117.

Odum, H. T. (1957). Trophic structure and productivity of Silver Springs, Florida. *Ecological Monographs*, *27*, 55–112.

Odum, H. T., & Odum, E. P. (1959). Principles and concepts pertaining to energy in ecological systems. In E. P. Odum & H. T. Odum (Eds.), *Fundamentals of ecology* (pp. 43–87). Philadelphia, PA: W. B. Saunders Company.

OECD. (1982). *Eutrophication of waters. monitoring assessment and control. Final report.* Paris, France: OECD.

O'Hara, S. L., Street-Perrott, F. A., & Burt, T. P. (1993). Accelerated soil erosion around a Mexican highland lake caused by prehispanic agriculture. *Nature, 362,* 48–51.

Oki, T., & Kanae, S. (2006). Global hydrological cycles and world water resources. *Science, 313,* 1068.

Olson, M. H., Hage, M. M., Binkley, M. D., & Binder, J. R. (2005). Impact of migratory snow geese on nitrogen and phosphorus dynamics in a freshwater reservoir. *Freshwater Biology, 50,* 882–890.

Olsson, H. (1991). Phosphatase activity in an acid, limed Swedish lake. In R. J. Chróst (Ed.), *Microbial enzymes in aquatic environments* (pp. 206–219). New York: Springer-Verlag.

Omernik, J. M. (1977). *Nonpoint source-stream nutrient level relationships: A nationwide study.* EPA-600/3-77-105, U.S. Environmental Protection Agency, Washington, DC.

Omernik, J. M. (1995). Ecoregions: A spatial framework for environmental management. In W. S. Davis & T. P. Simon (Eds.), *Biological assessment and criteria. Tools for water resource planning and decision making* (pp. 49–66). Boca Raton, FL: Lewis Publishers.

O'Neill, C. R., Jr. (1997). Economic impact of zebra mussels—results of the 1995 National Zebra Mussel Information Clearinghouse Study. *Great Lakes Research Review, 3,* 35–42.

Opperman, J., Galloway, G., Fargione, J., Mount, J., Richter, B., & Secchi, S. (2009). Sustainable floodplains through large-scale reconnection to rivers. *Science, 326,* 1487.

Oremland, R. S., & Stolz, J. F. (2003). The ecology of arsenic. *Science, 300,* 939–944.

O'Riordan, T. (1999). Economic challenges for lake management. *Hydrobiologia, 395/396,* 13–18.

Orr, C. H., Kroiss, S. J., Rogers, K. L., & Stanley, E. H. (2008). Downstream benthic responses to small dam removal in a cold-water stream. *River Research and Applications, 24,* 804–822.

Osenberg, C., Mittelbach, G., & Wainwright, P. (1992). Two-stage life histories in fish: The interaction between juvenile competition and adult performance. *Ecology,* 255–267.

Ottinger, R. L., Wooley, D. R., Robinson, N. A., Hodas, D. R., & Babb, S. E. (1990). *Environmental costs of electricity.* New York: Oceana Publications, Inc.

Pace, M. L., & Cole, J. J. (1994). Comparative and experimental approaches to top down and bottom-up regulation of bacteria. *Microbial Ecology, 28,* 181–193.

Pace, M. L., Cole, J. J., & Carpenter, S. R. (1998). Trophic cascades and compensation: differential responses of microzooplankton in whole-lake experiments. *Ecology, 79,* 138–152.

Pace, M. L., Cole, J. J., Carpenter, S. R., Kitchell, J. F., Hodgson, J. R., Bogert, M. C. V.d., et al. (2004). Whole-lake carbon-13 additions reveal terrestrial support of aquatic food webs. *Nature, 427,* 240–243.

Pace, N. R. (1997). A molecular view of microbial diversity and the biosphere. *Science, 276,* 734–740.

Paerl, H. W. (1990). Physiological ecology and regulation of N_2 fixation in natural waters. *Advances in Microbial Ecology, 11,* 305–344.

Paerl, H. W., & Pinckney, J. L. (1996). A mini-review of microbial consortia: Their roles in aquatic production and biogeochemical cycling. *Microbial Ecology, 31,* 225–247.

Page, L. M., & Burr, B. M. (1991). *A field guide to freshwater fishes, North America North of Mexico.* Boston: Houghton Mifflin Company.

Pajak, G. (1986). Development and structure of the Goczalkowice Reservoir ecosystem VIII. Phytoplankton. *Ekologia Polska, 34,* 397–413.

Palmer, M. A., Bernhardt, E. S., Allan, J. D., Lake, P. S., Alexander, G., Brooks, S., et al. (2005). Standards for ecologically successful river restoration. *Journal of Applied Ecology, 42,* 208–217.

Palmer, M. A., Covich, A. P., Finlay, B. J., Gibert, J., Hyde, K. D., Johnson, R. K., et al. (1997). Biodiversity and ecosystem processes in freshwater sediments. *Ambio, 26,* 571–577.

Palmer, M. A., & Poff, N. L. (1997). Heterogeneity in streams. The influence of environmental heterogeneity on patterns and processes in streams. *Journal of the North American Benthological Society, 16,* 169–173.

Palumbo, A. V., Mulholland, P. J., & Elwood, J. W. (1989). Epilithic microbial populations and leaf decomposition in acid-stressed streams. In S. S. Rao (Ed.), *Acid stress and aquatic microbial interactions* (pp. 70–88). Boca Raton, FL: CRC Press.

Pang, L., & Close, M. E. (1999). Attenuation and transport of atrazine and picloram in an alluvial gravel aquifer: a tracer test and batch study. *New Zealand Journal of Marine and Freshwater Research, 33,* 279–291.

Paragamian, V. I. (1987). Standing stocks of fish in some Iowa streams, with a comparison of channelized and natural stream reaches in the southern Iowa drift plain. *Proceedings of the Iowa Academy of Science, 94,* 128–134.

Parker, J. D., Caudill, C. C., & Hay, M. E. (2007). Beaver herbivory on aquatic plants. *Oecologia, 151,* 616–625.

Parris, M. J., & Cornelius, T. O. (2004). Fungal pathogen causes competitive and developmental stress in larval amphibian communities. *Ecology, 85,* 3385–3395.

Patrick, R. (1967). The effect of invasion rate, species pool, and size of area on the structure of diatom community. *Proceedings of the National Academy of Sciences,* United States of America, 58, 1335–1342.

Patrick, R., Reimer, C. W. (1966). The diatoms of the United States. Monographs of the Academy of Natural Sciences of Philadelphia, Philadelphia, PA.

Patrick, R., Reimer, C. W. (1975). The diatoms of the United States. Monographs of the Academy of Natural Sciences of Philadelphia, Philadelphia, PA.

Patten, B. C. (1993). Discussion: promoted coexistence through indirect effects: need for a new ecology of complexity. In H. Kawanabe, J. E. Cohen, & K. Iwasaki (Eds.), *Mutualism and community organization. Behavioural, theoretical, and food-web approaches.* New York: Oxford University Press.

Patten, D. T. (1998). Riparian ecosystems of semi-arid North America: diversity and human impacts. *Wetlands, 18,* 498–512.

Patterson, D. J. (1999). The diversity of eukaryotes. *American Naturalist, 154*(Suppl.), S96–S124.

Paul, M. J., Meyer, J. L., & Couch, C. A. (2006). Leaf breakdown in streams differing in catchment land use. *Freshwater Biology, 51,* 1684–1695.

Payne, W. J. (1981). *Denitrification.* New York: John Wiley & Sons.

Peary, J. A., & Castenholz, R. W. (1964). Temperature strains of a thermophilic blue-green alga. *Nature, 5/64,* 720–721.

Peckarsky, B. L. (1982). Aquatic insect predator-prey relations. *BioScience, 32,* 261–266.

Peckarsky, B. L., & Penton, M. A. (1989). Early warning lowers risk of stonefly predation for a vulnerable mayfly. *Oikos, 54,* 301–309.

Peckarsky, B. L., & Wilcox, R. S. (1989). Stonefly nymphs use hydrodynamic cues to discriminate between prey. *Oecologia, 79,* 265–270.

Pejler, B. (1995). Relation to habitat in rotifers. *Hydrobiologia, 313/314,* 267–278.

Pelton, D. K., Levine, S., & Braner, M. (1998). Measurements of phosphorus uptake by macrophytes and epiphytes from the LaPlatte River (VT) using ^{32}P in stream microcosms. *Freshwater Biology, 39,* 285–299.

Pennak, R. W. (1978). *Fresh-water invertebrates of the United States* (2nd ed.). New York: John Wiley & Sons.

Perry, M. (2008). Effects of environmental and occupational pesticide exposure on human sperm: a systematic review. *Human Reproduction Update, 14,* 233–242.

Perry, W. L., Lodge, D. M., & Lamberti, G. A. (1997). Impact of crayfish predation on exotic zebra mussels and native invertebrates in a lake-outlet stream. *Canadian Journal of Fisheries and Aquatic Sciences, 54,* 120–125.

Persson, A. (1997). Effects of fish predation and excretion on the configuration of aquatic food webs. *Oikos, 79,* 137–146.

Peterson, B. J. (1999). Stable isotopes as tracers of organic matter input and transfer in benthic food webs: a review. *Acta Oecologica, 20,* 479–487.

Peterson, B. J., Bahr, M., & Kling, G. W. (1997). A tracer investigation of nitrogen cycling in a pristine tundra river. *Canadian Journal of Fisheries and Aquatic Sciences, 54,* 2361–2367.

Peterson, B. J., Deegan, L. A., Helfrich, J., Hobbie, J. E., Hullar, M., Moller, B., et al. (1993). Biological response of a tundra river to fertilization. *Ecology, 74,* 653–672.

Peterson, B. J., & Fry, B. (1987). Stable isotopes in ecosystem studies. *Annual Review of Ecology and Systematics, 18,* 293–320.

Peterson, B. J., Wollheim, W. M., Mulholland, P. J., Webster, J. R., Meyer, J. L., Tank, J. L., et al. (2001). Control of nitrogen export from watersheds by headwater streams. *Science, 292,* 86–90.

Peterson, D. F., & Keller, A. A. (1990). Irrigation. In P. E. Waggoner (Ed.), *Climate change and U.S. water resources. Report of the American Association for the Advancement of Science panel on climatic variability, climate change and the planning and management of U.S. water resources* (pp. 269–306). New York: John Wiley and Sons.

Peterson, S. A., Sickle, J. V., Herlihy, A. T., & Hughes, R. M. (2007). Mercury concentration in fish from streams and rivers throughout the western United States. *Environmental Science and Technology, 41*, 58–65.

Petts, G., Maddock, I., Bickerton, M., & Ferguson, A. J. D. (1995). Linking hydrology and ecology: the scientific basis for river management. In D. M. Harper & A. J. D. Ferguson (Eds.), *The ecological basis for river management* (pp. 1–16). West Sussex, England: John Wiley & Sons, Ltd.

Phillips, G., & Lipton, J. (1995). Injury to aquatic resources caused by metals in Montana's Clark Fork River basin: historic perspective and overview. *Canadian Journal of Fisheries and Aquatic Sciences, 52*, 1990–1993.

Pickett, S. T. A., Kolasa, J., & Jones, C. G. (1994). *Ecological understanding*. San Diego, CA: Academic Press.

Pidwirny, M. (2006). *Fundamentals of physical geography* (2nd ed.). http://www.physicalgeography.net/.

Pielou, E. C. (1977). *Mathematical ecology*. New York: John Wiley & Sons.

Pijanowska, J. (1997). Alarm signals in *Daphnia? Oecologia, 112*, 12–16.

Pimentel, D., Acquay, H., Biltonen, M., Rice, P., Silva, M., Nelson, J., et al. (1992). Environmental and economic costs of pesticide use. *BioScience, 42*, 750–760.

Pimentel, D., Berger, B., Filiberto, D., Newton, M., Wolfe, B., Karabinakis, E., et al. (2004). Water resources: agricultural and environmental issues. *BioScience, 54*, 909–918.

Pimentel, D., Harvey, C., Resosudarmo, P., Sinclair, K., Kurz, D., McNair, M., et al. (1995). Environmental and economic costs of soil erosion and conservation benefits. *Science, 267*, 1117–1123.

Pimentel, D., Zuniga, R., & Morrison, D. (2005). Update on the environmental and economic costs associated with alien-invasive species in the United States. *Ecological Economics, 52*, 273–288.

Pimm, S. L. (1982). *Food webs*. London, UK: Chapman and Hall Ltd.

Pimm, S. L., Russell, G. J., Gittleman, J. L., & Brooks, T. M. (1995). The future of biodiversity. *Science, 269*, 347–350.

Pina, S., Creus, A., González, N., Gironés, R., Felip, M., & Sommaruga, R. (1998). Abundance, morphology and distribution of planktonic virus-like particles in two high-mountain lakes. *Journal of Plankton Research, 20*, 2413–2421.

Piper, G. L. (1996). Biological control of the wetlands weed purple loosestrife (*Lythrum salicaria*) in the Pacific northwestern United States. *Hydrobiologia, 340*, 291–294.

Pitcher, T. J., & Hart, P. J. B. (1982). *Fisheries ecology*. Westport, CT: The Avi Publishing Company, Inc.

Podolsky, R. D. (1994). Temperature and water viscosity: physiological versus mechanical effects on suspension feeding. *Science, 265*, 100–103.

Poff, N. L., Allan, J. D., Bain, M. B., Karr, J. R., Prestegaard, K. L., Richter, B. D., et al. (1997). The natural flow regime. *BioScience, 47*, 769–784.

Poff, N. L., Olden, J. D., Merritt, D. M., & Pepin, D. M. (2007). Homogenization of regional river dynamics by dams and global biodiversity implications. *Proceedings of the National Academy of Sciences of the United States of America, 104*, 5732–5737.

Poff, N. L., & Ward, J. V. (1989). Implications of streamflow variability and predictability for lotic community structure: A regional analysis of streamflow patterns. *Canadian Journal of Fisheries and Aquatic Sciences, 46*, 1805–1818.

Poinar, G. O., Jr. (1991). Nematoda and Nematomorpha. In J. H. Thorp & A. P. Covich (Eds.), *Ecology and classification of North American freshwater invertebrates* (pp. 249–284). San Diego, CA: Academic Press, Inc.

Polis, G. A., Anderson, W. B., & Holt, R. D. (1997). Toward an integration of landscape and food web ecology: the dynamics of spatially subsidized food webs. *Annual Review of Ecology and Systematics, 28*, 289–316.

Polis, G. A., Power, M. E., & Huxel, G. R. (Eds.). (2004). *Food webs at the landscape level*. Chicago: University of Chicago Press.

Pollard, A. I., & Reed, T. (2004). Benthic invertebrate assemblage change following dam removal in a Wisconsin stream. *Hydrobiologia, 513*, 51–58.

Pollock, M. M., Naiman, R. J., & Hanley, T. A. (1998). Plant species richness in riparian wetlands—a test of biodiversity theory. *Ecology, 79*, 94–105.

Popisil, P. (1994). The groundwater fauna of a Danube aquifer in the "Lobau" wetland in Vienna, Austria. In J. Gibert, D. L. Danielopol, & J. A. Stanford (Eds.), *Groundwater ecology* (pp. 347–390). San Diego, CA: Academic Press, Inc.

Por, F. D. (2007). Ophel: a groundwater biome based on chemoautotrophic resources: the global significance of the Ayyalon cave finds, Israel. *Hydrobiologia, 592,* 1–10.

Porter, K. G. (1973). Selective grazing and differential digestion of algae by zooplankton. *Nature, 244,* 179–180.

Porter, K. G. (1976). Enhancement of algal growth and productivity by grazing zooplankton. *Science, 192,* 1332–1334.

Porter, K. G. (1977). The plant-animal interface in freshwater ecosystems. *American Scientist, 65,* 159–170.

Porter, K. G., Bergstedt, A., & Freeman, M. C. (1999). The Okefenokee Swamp. Invertebrate communities and foodwebs. In D. P. Batzer, R. B. Rader, & S. A. Wissinger (Eds.), *Invertebrates in freshwater wetlands of North America: Ecology and management* (pp. 121–135). New York: John Wiley & Sons, Inc.

Porter, K. G., Gerritsen, J., & Orcutt, J. D., Jr. (1982). The effect of food concentration on swimming patterns, feeding behavior, ingestion, assimilation, and respiration by *Daphnia. Limnology and Oceanography, 27,* 935–949.

Porter, K. G., Sherr, E. B., Sherr, B. F., Pace, M., & Sanders, R. W. (1985). Protozoa in planktonic food webs. *Journal of Protozoology, 32,* 409–415.

Postel, S. (1996). Forging a sustainable water strategy. In *State of the world: A worldwatch institute report on progress toward a sustainable society.* New York: W. W. Norton & Co. (pp. 40–59).

Postel, S. L., Daily, G. C., & Ehrlich, P. R. (1996). Human appropriation of renewable fresh water. *Science, 271,* 785–788.

Pough, F. H., Andrews, R. M., Cadle, S. E., Crump, M. L., Savitzky, A. H., & Wells, K. D. (1998). *Herpetology.* Upper Saddle River, NJ: Prentice-Hall.

Pounds, J. A., Bustamante, M. R., Coloma, L. A., Consuegra, J. A., Fogden, M. P. L., Foster, P. N., et al. (2006). Widespread amphibian extinctions from epidemic disease driven by global warming. *Nature, 439,* 161–167.

Power, M. E. (1990a). Effects of fish in river food webs. *Science, 250,* 811–814.

Power, M. E. (1990b). Resource enhancement by indirect effects of grazers: Armored catfish, algae, and sediment. *Ecology, 71,* 897–904.

Power, M. E. (1992). Top-down and bottom-up forces in food webs: do plants have primacy? *Ecology, 73,* 733–746.

Power, M. E., & Matthews, W. J. (1983). Algae-grazing minnows (*Campostoma anomalum*), piscivorous bass (*Micropterus* spp.), and the distribution of attached algae in a small prairie-margin stream. *Oecologia, 60,* 328–332.

Power, M. E., Sun, A., Parker, G., Dietrich, W. E., & Wootton, J. T. (1995). Hydraulic food-chain models: an approach to the study of food-web dynamics in large rivers. *BioScience, 45,* 159–167.

Power, M. E., Tilman, D., Estes, J. A., Menge, B. A., Bond, W. J., Mills, L. S., et al. (1996). Challenges in the quest for keystones. *BioScience, 46,* 609–620.

Prepas, E. E., Kotak, B. G., Campbell, L. M., Evans, J. C., Hrudey, S. E., Holmes, C. F. B. (1997). Accumulation and elimination of cyanobacterial hepatotoxins by the freshwater clam *Anodonta grandis simpsoniana. Canadian Journal of Fisheries and Aquatic Sciences, 54,* 41–46.

Prescott, G. W. (1978). *How to know the freshwater Algae.* Dubuque, IA: Wm. C. Brown Co.

Prescott, G. W. (1982). *Algae of the western Great Lakes area.* Koenigstein, Germany: Otto Koeltz Science Publishers.

Preston, F. W. (1962). The canonical distribution of commonness and rarity: part I. *Ecology, 43,* 187–215.

Pringle, C. M. (1997). Exploring how disturbance is transmitted upstream: going against the flow. *Journal of the North American Benthological Society, 16,* 425–438.

Pringle, C. M., Hemphill, N., McDowell, W. H., Bednarek, A., & March, J. G. (1999). Linking species and ecosystems: different biotic assemblages cause interstream differences in organic matter. *Ecology, 80,* 1860–1872.

Priscu, J. C., Adams, E. A., Lyons, W. B., Voytek, M. A., Mogk, D. M., Brown, R. L., et al. (1999). Geomicrobiology of subglacial ice above Lake Vostok, Antarctica. *Science, 286,* 2141–2144.

Priscu, J. C., Fritsen, C. H., Adams, E. A., Giovannoni, S. J., Paerl, H. W., McKay, C. P., et al. (1998). Perennial Antarctic lake ice: an oasis for life in a polar desert. *Science, 280,* 1095–2098.

Proctor, V. W. (1959). Dispersal of fresh-water algae by migratory water birds. *Science, 130,* 623–624.

Proctor, V. W. (1966). Dispersal of desmids by waterbirds. *Phycologia, 5,* 227–232.

Proulx, M., Pick, F. R., Mazumdre, A., Hamilton, P. B., & Lean, D. R. S. (1996). Experimental evidence for interactive impacts of human activities on lake algal species richness. *Oikos, 76,* 191–195.

Pugh, P. J. A., & McInnes, S. J. (1998). The origin of arctic terrestrial and freshwater tardigrades. *Polar Biology, 19,* 177–182.

Purcell, L. M. (1977). Life at low Reynolds number. *American Journal of Physics, 45*, 3–11.

Pyke, G. H., Pulliam, H. R., & Charnov, E. L. (1977). Optimal foraging: a selective review of theory and tests. *The Quarterly Review of Biology, 52*.

Quinlan, E. L., Philips, E. J., Donnelly, K. A., Jett, C. H., Sleszynski, P., & Keller, S. (2008). Primary producers and nutrient loading in Silver Springs, FL, USA. *Aquatic Botany, 88*, 247–255.

Quinn, J. F., & Dunham, A. E. (1981). On hypothesis testing in ecology and evolution. *The American Naturalist, 122*, 602–617.

Rabalais, N. N., Turner, R. E., & Scavia, D. (2002). Beyond science into policy: Gulf of Mexico hypoxia and the Mississippi River. *BioScience, 52*, 129–142.

Rader, R. B. (1999). The Florida Everglades, natural variability, invertebrate diversity, and foodweb stability. In D. P. Batzer, R. B. Rader, & S. A. Wissinger (Eds.), *Invertebrates in freshwater wetlands of North America: Ecology and management* (pp. 25–54). New York: John Wiley & Sons, Inc.

Rader, R. B., & Belish, T. A. (1997a). Short-term effects of ambient and enhanced UV-B on moss (*Fontinalis neomexicana*) in a mountain stream. *Journal of Freshwater Ecology, 12*, 395–403.

Rader, R. B., & Belish, T. A. (1997b). Effects of ambient and enhanced UV-B radiation on periphyton in a mountain stream. *Journal of Freshwater Ecology, 12*, 615–628.

Rahel, F. J. (2000). Homogenization of fish faunas across the United States. *Science, 288*, 854–856.

Rahel, F. J., & Olden, J. D. (2008). Assessing the effects of climate change on aquatic invasive species. *Conservation Biology, 22*, 521–533.

Rainey, P. B., & Rainey, K. (2003). Evolution of cooperation and conflict in experimental bacterial populations. *Nature, 425*, 72–74.

Ramade, F. (1989). The pollution of the hydrosphere by global contaminants and its effects on aquatic ecosystems. In A. Boudou & F. Ribeyre (Eds.), *Aquatic ecotoxicology: Fundamental concepts and methodologies* (pp. 152–180). Boca Raton, FL: CRC Press.

Ranvestel, A. W., Lips, K. R., Pringle, C. M., Whiles, M. R., & Bixby, R. J. (2004). Neotropical tadpoles influence stream benthos: evidence for the ecological consequences of decline in amphibian populations. *Freshwater Biology, 49*, 274–285.

Räsänen, J., Kauppila, T., & Salonen, V. P. (2006). Sediment-based investigation of naturally or historically eutrophic lakes—implications for lake management. *Journal of Environmental Management, 79*, 253–265.

Raven, J. A. (1992). How benthic macroalgae cope with flowing freshwater: resource acquisition and retention. *Journal of Phycology, 28*, 133–146.

Reche, I., Pulido-Villena, E., Morales-Baquero, R., & Casamayor, E. O. (2005). Does ecosystem size determine aquatic bacterial richness? *Ecology, 86*, 1715–1722.

Redfield, A. C. (1958). The biological control of chemical factors in the environment. *American Scientist, 46*, 205–221.

Reemtsma, T., & These, A. (2005). Comparative investigation of low-molecular-weight fulvic acids of different origin by SEC-Q-TOF-MS: new insights into structure and formation. *Environmental Science and Technology, 39*, 3507–3512.

Regester, K. J., Whiles, M. R., & Lips, K. R. (2008). Variation in the trophic basis of production and energy flow associated with emergence of larval salamander assemblages from forest ponds. *Freshwater Biology, 53*, 1754–1767.

Reid, G. K., & Fichter, G. S. (1967). *Pond life, a guide to common plants and animals of North American ponds and lakes.* New York: Golden Press, Western Publishing Company Inc.

Reid, J. W. (1994). Latitudinal diversity patterns of continental benthic copepod species assemblages in the Americas. *Hydrobiologia, 292/293*, 341–349.

Rejmankova, E., Macek, P., & Epps, K. (2008). Wetland ecosystem changes after three years of phosphorus addition. *Wetlands, 28*, 914–927.

Relyea, R. (2005). The lethal impact of roundup on aquatic and terrestrial amphibians. *Ecological Applications, 15*, 1118–1124.

Resh, V. H., Brown, A. B., Covich, A. P., Gurtz, M. E., Li, H. W., Minshall, G. W., et al. (1988). The role of disturbance in stream ecology. *Journal of the North American Benthological Society, 7*, 433–455.

Resh, V. H., Leveque, C., & Statzner, B. (2004). Long-term, large-scale biomonitoring of the unknown: assessing the effects of insecticides to control river blindness (Onchocerciasis) in West Africa. *Annual Review of Entomology, 49*, 115–139.

Revsbech, N. P., & Jørgensen, B. B. (1986). Microelectrodes: their use in microbial ecology. In K. C. Marshall (Ed.), *Advances in Microbial Ecology: Vol. 9* (pp. 293–352). New York: Plenum.

Reynolds, C. S. (1984). *The ecology of freshwater phytoplankton.* Cambridge, UK: Cambridge University Press.

Reynolds, C. S. (1994). The role of fluid motion in the dynamics of phytoplankton in lakes and rivers. In P. S. Giller, A. G. Hildrew, & D. G. Raffaelli (Eds.), *Aquatic ecology. Scale, pattern and process* (pp. 141–188). Oxford, UK: Blackwell Scientific Publications.

Reynolds, J. B. (1996). Electrofishing. In B. R. Murphy & D. W. Willis (Eds.), *Fisheries techniques* (pp. 221–253). Bethesda, MD: American Fisheries Society.

Rheinheimer, G. (1991). *Aquatic microbiology* (4th ed.). Chichester, UK: John Wiley and Sons.

Ribblett, S. G., Palmer, M. A., & Coats, W. (2005). The importance of bacterivorous protists in the decomposition of stream leaf litter. *Freshwater Biology, 50,* 516–526.

Ricciardi, A., & MacIsaac, H. J. (2000). Recent mass invasion of the North American great lakes by Ponto-Caspian species. *Trends in Ecology and Evolution, 15,* 62–65.

Rice, E. L. (1984). *Allelopathy* (2nd ed.). Orlando, FL: Academic Press.

Rice, G., Stedman, K., Snyder, J., Wiedenheft, B., Willits, D., Brumfield, S., et al. (2001). Viruses from extreme thermal environments. *Proceedings of the National Academy of Sciences of the United States of America, 98,* 13341–13345.

Richardson, C. J., King, R. S., Qian, S. S., Vaithiyanthan, P., Qualls, R. G., & Stow, C. A. (2007). Estimating ecological thresholds for phosphorus in the Everglades. *Environmental Science and Technology, 41,* 8084–8091.

Richardson, C. J., & Schwegler, B. R. (1986). Algal bioassay and gross productivity experiments using sewage effluent in a Michigan wetland. *Water Resources Bulletin, 22,* 111–120.

Richardson, L. F. (1922). *Weather prediction by numerical processes.* Cambridge: Cambridge University Press.

Richardson, L. L., Aguilar, C., & Nealson, K. H. (1988). Manganese oxidation in pH and O_2 microenvironments produced by phytoplankton. *Limnology and Oceanography, 33,* 352–363.

Richardson, L. L., & Castenholz, R. W. (1987a). Diel vertical movements of the cyanobacterium *Oscillatoria terebriformis* in a sulfide-rich hot spring microbial mat. *Applied and Environmental Microbiology, 53,* 2142–2150.

Richardson, L. L., & Castenholz, R. W. (1987b). Enhanced survival of the cyanobacterium *Oscillatoria terebriformis* in darkness under anaerobic conditions. *Applied and Environmental Microbiology, 53,* 2151–2158.

Richter, J., Martin, L., & Beachy, C. K. (2009). Increased larval density induces accelerated metamorphosis independently of growth rate in the frog *Rana sphenocephala. Journal of Herpetology, 43,* 551–554.

Rickerl, D. H., Sancho, F. O., & Ananth, S. (1994). Vesicular-arbuscular endomycorrhizal colonization of wetland plants. *Journal of Environmental Quality, 23,* 913–916.

Ridge, I., Walters, J., & Street, M. (1999). Algal growth control by terrestrial leaf litter: a realistic tool? *Hydrobiologia, 395/396,* 173–180.

Riemer, D. N. (1984). *Introduction to freshwater vegetation.* Westport, CT: AVI Publishing.

Riera, J. L., Magnuson, J. J., Kratz, T. K., & Webster, K. E. (2000). A geomorphic template for the analysis of lake districts applied to the Northern Highland Lake District, Wisconsin, USA. *Freshwater Biology, 43,* 301–318.

Riess, W., Giere, O., Kohls, O., & Sarbu, S. M. (1999). Anoxic thermomineral cave waters and bacterial mats as habitat for freshwater nematodes. *Aquatic Microbial Ecology, 18,* 157–164.

Rigler, F. (1966). Radiobiological analysis of inorganic phosphorus in lakewater. *Verein Internationale Verein Limnologie, 16,* 465–470.

Ringelberg, J., & Gool, E. V. (1998). Do bacteria, not fish, produce "fish kairomone"? *Journal of Plankton Research, 20,* 1847–1852.

Ripley, E. A., Redmann, R. E., & Crowder, A. A. (1996). *Environmental effects of mining.* Delray Beach, FL: St. Lucie Press.

Ripple, W., & Beschta, R. (2004). Wolves, elk, willows, and trophic cascades in the upper Gallatin Range of Southwestern Montana, USA. *Forest Ecology and Management, 200,* 161–181.

Roback, S. S. (1974). Insects (Arthropoda: Insecta). In C. W. Hart, Jr. & S. L. H. Fuller (Eds.), *Pollution ecology of freshwater invertebrates* (pp. 313–376). New York: Academic Press.

Robarts, R. D., Donald, D. B., & Arts, M. T. (1995). Phytoplankton primary production of three temporary northern prairie wetlands. *Canadian Journal of Aquatic Sciences, 52,* 897–902.

Robb, G. A., & Robinson, J. D. F. (1995). Acid drainage from mines. *The Geographical Journal, 161,* 47–54.

Roberts, E. C., & Laybourn-Parry, J. (1999). Mixotrophic crytophytes and their predators in the dry valley lakes of Antarctica. *Freshwater Biology, 41,* 737–746.

Robinson, A., Fleishmann, A., McPerson, S., Heinrich, V., Gironella, E., & Pena, C. (2009). A spectacular new species of Nepenthes L. (Nepenthaceae) pitcher plant from central Palawan, Philippines. *Journal of the Linnean Society, 159,* 195–202.

Robinson, C., & Minshall, G. (1986). Effects of disturbance frequency on stream benthic community structure in relation to canopy cover and season. *Journal of the North American Benthological Society,* 237–248.

Robinson, C., & Uehlinger, U. (2008). Experimental floods cause ecosystem regime shift in a regulated river. *Ecological Applications, 18,* 511–526.

Robson, T. M., Pancott, V. A., Flint, S. D., Ballaré, C. L., Sala, O. E., Scopel, A. L., et al. (2003). Six years of solar UV-B manipulations affect growth of *Sphagnum* and vascular plants in Tierra del Fuego peatland. *New Phytologist, 160,* 379–389.

Rodas, V. L., & Costas, E. (1999). Preference of mice to consume *Microcystis aeruginosa* (toxin-producing cyanobacteria): a possible explanation for numerous fatalities of livestock and wildlife. *Research in Veterinary Science, 67,* 107–110.

Rodrigues, D. F., & Tiedje, J. M. (2008). Coping with our cold planet. *Applied and Environmental Microbiology, 74,* 1677–1686.

Roeselers, G., van Loosdrecht, M. C. M., & Muyzer, G. (2007). Heterotrophic pioneers facilitate phototrophic biofilm development. *Microbial Ecology, 54,* 578–585.

Rogers, P. (1986). Water: not as cheap as you think. *Technical Review,* 31–43.

Rogulj, B., Marmonier, P., Lattinger, R., & Danielopol, D. (1994). Fine-scale distribution of hypogean Ostracoda in the interstitial habitats of the Rivers Sava and Rhône. *Hydrobiologia, 287,* 19–28.

Root, R. B. (1967). The niche exploitation pattern of the blue-gray gnat-catcher. *Ecological Monographs, 37,* 317–350.

Rosemond, A. D., Mulholland, P. J., & Elwood, J. W. (1993). Top-down and bottom-up control of stream periphyton: Effect of nutrients and herbivores. *Ecology, 74,* 1264–1280.

Rosenzweig, M. L. (1995). *Species diversity in space and time.* Cambridge, UK: Cambridge University Press.

Rosenzweig, M. L. (1999). Heeding the warning in biodiversity's basic law. *Science, 284,* 276–277.

Rosenzweig, M. L. (2001). The four questions: What does the introduction of exotic species do to diversity? *Evolutionary Ecology Research, 3,* 361–367.

Rosgen, D. (2008). Discussion: Critical evaluation of how the Rosgen classification and associated natural channel design methods fail to integrate and quantify fluvial processes and channel responses by A. Simon, M. Doyle, M. Kondolf, F. D. Shields Jr., B. Rhoads, and M. McPhillips. *JAWRA Journal of the American Water Resources Association, 44,* 782–792.

Rosi-Marshall, E. J., Tank, J. L., Royer, T. V., Whiles, M. R., Evans-White, M., Chambers, C., et al. (2007). Toxins in transgenic crop byproducts may affect headwater stream ecosystems. *Proceedings of the National Academy of Sciences of the United States of America, 104,* 16204–16208.

Ross, D. H., & Wallace, J. B. (1981). Production of *Brachycentrus spinae* Ross (Trichoptera: Brachycentridae) and its role in seston dynamics of a southern Appalachian stream (USA). *Environmental Entomology, 10,* 240–246.

Roughgarden, J. (1989). The structure and assembly of communities. In J. Roughgarden, R. M. May, & S. A. Levin (Eds.), *Perspectives in ecological theory* (pp. 203–226). Princeton, NJ: Princeton University Press.

Rundle, H. D., Nagel, L., Boughman, J. W., & Schluter, D. (2000). Natural selection and parallel speciation in sympatric sticklebacks. *Science, 287,* 306–308.

Runkel, R. L. (2002). A new metric for determining the importance of transient storage. *Journal of the North American Benthological Society, 21,* 529–543.

Russell, D. F., Wilkens, L. A., & Moss, F. (1999). Use of behavioural stochastic resonance by paddle fish for feeding. *Nature, 402,* 291–294.

Russell, N. J., & Hamamoto, T. (1998). Psychrophiles. In K. Horikoshi & W. D. Grant (Eds.), *Extremophiles. Microbial life in extreme environments* (pp. 2–45). New York: Wiley-Liss, Inc.

Ruttner, F. (1963). *Fundamentals of limnology.* Toronto, Ontario, Canada: University of Toronto Press.

Ryding, S.-O., & Rast, W. (1989). *The control of Eutrophication of lakes and reservoirs.* Paris and United Kingdom: UNESCO and Parthenon.

Rytter, L., Arveby, A. S., & Granhall, U. (1991). Dinitrogen (C_2H_2) fixation in relation to nitrogen fertilization of grey alder [*Alnus incana* (L.) Moench.] plantations in a peat bog. *Biology and Fertility of Soils, 10,* 233–240.

Sahimi, M. (1995). *Porous media and fractured rock*. New York: Weinham.

Sakai, A., & Larcher, W. (1987). *Frost survival of plants*. Berlin, Germany: Springer-Verlag.

Salzet, M. (2001). Anticoagulants and inhibitors of platelet aggregation derived from leeches. *FEBS letters, 492*, 187–192.

Sanderson, S. L., Cech, J. J., Jr., & Patterson, M. R. (1991). Fluid dynamics in suspension-feeding blackfish. *Science, 251*, 1346–1348.

Sand-Jensen, K., Pedersen, N. L., & Sondergaard, M. (2007). Bacterial metabolism in small temperate streams under contemporary and future climates. *Freshwater Biology, 52*, 2340–2353.

Sand-Jensen, K., Riis, T., Markager, S., & Vincent, W. F. (1999). Slow growth and decomposition of mosses in Arctic lakes. *Canadian Journal of Fisheries and Aquatic Sciences, 56*, 388–393.

Sapp, J. (1991). Living together: Symbiosis and cytoplasmic inheritance. In L. Margulis & R. Fester (Eds.), *Symbiosis as a source of evolutionary innovation* (pp. 15–25). Cambridge, MA: MIT Press.

Sarbu, S. M., Kane, T. C., & Kinkle, B. K. (1996). A chemoautotrophically based cave ecosystem. *Science, 272*, 1953–1995.

Sarnelle, O., Cooper, S. D., Wiseman, S., & Mavuti, K. M. (1998). The relationship between nutrients and trophic-level biomass in turbid tropical ponds. *Freshwater Biology, 40*, 65–75.

Scavia, D., & Fahnenstiel, G. L. (1984). Small-scale nutrient patchiness: Some consequences and a new encounter mechanism. *Limnology and Oceanography, 29*, 785–793.

Schaake, J. C. (1990). From climate to flow. In P. E. Waggoner (Ed.), *Climate change and U.S. water resources* (pp. 177–206). New York: John Wiley & Sons.

Schaeffer, D. J., Malpas, P. B., & Barton, L. L. (1999). Risk assessment of microcystin in dietary *Aphanizomenon flos-aquae*. *Ecotoxicology and Environmental Safety, 44*, 73–80.

Schalla, R., & Walters, W. H. (1990). Rationale for the design of monitoring well screens and filter pack. In D. M. Nielsen & A. I. Johnson (Eds.), *Ground water and vadose zone monitoring* (pp. 64–75). Ann Arbor, MI: ASTM.

Scheffer, M. (1998). *Ecology of shallow lakes*. London: Chapman and Hall.

Scheffer, M., Carpenter, S., Foley, J. A., Folke, C., & Walker, B. (2001). Catastrophic shifts in ecosystems. *Nature, 413*, 591–596.

Schelske, C. L., & Stoermer, E. F. (1972). Phosphorus, silica, and eutrophication of Lake Michigan. *Limnology and Oceanography, Special Symposia, 1*, 157–171.

Schindler, D. E., Carpenter, S. R., Cole, J. J., Kitchell, J. F., & Pace, M. L. (1997). Influence of food web structure on carbon exchange between lakes and the atmosphere. *Science, 277*, 248–251.

Schindler, D. W. (1974). Eutrophication and recovery in experimental lakes: Implications for lake management. *Science, 184*, 897–899.

Schindler, D. W. (1998). Replications versus realism: The need for ecosystem-scale experiments. *Ecosystems, 1*, 323–334.

Schlesinger, W. H. (1997). *Biogeochemistry, an analysis of global change* (2nd ed.). San Diego: Academic Press.

Schloegel, L. M., Picco, A. M., Kilpatrick, A. M., Davies, A. J., Hyatt, A. D., & Daszak, P. (2009). Magnitude of the US trade in amphibians and presence of *Batrachochytrium dendrobatidis* and ranavirus infection in imported North American bullfrogs (*Rana catesbeiana*). *Biological Conservation, 142*, 1420–1426.

Schlumpf, M., Cotton, B., Conscience, M., Haller, V., Steinmann, B., & Lichtensteiger, W. (2001). *In vitro* and *in vivo* estrogenicity of UV screens. *Environmental Health Perspectives, 109*, 239–244.

Schmid, R. (2001). Recent advances in the description of the structure of water, the hydrophobic effect, and the like-dissolves-like rule. *Monatshefte für Chemie/Chemical Monthly, 132*, 1295–1326.

Schmitt, W. L. (1965). *Crustaceans*. Ann Arbor, MI: The University of Michigan Press.

Schneider, S., & Melzer, A. (2004). Sediment and water nutrient characteristics in patches of submerged macrophytes in running waters. *Hydrobiologia, 527*, 195–207.

Schoener, T. W. (1983). Field experiments on interspecific competition. *American Naturalist, 122*, 240–285.

Schoener, T. W. (1987). A brief history of optimal foraging ecology. In A. C. Kamil, J. R. Krebs, & H. R. Pulliam (Eds.), *Foraging behavior* (pp. 5–68). New York: Plenum Press.

Scholtz, G., Braband, A., Tolley, L., Reimann, A., Mittmann, B., Lukhaup, C., et al. (2003). Ecology—parthenogenesis in an outsider crayfish. *Nature, 421*, 806.

Schooler, S., McEvoy, P., & Coombs, E. (2006). Negative per capita effects of purple loosestrife and reed canary grass on plant diversity of wetland communities. *Diversity and Distributions, 12*, 351–363.

Schopf, J. W. (1993). Microfossils of the early Archaen apex chert: New evidence of the antiquity of life. *Science, 260*, 640–646.

Schrader, K. K., & Dennis, M. E. (2005). Cyanobacteria and earthy/musty compounds found in commercial catfish (*Ictalurus punctatus*) ponds in the Mississippi Delta and Mississippi-Alabama Blackland Prairie. *Water Research, 39*, 2807–2814.

Schulze, E.-D., & Mooney, H. A. (1994). Ecosystem function of biodiversity: A summary. In E.-D. Schulze & H. A. Mooney (Eds.), *Biodiversity and ecosystem function* (pp. 498–510). Berlin: Springer-Verlag.

Schutz, D., Taborsky, M., & Drapela, T. (2007). Air bells of water spiders are an extended phenotype modified in response to gas composition. *Journal of Experimental Zoology Part A: Ecological Genetics and Physiology*.

Schwarzenbach, R. P., Escher, B. I., Fenner, K., Hofstetter, T. B., Johnson, C. A., Gunten, U. V., et al. (2006). The challenge of micropollutants in aquatic systems. *Science, 313*, 1072.

Scott, J. T., Doyle, R. D., Back, J. A., & Dworkin, S. I. (2007). The role of N2 fixation in alleviating N limitation in wetland metaphyton: Enzymatic, iisotopic, and elemental evidence. *Biogeochemistry, 84*, 207–218.

Scott, W. (1923). The diurnal oxygen pulse in Eagle (Winona) Lake. In J. J. Davis (Ed.), *Proceedings of the Indiana Academy of Science* (pp. 311–314). Indianapolis, IN: Wm. B. Burford.

Scrimshaw, S., & Kerfoot, W. C. (1987). Chemical defenses of freshwater organisms: Beetles and bugs. In W. C. Kerfoot & A. Sih (Eds.), *Predation. Direct and indirect impacts on aquatic communities* (pp. 240–262). Hanover, NH: University Press of New England.

Sculthorpe, C. D. (1967). *The biology of aquatic vascular plants*. London, UK: Edward Arnold, Ltd.

Sedell, J. R., & Froggat, J. L. (1984). Importance of streamside forests to large rivers: the isolation of the Willamette River, Oregon, USA, from its floodplain by snagging and streamside forest removal. *Verein Internationale Verein Limnologie, 22*, 1828–1843.

Sedell, J. R., Richey, J. E., & Swanson, F. J. (1989). The river continuum concept: A basis for the expected ecosystem behavior of very large rivers. In D. P. Dodge (Ed.), *Proceedings of the International Large River Symposium*. Canadian Special Publications Fisheries and Aquatic Sciences.

Seehausen, O., & van Alphen, J. J. M. (1999). Can sympatric speciation by disruptive sexual selection explain rapid evolution of cichlid diversity in Lake Victoria? *Ecology Letters, 2*, 262–271.

Seehausen, O., van Alphen, J. J. M., & Witte, F. (1997). Cichlid fish diversity threatened by eutrophication that curbs sexual selection. *Science, 277*, 1808–1811.

Seely, C. J., & Lutnesky, M. M. F. (1998). Odour-induced antipredator behaviour of the water flea *Cariodaphnia reticulata*, in varying predator and prey densities. *Freshwater Biology, 40*, 17–24.

Seifert, R. P., & Seifert, F. H. (1976). A community matrix analysis of *Heliconia* insect communities. *The American Midland Naturalist, 110*, 461–483.

Sereda, J. M., Hudson, J. J., Taylor, W. D., & Demers, E. (2008). Fish as sources and sinks of nutrients in lakes. *Freshwater Biology, 53*, 278–289.

Shannon, J. P., Blinn, D. W., McKinney, T., Benenati, E. P., Wilson, K. P., & O'Brien, C. (2001). Aquatic food base response to the 1996 test flood below Glen Canyon Dam, Colorado River, Arizona. *Ecological Applications, 11*, 672–685.

Shapiro, J. (1979). The importance of trophic-level interactions to the abundance and species composition of algae in lakes. In J. Barica & L. R. Mur (Eds.), *Hypertrophic ecosystems developments in hydrobiology* (pp. 105–121). The Netherlands: Dr. W. Junk.

Shapiro, J. (1997). The role of carbon dioxide in the initiation and maintenance of blue-green dominance in lakes. *Freshwater Biology, 37*, 307–323.

Sharitz, R. R., & Batzer, D. P. (1999). An introduction to freshwater wetlands in North America and their invertebrates. In D. P. Batzer, R. B. Rader, & S. A. Wissinger (Eds.), *Invertebrates in freshwater wetlands of North America: ecology and management* (pp. 1–22). New York: John Wiley & Sons, Inc.

Sheath, R. G., & Müller, K. M. (1997). Distribution of stream macroalgae in four high arctic drainage basins. *Arctic, 50*, 355–364.

Sherbakov, D. Y. (1999). Molecular phylogenetic studies on the origin of biodiversity in Lake Baikal. *Trends in Ecology and Evolution, 14*, 92–95.

Sherbakov, D. Y., Kamaltynov, R. M., Ogarkov, O. B., & Verheyen, E. (1998). Patterns of evolutionary changes in Baikalian Gammarids inferred from DNA sequences (Crustacea, Amphipoda). *Molecular Phylogenetics and Evolution, 10*, 160–167.

Sherr, B. F., Sherr, E. B., & Fallon, R. D. (1987). Use of monodispersed fluorescently labeled bacteria to estimate in situ protozoan bactivory. *Applied and Environmental Microbiology, 53*, 958–965.

Sherr, E. B., & Sherr, B. F. (1994). Bacterivory and herbivory: Key roles of phagotrophic protists in pelagic food webs. *Microbial Ecology, 28*, 223–235.

Shotyk, W., Weiss, D., Appleby, P. G., Cheburkin, A. K., Frei, R., Gloor, M., et al. (1998). History of atmospheric lead deposition since 12,370 ^{14}C yr BP from a peat bog, Jura Mountains, Switzerland. *Science, 281*, 1635–1640.

Siegert, M. J., Kwok, R., Mayer, C., & Hubbard, B. (2000). Water exchange between the subglacial Lake Vostok and the overlying ice sheet. *Nature, 403*, 643–646.

Sigee, D. (2005). *Freshwater microbiology*. Wiley.

Simco, B. A., & Cross, F. B. (1966). Factors affecting growth and production of channel catfish, *Ictalurus punctatus*. *University of Kansas Museum of Natural History Publications, 17*, 193–256.

Simek, K., Babenzien, D., Bittl, T., Koschel, R., Macek, M., Nedoma, J., et al. (1998). Microbial food webs in an artificially divided acidic bog lake. *Internationale Revue der Hydrobiologie, 83*, 3–18.

Simis, S. G. H., Tijdens, M., Hoogveld, H. L., & Gons, H. J. (2005). Optical changes associated with cyanobacterial bloom termination by viral lysis. *Journal of Plankton Research, 27*, 937–949.

Simon, A., Doyle, M., Kondolf, M., Shields, F. D., Jr., Rhoads, B., & McPhillips, M. (2007). Critical evaluation of how the Rosgen classification and associated "natural channel design" methods fail to integrate and quantify fluvial processes and channel response. *Journal of the American Water Resources Association, 43*, 1117–1131.

Simon, A., Doyle, M., Kondolf, M., Shields, F. D., Jr., Rhoads, B., & Mcphillips, M. (2008). Reply to discussion by Dave Rosgen of critical evaluation of how the Rosgen classification and associated "natural channel design" methods fail to integrate and quantify fluvial processes and channel response. *Journal of the American Water Resources Association, 44*, 793–802.

Simonich, S. L., & Hites, R. A. (1995). Global distribution of persistent organochlorine compounds. *Science, 269*, 1851–1854.

Sinclair, J. L., & Ghiorse, W. C. (1989). Distribution of aerobic bacteria, protozoa, algae, and fungi in deep subsurface sediments. *Geomicrobiology Journal, 7*, 5–31.

Sinsabaugh, R. L., Repert, D., Weiland, T., Golladay, S. W., & Linkins, A. E. (1991). Exoenzyme accumulation in epilithic biofilms. *Hydrobiologia, 222*, 29–37.

Sinton, L. W., Finlay, R. K., Pang, L., & Scott, D. M. (1997). Transport of bacteria and bacteriophages in irrigated effluent into and through an alluvial gravel aquifer. *Water, Air, and Soil Pollution, 98*, 17–42.

Siver, P. A., Lord, W. D., & McCarthy, D. J. (1994). Forensic limnology: The use of freshwater algal community ecology to link suspects to an aquatic crime scene in southern New England. *Journal of Forensic Science, 39*, 847–853.

Skelly, D. K. (1997). Tadpole communities. *American Scientist, 85*, 36–45.

Skerratt, L. F., Berger, L., Speare, R., Cashins, S., McDonald, K. R., Phillott, A. D., et al. (2007). Spread of chytridiomycosis has caused the rapid global decline and extinction of frogs. *EcoHealth, 4*, 125–134.

Skubinna, J. P., Coon, T. G., & Batterson, T. R. (1995). Increased abundance and depth submersed macrophytes in response to decreased turbidity in Saginaw Bay, Lake Huron. *Journal of Great Lakes Research, 21*, 476–488.

Slavik, K., Peterson, B. J., Deegan, L. A., Bowden, W. B., Hershey, A. E., & Hobbie, J. E. (2004). Long-term responses of the Kuparuk River ecosystem to phosphorus fertilization. *Ecology, 85*, 939–954.

Slobodkin, L. E., & Bossert, P. E. (1991). The freshwater Cnidaria- or Coelenterates. In J. H. Thorp & A. P. Covich (Eds.), *Ecology and classification of North American freshwater invertebrates* (pp. 125–143). San Diego, CA: Academic Press, Inc.

Sloey, W. E., Spangler, F. L., & Fetter, C. W., Jr. (1978). Management of freshwater wetlands for nutrient assimilation. In R. E. Good, D. F. Whigham, & R. L. Simpson (Eds.), *Freshwater wetlands: ecological processes and management potential* (pp. 321–340). New York: Academic Press.

Smith, D. G. (2001). *Pennak's freshwater invertebrates of the United States, Porifera to Crustacea* (4th ed.). New York: John Wiley & Sons.

Smith, G. R., Rettig, J. E., Mittelbach, G. G., Valiulis, J. L., & Schaack, S. R. (1999). The effects of fish on assemblages of amphibians in ponds: A field experiment. *Freshwater Biology, 41*, 829–837.

Smith, I. M., & Cook, D. R. (1991). Water mites. In J. H. Thorp & A. P. Covich (Eds.), *Ecology and classification of North American freshwater invertebrates* (pp. 523–592). San Diego, CA: Academic Press, Inc.

Smith, J. P., Jr. (1977). *Vascular plant families.* Eureka, CA: Mad River Press, Inc.

Smith, K. G., Lips, K. R., & Chase, J. M. (2009). Selecting for extinction: Nonrandom disease-associated extinction homogenizes amphibian biotas. *Ecology Letters, 12,* 1–10.

Smith, R. A., Alexander, R. B., & Schwarz, G. E. (2003). Natural background concentrations of nutrients in streams and rivers of the conterminous United States. *Environmental Science and Technology, 37,* 2039–3047.

Smith, V. H. (1982). The nitrogen and phosphorus dependence of algal biomass in lakes: An empirical and theoretical analysis. *Limnology and Oceanography, 27,* 1101–1112.

Smolders, A., & Roelofs, J. G. M. (1993). Sulphate-mediated iron limitation and eutrophication in aquatic ecosystems. *Aquatic Botany, 46,* 247–253.

Smyth, J. D., & Smyth, M. M. (1980). *Frogs as host-parasite systems I.* Hong Kong: The MacMillan Press Ltd.

Snedcor, G. W., & Cochran, W. G. (1980). *Statistical methods* (7th ed.). Ames, IA: Iowa State University Press.

Snell, T. W. (1998). Chemical ecology of rotifers. *Hydrobiologia, 387/388,* 267–276.

Sobsey, M. D., & Shields, P. A. (1987). Survival and transport of viruses in soils: Model studies. In V. C. Rao & J. L. Melnick (Eds.), *Human viruses in sediments, sludges, and soils* (pp. 155–177). Boca Raton, FL: CRC Press.

Sokal, R. R., & Rohlf, F. J. (1981). *Biometry, the principles and practice of statistics.* New York: W. H. Freeman and Company.

Soltero, R. A., Sexton, L. M., Ashley, K. I., & McKee, K. O. (1994). Partial and full lift hypolimnetic aeration of Medical Lake, WA to improve water quality. *Water Research, 28,* 2297–2308.

Sommer, U. (1989). Toward a Darwinian ecology of plankton. In U. Sommer (Ed.), *Plankton ecology* (pp. 1–8). New York: Springer-Verlag.

Sommer, U. (1999). A comment on the proper use of nutrient ratios in microalgal ecology. *Archiv fur Hydrobiologie, 146,* 55–64.

Søndergaard, M. (1991). Phototrophic picoplankton in temperate lakes: Seasonal abundance and importance along a trophic gradient. *Internationale Revue der gesamten Hydrobiologie und Hydrographie, 76,* 505–522.

Søndergaard, M., & Laegaard, S. (1977). Vesicular-arbuscular mycorrhiza in some aquatic vascular plants. *Nature, 268,* 233.

Sonnenschein, C., & Soto, A. M. (1997). An updated review of environmental estrogen and androgen mimics and antagonists. *Journal of Steroid Biochemistry and Molecular Biology, 65,* 43–150.

Soranno, P. A., Webster, K. E., Riera, J. L., Kratz, T. K., Baron, J. S., Bukaveckas, P. A., et al. (1999). Spatial variation among lakes within landscapes: Ecological organization along lake chains. *Ecosystems, 2,* 395–410.

South, G. R., & Whittick, A. (1987). *Introduction to phycology.* Oxford, UK: Blackwell Scientific Publications.

Spackman, S. C., & Hughes, J. W. (1995). Assessment of minimum stream corridor width for biological conservation: species richness and distribution along mid-order streams in Vermont, USA. *Biological Conservation, 71,* 325–332.

Spadinger, R., & Maier, G. (1999). Selection and diel feeding of the freshwater jellyfish *Craspedacusta sowerbyi. Freshwater Biology, 41,* 567–573.

Sparling, D. W., & Fellers, G. M. (2009). Toxicity of two insecticides to California, USA, anurans and its relevance to declining amphibian populations. *Environmental Toxicology and Chemistry, 28,* 1696–1703.

Spencer, C. N., & Ellis, B. K. (1998). Role of nutrients and zooplankton in regulation of phytoplankton in Flathead Lake (Montana, USA), a large oligotrophic lake. *Freshwater Biology, 39,* 755–763.

Spencer, C. N., & King, D. L. (1984). Role of fish in regulation of plant and animal communities in eutrophic ponds. *Canadian Journal of Fisheries and Aquatic Sciences, 41,* 1851–1855.

Spencer, C. N., McClelland, B. R., & Stanford, J. A. (1991). Shrimp introduction, salmon collapse, and bald eagle displacement: Cascading interactions in the food web of a large aquatic ecosystem. *BioScience, 41,* 14–21.

Spigel, R. H., & Priscu, J. C. (1998). Physical limnology of the McMurdo dry valley lakes. In J. C. Priscu (Ed.), *Ecosystem dynamics in a polar desert* (pp. 153–186). Washington, DC: American Geophysical Union.

Sprung, M. (1993). The other life: An account of present knowledge of the larval phase of *Dreissena polymorpha.* In T. F. Nalepa & D. W. Schloesser (Eds.), *Zebra mussels. biology, impacts and control* (pp. 39–53). Boca Raton, FL: Lewis Publishers.

Srivastava, D. S. (2006). Habitat structure, trophic structure and ecosystem function: Interactive effects in a bromeliad-insect community. *Oecologia, 149,* 493–504.

Stahlschmidt-Allner, P., Allner, B., Römbke, J., & Knacker, T. (1997). Endocrine disrupters in the aquatic environment. *Environmental Science and Pollution Research, 4*, 155–162.

Stanford, J. A., & Gaufin, A. R. (1974). Hyporheic communities of two Montana rivers. *Science, 185*, 700–702.

Stanford, J. A., & Ward, J. V. (1988). The hyporheic habitat of river ecosystems. *Nature, 335*, 64–66.

Stanford, J. A., & Ward, J. V. (1993). An ecosystem perspective of alluvial rivers: Connectivity and the hyporheic corridor. *Journal of the North American Benthological Society, 12*, 48–60.

Stanley, E. H., Fisher, S. G., & Grimm, N. B. (1997). Ecosystem expansion and contraction in streams. *BioScience, 47*, 427–435.

Stanley, E. H., & Jones, J. B. (2000). Surface-subsurface interactions: Past, present, and future. In J. B. Jones & P. J. Mulholland (Eds.), *Streams and ground waters* (pp. 405–417). San Diego, CA: Academic Press.

Stanley, J. G., Miley, W. W., II, & Sutton, D. L. (1978). Reproductive requirements and likelihood for naturalization of escaped grass carp in the United States. *Transactions of the American Fisheries Society, 107*, 119–127.

Starmach, J. (1986). Development and structure of the Goczalkowice Reservoir ecosystem XV. *Ichthyofauna. Ekologia Polska, 34*, 515–521.

Stebbins, R. C. (2003). *A field guide to western reptiles and amphibians* (3rd ed.). Boston: Houghton Mifflin Company.

Stein, R. A., DeVries, D. R., & Dettmers, J. M. (1995). Food-web regulation by a planktivore: exploring the generality of the trophic cascade hypothesis. *Canadian Journal of Fisheries and Aquatic Sciences, 52*, 2518–2526.

Steinberg, C. E. W., & Geller, W. (1994). Biodiversity and interactions within pelagic nutrient cycling and productivity. In E.-D. Schulze & H. A. Mooney (Eds.), *Biodiversity and ecosystem function* (pp. 43–64). Berlin: Springer-Verlag.

Steinman, A. D. (1996). Effects of grazers on freshwater benthic algae. In J. R. Stevenson, M. L. Bothwell, & R. L. Lowe (Eds.), *Algal ecology: freshwater benthic ecosystems* (pp. 341–374). San Diego, CA: Academic Press.

Steinman, A. D., Conklin, J., Bohlen, P. J., & Uzarski, D. G. (2003). Influence of cattle grazing and pasture land use on macroinvertebrate communities in freshwater wetlands. *Wetlands, 23*, 877–889.

Steinmetz, J., Soluk, D., & Kohler, S. (2008). Facilitation between herons and smallmouth bass foraging on common prey. *Environmental Biology of Fishes, 81*, 51–61.

Stemberger, R., & Gilbert, J. J. (1985). Body size, food concentration, and population growth in planktonic rotifers. *Ecology, 66*, 1151–1159.

Stemberger, R. S., & Chen, C. Y. (1998). Fish tissue metals and zooplankton assemblages of northeastern US lakes. *Canadian Journal of Fisheries and Aquatic Sciences, 55*, 339–352.

Sterner, R. W. (1990). The ratio of nitrogen to phosphorus resupplied by herbivores: Zooplankton and the algal competitive arena. *The American Naturalist, 136*, 209–229.

Sterner, R. W. (1993). *Daphnia* growth on varying quality of *Scenedesmus*: Mineral limitation of zooplankton. *Ecology, 74*, 2351–2360.

Sterner, R. W., & Elser, J. J. (2002). *Ecological stoichiometry: The biology of elements from molecules to the biosphere*. Princeton, NJ: Princeton University Press.

Sterner, R. W., Elser, J. J., & Hessen, D. O. (1992). Stoichiometric relationships among producers, consumers and nutrient cycling in pelagic ecosystems. *Biogeochemistry, 17*, 49–67.

Sterner, R. W., & Hessen, D. O. (1994). Algal nutrient limitation and the nutrition of aquatic herbivores. *Annual Review of Ecology and Systematics, 25*, 1–29.

Stetter, K. O. (1998). Hyperthermophiles: Isolation, classification, and properties. In K. Horikoshi & W. D. Grant (Eds.), *Extremophiles. Microbial life in extreme environments* (pp. 1–24). New York: Wiley-Liss, Inc.

Stevens, L. E. (1995). Flow regulation, geomorphology, and Colorado River marsh development in the Grand Canyon, Arizona. *Ecological Applications, 5*, 1025–1039.

Stevens, L. E., Shannon, J. P., & Blinn, D. W. (1997). Colorado River benthic ecology in Grand Canyon, Arizona, USA: Dam, tributary and geomorphological influences. *Regulated Rivers: Research and Management, 13*, 129–149.

Stevens, M. H. H., & Cummins, K. W. (1999). Effects of long-term disturbance on riparian vegetation and in-stream characteristics. *Journal of Freshwater Ecology, 14*, 1–17.

Stevens, T. O. (1997). Subsurface microbiology and the evolution of the biosphere. In P. S. Amy & D. L. Haldeman (Eds.), *Microbiology of the terrestrial deep subsurface* (pp. 205–223). Boca Raton, FL: Lewis Publishers.

Stevens, T. O., & McKinley, J. P. (1995). Lithoautotrophic microbial ecosystems in deep basalt aquifers. *Science, 270*, 450–454.

Stevenson, R. J. (1996). The stimulation of drag and current. In R. J. Stevenson, M. L. Bothwell, & R. L. Lowe (Eds.), *Algal ecology: Freshwater benthic ecosystems* (pp. 321–340). San Diego, CA: Academic Press.

Stewart-Oaten, A., Bence, J. R., & Osenberg, C. W. (1992). Assessing effects of unreplicated perturbations: No simple solutions. *Ecology, 73*, 1396–1404.

Stewart-Oaten, A., Murdoch, W. W., & Parker, K. R. (1986). Environmental impact assessment: "pseudoreplication" in time? *Ecology, 67*, 929–940.

Stiassny, M., & Meyer, A. (1999). Cichlids of the rift lakes. *Scientific American, 280*, 64–69.

Stickney, R. R. (1994). *Principles of aquaculture*. New York: John Wiley & Sons, Inc.

Stockner, J. G., & MacIsaac, E. A. (1996). British Columbia lake enrichment programme: Two decades of habitat enhancement for Sockeye salmon. *Regulated Rivers: Research & Management, 12*, 547–561.

Stoddard, J. L., Jeffries, D. S., Lükewille, A., Clair, T. A., Dillon, P. J., Driscoll, C. T., et al. (1999). Regional trends in aquatic recovery from acidification in North America and Europe. *Nature, 401*, 575–578.

Stølum, H.-H. (1996). River meandering as a self-organization process. *Science, 271*, 1710–1713.

Stone, R. (2007). The last of the leviathans. *Science, 316*, 1684–1688.

Storey, K. B., & Storey, J. M. (1988). Freeze tolerance in animals. *Physiological Reviews, 68*, 27–84.

Strahler, A. N., & Strahler, A. H. (1979). *Elements of physical geography* (2nd ed.). New York: John Wiley and Sons.

Strauss, E. A., & Dodds, W. K. (1997). Influence of protozoa and nutrient availability on nitrification rates in subsurface sediments. *Microbial Ecology, 34*, 155–165.

Strauss, E. A., Dodds, W. K., & Edler, C. C. (1994). The impact of nutrient pulses on trophic interactions in a farm pond. *Journal of Freshwater Ecology, 9*, 217–228.

Strayer, D. L. (1991). Projected distribution of the zebra mussel, *Dreissena polymorpha*, in North America. *Canadian Journal of Fisheries and Aquatic Sciences, 48*, 1389–1395.

Strayer, D. L. (1994). Limits to biological distributions in groundwater. In J. Gibert, D. L. Danielopol, & J. A. Stanford (Eds.), *Groundwater ecology* (pp. 287–310). San Diego, CA: Academic Press, Inc.

Strayer, D. L. (1999). Effects of alien species on freshwater mollusks in North America. *Journal of the North American Benthological Society, 18*, 74–98.

Strayer, D. L. (2001). Endangered freshwater invertebrates. In S. A. Levin (Ed.), *Encyclopedia of biodiversity* (pp. 425–439). San Diego, CA: Academic Press.

Strayer, D. L. (2006). Challenges for freshwater invertebrate conservation. *Journal of the North American Benthological Society, 25*, 271–287.

Strayer, D. L., Caraco, N. F., Cole, J. J., Findlay, S., & Pace, M. L. (1999). Transformation of freshwater ecosystems by bivalves. *BioScience, 49*, 19–27.

Strayer, D. L., Downing, J. A., Haag, W. R., King, T. L., Layzer, J. B., Newton, T. J., et al. (2004). Changing perspectives on pearly mussels, North America's most imperiled animals. *BioScience, 54*, 429.

Strayer, D. L., & Hummon, W. D. (1991). Gastrotricha. In J. H. Thorp & A. P. Covich (Eds.), *Ecology and classification of North American freshwater invertebrates* (pp. 173–186). San Diego, CA: Academic Press, Inc.

Strayer, D. L., Smith, L. C., & Hunter, D. C. (1998). Effects of the zebra mussel (*Dreissena polymorpha*) invasion on the macrobenthos of the freshwater tidal Hudson River. *Canadian Journal of Zoology, 76*, 419–425.

Stream Solute Workshop. (1990). Concepts and methods for assessing solute dynamics in stream ecosystems. *Journal of the North American Benthological Society, 9*, 95–119.

Strickland, J. D. H., & Parsons, T. R. (1972). A practical handbook of sea-water analysis, 2nd ed. *Bulletin Fisheries Research Board of Canada, 167*.

Stuart, S. N., Chanson, J. S., Cox, N. A., Young, B. E., Rodrigues, A. S. L., Fischman, D. L., et al. (2004). Status and trends of amphibian declines and extinctions worldwide. *Science, 306*, 1783–1786.

Stübing, D., Hagen, W., & Schmidt, K. (2003). On the use of lipid biomarkers in marine food web analyses: An experimental case study on the Antarctic krill, *Euphausia superba*. *Limnology and Oceanography, 48*, 1685–1700.

Stumm, W., & Morgan, J. J. (1981). *Aquatic chemistry: An introduction emphasizing chemical equilibria in natural waters* (2nd ed.). New York: John Wiley & Sons.

Suberkropp, K. (1995). The influence of nutrients on fungal growth, productivity, and sporulation during leaf breakdown in streams. *Canadian Journal of Botany, 73*, S1361–S1369.

Suberkropp, K., & Chauvet, E. (1995). Regulation of leaf breakdown by fungi in streams: Influences of water chemistry. *Ecology, 76*, 1433–1445.

Suberkropp, K., & Weyers, H. (1996). Application of fungal and bacterial production methodologies to decomposing leaves in streams. *Applied and Environmental Microbiology, 62*, 1610–1615.

Sugiura, N., Iwami, N., Nishimura, Y. I. O., & Sudo, R. (1998). Significance of attached cyanobacteria relevant to the occurrence of musty odor in Lake Kasumigaura. *Water Research, 32*, 3549–3554.

Sushchik, N. N., Gladyshev, M. I., Moskvichova, A. V., Makhutova, O. N., & Kalachova, G. S. (2003). Comparison of fatty acid composition in major lipid classes of the dominant benthic invertebrates of the Yenisei River. *Comparative Biochemistry and Physiology, 134B*, 111–122.

Suttle, C. A., Chan, A. M., & Cottrell, M. T. (1990). Infection of phytoplankton by viruses and reduction of primary productivity. *Nature, 347*, 467–469.

Suttle, C. A., Stockner, J. G., Shortreed, K. S., & Harrison, P. J. (1988). Time-courses of size-fractionated phosphate uptake: Are larger cells better competitors for pulses of phosphate than smaller cells? *Oecologia, 74*, 571–576.

Swan, C. M., & Palmer, M. A. (2006). Composition of speciose leaf litter alters stream detritivore growth, feeding activity and leaf breakdown. *Oecologia, 147*, 469–478.

Syvitski, J. P. M., Vörösmarty, C. J., Kettner, A. J., & Green, P. (2005). Impact of humans on the flux of terrestrial sediment to the global coastal. *Science, 308*, 376.

Szuroczki, D., & Richardson, J. M. L. (2009). The role of trematode parasites in larval anuran communities: An aquatic ecologist's guide to the major players. *Oecologia, 161*, 371–385.

Tank, J. L., & Webster, J. R. (1998). Interaction of substrate and nutrient availability on wood biofilm processes in streams. *Ecology, 79*, 21268–22179.

Tank, S. E., & Schindler, D. W. (2004). The role of ultraviolet radiation in structuring epilithic algal communities in Rocky Mountain montane lakes: Evidence from pigments and taxonomy. *Canadian Journal of Fisheries and Aquatic Sciences, 61*, 1461–1474.

Tavares-Cromar, A. F., & Williams, D. D. (1996). The importance of temporal resolution in food web analysis: Evidence from a detritus-based stream. *Ecological Monographs, 66*, 91–113.

Taylor, B. W., Flecker, A. S., & Hall, R. O., Jr. (2006). Loss of a harvested fish species disrupts carbon flow in a diverse tropical river. *Science, 313*, 833–836.

Taylor, C. A., Warren, M. L., Jr., Fitzpatrick, J. F., Jr., Hobbs, H. H., III, Jezerinac, R. F., Pflieger, W. L., et al. (1996). Conservation status of crayfishes of the United States and Canada. *Fisheries, 21*, 25–38.

Taylor, F. J. R. (1999). Morphology (tabulation) and molecular evidence for dinoflagellate phylogeny reinforce each other. *Applied and Environmental Microbiology, 35*, 1–6.

Taylor, T. N., & Taylor, E. L. (1993). *The biology and evolution of fossil plants.* Englewood Cliffs, NJ: Prentice Hall.

Taylor, W. D., & Sanders, R. W. (1991). Protozoa. In J. H. Thorp & A. P. Covich (Eds.), *Ecology and classification of North American freshwater invertebrates* (pp. 37–93). San Diego, CA: Academic Press.

Telford, R. J., Vandvik, V., & Birks, H. J. B. (2006). Dispersal limitations matter for microbial morphospecies. *Science, 312*, 1015.

Templeton, R. G. (1995). *Freshwater fisheries management.* Osney Mead, Oxford: Fishing News Books.

Ternes, T. A. (1998). Occurrence of drugs in German sewage treatment plants and rivers. *Water Research, 32*, 3245–3260.

Tessier, A. J., & Consolatti, N. L. (1991). Resource quantity and offspring quality in *Daphnia. Ecology, 72*, 468–478.

Tezuka, Y. (1990). Bacterial regeneration of ammonium and phosphate as affected by the carbon: Nitrogen:phosphorus ratio of organic substrates. *Microbial Ecology, 19*, 227–238.

Thiébaut, G., & Muller, S. (1999). A macrophyte communities sequence as an indicator of eutrophication and acidification levels in weakly mineralised streams in north-eastern France. *Hydrobiologia, 410*, 17–24.

Thomas, E. P., Blinn, D. W., & Keim, P. (1998). Do xeric landscapes increase genetic divergence in aquatic ecosystems? *Freshwater Biology, 40*, 587–593.

Thomas, F., Schmidt-Rhaesa, A., Martin, G., Manu, C., Durand, P., & Renaud, F. (2002). Do hairworms (Nematomorpha) manipulate the water seeking behaviour of their terrestrial hosts? *Journal of Evolutionary Biology, 15*, 356–361.

Thomas, W. H., & Duval, B. (1995). Sierra Nevada, California, USA, snow-algae: Snow albedo changes, algal-bacterial inter-relationships, and ultraviolet radiation effects. *Arctic and Alpine Research, 27*, 389–399.

Thomson, J. R., Hart, D. D., Charles, D. F., Nightengale, T. L., & Winter, D. M. (2005). Effects of removal of a small dam on downstream macroinvertebrate and algal assemblages in a Pennsylvania stream. *Journal of the North American Benthological Society, 24*, 192–207.

Thorne, R. F. (1981). Are California vernal pools unique? Vernal pools and intermittent streams. In S. Jain & P. Moyle (Eds.), (pp. 1–8). Davis, CA: Institute of Ecology, University of California.

Thorp, J., & Delong, M. (1994). The riverine productivity model: An heuristic view of carbon sources and organic processing in large river ecosystems. *Oikos, 70*, 305–308.

Thorp, J. H., & Covich, A. P. (1991). *Ecology and classification of North American freshwater invertebrates*. San Diego, CA: Academic Press.

Thorp, J. H., & Covich, A. P. (1991). Introduction to freshwater invertebrates. In J. H. Thorp & A. P. Covich (Eds.), *Ecology and classification of North American freshwater invertebrates* (pp. 1–15). San Diego, CA: Academic Press.

Thorp, J. H., & Covich, A. P. (1991). An overview of freshwater habitats. In J. H. Thorp & A. P. Covich (Eds.), *Ecology and classification of North American freshwater invertebrates* (pp. 17–36). San Diego, CA: Academic Press.

Thorp, J. H., & Covich, A. P. (2001). *Ecology and classification of North American freshwater invertebrates* (2nd ed.). San Diego, CA: Academic Press.

Thorp, J. H., Thoms, M. C., & Delong, M. D. (2006). The riverine ecosystem synthesis: Biocomplexity in river networks across space and time. *River Research and Applications, 22*, 123–147.

Tijdens, M., van de Waal, D. B., Slovackova, H., Hoogveld, H. L., & Gons, H. J. (2008). Estimates of bacterial and phytoplankton mortality caused by viral lysis and microzooplankton grazing in a shallow eutrophic lake. *Freshwater Biology, 53*, 1126–1141.

Tilimanns, A. R., Wilson, A. E., Frances, R. P., & Sarnelle, O. (2008). Meta-analysis of cyanobacterial effects on zooplankton population growth rate: Species-specific responses. *Fundamental and Applied Limnology, 171*, 285–295.

Tilman, D. (1982). *Resource competition and community structure*. Princeton, NJ: Princeton University Press.

Tilman, D., Kilham, S. S., & Kilham, P. (1982). Phytoplankton community ecology: The role of limiting nutrients. *Annual Review of Ecology and Systematics, 13*, 349–372.

Tilman, D., Naeem, S., Knops, J., Reich, P., Siemann, E., Wedin, D., et al. (1997). Biodiversity and ecosystem properties. *Science, 278*, 1866–1867.

Tilzer, M. M., Gaedke, U., Schweizer, A., & Beese, B. (1991). Interannual variability of phytoplankton productivity and related parameters in Lake Constance: No response to decreased phosphorus loading? *Journal of Plankton Research, 13*, 755–777.

Timperman, J. (1969). Medico-legal problems in death by drowning: Its diagnosis by the diatom method. *Journal of Forensic Medicine, 16*, 45–73.

Tinbergen, L. (1951). *The study of instinct*. New York: Oxford University Press.

Tobert, H. A., Prior, S. A., Rogers, H. H., Schlesinger, W. H., Mullins, G. L., & Runion, G. B. (1996). Elevated atmospheric carbon dioxide in agroecosystems affects groundwater quality. *Journal of Environmental Quality, 25*, 720–726.

Tockner, K., Uehlinger, U., & Robinson, C. (2009). *Rivers of Europe*. Boston: Academic Press.

Tockner, K., & Ward, J. V. (1999). Biodiversity along riparian corridors. *Arch. Hydobiol. Suppl., 115*(3), 293–310.

Todd, D. K. (1970). *The water encyclopedia. A compendium of useful information on water resources*. Port Washington, NY: Water Information Center.

Tokmakoff, A. (2007). Shining light on the rapidly evolving structure of water. *Science, 317*, 54–55.

Tollrian, R., & Dodson, S. I. (1999). Inducible defenses in Cladocera: Constraints, costs and multipredator environments. In E. Tollrian & C. D. Harvell (Eds.), *The ecology and evolution of inducible defenses* (pp. 177–202). Princeton, NJ: Princeton University Press.

Tomassen, H. B., Smolders, A. J. P., Limpens, J., Lamers, L. P., & Roelofs, J. G. M. (2004). Expansion of invasive species on ombrotorphic bogs: Desiccation or high N deposition? *Journal of Applied Ecology, 41*, 139–150.

Tonn, W. M., & Magnuson, J. J. (1982). Patterns in the species composition and richness assemblages in northern Wisconsin lakes. *Ecology, 63*, 1149–1166.

Topping, D. J., Schmidt, J. C., & Vierra, L. E., Jr. (2003). Computation and analysis of the instantaneous-discharge for the Colorado River at Lees Ferry, Arizona: May 8, 1921 through September 30, 2000. GB1225.A6T67. *US Geological Survey Professional Paper, 1677*.

Torres-Ruiz, M., Wehr, J. D., & Perrone, A. A. (2007). Trophic relations in a stream food web: Importance of fatty acids for macroinvertebrate consumers. *Journal of the North American Benthological Society, 26*, 509–522.

Toth, L. A. (1996). Restoring the hydrogeomorphology of the channelized Kissimmee River. In A. Brookes & F. D. Shields Jr., (Eds.), *River channel restoration. Guiding principles for sustainable projects* (pp. 369–383). Chichester, England: John Wiley & Sons, Ltd.

Townsend, C. R. (1989). The patch dynamics concept of stream community ecology. *Journal of the North American Benthological Society, 8*, 36–50.

Townsend, C. R. (1996). Invasion biology and ecological impacts of brown trout *Salmo trutta* in New Zealand. *Biological Conservation, 78*, 13–22.

Traill, L. W., Bradshaw, C. J. A., & Brook, B. W. (2007). Minimum viable population size: A meta-analysis of 30 years of published estimates. *Biological Conservation, 139*, 159–166.

Trimble, S. W. (1999). Decreased rates of alluvial sediment storage in the Coon Creek Basin, Wisconsin, 1975–93. *Science, 285*, 1244–1246.

Triska, F. J., Sedell, J. R., Cromack, J. K., Gregory, S. V., & McCorison, F. M. (1984). Nitrogen budget for a small coniferous forest stream. *Ecological Monographs, 54*, 119–140.

Tsui, M. T. K., Finlay, J. C., & Nater, E. A. (2009). Mercury bioaccumulation in a stream network. *Environmental Science and Technology, 43*, 7016–7022.

Tuchman, N. C., Wahtera, K. A., Wetzel, R. G., Russo, N. M., Kilbane, G. M., Sasso, L. M., et al. (2003). Nutritional quality of leaf detritus altered by elevated atmospheric CO_2: Effect on development of mosquito larvae. *Freshwater Biology, 48*, 1432–1439.

Turner, M. A., Robinson, G. G. C., Townsend, B. E., Hann, B. J., & Amaral, J. A. (1995). Ecological effects of blooms of filamentous green algae in the littoral zone of an acid lake. *Canadian Journal of Fisheries and Aquatic Sciences, 52*, 2264–2275.

Turner, R. E., & Rabalais, N. N. (1994). Coastal eutrophication near the Mississippi River delta. *Nature, 368*, 619–621.

Turner, R. E., Rabalais, N. N., Justic, D., & Dortch, Q. (2003). Global patterns of dissolved N, P and Si in large rivers. *Biogeochemistry, 64*, 297–317.

Twolan-Strutt, L., & Keddy, P. A. (1996). Above- and belowground competition intensity in two contrasting wetland plant communities. *Ecology, 77*, 259–270.

United States Department of the Interior Fish and Wildlife Service, Department of Commerce, and Bureau of the Census. (1993). *National survey of fishing, hunting, and wildlife-associated recreation*. Washington, DC: US Government Printing Office.

United States Fish and Wildlife Service. (1981). *The Platte River ecology study special research report,* http://www.npwrc.usgs.gov/resource/othrdata/platteco/platteco.htm.

Untergasser, U. (1989). *Handbook of fish diseases*. Holbokon, NJ: T. F. H. Publications, Inc.

Urabe, J., Nakanishi, M., & Kawabata, K. (1995). Contributions of metazoan plankton to the cycling of nitrogen and phosphorus in Lake Biwa. *Limnology and Oceanography, 40*, 232–241.

Urban, N. R., & Eisenreich, S. J. (1988). Nitrogen cycling in a forested Minnesota bog. *Canadian Journal of Botany, 66*, 435–449.

US Air Force (1960). *Handbook of geophysics* (Revised ed.). New York: McMillan.

USEPA. (1985). *National water inventory, report to Congress. Based on reports submitted by states, tribes, commissions and the District of Columbia*. Washington, DC: United States Environmental Protection Agency.

USEPA. (1997). *EPA national water quality report*. United States Environmental Protection Agency, http://www.epa.gov/watrhome/resources.

USEPA. (2005). Bacillus thuringiensis *Cry3Bb1 protein and the genetic material necessary for its production (Vector ZMIR13L) in event MON 863 corn* & Bacillus thuringiensis *Cry1Ab Delta-endotoxin and the genetic material necessary for its production in corn (006430, 006484) fact sheet* EPA 730-F-05-001. Washington, DC: United States Environmental Protection Agency.

Vadeboncoeur, Y., Peterson, G., Zanden, M. J. V., & Kalff, J. (2008). Benthic algal production across lake size gradients: Interactions among morphometry, nutrients, and light. *Ecology, 89,* 2542–2552.

Vajda, A. M., Barber, L. B., Gray, J. L., Lopez, E. M., Woodling, J. D., & Norris, D. O. (2008). Reproductive disruption in fish downstream from an estrogenic wastewater effluent. *Environmental Science and Technology, 42,* 3407–3414.

Vallentyne, J. R. (1974). *The algal bowl, lakes and man.* Ottawa, Canada: Department of the Environment, Fisheries and Marine Service.

van der Valk, A. G. (2006). *The biology of freshwater wetlands.* New York: Oxford University Press.

Van der Zweerde, W. (1990). Biological control of aquatic weeds by means of phytophagous fish. In A. H. Pieterse & K. J. Murphy (Eds.), *Aquatic weeds. The ecology and management of nuisance aquatic vegetation* (pp. 201–221). New York: Oxford University Press.

Van Donk, E. (1989). The role of fungal parasites in phytoplankton succession. In U. Sommer (Ed.), *Plankton ecology: Succession in plankton communities* (pp. 171–194). New York: Springer-Verlag.

Van Duren, I. C., Boeye, D., & Grootjans, A. P. (1997). Nutrient limitations in an extant and drained poor fen: Implications for restoration. *Plant Ecology, 133,* 91–100.

van Wilgen, B. W., Cowling, R. M., & Burgers, C. J. (1996). Valuation of ecosystems services. *BioScience, 46,* 184–189.

Vander Zanden, M. J., & Fetzer, W. W. (2007). Global patterns of aquatic food chain length. *Oikos, 116,* 1378–1388.

Vanni, M. J. (2002). Nutrient cycling by animals in freshwater ecosystems. *Annual Review of Ecology and Systematics, 33,* 341–370.

Vanni, M. J., Andrews, J., Renwick, W., Gonzalez, M., & Noble, S. (2006). Nutrient and light limitation of reservoir phytoplankton in relation to storm-mediated pulses in stream discharge. *Archiv für Hydrobiologie, 167,* 421–445.

Vanni, M. J., Arend, K., Bremigan, M., Bunnell, D., Garvey, J., Gonzalez, M., et al. (2005). Linking landscapes and food webs: Effects of omnivorous fish and watersheds on reservoir ecosystems. *BioScience, 55,* 155–167.

Vanni, M. J., Bowling, A. M., Dickman, E. M., Hale, R. S., Higgins, K. A., Horgan, M. J., et al. (2006). Nutrient cycling by fish supports relatively more primary production as lake productivity increases. *Ecology, 87,* 1696–1709.

Vanni, M. J., & Layne, C. D. (1997). Nutrient recycling and herbivory as mechanisms in the "top-down" effect of fish on algae in lakes. *Ecology, 78,* 21–40.

Vanni, M. J., Layne, C. D., & Arnott, S. E. (1997). "Top-down" trophic interactions in lakes: Effects of fish on nutrient dynamics. *Ecology, 78,* 1–20.

Vannote, R. L., Minshall, G. W., Cummins, K. W., Sedell, J. R., & Cushing, C. E. (1980). The river continuum concept. *Canadian Journal of Fisheries and Aquatic Sciences, 37,* 130–137.

Vannote, R. L., & Sweeney, B. W. (1980). Geographic analysis of thermal equilibria: A conceptual model for evaluating the effects of natural and modified thermal regimes on aquatic insect communities. *American Naturalist, 115,* 667–695.

Vaughn, C. C., & Taylor, C. M. (1999). Impoundments and the decline of freshwater mussels: A case study of an extinction gradient. *Conservation Biology, 13,* 912–920.

Vaux, P. D., Paulson, L. J., Axler, R. P., & Leavitt, S. (1995). The water quality implications of artificially fertilizing a large desert reservoir for fisheries enhancement. *Water Environment Research, 67,* 189–200.

Vellieux, F., Madern, D., Zaccai, G., & Ebel, C. (2006). Molecular adaptation to salts. In C. Gerday & N. Glansdorff (Eds.), *Physiology and biochemistry of extremophiles* (pp. 240–253). Washington, DC: ASM Press.

Verduin, J. (1988). Chemical limnology. *Verhein Internationale Verein Limnologie, 23,* 103–105.

Verhoeven, J. T. A. (1986). Nutrient dynamics in minerotrophic peat mires. *Aquatic Botany, 25,* 117–137.

Verhoeven, J. T. A., Koerselman, W., & Meuleman, A. F. M. (1996). Nitrogen- or phosphorus-limited growth in herbaceous, wet vegetation: Relations with atmospheric inputs and management regimes. *Trends in Ecology and Evolution, 11,* 494–497.

Vighi, M., & Zanin, G. (1994). Agronomic and ecotoxicological aspects of herbicide contamination of groundwater in Italy. In L. Bergman & D. M. Pugh (Eds.), *Environmental toxicology, economics and institutions* (pp. 111–139). Dordrecht, The Netherlands: Kluwer Academic Publishers.

Vincent, W. F. (1988). *Microbial ecosystems of antarctica.* London, UK: Cambridge University Press.

Vinebrooke, R. D., & Leavitt, P. R. (1999). Differential responses of littoral communities to ultraviolet radiation in an alpine lake. *Ecology, 80,* 223–237.

Visser, P. M., Ibelings, B. W., Veer, B. V. D., Koedood, J., & Mur, L. R. (1996). Artificial mixing prevents nuisance blooms of the cyanobacterium *Microcystis* in Lake Nieuwe Meer, the Netherlands. *Freshwater Biology, 36,* 435–450.

Vitousek, P. M. (1994). Beyond global warming: Ecology and global change. *Ecology, 75,* 1861–1876.

Vitousek, P. M., Aber, J., Howarth, R. W., Likens, G. E., Matson, P. A., Schindler, D. W., et al. (1997). Human alteration of the global nitrogen cycle: Causes and consequences. *Issues in Ecology, 1,* 2–15.

Vitousek, P. M., D'Antonio, C. M., Loope, L. L., & Westbrooks, R. (1996). Biological invasions as global environmental change. *American Scientist, 84,* 468–478.

Vogel, S. (1994). *Life in moving fluids* (2nd ed.). Princeton, NJ: Princeton University Press.

Vogt, K. A., Gordon, J. C., Wargo, J. P., Vogt, D. J., Asbjornsen, H., Palmiotto, P. A., et al. (1997). *Ecosystems. Balancing science with management.* New York: Springer-Verlag.

Vollenweider, R. A. (1976). Advances in defining critical loading levels for phosphorus in lake eutrophication. *Memorie dell'Instituto Italiano di Idrobiologia, 33,* 53–83.

von Elert, E., & Franck, A. (1999). Colony formation in *Scenedesmus*: Grazer-mediated release and chemical features of the infochemical. *Journal of Plankton Research, 21,* 789–804.

Vorburger, C., & Ribi, G. (1999). Aggression and competition for shelter between a native and an introduced crayfish in Europe. *Freshwater Biology, 42,* 111–119.

Vörösmarty, C. J., Green, P., Salisbury, J., & Lammers, R. B. (2000). Global water resources: Vulnerability from climate change and population growth. *Science, 289,* 284–288.

Voshell, J. R., Jr., & Simmons, G. M., Jr. (1984). Colonization and succession of benthic macroinvertebrates in a new reservoir. *Hydrobiologia, 112,* 27–39.

Vrede, T., & Tranvik, L. J. (2006). Iron constraints on planktonic primary production in oligotrophic lakes. *Ecosystems, 9,* 1094–1105.

Vrijenhoek, R. C. (1998). Conservation genetics of freshwater fish. *Journal of Fish Biology, 53,* 394–412.

Vuori, K.-M., & Joensuu, I. (1996). Impact of forest drainage on the macroinvertebrates of a small boreal headwater stream: Do buffer zones protect lotic biodiversity? *Biological Conservation, 77,* 87–95.

Vymazal, J. (1995). *Algae and element cycling in wetlands.* Boca Raton, FL: CRC Press.

Wade, P. M. (1990). Physical control of aquatic weeds. In A. H. Pieterse & K. J. Murphy (Eds.), *Aquatic weeds: The ecology and management of nuisance aquatic vegetation* (pp. 93–135). New York: Oxford University Press.

Waggoner, P. E., & Schefter, J. (1990). Future water use in the present climate. In P. E. Waggoner (Ed.), *Climate change and US water resources. Report of the American association for the advancement of science panel on climatic variability, climate change and the planning and management of US water resources* (pp. 19–40). New York: John Wiley & Sons.

Wagner, R. (1991). The influence of the diel activity pattern of the larvae of *Sericostoma personatum* (Kirby and Spence) (Trichoptera) on organic matter distribution in stream-bed sediments—a laboratory study. *Hydrobiologia, 224,* 65–70.

Wake, D. B., & Vredenburg, V. T. (2008). Are we in the midst of the sixth mass extinction? A view from the world of amphibians. *Proceedings of the National Academy of Sciences of the United States of America, 105,* 11466–11473.

Walker, K. F., Boulton, A. M., Thoms, M. C., & Sheldon, F. (1994). Effects of water-level changes induced by weirs on the distribution of littoral plants along the River Murray, South Australia. *Australian Journal of Marine and Freshwater Research, 45,* 1421–1438.

Wallace, J. B., Cuffney, T. F., Webster, J. R., Lugthart, G. J., Chung, K., & Goldowitz, B. S. (1991). Export of fine organic particles from headwater streams: Effects of season, extreme discharges, and invertebrate manipulation. *Limnology and Oceanography, 36,* 670–682.

Wallace, J. B., Eggert, S. L., Meyer, J. L., & Webster, J. R. (1997). Multiple trophic levels of a forest stream linked to terrestrial litter inputs. *Science, 277,* 102–104.

Wallace, J. B., Eggert, S. L., Meyer, J. L., & Webster, J. R. (1999). Effects of resource limitation on a detrital-based ecosystem. *Ecological Monographs, 69,* 409–442.

Wallace, J. B., Hutchens, J. J., Jr., & Grubaugh, J. W. (2006). Transport and storage of FPOM. In F. R. Hauer & G. A. Lamberti (Eds.), *Methods in stream ecology* (pp. 249–272). New York: Academic Press.

Wallace, J. B., & Merritt, R. W. (1980). Filter-feeding ecology of aquatic insects. *Annual Review of Entomology, 25,* 103–132.

Wallace, J. B., & Webster, J. R. (1996). The role of macroinvertebrates in stream ecosystem function. *Annual Review of Entomology, 41*, 115–139.

Wallace, R. T., & Snell, T. W. (1991). Rotifera. In J. H. Thorp & A. P. Covich (Eds.), *Ecology and classification of North American freshwater invertebrates* (pp. 187–248). San Diego, CA: Academic Press, Inc.

Wallberg, P., Bergqvist, P.-A., & Andersson, A. (1997). Potential importance of protozoan grazing on the accumulation of polychlorinated biphenyls (PCBs) in the pelagic food web. *Hydrobiologia, 357*, 53–62.

Walsby, A. E. (1994). Gas vesicles. *Microbiological Reviews, 58*, 94–144.

Walter, R. C., & Merritts, D. J. (2008). Natural streams and the legacy of water-powered mills. *Science, 319*, 299–304.

Walters, D. M., Fritz, K. M., & Otter, R. R. (2008). The dark side of subsidies: Adult stream insects export organic contaminants to riparian predators. *Ecological Applications, 18*, 1835–1841.

Walther, D. A., & Whiles, M. R. (2008). Macroinvertebrate responses to constructed riffles in the Cache River, Illinois, USA. *Environmental Management, 41*, 516–527.

Wania, F., & Mackay, D. (1993). Global fractionation and cold condensation of low volatility organochlorine compounds in polar regions. *Ambio, 22*, 10–18.

Ward, A. K., Dahm, C. N., & Cummins, K. W. (1985). *Nostoc* (Cyanophyta) productivity in Oregon stream ecosystems: Invertebrate influences and differences between morphological types. *Journal of Phycology, 21*, 223–227.

Ward, B. B. (1996). Nitrification and denitrification: Probing the nitrogen cycle in aquatic environments. *Microbial Ecology, 32*, 247–261.

Ward, C. J., & Codd, G. A. (1999). Comparative toxicity of four microcystins of different hydrophobicities to the protozoan, *Tetrahymena pyriformis*. *Journal of Applied Microbiology, 86*, 874–882.

Ward, D. M., Weller, R., & Bateson, M. M. (1990). 16S rRNA sequences reveal numerous uncultured microorganisms in a natural community. *Nature, 345*, 63–65.

Ward, J. V. (1985). Thermal characteristics of running water. *Hydrobiologia, 125*, 31–46.

Ward, J. V., & Stanford, J. A. (1979). *The ecology of regulated rivers*. New York: Plenum Publishing Company.

Ward, J. V., & Stanford, J. A. (1983). The serial discontinuity concept of lotic ecosystems. In T. D. I. Fontaine & S. M. Bartell (Eds.), *Dynamics of lotic ecosystems* (pp. 29–42). Ann Arbor, MI: Ann Arbor Science Publishers.

Ward, J. V., & Voelz, N. J. (1994). Groundwater fauna of the South Platte River system, Colorado. In J. Gibert, D. L. Danielopol, & J. A. Stanford (Eds.), *Groundwater ecology* (pp. 391–423). San Diego, CA: Academic Press, Inc.

Ward-Perkins, J. B. (1970). Monterosi in the Etruscan and Roman periods. Ianula: An account of the history and development of the Lago di Monterosi, Latium, Italy. *Transactions of the American Philosophical Society, 60*, 10–16.

Warren, M. L., Jr., & Burr, B. M. (1994). Status of freshwater fishes of the United States. *Fisheries, 19*, 6–18.

Warren, M. L., Jr., Burr, B. M., Walsh, S. J., Bart, H. L., Cashner, R. C., Etnier, D. A., et al. (2000). Diversity, distribution, and conservation status of the native freshwater fishes of the southern United States. *Fisheries, 25*, 7–31.

Watanabe, H., & Tokuda, G. (2001). Animal cellulases. *Cellular and Molecular Life Sciences (CMLS), 58*, 1167–1178.

Waters, T. F. (1977). Secondary production in inland waters. *Advances in Ecological Research, 10*, 91–164.

Waters, T. F. (1995). *Sediment in streams, sources, biological effects, and control*. Bethesda, MD: American Fisheries Society.

Watson, S. B., Ridal, J., & Boyer, G. L. (2008). Taste and odor and cyanobacterial toxins: Impairment, prediction, and management in the Great Lakes. *Canadian Journal of Fisheries and Aquatic Sciences, 65*, 1779–1796.

Watson, V. (1989). Maximum levels of attached algae in the Clark Fork River. *Proceedings of the Montana Academy of Science, 49*, 27–35.

Weast, R. C. (1978). *CRC handbook of chemistry and physics*. West Palm Beach, FL: CRC Press.

Webster, J. R. (1975). *Potassium and calcium dynamics in stream ecosystems on three southern Appalachian watersheds of contrasting vegetation*. Dissertation. Athens, GA: University of Georgia.

Webster, J. R., Benfield, E. F., Ehrman, T. P., Schaeffer, M. A., Tank, J. L., Hutchens, J. J., et al. (1999). What happens to allochthonous material that falls into streams? A synthesis of new and published information from Coweeta. *Freshwater Biology, 41*, 687–705.

Webster, J. R., & Ehrman, R. P. (1996). Solute dynamics. In F. R. Hauer & G. A. Lamberti (Eds.), *Methods in stream ecology* (pp. 145–160). San Diego, CA: Academic Press.

Webster, K. E., Kratz, T. K., Bowser, C. J., Magnuson, J. J., & Rose, W. J. (1996). The influence of landscape position on lake chemical responses to drought in northern Wisconsin. *Limnology and Oceanography, 41*, 977–984.

Webster, K. E., Soranno, P. A., Baines, S. B., Kratz, T. K., Bowser, C. J., Dillon, P. J., et al. (2000). Structuring features of lake districts: Landscape controls on lake chemical responses to drought. *Freshwater Biology, 43*, 499–515.

Wehr, J., & Sheath, R. (2003). *Freshwater algae of North America: Ecology and classification.* San Diego, CA: Academic Press.

Welch, D. M., & Meselson, M. (2000). Evidence for the evolution of Bdelloid rotifers without sexual reproduction or genetic exchange. *Science, 288*, 1211–1215.

Welcomme, R. L. (1974). *Fisheries ecology of floodplain rivers.* London, UK: Longman.

Weldon, C., Du Preez, L. H., Hyatt, A. D., Muller, R., & Spears, R. (2004). Origin of the amphibian chytrid fungus. *Emerging Infectious Diseases, 10*, 2100–2105.

Weller, M. W. (1995). Use of two waterbird guilds as evaluation tools for the Kissimmee River restoration. *Restoration Ecology, 3*, 211–224.

Werner, E. E., & Hall, D. J. (1974). Optimal foraging and the size selection of prey by the bluegill sunfish (*Lepomis macrochirus*). *Ecology, 55*, 1042–1052.

Wernet, P., Nordlund, D., Cavalleri, U. B. M., Odelius, M., Ogasawara, H., Näslund, L. Å., et al. (2004). The structure of the first coordination shell in liquid water. *Science, 304*, 995–999.

Westermann, P. (1993). Wetland and swamp microbiology. In T. E. Ford (Ed.), *Aquatic microbiology: An ecological approach* (pp. 215–238). Oxford, UK: Blackwell Scientific Publications.

Wetzel, R. G. (1990). Reservoir ecosystems: conclusions and speculations. *Reservoir limnology: Ecological perspectives* (pp. 227–238).

Wetzel, R. G. (1983). *Limnology* (2nd ed.). Orlando, FL: Saunders College Publishing.

Wetzel, R. G. (2001). *Limnology* (3rd ed.). San Diego, CA: Academic Press.

Wetzel, R. G., & Likens, G. E. (1991). *Limnological analyses* (2nd ed.). New York: Springer-Verlag.

Whiles, M. R., Gladyshev, M. I., Peterson, S. D., Sushchik, N. N., Makhutova, O. N., Kalachova, G. S., et al. (In press). Use of fatty acid profiles to assess trophic relations of omnivorous tadpoles. *Freshwater Biology*.

Whiles, M. R., & Goldowitz, B. S. (2005). Macroinvertebrate communities in Central Platte River wetlands: Patterns across a hydrologic gradient. *Wetlands, 25*, 462–472.

Whiles, M. R., Goldowitz, B. S., & Charlton, R. E. (1999). Life history and production of a semi-terrestrial limnephilid caddis-fly in an intermittent Platte River wetland. *Journal of the North American Benthological Society, 18*, 533–544.

Whiles, M. R., Lips, K. R., Pringle, C. M., Kilham, S. S., Bixby, R. J., Brenes, R., et al. (2006). The effects of amphibian population declines on the structure and function of Neotropical stream ecosystems. *Frontiers in Ecology and the Environment, 4*, 27–34.

Whiles, M. R., Wallace, J. B., & Chung, K. (1993). The influence of *Lepidostoma* (Trichoptera: Lepidostomatidae) on recovery of leaf-litter processing in disturbed headwater streams. *American Midland Naturalist, 130*, 356–363.

White, D. C. (1995). Chemical ecology: Possible linkage between macro- and microbial ecology. *Oikos, 74*, 177–184.

White, D. S., & Hendricks, S. P. (2000). Lotic macrophytes and surface-subsurface exchange processes. In J. B. Jones & P. J. Mulholland (Eds.), *Streams and ground waters* (pp. 363–379). San Diego, CA: Academic Press.

White, I., Macdonald, B. C. T., Somerville, P. D., & Wasson, R. (2009). Evaluation of salt sources and loads in the upland areas of the Murray-Darling Basin, Australia. *Hydrological Processes, 23*, 2485–2495.

White, J. W. C., Ciais, P., Figge, R. A., Kenny, R., & Markgraf, V. (1994). A high-resolution record of atmospheric CO_2 content from carbon isotopes in peat. *Nature, 367*, 153–156.

White, P. S., & Pickett, S. T. A. (1985). Natural disturbance and patch dynamics: An introduction. In S. T. A. Pickett & P. S. White (Eds.), *The ecology of natural disturbance and patch dynamics* (pp. 3–13). Orlando, FL: Academic Press, Inc.

White, W. B., Culver, D. C., Herman, J. S., Kane, T. C., & Mylroie, J. E. (1995). Karst lands. *American Scientist, 83*, 450–459.

Whitehead, P. G., Wilby, R. L., Battarbee, R. W., Kernan, M., & Wade, A. J. (2009). A review of the potential impacts of climate change on surface water quality. *Hydrological Sciences, 54*, 101–123.

Whitfield, J. (2005). Biogeography: Is everything everywhere? *Science, 310*, 960–961.

Whitman, W. B., Coleman, D. C., & Wiebe, W. J. (1998). Prokaryotes: The unseen majority. *Proceedings of the National Academy of Science, 95*, 6578–6583.

Whittaker, R. H. (1975). *Communities and ecosystems* (2nd ed.). New York: Macmillan Publishing Co., Inc.

Wickstrom, C. E., & Castenholz, R. W. (1978). Association of *Pleurocapsa* and *Calothrix* (Cyanophyta) in a thermal stream. *Journal of Phycology, 14*, 84–88.

Wickstrom, C. E., & Castenholz, R. W. (1985). Dynamics of cyanobacterial and ostracod interactions in an Oregon hot spring. *Ecology, 66*, 1024–1041.

Wiegert, R., & Petersen, C. (1983). Energy transfer in insects. *Annual Review of Entomology, 28*, 455–486.

Wigand, C., Andersen, F. Ø., Christensen, K. K., Holmer, M., & Jensen, H. S. (1998). Endomycorrhizae of isoetids along a biogeochemical gradient. *Limnology and Oceanography, 43*, 508–515.

Wiggins, G. B. (1995). Trichoptera. In R. W. Merritt & K. W. Cummins (Eds.), *An introduction to the aquatic insects of North America* (pp. 271–347). Dubuque, IA: Kendall/Hunt Publishing Company.

Wigington, P. J., Baker, J., DeWalle, D., Kretser, W., Murdoch, P., Simonin, H., & Van Sickle, J. (1996). Episodic acidification of small streams in the northeastern United States: Ionic controls of episodes. *Ecological Applications, 6*, 389–407.

Wilber, C. G. (1983). *Turbidity in the aquatic environment: An environmental factor in fresh and oceanic waters*. Springfield, IL: Charles C. Thomas Publisher.

Wilbur, H. M. (1987). Regulation of structure in complex systems: Experimental temporary pond communities. *Ecology, 68*, 1437–1452.

Wilbur, H. M. (1997). Experimental ecology of food webs: Complex systems in temporary ponds. *Ecology, 78*, 2279–2302.

Wilhelm, F. M., Hudson, J. J., & Schindler, D. W. (1999). Contribution of *Gammarus lacustris* to phosphorus recycling in a fishless alpine lake. *Canadian Journal of Fisheries and Aquatic Sciences, 56*, 1679–1686.

Williams, D. D. (1987). *The ecology of temporary waters*. Portland. OR: Timber Press.

Williams, D. D. (1996). Environmental constraints in temporary fresh waters and their consequences for the insect fauna. *Journal of the North American Benthological Society, 15*, 634–650.

Williams, D. M., & Embley, T. M. (1996). Microbial diversity: Domains and kingdoms. *Annual Revue of Ecology and Systematics, 27*, 569–595.

Williams, W. D. (1993). Conservation of salt lakes. *Hydrobiologia, 267*, 291–306.

Williamson, C. E. (1991). Copepoda. In J. H. Thorp & A. P. Covich (Eds.), *Ecology and classification of North American freshwater invertebrates* (pp. 787–822). San Diego, CA: Academic Press, Inc.

Williamson, C. E., Zagarese, H. E., Schulze, P. C., Hargreaves, B. R., & Seva, J. (1994). The impact of short-term exposure to UV-B radiation on zooplankton communities in North Temperate Lakes. *Journal of Plankton Research, 16*, 205–218.

Willson, M. F., Gende, S. M., & Marston, B. H. (1998). Fishes and the forest. *BioScience, 48*, 455–462.

Wilson, L. G. (1990). Methods for sampling fluids in the vadose zone. In D. M. Nielsen & A. I. Johnson (Eds.), *Ground water and vadose zone monitoring* (pp. 7–24). Ann Arbor, MI: ASTM.

Wilson, M. A., & Carpenter, S. R. (1999). Economic valuation of freshwater ecosystem services in the United States: 1971–1997. *Ecological Applications, 9*, 772–783.

Wilson, S. C., Duarte-Davidson, R., & Jones, K. C. (1996). Screening the environmental fate of organic contaminants in sewage sludges applied to agricultural soils: 1. The potential for downward movement to groundwaters. *The Science of the Total Environment, 185*, 45–57.

Winemiller, K. O., & Rose, K. A. (1992). Patterns of life-history diversification in North-American fishes—implications for population regulation. *Canadian Journal of Fisheries and Aquatic Sciences, 49*, 2196–2218.

Wingham, D. J., Siegert, M. J., Shephard, A., & Muir, A. S. (2006). Rapid discharge connects Antarctic subglacial lakes. *Nature, 440*, 1033–1036.

Winterbourn, M. J. (1990). Interactions among nutrients, algae, and invertebrates in a New Zealand mountain stream. *Freshwater Biology, 23*, 463–474.

Winterbourn, M. J., Rounick, J. S., & Cowie, B. (1981). Are New Zealand stream ecosystems really different? *New Zealand Journal of Marine and Freshwater Research, 15*, 321–328.

Wissinger, S. A. (1999). Ecology of wetland invertebrates. Synthesis and applications for conservation and management. In D. P. Batzer, R. B. Rader, & S. A. Wissinger (Eds.), *Invertebrates in freshwater Wetlands of North America: Ecology and management* (pp. 1043–1086). New York: John Wiley & Sons, Inc.

Wissinger, S. A., Whiteman, H. H., Sparks, G. B., Rouse, G. L., & Brown, W. S. (1999). Foraging trade-offs along a predator-permanence gradient in subalpine wetlands. *Ecology, 80,* 2102–2116.

Wnorowski, A. (1992). Tastes and odors in the aquatic environment: A review. *Water SA, 18,* 203–214.

Woese, C. R., Kandler, O., & Wheelis, M. L. (1990). Towards a natural system of organisms: Proposal for the domains Archae, Bacteria, and Eukarya. *Proceedings of the National Academy of Science of the United States of America, 87,* 4576–4579.

Wolfe, B. E., Weishampel, P. A., & Klironomos, J. N. (2006). Arbuscular mycorrhizal fungi and water table affect wetland plant community composition. *Journal of Ecology, 94,* 905–914.

Wolfe, G. V., Steinke, M., & Kirst, G. O. (1997). Grazing-activated chemical defense in a unicellular marine alga. *Nature, 387,* 203–214.

Woltemade, C. J. (2000). Ability of restored wetlands to reduce nitrogen and phosphorus concentrations in agricultural drainage water. *Journal of Soil and Water Conservation, Third Quarter,* 303–309.

Wommack, K. E., Hill, R. T., Muller, T. A., & Colwell, R. R. (1996). Effects of sunlight on bacteriophage viability and structure. *Applied and Environmental Microbiology, 62,* 1342–1346.

Wong, M. K. M., Goh, T.-K., Hodgkiss, I. J., Hyde, K. D., Ranghoo, V. M., Tsui, C. K. M., et al. (1998). Role of fungi in freshwater ecosystems. *Biodiversity and Conservation, 7,* 1187–1206.

Wood, T. S. (1991). Bryozoans. In J. H. Thorp & A. P. Covich (Eds.), *Ecology and classification of North American freshwater invertebrates* (pp. 481–500). San Diego, CA: Academic Press, Inc.

Wootton, J. T. (1994). The nature and consequences of indirect effects in ecological communities. *Annual Review of Ecology and Systematics, 25,* 443–466.

World Health Organization Expert Committee. (1993). Geneva, Switzerland: World Health Organization.

Worthington, E. B. (1931). Vertical movements of fresh-water macroplankton. *Internationale Revue der gesamten Hydrobiologie und Hydrographie, 25,* 394–436.

Wrona, F. J., Prowse, T. D., Reist, J. D., Hobbie, J. E., Levesque, L. M. J., & Vincent, W. F. (2006). Climate change effects on aquatic biota, ecosystem structure and function. *Ambio, 35,* 359–369.

Wu, L., & Culver, D. A. (1991). Zooplankton grazing and phytoplankton abundance: An assessment before and after invasion of *Dreissena polymorpha. Journal of Great Lakes Research, 17,* 425–436.

Wurtsbaugh, W. A. (1992). Food-web modification by an invertebrate predator in the Great Salt Lake (USA). *Oecologia, 89,* 168–175.

Xenopoulos, M. A., Lodge, D. M., Alcamo, J., Marker, M., Schulze, K., & Van Vuuren, D. P. (2005). Scenarios of freshwater fish extinctions from climate change and water withdrawal. *Global Change Biology, 11,* 1557–1564.

Xenopoulos, M. A., & Schindler, D. W. (2003). Differential responses to UVR by bacterioplankton and phytoplankton from the surface and the base of the mixed layer. *Freshwater Biology, 48,* 108–122.

Xia, K., Bhandari, A., Das, K., & Pillar, G. (2005). Occurrence and fate of pharmaceuticals and personal care products (PPCPs) in biosolids. *Journal of Environmental Quality, 34,* 91–104.

Yamamoto, Y., Kouchiwa, T., Hodoki, Y., Hotta, K., Uchida, H., & Harada, K.-I. (1998). Distribution and identification of actinomycetes lysing cyanobacteria in a eutrophic lake. *Journal of Applied Phycology, 10,* 391–397.

Yan, N. D., Keller, W., Scully, N. M., Lean, D. R. S., & Dillon, P. J. (1996). Increased UV-B penetration in a lake owing to drought-induced acidification. *Nature, 381,* 141–143.

Yanagita, T. (1990). *Natural microbial communities. Ecological and physiological features.* Tokyo, Japan: Japan Scientific Societies Press.

Yang, Z., Kong, F. X., Shi, X. L., & Cao, H. S. (2006). Morphological response of *Microcystis aeruginosa* to grazing by different sorts of zooplankton. *Hydrobiologia, 563,* 225–230.

Yoch, D. C., Carraway, R. H., Friedman, R., & Kulkarni, N. (2001). Dimethylsulfide (DMS) production from dimethylsulfoniopropionate by freshwater river sediments: phylogeny of gram-positive DMS-producing isolates. *Fems Microbiology Ecology, 37,* 31–37.

Yoon, H.-S., & Golden, J. W. (1998). Heterocyst pattern formation controlled by a diffusible peptide. *Science, 30*, 935–938.

Yoon, I., Williams, R., Levine, E., Yoon, S., Dunne, J., & Martinez, N. (2004). Webs on the Web (WOW): 3D visualization of ecological networks on the WWW for collaborative research and education. *Proceedings of the IS&T/SPIE Symposium on Electronic Imaging, Visualization and Data Analysis, 5295*, 124–132.

Young, J. P. W. (1992). Phylogenetic classification of nitrogen-fixing organisms. In G. Stacey, R. H. Burris, & H. J. Evans (Eds.), *Biological nitrogen fixation*. New York: Chapman & Hall.

Young, P. (1996). Safe drinking water: A call for global action. *American Society of Microbiology News, 62*, 349–352.

Young, P. (1997). Major microbial diversity initiative recommended. *American Society of Microbiology News, 63*, 417–421.

Young, R. G., & Huryn, A. D. (1999). Effects of land use on stream metabolism and organic matter turnover. *Ecological Applications, 9*, 1359–1376.

Young, R. G., Matthaei, C. D., & Townsend, C. R. (2008). Organic matter breakdown and ecosystem metabolism: Functional indicators for assessing river ecosystem health. *Journal of the North American Benthological Society, 27*, 605–625.

Yuen, T. K., Hyde, K. D., & Hodgkiss, I. J. (1999). Interspecific interactions among tropical and subtropical freshwater fungi. *Microbial Ecology, 37*, 257–262.

Yurista, P. M. (2000). Cyclomorphosis in *Daphnia lumholtzi* induced by temperature. *Freshwater Biology, 43*, 207–213.

Zaret, T. M. (1980). *Predation and freshwater communities*. New Haven, CT: Yale University Press.

Zevenboom, W., Vaate, A. B. D., & Mur, L. R. (1982). Assessment of factors limiting growth rate of *Oscillatoria agardhii* in hypertrophic Lake Wolderqijd, 1978, by use of physiological indicators. *Limnology and Oceanography, 27*, 39–52.

Zimba, P. V. (1998). The use of nutrient enrichment bioassays to test for limiting factors affecting epiphytic growth in Lake Okeechobee, Florida: Confirmation of nitrogen and silica limitation. *Archiv für Hydrobiologie, 141*, 459–468.

Zlotnik, I., & Dubinsky, Z. (1989). The effect of light and temperature on DOC excretion by phytoplankton. *Limnology and Oceanography, 34*, 831–839.

Zwahlen, C., Hilbeck, A., Gugerli, P., & Nentwig, W. (2003). Degradation of the Cry1Ab protein within transgenic *Bacillus thuringiensis* corn tissue in the field. *Molecular Ecology, 12*, 765–775.

Zwart, G., Crump, B. C., Kamst-van Agterveld, M. P., Hagen, F., & Han, S.-K. (2002). Typical freshwater bacteria: an analysis of available 16S rRNA gene sequences from plankton of lakes and rivers. *Aquatic Microbial Ecology, 28*, 141–155.

Zweig, R. D. (1985). Freshwater aquaculture management for survival. *Ambio, 14*, 66–74.

Taxonomic Index

Note: f refers to figures and t refers to tables.

A

Abramis brama, 555–556, 566
Abyssogammarus sarmatus, 265f
Acanthomoeba, 210
Acanthomoeba castellani, 188t
Acer rubrum, 218t
Aedes albopictus, 550
Aedes sierrensis, 393
Aeolosoma, 231f, 236
Aeromonas, 540
Aeromonas salmonicida, 532t
African clawed frog (*Xenopus laevis*), 286
Agnetina capitata, 549f
Alces, 547–548
alder (*Alnus*), 350
alderflies (Megaloptera), 240t, 243
alewife (*Alosa*), 254f, 281t
algae
    Bacillariophycea (diatoms), 200t, 202, 204, 272f
    Charophyceae (stoneworts), 200t, 205–208
    Chlorophyceae (green algae), 200t, 205–208, 208f
    Chrysophyceae (golden algae), 199–202, 200t
    cyanobacteria, 195–198, 200t
    Dinophyceae, 200t, 202–205
    Euglenophyceae, 200t, 205
    Protozoa, 208–210, 209f
    Rhodophyceae (red algae), 199, 200t
Alligator mississippiensis, 103, 151
Alnus (alder), 350
Alosa (alewife), 254f
Alosa pseudoharengus (alewife), 281t
American bullfrog (*Lithobates catesbeianus*), 286
Amia, 254f
Amoeba, 209f, 210

amphibians, 255, 580
Amphipoda (scuds and sideswimmers), 250f, 251, 459
Anabaena, 196f, 197, 442–444
Anabaena azollae, 540
Anabaena flos-aquae, 541f
Anas acuta, 83f
Anatidae (duck), 547–548
Ancylobacter, 192–193t
Ancylostoma duodenale, 188t
Andreales, 213
Anguilla, 254f
Animalia, 176
animals, 221
    invertebrates, 181, 222–251
    vertebrates, 181, 252, 255, 256t
        amphibians, 255, 284–287
        birds, 255, 256t
        fishes, 252–255
        mammals, 255, 256t
        reptiles, 255, 256t
        tetrapods, 255
Annelida (segmented worms), 231f, 235–236
Anopheles, 244f
Anopheles claviger, 397f, 410
Antocha, 549f
ants (Hymenoptera), 240t, 245
Anura (frogs and toads), 255
    global decline, 284–287
Aphanizomenon, 196f, 197, 491–492
aphids (Homoptera), 242–243
Aquarius remigis (water striders), 25, 396, 575–576
Arachnida (water mites and spiders), 238–239
Arcella, 397f
Archaeoglobus, 192–193t
Ardea alba (great egret), 583
Ardea herodias (great blue heron), 583
Argyroneta, 239

arrow arum (*Peltrandra*), 216f, 217t
arrowhead (*Sagittaria*), 91t, 216f, 217t
Artemia franciscana (brine shrimp), 381
Artemia salina (brine shrimp), 248, 381
Arthrobotrys oligospora, 211f
Arthropoda, 238–251
Ascaris lumbricoides, 188t
Ascomycota (sac fungi), 211
Asellus, 251
Asiatic clam (*Corbicula fluminea*), 230–231, 282t
Asterionella, 203f, 366f, 532t
Atlantic white cedar (*Chamaecyparis thyoides*), 218t
Atya lanipes, 648
Avicennia (black mangrove), 94, 218t
Aythya fuligula, 600
Azolla (water velvet), 215f, 217t, 350, 540, 541f
Azotobacter, 192–193t

B

Bacillariophyceae (diatoms), 202
    colonization, 272f
    forensics, 203–204
    microbial diversity, 278
    pollution indicators, 204
    silicon, 363–364
Bacillus, 192–193t
Bacillus thuringiensis (*Bt*), 411–412
Bacteria, 174, 176, 180, 191–198
Baetis, 241f, 558
Balantidium coli, 188t, 210
bald eagle (*Haliaeetus leucocephalus*), 283–284
Basicladia, 208f
Basidiomycota (club fungi), 211
Basiliscus (lizards), 25
Bathybates, 266f

787

Subject Index

A

abiotic gradients, 672
abiotic oxidation, 358
absorbed light, 54
acetogenesis, 334f, 335t, 337
acid mine drainage, 418, 424
acid precipitation, 417–424
  biological effects, 418–424
  buffering capacity, 325–327, 418
  chemical effects, 293–294
  chemical treatments, 423
  sources and geography of, Color
    plate fig.9, 417–418
acidity, 213, 294, 335
  pH, 293–294
acidity titrations and CO_2, 327
acoustic Doppler velocity (ADV), 32
active filter feeders, 177, 529
acute exposure to toxins, 403
adaptations
  defenses, 553–557
  to extreme environments,
    377–380
  of fishes, 252, 614
  microbial, 530–531
  of predators, 557–560
adaptive radiation, 264, 613
additive toxic effects, 405
adhesion, 25
ADV (acoustic Doppler velocity), 32
advective transport/advection, 46,
  49, 135, 599–600
aeolian lakes, 151
aeration, hypolimnetic, 326f, 492
aerobic breakdown, 330–331, 331f
aerobic habitats, 302
aerobic respiration, 302, 333. *See
  also* oxygen
African Rift Lakes, 143–144,
  265–267
Agricultural Improvement and
  Reform Act, Federal (U.S.), 89

agriculture
  irrigation water, 9, 78–79
  land-use categories, 489f
  nitrate contamination, 356–357
  nutrient runoff, 480, 489,
    503–504
  organic pollutants, 410
  oxygen profiles, 317–318
  pollution, 410
  salt pollution, 430
  water uses, 5–6
  wetland use, 85–89, 97
akenites, 195
alarm chemicals, 554
Albert, Lake, 143–144
algae, 195–210
  acidification, 419–421
  blooms. *See* blooms
  defined, 199
  light absorption, Color plate fig.1,
    57–59, 199
  nutrient availability, 447t
  toxins, 197–198, 205
algal biomass, 481–484
alkalinity, 294. *See also* bicarbonate
  acidity/alkalinity titrations and
    CO_2, 327
alkalinity titrations and CO_2, 327
allelochemicals, 537
allelopathy, 537–538, 573, 576
Allen curve, 639
Allen paradox, 653
allochthonous production, 637–
  638, 650
allomones, 554
alpha (α) diversity, 263
Amazon River, 92t, 95–97, 109t,
  151, 614
ameboid protozoa, 210
amensalism, 179, 583–584
amictic lakes, 159
ammonia gas, 347

ammonification, 351
ammonium, 347, 351–352, 445,
  456–458, 460t
  decline of, 284–287
  amphibians, decline of, 284–287
Amur River, 109t
anadromous fishes, 253–255
anaerobic ammonium
  oxidation(anammox), 354
anammox (anaerobic ammonium
  oxidation), 354
anemones, 223–225
anoxia
  algal blooms, 312
  lake stratification, 160
anoxic (anaerobic) habitats, 302,
  312–319, 361
  carbon cycling in bogs, 336
  carbon oxidations, 333–334
  fish kills, 314–316, 476
  groundwaters, 317–318
  lakes, 313–316
  redox, 296–298
  sediment, 317
  streams, 315–317, 353–354
  wetlands, 315–316, 336, 337, 367
antagonism, 405
Antarctic Lakes, 142t, 150, 159, 376,
  387
anthropogenic effects
  demands for water, 5–8
  eutrophication (cultural), 470, 480
  extinctions, 283
  floods, 117
  global change, 68
  hydrological perturbations, 117,
    124–125
  nutrient pollution, 480, 481,
    489–492
  thermal pollution, 432–434
  toxic pollution. *See* toxins
  UV radiation, 434–435

795

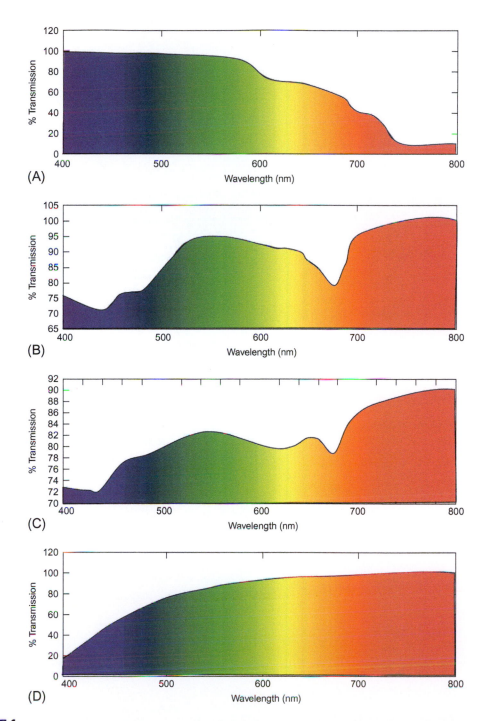

FIGURE 1

Percentage light transmission as a function of wavelength for (A) pure water; (B) a green alga, *Chorella*; (C) a cyanobacterium, *Microcystis*; and (D) humic substances. Note that water absorbs much of the red light, the green alga allows more green light through, the cyanobacterium lets a bit more blue light through, and the humic substances remove blue and green light. Thus, an oligotrophic lake has blue water, a lake with green algae appears green, a lake with cyanobacteria appears blue green, and a lake with high levels of dissolved humic substances appears reddish-brown. *(Data from Hakvoort, 1994; Wetzel, 2001).*

FIGURE 2
Contrasting color of lakes. A eutrophic pond in Oklahoma with a floating cyanobacterial bloom (left) and ultraoligotrophic Crater Lake (right).

FIGURE 3

Illustration of the appearance of colored objects at depth in lakes of different trophic status. Upper left, full light; upper right, a blue filter simulating light deep in an oligotrophic lake where white looks blue, and blue looks black. Lower left, a green filter as at moderate depth in a mesotrophic lake where white looks green and blue looks black. Lower right, a red filter as at shallow depth in a eutrophic lake where the contrast between red and white is strongly decreased.

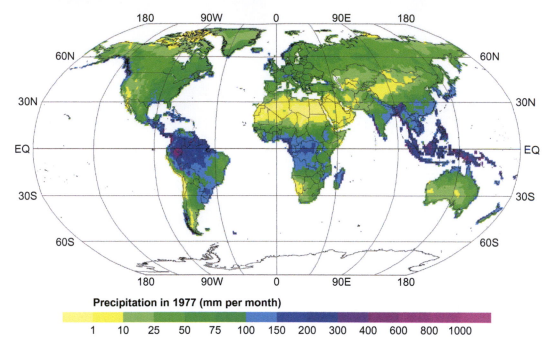

Precipitation in 1977 (mm per month)

1   10   25   50   75   100   150   200   300   400   600   800   1000

FIGURE 4

Global precipitation for 2006. *(From the Global Precipitation Climatology Centre).*

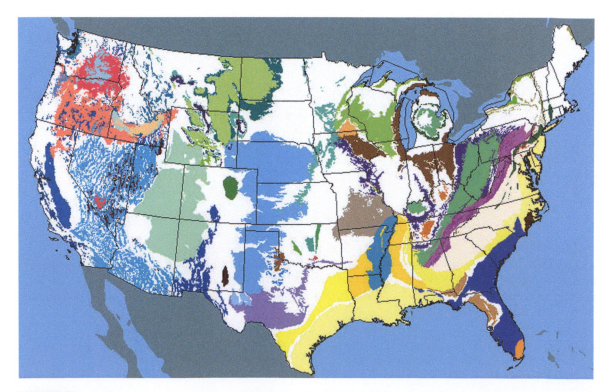

FIGURE 5

Major aquifers of the United States. Blue and turquoise areas are unconsolidated sand and gravel aquifers, yellow are semiconsolidated sand aquifers, green are sandstone aquifers, purple are sandstone and carbonate aquifers, brown and reddish brown are carbonate aquifers, red and pink are igneous and metamorphic rock aquifers. *(Image courtesy of the National Atlas, US Department of the Interior).*

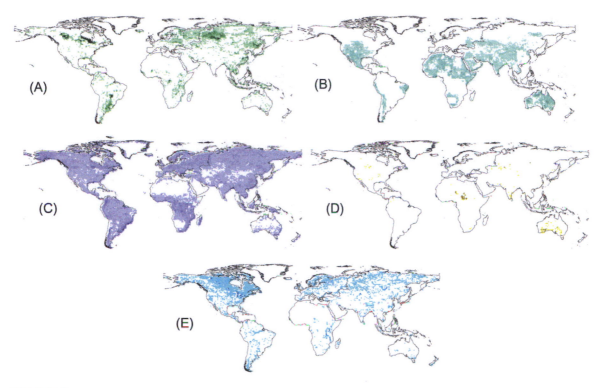

FIGURE 6

Global distribution of aquatic habitats. In all cases, increasingly dark colors indicate a greater density of the type of habitat. (A) Wetlands, (B) intermittent rivers, (C) perennial rivers, (D) intermittent lakes, (E) perennial lakes. *(Data from Cogley, 1994).*

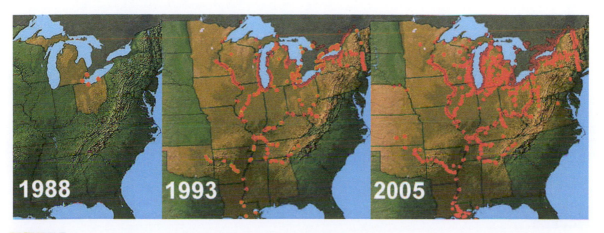

FIGURE 7

Zebra mussel spread in the United States from 1988 to 2005. Sightings or samplings are indicated by dots, and states with a brown tint are those where mussels have been found. *(Image courtesy of the US Geological Survey).*

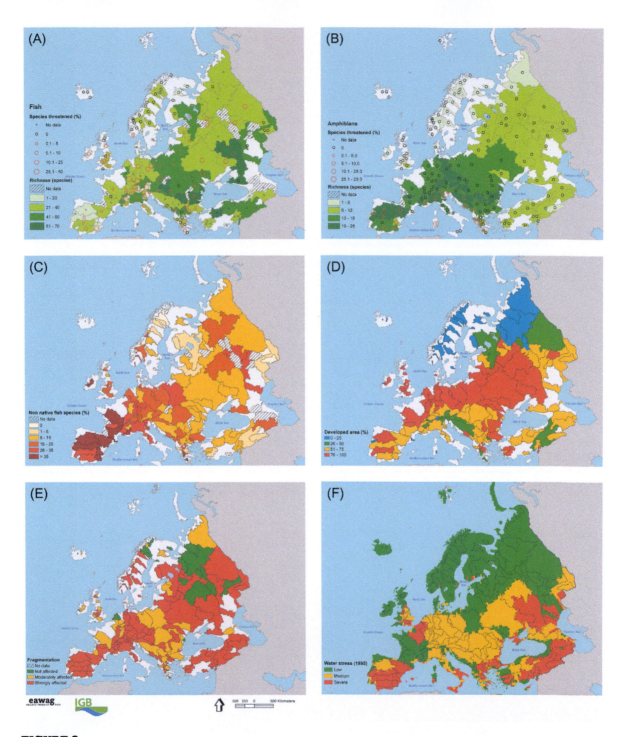

FIGURE 8

Percentage threatened species, numbers of native species, and stressors in Europe. (A) % fish threatened (larger red dots mean more threatened) and fish species (darker green areas have more native diversity), (B) % amphibians threatened (larger red dots mean more threatened) and amphibian species (darker green areas have more native diversity), (C) nonnative fishes (darker colors mean more species), (D) % developed areas, (E) fragmentation, and (F) degree of water stress. *(Image from Tockner et al., 2009).*

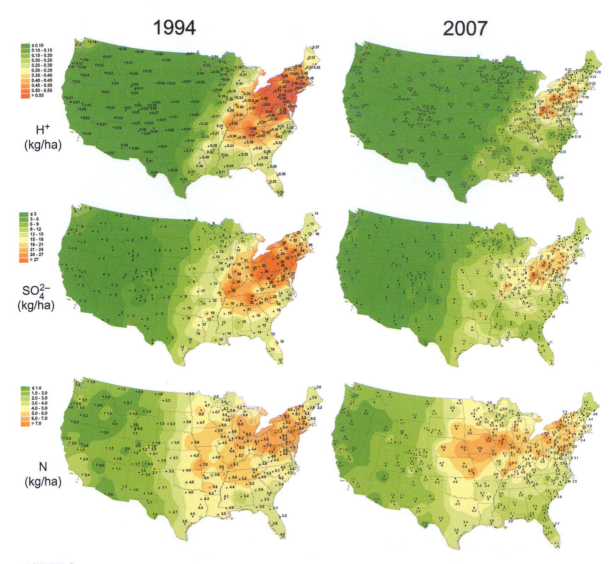

FIGURE 9

Atmospheric deposition of hydrogen ions, sulfate, and ammonium + nitrate in the United States from 1994 and 2007. In all images, low rates are in green, high rates are in red. *(Images courtesy of National Atmospheric Deposition Program (NRSP-3). 2009. NADP Program Office, Illinois State Water Survey, 2204 Griffith Dr., Champaign, IL 61820).*

FIGURE 10
Mesocosms made of 1 m diameter polyethylene bags to test the effects of nutrient limitation on uptake and remineralization rates of phosphorus in mesotrophic Milford Reservoir, Kansas.

FIGURE 11
Nutrient-diffusing substrata deployed in a stream (left) and after collection (right). The filters are placed on agar in containers with or without various nutrients and incubated in the stream channel for a few weeks. The filters are then removed from the tops of the containers and analyzed for metabolism (photosynthesis and respiration) and algal chlorophyll. In the assay to the right, there is a strong colimitation of N and P (the filters are much greener). *(Images courtesy of Laura Johnson)*.

FIGURE 12

A long-term leaf and wood exclusion experiment at Coweta Hydrologic Laboratory in the southern Appalachian Mountains under and above the netting. This ecosystem-scale experiment was designed to gauge the effects of excluding leaf and wood inputs into forested headwater streams and consisted of several hundred meters of mesh covering a headwater stream all the way up to the spring seep source. Also note the orange and gray plastic mesh to stop lateral leaf inputs. After wood removal, plastic pipes were added to simulate the physical effects of wood input without the carbon input. *(Images courtesy of S. L. Eggert).*

FIGURE 13

Scenes from the Pantanal illustrating various scales of observation. From top to bottom, a satellite image of the huge number of wetland pools, an aerial photo of a main channel and adjacent flooded areas, a pool with caiman, *Caiman yacre*, and a giant river otter, *Pteronura brasiliensis*. *(Images courtesy of Steve Hamilton).*